Unilateral Contact Problems

Variational Methods and Existence Theorems

T0139174

PURE AND APPLIED MATHEMATICS

A Program of Monographs, Textbooks, and Lecture Notes

EXECUTIVE EDITORS

Earl J. Taft
Rutgers University
New Brunswick, New Jersey

Zuhair Nashed
University of Central Florida
Orlando, Florida

EDITORIAL BOARD

M. S. Baouendi
University of California, San Diego

Jane Cronin
Rutgers University

Jack K. Hale
Georgia Institute of Technology

S. Kobayashi
University of California, Berkeley

Marvin Marcus
University of California, Santa Barbara

W. S. Masse
Yale University

Anil Nerode
Cornell University

Donald Passman
University of Wisconsin, Madison

Fred S. Roberts
Rutgers University

David L. Russell
Virginia Polytechnic Institute and State University

Walter Schempp
Universität Siegen

Mark Teply
University of Wisconsin, Milwaukee

MONOGRAPHS AND TEXTBOOKS IN PURE AND APPLIED MATHEMATICS

1. *K. Yano*, Integral Formulas in Riemannian Geometry (1970)
2. *S. Kobayashi*, Hyperbolic Manifolds and Holomorphic Mappings (1970)
3. *V. S. Vladimirov*, Equations of Mathematical Physics (A. Jeffrey, ed.; A. Littlewood, trans.) (1970)
4. *B. N. Pshenichnyi*, Necessary Conditions for an Extremum (L. Neustadt, translation ed.; K. Makowski, trans.) (1971)
5. *L. Narici et al.*, Functional Analysis and Valuation Theory (1971)
6. *S. S. Passman*, Infinite Group Rings (1971)
7. *L. Dornhoff*, Group Representation Theory. Part A: Ordinary Representation Theory. Part B: Modular Representation Theory (1971, 1972)
8. *W. Boothby and G. L. Weiss, eds.*, Symmetric Spaces (1972)
9. *Y. Matsushima*, Differentiable Manifolds (E. T. Kobayashi, trans.) (1972)
10. *L. E. Ward, Jr.*, Topology (1972)
11. *A. Babakhanian*, Cohomological Methods in Group Theory (1972)
12. *R. Gilmer*, Multiplicative Ideal Theory (1972)
13. *J. Yeh*, Stochastic Processes and the Wiener Integral (1973)
14. *J. Barros-Neto*, Introduction to the Theory of Distributions (1973)
15. *R. Larsen*, Functional Analysis (1973)
16. *K. Yano and S. Ishihara*, Tangent and Cotangent Bundles (1973)
17. *C. Procesi*, Rings with Polynomial Identities (1973)
18. *R. Hermann*, Geometry, Physics, and Systems (1973)
19. *N. R. Wallach*, Harmonic Analysis on Homogeneous Spaces (1973)
20. *J. Dieudonné*, Introduction to the Theory of Formal Groups (1973)
21. *I. Vaisman*, Cohomology and Differential Forms (1973)
22. *B.-Y. Chen*, Geometry of Submanifolds (1973)
23. *M. Marcus*, Finite Dimensional Multilinear Algebra (in two parts) (1973, 1975)
24. *R. Larsen*, Banach Algebras (1973)
25. *R. O. Kujala and A. L. Vitter, eds.*, Value Distribution Theory: Part A; Part B: Deficit and Bezout Estimates by Wilhelm Stoll (1973)
26. *K. B. Stolarsky*, Algebraic Numbers and Diophantine Approximation (1974)
27. *A. R. Magid*, The Separable Galois Theory of Commutative Rings (1974)
28. *B. R. McDonald*, Finite Rings with Identity (1974)
29. *J. Satake*, Linear Algebra (S. Koh et al., trans.) (1975)
30. *J. S. Golan*, Localization of Noncommutative Rings (1975)
31. *G. Klambauer*, Mathematical Analysis (1975)
32. *M. K. Agoston*, Algebraic Topology (1976)
33. *K. R. Goodearl*, Ring Theory (1976)
34. *L. E. Mansfield*, Linear Algebra with Geometric Applications (1976)
35. *N. J. Pullman*, Matrix Theory and Its Applications (1976)
36. *B. R. McDonald*, Geometric Algebra Over Local Rings (1976)

37. *C. W. Groetsch*, Generalized Inverses of Linear Operators (1977)
38. *J. E. Kuczkowski and J. L. Gersting*, Abstract Algebra (1977)
39. *C. O. Christenson and W. L. Voxman*, Aspects of Topology (1977)
40. *M. Nagata*, Field Theory (1977)
41. *R. L. Long*, Algebraic Number Theory (1977)
42. *W. F. Pfeffer*, Integrals and Measures (1977)
43. *R. L. Wheeden and A. Zygmund*, Measure and Integral (1977)
44. *J. H. Curtiss*, Introduction to Functions of a Complex Variable (1978)
45. *K. Hrbacek and T. Jech*, Introduction to Set Theory (1978)
46. *W. S. Massey*, Homology and Cohomology Theory (1978)
47. *M. Marcus*, Introduction to Modern Algebra (1978)
48. *E. C. Young*, Vector and Tensor Analysis (1978)
49. *S. B. Nadler, Jr.*, Hyperspaces of Sets (1978)
50. *S. K. Segal*, Topics in Group Rings (1978)
51. *A. C. M. van Rooij*, Non-Archimedean Functional Analysis (1978)
52. *L. Corwin and R. Szczarba*, Calculus in Vector Spaces (1979)
53. *C. Sadosky*, Interpolation of Operators and Singular Integrals (1979)
54. *J. Cronin*, Differential Equations (1980)
55. *C. W. Groetsch*, Elements of Applicable Functional Analysis (1980)
56. *I. Vaisman*, Foundations of Three-Dimensional Euclidean Geometry (1980)
57. *H. I. Freedan*, Deterministic Mathematical Models in Population Ecology (1980)
58. *S. B. Chae*, Lebesgue Integration (1980)
59. *C. S. Rees et al.*, Theory and Applications of Fourier Analysis (1981)
60. *L. Nachbin*, Introduction to Functional Analysis (R. M. Aron, trans.) (1981)
61. *G. Orzech and M. Orzech*, Plane Algebraic Curves (1981)
62. *R. Johnsonbaugh and W. E. Pfaffenberger*, Foundations of Mathematical Analysis (1981)
63. *W. L. Voxman and R. H. Goetschel*, Advanced Calculus (1981)
64. *L. J. Corwin and R. H. Szczarba*, Multivariable Calculus (1982)
65. *V. I. Istrăţescu*, Introduction to Linear Operator Theory (1981)
66. *R. D. Järvinen*, Finite and Infinite Dimensional Linear Spaces (1981)
67. *J. K. Beem and P. E. Ehrlich*, Global Lorentzian Geometry (1981)
68. *D. L. Armacost*, The Structure of Locally Compact Abelian Groups (1981)
69. *J. W. Brewer and M. K. Smith, eds.*, Emmy Noether: A Tribute (1981)
70. *K. H. Kim*, Boolean Matrix Theory and Applications (1982)
71. *T. W. Wieting*, The Mathematical Theory of Chromatic Plane Ornaments (1982)
72. *D. B.Gauld*, Differential Topology (1982)
73. *R. L. Faber*, Foundations of Euclidean and Non-Euclidean Geometry (1983)
74. *M. Carmeli*, Statistical Theory and Random Matrices (1983)
75. *J. H. Carruth et al.*, The Theory of Topological Semigroups (1983)
76. *R. L. Faber*, Differential Geometry and Relativity Theory (1983)
77. *S. Barnett*, Polynomials and Linear Control Systems (1983)

78. *G. Karpilovsky*, Commutative Group Algebras (1983)
79. *F. Van Oystaeyen and A. Verschoren*, Relative Invariants of Rings (1983)
80. *I. Vaisman*, A First Course in Differential Geometry (1984)
81. *G. W. Swan*, Applications of Optimal Control Theory in Biomedicine (1984)
82. *T. Petrie and J. D. Randall*, Transformation Groups on Manifolds (1984)
83. *K. Goebel and S. Reich*, Uniform Convexity, Hyperbolic Geometry, and Nonexpansive Mappings (1984)
84. *T. Albu and C. Nastasescu*, Relative Finiteness in Module Theory (1984)
85. *K. Hrbacek and T. Jech*, Introduction to Set Theory: Second Edition (1984)
86. *F. Van Oystaeyen and A. Verschoren*, Relative Invariants of Rings (1984)
87. *B. R. McDonald*, Linear Algebra Over Commutative Rings (1984)
88. *M. Namba*, Geometry of Projective Algebraic Curves (1984)
89. *G. F. Webb*, Theory of Nonlinear Age-Dependent Population Dynamics (1985)
90. *M. R. Bremner et al.*, Tables of Dominant Weight Multiplicities for Representations of Simple Lie Algebras (1985)
91. *A. E. Fekete*, Real Linear Algebra (1985)
92. *S. B. Chae*, Holomorphy and Calculus in Normed Spaces (1985)
93. *A. J. Jerri*, Introduction to Integral Equations with Applications (1985)
94. *G. Karpilovsky*, Projective Representations of Finite Groups (1985)
95. *L. Narici and E. Beckenstein*, Topological Vector Spaces (1985)
96. *J. Weeks*, The Shape of Space (1985)
97. *P. R. Gribik and K. O. Kortanek*, Extremal Methods of Operations Research (1985)
98. *J. A. Chao and W. A. Woyczynski, eds.*, Probability Theory and Harmonic Analysis (1986)
99. *G. D. Crown et al.*, Abstract Algebra (1986)
100. *J. H. Carruth et al.*, The Theory of Topological Semigroups, Volume 2 (1986)
101. *R. S. Doran and V. A. Belfi*, Characterizations of C*-Algebras (1986)
102. *M. W. Jeter*, Mathematical Programming (1986)
103. *M. Altman*, A Unified Theory of Nonlinear Operator and Evolution Equations with Applications (1986)
104. *A. Verschoren*, Relative Invariants of Sheaves (1987)
105. *R. A. Usmani*, Applied Linear Algebra (1987)
106. *P. Blass and J. Lang*, Zariski Surfaces and Differential Equations in Characteristic $p > 0$ (1987)
107. *J. A. Reneke et al.*, Structured Hereditary Systems (1987)
108. *H. Busemann and B. B. Phadke*, Spaces with Distinguished Geodesics (1987)
109. *R. Harte*, Invertibility and Singularity for Bounded Linear Operators (1988)
110. *G. S. Ladde et al.*, Oscillation Theory of Differential Equations with Deviating Arguments (1987)
111. *L. Dudkin et al.*, Iterative Aggregation Theory (1987)
112. *T. Okubo*, Differential Geometry (1987)

113. *D. L. Stancl and M. L. Stancl*, Real Analysis with Point-Set Topology (1987)
114. *T. C. Gard*, Introduction to Stochastic Differential Equations (1988)
115. *S. S. Abhyankar*, Enumerative Combinatorics of Young Tableaux (1988)
116. *H. Strade and R. Farnsteiner*, Modular Lie Algebras and Their Representations (1988)
117. *J. A. Huckaba*, Commutative Rings with Zero Divisors (1988)
118. *W. D. Wallis*, Combinatorial Designs (1988)
119. *W. Więsław*, Topological Fields (1988)
120. *G. Karpilovsky*, Field Theory (1988)
121. *S. Caenepeel and F. Van Oystaeyen*, Brauer Groups and the Cohomology of Graded Rings (1989)
122. *W. Kozlowski*, Modular Function Spaces (1988)
123. *E. Lowen-Colebunders*, Function Classes of Cauchy Continuous Maps (1989)
124. *M. Pavel*, Fundamentals of Pattern Recognition (1989)
125. *V. Lakshmikantham et al.*, Stability Analysis of Nonlinear Systems (1989)
126. *R. Sivaramakrishnan*, The Classical Theory of Arithmetic Functions (1989)
127. *N. A. Watson*, Parabolic Equations on an Infinite Strip (1989)
128. *K. J. Hastings*, Introduction to the Mathematics of Operations Research (1989)
129. *B. Fine*, Algebraic Theory of the Bianchi Groups (1989)
130. *D. N. Dikranjan et al.*, Topological Groups (1989)
131. *J. C. Morgan II*, Point Set Theory (1990)
132. *P. Biler and A. Witkowski*, Problems in Mathematical Analysis (1990)
133. *H. J. Sussmann*, Nonlinear Controllability and Optimal Control (1990)
134. *J.-P. Florens et al.*, Elements of Bayesian Statistics (1990)
135. *N. Shell*, Topological Fields and Near Valuations (1990)
136. *B. F. Doolin and C. F. Martin*, Introduction to Differential Geometry for Engineers (1990)
137. *S. S. Holland, Jr.*, Applied Analysis by the Hilbert Space Method (1990)
138. *J. Okniński*, Semigroup Algebras (1990)
139. *K. Zhu*, Operator Theory in Function Spaces (1990)
140. *G. B. Price*, An Introduction to Multicomplex Spaces and Functions (1991)
141. *R. B. Darst*, Introduction to Linear Programming (1991)
142. *P. L. Sachdev*, Nonlinear Ordinary Differential Equations and Their Applications (1991)
143. *T. Husain*, Orthogonal Schauder Bases (1991)
144. *J. Foran*, Fundamentals of Real Analysis (1991)
145. *W. C. Brown*, Matrices and Vector Spaces (1991)
146. *M. M. Rao and Z. D. Ren*, Theory of Orlicz Spaces (1991)
147. *J. S. Golan and T. Head*, Modules and the Structures of Rings (1991)
148. *C. Small*, Arithmetic of Finite Fields (1991)
149. *K. Yang*, Complex Algebraic Geometry (1991)
150. *D. G. Hoffman et al.*, Coding Theory (1991)
151. *M. O. González*, Classical Complex Analysis (1992)

152. *M. O. González*, Complex Analysis (1992)
153. *L. W. Baggett*, Functional Analysis (1992)
154. *M. Sniedovich*, Dynamic Programming (1992)
155. *R. P. Agarwal*, Difference Equations and Inequalities (1992)
156. *C. Brezinski*, Biorthogonality and Its Applications to Numerical Analysis (1992)
157. *C. Swartz*, An Introduction to Functional Analysis (1992)
158. *S. B. Nadler, Jr.*, Continuum Theory (1992)
159. *M. A. Al-Gwaiz*, Theory of Distributions (1992)
160. *E. Perry*, Geometry: Axiomatic Developments with Problem Solving (1992)
161. *E. Castillo and M. R. Ruiz-Cobo*, Functional Equations and Modelling in Science and Engineering (1992)
162. *A. J. Jerri*, Integral and Discrete Transforms with Applications and Error Analysis (1992)
163. *A. Charlier et al.*, Tensors and the Clifford Algebra (1992)
164. *P. Biler and T. Nadzieja*, Problems and Examples in Differential Equations (1992)
165. *E. Hansen*, Global Optimization Using Interval Analysis (1992)
166. *S. Guerre-Delabrière*, Classical Sequences in Banach Spaces (1992)
167. *Y. C. Wong*, Introductory Theory of Topological Vector Spaces (1992)
168. *S. H. Kulkarni and B. V. Limaye*, Real Function Algebras (1992)
169. *W. C. Brown*, Matrices Over Commutative Rings (1993)
170. *J. Loustau and M. Dillon*, Linear Geometry with Computer Graphics (1993)
171. *W. V. Petryshyn*, Approximation-Solvability of Nonlinear Functional and Differential Equations (1993)
172. *E. C. Young*, Vector and Tensor Analysis: Second Edition (1993)
173. *T. A. Bick*, Elementary Boundary Value Problems (1993)
174. *M. Pavel*, Fundamentals of Pattern Recognition: Second Edition (1993)
175. *S. A. Albeverio et al.*, Noncommutative Distributions (1993)
176. *W. Fulks*, Complex Variables (1993)
177. *M. M. Rao*, Conditional Measures and Applications (1993)
178. *A. Janicki and A. Weron*, Simulation and Chaotic Behavior of a-Stable Stochastic Processes (1994)
179. *P. Neittaanmäki and D. Tiba*, Optimal Control of Nonlinear Parabolic Systems (1994)
180. *J. Cronin*, Differential Equations: Introduction and Qualitative Theory, Second Edition (1994)
181. *S. Heikkilä and V. Lakshmikantham*, Monotone Iterative Techniques for Discontinuous Nonlinear Differential Equations (1994)
182. *X. Mao*, Exponential Stability of Stochastic Differential Equations (1994)
183. *B. S. Thomson*, Symmetric Properties of Real Functions (1994)
184. *J. E. Rubio*, Optimization and Nonstandard Analysis (1994)
185. *J. L. Bueso et al.*, Compatibility, Stability, and Sheaves (1995)
186. *A. N. Michel and K. Wang*, Qualitative Theory of Dynamical Systems (1995)
187. *M. R. Darnel*, Theory of Lattice-Ordered Groups (1995)

188. *Z. Naniewicz and P. D. Panagiotopoulos*, Mathematical Theory of Hemivariational Inequalities and Applications (1995)
189. *L. J. Corwin and R. H. Szczarba*, Calculus in Vector Spaces: Second Edition (1995)
190. *L. H. Erbe et al.*, Oscillation Theory for Functional Differential Equations (1995)
191. *S. Agaian et al.*, Binary Polynomial Transforms and Nonlinear Digital Filters (1995)
192. *M. I. Gil'*, Norm Estimations for Operation-Valued Functions and Applications (1995)
193. *P. A. Grillet*, Semigroups: An Introduction to the Structure Theory (1995)
194. *S. Kichenassamy*, Nonlinear Wave Equations (1996)
195. *V. F. Krotov*, Global Methods in Optimal Control Theory (1996)
196. *K. I. Beidar et al.*, Rings with Generalized Identities (1996)
197. *V. I. Arnautov et al.*, Introduction to the Theory of Topological Rings and Modules (1996)
198. *G. Sierksma*, Linear and Integer Programming (1996)
199. *R. Lasser*, Introduction to Fourier Series (1996)
200. *V. Sima*, Algorithms for Linear-Quadratic Optimization (1996)
201. *D. Redmond*, Number Theory (1996)
202. *J. K. Beem et al.*, Global Lorentzian Geometry: Second Edition (1996)
203. *M. Fontana et al.*, Prüfer Domains (1997)
204. *H. Tanabe*, Functional Analytic Methods for Partial Differential Equations (1997)
205. *C. Q. Zhang*, Integer Flows and Cycle Covers of Graphs (1997)
206. *E. Spiegel and C. J. O'Donnell*, Incidence Algebras (1997)
207. *B. Jakubczyk and W. Respondek*, Geometry of Feedback and Optimal Control (1998)
208. *T. W. Haynes et al.*, Fundamentals of Domination in Graphs (1998)
209. *T. W. Haynes et al.*, eds., Domination in Graphs: Advanced Topics (1998)
210. *L. A. D'Alotto et al.*, A Unified Signal Algebra Approach to Two-Dimensional Parallel Digital Signal Processing (1998)
211. *F. Halter-Koch*, Ideal Systems (1998)
212. *N. K. Govil et al.*, eds., Approximation Theory (1998)
213. *R. Cross*, Multivalued Linear Operators (1998)
214. *A. A. Martynyuk*, Stability by Liapunov's Matrix Function Method with Applications (1998)
215. *A. Favini and A. Yagi*, Degenerate Differential Equations in Banach Spaces (1999)
216. *A. Illanes and S. Nadler*, Jr., Hyperspaces: Fundamentals and Recent Advances (1999)
217. *G. Kato and D. Struppa*, Fundamentals of Algebraic Microlocal Analysis (1999)
218. *G. X.-Z. Yuan*, KKM Theory and Applications in Nonlinear Analysis (1999)
219. *D. Motreanu and N. H. Pavel*, Tangency, Flow Invariance for Differential Equations, and Optimization Problems (1999)
220. *K. Hrbacek and T. Jech*, Introduction to Set Theory, Third Edition (1999)
221. *G. E. Kolosov*, Optimal Design of Control Systems (1999)

222. *N. L. Johnson*, Subplane Covered Nets (2000)
223. *B. Fine and G. Rosenberger*, Algebraic Generalizations of Discrete Groups (1999)
224. *M. Väth, Volterra* and Integral Equations of Vector Functions (2000)
225. *S. S. Miller and P. T. Mocanu*, Differential Subordinations (2000)
226. *R. Li et al.*, Generalized Difference Methods for Differential Equations: Numerical Analysis of Finite Volume Methods (2000)
227. *H. Li and F. Van Oystaeyen*, A Primer of Algebraic Geometry (2000)
228. *R. P. Agarwal*, Difference Equations and Inequalities: Theory, Methods, and Applications, Second Edition (2000)
229. *A. B. Kharazishvili*, Strange Functions in Real Analysis (2000)
230. *J. M. Appell et al.*, Partial Integral Operators and Integro-Differential Equations (2000)
231. *A. I. Prilepko et al.*, Methods for Solving Inverse Problems in Mathematical Physics (2000)
232. *F. Van Oystaeyen*, Algebraic Geometry for Associative Algebras (2000)
233. *D. L. Jagerman*, Difference Equations with Applications to Queues (2000)
234. *D. R. Hankerson et al.*, Coding Theory and Cryptography: The Essentials, Second Edition, Revised and Expanded (2000)
235. *S. Dascalescu et al.*, Hopf Algebras: An Introduction (2001)
236. *R. Hagen et al.*, C*-Algebras and Numerical Analysis (2001)
237. *Y. Talpaert*, Differential Geometry: With Applications to Mechanics and Physics (2001)
238. *R. H. Villarreal*, Monomial Algebras (2001)
239. *A. N. Michel et al.*, Qualitative Theory of Dynamical Systems: Second Edition (2001)
240. *A. A. Samarskii*, The Theory of Difference Schemes (2001)
241. *J. Knopfmacher and W.-B. Zhang*, Number Theory Arising from Finite Fields (2001)
242. *S. Leader*, The Kurzweil-Henstock Integral and Its Differentials (2001)
243. *M. Biliotti et al.*, Foundations of Translation Planes (2001)
244. *A. N. Kochubei*, Pseudo-Differential Equations and Stochastics over Non-Archimedean Fields (2001)
245. *G. Sierksma*, Linear and Integer Programming: Second Edition (2002)
246. *A. A. Martynyuk*, Qualitative Methods in Nonlinear Dynamics: Novel Approaches to Liapunov's Matrix Functions (2002)
247. B. G. Pachpatte, Inequalities for Finite Difference Equations (2002)
248. *A. N. Michel and D. Liu*, Qualitative Analysis and Synthesis of Recurrent Neural Networks (2002)
249. *J. R. Weeks*, The Shape of Space: Second Edition (2002)
250. *M. M. Rao and Z. D. Ren*, Applications of Orlicz Spaces (2002)
251. *V. Lakshmikantham and D. Trigiante*, Theory of Difference Equations: Numerical Methods and Applications, Second Edition (2002)
252. *T. Albu*, Cogalois Theory (2003)
253. *A. Bezdek*, Discrete Geometry (2003)
254. *M. J. Corless and A. E. Frazho*, Linear Systems and Control: An Operator Perspective (2003)

255. *I. Graham and G. Kohr*, Geometric Function Theory in One and Higher Dimensions (2003)
256. *G. V. Demidenko and S. V. Uspenskii*, Partial Differential Equations and Systems Not Solvable with Respect to the Highest-Order Derivative (2003)
257. *A. Kelarev*, Graph Algebras and Automata (2003)
258. *A. H. Siddiqi*, Applied Functional Analysis: Numerical Methods, Wavelet Methods, and Image Processing (2004)
259. *F. W. Steutel and K. van Harn*, Infinite Divisibility of Probability Distributions on the Real Line (2004)
260. *G. S. Ladde and M. Sambandham*, Stochastic versus Deterministic Systems of Differential Equations (2004)
261. *B. J. Gardner and R. Wiegandt*, Radical Theory of Rings (2004)
262. *J. Haluska*, The Mathematical Theory of Tone Systems (2004)
263. *C. Menini and F. Van Oystaeyen*, Abstract Algebra: A Comprehensive Treatment (2004)
264. *E. Hansen and G. W. Walster*, Global Optimization Using Interval Analysis: Second Edition, Revised and Expanded (2004)
265. *M. M. Rao*, Measure Theory and Integration: Second Edition, Revised and Expanded (2004)
266. *W. J. Wickless*, A First Graduate Course in Abstract Algebra (2004)
267. *R. P. Agarwal, M. Bohner, and W-T Li*, Nonoscillation and Oscillation Theory for Functional Differential Equations (2004)
268. *J. Galambos and I. Simonelli*, Products of Random Variables: Applications to Problems of Physics and to Arithmetical Functions (2004)
269. *Walter Ferrer and Alvaro Rittatore*, Actions and Invariants of Algebraic Groups (2005)
270. *Christof Eck, Jiri Jarusek, and Miroslav Krbec*, Unilateral Contact Problems: Variational Methods and Existence Theorems (2005)

Unilateral Contact Problems

Variational Methods and Existence Theorems

Christof Eck
Universität Erlangen
Erlangen, Germany

Jiří Jarušek
Academy of Sciences of the Czech Republic
Prague, Czech Republic

Miroslav Krbec
Academy of Sciences of the Czech Republic
Prague, Czech Republic

CRC Press
Taylor & Francis Group
Boca Raton London New York

CRC Press is an imprint of the
Taylor & Francis Group, an **informa** business
A CHAPMAN & HALL BOOK

CRC Press
Taylor & Francis Group
6000 Broken Sound Parkway NW, Suite 300
Boca Raton, FL 33487-2742

First issued in paperback 2019

© 2005 by Taylor & Francis Group, LLC
CRC Press is an imprint of Taylor & Francis Group, an Informa business

No claim to original U.S. Government works

ISBN-13: 978-1-57444-629-6 (hbk)
ISBN-13: 978-0-367-39322-9 (pbk)

This book contains information obtained from authentic and highly regarded sources. Reasonable efforts have been made to publish reliable data and information, but the author and publisher cannot assume responsibility for the validity of all materials or the consequences of their use. The authors and publishers have attempted to trace the copyright holders of all material reproduced in this publication and apologize to copyright holders if permission to publish in this form has not been obtained. If any copyright material has not been acknowledged please write and let us know so we may rectify in any future reprint.

Except as permitted under U.S. Copyright Law, no part of this book may be reprinted, reproduced, transmitted, or utilized in any form by any electronic, mechanical, or other means, now known or hereafter invented, including photocopying, microfilming, and recording, or in any information storage or retrieval system, without written permission from the publishers.

For permission to photocopy or use material electronically from this work, please access www.copyright.com (http://www.copyright.com/) or contact the Copyright Clearance Center, Inc. (CCC), 222 Rosewood Drive, Danvers, MA 01923, 978-750-8400. CCC is a not-for-profit organization that provides licenses and registration for a variety of users. For organizations that have been granted a photocopy license by the CCC, a separate system of payment has been arranged.

Trademark Notice: Product or corporate names may be trademarks or registered trademarks, and are used only for identification and explanation without intent to infringe.

Library of Congress Card Number 2004064972

Library of Congress Cataloging-in-Publication Data

Eck, Christof.
Unilateral contact problems : variational methods and existence theorems / by Christof Eck, Jirí Janusĕk, and Miroslav Krbec.
p. cm.
Includes bibliographical references and index.
ISBN 1-57444-629-0 (alk. paper)
1. Differential equations, Partial. 2. Contact mechanics--Mathematical models. 3. Friction--Mathematical models. I. Janusĕk, Jirí. II. Krbec, Miroslav, 1950- III. Title.

QA377.E33 2005
531--dc22 2004064972

Visit the Taylor & Francis Web site at
http://www.taylorandfrancis.com

and the CRC Press Web site at
http://www.crcpress.com

Dedicated to the memory of Jindřich Nečas

Contents

Preface

The aim of this book is to present the contemporary state of the art for the mathematical analysis of contact problems with friction and a major part of the analysis for dynamic contact problems without friction. The main emphasis is put on the classical contact model, that is, a unilateral contact condition and the Coulomb friction law. The simplified normal-compliance approach—where a penetration of the bodies is allowed and the unilateral contact law is replaced by a constitutive law for the contact stress as function of the inter-penetration—is sometimes used as an approximation of the classical laws.

A significant part of the monograph emerged from research activities carried out by the authors during the past decade. This part deals with an approach that proved to be fruitful in many different situations. It starts from thin estimates of possible solutions and is based on an approximation of the problem by an auxiliary one, the proof of a moderate partial regularity of the solution to the approximate problem, and a passage to the original problem. The proof of the regularity of solutions often makes use of the shift technique developed by Fichera. This technique is perhaps not as well known in the literature as it deserves, but the authors believe that it is a very important tool, which may be useful not only for contact problems but also for other nonlinear problems with limited regularity. A careful attention is paid to quantification and precise results to get optimal bounds in sufficient conditions for existence theorems. A part of the content has been already published in several research articles.

Contact problems lie with the fields where the progress heavily depends on the interaction between "pure" and "applied" parts of mathematics. The theory especially relies on advanced results in various parts of functional analysis, in particular in the contemporary theory of function spaces; therefore we have included an exposition of relevant material.

The authors tried to give a self-contained survey, permitted by the extent. We hope that the book will be a useful tool for scientists working on the analysis of contact problems with or without friction and also for numerical analysts and engineers working on the numerical approximation of contact problems, and that it will also promote further research in this area.

During the research that led to this book and its preparation we appreciated help of many people. We wish to thank Wolfgang Wendland (Stuttgart), who initiated and continuously supported our cooperation, Michelle Schatzman (Lyon) for her careful proofreading and corrections of an essential part of the book, Tomáš Roubíček, Ivan Hlaváček, Jaroslav

Haslinger, (Prague), Hans Triebel and Hans-Jürgen Schmeisser (Jena) for valuable discussions, and of course many others colleagues and friends around the globe, too many to list without forgetting somebody. The support of the Grant Agency of the Academy of Sciences of the Czech Republic (project A 107 50 05), of the Grant Agency of the Czech Republic (project 101/01/1201) and of the DAAD and the DFG is gratefully acknowledged. Last but not least we appreciate very much the efficient collaboration with Kevin Sequiera, Jessica Vakili, and Tao Woolfe from Taylor and Francis Books.

We dedicate this book to the memory of Prof. Jindřich Nečas, one of the leading persons worldwide not only in this field, one of our prominent teachers, who belongs to initiators of the research in this area. His ideas and devotion to mathematics were often a key inspiration for mathematicians around him. He originated and/or substantially developed many of the techniques and procedures used in this book. Prof. Jindřich Nečas passed away last year and we wish to express here our deep gratitude for his constant support and encouragement.

Prague and Erlangen C.E., J.J., and M.K.

Chapter 1

Introduction

When one or several solid deformable bodies move, they may come into contact and forces are transmitted through their common area of contact. These forces have usually two components: the normal component prevents the inter-penetration of the bodies, and the tangential component is created by friction. A contact problem is described by a system of partial differential equations which describe the motion and deformation of the bodies, together with the boundary conditions modeling contact and friction forces.

Contact problems have applications to many fields of solid mechanics: in machine dynamics and manufacturing, contact problems arise when two parts hit one another; in many areas of mechanical engineering, a contact problem arises when cracks open and close. During an earthquake, contacts may take place between different parts of a building or different buildings, and between buildings and the earth. Due to the lower friction the solitary buildings usually survive an earthquake remarkably better than the buildings being in some contact with other ones. Therefore, neglecting contact and friction in a computer simulation of the motion of large structures during an earthquake may lead to false predictions. An earthquake itself can be seen as a gigantic contact problem between different lithospheric plates of the earth's surface, and indeed there are earthquake models which are precisely based on the build-up of tension in faults, followed by its release when an appropriate threshold is crossed, cf. e.g. [72]. Contact problems also occur in bio-mechanics; for instance a hip joint, be it natural or artificial, is subject to contact, and its wear depends on contact.

The normal contact forces are basically prescribed by the so-called *Signorini conditions,* and there is not much room for anything else, since they are geometrical and they just account for the non-interpenetration condition via a variational—or virtual work—condition. Friction conditions are quite another story: the phenomenologically (not mathematically) simplest and most popular conditions are given by the *Coulomb law of friction.* This law introduced in 1781 (see [37]) is also in good harmony with practical experience.

The existence of the static equilibrium of an elastic body in contact with an undeformable support and without friction has been proved by Fichera in 1964 [53]; the proof is based on the fact that this equilibrium is the minimum of a convex energy functional, and the solution of this problem led to the general concept of variational inequalities.

If the normal traction in the friction law of the Coulomb type is assumed to be a given function, then the problem is called a *contact problem with given friction*, or *the Tresca friction problem*. Duvaut and Lions [41] proved that the static contact problem with given friction is again a convex minimization problem, easily solvable by standard techniques. This fact was presented to a wider public in their monograph [42].

The analytical situation is much more complicated when Coulomb friction is included: convexity is lost, there are no good compactness properties, and therefore attempts to solve contact problems by standard tools from the analysis of optimization problems failed.

Panagiotopoulos [116] proposed an iterative algorithm for contact problems with Coulomb friction using the contact problem with given friction as an auxiliary problem. In each step of the iteration, he solves a contact problem with given friction, with normal force taken from the previous step, possibly by a suitable relaxation procedure.

A major progress is due to Nečas, Jarušek and Haslinger [113] who found a static equilibrium for an infinite elastic two-dimensional strip with Coulomb friction at the boundary. The proof used Tikhonov's fixed point theorem, showing that Panagiotopoulos' intuition can be turned into a mathematically sound proof. A crucial element of the proof was a regularity result obtained by the method of tangential translations (shifts), introduced to a wider public by Agmon, Douglis and Nirenberg [3] and into the mathematical mechanics by Fichera [54]. The result is valid only if the coefficient of friction is small enough and it turned out that the method of proof became the basis of many other results in frictional contact, including many of the results described in this book.

The result from [113] was extended to more general geometries, inhomogeneous materials, spatially varying material properties, semicoercive static problems without any Dirichlet boundary condition, and simple models of two-body contact problems in [73] and [74]. The papers [44] and [45] use the penalty method instead of a fixed point principle; they lead to essential generalizations of the requirements on the coefficient of friction. In particular, the coefficient of friction was allowed to depend on the solution; this dependence can be used to model the difference between the coefficients of friction of slip and friction of stick. Optimal bounds were derived for the "admissible magnitudes" of coefficients of friction with the help of special trace theorems for the solutions of the equations of elasticity. There are not only results pertaining to contact of a body with a rigid support, but also results for the contact between two bodies.

It is physically more realistic to consider dynamic than static models of contact problems. Here, additional complications arise, due to the hyperbolic nature of elastodynamics. The analysis is indeed so difficult that mathematicians often chose to replace the non-penetration constraint on displacements by an analogous constraint on velocities; this approximation is used in Duvaut and Lions' monograph [42] and it enabled them to solve contact problems in elastodynamics with a given time-independent friction force. This result has been extended by Jarušek to the case of a time-dependent friction force [79].

The first construction of a solution of a physically well posed dynamic Signorini problem was obtained by Amerio and Prouse [5] for a vibrating string. Despite considerable efforts by mathematicians, no significant results have been obtained for elasticity in dimensions larger than 1. We will describe in this book some important results from [127] and [15] in the case of the contact of a string with a concave obstacle and from [129] when the string may contact the obstacle at points, which may also be at the boundary. The situation is in fact quite different: for point-like obstacles, the conservation of the energy is a consequence of the equations. In the case of a continuous obstacle, it has to be imposed. Schatzman introduced a local conservation of energy principle and proved uniqueness under that condition [127].

Mathematicians from the Italian school studied the contact of two elastic strings and different types of response which are not completely elastic (which may occur e.g. for some gluing or semi-gluing supports), cf. e.g. [34], [33]. In higher space dimension, Kim has proved the existence of a weak solution of the wave equation with contact at the boundary; this solution satisfies an energy inequality [91]. Maruo has proved the existence result for unilateral contact problems where the space of admissible functions is a subset of the space of continuous functions [108]. This result, although disregarding the kind of response, suits for spatially polyharmonic problems, in particular for contact problems of plates. Lebeau and Schatzman prove the existence of a strong solution of the wave equation in a halfspace with contact at the boundary and conservation of energy [100]; the result essentially depends on this special geometric assumption.

Since the mathematical analysis of hyperbolic problems with contact at the boundary is so hard, an alternative has emerged: to consider viscoelastic materials instead of elastic materials, in the hope that the viscosity will have enough of a smoothing effect to allow for a solution of the contact problem. Existence results for contact with or without friction and different viscoelastic materials have been given in [76], [77], [78], [79] and [86].

Results on the frictional problem with unknown friction force are available only when the Signorini condition is formulated in terms of velocities rather than displacements. The first existence result in that case has been given in [82], and, as in the static case, it holds only for sufficiently small coefficients of friction. The paper [83] gives an explicit formula for this upper bound when the viscoelastic material has short-memory, i.e. it is a Kelvin-Voigt material; this bound depends only on the viscous part of the constitutive law. However, the upper bound of [83] is very restrictive and a better one has been found in [46] and [85] for an isotropic viscous material.

Another way out of the mathematical difficulties pertaining to contact problems has been proposed by some authors: non-local friction (starting from [40]) or normal compliance models. The former model compactifies the friction term by mollifying the normal force in it, while the tangential one remain mostly unchanged (despite no physical base for such different treatment of the components of the boundary traction has been presented). The latter model has received much attention and many papers on dynamic contact problems as well as some papers on static contact problems have used it: [11], [10], [93], [94], [106] and [107]. Both kinds of models involve one or several small parameters and can be physically justified with the help of a micro-scale analysis. However, the micro-scale analysis fixes the size of the small parameter to be of the order of magnitude of the roughness of the boundary, i.e. extremely small. We do not have any physical measurement of the size of these small parameters; therefore, we may well wonder how the solution depends on them. A formal passage to the limit in the normal compliance model gives a contact problem with Signorini contact condition and Coulomb friction law; therefore, the most important qualitative questions about the normal compliance models are essentially equivalent to the analysis of Coulomb friction: if the solution of a normal compliance model does not depend much on the small parameter involved in this model, it means that some kind of penalty approximation to Coulomb friction converges in practice.

One bad feature of normal compliance models is that the inter-penetration is not bounded; any modification that would enforce a bound on the inter-penetration seems to destroy the compactness of the penalty describing the compliance. Therefore such a modification leads to a much more difficult model than the originally formulated contact problem with Signorini condition and Coulomb friction law. Nevertheless, this approach, which has not been yet used in mathematical papers, could be an interesting way of describing the influence of the asperities.

In our opinion, and in view of all the difficulties of the other models, Signorini contact conditions together with Coulomb friction still remain the best models for contact with friction on a macroscopic scale.

Friction is a source of heat which cannot be always neglected. If the bodies have a large contact area, or the process is fast, the thermal effects become significant, and this is the reason why it is important to study the coupling of contact with the heat equation. The first results in this direction are presented for the normal compliance approach and in [47] with Signorini contact conditions and Coulomb friction law. There is a new difficulty coming in: we must cope with the quadratic growth of the heat source due to the combination of friction, deformation and viscosity. In [84], [48] and [81], this problem was handled by assuming that the heat diffusion coefficient grew fast enough with the temperature.

Many important problems remain unsolved. No satisfactorily solvable model is available for the dynamic contact of two viscoelastic bodies involving expected slip phenomena. Here the Signorini condition needs a suitable reformulation. Solving the dynamic frictional contact with Signorini conditions in displacements remains an important task. We do not expect uniqueness in any type of unilateral contact problem with Coulomb friction. A hint in the direction of non-uniqueness is the expected instability of friction with contact at the boundary. Until recently, there were only few published counterexamples, concerning a finite number of degrees of freedom and a quasi-static problem, cf. e.g. [95], [16]. In the last two years, counterexamples have been derived also for the infinite dimensional case. The most important ones are those by P. Hild [70], where it is shown that for the static version of the problem on a triangular domain in $\mathbb{R}^2$, a continuum of solutions of a certain type can be constructed provided the coefficient of friction is greater than 1. The magnitude of the coefficients of friction for which the existence of solutions is proved is every time below 1. Further important development in this field must be expected in the very close future. The most recent result we are informed about is [17], where two counterexamples of continuum of solutions for a static two-dimensional unilateral contact problem with Coulomb friction are considered: one for an unbounded and the other for a bounded domain. Both examples are constructed for any non-zero magnitude of the coefficient of friction. The authors expect a long and extensive discussion of mathematicians as well as of physicists and engineers after their publishing. The question of which of these solutions will turn out to be the most useful for practice and what additional features it possesses will naturally occur.

In this book, we present in a comprehensive manner the state of the art in the analysis of contact problems with Signorini conditions and Coulomb friction laws. The cases studied include the static, quasi-static and dynamic problems and diverse elastic and viscoelastic materials with rather general constitutive laws. The content is partly based on the authors' research of the last twenty years.

We overcome the above mentioned analytical difficulties with the help of a chain of auxiliary problems with simpler structure. The most useful are the contact problems with given friction, where the normal traction force in the friction law is assumed to be known, and the penalty approximation to the Signorini contact conditions. As mentioned above, the latter can be understood as an approximation of the contact law through a normal compliance model. The auxiliary models can be solved by standard techniques in functional analysis, and the original problem can be solved either by a fixed point argument in the case of given friction or by a limiting process in the case of the penalty approximation. However, these methods require additional information on the regularity of the traces on the part of the boundary where contact takes place: this is where the method of translations is employed.

The book is organized as follows: in Chapter 1, we give a short introduction to the theory of elasticity, we present the models of contact and friction and we give the mathematical formulation of the dynamic and static contact problems. Then we present a first existence proof using a fixed-point approach for a simplified static problem. The second chapter contains important definitions and results on function spaces, interpolation theorems, imbedding theorems, trace theorems, and fixed point theorems required later on. In Chapter 3, static and quasi-static problems are studied. Chapter 4 contains a summary of results on contact problems without friction or with given friction for various viscoelastic constitutive law and contact conditions formulated in velocities or in displacements. In Subsection 4.2.1, some contact problems for membranes are presented. The last chapter contains results on dynamic contact problems with contact conditions formulated in velocities and the Coulomb law of friction; we conclude with contact problems coupled with heat diffusion and generation of heat by friction.

1.1 Notations

We first introduce some important notations and conventions. Let Ω be a (bounded or unbounded) *domain* (an open connected set) in $\mathbb{R}^N$ with a boundary $\Gamma = \partial\Omega$. The vectors in $\mathbb{R}^m$ with $m \in \mathbb{N}$ and the m-dimensional vector fields are consistently denoted by bold letters. The components of a vector $\boldsymbol{u}$ are indicated by u_i. The scalar product in $\mathbb{R}^m$ will be consequently denoted by a dot and the Euclidean norm there by $|\cdot|$. For a matrix $\boldsymbol{M}$, the symbol $\boldsymbol{M}^\top$ denotes its transposition. For the unit matrix the symbol $\boldsymbol{Id}$ is used. For $y, z \in \mathbb{R} \cup \{\pm\infty\}$, $y < z$, the open interval is denoted by (y, z) and the closed one by $[y, z]$, with the usual extensions to intervals of mixed type. The symbol $\mathbb{R}_+$ stands for the interval $(0, +\infty)$ and $I_T = (0, T)$ for an appropriate time interval with $0 < T < +\infty$. If $u = u(t, \boldsymbol{x})$ is a function dependent on the time $t \in I_T$ and on the space variable $\boldsymbol{x} \in \Omega \subset \mathbb{R}^N$, then its (possibly weak) space derivatives are denoted by $\partial_i u \equiv \dfrac{\partial u}{\partial x_i}$, $i = 1, \ldots, N$, and the (possibly weak) time derivatives by dots, $\dot{u} = \dfrac{\partial u}{\partial t}$, $\ddot{u} = \dfrac{\partial^2 u}{\partial t^2}$, ... To be consistent in this notation, $\partial_z f$ will denote the partial derivative of a function f, smoothly dependent on a variable z, with respect to this variable. By $\dot{u}_+$ and $\dot{u}_-$ we denote the right and left time derivatives, respectively. We use the Einstein summation convention,

$$a_i b_i \equiv \sum_{i=1}^{r} a_i b_i,$$

with r denoting the range of the summation. Usually r is equal to the space dimension N. For two sets $M_1, M_2 \subset \mathbb{R}^N$ let

$$\operatorname{dist}(M_1, M_2) \equiv \inf_{\substack{\boldsymbol{x} \in M_1 \\ \boldsymbol{y} \in M_2}} |\boldsymbol{x} - \boldsymbol{y}|$$

be their distance. If $f : M \to \mathbb{R}$, $M \subset \mathbb{R}^m$, then $\operatorname{supp} f$, the closure of the set $\{\boldsymbol{x} \in M;\, f(\boldsymbol{x}) \neq 0\}$, is the support of f. In the sequel the bar over a subset of a topological space denotes its closure. For $z \in \mathbb{R}$, $[z]_+ = \max\{0, z\}$ is its *non-negative part* and $[z]_- = \max\{0, -z\}$ its *non-positive part*. For functions this relation is assumed to be pointwise. For two mappings $f : P \to Q$ and $g : Q \to R$, the symbol $g \circ f$ denotes their composition.

The notation of function spaces is the following: For an integer $\alpha \geq 0$ and a domain $\Omega \subset \mathbb{R}^N$ the space of functions $f : \Omega \to \mathbb{R}$ with bounded and continuous derivatives up to the order α is denoted by $C^\alpha(\Omega)$. It is endowed by the usual sup-norm. If α is not an integer, $\alpha = [\alpha] + \alpha'$ with $[\alpha] \in \mathbb{N}$ and $0 < \alpha' < 1$, then $C^\alpha(\Omega)$ is the space of functions in $C^{[\alpha]}(\Omega)$ whose derivatives of the order $[\alpha]$ are Hölder-continuous with the Hölder exponent α'. The space of bounded functions endowed with the sup norm is denoted by $B_0(M)$. Let $L_p(\Omega)$, $p \in [1, +\infty)$, denote the space of functions whose p-th power is integrable over Ω and $L_\infty(\Omega)$ be the space of essentially bounded functions on Ω; the definition has sense for any measurable $\Omega \subset \mathbb{R}^N$. In this introductory part of the book we shall use some basic spaces of Sobolev and Besov (Sobolev-Slobodetskii) type. We do not go into details here and treat them in a more general setting in Chapter 2. Some notation will be slightly different or more detailed in accordance with that usually used in the theory of function spaces. If $\alpha > 0$ is an integer and Ω a domain in $\mathbb{R}^N$, then $W_p^\alpha(\Omega)$ is the space of functions whose (possibly generalized) derivatives up to the order α belong to $L_p(\Omega)$. For a non-integer $\alpha > 0$ the symbol $W_p^\alpha(\Omega)$ denotes the corresponding *Sobolev-Slobodetskii* space; this is the subspace of $W_p^{[\alpha]}(\Omega)$ whose elements have the finite norm

$$\|u\|_{W_p^\alpha(\Omega)} \equiv \sqrt[p]{\|u\|^p_{W_p^{[\alpha]}(\Omega)} + \sum_{|\beta|=[\alpha]} \int_\Omega \int_\Omega \frac{\left|D^\beta u(\boldsymbol{x}) - D^\beta u(\boldsymbol{y})\right|^p}{|\boldsymbol{x} - \boldsymbol{y}|^{N+p\alpha'}}\, d\boldsymbol{x}\, d\boldsymbol{y}},$$

and the subspace is equipped with this norm. Here $\boldsymbol{\beta} = (\beta_1, \ldots, \beta_N) \in \mathbb{N}^N$ is a multiindex, $|\boldsymbol{\beta}| \equiv \sum_{i=1}^N |\beta_i|$ and $D^{\boldsymbol{\beta}} u \equiv \partial_1^{\beta_1} \cdots \partial_N^{\beta_N} u$. For $p = 2$ the space $W_2^\alpha(\Omega)$ will be denoted by $H^\alpha(\Omega)$. With $\mathring{W}_p^\alpha(\Omega)$ and $\mathring{H}^\alpha(\Omega)$ the corresponding spaces with vanishing traces are denoted. The Sobolev spaces with negative indices of differentiability are defined as the dual spaces $W_q^{-\alpha}(\Omega) = \big(\mathring{W}_p^\alpha(\Omega)\big)^*$ and $\mathring{W}_q^{-\alpha}(\Omega) = \big(W_p^\alpha(\Omega)\big)^*$ with p given by $\frac{1}{p} + \frac{1}{q} = 1$. The concept of vanishing traces and dual spaces is tackled in Section 2.8. At this moment let us just observe that integer values of $\alpha + \frac{1}{p}$ require a special attention. The corresponding definitions are also valid for $\mathring{H}^\alpha(\Omega)$ and $H^\alpha(\Omega)$, $\alpha < 0$. If the index of differentiability $\alpha = (\alpha_1, \alpha_2)$ in some of the definitions above has two components, then its first component refers to the time variable and the second one to the space variables. The spaces of functions defined on $M \subset \mathbb{R}^N$ with ranges in a Banach space B are denoted by $C^\alpha(M;B)$, $B_0(M;B)$, $L_p(M;B)$, $W_p^\alpha(M;B)$, $\mathring{W}_p^\alpha(M;B)$. The spaces of vector fields are sometimes also denoted by bold letters, $C^\alpha\big(\Omega;\mathbb{R}^N\big) = \boldsymbol{C}^\alpha(\Omega)$, $L_p\big(\Omega;\mathbb{R}^N\big) = \boldsymbol{L}_p(\Omega)$, etc. Let us observe that Ω will be often a *Lipschitz domain*, that is, the case when the boundary $\partial\Omega$ of Ω can be locally described as a graph of a Lipschitz continuous function and Ω lies on one side of $\partial\Omega$—see Section 1.6 for the formal definition and Subsection 2.7.1 in Chapter 2 (in particular Remark 2.7.2). For a Hilbert space $\mathscr{H}$ we denote the appropriate scalar product by $\langle \cdot, \cdot \rangle_{\mathscr{H}}$. Moreover, for a measurable set $M \subset \mathbb{R}^m$ or a measurable part M of the boundary $\Gamma = \partial\Omega$, $\langle \cdot, \cdot \rangle_M$ denotes the scalar product in $L_2(M)$ or in $\boldsymbol{L}_2(M)$. The unit normal vector of Γ at a point $\boldsymbol{x} \in \Gamma$ is denoted by $\boldsymbol{n}(\boldsymbol{x})$. The normal vector exists almost everywhere on Lipschitz boundaries and everywhere on C^1 boundaries; see also Section 1.7. If a vector field $\boldsymbol{u} = (u_1, \ldots, u_N)$ is defined on the boundary Γ of a domain Ω, then $u_n \equiv u_i n_i$ denotes its normal and $\boldsymbol{u_t} \equiv \boldsymbol{u} - u_n \boldsymbol{n}$ its tangential component. In order to simplify the notation, we use the same symbol for both a function defined on Ω and its trace defined on the boundary Γ. The symbol mes_Γ stands for the surface measure.

The standard notation for physical quantities used in this book is time variable t, space variable $\boldsymbol{x}$, displacement field $\boldsymbol{u}$, the stress tensor $\boldsymbol{\sigma}$ with components σ_{ij}, the strain tensor $\boldsymbol{\varepsilon}$ with components e_{ij} and components of the boundary traction $(\sigma_\Gamma)_i = \sigma_{ij} n_j$. The coefficient of friction will be denoted by $\mathfrak{F}$.

1.2 Linear elasticity

1.2.1 Deformation of bodies, the strain tensor

In this section we briefly tackle the mathematical description of deformations of solid bodies and the stresses created by deformations in linear elasticity. Readers familiar with linear elasticity may skip this section.

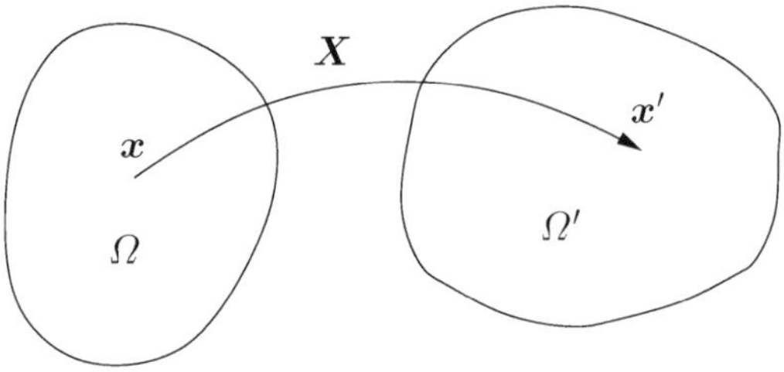

Figure 1.1: Deformation of a solid body.

Let us consider a body occupying in a reference configuration a domain $\Omega \subset \mathbb{R}^N$ of dimension $N = 2$ or $N = 3$. Every particle of the body is described by its position $\boldsymbol{x}$ in this reference configuration, and the body itself is identified with its reference configuration. A deformation of the body can be described by a one-to-one mapping

$$\boldsymbol{X} : \Omega \to \Omega'$$

from the reference configuration Ω to the deformed configuration Ω'. The function $\boldsymbol{X}$ maps any particle $x \in \Omega$ onto its position $\boldsymbol{X}(\boldsymbol{x}) \in \Omega'$ attained after the deformation. All information about the deformation is contained in the function $\boldsymbol{X}$. Instead of the map $\boldsymbol{X}$ we may also consider the *displacement field*

$$\boldsymbol{u}(\boldsymbol{x}) \equiv \boldsymbol{X}(\boldsymbol{x}) - \boldsymbol{x}.$$

We are interested in the local deformation occurring near a fixed material point $\boldsymbol{x} \in \Omega$. For simplicity we assume $\boldsymbol{X}$ to be smooth. As a first quantity measuring the local deformation we may consider the *deformation gradient*

$$\boldsymbol{D} \equiv \nabla \boldsymbol{X} = \{\partial_j \boldsymbol{X}_i\}_{i,j=1}^{N} = \boldsymbol{Id} + \nabla \boldsymbol{u}.$$

Some movements of a deformable body do not create deformations; these are translations and rotations of the body and any of their linear combinations. For translations the deformation gradient vanishes, but for rotations this is not true. Therefore, in order to have a suitable local measure for deformation, it is necessary to eliminate the contribution of rotations in the deformation gradient. This is done with the *polar decomposition*

$$\boldsymbol{D} = \boldsymbol{R}\boldsymbol{U}$$

with the *rotation tensor* $\boldsymbol{R}$ and the remainder $\boldsymbol{U}$ called *right stretch tensor*. The right stretch tensor is given by $\boldsymbol{U}^2 = \boldsymbol{D}^\top \boldsymbol{D}$. Alternatively we may use the decomposition

$$\boldsymbol{D} = \boldsymbol{V}\boldsymbol{R}$$

with the *left stretch tensor* $\boldsymbol{V}$ given by $\boldsymbol{V}^2 = \boldsymbol{D}\boldsymbol{D}^\top$. The right and left stretch tensors are suitable quantities to locally describe the deformation of a solid body; they are therefore used to formulate constitutive laws for different materials.

In linear elasticity the deformation gradient is assumed to be closed to the identity, and terms of order higher than one are neglected. In this case the right and left stretch tensors coincide,

$$\boldsymbol{U} = \boldsymbol{V} = \frac{1}{2}(\nabla \boldsymbol{X} + (\nabla \boldsymbol{X})^\top) = \boldsymbol{Id} + \frac{1}{2}(\nabla \boldsymbol{u} + (\nabla \boldsymbol{u})^\top).$$

The tensor

$$\varepsilon = \frac{1}{2}(\nabla \boldsymbol{u} + (\nabla \boldsymbol{u})^\top),$$

whose components are

$$e_{ij} = \frac{1}{2}(\partial_i u_j + \partial_j u_i), \quad i, j = 1, \dots, N,$$

is called the *linearized strain tensor*. It is the basic quantity used for formulation of constitutive laws in linear elasticity.

1.2.2 The stress tensor

Let us consider a $(N-1)$-dimensional hypersurface S dividing the body Ω into two parts Ω_1 and Ω_2; see Fig. 1.2. Then the part Ω_2 exerts upon Ω_1 a force along the common surface S. This force can be described by a force density $\sigma^{(1)}$ defined on the surface S. The value of this force density at a point $x \in S$ can be represented by a tensor $\boldsymbol{\sigma} \equiv \{\sigma_{ij}\}_{i,j=1}^{N}$, called the *stress tensor*, via

$$\sigma_i^{(1)}(\boldsymbol{x}) = \sigma_{ij}(\boldsymbol{x}) n_j^{(1)}(\boldsymbol{x}).$$

Here $\boldsymbol{n}^{(1)}$ denotes the normal vector of S pointing to the exterior of the domain $\Omega^{(1)}$. According to Newton's law "actio = reactio" the part Ω_2 exerts upon Ω_1 the same force with opposite direction. The corresponding force density is given by the same formula

$$\sigma_i^{(2)}(\boldsymbol{x}) = \sigma_{ij}(\boldsymbol{x}) n_j^{(2)}(\boldsymbol{x})$$

with the same stress tensor and the normal vector $\boldsymbol{n}^{(2)} = -\boldsymbol{n}^{(1)}$ pointing to the exterior of Ω^2. Note that all quantities here are calculated in the reference configuration and not in the deformed configuration. For material without internal rotations the stress tensor is symmetric, $\sigma_{ij} = \sigma_{ji}$; this is a consequence of the conservation of rotational momentum.

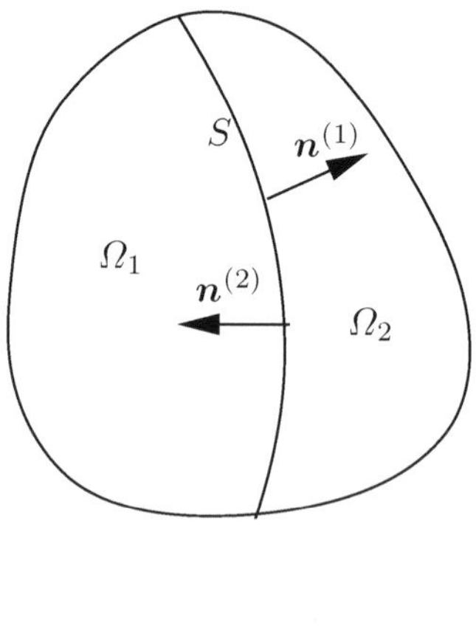

Figure 1.2: Definition of the stress tensor.

In order to model any given material, a constitutive law must be specified. Such laws are derived from experiments. For elastic materials the *constitutive law* is a relation between the strain tensor and the stress tensor. In linear elasticity this relation is assumed to be linear,

$$\sigma_{ij} = a_{ijk\ell} e_{k\ell}, \tag{1.2.1}$$

and the proportionality tensor $\{a_{ijk\ell}\}_{i,j,k,\ell=1}^N$ is called the *Hooke tensor.* Many materials of technical importance have a linear elastic behaviour in a wide range of loading; this is e.g. the case for most metals. Furthermore, the linear constitutive law can be considered as a linearization at a given configuration of a more general law. The Hooke tensor usually satisfies the conditions of *symmetry*

$$a_{ijk\ell} = a_{jik\ell} = a_{k\ell ij} \quad \text{for all } i, j, k, \ell = 1, \dots N, \tag{1.2.2}$$

it is elliptic,

$$a_{ijk\ell}\xi_{ij}\xi_{k\ell} \geq a_0 \xi_{ij}\xi_{ij} \tag{1.2.3}$$

for all symmetric tensors $\{\xi_{ij}\}_{i,j=1}^N$, i.e. satisfying $\xi_{ij} = \xi_{ji}$, and it is bounded,

$$a_{ijk\ell}\xi_{ij}\xi_{k\ell} \leq A_0 \xi_{ij}\xi_{ij} \tag{1.2.4}$$

for all symmetric $\{\xi_{ij}\}_{i,j=1}^N$. In a general situation the coefficients of the Hooke tensor may depend on the space variable $\boldsymbol{x}$; in this case the numbers a_0 and A_0 are assumed to be independent of $\boldsymbol{x}$. The tensor character of $\{a_{ijk\ell}\}$ implies that its eigenvalues do not depend on any isometry of the coordinate frame; in particular such a transformation does not change a_0 in (1.2.3) or A_0 in (1.2.4).

If the material's properties do not depend on the direction—the material is said to be *isotropic*—then the constitutive law is given by

$$\sigma_{ij} = \frac{E\nu}{(1+\nu)(1-2\nu)} \delta_{ij} \partial_k u_k + \frac{E}{1+\nu} e_{ij}; \tag{1.2.5}$$

the number E is called the *Young modulus* and the number ν is the *Poisson ratio.* If E and ν are constants, then the body is said to be *homogeneous.* Sometimes in the literature the coefficients in (1.2.5) are denoted by

$$\lambda = \frac{E\nu}{(1+\nu)(1-2\nu)} \quad \text{and} \quad \mu = \frac{E}{2+2\nu},$$

where λ and μ are the *Lamé constants.*

Besides elastic material we consider also some types of viscoelastic materials. A rather simple constitutive law of viscous type is given by the relation

$$\sigma_{ij} = \sigma_{ij}(\boldsymbol{u}, \dot{\boldsymbol{u}}) = a_{ijk\ell}^{(0)} e_{k\ell}(\boldsymbol{u}) + a_{ijk\ell}^{(1)} e_{k\ell}(\dot{\boldsymbol{u}}) \tag{1.2.6}$$

depending also on the *strain velocities* $e_{ij}(\dot{\boldsymbol{u}})$. Here $\{a^{(m)}_{ijk\ell}\}^N_{i,j,k,\ell=1}$, $m = 0, 1$, are tensors satisfying the above properties of symmetry, ellipticity and boundedness. Such a material is called *material with short memory.*

Another kind of viscosity involves *long time memory*. Its linearized form is as follows

$$\begin{aligned}\sigma_{ij} &= \sigma_{ij}(\boldsymbol{u}; t, \cdot)\\ &= a_{ijk\ell}e_{k\ell}(\boldsymbol{u}(t, \cdot)) + \int_0^t b_{ijk\ell}(t-s)e_{k\ell}(\boldsymbol{u}(s, \cdot))\, ds,\end{aligned} \tag{1.2.7}$$

where $b_{ijk\ell}$, the memory coefficients, satisfy a symmetry condition as in (1.2.2) and the coefficients

$$a^{\infty}_{ijk\ell} \equiv a_{ijk\ell} + \int_0^{+\infty} b_{ijk\ell}(s)\, ds, \quad i, j, k, \ell = 1, \ldots, N, \tag{1.2.8}$$

satisfy the conditions of ellipticity (1.2.3) and boundedness (1.2.4). Usually the memory coefficients behave like some negative powers of time in a neighbourhood of $t = 0$. Such memory is said to be singular. The integrability in (1.2.8) implies that the negative exponent is larger than -1. Later, we shall use in general a non-linear version of this concept.

1.2.3 The equations of linear elasticity

The differential equations of continuum mechanics are consequences of the physical conservation laws for mass, momentum, and energy. The equations of elasticity are derived from the energy balance. Let us consider an arbitrary time-independent subdomain $V \subset \Omega$ of the reference configuration of the elastic body. If $\boldsymbol{f}$ denotes the volume force acting inside the body (e.g. the gravitational force) and ρ is the (time-independent) density, then the conservation of momentum is formulated by

$$\frac{d}{dt}\left(\int_V \rho\, \dot{\boldsymbol{u}}(t, \boldsymbol{x})\, d\boldsymbol{x}\right) = \int_V \boldsymbol{f}\, d\boldsymbol{x} + \int_{\partial V} \sigma_{\partial V}(t, \boldsymbol{x})\, dt$$

with the boundary traction $\sigma_{\partial V}$ defined by the formula $(\sigma_{\partial V})_i = \sigma_{ij}n_j$. Using the Green formula this can be modified to

$$\int_V \rho\, \ddot{u}_i(t, \boldsymbol{x})\, d\boldsymbol{x} = \int_V f_i\, d\boldsymbol{x} + \int_V \partial_j \sigma_{ij}\, d\boldsymbol{x}, \quad i = 1, \ldots, N.$$

Here we assume $\boldsymbol{u}$ to be twice continuously differentiable. This formula is valid for every sufficiently smooth subdomain $V \subset \Omega$. Hence there holds the system of differential equations

$$\rho \ddot{u}_i - \partial_j \sigma_{ij}(\boldsymbol{u}) = f_i, \quad i = 1, \ldots, N. \tag{1.2.9}$$

Let us show that this is a hyperbolic system of equations of second order. It suffices to prove that the second part $-\partial_j \sigma_{ij}(\boldsymbol{u})$ is an elliptic operator. Let

$$A(\boldsymbol{u}, \boldsymbol{v}) \equiv \int_{\Omega} \sigma_{ij}(\boldsymbol{u}) e_{ij}(\boldsymbol{v})\, d\boldsymbol{x}$$

be the bilinear form of this operator. Due to the symmetry of the stress tensor the bilinear form is equivalent to $\int_{\Omega} \sigma_{ij}(\boldsymbol{u}) \partial_j v_i \, d\boldsymbol{x}$. Applying the Hooke law (1.2.1) yields

$$A(\boldsymbol{u}, \boldsymbol{v}) = \int_{\Omega} a_{ijk\ell} e_{k\ell}(\boldsymbol{u}) e_{ij}(\boldsymbol{v})\, d\boldsymbol{x}.$$

Since the Hooke tensor is bounded we immediately obtain

$$|A(\boldsymbol{u}, \boldsymbol{v})| \leq A_0 \|\nabla \boldsymbol{u}\|_{L_2(\Omega;\mathbb{R}^{N,N})} \|\nabla \boldsymbol{v}\|_{L_2(\Omega;\mathbb{R}^{N,N})}.$$

Due to the ellipticity condition (1.2.3) it holds

$$A(\boldsymbol{u}, \boldsymbol{u}) \geq \int_{\Omega} a_0 e_{ij}(\boldsymbol{u}) e_{ij}(\boldsymbol{u})\, d\boldsymbol{x}. \tag{1.2.10}$$

Hence A is elliptic, if

$$\mathfrak{E}_{\Omega} : \boldsymbol{u} \mapsto \int_{\Omega} e_{ij}(\boldsymbol{u}) e_{ij}(\boldsymbol{u})\, d\boldsymbol{x} \geq c_0 \|\boldsymbol{u}\|^2_{\boldsymbol{H}^1(\Omega)} \tag{1.2.11}$$

holds with a constant $c_0 > 0$. This important estimate is known as the *Korn inequality*. It will be proved in the next section.

1.2.4 The Korn inequality

The proof of relation (1.2.11) is performed in several steps. We start with the case of an infinite strip domain

$$Q_{(0,R)} := \mathbb{R}^{N-1} \times (0, R)$$

with $R > 0$. Let us denote the $\boldsymbol{H}^1(\Omega)$ seminorm by

$$\|\boldsymbol{u}\|'_{\boldsymbol{H}^1(\Omega)} = \|\nabla \boldsymbol{u}\|_{L_2(\Omega;\mathbb{R}^{N\times N})}.$$

Then the following relation is valid:

1.2.1 Theorem. *Let $\boldsymbol{u} \in \boldsymbol{H}^1(Q_{(0,R)})$ satisfy the boundary condition $\boldsymbol{u} = 0$ on $S_0 := \mathbb{R}^{N-1} \times \{0\}$. Then there holds the Korn inequality*

$$\mathfrak{E}_{\Omega}(\boldsymbol{u}) = \int_{\Omega} e_{ij}(\boldsymbol{u}) e_{ij}(\boldsymbol{u})\, d\boldsymbol{x} \geq \frac{1}{2} \|\boldsymbol{u}\|'^2_{\boldsymbol{H}^1(Q_{(0,R)})}. \tag{1.2.12}$$

Proof. In order to prove the result we extend the function u onto the strip $Q_{(-R,R)} := \mathbb{R}^{N-1} \times (-R, R)$ by

$$u_k(\boldsymbol{x}', -x_N) = -u_k(\boldsymbol{x}', x_N) \quad \text{for } k = 0, \dots, N-1,\ x_N > 0, \text{ and}$$
$$u_N(\boldsymbol{x}', -x_N) = u_N(\boldsymbol{x}', x_N) \quad \text{for } x_N > 0$$

with $\boldsymbol{x}' = (x_1, \dots, x_{N-1})^\top$. For this extension there holds

$$e_{ij}(\boldsymbol{u}; \boldsymbol{x}', x_N) e_{ij}(\boldsymbol{u}; \boldsymbol{x}', x_N) = e_{ij}(\boldsymbol{u}; \boldsymbol{x}', -x_N) e_{ij}(\boldsymbol{u}; \boldsymbol{x}', -x_N).$$

Consequently we have

$$\int_{Q_{(0,R)}} e_{ij}(\boldsymbol{u}) e_{ij}(\boldsymbol{u})\, d\boldsymbol{x} = \frac{1}{2} \int_{Q_{(-R,R)}} e_{ij}(\boldsymbol{u}) e_{ij}(\boldsymbol{u})\, d\boldsymbol{x}$$

and it suffices to prove the relation for the extended strip $Q_{(-R,R)}$. With the help of the partial Fourier transform

$$\mathscr{F}(u; \boldsymbol{\xi}', x_N) \equiv (2\pi)^{-(N-1)/2} \int_{\mathbb{R}^{N-1}} u(\boldsymbol{x}', x_N) \cdot e^{-i\boldsymbol{x}' \cdot \boldsymbol{\xi}'}\, d\boldsymbol{x}'$$

and the ($\boldsymbol{\xi}'$-dependent) Fourier coefficients

$$\mathfrak{a}_{k,m} = R^{-1/2} \int_{-R}^{R} \mathscr{F}(u_k; \boldsymbol{\xi}', x_N) \sin\Big(\frac{m\pi x_N}{R}\Big)\, dx_N,\ k = 1, \dots, N-1,$$
$$\mathfrak{b}_m = R^{-1/2} \int_{-R}^{R} \mathscr{F}(u_N; \boldsymbol{\xi}', x_N) \cos\Big(\frac{m\pi x_N}{R}\Big)\, dx_N,\ m \in \mathbb{N},$$
$$\mathfrak{b}_0 = (2R)^{-1/2} \int_{-R}^{R} \mathscr{F}(u_N; \boldsymbol{\xi}', x_N)\, dx_N,\ \mathfrak{a}_{k,0} = 0,\ k = 1, \dots, N-1,$$

we get

$$\begin{aligned} E_{(-R,R)}(\boldsymbol{u}) &= \int_{Q_{(-R,R)}} e_{ij}(\boldsymbol{u}) e_{ij}(\boldsymbol{u})\, d\boldsymbol{x} \\ &= \int_{Q_{(-R,R)}} \Bigg[\sum_{k=1}^{N} |\partial_k u_k|^2 + \frac{1}{2} \sum_{k=1}^{N-1} (\partial_k u_N + \partial_N u_k)^2 \\ &\quad + \frac{1}{4} \sum_{\substack{k,\ell=1, \\ k \neq \ell}}^{N-1} (\partial_k u_\ell + \partial_\ell u_k)^2 \Bigg]\, d\boldsymbol{x} \\ &= \int_{\mathbb{R}^{N-1}} \sum_{m=0}^{+\infty} \Bigg[\pi^2 R^{-2} m^2 |\mathfrak{b}_m|^2 + \sum_{k=1}^{N-1} \Big[\xi_k^2 |\mathfrak{a}_{k,m}|^2 \end{aligned}$$

$$+ \frac{1}{2}\big(\xi_k^2|\mathfrak{b}_m|^2 + \pi^2 R^{-2} m^2 |\mathfrak{a}_{k,m}|^2 + 2\pi R^{-1} m \operatorname{Re}\big(i\xi_k \mathfrak{b}_m \overline{\mathfrak{a}_{k,m}}\big)\big)\Big]$$

$$+ \frac{1}{4}\sum_{\substack{k,\ell=1,\\ k\neq\ell}}^{N-1} \big(\xi_k^2|\mathfrak{a}_{\ell,m}|^2 + \xi_\ell^2|\mathfrak{a}_{k,m}|^2 + 2\operatorname{Re}\big(\xi_k\xi_\ell i\mathfrak{a}_{k,m}\overline{i\mathfrak{a}_{\ell,m}}\big)\big)\Bigg] \cdot$$

$$\cdot\, d\boldsymbol{\xi}'\, dx_N .$$

Using the inequality

$$\frac{1}{2}\sum_{j=1}^{N}|z_j|^2 + \sum_{1\le j<k\le N}\operatorname{Re}\big(z_j\overline{z_k}\big) = \frac{1}{2}\Big|\sum_{j=1}^{N-1} z_j\Big|^2 \ge 0$$

for $z_j = i\xi_j\mathfrak{a}_{j,m}$, $j = 1,\dots,N-1$, and $z_N = \pi R^{-1} m b_m$ the theorem is proved. □

The just proved relation can be easily extended to halfspace domains of the type $Q = \mathbb{R}^{N-1}\times\mathbb{R}_+$ and to the whole space $\mathbb{R}^N$. In fact, the relation is valid for any width $R > 0$ of the strip. Considering an arbitrary field $\boldsymbol{u} \in \mathring{\boldsymbol{C}}^\infty(\mathbb{R}^N)$ for the whole space and $\boldsymbol{u} \in \boldsymbol{C}^\infty(Q)\cap\mathring{\boldsymbol{C}}^\infty(\mathbb{R}^N)$ for the halfspace, we can choose R such that $\sup \boldsymbol{u} \subset Q_{(0,R)}$ (after some possible translation of the support in the case of the full space). Then the Korn inequality for this strip yields the Korn inequality for the halfspace and for the whole domain. Since $\mathring{\boldsymbol{C}}^\infty(\mathbb{R}^N)$ is dense in $\boldsymbol{H}^1(\mathbb{R}^N)$ and $\boldsymbol{C}^\infty(Q)\cap\mathring{\boldsymbol{C}}^\infty(\mathbb{R}^N)$ is dense in $\boldsymbol{H}^1(Q)$, the inequality must be valid for the halfspace Q and the whole space $\mathbb{R}^N$ with the same constant. This yields

1.2.2 Corollary. *In the cases of a halfspace domain $Q = \mathbb{R}^{N-1}\times\mathbb{R}_+$ and of the whole space $Q = \mathbb{R}^N$ the Korn inequality is also valid,*

$$\int_Q e_{ij}(\boldsymbol{u})e_{ij}(\boldsymbol{u})\,d\boldsymbol{x} \ge \frac{1}{2}\|\boldsymbol{u}\|'^2_{\boldsymbol{H}^1(Q)} \quad \textit{for every } u \in \boldsymbol{H}^1(Q).$$

In order to extend the Korn inequality to sufficiently smooth general domains, we first formulate the coercivity of the strain tensor.

1.2.3 Theorem. *Let $\Omega \subset \mathbb{R}^N$ be a bounded domain with Lipschitz boundary $\Gamma = \partial\Omega$. Then there exists a constant c_0 such that for all $\boldsymbol{v} \in \boldsymbol{H}^1(\Omega)$,*

$$\int_\Omega \big(e_{ij}(\boldsymbol{v})e_{ij}(\boldsymbol{v}) + \boldsymbol{v}\cdot\boldsymbol{v}\big)\,d\boldsymbol{x} \ge c_0\|\boldsymbol{v}\|^2_{\boldsymbol{H}^1(\Omega)}. \tag{1.2.13}$$

1.2.4 Remark. If we introduce the space

$$\begin{aligned}\boldsymbol{E}(\Omega) &\equiv \big\{\boldsymbol{v}\in\boldsymbol{L}_2(\Omega);\ \|\boldsymbol{v}\|_{\boldsymbol{E}(\Omega)} < +\infty\big\},\\ &\text{where } \|\cdot\|_{\boldsymbol{E}(\Omega)} \equiv \sqrt{\|\cdot\|^2 + \mathfrak{E}_\Omega}\end{aligned} \tag{1.2.14}$$

endowed with the mentioned norm, Theorem 1.2.3 states the assumption under which $\boldsymbol{H}^1(\Omega)$ and $\boldsymbol{E}(\Omega)$ are equivalent.

Proof. The proof of this theorem for an infinite strip, a halfspace, and the whole space is obvious from the above Corollary. Its extension to a general bounded domain with a C^1 smooth boundary can be carried out with the help of the localization technique described in Section 1.6 below. The proof for domains with a Lipschitz boundary is based on the following lemma.

1.2.5 Lemma. *Let Ω be a domain satisfying the cone property (see definition in Chapter 2, p. 139); this is e.g. satisfied for $\Gamma \in C^{0,1}$. Then for a $v \in H^{-1}(\Omega)$ such that $\partial_i v \in H^{-1}(\Omega)$, $i = 1, \ldots, N$, the relation $v \in L_2(\Omega)$ is valid.*

The proof of this lemma is done for $\Omega = \mathbb{R}^N$ via the Fourier transform. For general domains it is based on the extension technique for domains with a Lipschitz boundary, described in Chapter 2 (see Subsection 2.8.2).

Since $\partial_j \partial_k v_i = \partial_j e_{ik}(\boldsymbol{v}) + \partial_k e_{ij}(\boldsymbol{v}) - \partial_i e_{jk}(\boldsymbol{v})$, Lemma 1.2.5 yields $\boldsymbol{v} \in \boldsymbol{H}^1(\Omega)$ whenever $\boldsymbol{v} \in \boldsymbol{E}(\Omega)$. The just proved surjectivity of the imbedding of $\boldsymbol{H}^1(\Omega)$ into $\boldsymbol{E}(\Omega)$ and the open mapping theorem yields the continuity of its inverse. □

For a bounded domain Ω, the kernel of the strain tensor $\boldsymbol{\varepsilon} : \boldsymbol{u} \mapsto \{e_{ij}(\boldsymbol{u})\}_{i,j=1}^N$ is given by the set

$$\ker(\boldsymbol{\varepsilon}) = \mathscr{R} \equiv \{\boldsymbol{x} \mapsto B\boldsymbol{x} + \boldsymbol{r}\,; B \in \mathbb{R}^{N\times N} \text{ is antisymmetric}, \boldsymbol{r} \in \mathbb{R}^N\}, \tag{1.2.15}$$

where "antisymmetric" means $B = -B^\top$. This fact can be easily verified with the help of the formula $\partial_j(B\boldsymbol{x} + \boldsymbol{r})_i = b_{ij}$ for $B = \{b_{ij}\}_{i,j=1}^N$, hence $e_{ij}(\boldsymbol{u}) = 0$ for every $\boldsymbol{u} \in \mathscr{R}$. Conversely, condition $\boldsymbol{\varepsilon}(\boldsymbol{u}) = 0$ is equivalent to $\partial_j u_i = -\partial_i u_j$ on Ω for all $i, j = 1, \ldots, N$. This yields $\partial_i \partial_j u_k = 0$, $i, j, k = 1, \ldots, N$, where the derivatives are considered in the distributional sense. In fact, $\partial_i \partial_k u_j = -\partial_i \partial_j u_k = -\partial_j \partial_i u_k = \partial_j \partial_k u_i$, where the second equality is obvious from the definition of the distributional derivatives (the test functions are smooth). From the same reason the first and the last term in the previous inequality equals to $\partial_k \partial_i u_j$, and $\partial_k \partial_j u_i$, respectively. However, simultaneously $\partial_k \partial_i u_j = -\partial_k \partial_j u_i$. Hence $\boldsymbol{u}$ is an affine field and has a representation $\boldsymbol{u}(\boldsymbol{x}) = B\boldsymbol{x} + \boldsymbol{r}$ with a matrix $B \in \mathbb{R}^{N\times N}$ and a vector $\boldsymbol{r} \in \mathbb{R}^N$. From $e_{ij}(\boldsymbol{u}) = \frac{1}{2}(b_{ij} + b_{ji}) = 0$ it follows $\boldsymbol{u} \in \mathscr{R}$.

We divide $\boldsymbol{H}^1(\Omega)$ into the two direct summands $\mathscr{R}$ and its (e.g. L_2-)orthogonal complement $\mathscr{Q}$. This decomposition also defines two continuous projections $\pi_{\mathscr{Q}} : \boldsymbol{H}^1(\Omega) \to \mathscr{Q}$ and $\pi_{\mathscr{R}} : \boldsymbol{H}^1(\Omega) \to \mathscr{R}$. The Korn inequality holds on $\mathscr{Q}$. Indeed, if this is not true, then there exists a sequence $\{\boldsymbol{v}_k\} \subset \mathscr{Q}$ such that $\varepsilon(\boldsymbol{v}_k) \to 0$ in $\boldsymbol{L}_2(\Omega)$ and $\|\boldsymbol{v}_k\|_{\boldsymbol{H}^1(\Omega)} = 1$. Due to the compact imbedding $\boldsymbol{H}^1(\Omega) \hookrightarrow \boldsymbol{L}_2(\Omega)$ we can extract a subsequence, denoted by $\boldsymbol{v}_k$ again, such that $\boldsymbol{v}_k \to \boldsymbol{v}$ in $\boldsymbol{L}_2(\Omega)$ and $\boldsymbol{v}_k \rightharpoonup \boldsymbol{v}$ in $\boldsymbol{H}^1(\Omega)$. Due to the convergence $\varepsilon(\boldsymbol{v}_k) \to 0$ there holds $\varepsilon(\boldsymbol{v}) = 0$. Passing to the limit $k \to \infty$ in (1.2.13) for $\boldsymbol{v}_k - \boldsymbol{v}$ and regarding $\dim \mathscr{R} < +\infty$, we can see $\boldsymbol{v}_k \to \boldsymbol{v}$ in $\boldsymbol{H}^1(\Omega)$. Hence $\boldsymbol{v} \in \mathscr{Q}$, $\|\boldsymbol{v}\|_{\boldsymbol{H}^1(\Omega)} = 1$ and $\varepsilon(\boldsymbol{v}) = 0$ which contradicts $\mathscr{Q} = \mathscr{R}^{\perp}$. Thus we have proved the following version of the Korn inequality:

1.2.6 Proposition. *Let $\Omega \subset \mathbb{R}^N$ be a bounded domain with Lipschitz boundary. Then*

$$\mathfrak{E}_{\Omega}(\boldsymbol{v}) \geq c_0 \|\pi_{\mathscr{Q}} \boldsymbol{v}\|^2_{\boldsymbol{H}^1(\Omega)}$$

for every $\boldsymbol{v} \in \boldsymbol{H}^1(\Omega)$.

An easy consequence of this result is

1.2.7 Corollary. *Let Ω be a bounded Lipschitz domain, an infinite strip, a halfspace or the full space. Let the space $\boldsymbol{V}$ be defined by $\boldsymbol{V} \equiv \{\boldsymbol{v} \in \boldsymbol{H}^1(\Omega)\,;\, \boldsymbol{v} = 0$ on $\Gamma_{\boldsymbol{U}}\}$ with a part $\Gamma_{\boldsymbol{U}}$ of the boundary having non-zero surface measure in the case of the bounded Lipschitz domain, $\boldsymbol{V} \equiv \{\boldsymbol{v} \in \boldsymbol{H}^1(\Omega)\,;\, \boldsymbol{v} = 0$ on $\mathbb{R}^{N-1} \times \{0\}\}$ for the strip $\Omega = Q_{(0,R)}$ or $\boldsymbol{V} = \boldsymbol{H}^1(\Omega)$ in the case of the halfspace and the full space. Then the Korn inequality*

$$\int_{\Omega} e_{ij}(\boldsymbol{u}) e_{ij}(\boldsymbol{u})\, dx \geq c_0 \|\boldsymbol{u}\|^2_{\boldsymbol{H}^1(\Omega)}$$

holds for every $\boldsymbol{u} \in \boldsymbol{V}$.

This Corollary follows from the fact that $\mathscr{R} \cap \boldsymbol{V}$ contains only the trivial function.

The set $\mathscr{R}$ is called *set of rigid motions*. Let us point out that it does not coincide with the set of translations and rotations of the domain Ω, as its name might suggest. In fact, it is the set of all derivatives

$$\boldsymbol{x} \mapsto \partial_{\varphi} A(\varphi)_{|\varphi=0}\, \boldsymbol{x}$$

of "real" rigid motions

$$\boldsymbol{x} \mapsto A(\varphi)\boldsymbol{x}$$

where $A : \mathbb{R} \to \mathbb{R}^{N,N}$ is a smooth mapping satisfying $A(0) = \boldsymbol{Id}$, $\det(A) = 1$ and $A(\varphi)A(\varphi)^{\top} = \boldsymbol{Id}$. This fact is a consequence of the linearization of the strain tensor.

1.2.5 Special cases

Let us assume that the thickness of the body is infinitely small, occupying the graph $\{(\boldsymbol{x}, \psi(\boldsymbol{x})); \boldsymbol{x} \in \Omega'\}$ in a reference configuration, where Ω' is a domain in $\mathbb{R}^{N-1}$, and ψ is some differentiable function on Ω'. For $N = 3$ this object is a membrane, for $N = 2$ it describes a string. Moreover, let the force act in the direction of the N-th axis. We expect that our object is deformable also only in that direction, i.e. u_N (further denoted by u) is the only possible non-zero component of the displacement. Its derivatives with respect to $x_i, i = 1, \ldots, N-1$, are the only possible non-zero components of the strain tensor. The components of the stress tensor are then limited to $\sigma_i(\boldsymbol{u}) \equiv \sigma_{iN}(\boldsymbol{u})$ and in the linear theory of elasticity $\sigma_i(u) = b_{ij}\partial_j u$, where $b_{ij} \equiv a_{iNjN}$, $i, j = 1, \ldots, N-1$. This concept can be extended to the mentioned viscous models. Here, however, some more general non-linear concepts are frequently used.

1.3 Formulation of contact problems

1.3.1 Contact and friction models

The phenomenologically most simple models for the description of contact and friction are the Signorini contact condition and the Coulomb law of friction. The Signorini condition models the exact non-penetrability of the bodies in contact with the following relations:

1. The normal displacement u_n of the boundary cannot be larger than the distance g_n between the bodies, that is, $u_n \leq g_n$.
2. by contact only compressive forces F_n can be transmitted, $F_n \leq 0$
3. forces can be transmitted only if there is contact, e.g. $F_n < 0 \Rightarrow u_n = g_n$.

The Coulomb law of friction was proposed very early in the year 1781(cf. [37]). It is based on the following observation: we consider a rigid block resting on a flat rigid plane. The block is pressed upon the plane with a normal force F_n (e.g. the gravity) and it is pulled in the horizontal direction. Then there is a threshold value such that the block starts to move only if the tangential force is bigger than this value. This threshold value is proportional to the normal force F_n with proportionality constant $\mathfrak{F}_1$ called coefficient of friction. The friction creates a force F_t opposite to the moving velocity with magnitude again proportional to the normal force, $F_t = \mathfrak{F}_2 F_n$. The proportionality constants $\mathfrak{F}_1$ and $\mathfrak{F}_2$ are usually different. However, throughout this book we either neglect this difference, $\mathfrak{F}_1 = \mathfrak{F}_2$, or we describe it approximately by a coefficient of friction depending on the sliding velocity $\dot{\boldsymbol{u}}_{\boldsymbol{t}}$, $\mathfrak{F} = \mathfrak{F}(\dot{\boldsymbol{u}}_{\boldsymbol{t}})$.

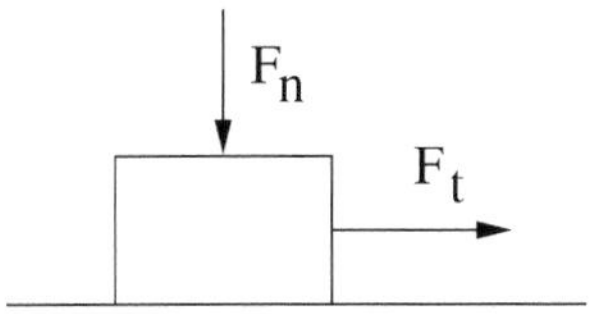

The contact and friction conditions, formulated here for rigid bodies, may be transferred to deformable bodies by replacing the forces F_n and $\boldsymbol{F_t}$ by force densities σ_n and $\boldsymbol{\sigma_t}$. The Signorini contact model is then given by the complementarity condition

$$u_n \leq g_n, \quad \sigma_n \leq 0, \quad \sigma_n(u_n - g_n) = 0. \tag{1.3.1}$$

The Coulomb friction law is defined by the relations

$$\dot{\boldsymbol{u}}_{\boldsymbol{t}} = 0 \Rightarrow |\boldsymbol{\sigma_t}| \leq \mathfrak{F}|\sigma_n|, \tag{1.3.2}$$

$$\dot{\boldsymbol{u}}_{\boldsymbol{t}} \neq 0 \Rightarrow \boldsymbol{\sigma_t} = -\mathfrak{F}|\sigma_n|\frac{\dot{\boldsymbol{u}}_{\boldsymbol{t}}}{|\dot{\boldsymbol{u}}_{\boldsymbol{t}}|} \tag{1.3.3}$$

with $\dot{\boldsymbol{u}}_{\boldsymbol{t}}$ denoting the (tangential) velocity.

Despite the simplicity of these models, the corresponding frictional contact problems are rather delicate. Due to the inequalities arising in the contact and friction laws, the problems have variational formulations in terms of *variational inequalities.* The functional describing the friction turns out to be neither monotone nor compact, hence the well established theory of variational inequalities cannot be applied to these problems directly. Moreover, the Euclidean norms make the friction non-smooth. These difficulties have led many researchers to propose different contact and friction models. Such simplifications include

- the concept of *non-local friction*, where the normal component of the contact traction in the Coulomb law is smoothed by convolution with a smooth kernel [40].
- The *normal compliance* contact model. There the exact non-penetration condition is relaxed and replaced by a (possibly) non-linear contact law [106]. The normal contact traction is given as a function of the magnitude of interpenetration of the bodies,

  $$\sigma_n(\boldsymbol{u}) = -\phi(u_n - g_n).$$

  This concept corresponds to a penalty approximation of the Signorini condition.

Both simplifications lead to contact problems with completely continuous friction laws. These problems can be analyzed with standard variational methods. The models are sometimes justified by an investigation of the micromechanics of the contact surfaces (although the smoothing of the normal boundary traction only while the tangential stress remains non-smoothed such as it is usually assumed in the first model does not seem to be fully physically verified). However, involving micromechanics means involving very small scales and parameters. In a physically realistic setting the diameter of the support of a suitable smoothing kernel in the *non-local friction* approach or the slope of the non-linear functional mapping the interpenetration to the boundary traction in the *normal compliance* model should have a magnitude of the size of the boundaries' microstructure. In the limit of vanishing microstructure parameters we again recover the Signorini condition and the Coulomb friction law. Therefore, in our opinion, the Signorini condition and the Coulomb law are still the most realistic models on a macroscopic level.

1.3.2 The dynamic contact problem

We give here the complete formulation of the dynamic contact problem with Coulomb friction for an elastic body and a rigid foundation. The boundary Γ of the body shall consist of three mutually disjoint parts $\Gamma_{\boldsymbol{U}}$, $\Gamma_{\boldsymbol{\tau}}$ and Γ_C. On $\Gamma_{\boldsymbol{U}}$ the boundary displacements are prescribed, on $\Gamma_{\boldsymbol{\tau}}$ the boundary tractions are given, and Γ_C is the potential contact part. Summarizing the models of the preceding sections, we arrive at the following problem: We search for a displacement field $\boldsymbol{u} : I_T \times \Omega \to \mathbb{R}^N$ with time interval $I_T = [0, T]$ such that the following relations are satisfied:

$$\ddot{\boldsymbol{u}}_i - \partial_j \sigma_{ij}(\boldsymbol{u}) = \boldsymbol{f}_i \text{ in } I_T \times \Omega, \tag{1.3.4}$$

$$\boldsymbol{u} = \boldsymbol{U} \text{ on } I_T \times \Gamma_{\boldsymbol{U}}, \tag{1.3.5}$$

$$\boldsymbol{\sigma}_\Gamma = \boldsymbol{\tau} \text{ on } I_T \times \Gamma_{\boldsymbol{\tau}}, \tag{1.3.6}$$

$$u_n \le g_n,\ \sigma_n \le 0,\ \sigma_n(u_n - g_n) = 0 \text{ on } I_T \times \Gamma_C, \tag{1.3.7}$$

$$\begin{aligned} \dot{\boldsymbol{u}}_{\boldsymbol{t}} = 0 \quad &\Rightarrow \quad |\boldsymbol{\sigma}_{\boldsymbol{t}}| \le \mathfrak{F}|\sigma_n|, \\ \dot{\boldsymbol{u}}_{\boldsymbol{t}} \ne 0 \quad &\Rightarrow \quad \boldsymbol{\sigma}_{\boldsymbol{t}} = -\mathfrak{F}|\sigma_n|\frac{\dot{\boldsymbol{u}}_{\boldsymbol{t}}}{|\dot{\boldsymbol{u}}_{\boldsymbol{t}}|} \end{aligned} \quad \text{on } I_T \times \Gamma_C, \tag{1.3.8}$$

$$\boldsymbol{u}(0, \boldsymbol{x}) = u_0, \quad \dot{\boldsymbol{u}}(0, \boldsymbol{x}) = u_1 \text{ for } \boldsymbol{x} \in \Omega. \tag{1.3.9}$$

This is the general formulation of a dynamic contact problem with Coulomb friction for an elastic body. Unfortunately, it is not possible to prove any result concerning the existence and uniqueness of solutions for this problem up to now. Even in the frictionless case there are only very few results about existence of solutions available [13]; all of them are valid for very special situations. In the present book we study the following situations:

- The static contact problem, derived from the dynamic problem by a time discretization, in Chapter 3.

- The dynamic problem for given normal force in the friction law. Here the term $\mathfrak{F}|\sigma_n|$ in the friction law (1.3.8) is assumed to be a given function G. This problem is not only of theoretical interest. If $\boldsymbol{u}_G$ denotes the solution of the problem with given friction G and this solution is unique, then an operator

  $$\Phi : G \to \mathfrak{F}|\sigma_n(\boldsymbol{u}_G)|$$

  is defined. The solution of the original contact problem is then given by a fixed point G of this operator and the corresponding displacement field $\boldsymbol{u}_G$. The first existence result for static contact problems with Coulomb friction, using such an approach, was proved in [113]. However, in the dynamic case the existence of a fixed point is still an open problem. The dynamic contact problem with given friction is treated in Chapter 4.

- Some contact problems for membranes in Section 4.3.

- The dynamic contact problem for viscoelastic material with contact condition formulated in displacement velocities,

  $$\dot{u}_n \leq 0, \quad \sigma_n \leq 0, \quad \sigma_n \dot{u}_n = 0. \tag{1.3.10}$$

  This condition can be interpreted as a first-order approximation (with respect to the time variable) of the original contact condition. It is physically realistic for a short time interval and for a vanishing initial gap $g_n = 0$ only. This problem is studied in Chapter 5. This chapter also contains results for thermo-viscoelastic contact problems.

1.4 Variational principles in mechanics

Let us briefly explain how the described models fit into the classical Lagrangean and Hamiltonian variational principles. A mechanical system is described in dependence of generalized coordinates $\boldsymbol{q}$ by a Lagrange function $L(\boldsymbol{q}, \dot{\boldsymbol{q}}, t)$ defined by difference of kinetic and potential energy,

$$L(\boldsymbol{q}, \dot{\boldsymbol{q}}, t) = E_{\text{kin}}(\boldsymbol{q}, \dot{\boldsymbol{q}}, t) - E_{\text{pot}}(\boldsymbol{q}, \dot{\boldsymbol{q}}, t).$$

The generalized coordinates are elements of a space of admissible coordinates V. According to the variational principle of Lagrange, the trajectory $\{\boldsymbol{q}(t)\,;\, t_0 < t < t_1\}$ between two known states $\boldsymbol{q}(t_0)$ and $\boldsymbol{q}(t_1)$ of the system is given by the solution of the optimization problem

$$\max_{\boldsymbol{q}\in C^1((t_0,t_1);V)} \mathscr{L}(t_0,t_1;\boldsymbol{q}) \quad \text{with } \mathscr{L}(t_0,t_1;\boldsymbol{q}) \equiv \int_{t_0}^{t_1} L(t,\boldsymbol{q}(t),\dot{\boldsymbol{q}}(t))\,dt.$$

The Euler-Lagrange equation

$$\partial_\varepsilon \mathscr{L}(t_0,t_1;\boldsymbol{q}+\varepsilon\boldsymbol{v})|_{\varepsilon=0} = 0 \tag{1.4.1}$$

of this problem is given by

$$\begin{aligned} 0 &= \int_{t_0}^{t_1} (\partial_{\boldsymbol{q}} L(t,\boldsymbol{q}(t),\dot{\boldsymbol{q}}(t))\boldsymbol{v}(t) + \partial_{\dot{\boldsymbol{q}}} L(t,\boldsymbol{q}(t),\dot{\boldsymbol{q}}(t))\dot{\boldsymbol{v}}(t))\,dt \\ &= \int_{t_0}^{t_1} \Big(\partial_{\boldsymbol{q}} L(t,\boldsymbol{q}(t),\dot{\boldsymbol{q}}(t)) - (\partial_{\dot{\boldsymbol{q}}} L(t,\boldsymbol{q}(t),\dot{\boldsymbol{q}}(t)))^{\bullet}\Big)\,\boldsymbol{v}(t)\,dt, \end{aligned}$$

where in the last step an integration by parts with respect to the time variable has been done. The equation is valid for all test functions from $\mathring{C}^1((t_0,t_1);V)$. This leads to the equation of motion of the Lagrangean formalism,

$$(\partial_{\dot{\boldsymbol{q}}} L)^{\cdot} = \partial_{\boldsymbol{q}} L. \tag{1.4.2}$$

The Hamiltonian of the physical system is defined as the dual of the Lagrange function with respect to the generalized velocities $\dot{\boldsymbol{q}}$,

$$H(\boldsymbol{q},\boldsymbol{p}) = \min_{\dot{\boldsymbol{q}}\in V} \big(\boldsymbol{p}\cdot\dot{\boldsymbol{q}} - L(\boldsymbol{q},\dot{\boldsymbol{q}})\big).$$

The necessary conditions for a minimum give the relation

$$\boldsymbol{p} = \partial_{\dot{\boldsymbol{q}}} L \tag{1.4.3}$$

for the generalized momentum $\boldsymbol{p}$. Differentiating the Hamiltonian $H(\boldsymbol{q},\boldsymbol{p}) = \boldsymbol{p}\cdot\dot{\boldsymbol{q}}(\boldsymbol{q},\boldsymbol{p}) - L(\boldsymbol{q},\dot{\boldsymbol{q}}(\boldsymbol{q},\boldsymbol{p}))$ with respect to $\boldsymbol{q}$ and $\boldsymbol{p}$ and using formula (1.4.3) yields $\partial_{\boldsymbol{q}} H = -\partial_{\boldsymbol{q}} L$ and $\partial_{\boldsymbol{p}} H = \dot{\boldsymbol{q}}$. In combination with (1.4.2) this leads to the equations of motion of the Hamiltonian formalism,

$$\dot{\boldsymbol{q}} = \partial_{\boldsymbol{p}} H \text{ and } \dot{\boldsymbol{p}} = -\partial_{\boldsymbol{q}} H. \tag{1.4.4}$$

For a linearly elastic body Ω subjected to a volume force $\boldsymbol{f}$ and to boundary tractions $\boldsymbol{\tau}$ the kinetic and potential energies are

$$\begin{aligned} E_{\text{kin}}(\boldsymbol{u}) &= \int_\Omega \tfrac{1}{2}\varrho|\dot{\boldsymbol{u}}|^2\,d\boldsymbol{x} \text{ and} \\ E_{\text{pot}}(\boldsymbol{u}) &= \int_\Omega \left(\tfrac{1}{2}a(\boldsymbol{u},\boldsymbol{u}) - \boldsymbol{f}\cdot\boldsymbol{u}\right)\,d\boldsymbol{x} - \int_\Gamma \boldsymbol{\tau}\cdot\boldsymbol{u}\,ds_{\boldsymbol{x}} \end{aligned}$$

with the bilinear form $a(\boldsymbol{u}, \boldsymbol{v}) = \sigma_{ij}(\boldsymbol{u})e_{ij}(\boldsymbol{v}) = a_{ijk\ell}e_{k\ell}(\boldsymbol{u})e_{ij}(\boldsymbol{v})$. Here, the displacement field $\boldsymbol{u}$ serves as generalized coordinates; it is an element of the space $V = \boldsymbol{H}^1(\Omega)$. The Lagrange functional, defined on $\boldsymbol{H}^1(\Omega) \times \boldsymbol{L}_2(\Omega)$, is given by

$$L(\boldsymbol{u}, \dot{\boldsymbol{u}}) = \int_\Omega \left(\tfrac{1}{2}\varrho|\dot{\boldsymbol{u}}|^2 - \tfrac{1}{2}a(\boldsymbol{u}, \boldsymbol{u}) + \boldsymbol{f} \cdot \boldsymbol{u}\right) d\boldsymbol{x} + \int_\Gamma \boldsymbol{\tau} \cdot \boldsymbol{u}\, ds_{\boldsymbol{x}}.$$

The derivatives $\partial_{\boldsymbol{u}} L$ and $\partial_{\dot{\boldsymbol{u}}} L$ are to be understood now in the sense of Fréchet derivatives; they are defined by

$$\partial_{\boldsymbol{u}} L : \boldsymbol{v} \mapsto \int_\Omega \left(-a(\boldsymbol{u}, \boldsymbol{v}) + \boldsymbol{f} \cdot \boldsymbol{v}\right) d\boldsymbol{x} + \int_\Gamma \boldsymbol{\tau} \cdot \boldsymbol{v}\, ds_{\boldsymbol{x}} \text{ and}$$

$$\partial_{\dot{\boldsymbol{u}}} L : \boldsymbol{v} \mapsto \int_\Omega \varrho\dot{\boldsymbol{u}} \cdot \boldsymbol{v}\, d\boldsymbol{x};$$

both derivatives are elements of $\boldsymbol{H}^1(\Omega)^*$. The Lagrangean equation of motion (1.4.2) is an operator equation in the space $\boldsymbol{H}^1(\Omega)^*$; it is equivalent to the variational equation

$$\int_{I_T} \int_\Omega \left((\varrho\dot{\boldsymbol{u}})^{\cdot} \cdot \boldsymbol{v} + a(\boldsymbol{u}, \boldsymbol{v}) - \boldsymbol{f} \cdot \boldsymbol{v}\right) d\boldsymbol{x}\, dt - \int_{I_T} \int_\Gamma \boldsymbol{\tau} \cdot \boldsymbol{v}\, ds_{\boldsymbol{x}}\, dt = 0.$$

This is the variational formulation of the problem

$$(\varrho\dot{u}_i)^{\cdot} - \partial_j\sigma_{ij}(\boldsymbol{u}) = f_i \text{ in } \Omega, \tag{1.4.5}$$

$$\boldsymbol{\sigma}_\Gamma(\boldsymbol{u}) = \boldsymbol{\tau} \text{ on } \Gamma. \tag{1.4.6}$$

The generalized momentum is given formally by the functional

$$\boldsymbol{p} : \boldsymbol{v} \mapsto \int_\Omega \varrho\dot{\boldsymbol{u}} \cdot \boldsymbol{v}\, d\boldsymbol{x}. \tag{1.4.7}$$

Using the usual identification $\boldsymbol{p} = \varrho\dot{\boldsymbol{u}} \in \boldsymbol{L}_2(\Omega)$, the Hamiltonian reads

$$H(\boldsymbol{u}, \boldsymbol{p}) = \int_\Omega \left(\tfrac{1}{2\varrho}|\boldsymbol{p}|^2 + \tfrac{1}{2}a(\boldsymbol{u}, \boldsymbol{u}) - \boldsymbol{f} \cdot \boldsymbol{u}\right) d\boldsymbol{x} - \int_\Gamma \boldsymbol{\tau} \cdot \boldsymbol{u}\, ds_{\boldsymbol{x}};$$

it coincides with the energy of the system. The Hamiltonian equation of motion (1.4.4) with $\dot{\boldsymbol{p}} : \boldsymbol{v} \mapsto \int_\Omega (\varrho\dot{\boldsymbol{u}})^{\cdot} \cdot \boldsymbol{v}\, d\boldsymbol{x}$ and

$$\partial_{\boldsymbol{u}} H : \boldsymbol{v} \mapsto \int_\Omega \left(a(\boldsymbol{u}, \boldsymbol{v}) - \boldsymbol{f} \cdot \boldsymbol{v}\right) d\boldsymbol{x} - \int_\Gamma \boldsymbol{\tau} \cdot \boldsymbol{v}\, ds_{\boldsymbol{x}} \tag{1.4.8}$$

leads again to the variational formulation of problem (1.4.5), (1.4.6).

In the case of contact without friction, the variational principles apply in a slightly different form. The kinetic and potential energy and the Lagrangean are defined as above, but in the set of admissible displacements $\mathscr{C} := \{\boldsymbol{v} \in \boldsymbol{H}^1(\Omega);\, v_n \leq g_n \text{ on } \Gamma_C\}$ we have an additional inequality constraint. Due to the convexity of this set the Euler-Lagrange equations (1.4.1) can be replaced by the inequality

$$\partial_\varepsilon \mathscr{L}(t_0, t_1; \boldsymbol{q} + \varepsilon(\boldsymbol{v} - \boldsymbol{q}))|_{\varepsilon=0} \leq 0. \tag{1.4.9}$$

This is equivalent to the variational inequality

$$\langle \partial_{\boldsymbol{u}} L, \boldsymbol{v} - \boldsymbol{u} \rangle_\Omega + \langle \partial_{\dot{\boldsymbol{u}}} L, \dot{\boldsymbol{v}} - \dot{\boldsymbol{u}} \rangle_\Omega \leq 0, \tag{1.4.10}$$

or

$$\int_\Omega (-a(\boldsymbol{u}, \boldsymbol{v} - \boldsymbol{u}) + \boldsymbol{f} \cdot (\boldsymbol{v} - \boldsymbol{u}))\, d\boldsymbol{x} + \int_\Gamma \boldsymbol{\tau} \cdot (\boldsymbol{v} - \boldsymbol{u})\, ds_{\boldsymbol{x}} \\ + \int_\Omega \varrho \dot{\boldsymbol{u}} \cdot (\dot{\boldsymbol{v}} - \dot{\boldsymbol{u}})\, d\boldsymbol{x} \leq 0,$$

which is indeed the variational formulation of the contact problem without friction.

The generalized momentum $\boldsymbol{p}$ and the Hamiltonian of the system are again the same as in the case of a boundary value problem without contact. The equation of motion in the Hamiltonian formalism is transformed into the inequality

$$\langle \dot{\boldsymbol{p}}, \boldsymbol{v} - \boldsymbol{u} \rangle_\Omega \geq - \langle \partial_{\boldsymbol{u}} H(\boldsymbol{u}, \boldsymbol{p}), \boldsymbol{v} - \boldsymbol{u} \rangle_\Omega\,,$$

as can be easily seen from (1.4.10) with $\boldsymbol{p} = \partial_{\dot{\boldsymbol{u}}} L$ and relation $\partial_{\boldsymbol{u}} H = -\partial_{\boldsymbol{u}} L$. Due to (1.4.7) and (1.4.8), this again yields the variational inequality of the contact problem without friction.

The contact problems to be studied later usually include Coulomb friction and viscous material laws. Due to the dissipative character of friction and viscosity, these problems cannot be written in a straightforward manner as Hamiltonian systems. In most cases to be studied, the contribution of friction and viscous energy dissipation will be considered as additional forces. In order to give a closed and consistent physical description of contact with friction and viscosity, it is necessary to consider also the heat equation because the dissipated energy is transformed to heat. In this case, the Lagrangean and Hamiltonian mechanics is no longer a suitable model; the right framework is thermodynamics. Contact problems including heat transfer will be studied at the end of Section 5, where also the thermodynamic consistency of the employed models is discussed.

1.5 The static contact problem

In the last part of this introduction we study the solvability of the static contact problem with a fixed point approach. This also motivates the necessity to use the mathematical tools presented in Chapter 2.

Usually static problems in continuum mechanics describe the equilibrium configuration which is attained by the solution of a certain problem with time-independent outer forces at the limit for infinite time. Often the static problem can be formulated as a minimization problem of a certain functional related with some kind of energy. However, for frictional contact problems the situation is not as simple because the friction leads to dissipation of energy. Here the static contact problem is obtained by approximating the time derivatives with finite differences. Let $0 = t^{(0)} < t^{(1)} < \ldots < t^{(K)} = T$ be some discrete time levels and let $u^{(\ell)}$ be an approximation for $u\big(t^{(\ell)}\big)$. In the simplest case we replace the time derivatives by backward differences,

$$\dot{\boldsymbol{u}} \to \frac{\Delta \boldsymbol{u}^{(\ell)}}{\Delta t^{(\ell)}} \text{ and } \ddot{\boldsymbol{u}} \to \frac{\dfrac{\Delta \boldsymbol{u}^{(\ell)}}{\Delta t^{(\ell)}} - \dfrac{\Delta \boldsymbol{u}^{(\ell-1)}}{\Delta t^{(\ell-1)}}}{\Delta t^{(\ell)}}$$

with the backward difference operator Δ defined by $\Delta \boldsymbol{u}^{(\ell)} = \boldsymbol{u}^{(\ell)} - \boldsymbol{u}^{(\ell-1)}$. This results in the sequence of problems

$$\frac{\Delta u_i^{(\ell)}}{\big(\Delta t^{(\ell)}\big)^2} - \partial_j \sigma_{ij}\big(\boldsymbol{u}^{(\ell)}\big) = f_i^{(\ell)} + \frac{\Delta u_i^{(\ell-1)}}{\Delta t^{(\ell)} \Delta t^{(\ell-1)}} \quad \text{in } \Omega, \tag{1.5.1}$$

$$\boldsymbol{u}^{(\ell)} = \boldsymbol{U}^{(\ell)} \quad \text{on } \Gamma_{\boldsymbol{U}}, \tag{1.5.2}$$

$$\boldsymbol{\sigma}_\Gamma^{(\ell)} = \boldsymbol{\tau}^{(\ell)} \quad \text{on } \Gamma_{\boldsymbol{\tau}}, \tag{1.5.3}$$

$$u_n^{(\ell)} \le g_n^{(\ell)}, \quad \sigma_n^{(\ell)} \le 0, \quad \sigma_n^{(\ell)}\big(u_n^{(\ell)} - g_n^{(\ell)}\big) = 0 \quad \text{on } \Gamma_C, \tag{1.5.4}$$

$$\left.\begin{aligned} \Delta \boldsymbol{u}_t^{(\ell)} = 0 &\;\Rightarrow\; \big|\boldsymbol{\sigma}_t^{(\ell)}\big| \le \mathfrak{F}\big|\sigma_n^{(\ell)}\big|, \\ \Delta \boldsymbol{u}_t^{(\ell)} \ne 0 &\;\Rightarrow\; \boldsymbol{\sigma}_t^{(\ell)} = -\mathfrak{F}\big|\sigma_n^{(\ell)}\big| \frac{\Delta \boldsymbol{u}_t^{(\ell)}}{\big|\Delta \boldsymbol{u}_t^{(\ell)}\big|} \end{aligned}\right\} \quad \text{on } \Gamma_C, \tag{1.5.5}$$

$$\boldsymbol{u}^{(0)} = \boldsymbol{u}_0, \quad \frac{\Delta \boldsymbol{u}^{(0)}}{\Delta t^{(0)}} = \boldsymbol{u}_1 \quad \text{in } \Omega. \tag{1.5.6}$$

In time step ℓ we consider this as a given problem for the displacement $\boldsymbol{u}^{(\ell)}$ with the functions $\boldsymbol{u}^{(\ell-1)}$ and $\Delta \boldsymbol{u}^{(\ell-1)}$ as part of the given data. This problem has the structure

$$d\,u_i - \partial_j \sigma_{ij}(\boldsymbol{u}) = f_i \quad \text{in } \Omega, \tag{1.5.7}$$

$$\boldsymbol{u} = \boldsymbol{U} \quad \text{on } \Gamma_{\boldsymbol{U}}, \tag{1.5.8}$$

$$\boldsymbol{\sigma}_\Gamma = \boldsymbol{\tau} \quad \text{on } \Gamma_{\boldsymbol{\tau}}, \tag{1.5.9}$$

$$u_n \le g_n, \quad \sigma_n \le 0, \quad \sigma_n(u_n - g_n) = 0 \quad \text{on } \Gamma_C, \tag{1.5.10}$$

$$\left.\begin{aligned} &\boldsymbol{u_t} - \boldsymbol{u_t}^{(\ell-1)} = 0 \Rightarrow |\boldsymbol{\sigma_t}| \le \mathfrak{F}|\sigma_n|, \\ &\boldsymbol{u_t} - \boldsymbol{u_t}^{(\ell-1)} \ne 0 \Rightarrow \boldsymbol{\sigma_t} = -\mathfrak{F}|\sigma_n| \frac{\boldsymbol{u_t} - \boldsymbol{u_t}^{(\ell-1)}}{\left|\boldsymbol{u_t} - \boldsymbol{u_t}^{(\ell-1)}\right|} \end{aligned}\right\} \quad \text{on } \Gamma_C, \tag{1.5.11}$$

where $d = \left(\Delta t^{(\ell)}\right)^{-2}$, $\boldsymbol{u}$ is the old $\boldsymbol{u}^{(\ell)}$, and the definitions of f_i, $\boldsymbol{U}$, $\boldsymbol{\tau}$ and g_n have been modified in an obvious way.

It is our aim to prove the existence of solutions to this problem. For simplicity of the presentation we assume $\boldsymbol{u_t}^{(\ell-1)} = \boldsymbol{0}$, as it is satisfied in the first step of the time discretization for initial datum $(\boldsymbol{u}_0)_{\boldsymbol{t}} = \boldsymbol{0}$. This restriction is not necessary for the mathematical analysis to be carried out, but the consideration of a non-zero $\boldsymbol{u_t}^{(\ell-1)} = 0$ leads to an enlargement of already large formulae without any substantial mathematical insight. All the proofs can be modified to include also the case of a non-zero, sufficiently smooth $\boldsymbol{u_t}^{(\ell-1)}$.

In order to derive the variational formulation of the problem it is necessary to find a variational formulation for the non-linear contact and friction boundary conditions. The Signorini contact condition is equivalent to

$$u_n \le g_n, \quad \sigma_n(v_n - u_n) \ge 0 \quad \text{for all } v_n \le g_n. \tag{1.5.12}$$

The proof is obvious. The Coulomb friction law admits the variational formulation

$$\boldsymbol{\sigma_t} \cdot (\boldsymbol{v_t} - \boldsymbol{u_t}) + \mathfrak{F}|\sigma_n|\big(|\boldsymbol{v_t}| - |\boldsymbol{u_t}|\big) \ge 0 \quad \forall \boldsymbol{v_t} \text{ orthogonal to } \boldsymbol{n}. \tag{1.5.13}$$

Indeed, assume that (1.5.11) is true (with $\boldsymbol{u_t}^{(\ell-1)} = \boldsymbol{0}$). Then, since $|\boldsymbol{\sigma_t}| \le \mathfrak{F}|\sigma_n|$, we have $\boldsymbol{\sigma_t} \cdot \boldsymbol{v_t} + \mathfrak{F}|\sigma_n||\boldsymbol{v_t}| \ge 0$. Moreover, $\boldsymbol{\sigma_t} \cdot \boldsymbol{u_t} + \mathfrak{F}|\sigma_n||\boldsymbol{u_t}| = 0$ because either $\boldsymbol{u_t} = 0$ or $\boldsymbol{\sigma_t} = -\mathfrak{F}|\sigma_n|\dfrac{\boldsymbol{u_t}}{|\boldsymbol{u_t}|}$. Hence (1.5.13) is proved. Now let (1.5.13) hold. Choosing $\boldsymbol{v_t} = -\boldsymbol{w_t} + \boldsymbol{u_t}$ for an arbitrary $\boldsymbol{w_t}$ perpendicular to $\boldsymbol{n}$ and using both the positivity of $\mathfrak{F}$ and the triangle inequality, we arrive at $\boldsymbol{\sigma_t} \cdot \boldsymbol{w_t} \le \mathfrak{F}|\sigma_n|\,|\boldsymbol{w_t}|$, hence $|\boldsymbol{\sigma_t}| \le \mathfrak{F}|\sigma_n|$. Choosing $\boldsymbol{v_t} = 0$ and $\boldsymbol{v_t} = 2\boldsymbol{u_t}$ yields $\boldsymbol{\sigma_t} \cdot \boldsymbol{u_t} + \mathfrak{F}|\sigma_n||\boldsymbol{u_t}| = 0$; hence either $\boldsymbol{u_t} = 0$ or, if this is not true, $\boldsymbol{\sigma_t} = -\mathfrak{F}|\sigma_n|\dfrac{\boldsymbol{u_t}}{|\boldsymbol{u_t}|}$. Therefore the classical friction law (1.5.11) is verified.

In order to derive a variational formulation of the contact problem, we multiply the differential equations (1.5.7) with a test function of the form $v_i - u_i$, we sum up the results over $i = 1, \dots, N$ and integrate by parts,

$$\int_\Omega \partial_j \sigma_{ij}(\boldsymbol{u})\, v_i \, d\boldsymbol{x} = \int_\Gamma \boldsymbol{\sigma}_\Gamma(\boldsymbol{u}) \cdot \boldsymbol{v}\, ds_{\boldsymbol{x}} - \int_\Omega \sigma_{ij}(\boldsymbol{u}) e_{ij}(\boldsymbol{v})\, d\boldsymbol{x}.$$

Inserting the variational formulations of the contact and friction conditions we obtain

$$A(\boldsymbol{u}, \boldsymbol{v}-\boldsymbol{u}) + \int_{\Gamma_C} \mathfrak{F} |\sigma_n(\boldsymbol{u})| \big(|\boldsymbol{v_t}| - |\boldsymbol{u_t}|\big)\, ds_{\boldsymbol{x}} \geq \mathscr{L}(\boldsymbol{v}-\boldsymbol{u}) \tag{1.5.14}$$

with the bilinear form

$$A(\boldsymbol{u}, \boldsymbol{v}) = \int_\Omega \big(d\,\boldsymbol{u} \cdot (\boldsymbol{v}-\boldsymbol{u}) + a(\boldsymbol{u}, \boldsymbol{v}-\boldsymbol{u})\big)\, d\boldsymbol{x},$$

where

$$a(\boldsymbol{u}, \boldsymbol{v}) \equiv \sigma_{ij}(\boldsymbol{u}) e_{ij}(\boldsymbol{v}) = a_{ijk\ell}\, e_{k\ell}(\boldsymbol{u}) e_{ij}(\boldsymbol{v})$$

and the linear functional of the given outer forces

$$\mathscr{L}(\boldsymbol{v}) \equiv \int_\Omega \boldsymbol{f} \cdot \boldsymbol{v}\, d\boldsymbol{x} + \int_{\Gamma_\tau} \boldsymbol{\tau} \cdot \boldsymbol{v}\, ds_{\boldsymbol{x}}.$$

The variational equation is valid for all functions v in the cone

$$\mathscr{C} := \big\{\boldsymbol{v} \in \boldsymbol{H}^1(\Omega);\ \boldsymbol{v} = \boldsymbol{U} \text{ on } \Gamma_U \text{ and } v_n \leq g_n \text{ on } \Gamma_C\big\}$$

with the vertex at $\boldsymbol{U}$ (we assume $\boldsymbol{U} \in \boldsymbol{H}^1(\Omega)$) and the solution $\boldsymbol{u}$ is searched for in the same cone.

Any twice differentiable weak solution $\boldsymbol{u}$ of (1.5.14) is also a solution of the classical formulation (1.5.7)–(1.5.11) (with $\boldsymbol{u}_t^{(\ell-1)} = \boldsymbol{0}$). Indeed, by putting $\boldsymbol{v} = \boldsymbol{w} + \boldsymbol{u}$ with $\boldsymbol{w} \in \mathring{\boldsymbol{H}}^1(\Omega)$ and using the Green formula, equation (1.5.7) is derived. The condition (1.5.8) is just required for the solution of (1.5.14). Then, by putting a general $\boldsymbol{v} \in \mathscr{C}$ satisfying first $\boldsymbol{v} = \boldsymbol{0}$ on Γ_C, then $\boldsymbol{v_t} = \boldsymbol{0}$ on Γ_C and finally without any further condition and using the weak formulations of the contact condition (1.5.12) and of the friction law (1.5.13) we successively verify the validity of the remaining boundary conditions (1.5.9), (1.5.10), (1.5.11).

The weak formulation (1.5.14) of the contact problem is a *variational inequality*. There exists an extensive theory on variational inequalities, see e.g. the monographs [19], [42], [92]. The existence results developed there are mainly based on the equivalence of many variational inequalities to certain non-smooth optimization problems. However, in our case it is not possible to reformulate (1.5.14) as an optimization problem; therefore these known results cannot be applied directly. In order to prove the solvability of the problem we use an approximation by a sequence of auxiliary problems with simpler structure. Here we employ the historically older approach using the contact problem with given friction. In this problem the contact stress $\mathfrak{F}|\sigma_n|$ in the friction functional is assumed to be a known non-negative functional G. Then (1.5.14) has the form: *Find a function $\boldsymbol{u} \in \mathscr{C}$ such that for all $\boldsymbol{v} \in \mathscr{C}$*

$$A(\boldsymbol{u}, \boldsymbol{v} - \boldsymbol{u}) + \int_{\Gamma_C} G\big(|\boldsymbol{v}_t| - |\boldsymbol{u}_t|\big)\, ds_{\boldsymbol{x}} \geq \mathscr{L}(\boldsymbol{v} - \boldsymbol{u}). \tag{1.5.15}$$

This is equivalent to the optimization problem

$$\boldsymbol{u} = \operatorname*{Argmin}_{\boldsymbol{v} \in \mathscr{C}} (J_0(\boldsymbol{v}) + j_G(\boldsymbol{v})) \tag{1.5.16}$$

with the functionals

$$\begin{aligned} J_0 &: \boldsymbol{v} \mapsto \tfrac{1}{2} A(\boldsymbol{v}, \boldsymbol{v}) - \mathscr{L}(\boldsymbol{v}), \\ j_G &: \boldsymbol{v} \mapsto \int_{\Gamma_C} G|\boldsymbol{v}_t|\, ds_{\boldsymbol{x}}. \end{aligned}$$

The set of admissible functions $\mathscr{C}$ is a convex cone and the functional $J_G = J_0 + j_G$ to be optimized is convex and continuous. The functional J_0 is affine while j_G is non-smooth. The optimality condition of problem (1.5.15) is given by variational inequality (1.5.14). In fact, for a convex continuous functional $F : X \to \mathbb{R}$ defined on a convex subset C of a Banach space X, the inequality $\big(F(u + \varepsilon h) - F(u)\big)/\varepsilon \leq F(u + h) - F(u)$ holds for any u, $u + h \in C$ and $\varepsilon \in (0, 1)$; hence the function $\varepsilon \mapsto \big(F(u + \varepsilon h) - F(u)\big)/\varepsilon$ is non-decreasing on $(0, 1)$. In particular, such functionals are directionally differentiable in such directions with the directional derivative

$$DF(u; h) \equiv \lim_{\varepsilon > 0} \frac{F(u + \varepsilon h) - F(u)}{\varepsilon} = \inf_{\varepsilon > 0} \frac{F(u + \varepsilon h) - F(u)}{\varepsilon}.$$

This implies in particular

$$\lim_{\varepsilon > 0} \frac{F(u + \varepsilon h) - F(u)}{\varepsilon} \leq F(u + h) - F(u). \tag{1.5.17}$$

In our case (1.5.17) applied to $F = j_G$ and $h = \boldsymbol{v} - \boldsymbol{u}$ easily yields the inequality

$$\begin{aligned} &\lim_{\varepsilon \to 0} \tfrac{1}{\varepsilon} (J_G(\boldsymbol{u} + \varepsilon(\boldsymbol{v} - \boldsymbol{u})) - J_G(\boldsymbol{u})) \\ &\quad \leq \langle DJ_0(\boldsymbol{u}), \boldsymbol{v} - \boldsymbol{u} \rangle + j_G(\boldsymbol{v}) - j_G(\boldsymbol{u}) \ \text{ for all } \boldsymbol{v} \in \mathscr{C}, \end{aligned} \tag{1.5.18}$$

with the Fréchet derivative $\langle DJ_0(\boldsymbol{u}), \boldsymbol{v}\rangle = A(\boldsymbol{u}, \boldsymbol{v}) - \mathscr{L}(\boldsymbol{v})$. If $\boldsymbol{u}$ is a solution of the optimization problem, then the left hand side of (1.5.18) is nonnegative for every $\boldsymbol{v} \in \mathscr{C}$; hence $\boldsymbol{u}$ is also a solution of the variational inequality. On the other hand, if $\boldsymbol{u}$ is a solution of the variational inequality, then (1.5.17) applied to J_0 and to $h = \boldsymbol{v} - \boldsymbol{u}$ shows

$$\langle DJ_0(\boldsymbol{u}), \boldsymbol{v} - \boldsymbol{u}\rangle \leq J_0(\boldsymbol{v}) - J_0(\boldsymbol{u})$$

and (1.5.15) therefore implies $J_0(\boldsymbol{v}) + j_G(\boldsymbol{v}) \geq J_0(\boldsymbol{u}) + j_G(\boldsymbol{u})$. This proves the equivalence of the formulations (1.5.15) and (1.5.16).

It is now rather easy to prove the existence and uniqueness of solutions to the optimization problem (1.5.16). The necessary assumptions are as follows.

1.5.1 Assumption. *Let Ω be a bounded domain with a Lipschitz boundary Γ consisting of three measurable (with respect to the surface measure) and mutually disjoint parts Γ_U, Γ_τ and Γ_C. Let $\mathscr{L} \in \boldsymbol{H}_0^1(\Omega)^*$ with $\boldsymbol{H}_0^1(\Omega) := \{\boldsymbol{v} \in \boldsymbol{H}^1(\Omega);\ \boldsymbol{v} = 0 \text{ on } \Gamma_U\}$ and let $\boldsymbol{U} \in \boldsymbol{H}^1(\Omega)$ such that $\boldsymbol{U}|_{\Gamma_C} = 0$. The functional G is assumed to be an element of the dual cone*

$$\mathscr{C}^* := \big\{G \in H^{-1/2}(\Gamma_C);\ \langle G, v\rangle_{\Gamma_C} \geq 0 \text{ for all } v \in H^{1/2}(\Gamma_C) \text{ with } v \geq 0\big\}. \tag{1.5.19}$$

We assume either $d = d(\boldsymbol{x}) \geq \alpha_0 > 0$ for all $\boldsymbol{x} \in \Omega$ or $\operatorname{mes}_\Gamma \Gamma_U > 0$. The coefficients $a_{ijk\ell}$ of the constitutive relation are bounded, symmetric, elliptic (in the sense described above), and $d \in L_\infty(\Omega)$.

Under these assumptions the following Theorem holds true.

1.5.2 Theorem. *The contact problem with given friction has a unique solution $\boldsymbol{u} = \boldsymbol{u}_G \in \mathscr{C}$. The solution satisfies the* a priori *estimate*

$$\|\boldsymbol{u}\|_{\boldsymbol{H}^1(\Omega)} \leq c_0 \quad \text{and} \quad \|\sigma_n(\boldsymbol{u})\|_{H^{-1/2}(\Gamma)} \leq C_0 \tag{1.5.20}$$

with constants c_0, C_0 independent of the given friction force G. The solution $\boldsymbol{u}$ depends continuously on the given friction: if $\{G_k\}_{k=1}^\infty$, $G_k \in \mathscr{C}^*$, is a sequence with a strong $H^{-1/2}(\Gamma_C)$-limit G_0, then the corresponding sequence $\{\boldsymbol{u}_k\}_{k=1}^\infty$ of solutions of problem (1.5.15) converges strongly in $\boldsymbol{H}^1(\Omega)$ to the solution $\boldsymbol{u}_0$ of the contact problem with given friction force G_0.

Here and in the sequel the strong convergence denotes the convergence in the appropriate norm topology.

This existence and uniqueness result follows from the standard optimization theorem

1.5.3 Theorem. *Let X be a reflexive Banach space, let $C \subset X$ be a convex closed subset and let $F : X \to \mathbb{R}$ be a convex, weakly lower semicontinuous functional satisfying the property of coercivity*

$$\lim_{\substack{x \in C \\ \|x\|_X \to +\infty}} F(x) = +\infty.$$

Then there exists a solution x_0 of the optimization problem

$$\min_{x \in C} F(x).$$

If F is strictly convex, then this solution is unique.

Proof. Due to the coercivity every minimizing sequence of the problem is bounded in X. Since X is reflexive, every bounded subset is sequentially weakly compact. It is therefore possible to find a weakly convergent subsequence whose limit x_0 is a solution of the problem. For details see e.g. [51]. □

We apply this theorem to the functional J_0 defined on the Banach space $X = \boldsymbol{H}^1(\Omega)$ and the convex subset $\mathscr{C} \subset X$. It is easy to see that $\mathscr{C}$ is closed in $\boldsymbol{H}^1(\Omega)$. Due to the Korn inequality used for $\boldsymbol{u} - \boldsymbol{U}$ and Assumption 1.5.1 we have

$$J(\boldsymbol{u}) \geq c_0 \|\boldsymbol{u}\|^2_{\boldsymbol{H}^1(\Omega)} - \left(c_1 \|\boldsymbol{U}\|_{\boldsymbol{H}^1(\Omega)} + \|\mathscr{L}\|_{\boldsymbol{H}^1_0(\Omega)^*} \right) \|\boldsymbol{u}\|_{\boldsymbol{H}^1(\Omega)} \\ - c_2 \|\boldsymbol{U}\|^2_{H^1(\Omega)},$$

with suitable constants c_1, c_2 and this proves the coercivity of J. The strict convexity of J again follows from the Korn inequality. Moreover, J is continuous on $\boldsymbol{H}^1(\Omega)$, i.e it is also weakly lower semicontinuous. Hence existence and uniqueness of a solution are verified.

The *a priori* estimate (1.5.20) is proved by taking a fixed test function $\boldsymbol{v} \in \mathscr{C}$ with $\boldsymbol{v_t}|_{\Gamma_C} = 0$ in variational inequality (1.5.15), using the ellipticity of the bilinear form and the relation $\int_{\Gamma_C} G|\boldsymbol{u_t}|\, ds_{\boldsymbol{x}} \geq 0$ valid for $G \in \mathscr{C}^*$. The second estimate in (1.5.20) is a consequence of the first one and the definition of the normal stress by the Green formula

$$\langle \sigma_n(\boldsymbol{u}), v_n \rangle \equiv \int_\Omega \big(d\boldsymbol{u} \cdot \boldsymbol{v} + a(\boldsymbol{u}, \boldsymbol{v}) - \boldsymbol{f} \cdot \boldsymbol{v} \big)\, d\boldsymbol{x} \tag{1.5.21}$$

for any $\boldsymbol{v} \in \boldsymbol{H}^1(\Omega)$ with $\boldsymbol{v_t} = 0$ on Γ.

The continuous dependence of the solution $\boldsymbol{u}$ on G can be verified in the standard way. We take the problem with given friction for two different friction forces G_1 and G_2, denote their solutions by $\boldsymbol{u}_1$ and $\boldsymbol{u}_2$ and insert

$\boldsymbol{v} = \boldsymbol{u}_2$ and $\boldsymbol{v} = \boldsymbol{u}_1$ into the variational inequality (1.5.15) for $G = G_1$ and $G = G_2$, respectively. Then, after adding both inequalities we obtain with help of the Korn inequality (1.2.13)

$$\|\boldsymbol{u}_1 - \boldsymbol{u}_2\|^2_{\boldsymbol{H}^1(\Omega)} \leq \text{const.}\|G_1 - G_2\|_{H^{-1/2}(\Gamma_C)} \cdot \|\,|(\boldsymbol{u}_1)_{\boldsymbol{t}}| - |(\boldsymbol{u}_2)_{\boldsymbol{t}}|\,\|_{\boldsymbol{H}^{1/2}(\Gamma_C)}. \tag{1.5.22}$$

Here, for the second factor only the estimate $\|\,|(\boldsymbol{u}_1)_{\boldsymbol{t}}| - |(\boldsymbol{u}_2)_{\boldsymbol{t}}|\,\|_{\boldsymbol{H}^{1/2}(\Gamma_C)} \leq \|\boldsymbol{u}_1\|_{\boldsymbol{H}^{1/2}(\Gamma_C)} + \|\boldsymbol{u}_2\|_{\boldsymbol{H}^{1/2}(\Gamma_C)}$ is available. This is a consequence of the inequality

$$\|\,|\boldsymbol{v}|\,\|_{H^\beta(M)} \leq \|\boldsymbol{v}\|_{H^\beta(M)}. \tag{1.5.23}$$

The proof for $\beta = 0$ is obvious; for $\beta \in (0,1)$ it is based on the triangle inequality in $\mathbb{R}^N$. Let us remark that an estimate of the type $\|\,|v| - |w|\,\|_{H^\beta(\Gamma_C)} \leq c\|v - w\|_{H^\beta(\Gamma_C)}$ is in general *not valid.* This can be seen e.g. by considering the functions $v_k = \frac{1}{2}(1 + \sin(kx))$ and $w_k = v_k - 1$ on the domain $(0, 2\pi)$. These functions satisfy $|v_k| - |w_k| = 2v_k - 1$ and $v_k - w_k = 1$; as a consequence there holds $\|v_k - w_k\|_{H^\beta(0,2\pi)} = \sqrt{2\pi}$, but $\|\,|v_k| - |w_k|\,\|_{H^\beta(0,2\pi)} \to +\infty$ for $k \to \infty$ and fixed $\beta \in (0,1)$.

Using a standard trace theorem and the *a priori* estimate (1.5.20) for $\boldsymbol{u}_1$ and $\boldsymbol{u}_2$ yields the 1/2-Hölder continuity of the operator $G \mapsto \boldsymbol{u} = \boldsymbol{u}_G$. We remark that (1.5.23) holds also for $\beta = 1$, but the definition of the space needs more assumptions about the set M; it is valid e.g. for an open set with a finite number of components having a Lipschitz boundary and non-zero mutual distances of the components, or a C^1-surface of dimension $N - 1$ in $\mathbb{R}^N$.

Relation (1.5.23) does not hold in general for $\beta \in (-1, 0)$ even if $v \in L_2(M)$. Indeed for $M = (0,1)$ and $v_k = \operatorname{sign}\sin(kx)$ the convergence $\|v_k\|_{H^\beta(M)} \to 0$ for $k \to +\infty$ holds while $\||v_k|\|_{H^\beta(M)} = 1$ for all $k \in \mathbb{N}$.

In the sequel we need the following

1.5.4 Lemma. *Let $M \subset \mathbb{R}^m$ be open and bounded or let M be an open bounded subset of a manifold $\Gamma \subset \mathbb{R}^{m+k}$ of the class $C^{1+\beta}$ having the dimension m for some integer $k > 0$. We assume that M consists of a finite number of components having a Lipschitz boundary or* $\operatorname{supp} F \subset M$. *Let*

$$\begin{aligned} &\alpha = 0 \text{ and } F \in L_\infty(M) \text{ or} \\ &\alpha \in (0, \min(1,\beta)) \text{ and } F \in L_\infty(M) \cap W_p^{\alpha+\varepsilon}(M) \\ &\qquad \text{for } \varepsilon > 0 \text{ arbitrarily small.} \end{aligned} \tag{1.5.24}$$

Here, the index p satisfies $p \geq m/\alpha$ for $\alpha < m/2$, $p > 2$ for $\alpha = m/2$ and $p \geq 2$ for $\alpha > m/2$. Then, if $g \in H^{-\alpha}(M)$ is non-positive or non-negative in the dual sense, there holds

$$\|Fg\|_{H^{-\alpha}(M)} \leq \|F\|_{L_\infty(M)} \|g\|_{H^{-\alpha}(M)}. \tag{1.5.25}$$

If g is a general element of $H^{-\alpha}(M)$ and $F \in L_\infty(M) \cap W_p^{\alpha+\varepsilon}(M)$ with p satisfying the requirements mentioned above and arbitrarily small $\varepsilon > 0$, then

$$\|Fg\|_{H^{-\alpha}(M)} \leq \left(\|F\|_{L_\infty(M)} + \text{const.} \|F\|_{W_p^{\alpha+\varepsilon}(M)} \right) \|g\|_{H^{-\alpha}(M)}. \tag{1.5.26}$$

Proof. Let us first consider the case of M being composed of a finite number of domains with Lipschitz boundary. The definition of the dual norm, estimate (1.5.23) and Hölder's inequality show the validity of (1.5.25) for $g \in L_2(M)$ or $\alpha = 0$. In the latter case the assumption on the sign of g is redundant. For $F \in W_p^{\alpha+\varepsilon}(M) \cap L_\infty(M)$ with $p \in (1, \infty)$, any small $\varepsilon > 0$ and any $\boldsymbol{v} \in H^\alpha(M)$ there holds

$$\|Fv\|_{H^\alpha(M)} \leq \|F\|_{L_\infty(M)} \|v\|_{H^\alpha(M)} + \|F\|_{W_{2q}^{\alpha+\varepsilon}(M)} \|v\|_{L_{2q'}(M)} \tag{1.5.27}$$

with indices $q, q' \geq 1$ satisfying $2q \leq p$ and $q = q'/(q'-1)$. Then the imbedding $H^\alpha(M) \hookrightarrow L_{2q'}(M)$ is employed; it is valid for $q' \leq m/(m-2\alpha)$ if $\alpha < m/2$, for any finite $q' \geq 1$ if $\alpha = m/2$ and for $q' = +\infty$ if $\alpha > m/2$. This determines the value of the index p.

If M is not regular in the above mentioned sense, but $\operatorname{supp} F \subset M$, then there exists a regular set M' such that $\operatorname{supp} F \subset M'$. In this case the proof is done for the set M', and the result is also valid for the original set M because $F = 0$ on $M \setminus M'$.

Up to now, estimate (1.5.25) has been proved for the case $g \in L_2(M)$ only. However, due to (1.5.26) the operator $g \mapsto Fg$ is continuous on $H^{-\alpha}(M)$ for F satisfying the assumptions of the Lemma. Due to the density of non-negative (non-positive) elements of $L_2(M)$ in a set of those from $H^{-\alpha}(M)$ which is a consequence of the mollifier technique (cf. (2.2)), the assertion (1.5.25) is also valid for every non-negative (non-positive) $g \in H^{-\alpha}(M)$ and the Lemma is proved. □

1.5.5 Remark. If the boundary of M is at least $C^{1+\beta}$-smooth for $\beta > \alpha$ or if $\operatorname{supp} F \subset M$ holds, then $L_\infty(M) \cap W_p^{\alpha+\varepsilon}(M)$ is dense in $C^0(\overline{M})$. Hence in these cases the Lemma is also valid for $F \in C^0(\overline{M})$.

A solution of the original contact problem with Coulomb friction is characterized by a fixed point of the operator

$$\Phi : \mathscr{C}^* \ni G \to \mathfrak{F}\sigma_n(\boldsymbol{u}_G) \in \mathscr{C}^*.$$

The existence of a fixed point is proved by Tikhonov's theorem.

1.5.6 Theorem. *Let $\mathscr{X}$ be a locally convex space and let $\mathscr{S} \subset \mathscr{X}$ be a convex compact subset. Then any continuous mapping $A : \mathscr{S} \to \mathscr{S}$ has a fixed point in $\mathscr{S}$.*

In order to apply this Theorem to the contact problem, we choose the space $\mathscr{X} \equiv H^{-1/2}(\Gamma_C)$, endowed with the weak topology. We shall prove that the operator Φ maps a suitable compact subset $\mathscr{S} \subset \mathscr{X}$ continuously into itself. Using the well-known weak compactness of closed, convex and bounded sets in reflexive Banach spaces we may choose $\mathscr{S}$ as

$$\mathscr{S} := \{G \in \mathscr{C}^*;\ \|G\|_{H^{-\alpha}(\Gamma_C)} \leq R\}$$

with suitable values $\alpha < 1/2$ and $R < +\infty$. Then it is necessary to prove:

$$\|G\|_{H^{-\alpha}(\Gamma_C)} \leq R \Rightarrow \|\Phi(G)\|_{H^{-\alpha}(\Gamma_C)} \leq R.$$

This *regularity result* for the solution of the contact problem will be proved under the following

1.5.7 Assumption. *In addition to Assumption 1.5.1 let the following conditions be satisfied for an index $\alpha \in (0, 1/2)$: the contact part of the boundary is of the class $C^{2+\beta}$ with $\beta > 2\alpha$, the coefficients d and $a_{ijk\ell}$ of the bilinear form are of the class $C^{\beta'}$ with $\beta' \in (\alpha, 1]$, the coefficient of friction is bounded on Γ_C by the constant $C_{\mathfrak{F}}$ defined below and its support is contained in a set $\Gamma_{\mathfrak{F}} \subset \Gamma_C$ having a positive distance to $\Gamma \setminus \Gamma_C$. The given data satisfy $f \in \mathring{\boldsymbol{H}}^{-1+\alpha}(\Omega_C)$ with a subdomain $\Omega_C \Subset \Omega$ such that $\Gamma_C \subset \partial\Omega_C$ and $g_n \in H^{1/2+\alpha}(\Gamma_C)$.*

1.5.8 Theorem. *Let the condition of Assumption 1.5.7 be true. Then the normal boundary stress $\sigma_n(\boldsymbol{u}_G)$ to the solution $\boldsymbol{u}_G$ of the problem with given friction $G \in H^{-1/2+\alpha}(\Gamma_C)$ satisfies the* a priori *estimates*

$$\|\sigma_n(\boldsymbol{u}_G)\|_{\boldsymbol{H}^{-1/2+\alpha}(\Gamma_{\mathfrak{F}})} \leq (1+\varepsilon)C_0\|G\|_{H^{-1/2+\alpha}(\Gamma_C)} + c_1(\varepsilon), \qquad (1.5.28)$$

where $\varepsilon > 0$ can be arbitrarily small. The constant C_0 depends on the geometry of the domain and on the bilinear form a only, while the constant c_1 depends on ε and on the given data f, $\boldsymbol{\tau}$ and $\boldsymbol{U}$.

The *proof* of this theorem will be given in Subsection 1.7.2 for the case of a halfspace domain and in Subsection 1.7.3 for the general case.

Combination of the estimates (1.5.28) and (1.5.25) with the non-negativity of both $\mathfrak{F}$ and $\sigma_n(\boldsymbol{u}_G)$ yields:

$$\begin{aligned}\|\Phi(G)\|_{H^{-1/2+\alpha}(\Gamma_{\mathfrak{F}})} &= \|\mathfrak{F}\sigma_n(\boldsymbol{u}_G)\|_{H^{-1/2+\alpha}(\Gamma_{\mathfrak{F}})}\\ &\leq \|\mathfrak{F}\|_{L_\infty(\Gamma_C)}\left(C_0(1+\varepsilon)\|G\|_{H^{-1/2+\alpha}(\Gamma_{\mathfrak{F}})} + c_1(\varepsilon)\right).\end{aligned}$$

If

$$\|\mathfrak{F}\|_{L_\infty(\Gamma_C)} < C_{\mathfrak{F}} \equiv C_0^{-1}, \tag{1.5.29}$$

then there is a $R > 0$ such that $\|G\|_{H^{-1/2+\alpha}(\Gamma_{\mathfrak{F}})} \leq R$ implies

$$\|\Phi(G)\|_{H^{-1/2+\alpha}(\Gamma_{\mathfrak{F}})} \leq R.$$

Moreover, if

$$\begin{aligned}\mathfrak{F} \in W_p^\lambda(\Gamma_C) \text{ with } \lambda > \tfrac{1}{2} - \alpha \text{ and}\\ p \text{ given in Lemma 1.5.4,}\end{aligned} \tag{1.5.30}$$

then the operator Φ is weakly continuous in $\mathscr{S}$. In order to verify this, let us consider a sequence $\{G_k\}$ in $\mathscr{C}^* \cap H^{-1/2+\alpha}(\Gamma_{\mathfrak{F}})$ tending weakly to some limit G_0. These functionals are extended by 0 to $H^{-1/2+\alpha}(\Gamma_C)$, i.e. by the identity $\langle G_k, v\rangle_{\Gamma_C} = \langle G_k, v\rangle_{\Gamma_{\mathfrak{F}}}$ for any $k \in \{0\} \cup \mathbb{N}$ and $v \in H^{-1/2+\alpha}(\Gamma_C)$. Then, due to (1.5.21) and (1.5.22) there holds $\sigma_n(\boldsymbol{u}_{G_k}) \to \sigma_n(\boldsymbol{u}_G)$ in $H^{-1/2}(\Gamma_C)$. Any in $H^{-1/2+\alpha}(\Gamma_{\mathfrak{F}})$ weakly convergent subsequence of $\sigma_n(\boldsymbol{u}_{G_k})$ has the same limit; consequently the operator $G \mapsto \sigma_n(\boldsymbol{u}_G)$ is weakly continuous. Lemma 1.5.4 ensures the weak continuity of the linear operator $g \mapsto \mathfrak{F}g$ on $H^{-1/2+\alpha}(\Gamma_{\mathfrak{F}})$, and hence the operator Φ is weakly continuous.

With these facts the assumptions of Tikhonov's fixed point theorem are satisfied and the existence of a solution to the static contact problem with given friction is proved. The result is summarized in the following

1.5.9 Theorem. *Let the Assumptions 1.5.1, 1.5.7 and relations* (1.5.29) *and* (1.5.30) *be valid. Then the contact problem with Coulomb friction has a solution.*

We remark that the condition $\mathfrak{F} \in C^\varepsilon(\Gamma_C)$ for an arbitrarily small $\varepsilon > 0$ is sufficient to fulfill the conditions on $\mathfrak{F}$ in Assumption 1.5.1.

1.6 Geometry of domains

We now give a more detailed description of the geometrical properties of the boundary of a domain Ω and its relation to the definition of norms. Let $B_{N-1}(a) := \{x \in \mathbb{R}^{N-1}\,;\, |x| < a\}$ denote the $(N-1)$-dimensional ball of radius a and for $\boldsymbol{x} \in \mathbb{R}^N$ let $\boldsymbol{x}' \equiv (x_1, \ldots, x_{N-1})^\top$. A bounded domain Ω is said to have a *Lipschitz boundary* Γ or to be of the class $C^{0,1}$, if for any $\boldsymbol{x}_0 \in \Gamma$ there exists a transformation $\mathscr{O}_{\boldsymbol{x}_0}$ composed of a rotation and a translation, taking $\boldsymbol{x}_0$ to the origin, and a map $\Psi_{\boldsymbol{x}_0} : \boldsymbol{x} \mapsto (\boldsymbol{x}', x_N - \psi_{\boldsymbol{x}_0}(\boldsymbol{x}'))$

with Lipschitz function $\psi_{\boldsymbol{x}_0} : B_{N-1}(a) \to \mathbb{R}$ satisfying $\psi_{\boldsymbol{x}_0}(0) = 0$ such that

$$B_{N-1}(a) \times (-b, 0) \subset \Psi_{\boldsymbol{x}_0} \circ \mathscr{O}_{\boldsymbol{x}_0}(\mathbb{R}^N \setminus \Omega) \quad \text{and}$$
$$B_{N-1}(a) \times (0, b) \subset \Psi_{\boldsymbol{x}_0} \circ \mathscr{O}_{\boldsymbol{x}_0}(\Omega)$$

for suitable parameters $a, b > 0$. In other words, after a suitable rotation and translation $\mathscr{O}_{\boldsymbol{x}_0}$ of the domain, the boundary can be represented locally as a graph of the Lipschitz function $\psi_{\boldsymbol{x}_0}$ and the domain is locally on one side of the boundary.

Analogously, a boundary of the class $C^{k,1}$ for any positive integer k is given by the same definition, if $\psi_{\boldsymbol{x}_0} \in C^{k,1}(B_{N-1}(a))$, and boundaries of the class $C^k, k \geq 0$ (C^k-boundaries) are given, if all $\psi_{\boldsymbol{x}_0}$ are C^k-functions.

For Γ of the class C^k, $k \geq 1$, the normal vector is defined everywhere on Γ while for a Lipschitz boundary the normal vector is defined for almost every point of the boundary. If the normal vector is defined in $\boldsymbol{x}_0$, then the transformation $\mathscr{O}_{\boldsymbol{x}_0}$ can be chosen such that it maps the tangent hyperplane to Γ at $\boldsymbol{x}_0$ onto the hyperplane $\{(\boldsymbol{x}', 0);\ \boldsymbol{x}' \in \mathbb{R}^{N-1}\}$.

The composition $\mathfrak{R}_{\boldsymbol{x}_0} \equiv \Psi_{\boldsymbol{x}_0} \circ \mathscr{O}_{\boldsymbol{x}_0}$ is called a *local rectification* of Ω in the neighbourhood of $\boldsymbol{x}_0$.

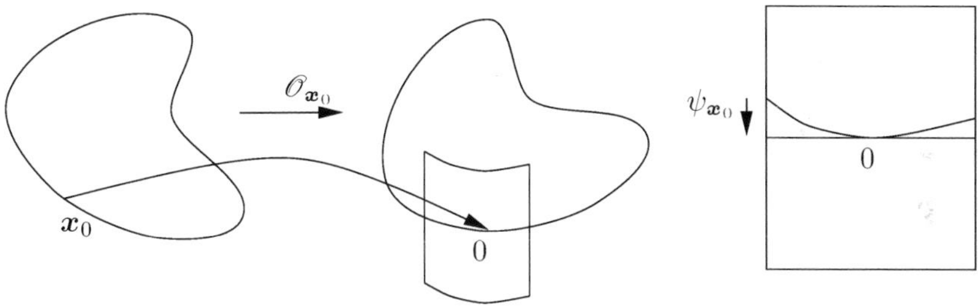

Figure 1.3: Local rectification of the domain.

If Ω is a bounded domain with a Lipschitz boundary, then its boundary is compact and for any $\delta > 0$ there can be constructed a finite covering $\mathscr{U} \equiv \{U_i; i \in \mathscr{I}_\delta\}$ of Γ of the type $U_i \equiv \mathfrak{R}_i^{-1}(B_{N-1}(a_i) \times (-b_i, b_i))$, $i \in \mathscr{I}_\delta \equiv \{1, \ldots, n(\delta)\}$ with suitable $a_i, b_i > 0$ such that $\operatorname{diam} U_i < \delta$. Here, $\mathfrak{R}_i$ denotes a local rectification in a neighbourhood of a point $x_i \in \Gamma$. Moreover, there is a $r \equiv r(\delta)$ such that for all $x \in \Gamma$ there exists a $i \in \mathscr{I}_\delta$ with $\operatorname{dist}(x, \mathbb{R}^N \setminus U_i) \geq 2r(\delta)$. As a consequence, there exists a sufficiently smooth partition of unity on a r-neighbourhood of Γ subordinate to the covering $\mathscr{U}$. This is a system of non-negative functions ϱ_i, $i \in \mathscr{I}_\delta$ of a class C^k for a suitable $k > 0$ such that $\operatorname{supp} \varrho_i \subset U_i$, $\varrho = \sum\limits_{i=1}^{n(\delta)} \varrho_i = 1$ on

$\{x \in \mathbb{R}^N; \operatorname{dist}(x, \Gamma) \leq \rho\}$ and $\varrho(\mathbb{R}^N) \subset [0,1]$. Then, adding the function $\varrho_{n(\delta)+1}$ defined by $1 - \varrho$ on Ω and by 0 outside completes a partition of unity $\mathfrak{P} \equiv \{\varrho_i, i = 1, \ldots, n(\delta)+1\}$ on $\overline{\Omega}$.

For any Sobolev or Sobolev-Slobodetskii space $W_p^k(\Omega)$ and the finite partition of unity $\mathfrak{P}$ defined above, the norms $\|\cdot\|_{W_p^k(\Omega)}$ and $\|u\|_{k,p,\Omega,\mathfrak{P}} \equiv \sum_{i=1}^{n(\delta)+1} \|\varrho_i u\|_{W_p^k(\Omega)}$ are equivalent. Indeed, by the triangle inequality the latter norm is greater than the former one. If the norms are not equivalent, then there exists a sequence u_n such that $\|u_n\|_{k,p,\Omega,\mathfrak{P}} = 1$ and $u_n \to 0$ in $W_p^k(\Omega)$. This leads to an obvious contradiction to the continuity of the operators $u \mapsto \varrho_i u$ on $W_p^k(\Omega)$, $i = 1, \ldots, n(\delta)+1$. Moreover, if Ω is bounded and its boundary Γ is at least of the class C^k, then we can use the local rectification for all elements of the constructed partition except the last one and extend the functions $(\varrho_i u) \circ \mathfrak{R}_i^{-1}$, $i = 1, \ldots, n(\delta)$, by 0 to the halfspace $O \equiv \mathbb{R}^{N-1} \times \mathbb{R}_+$ while the last element $\varrho_{n(\delta)+1} u$ can be extended by 0 to $\mathbb{R}^N$. The original $W_p^k(\Omega)$-norm is then equivalent to the norm

$$u \mapsto \sum_{i=1}^{n(\delta)} \|(\varrho_i u) \circ \mathfrak{R}_i^{-1}\|_{W_k^p(O)} + \|\varrho_{n(\delta)+1} u\|_{W_p^k(\mathbb{R}^N)}.$$

The above described procedure enables us to extend functions from $W_p^k(\Omega)$ to $W_p^k(\mathbb{R}^N)$ for a bounded domain Ω with a boundary of the class at least C^k, i.e. to define a bounded linear operator $\mathscr{E} : W_p^k(\Omega) \to W_p^k(\mathbb{R}^N)$ such that $\mathscr{E}u|_\Omega = u$ for all $u \in W_p^k(\Omega)$. Indeed it suffices to define $\mathscr{E}$ for $\Omega = O$ and for k integer due to the interpolation results of the next Chapter. This extension is easy: for $\boldsymbol{x}' \in \mathbb{R}^{N-1}$ and $x_N \leq 0$ we take $u(\boldsymbol{x}', x_N) = \sum_{j=1}^{k} c_j u(\boldsymbol{x}', -j x_N)$, where the constants c_j are defined in such a way that the extension to $C^{k-1}(\mathbb{R}^N)$ works for $u \in C^{k-1}(O)$ (for details see the next chapter).

With this technique we prove Theorem 1.2.3 for $\Gamma \in C^1$. The "localized" version $\|\cdot\|_{\boldsymbol{E},\Omega,\mathfrak{P}}$ of the norm $\|\cdot\|_{\boldsymbol{E}(\Omega)}$ introduced in (1.2.14) is well defined and equivalent to $\|\cdot\|_{\boldsymbol{E}(\Omega)}$; this is proved in the same way as the corresponding equivalence for the norm $W_p^k(\Omega)$ above. The local rectification leads to the estimate $E_O(\varrho_i \boldsymbol{u}) \leq E_\Omega(\varrho_i \boldsymbol{u}) + c m_\Gamma^2(\delta) \|\nabla(\varrho_i \boldsymbol{u})\|^2_{L_2(O;R^{N\times N})}$ with the modulus of continuity m_Γ of the gradient of the boundary description $\psi_i \equiv \psi_{x_i}$ and a constant c independent of $i, \boldsymbol{u}, \varrho_i$ and δ. Since $m_\Gamma(\delta) \to 0$ for $\delta \to 0$, Corollary 1.2.2 yields $(1/2-\varepsilon)\|\nabla(\varrho_i \boldsymbol{u})\|^2_{L_2(O;\mathbb{R}^{N\times N})} \leq \mathfrak{E}_\Omega(\varrho_i \boldsymbol{u})$ for any $\varepsilon \in (0, 1/2)$, if $\delta = \delta(\varepsilon)$ is taken sufficiently small. This and the above proved equivalences of the appropriate norms lead to the desired result.

1.7 The method of tangential translations

From the example mentioned in Section 1.5 it is seen that the regularity of solutions close to the boundary is a crucial part of the existence proof. This will be true for the majority of examples studied in this book. In the present section, the proof of Theorem 1.5.8 is carried out. This proof is intended to introduce—at an easy example—two important techniques, namely the localization technique and the technique of local tangential translations in arguments of the employed functions, which we call *translation method* or *shift technique*. These techniques are also needed in most of the problems studied later.

1.7.1 Preliminaries

For an introduction into the basic principle of the shift technique, let us consider the Sobolev-Slobodetskii space of order α (the Hilbert case and on the full space $\mathbb{R}^N$),

$$\|u\|^2_{H^\alpha(\mathbb{R}^N)} = \|u\|^2_{L_2(\mathbb{R}^N)} + \int_{\mathbb{R}^N} \frac{\|u(\cdot + \boldsymbol{h}) - u(\cdot)\|^2_{L_2(\mathbb{R}^N)}}{|\boldsymbol{h}|^{N+2\alpha}}\, d\boldsymbol{h} \tag{1.7.1}$$

(see Chapter 2, Section 2.4 for more general cases). This norm essentially represents a higher-order Sobolev norm (the H^α-norm) by taking a lower-order norm (the L_2-norm) of a difference of shifted function and non-shifted function to some power, multiplying the result by another power of the shift parameter $\boldsymbol{h}$ and integrating with respect to $\boldsymbol{h}$. This kind of representation is very well suited for variational formulations of linear and quasi-linear partial differential equations: we are able to shift the arguments of functions, to use those shifted functions as test functions and to shift the arguments (or, more precisely, the variables of integration) of variational formulations. Usually, the lower-order Sobolev norm is not the L_2-norm (as in the example above); in most cases it is the (possibly localized) energy norm of the problem.

In the calculations it is often advantageous to employ also the definition of Sobolev and Sobolev-Slobodetskii norms on $\mathbb{R}^N$ via the Fourier transform

$$\widehat{u}(\boldsymbol{\xi}) \equiv \mathscr{F}(u;\boldsymbol{\xi}) = \frac{1}{(2\pi)^{N/2}} \int_{\mathbb{R}^N} u(\boldsymbol{x}) e^{-i\boldsymbol{x}\cdot\boldsymbol{\xi}}\, d\boldsymbol{x},$$

that is,

$$\|u\|^2_{H^\alpha(\mathbb{R}^N)} = \int_{\mathbb{R}^N} |\mathscr{F}(u;\boldsymbol{\xi})|^2 (1 + c|\boldsymbol{\xi}|^{2\alpha})\, d\boldsymbol{\xi}.$$

The well-known Plancherel identity $\|f\|_{L_2(\mathbb{R}^N)} = \|\mathscr{F}(f)\|_{L_2(\mathbb{R}^N)}$ shows $c = 1$ for non-negative integer α. Moreover, for $0 < \alpha < 1$,

$$\begin{aligned} \|u\|'^2_{H^\alpha(\mathbb{R}^N)} &\equiv \int_{\mathbb{R}^N} \frac{\|u(\cdot + \boldsymbol{h}) - u(\cdot)\|^2_{L_2(\mathbb{R}^N)}}{|\boldsymbol{h}|^{N+2\alpha}}\, d\boldsymbol{h} \\ &= c_N(\alpha) \int_{\mathbb{R}^N} |\widehat{u}(\boldsymbol{\xi})|^2 |\boldsymbol{\xi}|^{2\alpha}\, d\boldsymbol{\xi} \end{aligned} \tag{1.7.2}$$

with constants $c_N(\alpha)$ given by

$$\begin{aligned} c_1(\alpha) &= 2^{2-2\alpha} \int_{\mathbb{R}} \frac{\sin^2 t}{|t|^{1+2\alpha}}\, dt \text{ and} \\ c_N(\alpha) &= c_1(\alpha) \int_{\mathbb{R}^{N-1}} (1 + |s|^2)^{-N/2-\alpha}\, ds, \ N \geq 2. \end{aligned}$$

These constants are calculated from the relation

$$\int_{\mathbb{R}^N} \frac{\left|e^{i\boldsymbol{h}\cdot\boldsymbol{\xi}} - 1\right|^2}{|\boldsymbol{h}|^{N+2\alpha}}\, d\boldsymbol{h} = c_N(\alpha)|\boldsymbol{\xi}|^{2\alpha}.$$

Observe that $e^{i\boldsymbol{h}\cdot\boldsymbol{\xi}} - 1$ comes from $\mathscr{F}(u(\cdot + \boldsymbol{h}) - u(\cdot); \boldsymbol{\xi}) = \left(e^{i\boldsymbol{h}\cdot\boldsymbol{\xi}} - 1\right)\mathscr{F}(u; \boldsymbol{\xi})$. Hence the c in (1.7.1) equals to $c_N(\alpha)$ for $\alpha \in (0, 1)$.

For $\Omega = \mathbb{R}^N$ the relation (1.7.1) leads to the expression for the negative norms

$$\|u\|^2_{H^{-\alpha}(\mathbb{R}^N)} \equiv \int_{\mathbb{R}^N} |\widehat{u}(\boldsymbol{\xi})|^2 \left(1 + c|\boldsymbol{\xi}|^{2\alpha}\right)^{-1} d\boldsymbol{\xi}, \tag{1.7.3}$$

where $c = 1$ for integer α and $c = c_N(\alpha)$ for $\alpha \in (0, 1)$. From (1.7.1) and (1.7.3) there follows the continuous imbedding $H^\alpha(\mathbb{R}^N) \hookrightarrow H^\beta(\mathbb{R}^N)$ for $\beta < \alpha$. The same can be proved for the case of a halfspace O because there exists an extension $\mathscr{E}$ of $H^\alpha(O)$ to $H^\alpha(\mathbb{R}^N)$ such that the norms $\|\mathscr{E}(\cdot)\|_{H^\alpha(\mathbb{R}^N)}$ and $\|\cdot\|_{H^\alpha(O)}$ are equivalent.

For a function f defined on $\mathbb{R}^N$ and any $\boldsymbol{h} \in \mathbb{R}^N$ we introduce the *translation (shift) operator* and the *difference operator*

$$S_{\boldsymbol{h}} f \equiv f_{-\boldsymbol{h}} : \boldsymbol{x} \mapsto f(\boldsymbol{x} + \boldsymbol{h}),\ \boldsymbol{x} \in \mathbb{R}^N \tag{1.7.4}$$

$$\Delta_{\boldsymbol{h}} f : \boldsymbol{x} \mapsto f(\boldsymbol{x} + \boldsymbol{h}) - f(\boldsymbol{x}),\ \boldsymbol{x} \in \mathbb{R}^N. \tag{1.7.5}$$

The same notation is employed for f defined in $\mathbb{R}^{N-1} \times I$ for any interval $I \subset \mathbb{R}$ (possibly degenerated into a point) with $\boldsymbol{h} = (\boldsymbol{h}', 0)$ for any $\boldsymbol{h}' \in \mathbb{R}^{N-1}$.

Relation (1.7.2) can be easily generalized to the case of an arbitrary $H^{\beta}(\mathbb{R}^N)$-norm instead of the $L_2(\mathbb{R}^N)$-norm with $\alpha, \beta > 0$ and $\alpha + \beta < 1$,

$$\int_{\mathbb{R}^N} \frac{\|\Delta_{\boldsymbol{h}} u\|'^2_{H^{\beta}(\mathbb{R}^N)}}{|\boldsymbol{h}|^{N+2\alpha}} \, d\boldsymbol{h} = d_N(\alpha,\beta) \|u\|'^2_{H^{\alpha+\beta}(\mathbb{R}^N)}, \qquad \text{with } d_N(\alpha,\beta) \equiv \frac{c_N(\alpha) c_N(\beta)}{c_N(\alpha+\beta)}. \tag{1.7.6}$$

There exists also a version of this relation for negative Sobolev-Slobodetskii indices,

$$\int_{\mathbb{R}^N} \frac{\|\Delta_{\boldsymbol{h}} u\|^2_{H^{-\beta}(\mathbb{R}^N)}}{|\boldsymbol{h}|^{N+2\alpha}} \, d\boldsymbol{h} = d^*_N(\alpha,\beta) \|u\|^2_{H^{-\beta+\alpha}(\mathbb{R}^N)} + R_{\alpha,\beta}(u) \tag{1.7.7}$$

for $0 \le \alpha \le \beta$ with

$$d^*_N(\alpha,\beta) \equiv \frac{c_N(\alpha) c_N(\beta - \alpha)}{c_N(\beta)} \tag{1.7.8}$$

and a lower-order remainder $R_{\alpha,\beta}(u)$ satisfying

$$|R_{\alpha,\beta}(u)| \le c(\alpha,\beta) \|u\|^2_{H^{2(\alpha-\beta)}(\mathbb{R}^N)}, \tag{1.7.9}$$

where $c(\alpha,\beta)$ depends only on α, β and on N. Moreover, there holds $d^*_N(\alpha,\alpha) = 1$ and $R_{\alpha,\alpha}(u) = \|u\|^2_{H^{-\alpha}(\Omega)}$. These renormation formulae are valid for the whole space. In the case of a halfspace $\mathbb{R}^{N-1} \times \mathbb{R}^+$ similar relations are valid for the tangential variables $(\boldsymbol{x}', 0)^\top$, where $\boldsymbol{x}' = (x_1, \ldots, x_{N-1})^\top$. For example, instead of relation (1.7.6) we have

$$\int_{\mathbb{R}^{N-1}} \frac{\|\Delta_{\boldsymbol{h}} u\|'^2_{H^{\beta}(\mathbb{R}^N)}}{|\boldsymbol{h}'|^{N-1+2\alpha}} \, d\boldsymbol{h}' = d_{N-1}(\alpha,\beta) \|u\|'^2_{H^{\alpha+\beta,\alpha}(\mathbb{R}^N)}$$

with $\boldsymbol{h} = (\boldsymbol{h}', 0)$, where the first index in the anisotropic Sobolev-Slobodetskii norm corresponds to the variables $\boldsymbol{x}'$ and the second one to x_N.

All the mentioned relations will be extended in Chapter 2 together with many more important relations between different definitions of norms.

1.7.2 Shift technique for a halfspace

In order to illustrate the shift technique, let us consider problem (1.5.15) for the case of a halfspace domain $\Omega = \mathbb{R}^{N-1} \times (0, +\infty)$ with its whole boundary $\Gamma = \Gamma_C = \mathbb{R}^{N-1} \times \{0\}$ as potential contact boundary. We assume that the support of the coefficient of friction $\mathfrak{F}$ is a subset of an open bounded ball in Γ and the friction force G has its support again inside this ball. The volume force $\boldsymbol{f}$ is also assumed to have a bounded support. In this case, the set of admissible functions is $\mathscr{C} = \{\boldsymbol{v} \in \boldsymbol{H}^1(\Omega);\ v_n \le g_n \text{ on } \Gamma\}$. Moreover, for simplicity of the presentation the coefficients $a_{ijk\ell}$ and d of the bilinear form are assumed to be constant.

The solvability of the problem on such a domain Ω can be proved as in Theorem 1.5.2. In order to avoid redundant complexities we assume $d > 0$ which easily ensures the validity of the *a priori* estimate (1.5.20).

For the halfspace domain we apply the shift operator defined in (1.7.4) to a function $\boldsymbol{v}$ defined on Ω in any direction $\boldsymbol{h} = (\boldsymbol{h}', 0)$ with $\boldsymbol{h}' \in \mathbb{R}^{N-1}$ tangential to the boundary. Inserting the test function $\boldsymbol{v} = \boldsymbol{u}_{-\boldsymbol{h}}$ into inequality (1.5.15) we obtain

$$\begin{aligned}\int_\Omega \big(d\boldsymbol{u}\cdot(\boldsymbol{u}_{-\boldsymbol{h}} - \boldsymbol{u}) + a(\boldsymbol{u}, \boldsymbol{u}_{-\boldsymbol{h}} - \boldsymbol{u})\big)\,d\boldsymbol{x}\qquad&\\ + \int_\Gamma G\big(|(\boldsymbol{u}_{-\boldsymbol{h}})_t| - |\boldsymbol{u}_t|\big)\,ds_{\boldsymbol{x}} \ge \int_\Omega \boldsymbol{f}\cdot(\boldsymbol{u}_{-\boldsymbol{h}} - \boldsymbol{u})\,d\boldsymbol{x}.&\end{aligned} \tag{1.7.10}$$

Then, the integration variable $\boldsymbol{x}$ of inequality (1.5.15) is shifted. This corresponds to a transformation of variables applied in all integrals, giving inequality

$$\begin{aligned}&\int_\Omega \big(d\boldsymbol{u}_{-\boldsymbol{h}}\cdot(\boldsymbol{v}_{-\boldsymbol{h}} - \boldsymbol{u}_{-\boldsymbol{h}}) + a(\boldsymbol{u}_{-\boldsymbol{h}}, \boldsymbol{v}_{-\boldsymbol{h}} - \boldsymbol{u}_{-\boldsymbol{h}})\big)\,d\boldsymbol{x}\\ &\qquad + \int_\Gamma G_{-\boldsymbol{h}}\big(|(\boldsymbol{v}_t)_{-\boldsymbol{h}}| - |(\boldsymbol{u}_t)_{-\boldsymbol{h}}|\big)\,ds_{\boldsymbol{x}}\\ &\quad \ge \int_\Omega \boldsymbol{f}_{-\boldsymbol{h}}\cdot(\boldsymbol{v}_{-\boldsymbol{h}} - \boldsymbol{u}_{-\boldsymbol{h}})\,d\boldsymbol{x}.\end{aligned} \tag{1.7.11}$$

Employing the test function $\boldsymbol{v}_{-\boldsymbol{h}} \equiv \boldsymbol{u}$ in the previous inequality, adding the result to (1.7.10) and multiplying the resulting inequality by -1 gives

$$\begin{aligned}\int_\Omega \big(d\,\Delta_{\boldsymbol{h}}\boldsymbol{u}\cdot\Delta_{\boldsymbol{h}}\boldsymbol{u} + a(\Delta_{\boldsymbol{h}}\boldsymbol{u}, \Delta_{\boldsymbol{h}}\boldsymbol{u})\big)\,d\boldsymbol{x} + \int_\Gamma \Delta_{\boldsymbol{h}}G\,\Delta_{\boldsymbol{h}}|\boldsymbol{u}_t|\,ds_{\boldsymbol{x}}&\\ \le \int_\Omega \Delta_{\boldsymbol{h}}\boldsymbol{f}\cdot\Delta_{\boldsymbol{h}}\boldsymbol{u}\,d\boldsymbol{x}.&\end{aligned}$$

This inequality is multiplied by $|\boldsymbol{h}'|^{-(N-1+2\alpha)}$ with $0 < \alpha < 1$ and then integrated with respect to $\boldsymbol{h}' \in \mathbb{R}^{N-1}$. Then for the expression

$$\mathscr{A}(\boldsymbol{u}) \equiv \left(\int_{\mathbb{R}^{N-1}} \frac{\|\Delta_{\boldsymbol{h}}\boldsymbol{u}\|_A^2}{|\boldsymbol{h}'|^{N-1+2\alpha}}\,d\boldsymbol{h}'\right)^{1/2}$$

with the energy norm $\|\boldsymbol{u}\|_A \equiv \sqrt{\int_\Omega a(\boldsymbol{u},\boldsymbol{u})\,d\boldsymbol{x}}$ the inequality

$$\begin{aligned}\mathscr{A}^2(\boldsymbol{u}) \leq &- \int_{\mathbb{R}^{N-1}} \int_\Gamma |\boldsymbol{h}'|^{-(N-1+2\alpha)} \Delta_{\boldsymbol{h}} G \Delta_{\boldsymbol{h}} |\boldsymbol{u_t}|\, ds_{\boldsymbol{x}}\, d\boldsymbol{h}' \\ &+ \int_{\mathbb{R}^{N-1}} |\boldsymbol{h}'|^{-(N-1+2\alpha)} \|\Delta_{\boldsymbol{h}} \boldsymbol{f}\|_{\mathring{\boldsymbol{H}}^{-1}(\Omega)} \|\Delta_{\boldsymbol{h}} \boldsymbol{u}\|_{\boldsymbol{H}^1(\Omega)}\, d\boldsymbol{h}'\end{aligned} \tag{1.7.12}$$

is derived, if we neglect the positive term containing $d\Delta_{\boldsymbol{h}}\boldsymbol{u} \cdot \Delta_{\boldsymbol{h}}\boldsymbol{u}$.

In order to estimate the friction term in (1.7.12) we employ the $(N-1)$-dimensional Fourier transform. It holds

$$\begin{aligned}&- \int_{\mathbb{R}^{N-1}} \int_\Gamma |\boldsymbol{h}'|^{-(N-1+2\alpha)} \Delta_{\boldsymbol{h}} G \Delta_{\boldsymbol{h}} |\boldsymbol{u_t}|\, ds_{\boldsymbol{x}}\, d\boldsymbol{h}' \\ &= -\mathrm{Re} \int_{\mathbb{R}^{N-1}} c_{N-1}(\alpha) |\boldsymbol{\xi}|^{2\alpha} \mathfrak{F}(G) \overline{\mathfrak{F}(|\boldsymbol{u_t}|)}\, d\boldsymbol{\xi} \\ &\leq c_{N-1}(\alpha) \|G\|_{H^{-(1/2-\alpha)}(\Gamma_C)} \cdot \\ &\quad \cdot \sqrt{\int_{R^{N-1}} |\mathfrak{F}(|\boldsymbol{u_t}|)|^2 |\boldsymbol{\xi}|^{4\alpha} \left(1 + c_{N-1}(\tfrac{1}{2} - \alpha)|\boldsymbol{\xi}|^{1-2\alpha}\right) d\boldsymbol{\xi}} \\ &\leq c_{N-1}(\alpha)(1+\varepsilon) \|G\|_{H^{-(1/2-\alpha)}(\Gamma_C)} \cdot \\ &\quad \cdot \left[\sqrt{\frac{c_{N-1}(\frac{1}{2}-\alpha)}{c_{N-1}(\frac{1}{2}+\alpha)}} \|\boldsymbol{u_t}\|'_{H^{1/2+\alpha}(\Gamma_C)} + c(\alpha,\varepsilon) \|\boldsymbol{u_t}\|'_{H^{1/2}(\Gamma_C)} \right]\end{aligned} \tag{1.7.13}$$

with $\varepsilon > 0$ arbitrarily small. The last row of (1.7.13) is obvious for $\alpha \in (0, 1/4]$ with $\varepsilon = 0$ and the estimate (1.5.23) for the Euclidean norm. For $1/4 < \alpha < 1/2$ the term $|\boldsymbol{\xi}|^{4\alpha}$ is estimated with the help of Hölder's inequality $yz \leq p^{-1}y^p + q^{-1}z^q$ applied for $p = 2\alpha/(4\alpha - 1)$, $q = 2\alpha/(1-2\alpha)$, $y = (p\varepsilon|\boldsymbol{\xi}|^{1+2\alpha})^{1/p}$ and $z = (p\varepsilon)^{-1/p}|\boldsymbol{\xi}|^{1/q}$. For $\alpha = 1/2$ the second row of (1.7.13) has the form $\|G\|_{L_2(\Gamma_C)} \left(\int_\Omega |\mathfrak{F}(|\boldsymbol{u_t}|)|^2 |\boldsymbol{\xi}|^2 d\boldsymbol{\xi}\right)^{1/2}$ which is bounded by $\|G\|_{L_2(\Gamma_C)} \|\boldsymbol{u_t}\|_{H^1(\Gamma_C)}$.

The inequality (1.5.23), valid for $0 \leq \beta \leq 1$ only, limits the regularity gain α to $\alpha \leq 1/2$. This limitation of regularity of solutions is a consequence of the non-smooth character of the Coulomb friction which we will face in many problems studied in the next chapters.

The trace theorem for halfspaces and the Korn inequality (1.2.11) yield

$$\|\boldsymbol{v_t}\|^2_{\boldsymbol{H}^{1/2}(\Gamma)} \leq C_1^{(0)} \|\boldsymbol{v}\|_A^2 + c_1 \|\boldsymbol{v}\|^2_{\boldsymbol{L}_2(\Omega)} \tag{1.7.14}$$

for all $\boldsymbol{v} \in \boldsymbol{H}^1(\Omega)$ with a specific constant $C_1^{(0)}$. This gives

$$\|\boldsymbol{u_t}\|'^2_{H^{1/2+\alpha}(\Gamma_C)} = d_{N-1}^{-1}\left(\alpha, \tfrac{1}{2}\right) \int_{\mathbb{R}^{N-1}} \frac{\|\Delta_{\boldsymbol{h}} \boldsymbol{u_t}\|'^2_{H^{1/2}(\Gamma)}}{|\boldsymbol{h}'|^{N-1+2\alpha}}\, d\boldsymbol{h}'$$

$$\leq d_{N-1}^{-1}\left(\alpha, \tfrac{1}{2}\right) C_1^{(0)} \mathscr{A}^2(\boldsymbol{u}) + c\|\boldsymbol{u}\|^2_{H^1(\Omega)}.$$

Using the inequalities

$$\int_{\mathbb{R}^{N-1}} \frac{\|\Delta_{\boldsymbol{h}} \boldsymbol{f}\|^2_{\mathring{\boldsymbol{H}}^{-1}(\Omega)}}{|\boldsymbol{h}'|^{N-1+2\alpha}}\, d\boldsymbol{h}' \leq c\|\boldsymbol{f}\|^2_{\mathring{\boldsymbol{H}}^{-1+\alpha}(\Omega)} \qquad \text{and}$$

$$\int_{\mathbb{R}^{N-1}} \frac{\|\Delta_{\boldsymbol{h}} \boldsymbol{u}\|^2_{\boldsymbol{H}^1(\Omega)}}{|\boldsymbol{h}'|^{N-1+2\alpha}}\, d\boldsymbol{h}' \leq c_1 \mathscr{A}^2(\boldsymbol{u}) + c_2\|\boldsymbol{u}\|^2_{\boldsymbol{H}^\alpha(\Omega)},$$

the estimates for the lower order norms by

$$\int_{\mathbb{R}^{N-1}} \frac{\|\Delta_{\boldsymbol{h}} \boldsymbol{u}\|^2_{\boldsymbol{L}_2(M)}}{|\boldsymbol{h}'|^{N-1+2\alpha}}\, d\boldsymbol{h}' \ \leq \ c_1\|\boldsymbol{u}\|^2_{\boldsymbol{H}^\alpha(M)}$$

for $M = \Omega$ and $M = \Gamma$ and the relation $d_{N-1}^{-1}(\alpha, \frac{1}{2}) \dfrac{c_{N-1}(\frac{1}{2} - \alpha)}{c_{N-1}(\frac{1}{2} + \alpha)}\, c_{N-1}^2(\alpha) = d_{N-1}^*(\alpha, \frac{1}{2})$, we obtain

$$\mathscr{A}^2(\boldsymbol{u}) \leq (1+\varepsilon)\sqrt{d_{N-1}^*(\alpha, \tfrac{1}{2}) C_1^{(0)}}\, \|G\|_{H^{1/2-\alpha}(\Gamma)} \mathscr{A}(\boldsymbol{u}) + c(\varepsilon)$$

with $\varepsilon > 0$ arbitrarily small and c dependent on $\|\boldsymbol{f}\|_{\mathring{\boldsymbol{H}}^{-1+\alpha}(\Omega)}$ only (due to the validity of the appropriate version of the *a priori* estimate (1.5.20)). From this the inequality

$$\mathscr{A}(\boldsymbol{u}) \leq (1+\varepsilon)\sqrt{d_{N-1}^*(\alpha, \tfrac{1}{2}) C_1^{(0)}}\, \|G\|_{H^{1/2-\alpha}(\Gamma)} + c(\varepsilon) \tag{1.7.15}$$

follows with $\varepsilon > 0$ arbitrarily small and $c(\varepsilon)$ dependent on $\|\boldsymbol{f}\|_{\mathring{\boldsymbol{H}}^{-1+\alpha}(\Omega)}$.

In order to verify the estimate of the boundary traction 1.5.28 the shift technique is applied to the Green formula

$$\langle \sigma_n(\boldsymbol{u}), v_n \rangle \equiv \int_\Omega \left(d\,\boldsymbol{u} \cdot \boldsymbol{v} + a(\boldsymbol{u}, \boldsymbol{v}) - \boldsymbol{f} \cdot v \right) d\boldsymbol{x} \tag{1.7.16}$$

valid for all $\boldsymbol{v} \in \boldsymbol{H}^1(\Omega)$ with $\boldsymbol{v_t} = (v_1, \ldots, v_{N-1}, 0) = 0$ on Γ. Let $\boldsymbol{w} \in \boldsymbol{H}^1(\Omega)$ with $\boldsymbol{w_t} = 0$ be a fixed test function. We use the test function

$\boldsymbol{v} = \boldsymbol{w}_{-\boldsymbol{h}} - \boldsymbol{w}$ in equation (1.7.16), we use $\boldsymbol{v}_{-\boldsymbol{h}} \equiv \boldsymbol{w} - \boldsymbol{w}_{-\boldsymbol{h}}$ in the shifted Green formula

$$\langle (\sigma_n(\boldsymbol{u}))_{-\boldsymbol{h}}, (v_n)_{-\boldsymbol{h}} \rangle \equiv \int_{\Omega} \big(d\boldsymbol{u}_{-\boldsymbol{h}} \cdot \boldsymbol{v}_{-\boldsymbol{h}} + a(\boldsymbol{u}_{-\boldsymbol{h}}, \boldsymbol{v}_{-\boldsymbol{h}}) \\ - \boldsymbol{f}_{-\boldsymbol{h}} \cdot \boldsymbol{v}_{-\boldsymbol{h}} \big)\, dx, \tag{1.7.17}$$

we add both results, multiply by $|\boldsymbol{h}|^{-(N-1+2\alpha)}$ and integrate with respect to $\boldsymbol{h}' \in \mathbb{R}^{N-1}$. Then the estimate

$$\int_{\mathbb{R}^{N-1}} \frac{\langle \Delta_{\boldsymbol{h}}(\sigma_n(\boldsymbol{u})), \Delta_{\boldsymbol{h}} w_n \rangle}{|\boldsymbol{h}'|^{N-1+2\alpha}}\, d\boldsymbol{h}' \\ \leq \mathscr{A}(\boldsymbol{u})\mathscr{A}(\boldsymbol{w}) + c_1 \|\boldsymbol{u}\|_{\boldsymbol{H}^{\alpha}(\Omega)} \|\boldsymbol{w}\|_{\boldsymbol{H}^{\alpha}(\Omega)} \\ + c_2 \|\boldsymbol{f}\|_{\mathring{\boldsymbol{H}}^{-1+\alpha}(\Omega)} \big(\mathscr{A}(\boldsymbol{w}) + \|\boldsymbol{w}\|_{\boldsymbol{H}^{\alpha}(\Omega)} \big) \tag{1.7.18}$$

is derived. The left hand side of this relation has the following representation in terms of Fourier transformed functions:

$$\mathrm{Re}\left(\int_{\mathbb{R}^{N-1}} \mathfrak{F}(\sigma_n(\boldsymbol{u}); \boldsymbol{\xi}) \overline{\mathfrak{F}(w_n; \boldsymbol{\xi})} c_{N-1}(\alpha) |\boldsymbol{\xi}|^{2\alpha}\, d\boldsymbol{\xi} \right). \tag{1.7.19}$$

The function $\boldsymbol{w}$ is chosen in such a way that

$$\mathfrak{F}(w_n; \boldsymbol{\xi}) \equiv \mathfrak{F}(\sigma_n(\boldsymbol{u}); \boldsymbol{\xi}) \big(1 + c_{N-1}(\tfrac{1}{2})|\boldsymbol{\xi}|\big)^{-1}. \tag{1.7.20}$$

From the equation $\dfrac{c_{N-1}(\frac{1}{2})|\boldsymbol{\xi}|}{1 + c_{N-1}(\frac{1}{2})|\boldsymbol{\xi}|} = 1 - \dfrac{1}{1 + c_{N-1}(\frac{1}{2})|\boldsymbol{\xi}|}$ and the inequality

$$\frac{c_{N-1}(\alpha)|\boldsymbol{\xi}|^{2\alpha}}{1 + c_{N-1}(\frac{1}{2})|\boldsymbol{\xi}|} \geq \frac{c_{N-1}(\alpha) c_{N-1}(\frac{1}{2} - \alpha)}{c_{N-1}(\frac{1}{2})} \frac{1}{1 + c_{N-1}(\frac{1}{2} - \alpha)|\boldsymbol{\xi}|^{1-2\alpha}} \\ - c \frac{1}{1 + |\boldsymbol{\xi}|^{2-4\alpha}}$$

there follows an estimate of expression (1.7.19) from below by

$$\|\sigma_n(\boldsymbol{u})\|^2_{L_2(\Gamma)} - \|\sigma_n(\boldsymbol{u})\|^2_{H^{-1/2}(\Gamma)}$$

for $\alpha = \frac{1}{2}$ and by

$$d_{N-1}(\alpha, \tfrac{1}{2} - \alpha) \|\sigma_n(\boldsymbol{u})\|^2_{H^{-(1/2-\alpha)}(\Gamma)} - c_1 \|\sigma_n(\boldsymbol{u})\|^2_{H^{-2(1/2-\alpha)}(\Gamma)}$$

for $\alpha < \frac{1}{2}$. In the latter case the second term can be further estimated using

$$\|\sigma_n(\boldsymbol{u})\|^2_{H^{-2(1/2-\alpha)}(\Gamma)} \leq \varepsilon \|\sigma_n(\boldsymbol{u})\|^2_{H^{-(1/2-\alpha)}(\Gamma)} + c(\varepsilon) \|\sigma_n(\boldsymbol{u})\|^2_{H^{-1/2}(\Gamma)}$$

with $\varepsilon > 0$ arbitrarily small at the expense of constant $c(\varepsilon)$. Moreover, there holds

$$\|w_n\|^2_{H^{1/2}(\Gamma)} = \|\sigma_n(\boldsymbol{u})\|^2_{H^{-1/2}(\Gamma)} \text{ and}$$

$$\int_{\mathbb{R}^{N-1}} \frac{\|\Delta_{\boldsymbol{h}} w_n\|'^2_{H^{1/2}(\Gamma)}}{|\boldsymbol{h}'|^{N-1+2\alpha}} \, d\boldsymbol{h}' \leq (1+\varepsilon) d_{N-1}(\alpha, \tfrac{1}{2} - \alpha) \|\sigma_n\|^2_{H^{-(1/2-\alpha)}(\Gamma)} + c_2(\varepsilon) \|\sigma_n\|^2_{H^{-1/2}(\Gamma)},$$

the latter inequality follows from

$$\frac{c_{N-1}(\alpha)|\boldsymbol{\xi}|^{2\alpha}}{1 + c_{N-1}(\frac{1}{2})|\boldsymbol{\xi}|} \leq \frac{c_{N-1}(\alpha) c_{N-1}(\frac{1}{2} - \alpha)}{c_{N-1}(\frac{1}{2})} \frac{1}{1 + c_{N-1}(\frac{1}{2} - \alpha)|\boldsymbol{\xi}|^{1-2\alpha}} + c \frac{1}{1 + |\boldsymbol{\xi}|^{2-4\alpha}}.$$

There exists a linear continuous extension operator $\boldsymbol{H}^{1/2}(\Gamma) \to \boldsymbol{H}^1(O)$ such that the inequality

$$\|\boldsymbol{v}\|^2_A \leq C_0^{(2)} \|v_n\|^2_{H^{1/2}(\Gamma)} \tag{1.7.21}$$

is valid for any $\boldsymbol{v} = (0, \ldots, 0, v_n)^\top \in \boldsymbol{H}^{1/2}(\Gamma)$. This is possible due to the trace theorem and the boundedness of the bilinear form a. Employing all these relations and the extension operator to $v_n = w_n$ it can be proved from (1.7.18) that

$$\|\sigma_n(\boldsymbol{u})\|_{H^{-(1/2-\alpha)}(\Gamma)} \leq (1+\varepsilon) d_{N-1}^{-1/2}\left(\alpha, \tfrac{1}{2} - \alpha\right) \sqrt{C_0^{(2)} \mathscr{A}(\boldsymbol{u})} + c_1(\varepsilon) \sqrt{\mathscr{A}(\boldsymbol{u})} + c_2(\varepsilon) \tag{1.7.22}$$

with $\varepsilon > 0$ arbitrarily small. Hence for $\Omega = \mathbb{R}^{N-1} \times \mathbb{R}_+$ Theorem 1.5.8 is proved with constant $C_0 = \sqrt{C_0^{(1)} C_0^{(2)}}$.

The calculation of the constants $C_0^{(1)}, C_0^{(2)}$, which are crucial for the existence of solutions to the contact problems with Coulomb friction, should be performed in some optimal way. Since such a calculation may differ for different forms of the Hooke tensor, we omit it here. However, if they are calculated for the halfspace domain, their value is also suitable for more general domains as it will be shown below.

1.7.3 Shift technique with local rectification

The shift technique for a general domain will be studied for the case of a vanishing initial gap $g_n = 0$ on Γ_C. This is no restriction; if the initial gap does not vanish, we represent the unknown function $\boldsymbol{u}$ by $\widetilde{\boldsymbol{u}} + \boldsymbol{w}$, where $\boldsymbol{w}$ satisfies $w_n = g_n$ and $\boldsymbol{w_t} = 0$ on Γ_C. The new function $\widetilde{\boldsymbol{u}}$ then satisfies the variational inequality (1.5.15) with the modified data $\widetilde{\boldsymbol{U}} = \boldsymbol{U} - \boldsymbol{w}$, $\widetilde{g}_n = 0$ and $\widetilde{\mathscr{L}}(\boldsymbol{v}) \equiv \mathscr{L}(\boldsymbol{v}) - A(\boldsymbol{w}, \boldsymbol{v})$, and the regularity proof is carried out for this modified version. Due to the requirements $g_n \in H^{1/2+\alpha}(\Gamma_C)$ and $\Gamma_C \in C^2$ the function $\boldsymbol{w}$ can be chosen from $H^{1+\alpha}(\Omega_C)$, and then there still holds $\widetilde{\mathscr{L}} \in \mathring{\boldsymbol{H}}^{-1+\alpha}(\Omega_C)$.

The application of the shift technique to a general domain requires a local rectification of the boundary in order to transform the problem (locally) onto a halfspace. This is described here as the example of problem (1.5.15) serving again as a model problem. In the description of the rectification, the notions of Section 1.6 will be employed. The requirements for the smoothness of the boundary are $\Gamma_C \in C^{2+\beta}$ with $\beta > 0$ to be specified below.

For a parameter $\delta > 0$—which is chosen sufficiently small as described below—let $\mathscr{U}_\delta := \{U_i; i \in \mathscr{I}_\delta\}$ denote a finite covering of $\Gamma_{\mathfrak{F}}$ of the type described in Section 1.6, with the local rectification $\mathfrak{R}_i = \Psi_i \circ \mathscr{O}_i$. Since $\Gamma_{\mathfrak{F}}$ has a positive distance to $\Gamma \setminus \Gamma_C$ we may assume $U_i \cap \Gamma \setminus \Gamma_C = \emptyset$. For simplicity of the presentation we assume furthermore $U_i = \mathfrak{R}_i^{-1}\big(B_{N-1}(\delta) \times (-\delta, \delta)\big)$. We recall that $\mathscr{O}_i$ is a combination of a rotation and a translation mapping the tangent plane to Γ at the point $\boldsymbol{x}_i \equiv \mathscr{O}_i(0) \in \Gamma_{\mathfrak{F}}$ onto the hyperplane $\{(\boldsymbol{y}', 0);\ \boldsymbol{y}' \in \mathbb{R}^{N-1}\}$, and Ψ_i denotes a straightening of the graph of the function ψ_i locally describing the boundary. Then, for ψ_i there holds $\nabla\psi_i(\boldsymbol{0}) = \boldsymbol{0}$,

$$|\nabla\psi_i(\boldsymbol{x})| \leq c\delta \quad \text{for all } \boldsymbol{x} \in B_{N-1}(\delta) \tag{1.7.23}$$

and there exists a $\delta_0 > 0$ such that the constant c is independent of the choice of δ for $\delta \leq \delta_0$ and of $i \in \mathscr{I}_\delta$. Let

$$\mathfrak{P}_\delta \equiv \{\varrho_i;\ i \in \mathscr{I}_\delta\} \tag{1.7.24}$$

be a partition of unity on $\Gamma_{\mathfrak{F}}$ subordinate to the covering $\mathscr{U}_\delta$.

We fix an index $i \in \mathscr{I}_\delta$, consider the corresponding cutoff function ϱ_i, and denote its support in Γ by Γ_ϱ. If δ is sufficiently small, then the set

$$\Omega_\varrho := \{\mathscr{O}_i^{-1}(\boldsymbol{y}', \psi(\boldsymbol{y}') + y_N);\ \boldsymbol{y}' \in B_{N-1}(\delta),\ y_N \in (0, \delta)\}$$

describes the intersection of Ω with a neighbourhood of $\boldsymbol{x} = \boldsymbol{x}_i$ in $\mathbb{R}^N$. Let us further denote $\Gamma_\varrho = \Gamma \cap \partial\Omega_\varrho$. For simplicity of the notation, we omit the index i in ϱ_i, $\mathscr{O}_i$, Ψ_i and ψ_i, and we recall the definition of the local rectification map $\mathfrak{R} \equiv \Psi \circ \mathscr{O}$ of Ω.

Inserting the test function $\boldsymbol{u} + \varrho(\boldsymbol{v} - \varrho\boldsymbol{u})$ into variational inequality (1.5.15) and shifting the cutoff function ϱ from the right hand side of each form to the left hand side yields

$$\begin{aligned} &\int_{\Omega_\varrho} \left(d\,\varrho\boldsymbol{u}\cdot(\boldsymbol{v}-\varrho\boldsymbol{u}) + a(\varrho\boldsymbol{u},\boldsymbol{v}-\varrho\boldsymbol{u}) + b(\boldsymbol{u},\boldsymbol{v}-\varrho\boldsymbol{u})\right)dx \\ &\quad + \int_{\Gamma_\varrho} \varrho G\left(|\boldsymbol{v_t}| - |(\varrho\boldsymbol{u})_{\boldsymbol{t}}|\right) ds_{\boldsymbol{x}} \geq \int_{\Omega_\varrho} \varrho\boldsymbol{f}\cdot(\boldsymbol{v}-\varrho\boldsymbol{u})\,dx \end{aligned} \tag{1.7.25}$$

with the bilinear form

$$\begin{aligned} b(\boldsymbol{u},\boldsymbol{v}) &= a(\boldsymbol{u},\varrho\boldsymbol{v}) - a(\varrho\boldsymbol{u},\boldsymbol{v}) \\ &= \int_{\Omega_\varrho} a_{ijk\ell}\left(e_{ij}(\boldsymbol{u})\partial_\ell\varrho\, v_k - \partial_j\varrho u_i\, e_{k\ell}(\boldsymbol{v})\right) d\boldsymbol{x}. \end{aligned} \tag{1.7.26}$$

Here the estimate $|\boldsymbol{u_t} + \varrho(\boldsymbol{v_t} - \varrho\boldsymbol{u_t})| - |\boldsymbol{u_t}| \leq \varrho|\boldsymbol{v_t}| - \varrho^2|\boldsymbol{u_t}|$ and the non-negativity of G is essential. Performing the change of variables

$$\boldsymbol{x} = \mathfrak{Q}(\boldsymbol{y}) = \mathfrak{R}^{-1}(\boldsymbol{y}),\ \boldsymbol{y} = (\boldsymbol{y}', y_N) \in O_\delta := B_{N-1}(0,\delta)\times(0,\delta), \tag{1.7.27}$$

this variational inequality is transformed onto the domain O_δ. Due to the special form of the function $\Psi : (\boldsymbol{x}', x_N) \mapsto (\boldsymbol{x}', x_N - \psi(\boldsymbol{x}'))$ its inverse is given by $(\boldsymbol{x}', x_N) \mapsto (\boldsymbol{x}', x_N + \psi(\boldsymbol{x}'))$ and we have

$$\partial_{y_\ell}\Psi_k^{-1} = \delta_{k\ell} + \delta_{kN}\partial_{y_\ell}\psi$$

valid for $k,\ell = 1,\ldots,N$ with the convention $\partial_{y_N}\psi = 0$. Hence, the determinant of the Jacobian of the parametrization (1.7.27) equals to 1 and the corresponding density of surface measure is equal to $J \equiv \sqrt{1+|\nabla\psi|^2}$. The latter expression is a consequence of

$$\left\{\partial_i\Psi^{-1}\cdot\partial_j\Psi^{-1}\right\}_{i,j=1}^{N} = \boldsymbol{Id} + \nabla\psi(\nabla\psi)^\top$$

and the formula $\det(\boldsymbol{Id} + \boldsymbol{z}\boldsymbol{z}^\top) = 1 + |\boldsymbol{z}|^2$ valid for a vector $\boldsymbol{z} \in \mathbb{R}^N$.

For any function $\varphi : \Omega_\varrho \to \mathbb{R}$ we denote the transformed function by $\widetilde{\varphi} \equiv \varphi \circ \mathfrak{Q}$. Then there holds the local representation $\varphi(\boldsymbol{x}) = \widetilde{\varphi}(\mathfrak{R}(\boldsymbol{x}))$ with $\mathfrak{R} : \boldsymbol{x} \mapsto \Psi \circ \mathscr{O}(\boldsymbol{x})$, cf. Subsection 1.6. For its derivatives there holds

$$\partial_j\varphi = \sum_{k,\ell=1}^{N} \partial_k\widetilde{\varphi}\,\mathscr{O}_{\ell j}\partial_\ell\Psi_k$$

with $\{\mathscr{O}_{k\ell}\} = \{\partial_\ell \mathscr{O}_k\}$ denoting the (constant) gradient of the transformation $\mathscr{O}$. Hence, the transformation of the bilinear form is

$$\begin{aligned} a(\boldsymbol{u},\boldsymbol{v}) &= \sum_{i,j,k,\ell=1}^{N} a_{ijk\ell}\partial_j u_i \partial_\ell v_k \\ &= \sum_{i,j,k,\ell,r,s=1}^{N} a_{ijk\ell}\mathscr{O}_{rj}\mathscr{O}_{s\ell}\partial_r \widetilde{u}_i \partial_s \widetilde{v}_k \\ &\quad - \sum_{i,j,k,\ell,r,s=1}^{N} a_{ijk\ell}\mathscr{O}_{rj}\partial_r\psi\mathscr{O}_{s\ell}\partial_s\psi\partial_N\widetilde{u}_i\partial_N\widetilde{v}_k \\ &= \widetilde{a}(\widetilde{\boldsymbol{u}},\widetilde{\boldsymbol{v}}) - b_0(\widetilde{\boldsymbol{u}},\widetilde{\boldsymbol{v}}) \end{aligned} \tag{1.7.28}$$

with bilinear form $\widetilde{a}(\cdot,\cdot)$ having the coefficients

$$\widetilde{a}_{ijk\ell} = \sum_{r,s=1}^{N} a_{irks}\mathscr{O}_{jr}\mathscr{O}_{\ell s} \tag{1.7.29}$$

and with a small remainder $b_0(\cdot,\cdot)$ satisfying

$$|b_0(\widetilde{\boldsymbol{u}},\widetilde{\boldsymbol{v}})| \leq C\delta|\nabla\widetilde{\boldsymbol{u}}||\nabla\widetilde{\boldsymbol{v}}|. \tag{1.7.30}$$

For homogeneous, isotropic material the coefficients $a_{ijk\ell}$ remain the same, $\widetilde{a}_{ijk\ell} = a_{ijk\ell}$. In the case of anisotropic material, the coefficients change, but the constants a_0 and A_0 from the ellipticity and boundedness conditions do not change. In a similar way $b(\boldsymbol{u},\boldsymbol{v})$ is transformed to $\widetilde{b}(\widetilde{\boldsymbol{u}},\widetilde{\boldsymbol{v}})$ satisfying

$$|\widetilde{b}(\widetilde{\boldsymbol{u}},\widetilde{\boldsymbol{v}})| \leq c\big(\|\widetilde{\boldsymbol{u}}\|_{\boldsymbol{L}_2(O_\delta)}\|\widetilde{\boldsymbol{v}}\|_{\boldsymbol{H}^1(O_\delta)} + \|\widetilde{\boldsymbol{u}}\|_{\boldsymbol{H}^1(O_\delta)}\|\widetilde{\boldsymbol{v}}\|_{\boldsymbol{L}_2(O_\delta)}\big). \tag{1.7.31}$$

As a consequence, we obtain the transformed variational inequality

$$\begin{aligned} &\int_{O_\delta} \big(d\,\widetilde{\varrho\boldsymbol{u}}\cdot(\widetilde{\boldsymbol{v}} - \widetilde{\varrho\boldsymbol{u}}) + \widetilde{a}(\widetilde{\varrho\boldsymbol{u}},\widetilde{\boldsymbol{v}} - \widetilde{\varrho\boldsymbol{u}}) + \widetilde{b}(\widetilde{\boldsymbol{u}},\widetilde{\boldsymbol{v}} - \widetilde{\varrho\boldsymbol{u}}) \\ &\qquad + b_0(\widetilde{\varrho\boldsymbol{u}},\widetilde{\boldsymbol{v}} - \widetilde{\varrho\boldsymbol{u}})\big)\,d\boldsymbol{x} + \int_{S_\delta} J\widetilde{\varrho G}\,(|\widetilde{\boldsymbol{v}}_{\widetilde{\boldsymbol{t}}}| - |(\widetilde{\varrho\boldsymbol{u}})_{\widetilde{\boldsymbol{t}}}|)\,ds_{\boldsymbol{x}} \\ &\qquad \geq \int_{S_\delta} \widetilde{\varrho\boldsymbol{f}}\cdot(\widetilde{\boldsymbol{v}} - \widetilde{\varrho\boldsymbol{u}})\,d\boldsymbol{x}. \end{aligned} \tag{1.7.32}$$

Here, the remaining part of the boundary is denoted by $S_\delta := B_{N-1}(\delta) \times \{0\}$. The tilde on the index $\boldsymbol{t}$ of the tangential component shall indicate that it is tangential to the old boundary and not to the boundary of the infinite halfspace. Now, all the integrands in this variational formulation can be extended by 0 to the halfspace $O \equiv \mathbb{R}^{N-1} \times \mathbb{R}_+$ because in every term we have the cutoff function ϱ or one of its derivatives as a multiplicative factor. In order to be formally precise, we extend $\widetilde{\varrho \boldsymbol{u}}$, $\widetilde{\varrho G}$ and $\widetilde{\varrho \boldsymbol{f}}$ by 0, the function $\widetilde{\boldsymbol{u}}$ in such a way that $\|\widetilde{\boldsymbol{u}}\|_{\boldsymbol{H}^1(O)} \leq c\|\widetilde{\boldsymbol{u}}\|_{\boldsymbol{H}^1(O_\delta)}$ holds. The coefficients of the bilinear form a are extended such that they remain constant in the case of homogeneous anisotropic material or that they remain Lipschitz and the constants a_0 and A_0 of ellipticity and boundedness are valid for the extended forms, too. The remaining bilinear forms $\widetilde{b}$ and b_0 are extended such that their coefficients remain bounded and Lipschitz and the relations (1.7.30) and (1.7.31) are still valid. Changing the notation of the localized and transformed function(al)s to $\boldsymbol{u}$ for $\widetilde{\varrho \boldsymbol{u}}$, G for $J\widetilde{\varrho G}$, $\boldsymbol{f}$ for $\widetilde{\varrho \boldsymbol{f}}$ and omitting the tildes at the forms a, b, b_0 and the index t for tangential components we obtain a variational formulation on the halfspace,

$$\int_O \big(d\boldsymbol{u} \cdot (\boldsymbol{v} - \boldsymbol{u}) + a(\boldsymbol{u}, \boldsymbol{v} - \boldsymbol{u}) + b_0(\boldsymbol{u}, \boldsymbol{v} - \boldsymbol{u}) + b(\widetilde{\boldsymbol{u}}, \boldsymbol{v} - \boldsymbol{u})\big)\, dx$$

$$+ \int_{\mathbb{R}^{N-1}} G\left(|\boldsymbol{v_t}| - |\boldsymbol{u_t}|\right) ds_{\boldsymbol{x}} \geq \int_O \boldsymbol{f} \cdot (\boldsymbol{v} - \boldsymbol{u})\, dx. \tag{1.7.33}$$

In this formula, the tilde on $\boldsymbol{u}$ in $b(\widetilde{\boldsymbol{u}}, \boldsymbol{v} - \boldsymbol{u})$ remains in order to indicate that in (1.7.32) the first argument of b is $\widetilde{\boldsymbol{u}}$ and not $\widetilde{\varrho \boldsymbol{u}}$.

The transformation of the Green formula after the application of the local rectification has the form

$$\int_{\partial O} \sigma_n v_n \, ds_{\boldsymbol{x}} = \int_\Omega \big(d\boldsymbol{u} \cdot \boldsymbol{v} + a(\boldsymbol{u}, \boldsymbol{v}) + b_0(\boldsymbol{u}, \boldsymbol{v}) + b(\widetilde{\boldsymbol{u}}, \boldsymbol{v}) - \boldsymbol{f} \cdot \boldsymbol{v}\big) dx,$$

where σ_n denotes the extended version of $J\widetilde{\varrho \sigma_n}$. This formula is valid for functions $\boldsymbol{v}$ satisfying $\boldsymbol{v_t} = 0$, where $\boldsymbol{v_t}$ is still the component tangential to the old boundary.

Based on these two relations, the shift technique can be performed similarly to the preceding section. There are basically three different modifications necessary: first, the additional perturbed forms b and b_0 must be estimated, second, the dependence of the coefficients of a on the space variable must be taken into account and third, we must cope with the different orientation of the normal vector at different points of the boundary. In order to tackle the last task, we define the rotation operator $\mathfrak{T}_{-\boldsymbol{h}}(\boldsymbol{x})$ transforming the unit outer normal vector $\boldsymbol{n}_{-\boldsymbol{h}}(\boldsymbol{x}) = \boldsymbol{n}(\boldsymbol{x} + \boldsymbol{h})$ to $\boldsymbol{n}(\boldsymbol{x})$ with axis of rotation perpendicular to both vectors. Such a rotation transforms the tangential hyperplane at $\boldsymbol{x} + \boldsymbol{h}$ to that at $\boldsymbol{x}$. If $\Gamma_{\mathfrak{F}} \in C^{2+\beta}$ holds, then the normal vector $\boldsymbol{n}$ is in $C^{1+\beta}(\partial O)$ (after the straightening of the boundary), and there holds $|\boldsymbol{n} - \boldsymbol{n}_{-\boldsymbol{h}}| \leq C_1|\boldsymbol{h}'|$ and $|\partial_j(\boldsymbol{n} - \boldsymbol{n}_{-\boldsymbol{h}})| \leq C_2|\boldsymbol{h}'|^\beta$ for $j = 1, \ldots, N-1$. Using these estimates it is possible to prove $|\mathfrak{T}_{-\boldsymbol{h}} - \boldsymbol{Id}| \leq C_3|\boldsymbol{h}'|$ and

$|\partial_j \mathfrak{T}_{-\boldsymbol{h}}| \le C_4|\boldsymbol{h}'|^{\beta}$ for $j = 1, \ldots, N-1$. Details can be found in Lemma 3.2 of [73], where the procedure is studied for space dimension $N = 3$.

In the shift technique it is necessary to keep the test functions in the cone $\mathscr{C}$ by satisfying $\boldsymbol{v}_n \le g_n$. This is not trivial because the shifted function $\boldsymbol{u}_{-\boldsymbol{h}}$ satisfies $(u_n)_{-\boldsymbol{h}} = \boldsymbol{u}_{-\boldsymbol{h}} \cdot \boldsymbol{n}_{-\boldsymbol{h}} \le 0$, but in general not $(\boldsymbol{u}_{-\boldsymbol{h}})_n = \boldsymbol{u}_{-\boldsymbol{h}} \cdot \boldsymbol{n} \le 0$ due to the change of the normal vector. Therefore, in the non-shifted variational inequality we insert the test function $\boldsymbol{v} = \mathfrak{T}_{-\boldsymbol{h}}\boldsymbol{u}_{-\boldsymbol{h}}$ instead of $\boldsymbol{u}_{-\boldsymbol{h}}$ while for the shifted inequality we insert $\boldsymbol{v}_{-\boldsymbol{h}} = (\mathfrak{T}_{\boldsymbol{h}})_{-\boldsymbol{h}}\boldsymbol{u}$. Here, $(\mathfrak{T}_{\boldsymbol{h}})_{-\boldsymbol{h}}(\boldsymbol{x}) = \mathfrak{T}_{\boldsymbol{h}}(\boldsymbol{x} + \boldsymbol{h})$ maps $\boldsymbol{n}(\boldsymbol{x})$ to $\boldsymbol{n}(\boldsymbol{x} + \boldsymbol{h})$.

For a space-dependent bilinear form the shift technique yields the term

$$\begin{aligned} &a_{-\boldsymbol{h}}(\boldsymbol{u}_{-\boldsymbol{h}}, (\mathfrak{T}_{\boldsymbol{h}})_{-\boldsymbol{h}}\boldsymbol{u} - \boldsymbol{u}_{-\boldsymbol{h}}) + a(\boldsymbol{u}, \mathfrak{T}_{-\boldsymbol{h}}\boldsymbol{u}_{-\boldsymbol{h}} - \boldsymbol{u}) \\ &\quad = -a(\Delta_{\boldsymbol{h}}\boldsymbol{u}, \Delta_{\boldsymbol{h}}\boldsymbol{u}) + (a - a_{-\boldsymbol{h}})(\boldsymbol{u}_{-\boldsymbol{h}}, \Delta_{\boldsymbol{h}}\boldsymbol{u}) + R_{\mathfrak{T},a}(\boldsymbol{u}) \end{aligned} \tag{1.7.34}$$

with remainder

$$R_{\mathfrak{T},a}(\boldsymbol{u}) \equiv a_{-\boldsymbol{h}}(\boldsymbol{u}_{-\boldsymbol{h}}, (\mathfrak{T}_{\boldsymbol{h}})_{-\boldsymbol{h}}\boldsymbol{u} - \boldsymbol{u}) + a(\boldsymbol{u}, \mathfrak{T}_{-\boldsymbol{h}}\boldsymbol{u}_{-\boldsymbol{h}} - \boldsymbol{u}_{-\boldsymbol{h}}).$$

The first term in (1.7.34) gives—after the multiplication by $|\boldsymbol{h}'|^{-N+1-2\alpha}$ and after the integration—the expression $-\mathscr{A}(\boldsymbol{u})$. The second term can be estimated due to the Hölder continuity of the coefficients by $C|\boldsymbol{h}'|^{\beta}|\nabla \boldsymbol{u}_{-\boldsymbol{h}}| \cdot |\nabla(\boldsymbol{u} - \boldsymbol{u}_{-\boldsymbol{h}})|$, and we get

$$\begin{aligned} &\left| \int_{\mathbb{R}^{N-1}} \int_O \frac{(a - a_{-\boldsymbol{h}})(\boldsymbol{u}_{-\boldsymbol{h}}, \Delta_{\boldsymbol{h}}\boldsymbol{u})}{|\boldsymbol{h}'|^{N-1+2\alpha}}\, dx\, dh' \right| \\ &\quad \le c_1 \int_{|\boldsymbol{h}'|<1} \int_O \frac{|\nabla \boldsymbol{u}_{-\boldsymbol{h}}||\nabla(\boldsymbol{u} - \boldsymbol{u}_{-\boldsymbol{h}})|}{|\boldsymbol{h}'|^{N-1-\beta'+2\alpha}}\, dx\, dh' + c_2\|\boldsymbol{u}\|^2_{\boldsymbol{H}^1(O)} \\ &\quad \le c_3\|\boldsymbol{u}\|_{\boldsymbol{H}^1(O)}\mathscr{A}(\boldsymbol{u}) + c_2\|\boldsymbol{u}\|^2_{\boldsymbol{H}^1(O)}. \end{aligned}$$

Here, a split of the integral with respect to $\boldsymbol{h}$ for $|\boldsymbol{h}'| < 1$ and $|\boldsymbol{h}'| > 1$ is employed. For the first integral Hölder's inequality and the condition $\beta' > \alpha$ is used. The integral over the remainder $R_{\mathfrak{T},a}(\boldsymbol{u})$ for $\alpha < 1/2$ can be expressed as

$$\begin{aligned} \int_O R_{\mathfrak{T},a}(\boldsymbol{u})\, dx &= \int_O a_{-\boldsymbol{h}}\big(\boldsymbol{u}_{-\boldsymbol{h}}, ((\mathfrak{T}_{\boldsymbol{h}})_{-\boldsymbol{h}} - \boldsymbol{Id})\boldsymbol{u}\big)\, dx \\ &\quad + \int_O a\big(\boldsymbol{u}_{-\boldsymbol{h}}, (\mathfrak{T}_{-\boldsymbol{h}} - \boldsymbol{Id})\boldsymbol{u}_{-\boldsymbol{h}}\big)\, dx. \end{aligned}$$

After the multiplication by $|\boldsymbol{h}'|^{N-1-2\alpha}$ and integration in $\boldsymbol{h}$ over $\mathbb{R}^{N-1}$ we employ the appropriate smoothness of Γ_C resulting in the "Hölder" continuity $\|\mathfrak{T}_{\pm h} - \boldsymbol{Id}\|_{C^1} \le |\boldsymbol{h}'|^{\beta}$ and the fact $\int_{\mathbb{R}^{N-1}} |\boldsymbol{h}'|^{1+\varepsilon-N} d\boldsymbol{h}' < +\infty$

for $\varepsilon = \min(1,\beta) - 2\alpha > 0$ to estimate $R_{\mathfrak{T}}$ by means of the *a priori* estimate (1.5.20). For $\alpha = 1/2$ it can be estimated by

$$\int_O R_{\mathfrak{T},a}(\boldsymbol{u})\,dx = \int_O \big(a(\boldsymbol{u}, \mathfrak{T}_{\boldsymbol{h}}\boldsymbol{u}_{\boldsymbol{h}} - \boldsymbol{u}_{\boldsymbol{h}}) + a(\boldsymbol{u}, \mathfrak{T}_{-\boldsymbol{h}}\boldsymbol{u}_{-\boldsymbol{h}} - \boldsymbol{u}_{-\boldsymbol{h}})\big)\,dx$$

$$= \int_O \big(a(\boldsymbol{u}, (\mathfrak{T}_{-\boldsymbol{h}} - \boldsymbol{Id})(\boldsymbol{u}_{-\boldsymbol{h}} - \boldsymbol{u})) + a(\boldsymbol{u}, (\mathfrak{T}_{\boldsymbol{h}} - \boldsymbol{Id})(\boldsymbol{u}_{\boldsymbol{h}} - \boldsymbol{u}))$$

$$+\, a(\boldsymbol{u}, (\mathfrak{T}_{\boldsymbol{h}} + \mathfrak{T}_{-\boldsymbol{h}} - 2\,\boldsymbol{Id})\boldsymbol{u})\big)\,dx,$$

where a backward shift in the arguments of the first bilinear form was applied. Due to the properties of $(\boldsymbol{Id} - \mathfrak{T}_{\pm\boldsymbol{h}})$, the first two terms can be estimated just as the differences of the coefficients of the form a above. In the last term we have the second difference of the appropriate transformation which satisfies $\|\mathfrak{T}_{\boldsymbol{h}} + \mathfrak{T}_{-\boldsymbol{h}} - 2\,\boldsymbol{Id}\,\|_{C^1} \le c|\boldsymbol{h}'|^{1+\beta}$ if $\Gamma_C \in C^{3+\beta}$. With this rotation the "direct" rectification technique has quite a high requirement to the smoothness of the boundary.

A similar technique is also applied to the term $b_0(\boldsymbol{u},\boldsymbol{v})$. Here, the first term is bounded by

$$\left|\int_{\mathbb{R}^{N-1}}\int_O \frac{b_0(\Delta_{\boldsymbol{h}}\boldsymbol{u}, \Delta_{\boldsymbol{h}}\boldsymbol{u})}{|\boldsymbol{h}'|^{N-1+2\alpha}}\,dx\,d\boldsymbol{h}'\right| \le c_1\delta\mathscr{A}^2(\boldsymbol{u})$$

while the second one is estimated by

$$\int_{\mathbb{R}^{N-1}}\int_O \left|\frac{(b_0 - (b_0)_{-\boldsymbol{h}})(\boldsymbol{u}_{-\boldsymbol{h}}, \Delta_{\boldsymbol{h}}\boldsymbol{u})}{|\boldsymbol{h}'|^{N-1+2\alpha}}\right| dx\,d\boldsymbol{h}' \le c_2\|\boldsymbol{u}\|_{\boldsymbol{H}^1(O)}\mathscr{A}(\boldsymbol{u})$$

$$+\, c_3\|\boldsymbol{u}\|^2_{\boldsymbol{H}^1(O)}$$

and the remainder $R_{\mathfrak{T},b_0}(\boldsymbol{u})$ is estimated similarly to $R_{\mathfrak{T}}(a)$. For the terms of b we use the Plancherel identity. Employing the representation

$$b(\tilde{\boldsymbol{u}}, \boldsymbol{v}) = b^{(1)}_{ijk}\partial_i\tilde{u}_j v_k + b^{(2)}_{ijk}\tilde{u}_i\partial_j v_k$$

with C^β-coefficients $b^{(\ell)}_{ijk}$ we get

$$\left|\int_{\mathbb{R}^{N-1}}\int_O \frac{b_{-\boldsymbol{h}}(\tilde{\boldsymbol{u}}_{-\boldsymbol{h}}, \Delta_{\boldsymbol{h}}\boldsymbol{u}) + b(\tilde{\boldsymbol{u}}, -\Delta_{\boldsymbol{h}}\boldsymbol{u})}{|\boldsymbol{h}'|^{N-1+2\alpha}}\,dx\,d\boldsymbol{h}'\right|$$

$$\le c_1\left|\int_{\mathbb{R}^{N-1}}\int_O \frac{\Delta_{\boldsymbol{h}}(b^{(1)}_{ijk}\partial_i\tilde{u}_j)\Delta_{\boldsymbol{h}}u_k + (\Delta_{\boldsymbol{h}}b^{(2)}_{ijk}\tilde{u}_i)\Delta_{\boldsymbol{h}}(\partial_j u_k)}{|\boldsymbol{h}'|^{N-1+2\alpha}}\,dx\,d\boldsymbol{h}'\right|$$

$$\le c_2\left|\int_{\mathbb{R}^{N-1}}\int_{\mathbb{R}_+}\Big(\mathscr{F}(b^{(1)}_{ijk}\partial_i\tilde{u}_j;\boldsymbol{\xi}', x_N)\overline{\mathscr{F}(u_k;\boldsymbol{\xi}', x_N)}\right.$$

$$+ \mathscr{F}\left(b^{(2)}_{ijk}\tilde{u}_i;\boldsymbol{\xi}',x_N\right)\overline{\mathscr{F}\left(\partial_j u_k;\boldsymbol{\xi}',x_N\right)}\Big)|\boldsymbol{\xi}'|^{2\alpha}\,dx_N\,d\boldsymbol{\xi}'\Big|$$

$$\le c_3\|\tilde{\boldsymbol{u}}\|_{H^1(O)}\|\boldsymbol{u}\|_{H^{2\alpha}(O)}.$$

Of course, the remainder $R_{\mathfrak{T},b}(\boldsymbol{u})$ is again estimated as that for the norm a.

The estimate of the friction term is similar to that in the preceding subsection. Due to the employed rotation operators the test function for the non-shifted inequality $\boldsymbol{v} = \mathfrak{T}_{-\boldsymbol{h}}\boldsymbol{u}_{-\boldsymbol{h}}$ satisfies $\boldsymbol{v}_{\tilde{\boldsymbol{t}}} = (\boldsymbol{u}_{\tilde{\boldsymbol{t}}})_{-\boldsymbol{h}}$ and that for the shifted inequality $\boldsymbol{v}_{-\boldsymbol{h}} = (\mathfrak{T}_{\boldsymbol{h}})_{-\boldsymbol{h}}u$ satisfies $(\boldsymbol{v}_{\tilde{\boldsymbol{t}}})_{-\boldsymbol{h}} = \boldsymbol{u}_{\tilde{\boldsymbol{t}}}$. Hence the contribution of the friction term is given by

$$-\Delta_{\boldsymbol{h}}G\Delta_{\boldsymbol{h}}|\boldsymbol{u}_{\tilde{\boldsymbol{t}}}|.$$

This term can be estimated by the right hand side of formula (1.7.13) with $\boldsymbol{u}_{\boldsymbol{t}}$ there replaced by $\boldsymbol{u}_{\tilde{\boldsymbol{t}}}$. However, due to the smoothness of Γ_C and the usual trace theorem there holds

$$\|\boldsymbol{u}_{\boldsymbol{t}} - \boldsymbol{u}_{\tilde{\boldsymbol{t}}}\|_{\boldsymbol{H}^{1/2+\alpha}(\mathbb{R}^{N-1})} \le c_1\|\boldsymbol{u}\|^2_{\boldsymbol{L}_2(\mathbb{R}^{N-1})} + c_2\delta\mathscr{A}(\boldsymbol{u}).$$

Hence the friction term can be estimated by

$$c_{N-1}(\alpha)(1+\varepsilon+c_1\delta)\|G\|_{H^{-(1/2-\alpha)}(\mathbb{R}^{N-1})}\cdot$$
$$\left[\sqrt{\frac{c_{N-1}(\frac{1}{2}-\alpha)}{c_{N-1}(\frac{1}{2}+\alpha)}}\|\boldsymbol{u}_{\boldsymbol{t}}\|_{\boldsymbol{H}^{1/2+\alpha}(\mathbb{R}^{N-1})} + c_2(\varepsilon)\|\boldsymbol{u}_{\boldsymbol{t}}\|_{\boldsymbol{H}^{1/2}(\mathbb{R}^{N-1})}\right] \quad (1.7.35)$$
$$+ c_3\delta\|G\|_{H^{-(1/2+\alpha)}(\mathbb{R}^{N-1})}\mathscr{A}(\boldsymbol{u})$$

with $\varepsilon > 0$ arbitrarily small and subscript $\boldsymbol{t}$ denoting the tangential component with respect to the boundary of the halfspace O. The result of the shift technique is

$$\mathscr{A}^2(\boldsymbol{u}) \le (1+\varepsilon+c_1\delta)\left(\sqrt{d^*_{N-1}\left(\alpha,\tfrac{1}{2}\right)C_0^{(1)}}\|G\|_{H^{-1/2+\alpha}(\mathbb{R}^{N-1})}\mathscr{A}(\boldsymbol{u})\right)$$
$$+ c_2\delta\mathscr{A}^2(\boldsymbol{u}) + c_3\|G\|_{H^{-1/2}(\mathbb{R}^{N-1})} + c_4\mathscr{A}(\boldsymbol{u}) + c_5 \quad (1.7.36)$$

with an arbitrarily small $\varepsilon > 0$.

The estimate for the normal traction σ_n (or, to be more precise, its locally rectified form) is done as in the case of the halfspace, with similar modifications as above. It is based on the rectified Green formula

$$\langle\sigma_{\tilde{n}}(\boldsymbol{u}),\boldsymbol{v}_{\tilde{n}}\rangle_{\partial O} = -\int_O \left(a(\boldsymbol{u},\boldsymbol{v}) + b(\tilde{\boldsymbol{u}},\boldsymbol{v}) + b_0(\boldsymbol{u},\boldsymbol{v}) - \boldsymbol{f}\cdot\boldsymbol{v}\right)dx$$

valid for all functions $\boldsymbol{v} \in \boldsymbol{H}^1(O)$ satisfying $v_{\tilde{t}} = 0$. Let us define a (scalar) function w as in formula (1.7.20), and let us denote by $\mathscr{E}$ an extension operator from $H^{1/2}(\partial O)$ to $\boldsymbol{H}^1(O)$ satisfying

$$\|\mathscr{E}\boldsymbol{v}\|_A \leq C_0^{(2)} \|\boldsymbol{v}\|_{\boldsymbol{H}^{1/2}(\partial O)} \tag{1.7.37}$$

with a certain constant $C_0^{(2)}$. Then the shift technique is applied to the Green formula with test functions $\boldsymbol{v} = \mathscr{E}\big((w - w_{-\boldsymbol{h}})\tilde{\boldsymbol{n}}\big)$ for the non-shifted equation and $\boldsymbol{v}_{-\boldsymbol{h}} = \mathscr{E}\big((w_{-\boldsymbol{h}} - w)\tilde{\boldsymbol{n}}_{-\boldsymbol{h}}\big)$ for the shifted equation. Then, on the left hand side of the resulting inequality we get the term

$$\langle \Delta_{\boldsymbol{h}} G, \Delta_{\boldsymbol{h}} w\rangle_{\Gamma_C}.$$

The contribution of the bilinear form a is decomposed into

$$\begin{aligned} a(\boldsymbol{u},\boldsymbol{v}) - a_{-\boldsymbol{h}}(\boldsymbol{u}_{-\boldsymbol{h}},\boldsymbol{v}_{-\boldsymbol{h}}) &= -a(\Delta_{\boldsymbol{h}}\boldsymbol{u}, \mathscr{E}(\Delta_{\boldsymbol{h}} w\, \tilde{\boldsymbol{n}})) \\ &+ (a_{-\boldsymbol{h}} - a)(\boldsymbol{u}_{-\boldsymbol{h}}, \mathscr{E}(\Delta_{\boldsymbol{h}} w\, \tilde{\boldsymbol{n}})) + a_{-\boldsymbol{h}}(\boldsymbol{u}_{-\boldsymbol{h}}, \mathscr{E}(\Delta_{\boldsymbol{h}} w(\tilde{\boldsymbol{n}}_{-\boldsymbol{h}} - \tilde{\boldsymbol{n}})). \end{aligned}$$

An estimate of the second term on the right hand side of this relation using the C^β-Hölder continuity of the coefficients of a gives

$$\begin{aligned} &\int_{\mathbb{R}^{N-1}} \int_O \frac{(a_{-\boldsymbol{h}} - a)(\boldsymbol{u}_{-\boldsymbol{h}}, \mathscr{E}(\Delta_{\boldsymbol{h}} w\, \tilde{\boldsymbol{n}}))}{|\boldsymbol{h}'|^{N-1+2\alpha}}\, d\boldsymbol{x}\, d\boldsymbol{h}' \\ &\quad \leq c_1 \|\boldsymbol{u}\|_{\boldsymbol{H}^1(O)} \|w\|_{H^{1/2+\alpha}(\partial O)}. \end{aligned}$$

For the estimate of the last term we employ the relation

$$\begin{aligned} \|\Delta_{\boldsymbol{h}} w(\tilde{\boldsymbol{n}}_{-\boldsymbol{h}} - \tilde{\boldsymbol{n}})\|_{\boldsymbol{H}^{1/2}(\partial O)} \leq & c_1 |\boldsymbol{h}'| \|\Delta_{\boldsymbol{h}} w\|_{H^{1/2}(\partial O)} \\ & + c_2 |h|^\beta \|\Delta_{\boldsymbol{h}} w\|_{L_2(\partial O)}. \end{aligned}$$

This results in an estimate of the corresponding integral by

$$\begin{aligned} &\int_{\mathbb{R}^{N-1}} \int_O \frac{a_{-\boldsymbol{h}}(\boldsymbol{u}_{-\boldsymbol{h}}, \mathscr{E}(\Delta_{\boldsymbol{h}} w(\tilde{\boldsymbol{n}}_{-\boldsymbol{h}} - \tilde{\boldsymbol{n}})))}{|\boldsymbol{h}'|^{N-1-2\alpha}}\, d\boldsymbol{x}\, d\boldsymbol{h}' \\ &\quad \leq c \|\boldsymbol{u}\|_{\boldsymbol{H}^1(O)} \|w\|_{H^{1/2+\alpha}(\partial O)}. \end{aligned}$$

Using similar estimates for the remaining linear and bilinear forms and the techniques of the preceding subsection, the following inequality is derived:

$$\begin{aligned} \|\sigma_{\tilde{\boldsymbol{n}}}\|_{H^{-(1/2-\alpha)}(\mathbb{R}^{N-1})} \leq (1+\varepsilon) d_{N-1}^{-1/2}\big(\alpha, \tfrac{1}{2} - \alpha\big) \sqrt{C_0^{(2)}}\, \mathscr{A}(\boldsymbol{u}) \\ + c_1 \delta \mathscr{A}(\boldsymbol{u}) + c_2(\varepsilon) \sqrt{\mathscr{A}(\boldsymbol{u})} + c_3(\varepsilon) \end{aligned}$$

with ε arbitrarily small.

The constant $C_0^{(2)}$ in relation (1.7.37) must be valid for functions $w\,\tilde{\boldsymbol{n}}$ having only a component in direction of the normal to the original boundary. However, in the calculation of the coefficient $C_0^{(2)}$ it is convenient to use functions of the type $w\boldsymbol{n}$ with normal $\boldsymbol{n}$ to the boundary of the halfspace. This is possible due to the estimate

$$\|w(\tilde{\boldsymbol{n}}-\boldsymbol{n})\|_{\boldsymbol{H}^\gamma(\partial O)} \le c\delta\|w\|_{H^\gamma(\partial O)} + \tilde{c}\|w\|_{L_2(\partial O)}. \tag{1.7.38}$$

In fact, there holds

$$\begin{aligned}\|\mathscr{E}w\tilde{\boldsymbol{n}}\|_A &\le \|\mathscr{E}w\boldsymbol{n}\|_A + \|\mathscr{E}w(\tilde{\boldsymbol{n}}-\boldsymbol{n})\|_A \\ &\le C_0^{(2)}\|w\|_{H^{1/2}(\partial O)} + c_1\|w(\tilde{\boldsymbol{n}}-\boldsymbol{n})\|_{\boldsymbol{H}^{1/2}(\partial O)} \\ &\le \big(C_0^{(2)} + c_2\delta\big)\|w\|_{H^{1/2}(\partial O)} + \tilde{c}\|w\|_{L_2(\partial O)}.\end{aligned}$$

In order to prove Theorem 1.5.8, it is necessary to consider the whole picture consisting of the locally rectified versions of the problem for all cutoff functions of the partition of unity. Therefore, the new norms

$$\mathfrak{A}(\boldsymbol{u}) \equiv \left(\sum_{i=1}^{I_\delta} \mathscr{A}^2(\widetilde{\varrho\boldsymbol{u}})\right)^{1/2}$$

and

$$\|G\|_{-1/2+\alpha,\Gamma_{\mathfrak{F}},\mathfrak{P}} \equiv \left(\sum_{i=1}^{I_\delta} \|J\widetilde{\varrho G}\|^2_{H^{-1/2+\alpha}(\mathbb{R}^{N-1})}\right)^{1/2}$$

are considered. These norms are equivalent to the $\boldsymbol{H}^{1/2+\alpha}(\Gamma_{\mathfrak{F}})$-norm of $\boldsymbol{u}$ and to the $H^{-(1/2-\alpha)}(\Gamma_{\mathfrak{F}})$-norm of G. The constants in this equivalence depend on the employed partition of unity. From the local estimates we derive

$$\begin{aligned}\mathfrak{A}^2(\boldsymbol{u}) \le d_{N-1}^*\left(\alpha,\tfrac{1}{2}\right)\sqrt{C_0^{(1)}}(1+\varepsilon+c_1\delta)\|G\|_{-1/2+\alpha,\Gamma_{\mathfrak{F}},\mathfrak{P}}\mathfrak{A}(\boldsymbol{u}) \\ + c_2\delta\mathfrak{A}^2(\boldsymbol{u}) + c_3(\varepsilon)\mathfrak{A}(\boldsymbol{u}) + c_4(\varepsilon)\end{aligned}$$

and

$$\begin{aligned}\|\sigma_n\|_{-1/2+\alpha,\Gamma_{\mathfrak{F}},\mathfrak{P}} \le &(1+\varepsilon)d_{N-1}^{-1/2}\left(\alpha,\tfrac{1}{2}-\alpha\right)\sqrt{C_0^{(2)}}\mathfrak{A}(\boldsymbol{u}) \\ &+ c_1\delta\mathfrak{A}(\boldsymbol{u}) + c_2(\varepsilon)\sqrt{\mathfrak{A}(\boldsymbol{u})} + c_3(\varepsilon)\end{aligned}$$

with $\varepsilon > 0$ arbitrarily small. The combination of these estimates proves Theorem 1.5.8. There, the constant C_0 is given by $C_0 = \sqrt{C_0^{(1)}C_0^{(2)}}$, if the norms $\|\cdot\|_{H^{1/2+\alpha}(\Gamma_C)}$ and $\|\cdot\|_{H^{1/2+\alpha}(\Gamma_C)}$ are replaced by the new norms introduced above. The assertion is also valid for the original norms, but in this case the constant C_0 is different.

Hence again, if the coefficient of friction satisfies the condition

$$\|\mathfrak{F}\|_{L_\infty(\Gamma)} < C_{\mathfrak{F}} \text{ with } C_{\mathfrak{F}} = \left(C_0^{(1)} C_0^{(2)}\right)^{-1/2},$$

and $\mathfrak{F}$ satisfies (1.5.30) for a suitable $\alpha > 0$ then with an appropriate choice of ε and δ (implying an appropriate choice of the partition of unity), the Tikhonov fixed point theorem proves the solvability of the contact problem with friction.

1.7.1 Remark. The described combination of the direct rectification of the boundary with the fixed point approach leads to quite excessive requirements on the smoothness of the contact part of the boundary because of the necessity to employ the rotation maps $\mathfrak{T}_{-\boldsymbol{h}}$. In Chapter 3 the penalization of the non-penetrability condition leads to a Newton-type boundary value condition in the auxiliary problems instead of the Signorini condition on Γ_C. For such a condition we can use $\boldsymbol{v} = \boldsymbol{u}_{-\boldsymbol{h}}$, $\boldsymbol{v}_{-\boldsymbol{h}} = \boldsymbol{u}$ as the test function in the original and shifted boundary value problem, respectively. The remainders in the estimates of the bilinear forms cancel. This enables requiring $\Gamma \in C^{1+\beta}$ only. If this approach is applied to the contact problem with given friction, then the friction term has the form

$$G\big(|\pi_{\boldsymbol{t}}\boldsymbol{u}_{-\boldsymbol{h}}| - |\pi_{\boldsymbol{t}}\boldsymbol{u}|\big) + G_{-\boldsymbol{h}}\big(|(\pi_{\boldsymbol{t}})_{-\boldsymbol{h}}\boldsymbol{u}| - |(\pi_{\boldsymbol{t}}\boldsymbol{u})_{-\boldsymbol{h}}|\big), \tag{1.7.39}$$

where G represents the product of the localized penalty term with the coefficient of friction and $\pi_{\boldsymbol{t}}$ denotes the projection onto the tangential hyperplane. Decomposing the term (1.7.39) into

$$\begin{aligned}(G_{-\boldsymbol{h}} - G)\big(|\pi_{\boldsymbol{t}}\boldsymbol{u}| - |(\pi_{\boldsymbol{t}}\boldsymbol{u})_{-\boldsymbol{h}}|\big) + G_{-\boldsymbol{h}}\big(|(\pi_{\boldsymbol{t}})_{-\boldsymbol{h}}\boldsymbol{u}| - |\pi_{\boldsymbol{t}}\boldsymbol{u}|\big)\\ + G\big(|\pi_{\boldsymbol{t}}\boldsymbol{u}_{-\boldsymbol{h}}| - |(\pi_{\boldsymbol{t}})_{-\boldsymbol{h}}\boldsymbol{u}_{-\boldsymbol{h}}|\big)\end{aligned} \tag{1.7.40}$$

we can estimate the first term as above via (1.7.13). However, it is not clear how to rewrite the remainder into a form containing products of differences or differences of the second order which can be estimated. The problem consists in the impossibility to estimate $\int_{|\boldsymbol{h}'|\le h_0} |\boldsymbol{h}'|^{-N} \| |(\pi_{\boldsymbol{t}})_{-\boldsymbol{h}}\boldsymbol{u}| + |(\pi_{\boldsymbol{t}})_{\boldsymbol{h}}\boldsymbol{u}| - 2|\pi_{\boldsymbol{t}}\boldsymbol{u}| \|_{\boldsymbol{H}^{1/2}(\Gamma_C)} \, d\boldsymbol{h}'$ for any $h_0 > 0$. Hence this technique is suitable only if the gain of regularity α for the solution of the contact problem in the case of a general domain is limited to $\alpha < 1/2$. If the coefficient of friction is assumed to depend on the solution $\boldsymbol{u}$, then this small loss of regularity is restrictive, due to the additional requirements for F in Lemma 1.5.4.

However, there exists another kind of local representation of the boundary, where the normal and tangential components of the original boundary are transformed onto the normal and tangential components with respect to the boundary of the halfspace without using the artificial local rotations. This representation uses curvilinear coordinates. With this approach it is possible to obtain the regularity result with $\alpha = 1/2$ also in the case of a general domain, but at the price of stronger assumptions concerning the smoothness of the domain. The result is presented in the next Theorem.

1.7.2 Theorem. *Let the Assumptions 1.5.7 be valid for $\alpha = 1/2$, and let moreover $\Gamma_C \in C^{2+\beta}$ for some $\beta > 1/2$. Then the solution of the penalized contact problem satisfies the* a priori *estimate* (1.5.28) *uniformly with respect to ε.*

Proof. We start from a local parametrization of the boundary by means of a $C^{2+\beta}$-smooth function,

$$\boldsymbol{x}_\Gamma : B_{N-1}(\delta) \ni \boldsymbol{y}' \mapsto \boldsymbol{x}_\Gamma(\boldsymbol{y}') \in \Gamma.$$

Let us denote the image of this parametrization by $\Gamma_\delta \equiv \boldsymbol{x}_\Gamma(B_{N-1}(\delta))$. From the parametrization of the boundary we construct a parametrization

$$\boldsymbol{x} = \mathfrak{Q}(\boldsymbol{y}) = \boldsymbol{x}_\Gamma(\boldsymbol{y}') + y_N \boldsymbol{n}(\boldsymbol{x}_\Gamma(\boldsymbol{y}')) \qquad \text{for } \boldsymbol{y} = (\boldsymbol{y}', y_N) \in B_{N-1}(\delta) \times (0, \delta)$$

of some local neighbourhood $\Omega_\delta \subset \Omega$ of the boundary part Γ_δ. If δ is sufficiently small, this is a one-to-one mapping of the parameter domain $B_{N-1}(\delta) \times (0, \delta)$ onto Ω_δ; in fact δ must be smaller than the inverse of the curvature's maximum.

The partial derivatives $\mathfrak{g}_j \equiv \partial_{y_j} \boldsymbol{x}$, $j = 1, \ldots, N$ of the parametrization are called *covariant basis vectors* in differential geometry. On the boundary we have $y_N = 0$, hence $\mathfrak{g}_j = \partial_{y_j} \boldsymbol{x}_\Gamma$ for $j = 1, \ldots, N-1$ and $\mathfrak{g}_N = \boldsymbol{n}$. The tensor $\{\mathfrak{g}_{ij}\}_{i,j=1}^N$ defined by $\mathfrak{g}_{ij} = \mathfrak{g}_i \cdot \mathfrak{g}_j$ is the *covariant metric tensor*. Without loss of generality we may assume $\mathfrak{g}_{ij}(\boldsymbol{y} = \boldsymbol{0}) = \delta_{ij}$; this can be always achieved by an additional linear transformation of the parametrization. Then the determinant of the Jacobian of this parametrization has the form

$$J_\Omega = |\det(\mathfrak{g}_1, \ldots, \mathfrak{g}_N)| = 1 + J_{\Omega,R}$$

with a remainder $J_{\Omega,R} \in C^{1,\beta}$ satisfying $J_{\Omega,R}(0) = 0$, and the density of surface measure is

$$J_\Gamma \equiv |\det\{\mathfrak{g}_{ij}\}_{i,j=1}^{N-1}| = J_\Gamma = 1 + J_{\Gamma,R}$$

with a $C^{1,\beta}$-smooth function $J_{\Gamma,R}$ satisfying $J_{\Gamma,R}(0) = 0$.

A crucial step is the representation of vector fields. This is done by means of a local basis consisting of the *contravariant basis vectors* $\mathfrak{g}^1, \ldots, \mathfrak{g}^N$ which represent the biorthogonal system to the covariant vectors $\mathfrak{g}_i$. In terms of the mapping $\boldsymbol{y} \mapsto \boldsymbol{x} = \mathfrak{Q}(\boldsymbol{y})$ and its inverse $\boldsymbol{x} \mapsto \boldsymbol{y} = \mathfrak{Q}^{-1}(\boldsymbol{x})$ the contravariant basis is given by $\mathfrak{g}^i = \left\{\partial_{x_\ell} y_i\right\}_{\ell=1}^N$. A vector field $\boldsymbol{u}$ may be represented by

$$\boldsymbol{u} = u_i e_i = \overline{u}_i \mathfrak{g}^i$$

with cartesian components u_i and covariant components $\overline{u}_i$. The cartesian components can be expressed in terms of the covariant ones by

$$u_i = \overline{u}_j \langle \mathfrak{g}^j, e_i \rangle = \overline{u}_j \partial_{x_i} y_j.$$

The big advantage of the representation in terms of covariant components is the transformation of normal and tangential components on the boundary; it holds

$$u_n = \overline{u}_N \text{ and } \boldsymbol{u_t} = \sum_{i=1}^{N-1} \overline{u}_i \mathfrak{g}^i.$$

In other words, the normal component u_n with respect to the normal of the domain is transformed onto the normal component with respect to the halfspace, and analogously for the tangential components. The derivatives of cartesian components with respect to the old space variables are transformed according to

$$\frac{\partial u_i}{\partial x_j} = \frac{\partial}{\partial x_j}\left(\overline{u}_\ell \frac{\partial y_\ell}{\partial x_i}\right) = \frac{\partial \overline{u}_\ell}{\partial y_k}\frac{\partial y_k}{\partial x_j}\frac{\partial y_\ell}{\partial x_i} + \overline{u}_\ell \frac{\partial^2 y_\ell}{\partial x_i \partial x_j}. \tag{1.7.41}$$

Since the covariant basis vectors $\mathfrak{g}_i$, $i = 1, \ldots, N$, are $C^{1+\beta}$-smooth and orthogonal at $y = 0$ there holds

$$\frac{\partial y_i}{\partial x_j} = \delta_{ij} + R_{ij} \tag{1.7.42}$$

with remainder $R_{ij} \in C^\beta(\Omega)$ satisfying $R_{ij}(\mathbf{0}) = \mathbf{0}$. In particular there follows

$$|R_{ij}(\boldsymbol{y})| \leq c_1 \delta$$

for $\boldsymbol{y} \in B_{N-1}(\delta) \times (0, \delta)$. Therefore the bilinear form is transformed according to

$$
\begin{aligned}
a(\boldsymbol{u}, \boldsymbol{v}) &= a_{ijk\ell} \frac{\partial u_j}{\partial x_i} \frac{\partial v_\ell}{\partial x_k} \\
&= a_{ijk\ell} \left(\frac{\partial \overline{u}_m}{\partial y_q} \frac{\partial y_q}{\partial x_i} \frac{\partial y_m}{\partial x_j} + \overline{u}_m \frac{\partial^2 y_m}{\partial x_i x_j} \right) \\
&\quad \cdot \left(\frac{\partial \overline{v}_r}{\partial y_s} \frac{\partial y_s}{\partial x_k} \frac{\partial y_r}{\partial x_\ell} + \overline{v}_r \frac{\partial^2 y_r}{\partial x_k x_\ell} \right) \\
&= a_{ijk\ell} \frac{\partial \overline{u}_j}{\partial x_i} \frac{\partial \overline{v}_\ell}{\partial x_k} + b_0(\overline{u}, \overline{v}) + b_1(\overline{u}, \overline{v})
\end{aligned}
$$

with perturbations b_0, b_1 satisfying

$$
\begin{aligned}
&|b_0(\boldsymbol{u}, \boldsymbol{v})| \leq c_1 \delta |\nabla \boldsymbol{u}||\nabla \boldsymbol{v}| \text{ and} \\
&|b_1(\boldsymbol{u}, \boldsymbol{v})| \leq c_2 (|\nabla \boldsymbol{u}||\boldsymbol{v}| + |\nabla \boldsymbol{v}||\boldsymbol{u}| + |\boldsymbol{u}||\boldsymbol{v}|).
\end{aligned}
$$

The shift technique is done with this new rectification. As in Subsection 1.7.3 we start from a finite covering $\mathscr{U} = \{U_i, i \in \mathscr{I}_\delta\}$, where now the U_i are images of the local curvilinear representation of the type $B_{N-1}(\delta) \times (0, \delta) \ni \boldsymbol{y} \mapsto \boldsymbol{x} = \mathfrak{Q}_i(\boldsymbol{y})$ described above, and a subordinated partition of unity $\mathfrak{P}_\delta$ consisting of smooth cutoff functions ϱ_i, $i = 1, \ldots, \mathscr{I}_\delta$. For a fixed index i—which is omitted for simplicity of the notation—the localization can be performed in the usual way, leading to inequality (1.7.25). Then, by the transform to the new coordinates $\boldsymbol{y} = \mathfrak{Q}_i(\boldsymbol{x})$, a variational inequality of the type (1.7.32) is derived for the vector $\tilde{\boldsymbol{u}} = (\overline{u}_i, \ldots, \overline{u}_N)^\top$ of the transformed covariant components. Here, the transformed bilinear form can be written as

$$
\int_{B_{N-1}(\delta) \times (0,\delta)} J_\Omega \big(a(\overline{\boldsymbol{u}}, \overline{\boldsymbol{v}}) + b_0(\overline{\boldsymbol{u}}, \overline{\boldsymbol{v}}) + b_1(\overline{\boldsymbol{u}}, \overline{\boldsymbol{v}}) + b(\overline{\boldsymbol{u}}, \overline{\boldsymbol{v}}) \big) \, d\boldsymbol{x}
$$

with b denoting the transformed version of the perturbation caused by the localization. The bilinear forms b_1 and b have the same properties. They can be merged into one form denoted again by b. The products of bilinear forms with the perturbation of the Jacobian $J_{\Omega,R}$ will be included into the bilinear form b_0. Hence the transformed bilinear form has exactly the same properties as in the case of the "direct" rectification. After the standard extension of the variational inequality onto the halfspace, the rectified variational inequality is the same as (1.7.33), with the important difference that now the index $\boldsymbol{t}$ represents the tangential component $\overline{\boldsymbol{u}}_{\boldsymbol{t}} = (\overline{u}_1, \ldots, \overline{u}_{N-1}, 0)^\top$ with respect to the straightened boundary. Hence the shift technique can be performed as above; and the corresponding term arising from the friction functional can be simply estimated as in (1.7.13) by

$$\int_{\mathbb{R}^{N-1}} \int_{\mathbb{R}^{N-1}\times\{0\}} |\boldsymbol{h}'|^{-(N-1+2\alpha)} \Delta_{\boldsymbol{h}} G \, \Delta_{\boldsymbol{h}} |\boldsymbol{u_t}| \, ds_{\boldsymbol{x}} \, d\boldsymbol{h}'$$
$$\leq c_{N-1}(\alpha)(1+\varepsilon)\|G\|_{H^{1/2-\alpha}(\Gamma_C)}$$
$$\cdot \left[\sqrt{\frac{c_{N-1}(\frac{1}{2}-\alpha)}{c_{N-1}(\frac{1}{2}+\alpha)}} \|\boldsymbol{u_t}\|'_{\boldsymbol{H}^{1/2+\alpha}(\Gamma_C)} + c\|\boldsymbol{u_t}\|'_{\boldsymbol{H}^{1/2}(\Gamma_C)} \right] .$$

This estimate is also valid for $\alpha = \frac{1}{2}$. Hence the proof can be finished as described above. □

The authors hope that the difficulties arising just at the first look to the contact problems with Coulomb friction have convinced the readers that a deeper understanding of different aspects of these problems needs a more comprehensive knowledge of the theoretical background. This will be done in the next chapter.

Chapter 2

Background

In this chapter we supply the readers with the necessary knowledge to follow the proofs of the results throughout the book. Two areas should be well understood for this purpose, namely:

- *Fixed point theory.* Schauder's or Tikhonov's fixed point theorem are the main ingredients of most existence theorems in this book.
- *Theory of suitable spaces of functions with "fractional-order" derivatives on domains.* The non-smooth character of our problems requires getting the best possible estimates, employing efficiently every smoothness occurring in the assumptions. Hence imbedding, interpolation and trace theorems for such spaces together with the local rectification and extension techniques are the inevitable base for the results presented in our book.

The authors feel the lack of a suitable (series of) monograph(s), suitable for a general reader, on which a research requiring appropriate deeper knowledge can be well based. There is a certain difference between the language and aims in functional analysis or function space theory and those in PDEs theory. Their interplay, however, has always been of great mutual benefit and the authors' effort was that of builders of bridges, connecting advanced techniques and results from a variety of fields. Since rather extensive recent or contemporary theories are touched upon here, this chapter should be understood more as a certain vademecum and a series of hints how to build them for such purpose with references to special literature for those who need to learn the details.

2.1 Fixed point theorems

We prove Brouwer's and Tikhonov's fixed point theorems in this section. Among several standard approaches to Brouwer's theorems we have chosen the proof based on the Brouwer degree.

We shall not need much of special notation in this section. If $\Omega \subset \mathbb{R}^N$ is a domain, then $C(\Omega)$ and $C(\overline{\Omega})$ denote the space of functions continuous in Ω and in $\overline{\Omega}$, respectively. Similarly, $C^k(\Omega)$ and $C^k(\overline{\Omega})$, $k = 1, \dots,$ have their usual meaning.

2.1.1 Covering theorems

We start with several fundamental theorems. First we present Besicovitch' covering theorem and Morse's theorems, and then (a generalization of) Sard's theorem on critical points. Besicovitch' theorem is one of the basic building blocks of a large area of analysis and its various versions can be found in many monographs (see e.g. [63], Chapter 1, [58], [143], Chapter 1, etc.).

In what follows, a *closed cube* will always be a *closed cube with edges parallel to the coordinate axes.* Analogous meaning will have an *open cube.*

2.1.1 Theorem (Besicovitch' covering theorem). *Let $A \subset \mathbb{R}^N$. Suppose that for every $x \in A$ there is given a closed cube $Q(x)$ centered at x. If A is unbounded, suppose additionally that* $\sup_{x\in A} \operatorname{diam} Q(x) < \infty$. *Then there exists a sequence $\{x_k\}_{k\in\mathbb{N}}$ of points in A such that*

(i) $A \subset \bigcup_{k\in\mathbb{N}} Q(x_k)$

(ii) *the sequence $\{Q_k\}_{k\in\mathbb{N}}$ is uniformly locally finite covering of $Q(x)$, i.e. there exists a constant θ_N depending only on N such that*

$$\sum_{k\in\mathbb{N}} \chi_{Q_k}(y) \leq \theta_N, \qquad x \in \mathbb{R}^N$$

(iii) *there exists a number ξ_N depending only on N such that the sequence $\{Q_k\}_{k\in\mathbb{N}}$ can be split to ξ_N subsequences, each of them consisting of cubes with non-intersecting interiors.*

Proof. Step 1. First we shall consider a simplified situation, when the principle of the proof is very transparent. Let us suppose that A is bounded and that we are given a sequence of closed cubes $\{Q(k)\}_{k\in\mathbb{N}}$ centered at the origin such that $\lim_{k\to+\infty} \operatorname{diam} Q(k) = 0$. For every $x \in A$ choose an arbitrary natural number $i(x)$ and put $Q(x) = x + Q(i(x))$. We shall construct an enumerable covering as follows. We choose x_1 such that $Q(x_1)$ has maximal diameter. If $x_1, \dots, x_{m-1}$ are chosen, let x_{m+1} be a point outside $\bigcup_{j=1}^m Q(x_j)$ such that $\operatorname{diam} Q(x_{m+1})$ is maximal among diameters of the remaining cubes. We get a sequence of cubes $Q_k \equiv Q(x_k)$, possibly finite. If $k_1 \neq k_2$, then $x_{k_1} \notin Q_{k_2}$. Indeed, $k_1 > k_2$ and $x_{k_1} \in Q_{k_2}$ at the same time is impossible since centers of "younger" cubes lie outside the "older" cubes. Similarly $k_1 \leq k_2$ and $x_{k_1} \in Q_{k_2}$ at the same time leads to a contradiction since then x_{k_2} lies outside the cube Q_{k_1} and therefore the cube Q_{k_2} would be bigger than Q_{k_1}. But the sequence $\{\operatorname{diam} Q_k\}$ is non-increasing thanks to the construction of $\{Q_k\}$. Therefore the cubes $x_k + 2^{-1}Q_k$ are disjoint. Hence the sequence $\{\operatorname{diam} Q_k\}$ is either finite or it tends to zero.

If the sequence $\{Q_k\}$ is finite, then it is plainly a covering of A. If it is infinite, then $A \setminus \bigcup_{k=1}^{+\infty} Q_k$ must be empty; if not, then it contains some $x \in A$ and there exists k_0 such that $\operatorname{diam} Q(x) > \operatorname{diam} Q_{k_0}$ and this contradicts the maximality of diameters of the cubes that we have chosen at each step of the construction; there would be at least one better choice at the k_0-th step, namely the cube $Q(x)$.

For $z \in A$ consider 2^N hyperplanes containing z and parallel with coordinate axes. They divide $\mathbb{R}^N$ into 2^N connected infinite paralellepipeds. For each of them there is at most one x_j such that $z \in Q_j$. Indeed, if $x \in Q_i \cap Q_j$, then the bigger of these cubes should contain the center of the smaller cube, which is impossible as we have seen above. Hence no point of A can belong to more than 2^N cubes from $\{Q_k\}$.

It remains to prove the property (iii). Fix some Q_j. According to the above argument each of the 2^N corners of Q_j belongs to at most 2^N among the remaining cubes of the covering. Note that if $i < j$, then the diameter of Q_i is not smaller than the diameter of Q_j, hence Q_i must contain at least one vertex of Q_j. There are at most 4^N cubes altogether that contain at least one vertex of Q_j. Consequently, the number of the "older" cubes in the sequence, which have a non-empty intersection with Q_j, does not exceed 4^N. This actually suggests how to divide the whole sequence into families of disjoint cubes. Indeed, start with Q_1 and put it into the first family, put Q_2 into the second family, and so on, (4^N+1)-times, independently of whether they have empty or non-empty intersection with their predecessors. The (4^N+2)nd cube in the sequence must have an empty intersection with at least one of the older cubes, say, with Q_i. We put Q_{4^N+2} into the i-th family and continue this way. As a result the whole sequence $\{Q_k\}$ is split into 4^N+1 families of disjoint cubes.

Step 2. Let the assumptions of the theorem be fulfilled and suppose additionally that A is bounded. If $s_0 = \sup_{x\in A} \operatorname{diam}(Q(x)) = +\infty$, then there is nothing to prove. Hence assume that s_0 is finite and construct a sequence $\{Q_k\}$ as in Step 1, with the only difference that at the beginning we choose a cube whose diameter is bigger than $s_0/2$ and at every next stage we select a cube, whose diameter is at least half of the supremum of diameters of the remaining cubes. Now the situation is different than that of Step 1. If $i > j$, then $x_i \notin Q_j$ by the construction. If $i < j$, then x_i can be in Q_j since neither Q_j nor Q_i is necessarily maximal. Nevertheless, we always have $3^{-1}Q_i \cap 3^{-1}Q_j = \emptyset$ for $i \neq j$, where $3^{-1}Q_k$ denotes the three times squeezed Q_k, that is, the cube concentric with Q_k, and whose sidelength is one third of that of Q_k. Namely, if $i > j$, then $x_i \notin Q_j$ and at the same time $\operatorname{diam} Q_j < 2^{-1} \operatorname{diam} Q_i$. Hence if $3^{-1}Q_i \cap 3^{-1}Q_j \neq \emptyset$, then Q_j would contain x_i (draw a picture). The role of i and j was symmetric hence the three times squeezed cubes do not intersect. Since A is bounded the sequence $\{Q_i\}$ is either finite or $\operatorname{diam} Q_i \to 0$ as $i \to +\infty$.

Now the covering property follows. If there exists $x \in A \setminus \bigcup Q_k$, then there is i_0 such that $\operatorname{diam} Q(x) > 2^{-1} \sup_{i \geq i_0} \operatorname{diam} Q_i$. Hence at the i_0-th step at the latest we should have chosen $Q(x)$ (or other cube whose diameter has the same property) instead of Q_{i_0+1}. This is a contradiction.

We omit details of proofs of the assertions (ii) and (iii).

Step 3. If A is unbounded and $\sup_{x \in A} \operatorname{diam} Q(x) = S < \infty$, then we cut $\mathbb{R}^N$ into cubes $\{K_i\}$ whose diameter equals S and consider $A \cap K_i$ instead of A. As a result we get countable families $\{Q^{(i)_k}\}_k$, where arbitrary cubes $Q_k^{(i+3)}$ and $Q_{k'}^{(i)}$ from the families corresponding to $A \cap K_{i+3}$ and $A \cap K_i$, respectively, do not intersect. □

2.1.2 Theorem (Sard). *Let $A \subset \mathbb{R}^N$ be bounded and for each $x \in A$ let there be given a set $H(x)$ possessing the following properties:*
(i) *there exists $M > 0$ independent of x such that for every $x \in A$ there are two closed balls $B(x, r(x))$ and $B(x, Mr(x))$ centered at x such that $b(x, r(x)) \subset H(x) \subset B(x, Mr(x))$,*
(ii) *for every $x \in A$ and every $y \in H(x)$ the set $H(x)$ contains the convex hull of $\{y\} \cup B(x, r(x))$. Then the conclusions of Theorem 2.1.1 are true.*

We refer to [63] for the proof, references, and further discussion of covering theorems of this type. In particular the same key idea works in the two preceding theorems, and at the same time Sard's theorem is more general. Covering theorems play a basic role in the theory of classical operators on Euclidean (and more general) spaces and the proofs can be found in many other well-known monographs—see e.g. [143], [126], [58]. General covering theorems on spaces with non-Euclidean geometry (on spaces of homogeneous type) can be found in [59].

2.1.3 Remark. Besicovitch' covering theorem holds also for (centered) balls replacing cubes and it is easy to realize that the boundedness of A is not essential if $\sup_{x \in A} \operatorname{diam} Q(x) < \infty$. The cubes $Q(x)$ need not be closed, either.

Let $A \subset \mathbb{R}^N$ and let (X, μ) be a set endowed with an outer measure. Let $f : A \to X$ be an arbitrary function. A point $x \in \Omega$ is a *critical point of f* if there exists a sequence $\{Q_k(x)\}$ of open cubes centered at x and such that $\lim_{k \to +\infty} \operatorname{diam}(Q_k(x)) = 0$ and $\lim_{k \to +\infty} \mu(f(Q_k(x)))\big(|Q_k(x)|\big)^{-1} = 0$, where $|Q_k(x)|$ denotes the Lebesgue measure of $Q_k(x)$. Next theorem is due to de Guzmán [64] and it generalizes Sard's theorem on critical points.

2.1.4 Theorem (de Guzmán). *Let Ω be an open subset of $\mathbb{R}^N$ and let (X, μ) be a space with a σ-finite outer measure. Let C be the set of all critical points of an arbitrary $f : \Omega \to X$. Then $\mu(f(C)) = 0$.*

Proof. It suffices to prove the theorem for an arbitrary $S \subset C$ with a finite measure. Let $\varepsilon > 0$ and choose an open set G containing S. For every $x \in S$ let $Q(x) \subset G$ be an open cube centered at x and such that $\mu(f(Q(x))) < \varepsilon|Q(x)|$. According to Besicovitch' theorem (see also Remark 2.1.3) there exists a sequence $\{Q_k\} = \{Q(x_k)\}$, extracted from our family $\{Q(x)\}$, satisfying $\sum_{k\in\mathbb{N}} \chi_{Q_k} \le \theta_N$ and $\mu(f(Q_k)) < \varepsilon|Q_k|$, where θ_N depends only on N. Hence

$$\begin{aligned}\mu(f(S)) &\le \mu\Big(f\Big(\bigcup_{k\in\mathbb{N}} Q_k\Big)\Big) = \mu\Big(\bigcup_{k\in\mathbb{N}} f(Q_k)\Big) \le \sum_{k\in\mathbb{N}} \mu(f(Q_k)) \\ &\le \varepsilon \sum_{k\in\mathbb{N}} |Q_k| = \varepsilon \int_{\bigcup_{k\in\mathbb{N}} Q_k} \Big(\sum_{k\in\mathbb{N}} \chi_{Q_k}(y)\, dy\Big) \\ &\le \varepsilon\theta_N \Big|\bigcup_{k\in\mathbb{N}} Q_k\Big| \le \varepsilon\theta_N |G|.\end{aligned}$$

Since ε was arbitrary the proof is complete. □

Let $f \in C^1(\mathbb{R}^N)$ and denote by J_f the Jacobian of f. If $\det J_f(x) = 0$, then x is the critical point in the above sense. This immediately implies the following

2.1.5 Corollary (Sard's theorem). *Let $f \in C^1(\mathbb{R}^N)$ and let C be the set of those $x \in \mathbb{R}^N$, where J_f, the Jacobian determinant of f, vanishes. Then $|f(C)| = 0$.*

2.1.2 Fixed point principles

In the remainder of this section Ω will be an open and bounded subset of $\mathbb{R}^N$. Let $f : \Omega \to \mathbb{R}^N$, $f \in C(\overline{\Omega}) \cap C^1(\Omega)$ and denote by J_f the Jacobian determinant of f. For $p \in \mathbb{R}^N$, a point $x \in f^{-1}(p) = \{x \in \Omega : f(x) = p\}$ is called *regular* if $J_f(x) \neq 0$. Observe that if $p \in f(\overline{\Omega}) \setminus f(\partial\Omega)$ and if $f^{-1}(p)$ consists of regular points, then $f^{-1}(p)$ is finite. Indeed, $f^{-1}(p)$ is closed and at each of its condensation points one can use the implicit function theorem. Hence assuming that $f^{-1}(p)$ is infinite would lead to a contradiction.

2.1.6 Definition. Let $f \in C^1(\Omega) \cap C(\overline{\Omega})$ and $p \in \mathbb{R}^N \setminus f(\partial\Omega)$. Suppose that f^{-1} is non-empty and contains only regular points. Then the *degree of the mapping f with respect to Ω and p* (or shortly the degree of f) is defined as

$$d[f; \Omega, p] = \sum_{x\in f^{-1}(p)} \operatorname{sign} J_f(x).$$

Formally we put $d[f; \Omega, p] = 0$ if $f^{-1}(p) = \emptyset$.

2.1.7 Remark. Note that it is easy to find e.g. the degree of the identical mapping id : $\Omega \to \Omega$. Indeed, we have $d[id; \Omega, p] = 1$ if $p \in \Omega$ and $d[id; \Omega, p] = 0$ if $p \notin \Omega$. This is straightforward. Nevertheless, it is a very non-trivial problem for general f and the general formula is not very helpful. Therefore we shall describe an alternative construction of the degree and an effective way how to calculate it.

Let $r_0 > 0$ and let φ be a *mollifier* supported in $[0, r_0]$, that is, a function in $\mathscr{D}(\mathbb{R})$, vanishing in $(-\infty, 0] \cup [r_0, +\infty)$ and such that $\|\varphi\|_1 = 1$. Let $f \in C^1(\Omega) \cap C(\overline{\Omega})$ and suppose that for some $p \in \mathbb{R}^N \setminus f(\partial\Omega)$ the set $f^{-1}(p)$ consists only of regular points of f. Then f is a diffeomorphism in a neighbourhood of every point in $f^{-1}(p)$, say, on $U(x_k)$, where k runs over some finite set K of indices. Put $\psi(x) = f(x) - p$. Plainly there is $\eta > 0$ such that $B(0, \eta) \subset \bigcup_{k \in K} \psi(U(x_k))$. Hence if $x \in \overline{\Omega} \setminus \bigcup_{k \in K} U(x_k)$, then $|\psi(x)| \geq \delta$.

The construction described in the sequel, based on the concept of a degree of a mapping, follows Nagumo [112] and Heinz [68].

2.1.8 Lemma. *Let $f \in C^1(\Omega) \cap C(\overline{\Omega})$ and suppose that for some $p \in \mathbb{R}^N \setminus f(\partial\Omega)$ the set $f^{-1}(p)$ has (finite) cardinality K and contains only regular points of f. Let φ be a mollifier living in $[0, \min(\delta, \eta)]$ with δ and η having the above described properties. Then*

$$\int_\Omega \varphi(|f(x) - p|) J_f(x)\, dx = \sum_{k \in K} \operatorname{sign} J_f(x_k) = d[f; \Omega, p]. \tag{2.1.1}$$

Proof. The left hand side of (2.1.1) equals to

$$\begin{aligned}
\int_{\bigcup_{k \in K} U(x_k)} \varphi(|\psi(x)|) J_f(x)\, dx &= \sum_{k \in K} \int_{U(x_k)} \varphi(|\psi(x)|) J_f(x)\, dx \\
&= \sum_{k \in K} \operatorname{sign} J_\psi(x_k) \int_{U(x_k)} \varphi(|\psi(x)|) |J_f(x)|\, dx \\
&= \sum_{k \in K} \operatorname{sign} J_f(x_k) \int_{\psi(U(x_k))} \varphi(|\psi(y)|)\, dy
\end{aligned}$$

and with the help of the change of variables $y = \psi(x)$

$$= \sum_{k \in K} \operatorname{sign} J_f(x_k).$$

□

It follows that the value of the integral on the left hand side of (2.1.1) is independent of a particular choice of a mollifier. We are going to show that this assertion remains to be valid even for ψ supported in the interval $[0, \min_{x\in\partial\Omega} |f(x) - p|]$. This will follow from the next lemma.

2.1.9 Lemma. *Let* $\omega : \overline{\Omega} \to \mathbb{R}^N$, $\omega \in C^1(\Omega) \cap C(\overline{\Omega})$. *Let*

$$0 < \varepsilon < \inf_{x\in\partial\Omega} |\omega(x)|. \tag{2.1.2}$$

Suppose that φ *is a continuous real function supported in* $[0, \varepsilon]$ *satisfying*

$$\int_0^{+\infty} r^{N-1}\varphi(r)\, dr = 0.$$

Then

$$\int_\Omega \varphi(|\omega(x)|) J_\omega(x)\, dx = 0. \tag{2.1.3}$$

Proof. Thanks to the Weierstrass approximation theorem we can suppose that $\omega \in C^2(\Omega) \cap C(\overline{\Omega})$. Indeed, ω can be uniformly approximated by polynomials, which can be additionally assumed to satisfy (2.1.2). Define

$$\psi(x) = \begin{cases} x^{-N} \displaystyle\int_0^x t^{N-1}\varphi(t)\, dt & \text{if } x > 0, \\ 0 & \text{if } x = 0. \end{cases}$$

Plainly $\psi \in C^1(0, +\infty)$, $\operatorname{supp}\psi \subset [0, \varepsilon]$ and

$$x\psi'(x) + N\psi(x) = \varphi(x). \tag{2.1.4}$$

Put

$$g_i(y) = \psi(|y|)y_i, \qquad y \in \mathbb{R}^N,\ i - 1, \dots, N.$$

Then trivially $g_i \in C(\mathbb{R}^N) \cap C^1(\mathbb{R}^N \setminus \{0\})$. Further, $\partial g_i(0)/\partial y_j = 0$ and it is not difficult to show that $\partial g_i(y)/\partial y_j$ is continuous at the origin, hence $g_i \in C^1(\mathbb{R}^N)$. It follows from the definition of ψ and properties of ω that $g_i(\omega(\cdot))$ vanishes in some neighbourhood of $\partial\Omega$.

Denote by $A_{ij}(x)$ the co-factors of the element $\partial\omega_i(x)/\partial x_j$ in the Jacobian of ω (i.e. the subdeterminant of order $N-1$, resulting from omitting the i-th row and the j-th column in the original matrix multiplied by $(-1)^{i+j}$). Since $\omega \in C^2(\mathbb{R}^N)$ we have $A_{ij} \in C^1(\Omega)$ and

$$\sum_{j=1}^N \partial A_{ij}(x)/\partial x_j = 0 \tag{2.1.5}$$

for every $x \in \Omega$ and all $i = 1, \ldots, N$. A formal proof would be rather lengthy, but it suffices to realize what the operation $\partial/\partial x_i$ makes with A_{ij}. Every product in A_{ij} (with $N-1$ terms) gives in turn $N-1$ products with exactly one occurrence of the second derivative. Since derivatives of second order are interchangeable here, it follows easily from the definition of the determinant and calculus rules for determinants (change of sign when interchanging position of neighbouring rows or columns in accordance with the change of the sign of the index permutation in question) that for every term in (2.1.5) after differentiation $\partial/\partial x_i$ applied to $A_{ij}(x)$ there is a counterpart with opposite sign resulting from taking $\partial/\partial x_j$ of a suitable term and hence the total result is zero.

Choose $x \in \Omega$ such that $\omega(x) \neq 0$ and denote by δ_{ik} the Kronecker symbol, that is $\delta_{ik} = 0$ if $i \neq k$ and $\delta_{ik} = 0$ if $i = k$. In view of (2.1.5) and (2.1.4) we have

$$\begin{aligned} &\operatorname{div} \sum_{i=1}^{N} A_{ij}(x) g_i(\omega(x)) \\ &\quad = \sum_{i=1}^{N} g_i(\omega(x)) \sum_{j=1}^{N} \frac{\partial A_{ij}(x)}{\partial x_j} + \sum_{i,j,k=1}^{N} A_{ij}(x) \frac{\partial g_i(\omega(x))}{\partial y_k} \frac{\partial \omega_k(x)}{\partial x_j} \\ &\quad = \sum_{i,k=1}^{N} \frac{\partial g_i(\omega(x))}{\partial y_k} \cdot \delta_{ik} = J_\omega(x) \sum_{\ell=1}^{N} \frac{\partial g_\ell(\omega(x))}{\partial y_\ell} \\ &\quad = J_\omega(x) \left[N\psi(|\omega(x)|) + |\omega(x)|\psi'(|\omega(x)|) \right] = \varphi(|\omega(x)|) J_{\omega(x)}. \end{aligned}$$

If $\omega(x) = 0$, then

$$\operatorname{div} \sum_{i=1}^{N} A_{ij}(x) g_i(\omega(x)) = \varphi(|\omega(x)|) J_{\omega(x)}$$

holds trivially. Let $F = (F_j)$, where for $j = 1, \ldots, N$,

$$F_j(x) = \begin{cases} \displaystyle\sum_{i=1}^{N} A_{ij}(x) g_i(\omega(x)), & \text{if } x \in \Omega, \\ 0 & \text{if } x \in \mathbb{R}^N \setminus \Omega, \end{cases}$$

and let Q be a cube containing Ω. Then $F_j \in C^1(Q) \cap C(\overline{Q})$ and we have

$$\int_\Omega \varphi(|\omega(x)|) J_\omega(x)\, dx = \int_\Omega \operatorname{div}\Big(\sum_{i=1}^N A_{ij}(x) g_i(\omega(x))\Big)\, dx$$
$$= \sum_{j=1}^N \int_Q \operatorname{div} F(x)\, dx = 0.$$

The proof is complete. □

The next step is to show that the property (2.1.3) is independent of a particular choice of the mollifier φ. Specifically we have

2.1.10 Lemma. *Let $\omega \in C^1(\Omega) \cap C(\overline{\Omega})$ and suppose that φ_1 and φ_2 are mollifiers supported in $[0, \varepsilon]$, where $0 < \varepsilon < \inf_{x \in \partial\Omega} |\omega(x)|$. Then*

$$\int_\Omega \varphi_1(|\omega(x)|) J_\omega(x)\, dx = \int_\Omega \varphi_2(|\omega(x)|) J_\omega(x)\, dx.$$

The proof is easy (based on the previous Lemma 2.1.9) and we leave it to the reader.

Now we need to show that the degree with respect to a point p is locally constant for functions close in the $C(\Omega)$ norm, but sufficiently far from p on $\partial\Omega$.

2.1.11 Lemma. *Let $p \in \mathbb{R}^N$ and $f_i \in C^1(\Omega) \cap C(\overline{\Omega})$, $i = 1, 2$. Let us assume that for some $\varepsilon > 0$,*

$$|f_i(x) - p| \geq 7\varepsilon, \qquad i = 1, 2,\ x \in \partial\Omega,$$

and

$$|f_1(x) - f_2(x)| < \varepsilon, \qquad x \in \overline{\Omega}.$$

Further, let all points in $(f_i)^{-1}(p)$ be regular, $i = 1, 2$. Then

$$d[f_1; \Omega, p] = d[f_2; \Omega, p].$$

Proof. Put $\psi_i(x) = f_i(x) - p$, $x \in \overline{\Omega}$, $i = 1, 2$, and fix a C^1-function $\gamma\colon [0, +\infty) \to [0, 1]$ such that $\gamma(r) = 1$ if $0 \leq r \leq 2\varepsilon$, and $\gamma(r) = 0$ if $3\varepsilon \leq r$. Let

$$\psi_3(x) = \big(1 - \gamma(|\psi_1(x)|)\big)\psi_1(x) + \gamma(|\psi_1(x)|)\psi_2(x).$$

If $x \in \partial\Omega$, then

$$|\psi_3(x)| \geq |\psi_1(1)| - \gamma(|\psi_1(x)|)(|\psi_1(x) - \psi_2(x)|) \geq 7\varepsilon - \varepsilon = 6\varepsilon,$$

and plainly $\psi_3(x) = \psi_1(x)$ if $|\psi_1(x)| > 3\varepsilon$, whereas $\psi_3(x) = \psi_2(x)$ if $|\psi_1(x)| < 2\varepsilon$. Let φ_1 and φ_2 be regularizators supported in $(4\varepsilon, 5\varepsilon)$ and in $(0, \varepsilon)$, respectively. Then, for all $x \in \Omega$,

$$\varphi_1(|\psi_3(x)|)J_{\psi_3}(x) = \varphi_1(|\psi_1(x)|)J_{\psi_1}(x)$$

and

$$\varphi_2(|\psi_3(x)|)J_{\psi_3}(x) = \varphi_2(|\psi_2(x)|)J_{\psi_1}(x). \tag{2.1.6}$$

Since

$$\int_\Omega \varphi_1(|\psi_3(x)|)J_{\psi_3(x)}\,dx = \int_\Omega \varphi_2(|\psi_3(x)|)J_{\psi_3(x)}\,dx$$

(see Lemma 2.1.9) we can integrate the identities in (2.1.6) to get

$$\int_\Omega \varphi_1(|\psi_1(x)|)J_{\psi_1(x)}\,dx = \int_\Omega \varphi_2(|\psi_2(x)|)J_{\psi_2(x)}\,dx$$

which completes the proof in view of Lemma 2.1.10. □

2.1.12 Lemma. *Let $f_1, f_2 \in C^1(\Omega) \cap C(\overline{\Omega})$, and $p_1, p_2 \in \mathbb{R}^N$. Let $\varepsilon > 0$ and*

$$\begin{aligned} |f_i(x) - p_i| &\geq 7\varepsilon, \qquad && x \in \partial\Omega,\ i = 1, 2, \\ |f_1(x) - f_2(x)| &< \varepsilon, && x \in \partial\Omega,\ i = 1, 2, \\ |p_1 - p_2| &< \varepsilon. \end{aligned}$$

Assume that all points in $(f_i)^{-1}(p_j)$, $i, j = 1, 2$, are regular. Then

$$d[f_1; \Omega, p_1] = d[f_2; \Omega, p_2].$$

Proof. According to Lemma 2.1.11, $d[f_1; \Omega, p_1] = d[f_2; \Omega, p_1]$. At the same time $d[f_2; \Omega, p_1] = d[f_2 + (p_1 - p_2); \Omega, p_1]$. But the last quantity is nothing but $d[f_2; \Omega, p_2]$ by the definition of the degree. □

For completeness we state a variant of Lemma 2.1.11 for an arbitrary $p \in \mathbb{R}^N$.

2.1.13 Lemma. *Let $f_1, f_2 \in C^1(\Omega) \cap C(\overline{\Omega})$ and $p \in \mathbb{R}^N$. Let $\varepsilon > 0$ be such that*

$$|f_i(x) - p| \geq 8\varepsilon, \qquad x \in \partial\Omega,$$
$$|f_1(x) - f_2(x)| < \varepsilon, \qquad x \in \overline{\Omega}.$$

Then $d[f_1; \Omega, p] = d[f_2; \Omega, p]$.

Proof. It suffices to realize that we can approach p by a sequence of points $\{p_k\}$ such that both sets $(f_i)^{-1}(p)$ contain only regular points. Then Lemma 2.1.11 implies that the numbers $d[f_i; \Omega, p_k]$, $i = 1, 2$, are the same for sufficiently large k. □

Next step in the construction of the degree of a mapping f is to consider non-regular points $p \notin f(\partial\Omega)$. Invoking Sard's theorem (see Theorem 2.1.5) we see that the image $f(C)$ of the set C of non-regular points of f has zero Lebesgue measure and therefore for every $p \in f(C)$ there exists a sequence of *regular* points $p_k \notin f(\partial\Omega)$ tending to p as $k \to +\infty$. We have proved in the previous lemmas that $d[f; \Omega, p_k]$ is independent of a particular choice of p_k provided p_k are sufficiently close to each other. Therefore, $d[f; \Omega, p_k]$ is constant provided k is large enough. Furthermore, if two mappings are close enough to each other, they have the same degree with respect to regular points (Lemma 2.1.11). Combining these considerations with the Weierstrass approximation theorem, we can get rid of the assumption $f \in C^1(\Omega)$ and arrive at a definition of the degree for continuous functions: For $f \in C(\overline{\Omega})$ and $p \notin f(\partial\Omega)$ the number $d[f; \Omega, p]$ is defined as $\lim_{k\to+\infty} d[f_k; \Omega, p_k]$.

2.1.14 Remark. The whole procedure we went through, starting with Definition 2.1.6, shows that it is practically impossible to calculate the degree of a particular mapping with respect to some point.

Hence of great importance are some clever tricks. One of them is actually the proof of Brouwer's fixed point theorem.

We start with two consequences of preceding lemmas.

2.1.15 Theorem. *Let Ω be a bounded, open, non-empty subset of $\mathbb{R}^N$, $f \in C(\overline{\Omega})$, and $p \in \mathbb{R}^N \setminus f(\partial\Omega)$. Suppose that $d[f; \Omega, p] \neq 0$. Then there exists $x_0 \in \Omega$ such that $f(x_0) = p$.*

Proof. Let f_k be a sequence of C^1-functions uniformly converging to f and let $\{p_k\}$ be a sequence tending to p such that p_k is a regular point of f_k, $k \in N$. There exists k_0 such that $d[f_k; \Omega, p_k] = d[f; \Omega, p] \neq 0$ for all $k \geq k_0$. Hence for each $k \geq k_0$ there exists $x_k \in \Omega$ such that $f_k(x_k) = p_k$. Since $\overline{\Omega}$ is a compact set we can suppose without loss of generality that

$\lim_{k\to+\infty} x_k$ exists. Denote this limit by x_0. By continuity argument, we have $f(x_0) = p$. □

2.1.16 Theorem (homotopy theorem). *Let Ω be a bounded, open, non-empty subset of $\mathbb{R}^N$. Let $h : \overline{\Omega} \times [0,1] \to \mathbb{R}^N$ be continuous. Let $p \in \mathbb{R}^N$ and suppose that $h(x,t) \neq p$ for every $t \in [0,1]$ and for every $x \in \partial\Omega$. Then the function $t \mapsto d[h(t,\cdot);\Omega,p]$ is constant on $[0,1]$.*

Proof. We shall verify the assumptions of Lemma 2.1.13. Choose $\varepsilon > 0$ such that $|h(x,t) - p| \geq 9\varepsilon$ for all $x \in \partial\Omega$ and all $t \in [0,1]$. The mapping h is uniformly continuous, therefore $|h(x,t_1) - h(x,t_2)| < \varepsilon/3$ for all $x \in \overline{\Omega}$ and all $t_1, t_2 \in [0,1]$ that are sufficiently close, say, $|t_1 - t_2| \leq \delta$. Fix such t_1, t_2 and choose sequences $\{f_k^j\}_{k\geq 1}$, $j = 1,2$, of C^1-functions such that $f_k^j \to h(\cdot,t_j)$ uniformly on $\overline{\Omega}$, $j = 1,2$. For large k's,

$$\begin{aligned} |f_k^j - h(x,t_j)| &< \varepsilon/3, \qquad && x \in \overline{\Omega}, \\ |f_k^j - p| &\geq 8\varepsilon, && x \in \partial\Omega, \\ |f_k^1(x) - f_k^2(x)| &< \varepsilon, && x \in \overline{\Omega}. \end{aligned}$$

Hence Lemma 2.1.13 implies that

$$d[h(\cdot,t_1);\Omega,p] = d[f_k^1;\Omega,p] = d[f_k^2);\Omega,p] = d[h(\cdot,t_2);\Omega,p]$$

and we see that $d[h(\cdot,t_1);\Omega,p]$ is locally constant on $[0,1]$. Since $[0,1]$ is compact we are done. □

We come to the highlights of this section.

2.1.17 Theorem (Brouwer's fixed point theorem in $\mathbb{R}^N$). *Let f be a continuous mapping from the closed unit ball of $\mathbb{R}^N$ into itself. Then f has a fixed point.*

Proof. Denote by $B(1)$ the closed unit ball in $\mathbb{R}^N$. Let us suppose that $x - f(x) \neq 0$ for all $x \in \partial B(1)$; otherwise there is nothing to prove. Put $h(x,t) = x - tf(x)$, $x \in B(1)$, $t \in [0,1]$. Then

$$1 = d[\mathrm{id}; B(1), 0] = d[\mathrm{id} - f; B(1), 0]$$

and the claim follows from Theorem 2.1.15. □

2.1.18 Remark. Theorem 2.1.17 holds also under the weaker assumption $f(\partial B(1)) \subset B(1)$. This is clear from the proof.

2.1.19 Remark. Since all finite dimensional locally convex spaces are isomorphic to some Euclidean space, in particular, their topology is metrizable, we get the analogous fixed point theorem in such spaces for mappings of all sets that are homeomorphic to (homeomorphic copies of) the closed unit ball in $\mathbb{R}^N$. Hence we may ask, how these copies in $\mathbb{R}^N$ look like. If $K \subset \mathbb{R}^N$ is convex, closed, bounded, and with non-empty interior, then it suffices to move K in order that the origin lies in the interior of K and to consider points on rays through the origin to construct a homeomorphism between K and $B(1)$. On the other hand, all homeomorphic copies of $B(1)$ must have the properties that the above set K enjoys. Hence we see that the isomorphic copies of $B(1) \subset \mathbb{R}^N$ are exactly all the subsets of $\mathbb{R}^N$, which are convex, closed, bounded, and with non-empty interior. Next, if $K \neq \emptyset$, $K \subset \mathbb{R}^N$ is convex, closed, bounded, and its interior is empty, then there exist at most $N-1$ linearly independent points in K. This can be easily seen by an induction argument. If ℓ is the maximum of cardinalities of all linearly independent subsets of K, then K is a subset of a subspace $L \subset \mathbb{R}^N$, the dimension of L is of course ℓ, and K has a non-empty interior with respect to L. In this way the situation is reduced to the first case. We shall formulate the consequence separately as next Theorem.

2.1.20 Theorem. *Let X be a finitely dimensional linear space and let $K \subset X$ be non-empty, closed, and convex. Let $T : K \to K$ and suppose that $T(\partial K) \subset K$. Then there exists $x_0 \in K$ such that $T(x_0) = x_0$.*

Brouwer's fixed point theorem or its above mentioned generalization does not hold in infinite dimensional Banach spaces (see [88]). Nevertheless, there is a class of mappings and sets in them, namely compact sets and completely continuous mappings, which permits reduction to finite dimension and an application of Brouwer's theorem to get approximations, which converge to the desired fixed point. We will formulate and prove Schauder's fixed point theorem. First, however, we shall need an easy generalization of Theorem 2.1.17

Recall that a subset M of a normed linear space X is *totally bounded* if for every $\varepsilon > 0$ there exist finitely many points, say, $x_1, x_2, \dots, x_{n(\varepsilon)}$ such that $\min_{1 \le j \le n(\varepsilon)} \operatorname{dist}(x, x_j) < \varepsilon$ for all $x \in M$ (a finite ε-net). A mapping T from a linear normed space X into a linear normed space Y is *completely continuous* if for every bounded $M \subset X$ the set $T(M)$ is totally bounded in Y.

We are in a position to prove a fixed point theorem in infinite dimensional Banach space, known also as the Schauder principle [130].

2.1.21 Theorem (Schauder's fixed point theorem). *Let X be a Banach space, let $\emptyset \neq K \subset X$ be bounded, closed, and convex. Let $T : K \to X$*

be completely continuous and assume that $T(K) \subset K$. Then there exists $x_0 \in K$ such that $Tx_0 = x_0$.

Proof. Let $\varepsilon > 0$ and let $V_n = \{v_1, \ldots, v_{n(\ell)}\}$ be a $1/n$-net in X. For $i = 1, \ldots, n(\ell)$ let us define

$$m_i(x) = \begin{cases} 0 & \text{if } \|x - v_i\| > 1/n, \\ \varepsilon - \|x - v_i\| & \text{if } \|x - v_i\| \leq 1/n, \end{cases}$$

and

$$F_n(x) = \left(\sum_{i=1}^{n(\ell)} m_i(x)\right)^{-1} \left(\sum_{i=1}^{n(\ell)} m_i(x) v_i\right).$$

Then we plainly have $\|x - F_n(x)\| \leq 1/n$ for all $x \in X$. Put $T_n = F_n \circ T$, denote by W_n the linear hull of V_n. Then $T_n(K \cap W_n) \subset K \cap W_n$ because F_n is a convex linear combination of elements in $K \cap W_n$. Hence for each $n \in \mathbb{N}$ there is $x_n \in K \cap W_n$ such that $x_n = T_n(x_n)$. Since $\{x_n\}_{n \geq 1}$ is a bounded sequence the set $\{T(x_n)\}$ is totally bounded. Consequently, we can find a subsequence of $\{Tx_n\}_{n \geq 1}$, which we denote in the same way, converging to some element $z \in K$. We have

$$\|T(x_n) - x_n\| = \|T(x_n) - T_n(x_n)\| \leq \frac{1}{n}.$$

By continuity of T, if $x_0 = \lim_{n \to +\infty} x_n$, then $x_0 = T(x_0)$. □

Schauder's fixed point theorem was generalized by Tikhonov [146] to the case of continuous mappings of convex compact subsets of locally convex spaces into itself.

Recall that a *topological linear space* is a linear space X (over real or complex numbers) endowed with a topology, for which addition and multiplication by scalars are continuous. Thus, the topology in X can be given by the family of neighbourhoods of the origin. If there exists a fundamental system, say $\mathscr{U}$, of convex neighbourhoods of zero (that is, for every neighbourhood V of zero in X there is $U \in \mathscr{U}$ such that $U \subset V$), then the space X is called *locally convex*.

Observe that one of the ideas in the proofs of the previous theorems was to choose sufficiently fine ε-nets. If we do not have a metric, we may still use special covering refinements. Let X be a locally convex topological linear space and let $\{W_\varkappa\}_\varkappa$ be a family of subsets of X. A family $\{U_\kappa\}_\kappa$ is called a *double refinement of* $\{W_\varkappa\}_\varkappa$ if for every U_κ there exists $W_\varkappa \supset U_\kappa$ such that every U_{κ_1} having a non-empty intersection with U_κ is also a subset of $W_\varkappa$.

2.1.22 Lemma. *Let F be a compact subset of a locally convex topological linear space X and $\{W_\varkappa\}_\varkappa$ a family of open sets covering F. Then there exists a double refinement of $\{W_\varkappa\}_\varkappa$.*

Proof. By compactness we can choose W_j, $j = 1, \dots, n$, covering F. For every $x \in F$ let $W^*(x)$ be a neighbourhood of x, whose closure is contained in some of the sets W_j. The family $\{W^*(x)\}_{x \in F}$ is a covering of F, hence we can pass to some of its finite part, say, W_k^*, $k = 1, \dots, m$. For every $x \in F$, put

$$U^1(x) = \bigcap_{x \in W_k^*} W_k^*, \qquad U^2(x) = \bigcap_{x \in X \setminus \overline{W_k^*}} W_k^*, \qquad U^3(x) = \bigcap_{x \in W_j} W_j,$$

and put $U(x) = U^1(x) \cap U^2(x) \cap U^3(x)$, $x \in F$. Assume $U(x) \cap U'(x') \neq \emptyset$. Choose some $W_k^* \supset U(x)$ and then select some $W_i \supset \overline{W_k^*}$. Hence W_i is one of the sets in the intersection defining $U^3(x)$. Then x' must be in $\overline{W_k^*}$. For, if $x' \in F \setminus \overline{W_k^*}$, then according to the definition $U'(x') \subset F \setminus \overline{W_k^*}$ and $U'(x')$ would have not a common point with $U(x)$. Thus $x' \in \overline{W_k^*} \subset W_i$ and we conclude that $U'(x') \subset W_i$. Since the family $\{U(x)\}_{x \in F}$ is finite we are done. □

2.1.23 Theorem (Tikhonov's fixed point theorem). *Let F be a convex and compact subset of a locally convex topological linear space X and $f : F \to F$ be continuous. Then f has a fixed point.*

Proof. Let us assume that f has no fixed point. For every $x \in F$ choose disjoint convex neighbourhoods $W(x)$ and $W'(f(x))$. By continuity of f we can suppose that $f(U(x)) \subset W'(f(x))$. According to Lemma 2.1.22 there exists a finite family $\{U_i\}_{i=1}^r$, a double refinement of $\{W(x)\}_{x \in F}$. Further, for every $x \in F$, let $V(x)$ be a neighbourhood of x such that $f(V(x))$ is a subset of some U_i. Let $x_i \in U_i$, $i = 1, \dots, r$, fix some $V(x)$ and denote it with V_i. Let L be a convex hull of $x_1, \dots, x_r$. Since the linear hull of these points is a finite dimensional linear space, that is, a homeomorphic copy of a Euclidean space $\mathbb{R}^s$ with $s \leq r$, the set L is a homeomorphic copy of a simplex in $\mathbb{R}^s$. Let us divide L into a family $\{S_\ell\}$ of smaller simplexes if necessary in such a way that every S_ℓ is a subset of some V_i. Hence $f(S_\ell)$ is a subset of some U_i. Let $y_1, \dots, y_m$ be the extremal points (i.e. the vertices) of all S_ℓ. If $f(y_i) \in U_k$, then we put

$$\varphi(y_i) = x_{k(i)}, \qquad i = 1, \dots, m,$$

and we extend φ to S_ℓ by linear interpolation. By convexity, φ takes L into L and by Brouwer's fixed point theorem there exists $x_0 \in L$ such that $\varphi(x_0) = x_0$.

Consider the finite covering $\{W_i\}$ from Lemma 2.1.22. We claim that for every $x \in L$ the images $\varphi(x)$ and $f(x)$ are in the same W_i. This will be plainly a contradiction since according to our construction of the neighbourhoods W_i the points $x_0 = \varphi(x_0)$ and $f(x_0)$ cannot belong to the same W_i. To this end choose $x \in L$, denote by S the corresponding subsimplex containing x and denote by $y_1, \ldots, y_{n(S)}$ its vertices. Then there exists $i_0 \in \{1, \ldots, r\}$ so that $f(S) \subset U_{i_0}$. According to the definition of φ we have also $\varphi(y_i), f(y_i) \in U_i$. Thus $f(y_i) \in U_{i_0} \cap U_i$, i.e., the intersection of U_i with U_{i_0} is non-empty, hence all U_i, having the double refinement property, must be contained in one of the sets W_k, say, W_{k_0}, which corresponds to U_{i_0}. By convexity argument, $\varphi(x) \in W_{k_0}$. Consequently also $f(x) \in U_{i_0} \subset W_{k_0}$ and $\varphi(x) \in W_{k_0}$. The theorem is proved. □

2.2 Some general remarks

The theory exposed in the book heavily relies on properties of various function spaces. Problems that we study here belong to the expanding class of the PDE problems, requiring spaces more general than the traditional Sobolev spaces. In this chapter we present some basic properties of spaces used in the sequel. We regret that it is an impossible task to give a self-contained account even for employed clones of Sobolev spaces, which are the main analytic tool for our considerations. We give proofs always when it is possible with respect to the limited extent of the book. Our intention is to provide basic orientation in the theory in order that the reader not familiar with the contemporary function spaces theory and its methods gets some useful knowledge for reading specialized literature.

Today, the theory of function spaces is a rather wide area of analysis with no clear boundaries; it uses a broad spectrum of methods and tools: classical and functional analysis, interpolation theory, Fourier analysis and more. It should be observed that the basic mighty impetus came from the PDE theory with the concept of generalized derivatives. The contemporary state of the function spaces theory is not purely the result of this basic PDE feedback. At one hand, it provides a very general theory, in particular that based on decompositions, theorems of Littlewood-Paley and Nikolskii type, together with deep links to harmonic analysis; on the other hand, it seems that some cases needed in applications are not yet sufficiently covered by the general theory. The most basic references for the theory presented in this chapter are: The books by Peetre [119], by Bergh and Löfström [23] and Triebel [148] for the interpolation theory and the decomposition approach to spaces of Sobolev and Besov type. The reader will find also very useful the book by Bennett and Sharpley [22] in connection with the interpolation of Besov spaces in a more classical spirit. Stein's book [143] contains the classical part on singular integral operators, Riesz and Bessel potentials and potential Sobolev spaces. For a first encounter with basic function spaces, classical interpolation theorems and harmonic analysis operators the book by Sadosky [126] can be also recommended. Then there are the famous Zygmund's books [158]. Anisotropic spaces, in particular, the vector-valued Sobolev and Besov spaces, will be needed in the form of mappings from an interval into a space of Sobolev type. They fall into the class of so called spaces with dominating smoothness properties. Owing to their importance for the PDEs they have been studied in a classical setting, in particular in the Hilbert case (see, e.g. [102]) and they have been investigated by many authors. The unified theory in the Fourier analysis framework of these spaces is due to Schmeisser [131], [132], [133]; see also the book by Schmeisser and Triebel [135]. Finally we refer to the deep theory of anisotropic spaces developed in the monograph by Besov, Il'in and Nikol'skii [26].

When our spaces have a Hilbert structure, we may rely on the simpler properties of Fourier transform and of Bessel kernels. The general case is, however, more complicated and it uses deeper facts.

Fourier analysis naturally requires spaces defined on the whole of $\mathbb{R}^N$. From a purely theoretical point of view, this is only a technical problem to pass to spaces on domains with the extension property with respect to spaces under consideration. This allows restricting our attention to spaces defined on $\mathbb{R}^N$ ($N \geq 1$) instead on a domain since in this case the corresponding space on the domain can be understood as a factorspace modulo equality on the given domain. On the other hand, the extension theorems do not belong to the easiest part of the function spaces theory. Nevertheless, the machinery of the localization technique is well understood now and has become one of the standard tools. The reader will meet localization arguments throughout the book quite frequently and sometimes we omit the lengthy technical details.

We shall introduce the basic spaces and nail down the notation. Let $\Omega \subset \mathbb{R}^N$ be measurable. Let $1 \leq p \leq \infty$. Then the *Lebesgue space* $L_p(\Omega)$ is defined as the space of all measurable functions on Ω with the finite norm $\|f\|_p = \|f\|_{L_p(\Omega)} = \left(\int_\Omega |f(x)|^p \, dx\right)^{1/p}$ if $p < \infty$, and $\|f\|_\infty = \|f\|_{L_\infty(\Omega)} = \operatorname*{ess\,sup}_{x\in\Omega} |f(x)|$ for $p = \infty$. It is well known that for $1 < p < \infty$, the $L_p(\Omega)$ spaces are separable and reflexive Banach spaces; for $p = 2$ we get a Hilbert space with inner product $\int_\Omega f(x)g(x)\, dx$. The space $L_1(\Omega)$ is separable, whereas $L_\infty(\Omega)$ is *not*: consider e.g. the set of characteristic functions of all balls centered at a point of density of Ω. If $1 \leq p \leq \infty$, then $p' = p/(p-1)$ will denote the *conjugate exponent* to p; we put $p' = \infty$ if $p = 1$, and $p' = 1$ if $p = \infty$. If $1 < p < \infty$, then $(L_p(\Omega))'$, the dual space to $L_p(\Omega)$, is equal to $L_{p'}(\Omega)$ and every bounded linear functional F on $L_p(\Omega)$ has a unique representation in terms of the duality mapping $f \mapsto < f, g >= \int_\Omega f(x)g(x)\,dx$ for a (unique) $g \in L_{p'}$. At the endpoints of the scale, $p = 1$ or $p = \infty$, the duality relation gives only the norm, namely, $\|f\|_p = \sup\left\{\int_\Omega f(x)g(x)\,dx;\ \|g\|_{p'} \leq 1\right\}$.

By the symbol $C^\alpha(\Omega)$, $0 < \alpha \leq 1$, we shall denote the (real-valued) *Hölder spaces*; for $\alpha = 1$ we usually talk about the spaces of Lipschitz continuous functions.

We shall suppose that the reader is familiar with the construction and properties of the Bochner integral of vector-valued functions. Relevant properties of real- (or complex-) valued functions carry over to the vector case. Considering functions with values in a Banach space X that are Bochner integrable with the p-th power we get the *vector-valued spaces* $L_p(X) = L_p(\Omega, X)$, that is, the space of all functions $f : \Omega \to X$ with the finite norm $\|f\|_p = \|f\|_{L_p(\Omega,X)}$ (with the corresponding change if $p = 1$ or $p = \infty$). By $C^\alpha(X)$ or $C^\alpha(\Omega, X)$ we shall denote the space of vector-valued Hölder continuous functions. We shall frequently work with $X = \mathbb{R}^N$, the N-dimensional Euclidean spaces. Then the above spaces (and also spaces of Sobolev type and so on) will often be denoted by bold letters. As to spaces of vector-valued functions we refer to [39], [7], [8], [134].

One of the main tools is the Fourier transform. We shall use the formulae

$$\mathscr{F}(f;\xi) = \frac{1}{(2\pi)^{N/2}} \int_{\mathbb{R}^N} e^{-ix\cdot\xi} f(x)\,dx, \qquad \xi \in \mathbb{R}^N,$$

and

$$\mathscr{F}^{-1}(f;x) = \frac{1}{(2\pi)^{N/2}} \int_{\mathbb{R}^N} e^{ix\cdot\xi} f(\xi)\,d\xi, \qquad \xi \in \mathbb{R}^N,$$

for $f \in \mathscr{S}(\mathbb{R}^N)$, the space of rapidly decreasing infinitely differentiable functions, where $x \cdot \xi$ is the scalar product of x and ξ in $\mathbb{R}^N$. Observe that sometimes another multiplicative constant instead of $(2\pi)^{-N/2}$ occurs in front of the integrals above and/or $x \cdot \xi$ is replaced by $2\pi x \cdot \xi$. Usage changes from author to author and the major intrinsic reason is aesthetically pleasing formulae, depending of course on the sort of objectives we are concerned with.

For the reader's convenience we recall *Hölder's inequality*, linking the norms in dual spaces L_p and $L_{p'}$ ($p' = p/(p-1)$ if $1 < p < \infty$, $p' = 1$ if $p = \infty$, and $p' = \infty$ if $p = 1$),

$$\int_\Omega f(x)g(x)\,dx \leq \|f\|_p\,\|g\|_{p'}$$

whenever the right hand side has sense.

We recall *Minkowski's inequality*: If $1 \le p \le \infty$, $\Omega_1, \Omega_2 \subset \mathbb{R}^N$ are measurable, and f is measurable and non-negative on $\Omega_1 \times \Omega_2$, then

$$\left(\int_{\Omega_1}\left(\int_{\Omega_2} f(x,y)\,dx\right)^p dy\right)^{1/p} \le \int_{\Omega_2}\left(\int_{\Omega_1} f(x,y)^p\,dy\right)^{1/p} dx.$$

The well-known (triangle) inequality $\|f+g\|_p \le \|f\|_p + \|g\|_p$ is usually also called Minkowski's inequality.

Third, for reader's convenience, we present the classical Hardy inequality. We refer to [65], [114], [99], Chapter 6, for the proof.

2.2.1 Proposition (Hardy's inequality). *Suppose that $1 \le p < \infty$, $s \ge 0$ and let $f \ge 0$ be measurable function on $(0, +\infty)$. Put $F(x) = x^{-1}\int_0^x f(t)\,dt$. Then*

$$\int_0^{+\infty} F(x)^p x^{p-s-1}\,dx \le \frac{p}{s}\int_0^{+\infty} f(t)^p t^{p-s-1}\,dt. \tag{2.2.1}$$

We shall occasionally use the space $\mathscr{D}(\mathbb{R}^N)$ of all C^∞ functions (real- or complex-valued) with compact support in $\mathbb{R}^N$ as well as the space $\mathscr{D}'(\mathbb{R}^N)$, the dual of $\mathscr{D}(\mathbb{R}^N)$, whose elements are known as *distributions. Tempered distributions* are, however, more useful in the theory of function spaces.

Let X be a Banach space. Following the common usage we denote by $\mathscr{S}(\mathbb{R}^N, X) = \mathscr{S}(X)$ the space of all rapidly decreasing infinitely differentiable functions on $\mathbb{R}^N$ with the inductive limit topology given by the seminorms

$$\|\varphi\|_{k,\ell} = \sup_{x\in\mathbb{R}^N}\max_{|\alpha|\le\ell}\left(1+|x|^k\right)\|D^\alpha\varphi(x)\|_X, \tag{2.2.2}$$

where k and ℓ run over all non-negative integers.

Given Banach spaces X and Y, a linear mapping $f : \mathscr{S}(\mathbb{R}^N, X) \to Y$ is said to be a *tempered distribution* with values in Y if for every non-negative integers k and ℓ there exist $c > 0$ such that

$$\|f(\varphi)\|_Y \le c\|\varphi\|_{k,\ell}$$

for every $\varphi \in \mathscr{S}(\mathbb{R}^N, X)$. We shall write $f \in \mathscr{S}'(\mathbb{R}^N, X, Y)$ or more simply $f \in \mathscr{S}'(X, Y)$. The space $\mathscr{S}'(X, Y)$ of tempered distributions is a vector space, complete with respect to the weak topology induced by the pairing $(\mathscr{S}(\mathbb{R}^N, X), \mathscr{S}'(X, Y))$.

If $X = Y$, we shall use the notation $\mathscr{S}'(X)$ instead of $\mathscr{S}'(X, X)$. In the important particular case when $X = Y = \mathbb{R}^1$ (or $\mathbb{C}$), the formula (2.2.2) becomes

$$\|\varphi\|_{k,\ell} = \sup_{x \in \mathbb{R}^N} \max_{|\alpha| \le \ell} \left(1 + |x|^k\right) |D^\alpha \varphi(x)|,$$

and the spaces $\mathscr{S}(\mathbb{R}^N, X)$ and $\mathscr{S}'(\mathbb{R}^N, X, Y)$ are usually denoted by $\mathscr{S}(\mathbb{R}^N)$ and $\mathscr{S}'(\mathbb{R}^N)$, respectively. This is slightly abusing as the reader has certainly observed since it makes no difference between mappings into $\mathbb{R}^1$ and $\mathbb{R}^N$, nevertheless, the notation $\mathscr{S}(\mathbb{R}^N)$ and $\mathscr{S}'(\mathbb{R}^N)$ is widely used in the "classical" situation with real- or complex-valued tempered distributions so that confusion with the more general situation can hardly occur.

Let X be a Banach space, let $f \in \mathscr{S}'(X)$ and $\alpha = (\alpha_1, \dots, \alpha_n)$ be a multiindex. Then the *weak derivative* (or the *distributional derivative*) of order α of a tempered distribution f is the tempered distribution defined by

$$D^\alpha f(\varphi) = (-1)^\alpha f(D^\alpha \varphi), \qquad \varphi \in \mathscr{S}(\mathbb{R}^N).$$

We shall not pursue the properties of $\mathscr{S}(\mathbb{R}^N)$ in detail, in particular the definition of the topology in $\mathscr{S}(\mathbb{R}^N)$ and convergence in it. There are many excellent available expositions of the subject.

Later on we shall frequently work with convolutions. Let us recall that for $f_1 \in L_1(\mathbb{R}^N)$ and $f_2 \in L_p(\mathbb{R}^N)$, where $1 \le p \le \infty$, the *convolution* is defined as

$$f_1 * f_2(x) = \int_{\mathbb{R}^N} f_1(y) f_2(x - y)\, dy, \qquad x \in \mathbb{R}^N. \tag{2.2.3}$$

This definition makes sense in view of Minkowski's inequality and $f_1 * g_2 \in L_p(\mathbb{R}^N)$. Similarly, Hölder's inequality implies that if $f_1 \in L_p(\mathbb{R}^N)$ and $f_2 \in L_{p'}$ with $p' = p/(p-1)$, we have $f_1 * f_2 \in L_\infty(\mathbb{R}^N)$.

Plainly if $f_1, f_2 \in \mathscr{S}(\mathbb{R}^N)$, then $f_1 * f_2 \in \mathscr{S}(\mathbb{R}^N)$.

The definition in (2.2.3) is meaningful also in other important cases. For instance, if both f_1 and f_2 have compact supports in $\mathbb{R}^N$, and $f_i \in L_{p_i}$, $i = 1, 2$, then extending f_1 and f_2 by zero to the whole of $\mathbb{R}^N$, the convolution $f_1 * f_2$ is in $L_{\max\{p_1, p_2\}}$. This follows easily on applying Minkowski's and Hölder's inequalities.

The concept of the convolution can be extended to a pair $f \in \mathscr{S}'(\mathbb{R}^N)$ and $\varphi \in \mathscr{S}(\mathbb{R}^N)$; we put

$$f * \varphi(x) = f(\varphi(x - \cdot)), \qquad x \in \mathbb{R}^N. \tag{2.2.4}$$

The last definition can be further extended in an obvious way to convolutions of tempered distributions and functions with values in any normed linear space E provided $\mathscr{S}(\mathbb{R}^N)$ is dense in E.

More generally, let X be a Banach space. Given $f \in \mathscr{S}'(\mathbb{R}^N, X)$ and $\varphi \in \mathscr{S}(\mathbb{R}^N)$, we define the convolution $f * \varphi$ as in (2.2.4).

If f_1 and f_2 are functions in $L_p(\mathbb{R}^N)$ and $L_{p'}(\mathbb{R}^N)$, respectively, where p and p' are conjugate exponents, then $f_1 * f_2$ is essentially bounded.

If $f_1 \in L_p(\mathbb{R}^N)$ and $f_2 \in L_r(\mathbb{R}^N)$, where $1 < p < r' < \infty$, then $f_1 * f_2 \in L_q\mathbb{R}^N$, where $1/q = 1/p - 1/r'$ (Young's inequality, see Theorem 2.3.16).

A *standard regularizator* or a *standard mollifier* is a non-negative function $\varphi \in \mathscr{D}(\mathbb{R}^N)$, supported in $[-1,1]^N$, and satisfying $\|\varphi\|_1 = 1$. A usual example of such a function is $\varphi(x) = \varphi_1(x)/\|\varphi\|_1$, where $\varphi_1(x) = \exp(-(1-|x|^2))$ if $|x| \le 1$ and 0 otherwise. The family $\{\varphi_\varepsilon\}_{\varepsilon>0}$, where $\varphi_\varepsilon(x) = \varepsilon^{-N}\varphi(x/\varepsilon)$, $x \in \mathbb{R}^N$, is called an *approximation of unity*, or *resolution of unity*; sometimes we speak about a *smooth decomposition* (or *resolution*) *of unity*. The convolutions $\varphi_\varepsilon * f$, where $f \in L_{1,\mathrm{loc}}$ are one of the standard tools (not only) in the theory of function spaces. It is well known that $\varphi_\varepsilon * f$ is a C^∞ approximation of f in many important spaces, particularly in L_p and in Sobolev spaces W_p^k, $1 \le p < \infty$. If this convolution has sense, the points in $\operatorname{supp}(\varphi_\varepsilon * f)$ have distance from $\operatorname{supp} f$ not bigger than ε and this is especially important in applications. We refer e.g. to [99], Chapter 2, for detailed proofs.

Let $f : \Omega \to \mathbb{R}^1$ be measurable. Then we define the *distribution function* of f as

$$m(f,\lambda) = \operatorname{mes}\{x \in \Omega;\ |f(x)| > \lambda\}, \qquad \lambda > 0.$$

The function $m(f,\cdot)$ is measurable, non-increasing, and it is continuous from the right on $(0,+\infty)$. The *non-increasing rearrangement* of f is the generalized inverse of $m(f.,)$, that is,

$$f^*(t) = \inf\{\lambda;\ m(f,\lambda) \le t\}, \qquad t > 0.$$

The function f^* is non-increasing and continuous from the right on $(0,+\infty)$ as well and it is a useful exercise for a reader to draw a picture showing the behaviour of g^* if $g(t) = \sum_j c_j \chi_{I_j}(t)$, $t \in \mathbb{R}^1$, where $\{I_j\}$ is a finite sequence of disjoint intervals in $\mathbb{R}^1$: The function g^* equals c_i on intervals of length $\operatorname{mes} I_j$ and, starting from the origin, the values c_j are in non-increasing order. Closely related is the concept of a *spherical radially non-increasing rearrangement* (sometimes called a *spherical symmetrization*): instead of a rearrangement on $(0,+\infty)$ we consider a rearrangement $f^\#$ on

$\mathbb{R}^N$, defined by $f^{\#}(x) = f^*(\omega_N |x|^N)$, $x \in \mathbb{R}^N$, where ω_N is the volume of the unit ball in $\mathbb{R}^n$. Plainly $\|f^{\#}\|_{L_q(\mathbb{R}^N)} = \|f^*\|_{L_q(0,+\infty)}$ for all q, $1 \le q \le \infty$. The idea of symmetrization of a function goes back to the end of the 19th century and works by Steiner and Schwartz. The rearrangements were systematically studied by Hardy and Littlewood (see [65]) and since then they have become an essential tool in the study of function spaces. We refer e.g. to Kawohl's book [90] for an account of various types of rearrangements and further references. We recall an alternative formula (see [96], [55]), giving another nice geometrical idea of what happens when we rearrange a function, namely,

$$f^*(t) = \sup_{\operatorname{mes} E = t} \inf_{x \in E} |f(x)|, \qquad t > 0.$$

Rearrangements have been widely used for proofs of imbedding theorems for Sobolev spaces and their connections with isoperimetric formula and geometric measure theory have been clarified. Let us recall the one of the (highly non-trivial) key facts, namely, that $\|\Phi(\nabla u^{\#})\|_1 \le \|\Phi(\nabla u)\|_1$ for every Lipschitz continuous u, and convex non-decreasing Φ such that $\Phi(0) = 0$. From the rich literature on this subject we refer to the survey paper by Talenti [144].

Sometimes it may be useful to work with the integral average

$$f^{**}(t) = \frac{1}{t} \int_0^t f^*(\tau)\, d\tau.$$

The main property linking f and f^* is their *equimeasurability*, i.e. we have $m(f^*, t) = m(f, t)$, $t > 0$. By Fubini's theorem this implies that $\|f^*\|_{L_p(0,+\infty)} = \|f\|_{L_p(\Omega)}$, $1 \le p < \infty$, and $\|f\|_{L_\infty(\Omega)} = \|f^*\|_{L_\infty(0,+\infty)}$. The *Marcinkiewicz* (or the *weak* L_p)*-space* $L_p^*(\Omega)$ is defined as the space of all measurable functions f on Ω with the finite norm

$$\|f\|_{L_p^*(\Omega)} = \sup_{\lambda > 0} \lambda m(f, \lambda)^{1/p} = \sup_{t > 0} t^{1/p} f^*(t), \qquad 1 \le p < \infty, \quad (2.2.5)$$

and/or

$$\|f\|_{L_\infty^*(\Omega)} = \sup_{t > 0} f^*(t).$$

It is easy to see that $L_p(\Omega) \hookrightarrow L_p^*(\Omega)$ and $L_\infty^*(\Omega) = L_\infty(\Omega)$. Moreover, if $p > 1$, then the weak L_p-spaces can alternatively be defined with use of f^{**} instead of f^*. One estimate is easy—it follows from the pointwise estimate $f^{**} \ge f^*$, which is moreover independent of p. The opposite inequality holds (for integrals) in view of Hardy's inequality (see (2.2.1)) and it explains the restriction $p > 1$ (the constant here is equivalent to $(p-1)^{-1}$ as $p \to 1$). The case $p = 1$ will play no role for us; however, it is very much worthy of interest. It can be shown that f^{**} is equivalent to the non-increasing rearrangement of the maximal operator Mf of the function f (Herz' theorem, see e.g. [22], Chapter 3). Recall that the maximal operator is defined, for a locally integrable f, as

$$Mf(x) = \sup \frac{1}{\operatorname{mes} Q} \int_Q |f(y)| \, dy, \qquad x \in \mathbb{R}^N, \tag{2.2.6}$$

where the supremum is taken over all cubes containing x and with edges parallel with coordinate axes. A certain $\mathbb{R}^N$ analog to Hardy's inequality is the maximal inequality, saying, that for $1 < p \le \infty$, the function Mf is in $L_p(\mathbb{R}^N)$ if and only if $f \in L_p(\mathbb{R}^N)$. If $f \in L_1(\mathbb{R}^N)$, then, in general, $Mf \notin L_1(\mathbb{R}^N)$. On the other hand, $f \in L_1(\mathbb{R}^N)$ implies $Mf \in L_1^*$. (Together with the (trivial) boundedness of M in $L_\infty(\mathbb{R}^N)$ this implies the maximal inequality in view of the Marcinkiewicz interpolation theorem; see Theorem 2.3.2 in Section 2.3.) A further analysis reveals that Mf is integrable over the set $\{Mf > 1\}$ if and only if $|f| \log^+ |f|$ is integrable (Stein's theorem). This is closely connected with the equivalence $(Mf)^*$ and f^{**} (Herz' theorem). We refer e.g. to [63], [22], [58]. We also note that this is naturally linked with deep differentiation properties of real functions of several variables with respect to various differentiation bases, depending heavily on the geometric properties of elements of the basis in question (see e.g. [63]). Nevertheless, sticking even to nice bases as in (2.2.6), the behaviour of many important operators, including the maximal operator or the integral average as in Hardy's inequality, changes dramatically as $p \to 1$ and it turns out that L_1 is not the "best" endpoint of the L_p scale. On the other hand the space L_1 and Sobolev spaces based on it play an important role in PDEs; hence this is a somewhat unpleasant situation. The situation on the other end of the L_p scale is not complicated in this way; however, it sometimes is possible to replace L_∞ by other spaces, with weaker topology, particularly in interpolation arguments—the BMO space can serve to this purpose, for instance. Nevertheless, both the spaces L_∞ and the BMO are not spaces "good" for the general Fourier analysis approach. (For including of the bmo spaces, an $\mathbb{R}^N$ variant of BMO, see e.g. [149].)

An interested reader will find the exposition of the more general concept of the *Lorentz space* $L_{pq}(\Omega)$ in many well-known books, e.g. [29], [23], Chapters 1 and 5, [22], Chapter 4.

Relevant elementary properties carry over to the case of Bochner integrable vector-valued functions.

A bounded (sub-)linear mapping $T : L_p(\Omega_1) \to L_q(\Omega_2)$ is said to be *of type* (p, q). The number $\|T\| = \sup\{\|Tf\|_q;\ \|f\|_p \le 1\}$ is the norm of T.

If the target space is replaced with $L_q^*(\Omega_2)$, then T is said to be *of weak type* (p, q). Explicitly, this means that there exists $c > 0$ such that for all $f \in L_p(\Omega_1)$ and all $\lambda > 0$,

$$m(Tf, \lambda) \le c^q \lambda^{-q} \|f\|_p^q \tag{2.2.7}$$

(which is equivalent to $\|Tf\|_{L_q^*(\Omega)} \le c\|f\|_{L_p(\Omega_1)}$). The best c in (2.2.7) is the norm of T.

The importance of the Marcinkiewicz spaces in our context consists in the fact that they are natural target spaces for integral operators shifting the smoothness properties of Lebesgue spaces.

We shall suppose that the reader is familiar with the elements of the theory of Sobolev spaces of integer order on a domain in $\mathbb{R}^N$. Let us just recall that they can be defined in several standard equivalent ways, namely, as the spaces with distributional derivatives, as a completion of functions smooth in the domain in question (the celebrated Meyers-Serrin theorem [111] gives their coincidence), or as the so called Beppo-Levi spaces. We refer to well-known expository texts [1], [99].

We also point out another essential topic in the theory of the Sobolev type spaces. One often meets density arguments in proofs and the spaces can be considered on various domains or on the whole of $\mathbb{R}^N$. Particularly spaces on $\mathbb{R}^N$ permit the Fourier analytic approach, embracing very general scales of spaces (potential Sobolev spaces, Besov spaces, Lizorkin-Triebel spaces). According to the Meyers-Serrin theorem one can approximate functions in Sobolev spaces in a domain, say Ω, by functions in $C^\infty(\Omega)$; however, the situation changes dramatically when the latter space is replaced by $C^\infty(\overline{\Omega})$. Analogous problems appear when considering imbeddings, extensions, and traces and it turns out that the geometric properties of the boundary play the major role here. We touch upon these problems (at least partly) in Subsections 2.7.1 and 2.7.3 (see also Section 1.6 for an introduction).

In this connection let us again recall the concept of a *mollifier* (or of a *regularizator*) in more details. Consider, for simplicity, $f \in L_p(\mathbb{R}^N)$. Then there exists $f_0 \in L_p(\mathbb{R}^N)$ with a compact support such that $\|f - f_0\|_p$ is arbitrarily small. Let

$$\psi(x) = \begin{cases} \exp(-1/(1-|x|^2)), & |x| \le 1, \\ 0, & \text{otherwise}, \end{cases}$$

and put $\varphi(x) = \psi(x) \left(\|\psi\|_1\right)^{-1}$. The function φ is usually called the *standard mollifier* or briefly the *mollifier*. For $x \in \mathbb{R}^N$, $\varepsilon > 0$, let us define the family

$$\varphi_\varepsilon(x) = \varepsilon^{-N} \varphi(x/\varepsilon). \tag{2.2.8}$$

Then the functions φ_ε are in $C_0^\infty(\mathbb{R}^N)$, $\operatorname{supp}\varphi_\varepsilon \subset \{|x| \le \varepsilon\}$, and $\|\varphi_\varepsilon\|_1 = 1$ for all $\varepsilon > 0$. Consider functions $f_\varepsilon(x) = f * \varphi_\varepsilon(x)$. It is not difficult to show that $f_\varepsilon \in C_0^\infty(\mathbb{R}^N)$ and that $f_\varepsilon \to f$ in $L_p(\mathbb{R}^N)$ as $\varepsilon \to 0$ (using Minkowski's inequality and the L_p-mean continuity of f). Similarly, if f has a distributional derivative $D^\beta f \in L_p(\mathbb{R}^N)$, then $D^\beta f_\varepsilon \to D^\beta$ in $L_p(\mathbb{R}^N)$. Moreover, if $f \in W_p^k(\mathbb{R}^N)$, that is, when $D^\alpha f \in L_p(\mathbb{R}^N)$ for all $|\alpha| \le k$, then first one can approximate f by a function $\widetilde{f}$ with a compact support and close to f in the corresponding Sobolev space norm and then consider $\widetilde{f} * \varphi_\varepsilon$ to prove the density of $C_0^\infty(\mathbb{R}^N)$ in the Sobolev spaces $W_p^k(\mathbb{R}^N)$. If Ω is a domain in $\mathbb{R}^N$, $N \ge 2$, with, say, a Lipschitz boundary $\partial\Omega$, then the procedure is only formally more complicated: After writing f as $\sum f_i$, $f_i \in W_p^k(\Omega)$, using a suitable decomposition of unity and after a rotation if necessary we arrive at the situation when either f_i has a compact support contained in Ω or f_i is non-zero in a domain say Ω_i and $\partial\Omega \cap \partial\Omega_i$ is non-empty and it is a graph of a Lipschitz function. Then consider the function $f_{ih}(x) = f_i(x+h)$, $x \in \Omega$ and h sufficiently small. One can show easily that $f_{ih} \to f_i$ in $W_p^k(\Omega)$ as $h \to 0$ and then it suffices to approximate f_{ih} by $f_{ih} * \varphi_\varepsilon$ to show that $C^\infty(\overline{\Omega})$ is dense in $W_p^k(\Omega)$. These considerations are of course much easier when $N = 1$.

2.3 Crash course in interpolation

We shall recall several basic facts from the interpolation theory. The beginning of the interpolation theory goes back to theorems on bilinear forms (Schur, M. Riesz) and the role of model cases is usually played by the Riesz-Thorin theorem (the complex interpolation), hanging on the Hadamard principle, and the Marcinkiewicz theorem (the real interpolation). The theory underwent the major development in 1950s and 1960s in close relations with the function spaces theory and since then it has been constantly an area of an intensive research.

The basic problem in the interpolation theory is the following: Given a linear operator T acting continuously between two couples of normed linear spaces, say, $T : X_j \to Y_j$, $j = 1, 2$, where (X_1, X_2) and (Y_1, Y_2) are *compatible* (or a *compatible couple of spaces*), that is, both two last pairs of spaces are subspaces of some Hausdorff topological linear space, we look for spaces X and Y such that $X_1 \cap X_2 \subset X \subset X_1 + X_2$ and $Y_1 \cap Y_2 \subset Y \subset Y_1 + Y_2$ and such that $T : X \to Y$ is continuous. The spaces $X_1 \cap X_2$ and $X_1 + X_2$ are endowed with the standard norms $\max\{\|x\|_{X_1}, \|x\|_{X_2}\}$ and $\inf\{\|x_1\|_{X_1} + \|x_2\|_{X_2};\ x = x_1 + x_2,\ x_j \in X_j,\ j = 1, 2\}$, resp. If $X_1 \cap X_2 \subset X \subset X_1 + X_2$, then the space X is usually called *intermediate* with respect to the couple (X_1, X_2). According to the kind of quantitative relations between the norms $\|T\|_{X_j\to Y_j}$, $j=1,2$, and $\|T\|_{X\to Y}$ one can introduce more special concepts of interpolation spaces. If $\|T\|_{X\to Y}\le C\max_{j=1,2}\|T\|_{X_j\to Y_j}$, then X and Y are said to be *uniform interpolation spaces*; if $C=1$ here, we talk about *exact interpolation spaces*. If $\|T\|_{X\to Y}\le C\|T\|^{1-\theta}_{X_1\to Y_1}\|T\|^{\theta}_{X_2\to Y_2}$, then X and Y are said to be *interpolation spaces of exponent* θ; if $C=1$, they are usually called *exact interpolation spaces of exponent* θ. On this level it is sometimes useful to consider the *category* of normed linear spaces imbedded into some (fixed) Hausdorff topological linear space with bounded (sub-)linear mappings as *morphisms* in this category. Then one can introduce the *interpolation functor* as a mapping taking couples of spaces in a given category to another spaces, which has interpolation properties described above; then one speaks e.g. about exact interpolation functors etc. We shall not go into further details and refer to [23], Chapter 2 or [148], Chapter 1. The general theory of interpolation spaces together with a description of numerous interpolation methods is sufficiently exposed in well-known monographs (see e.g. [23], [148]). A reader who would like to go to the modern roots of the theory and to original works will find also many references there. For all of them let us recall Peetre's book [119].

2.3.1 Classical theorems

Essentially, we shall need basic interpolation facts for the L_p spaces, but our considerations include also vector-valued spaces; hence we shall also briefly describe the complex interpolation method. First of all, however, we state two classical interpolation theorems (see e.g. [23], Chapter 1, [126], Chapter 4, [148], Chapter 1).

2.3.1 Theorem (Riesz-Thorin convexity theorem). *Let $\Omega\subset\mathbb{R}^N$ be measurable. Let $1\le p_i\le\infty$, $1\le q_i\le\infty$, $i=1,2$ and suppose that T is a (sub-)linear operator*

$$T: L_{p_1}(\Omega)\to L_{q_1}(\Omega),\quad \textit{and}\quad T: L_{p_2}(\Omega)\to L_{q_2}(\Omega),$$

with norms M_1 and M_2, resp. Let $0\le\theta\le 1$,

$$\frac{1}{p}=\frac{1-\theta}{p_1}+\frac{\theta}{p_2},\qquad \textit{and}\qquad \frac{1}{q}=\frac{1-\theta}{q_1}+\frac{\theta}{q_2} \tag{2.3.1}$$

(formally we put $a/\infty=0$ for any real a). Then T is a bounded mapping from $L_p(\Omega)$ into $L_q(\Omega)$ with norm less or equal to $M_1^{1-\theta}M_2^{\theta}$.

Observe that in connection with the relations (2.3.1) it is useful to consider the norm of T as a function of $1/p$. Then the theorem states that the quantity $M=M(\theta)$ is a logarithmically convex function of $1/p$.

Recall that the (Thorin's) proof of the above theorem hangs on the Hadamard three lines (or equivalently—via conformal mapping—three circles) principle: If F is analytic on the open strip $0 < \operatorname{Re} z < 1$ in the complex plane, bounded and continuous on its closure, and if $|F(k-1+it)| \leq M_k$, $k = 1, 2$, $t \in \mathbb{R}^1$, then $|F(\theta + it)| \leq M_1^{1-\theta} M_2^{\theta}$ for all $0 < \theta < 1$ and all $t \in \mathbb{R}^1$.

More generally, one can consider L_p spaces on spaces with positive measure and consisting of vector-valued functions; we shall do this in the context of general interpolation methods (see below).

Let us recall another classical interpolation principle, belonging to the standard tools in many areas of analysis. Recall the notation $L_p^*(\Omega)$ for the weak Lebesgue space $L_p(\Omega)$ (see (2.2.5) on p. 82).

2.3.2 Theorem (Marcinkiewicz interpolation theorem). *Let $\Omega \subset \mathbb{R}^N$ be measurable. Let $1 \leq p_i \leq \infty$, $1 \leq q_i \leq \infty$, $i = 1, 2$ and suppose that T is a (sub-)linear operator*

$$T : L_{p_1}(\Omega) \to L_{q_1}^*(\Omega), \quad \text{and} \quad T : L_{p_2}(\Omega) \to L_{q_2}^*(\Omega),$$

with norms M_1 and M_2, resp. Let $0 \leq \theta \leq 1$,

$$\frac{1}{p} = \frac{1-\theta}{p_1} + \frac{\theta}{p_2}, \quad \text{and} \quad \frac{1}{q} = \frac{1-\theta}{q_1} + \frac{\theta}{q_2} \tag{2.3.2}$$

(again, we formally put $a/\infty = 0$ for any real a) and suppose that $p \leq q$. Then T is a bounded mapping from $L_p(\Omega)$ into $L_q(\Omega)$ with norm less or equal to $C(\theta) M_1^{1-\theta} M_2^{\theta}$.

The behaviour of the constant $C(\theta)$ above can be estimated—it generally blows-up as $p \to p_j$, $j = 1, 2$ (see e.g. [147], Chapter 4).

The theorem holds also for vector-valued functions. Observe that the major difference between these two classical theorems is the boundedness of the operator in question between different pairs of spaces. Plainly, L_p is imbedded into L_p^* and the latter space is effectively larger, and hence the assumptions in the Marcinkiewicz theorem are weaker; the price for that is the necessary requirement $p \leq q$. Many classical operators as the Riesz operator and various kinds of maximal operators fit into this scheme; we shall see a prominent example based on Young's inequality in the following.

We shall give a brief account of the abstract interpolation methods in the next subsections.

2.3.2 The complex interpolation method

We give only a very brief description of the complex interpolation method and several basic properties since we shall mainly use the real interpolation in the following. Observe that the abstract construction of complex interpolation spaces is actually based on the main tool in the proof of the Riesz-Thorin theorem, namely, on the Hadamard principle. Of course, the Riesz-Thorin theorem becomes a special important claim of the complex interpolation. We briefly describe the definition of the complex interpolation scale. Let $\overline{X} = (X_1, X_2)$ be an admissible couple of normed linear spaces. Denote by $\mathcal{F} = \mathcal{F}(\overline{X})$ the linear set of all bounded and continuous functions $f : \{z \in \mathbb{C};\ 0 \leq \operatorname{Re} z \leq 1\} \to X_1 + X_2$, analytic in $\{z \in \mathbb{C};\ 0 < \operatorname{Re} z < 1\}$ and such that the functions $t \mapsto f(j-1+it)$, $t \in \mathbb{R}^1$, are continuous into X_j, and satisfy $\lim_{|t| \to +\infty} f(j-1+it) = 0$, $j = 1, 2$. The set $\mathcal{F}(\overline{X})$ endowed with the norm $\max_{j=1,2} \sup_{t \in \mathbb{R}^1} |f(j-1+it)|$ becomes a Banach space (see [23], Chapter 4, [148], Chapter 1). For $0 < \theta < 1$, we define the space

$$[X_1, X_2]_\theta = \{x \in X_1 + X_2;\ \text{there exists } f \in \mathcal{F} \text{ such that } f(\theta) = x\}$$

and provide it with the norm

$$\|x\|_{[X_1, X_2]_\theta} = \inf_{f(\theta)=x,\, f \in \mathcal{F}} \Big(\max_{j=0,1,\, t \in \mathbb{R}^N} |f(j-1+it)| \Big).$$

As a result we obtain a scale of Banach spaces $[X_1, X_2]_\theta$, $0 < \theta < 1$, which are exact interpolation spaces of exponent θ with respect to the couple (X_1, X_2), that is, the corresponding quantitative relation for the norms of linear operators acting on complex interpolation spaces is precisely the same as in the Riesz-Thorin theorem recalled above. This is of course not a pure coincidence; interpolating L_p spaces with the complex method yields an L_p space "between" the given couple. Moreover, the complex interpolation method enjoys the stability property, that is, if we interpolate two compatible Banach spaces, using different indices and if we repeat the interpolation process with these results of interpolation, that is, if we reiterate the interpolation, we arrive at a space, which can be constructed by means of the same method from the couple of the spaces at the beginning. More precisely:

2.3.3 Theorem (stability theorem). *Let* $\overline{X} = (X_1, X_2)$ *be an admissible couple of Banach spaces and let* $Y_j = [X_1, X_2]_{\theta_j}$, $0 < \theta_j < 1$, $j = 1, 2$. *Suppose that* $X_1 \cap X_2$ *is dense in* X_1, X_2, *and* $Y_1 \cap Y_2$. *Then* $[Y_1, Y_2]_\eta = [X_1, X_2]_\theta$ *for any* $\theta = (1-\eta)\theta_1 + \eta\theta_2$, *where* $0 \leq \eta \leq 1$.

The Riesz-Thorin theorem appears in the following form:

2.3.4 Theorem. *Let* $\Omega \subset \mathbb{R}^N$ *be measurable. Let* $1 \leq p_j \leq \infty$, $j = 1, 2$, *and* $0 < \theta < 1$. *Put* $1/p = (1-\theta)/p_1 + \theta/p_2$. *Then* $[L_{p_1}(\Omega), L_{p_2}(\Omega)]_\theta = L_p(\Omega)$ *with equal norms.*

Proofs of both theorems can be found in well-known monographs (see e.g. [22], Chapter 4, [23], Chapter 5, [148], Chapter 1).

Observe that the Riesz-Thorin theorem holds also for vector-valued L_p spaces on sets with positive measure (see again [23], Chapter 5). We shall present a more general theorem on complex interpolation of vector-valued functions. Let (M, μ) be a space with a positive measure μ and X a Banach space. For $1 \leq p < \infty$, denote by $L_p(X)$ the space of all functions $f : M \to X$ such that $\|f\|_p = \left(\int_M \|f(x)\|_X^p \, d\mu\right)^{1/p} < \infty$.

2.3.5 Theorem. *Let (X_1, X_2) be a compatible couple of Banach spaces and assume that $1 \leq p_1, p_2 < \infty$, $0 < \theta < 1$, and $1/p = (1-\theta)/p_1 + \theta/p_2$. Then*

$$[L_{p_1}(X_1), L_{p_2}(X_2)]_\theta = L_p([X_1, X_2]_\theta) \tag{2.3.3}$$

with equivalent norms.

The *proof* of the last theorem can be found e.g. in [23], Chapter 5.

2.3.3 Fourier multipliers

Proofs of estimates in Sobolev spaces in the Hilbert case $p = 2$ are usually much easier due to the nice behaviour of the Fourier transform on L_2. The general case is much harder and a standard approach includes use of multipliers. Let us observe that in the general theory of function spaces it is just the "sufficient supply" of multipliers for them, which is essential for their properties (cf. [148], [149], etc.).

We shall recall the concept of a Fourier multiplier and formulate a multiplier theorem of Mikhlin type. The proof is not easy and it would go necessarily far beyond the purpose of this book; we refer to the rich existing literature ([119], [23], Chapter 6, [148], [149]). Let us recall our notation: If H is a Hilbert space, then $\mathscr{S}(H) = \mathscr{S}(\mathbb{R}^N, H)$ is the space of all infinitely differentiable, rapidly decreasing functions from $\mathbb{R}^N$ into H with the topology of seminorms given as in (2.2.2) with $X = H$. Given Hilbert spaces H_1 and H_2, let $\mathscr{S}'(\mathbb{R}^N, H_1, H_2) = \mathscr{S}'(H_1, H_2)$ be the space of all bounded linear mappings from $\mathscr{S}(\mathbb{R}^N, H_1)$ into H_2.

2.3.6 Definition. A tempered distribution $\rho \in \mathscr{S}'(\mathbb{R}^N, H_1, H_2)$ is the *Fourier multiplier* for the couple $(L_p(H_1), L_p(H_2))$ if $\mathscr{F}^{-1}(\rho) * f \in L_p(H_2)$ for all $f \in \mathscr{S}(\mathbb{R}^N, H_1)$ and

$$\sup_{\|f\|_{L_p(H_2)} \leq 1} \|\mathscr{F}^{-1}(\rho) * f\|_{L_p(H_1)} < +\infty.$$

The sup on the left hand side is the *norm of the Fourier multiplier* ϱ, denoted with $\|\varrho\|_{M_p(H_1,H_2)}$. The space of all Fourier multipliers for the couple $(L_p(H_1), L_p(H_2))$ will be denoted by $M_p(H_1, H_2)$. If $H_1 = H_2 = \mathbb{R}^N$, we shall write $M_p(\mathbb{R}^N)$ or simply M_p and $\|\varrho\|_{M_p(\mathbb{R}^N)}$ or $\|\varrho\|_{M_p}$ if the dimension is not essential in the given context.

2.3.7 Theorem (Mikhlin's multiplier theorem). *Let H_1 and H_2 be Hilbert spaces and assume that there is a positive constant M such that a mapping $\rho : \mathbb{R}^N \to L(\mathscr{S}(H_1), H_2)$ satisfies the condition*

$$|\xi|^\alpha \|D^\alpha \rho(\xi)\|_{L(\mathscr{S}(H_1),H_2)} \le M \tag{2.3.4}$$

for all $\xi \in \mathbb{R}^N$ and all $|\alpha| \le K$, where K is an integer bigger than $N/2$. Then ρ is a multiplier for the couple $(L_p(H_1), L_p(H_2)$ for every $1 < p < \infty$.

Note that if $H_1 = H_2 = \mathbb{R}^1$, then the condition (2.3.4) reduces to $|\xi|^\alpha |D^\alpha \rho(\xi)| \le$ const. for all $|\alpha| \le K$, where K is an integer, $K > N/2$.

In the following we shall also need other sufficient conditions guaranteeing that a given function is a Fourier multiplier. We shall restrict ourselves to the case of real (or complex) valued L_p spaces; hence our multipliers will be elements of $\mathscr{S}'$. For brevity, we shall use the notation M_p for the space of multipliers in L_p. Occasionally, we shall write $M_p(\mathbb{R}^N)$ to emphasize the dimension.

2.3.8 Lemma. *Let ϱ be a multiplier in L_p for some $1 \le p \le \infty$ with the norm $\|\varrho\|_{M_p} = \sup_{f \ne 0} \|\mathscr{F}^{-1}\varrho * f\|_p \|f\|_p^{-1}$. Then for every $\alpha \in \mathbb{R}^1 \setminus \{0\}$ the function $\varrho_\alpha(x) = \varrho(\alpha x)$, $x \in \mathbb{R}^N$, is a multiplier in L_p and $\|\varrho_\alpha\| = \|\varrho\|_{M_p}$.*

Proof. We have

$$\begin{aligned}
\mathscr{F}^{-1}(\varrho(\alpha x); y) &= \frac{1}{(2\pi)^{N/2}} \int_{\mathbb{R}^N} \varrho_\alpha(x) e^{ix\cdot y}\, dx \\
&= \frac{\alpha^{-N}}{(2\pi)^{N/2}} \int_{\mathbb{R}^N} \varrho(\eta) e^{i\eta\cdot y/\alpha}\, dy \\
&= \alpha^{-N} \mathscr{F}^{-1}(\varrho; y/\alpha).
\end{aligned}$$

Further,

$$\begin{aligned}
\|\mathscr{F}^{-1}(\varrho_\alpha) * f\|_p &= \alpha^{-N} \left(\int_{\mathbb{R}^N} \left| \int_{\mathbb{R}^N} \mathscr{F}^{-1}(\varrho; y/\alpha) f(x-y)\, dy \right|^p dx \right)^{1/p} \\
&= \left(\int_{\mathbb{R}^N} \left| \int_{\mathbb{R}^N} \mathscr{F}^{-1}(\varrho; \eta) f(x - \alpha\eta)\, d\eta \right|^p dx \right)^{1/p} \\
&= \left(\int_{\mathbb{R}^N} \left| \int_{\mathbb{R}^N} \mathscr{F}^{-1}(\varrho; \eta) f(\alpha(y - x))\, dx \right|^p d\eta \right)^{1/p} \\
&= \alpha^{N/p} \|f(\alpha\cdot)\|_p \, \|\varrho\|_{M_p}.
\end{aligned}$$

Since $\|f(\alpha\,\cdot)\|_p \alpha^{N/p} = \|f\|_p$ we are done. $\square$

2.3.9 Lemma. *Let $\varrho \in M_p(\mathbb{R}^1)$ for some $1 \le p \le \infty$ and let $\xi \in \mathbb{R}^N$. Then the function $\varrho_y = \varrho(\xi \cdot y)$ is in $M_p(\mathbb{R}^N)$ with the same M_p norm.*

Proof. It is a matter of routine to verify that M_p and its norm are invariant with respect to surjective linear transformations in $\mathbb{R}^N$. (One can go along the lines of the proof of the previous lemma, for instance.) Hence we can suppose from the beginning that $\xi \cdot y = y_1$. Denote by ϱ_ξ^* our ϱ_ξ in this new coordinate system. Then

$$\varrho_\xi^*(y) = \varrho(y_1) \otimes 1 \otimes \cdots \otimes 1.$$

Let δ denote the Dirac distribution. Then

$$\begin{aligned}\|\mathscr{F}^{-1}(\varrho_\xi^*) * f\|_p &= \|(\mathscr{F}^{-1}(\varrho \otimes \delta \otimes \cdots \otimes \delta) * f\|_p \\ &\le \|\varrho\|_{M_p}\|f\|_p,\end{aligned}$$

which gives $\|\varrho_\xi\|_{M_p(\mathbb{R}^N)} = \|\varrho_\xi^*\| \le \|\varrho\|_{M_p(\mathbb{R}^1)}$. The equality follows on taking functions of the form $f_1(x_1)f_2(x_2, \ldots, x_N)$. $\square$

Next theorem gives conditions in terms of square integrability of a function and its derivatives of sufficiently high orders.

2.3.10 Theorem. *Let m be an integer, $m > N/2$. Assume that $\varrho \in L_2$ and that $D^\alpha \varrho \in L_2(\mathbb{R}^N)$ for all $|\alpha| = m$. Then ϱ is a multiplier in $L_p(\mathbb{R}^N)$ for all $1 \le p \le \infty$ and*

$$\|\varrho\|_{M_p} \le c\|\varrho\|_2^{1-N/(2m)} \left(\sup_{|\alpha|=m} \|D^\alpha \varrho\|_2 \right)^{N/(2m)} . \tag{2.3.5}$$

Proof. First we show that $M_1 \subset M_p$ for $1 \le p \le 2$. Together with the (trivial) equality $M_p = M_{p'}$, where $p' = p/(p-1)$, this will imply $M_1 \subset M_p$ for all $1 \le p \le \infty$. Observe that $M_2 = L_\infty$. This immediately follows from Parseval's formula because

$$\|\mathscr{F}^{-1}(\varrho) * f\|_2 = \|\varrho \mathscr{F}(f)\|_2$$

and $\|f\|_2 = \|\mathscr{F}(f)\|_2$. Hence if $\varrho \in M_1 \cap M_p = M_1 \cap M_{p'}$ and $1 \le p \le 2$, then we have, by virtue of Riesz-Thorin theorem (see Theorem 2.3.1),

$$\|\varrho\|_{M_p} \le \|\varrho\|_{M_1}^{2-p} \|\varrho\|_{M_p}^{p-1}$$

since $1/p = 1 - \theta + \theta/p'$ for $\theta = p - 1$; hence $1 - \theta = 2 - p$ and $\theta = p - 1$. From this we immediately get the desired imbedding of M_1 into M_p.

Hence it will be sufficient to prove (2.3.5) for $p = 1$. According to the definition, $M_1 = M_\infty$ is the space of Fourier transforms of bounded measures in $\mathbb{R}^N$ and

$$\|\varrho\|_{M_1} = \|\varrho\|_{M_\infty} = \int_{\mathbb{R}^N} |\mathscr{F}^{-1}(\varrho; x)|\, dx.$$

This follows from the simple estimate

$$|\mathscr{F}^{-1}(\varrho) * f(0)| \le \|\varrho\|_{M_\infty} \|f\|_\infty, \qquad f \in \mathscr{S}.$$

Fix some $t > 0$. By Hölder's inequality and Parseval's formula,

$$\begin{aligned}\int_{|x|<t} |\mathscr{F}^{-1}(\varrho; x)|\, dx &\le c_1 t^{N/2} \Big(\int_{|x|<t} |\mathscr{F}^{-1}(\varrho; x)|^2\, dx \Big)^{1/2} \\ &\le c_1 t^{N/2} \|\varrho\|_2.\end{aligned}$$

At the same time, using Hölder's inequality once more,

$$\begin{aligned}\int_{|x|\ge t} |\mathscr{F}^{-1}(\varrho; x)|\, dx &\le \int_{|x|\ge t} |x|^{-m} |x|^m |\mathscr{F}^{-1}(\varrho; x)|\, dx \\ &\le \Big(|x|^{-2m}\, dx \Big)^{1/2} \Big(\int_{\mathbb{R}^N} |x|^{2m} |\mathscr{F}^{-1}(\varrho; x)|^2 \Big)^{1/2} \\ &\le c_2 t^{N/2-m} \Big(\sum_{|\alpha|=m} \int_{\mathbb{R}^N} |D^\alpha \varrho(x)|^2 \Big)^{1/2} \\ &\le c_3 t^{N/2-m} \max_{|\alpha|=m} \|D^\alpha \varrho\|_2.\end{aligned}$$

On choosing t such that $c_1 t^{N/2} \|\varrho\|_2 = c_3 t^{N/2-m} \max_{|\alpha|=m} \|D^\alpha \varrho\|_2$, that is,

$$t = c_4 \|\varrho\|_2^{-1/m} \max_{|\alpha|=m} \|D^\alpha \varrho\|_2^{1/m}$$

for a suitable c_4 independent of t, we conclude that (2.3.5) holds with $\|\varrho\|_{M_1}$ on the left hand side and the proof is complete. □

2.3.11 Remark. It follows from the proof of the previous theorem that M_p is an algebra. Furthermore, it is not difficult to prove that M_p is complete, hence, M_p is even a Banach algebra.

We still have to go through several interpolation technicalities useful in the sequel. Let X be a Banach space. For $1 \le p < \infty$ and $s \in \mathbb{R}^1$, let ℓ_p^s denote the space of all sequences $a = \{a_k\}_{k\ge 0} \subset X$ with the finite norm

$$\|\{a\}\|_{\ell_p^s(X)} = \Big(\sum_{k=0}^{+\infty} (2^{ks} \|a_k\|_X)^p \Big)^{1/p}.$$

2.3.12 Proposition. *Let $0 < \theta < 1$, $s_1, s_2 \in \mathbb{R}^1$, $1 \leq p_1, p_2 < \infty$ and let X_1 and X_2 be Banach spaces. Then, with equal norms,*

$$[\ell_{p_1}^{s_1}(X_1), \ell_{p_2}^{s_2}(X_2)]_\theta = \ell_p^s([X_1, X_2]_\theta),$$

where $s = (1-\theta)s_1 + \theta s_2$ and $1/p = (1-\theta)/p_1 + \theta/p_2$.

Proof. If $f \in \mathcal{F}(\ell_{p_1}^{s_1}(X_1), \ell_{p_2}^{s_2}(X_2))$ with $f(z) = \{f_k(z)\}_{k \geq 0}$, then put

$$\widetilde{f}_k(z) = 2^{s_1(1-z)k}\, 2^{s_2 zk} f_k(z).$$

The mapping $f \mapsto \widetilde{f}$ is an isometry between the spaces $\mathcal{F}(\ell_{p_1}^{s_1}(X_1), \ell_{p_2}^{s_2}(X_2))$ and $\mathcal{F}(\ell_{p_1}(X_1), \ell_{p_2}(X_2))$. Realizing that Theorem 2.3.5 holds for discrete measures as well, we are done. □

2.3.4 The real interpolation method

We shall turn our attention to the real interpolation method. Let (X_1, X_2) be a compatible couple of normed linear spaces. The (Peetre's) K-*functional* is the real valued function $K(t, x)$ defined on $(0, +\infty) \times (X_1 + X_2)$ by

$$\begin{aligned} K(t,x) &= K(t, x, X_1, X_2) \\ &= \inf\{\|x_1\|_{X_1} + t\|x_2\|_{X_2};\ x = x_0 + x_1,\ x_j \in X_j,\ j = 0, 1\}. \end{aligned}$$

For $0 < \theta < 1$ and $1 \leq p < \infty$ we define the space $(X_1, X_2)_{\theta,p}$ as the linear set of all $x \in X_1 + X_2$ with the finite norm

$$\|x\|_{\theta,p} = \left(\int_0^{+\infty} \left[t^{-\theta} K(t,x)\right]^p \frac{dt}{t} \right)^{1/p}.$$

For $0 < \theta < 1$, the norm in $(X_1, X_2)_{\theta,\infty}$ is given by

$$\|x\|_{\theta,\infty} = \sup_{t>0} t^{-\theta} K(t,x).$$

The result of this construction is again a scale of interpolation spaces between $X_1 \cap X_2$ and $X_1 + X_2$. The same relation for norms of a linear operator holds as in the case of the complex interpolation method, namely, the real interpolation spaces $(X_1, X_2)_{\theta,p}$ are exact of exponent θ, for all $1 \leq p \leq \infty$.

The real interpolation method is stable, too. We have (for the proof see e.g. [22], Chapter 5, [23], Chapter 3, [148], Chapter 1) the following reiteration theorem.

2.3.13 Theorem (reiteration theorem). *Let* (X_1, X_2) *be a compatible couple of Banach spaces and* $0 \le \theta_j \le 1$, $j = 1, 2$, $\theta_1 \ne \theta_2$. *Put* $Y_j = (X_1, X_2)_{\theta_j, q_j}$, *where* $1 \le q_j \le \infty$, $j = 1, 2$. *Let* $0 < \eta < 1$ *and* $\theta = (1 - \eta)\theta_1 + \eta\theta_2$. *Then, for every* $1 \le q \le \infty$, *it is* $(Y_1, Y_2)_{\eta,q} = (X_1, X_2)_{\theta,q}$ *with equivalent norms. More generally, if* Z_j *are Banach spaces,* $X_1 \cap X_2 \subset Z_j \subset X_1 + X_2$, $j = 1, 2$, *which are of class* θ_j, *that is, if* $(X_1, X_2)_{\theta_j,1} \hookrightarrow Z_j \hookrightarrow (X_1, X_2)_{\theta_j,\infty}$, *then the same reiteration formula holds with* Z_j *replacing* Y_j, $j = 1, 2$.

On the general level we shall state yet a theorem on interpolation of dual spaces. In particular, this will make possible later to derive directly interpolation formulae for Sobolev-Slobodetskii spaces with negative smoothness.

2.3.14 Theorem (duality theorem). *Let* (X_1, X_2) *be a compatible couple of Banach spaces. Assume that* $X_1 \cap X_2$ *is dense both in* X_1 *and* X_2. *Let* $0 < \theta < 1$, $1 \le p \le \infty$, *and* $p' = p/(p-1)$ *(we put formally* $1/\infty = 0$*). Then, with equivalent norms,*

$$\big((X_1, X_2)_{\theta,p}\big)' = (X_1', X_2')_{\theta,p'}.$$

The trace method

Later we shall tackle the problem of traces, which is very natural to be considered in the framework of the interpolation trace method. We shall proceed briefly—the methods in the following are in details described e.g. in [23], Chapter 3, [148], Chapter 1. Let $\overline{X} = (X_1, X_2)$ be a compatible couple of Banach spaces, $\overline{p} = (p_1, p_2)$, with $1 \le p_1, p_2 \le \infty$, $0 < \theta < 1$. If $u : (0, +\infty) \to X$, where X is a Banach space, let u' denote the distributional derivative of u. Let $m \in \mathbb{N}$, $0 < \theta < 1$, $\overline{p} = (p_1, p_2)$, where $1 \le p_i \le \infty$, $i = 1, 2$, and let $\overline{X} = (X_1, X_2)$ be a compatible couple of Banach spaces. We define the space $V^m = V^m(\overline{X}, \overline{p}, \theta)$ as the space of all functions $u : (0, +\infty) \to X_1 + X_2$ such that u is locally X_1-integrable and $u^{(m)}$ is locally X_2-integrable, with the finite norm

$$\|u\|_{V^m} = \max\left(\|t^\theta u\|_{L_{p_1}(X_1, dt/t)}, \|t^{\theta-1} u^{(m)}\|_{L_{p_2}(X_2, dt/t)}\right). \tag{2.3.6}$$

An element $x \in X_1 + X_2$ is said to be the *trace of a function* if there exists $u \in V^m$ such that $x = \lim_{t \to 0} u(t)$ in $X_1 + X_2$. We shall use the notation $\operatorname{Tr} u = x$. We denote by T^m the factorspace of all such limits equipped with the factornorm

$$\|x\|_{T^m} = \inf_{\operatorname{Tr} u = x} \|u\|_{V^m}.$$

The spaces V^m and T^m are Banach spaces and their crucial property is their coincidence with the real interpolation spaces obtained by the K-method:

2.3.15 Theorem. *Let $\overline{X} = (X_1, X_2)$ be a compatible couple of Banach spaces, let $0 < \theta < 1$, $1 \le p_i < \infty$, $i = 1, 2$, and $1/p = (1-\theta)/p_1 + \theta/p_2$. Then $T^m(\overline{X}) = (X_1, X_2)_{\theta,p}$.*

The proof of the above theorem would occupy a lot of space; we refer to [23], Chapter 3, [148], Chapter 1.

Semigroups and interpolation

Let us recall the concept of a *equibounded strongly continuous semigroup.* Let X be a Banach space and let $\{G(t)\}_{t\ge 0}$ be a one-parameter family of bounded linear operators mapping X into itself, possessing the following properties: (i) $G(t)G(s) = G(s+t)$ for all $s, t \ge 0$, $G(0) = I$, (ii) $\lim_{s\to t} G(s)(x) = G(t)(x)$ in X for all $t > 0$, (iii) $\lim_{t\to 0} G(t)x = x$ for all $x \in X$, (iv) there exists $C > 0$ such that $\|G(t)x\| \le C\|x\|$ holds for all $x \in X$ and all $t > 0$. Given a semigroup X with these properties we can define the *infinitesimal generator of a semigroup* of X as the mapping $\Lambda x = \lim_{t\to 0} t^{-1}(G(t) - I)$ defined on $\mathcal{D}(\Lambda)$, consisting of those $x \in X$, for which this limit exists. It is well known that Λ is a closed operator, $\mathcal{D}(\Lambda)$ is dense in X, Λ commutes with $G(t)$ for all $t > 0$. The space $\mathcal{D}(\Lambda)$ becomes a Banach space when equipped with the norm $\|x\|_X + \|\Lambda x\|_X$. We refer to [29], [154].

We shall now briefly describe the specific role of semigroups in real interpolation.

Given semigroups $\{G_j(t)\}_{t\ge 0}$ on a Banach space X, $j = 1, \dots, N$, such that $G_j(t)G_k(s) = G_k(s)G_j(t)$, $j, k = 1, \dots, N$ with corresponding infinitesimal operators Λ_j, let, for m any positive integer, Λ_j^m denote the m-th power of Λ_j. The space $\bigcap_{j=1}^N \mathcal{D}(\Lambda_j)$ endowed with the norm $\|x\|_X + \max_{1\le j\le N} \|\Lambda_j^m x\|_X$ is a Banach space. Put

$$K^m = \bigcap_{|\alpha|\le m} \mathcal{D}(\Lambda_1^{\alpha_1} \dots \Lambda_N^{\alpha_N})$$

and define the norm in K^m by

$$\|x\|_{K^m} = \sum_{|\alpha|\le m} \|\Lambda_1^{\alpha_1} \dots \Lambda_N^{\alpha_N} x\|_X,$$

where $\alpha = (\alpha_1, \dots, \alpha_N)$ stands for N-dimensional multiindices. Then (see e.g. [148], Chapter 1), for $0 < \theta < 1$ and $1 \leq q \leq \infty$,

$$(X, K^m)_{\theta,q} = \left(X, \bigcap_{j=1}^{N} \mathcal{D}(A_j^m) \right)_{\theta,q} . \tag{2.3.7}$$

If k is an integer, $k < s = \theta m$, and $\ell > s - k$, then the norm in $(X, K^m)_{\theta,q}$ is equivalent to

$$\|x\|_X + \sum_{j=1}^{N} \left(\int_0^{+\infty} t^{-(s-k)q} \|(G_j(t) - I)^\ell A_j^k x\|_X^q \frac{dt}{t} \right)^{1/q} \tag{2.3.8}$$

or to

$$\|x\|_X + \sum_{j=1}^{N} \left(\int_0^{+\infty} \cdots \int_0^{+\infty} |h|^{-(s-k)q} \left\| \left(\prod_{j=1}^{N} G_r(h_r) - I \right)^\ell \prod_{j=1}^{N} A_j^{\alpha_j} x \right\|_X^q \frac{dh_1 \ldots dh_N}{|h|^N} \right)^{1/q} \tag{2.3.9}$$

(appropriately modified if $q = \infty$).

Observe that, generally, $K^m \neq \bigcap_{j=1}^N \mathcal{D}(A_j^m)$ (see e.g., [148], Chapter 1).

The above abstract scheme is tailored to fit the situation when the semigroups $\{G_j(t)\}$ are the translation operators

$$\begin{aligned}(G_j(t)f)(x) = &f(x_1, \ldots, x_{j-1}, x_j + t, x_{j+1}, \ldots, x_N), \\ &t \geq 0,\ j = 1, \ldots, N,\end{aligned} \tag{2.3.10}$$

acting on $f \in L_p(X)$, where X is a Banach space. It is a matter of simple routine to show that

$$(A_j f)(\varphi) = -\int_{\mathbb{R}^N} \frac{\partial \varphi(x)}{\partial x} dx = \frac{\partial f}{\partial x_j}(\varphi), \qquad j = 1, \ldots, N,$$

for all test functions φ and that

$$\mathcal{D}(A_j) = \{f \in L_p(X);\ \partial f / \partial x_j \in L_p(X)\}, \qquad j = 1, \ldots, N,$$

more generally, that

$$A_1^{\alpha_1} \ldots A_n^{\alpha_n} f = D^\alpha f$$

for $f \in K^m$ and $|\alpha| \leq m$. The space $K^m = \bigcap_{j=1}^N \mathcal{D}(A_j^m)$ can be shown to be the Sobolev space $W_p^m(X)$ up to equivalence of norms.

Of importance and interest for us is of course the basic question about real interpolation of L_p spaces. We recall the remarkable formula for the K-functional in L_p spaces (due to Holmstedt; see e.g. [23], Chapter 3), where the functions can be vector-valued:

$$K(t, f, L_{p_1}, L_{p_2}) \sim \left(\int_0^{t^\alpha} [f^*(s)]^{p_1}\, ds\right)^{1/p_1} + \left(\int_{t^\alpha}^{+\infty} [f^*(s)]^{p_2}\, ds\right)^{1/p_2}$$

with $0 < p_1 < p_2$, $1/\alpha = 1/p_1 - 1/p_2$. In particular, there is a simple and elegant formula $K(t, f, L_1, L_\infty) = \int_0^t f^*(s)\, ds$, due to Calderón (see e.g. [22]). Together with the reiteration formula for the real interpolation method (cf. [23], Chapter 3, [148], Chapter 1) this yields

$$(L_{p_1}, L_{p_2})_{\theta,p} = L_p \quad \text{and} \quad (L_{p_1}, L_{p_2})_{\theta,\infty} = L_p^*$$

with $1/p = (1-\theta)/p_1 + \theta/p_2$.

Replacing the interpolation parameter p above with an arbitrary $q \in [1, \infty)$, one has to step out of the scale of the Lebesgue and Marcinkiewicz spaces and to consider the Lorentz spaces. The last scale is then closed with respect to the real interpolation. This, however, goes beyond our needs here and we shall not deal with it. A detailed account of the Lorentz spaces can be found e.g. in [22].

We shall illustrate the use of the real interpolation method to prove a theorem on convolution, which gives one of the possibilities of proving Sobolev imbedding theorems for potential type spaces.

2.3.16 Theorem (Young's inequality). *Let $K \in L_\gamma(\mathbb{R}^N)$ for some $\gamma \in (1, \infty)$ and $f \in L_p(\mathbb{R}^N)$, where $1 < p < \gamma/(\gamma - 1)$. Then $K * f \in L_q(\mathbb{R}^N)$ for $1/q = 1/p - (\gamma - 1)/\gamma$ and $\|K * f\|_q \le \|K\|_\gamma \, \|f\|_{L_p(\mathbb{R}^N)}$.*

Proof. The proof of Young's inequality is one of the standard applications, demonstrating the power of the Marcinkiewicz interpolation theorem. By Minkowski's inequality it is very easy to show that the mapping $\mathcal{K} : f \mapsto K * f$ maps L_1 into L_γ with the norm $\|K\|_\gamma$ and, further, that it maps $L_{\gamma/(\gamma-1)}$ into L_∞ with the same estimate for the norm. Consequently, interpolating $(L_1, L_{\gamma/(\gamma-1)})$ and the target couple (L_γ, L_∞) one arrives at the desired inequality. □

A deeper analysis shows that the assumptions on the kernel K can be weakened: it suffices to assume that $K \in L_\gamma^*(\mathbb{R}^N)$ for some $\gamma \in (1, \infty)$. This is for instance the case with the Riesz kernel $K_\gamma(x) \equiv |x|^{\gamma - N}$, $0 < \gamma < N$. In particular, for $\gamma = 1$, we get the Sobolev imbedding theorem because $|f(x)| \le \text{const.}(K_1 * (\nabla f))(x)$, $x \in \mathbb{R}^N$, for f smooth.

Note that various generalizations of Young's inequality can be found in [101].

Interpolation of compact operators

It is natural to ask how the compactness properties of operators behave with respect to the interpolation. At the same time it is important for applications. The answer is sufficiently satisfactory for our purposes. We start with a general concept of $K(\theta)$ and $J(\theta)$ classes (cf. also Theorem 2.3.13).

2.3.17 Definition. Let (X_1, X_2) be an interpolation couple and $0 < \theta < 1$.

A Banach space X is said to be of *class* $J(\theta, X_1, X_2)$ (or shortly of class $J(\theta)$) if $(X_1, X_2)_{\theta,1} \hookrightarrow X \hookrightarrow X_1 + X_2$.

A Banach space X is said to be of *class* $K(\theta, X_1, X_2)$ (or shortly of class $K(\theta)$) if $X_1 \cap X_2 \hookrightarrow X \hookrightarrow (X_1, X_2)_{\theta,\infty}$.

2.3.18 Remark. It is not difficult to show that for $X_1 \cap X_2 \hookrightarrow X \hookrightarrow X_1 + X_2$, the space X is of class $K(\theta)$ if and only if there is $c > 0$ such that $K(t, x) \le ct^\theta \|x\|_X$ for all $t > 0$ and all $x \in X$ (where $K(t, x)$ is the Peetre functional). The membership in $J(\theta)$ can be shown equivalent to the inequality $\|x\|_X \le c\|x\|_{X_1}^{1-\theta} \|x\|_{X_2}^\theta$ for all $x \in X_1 \cap X_2$ and some c independent of x. Also, X is of class $J(\theta)$ if and only if there is $c > 0$ such that $\|x\|_X \le c \max(t^{-\theta}\|x\|_{X_1}, t^{1-\theta}\|x\|_{X_2})$ for all $x \in X_1 \cap X_2$ and all $t > 0$. (See [23], Chapter 3 or [148], Chapter 1 for details.)

Note that in the interpolation theory one encounters the functional

$$J(t, x) = J(t, x, X_1, X_2) = \max(\|x\|_{X_1}, t\|x\|_{X_2}), \qquad x \in X_1 \cap X_2,\ t > 0.$$

Hence X is of class $J(\theta)$ if and only if there is $c > 0$ such that $\|x\|_X \le t^{-\theta} J(t, x)$ for all $x \in X_1 \cap X_2$ and all $t > 0$.

Recall the notation $(X_1, X_2)_{\theta,q}$ and $[X_1, X_2]_\theta$ for the result of the real and complex interpolation (see p. 93 and p. 88, respectively). Without proof (we refer again to [23], Chapter 3 or [148], Chapter 1) we state

2.3.19 Theorem. *Let (X_1, X_2) be an interpolation couple, $0 < \theta < 1$, and $1 \le q \le \infty$. Then both $(X_1, X_2)_{\theta,q}$ and $[X_1, X_2]_\theta$ are of class $K(\theta)$ and $J(\theta)$.*

Now we state

2.3.20 Theorem. *Let (X_1, X_2) be an interpolation couple and let Y be a Banach space. Then the following implications hold:*

(i) *Let T be a linear operator, and let us assume that $T : X_1 \to Y$ is bounded, $T : X_2 \to Y$ is compact, and that X is of class $K(\theta, X_1, X_2)$. Then T takes X into Y compactly.*

(ii) *Let T be a linear operator, and let us assume that $T : Y \to X_1$ is bounded, $T : Y \to X_2$ is compact, and that X is of class $J(\theta, X_1, X_2)$. Then T takes Y into X compactly.*

We refer to [23], Chapter 3 and [148], Chapter 1 for further references concerning especially the more general situation with another interpolation couple (Y_1, Y_2) instead of the (trivial) couple (Y, Y).

An easy consequence of the last theorem and Theorem 2.3.13 is

2.3.21 Corollary. *Let $X_1 \hookrightarrow X_2$ be Banach spaces and suppose that X_1 is compactly imbedded into X_2. Let $0 < \theta_1 < \theta_2 < 1$ and $1 \le p_1, p_2 \le \infty$. Then $(X_1, X_2)_{\theta_2, p_2}$ is compactly imbedded into $(X_1, X_2)_{\theta_1, p_1}$.*

A bit more general theorem can be proved:

2.3.22 Theorem. *Let $X_1 \hookrightarrow X_2$ be Banach spaces and suppose that X_1 is compactly imbedded into X_2. Let $0 < \theta_1 < \theta_2 < 1$ and Banach spaces Y_{θ_1} and Y_{θ_2} of class $K(\theta_1, X_1, X_2)$ and of class $J(\theta_2, X_1, X_2)$, respectively. Then Y_θ is compactly imbedded into Y_{θ_2}. The same conclusion is true in the case when $Y_{\theta_1} = X_1$ or $Y_{\theta_2} = X_2$.*

2.4 Besov and Lizorkin-Triebel spaces

2.4.1 The "classical" definition

The next natural question after identifying the interpolation spaces between the Lebesgue spaces and/or Marcinkiewicz spaces is what happens between Sobolev spaces. We shall assume that the reader is familiar with the standard concept of a generalized (or distributional) derivative and with *Sobolev spaces* of integer order. In the Section 2.5 we shall see that the scale of potential Sobolev spaces is closed under complex interpolation method. Nevertheless, as well known, the potential Sobolev spaces coincide with the "classical" Sobolev spaces only for integer derivatives. In general, imbeddings between general Besov and Lizorkin-Triebel spaces imply interpolation estimates for Sobolev-Slobodetskii spaces via estimates for potential Sobolev spaces when $p \ge 2$. This is clearly not enough for some applications. Looking at the historical development, Besov spaces naturally appear in the study of the trace problem. It is easy to show that given $f \in C^\infty(\overline{\Omega}) \cap W_p^k(\Omega)$ (with Ω sufficiently smooth) we have $f \in L_p(\partial\Omega)$ and the operator of this restriction is bounded, $1 \le p < \infty$. Hence this restriction can be uniquely extended to $W_p^k(\Omega)$, calling it a *trace operator*. Nevertheless, one would like to have a complete description of the target space, i.e. to have also an extension from the boundary. Here, the *Sobolev-Slobodetskii spaces* come into play. We refer to key papers by Gagliardo [56], Slobodetskii [138], [139], and Besov [24], [25]. The Sobolev-Slobodetskii spaces fall into the more general class of Besov spaces; however, because of their importance for the trace theory we start with the definition due to Slobodetskii.

2.4.1 Definition. Let $s > 0$ be a non-integer and denote with $[s]$ the integer part of s. Let $1 \le p \le \infty$ and Ω be a domain in $\mathbb{R}^N$. Then the *Sobolev-Slobodetskii space* $W_p^s(\Omega)$ is defined as the linear space of all functions $f \in L_p(\mathbb{R}^N)$ with

$$\|f\|_{W_p^s(\mathbb{R}^N)} = \|f\|_{L_p(\mathbb{R}^N)} + \sum_{|\alpha|=[s]} \left(\int_\Omega \int_\Omega \frac{|D^\alpha f(x) - D^\alpha f(y)|^p}{|x-y|^{(s-[s])p+N}} \, dx dy \right)^{1/p} < +\infty$$

(with the appropriate modification if $p = \infty$).

It was proved in papers mentioned above that for $s > 1/p$, $1 < p < \infty$, and $k \ge 1$ an integer, the space of traces of functions from $W_p^k(\Omega)$ (Ω sufficiently smooth) is exactly $W_p^{k-1/p}(\partial\Omega)$. (The situation for $p = 1$ is in some sense singular—the trace space for $W_p^1(\Omega)$ is $L_1(\Omega)$ as proved by Gagliardo [56].) Note that, for Ω sufficiently smooth, after standard decomposition and localization the situation can be transferred to functions defined in $\mathbb{R}^N$, considering their traces on $\mathbb{R}^{N-1}$. In the sequel we shall mostly handle this basic $\mathbb{R}^N$ setting. As observed above the Sobolev-Slobodetskii spaces W_p^s are special cases of more general Besov spaces $B_{p\,q}^s$ (it is $W_p^s = B_{p\,p}^s$ for non-integer $s > 0$), which are, for $s > 1/p$, "closed" with respect to the trace operator, i.e., the trace of a function in a Besov space belongs to another Besov space on the boundary. The same holds even for traces on manifolds of lower dimension than $N - 1$ and also permits generalizations to certain kinds of anisotropic spaces. There is a vast literature on this and related subjects and we refer to [26] for a detailed account.

Now the reader should look at the definition of a general Besov space below. It is not difficult to see that working either with differences or with moduli of continuity leads to the same result. On the other hand, the general definition gives a certain freedom for the choice of order of differences (or moduli of continuity). That the results are equivalent norms is not immediately clear and it will be discussed in detail in the sequel.

Let us point out that the developed theory and existing approaches to Besov spaces allow handling them for a general scope of parameters s, p, and q and there is no particular reason for restricting our attention artificially to the case $p = q$. Hence we introduce now the concept of the Besov spaces $B_{p\,q}^s$. We shall start with a "more classical" definition of the Besov spaces. This also corresponds to the historical development of the topic. Later, we describe the Fourier analysis approach, using appropriate decompositions. At the first look the latter approach is more complicated. However, this technique is very powerful and the way to the desired goal is often more straightforward. Also, these tools are sometimes more appropriate for description of more complicated spaces with some anisotropic properties.

Let us introduce the following notation. For $h \in \mathbb{R}^N$ let T_h denote the translation operator defined by $T_h f(x) = f(x+h)$. Let

$$\Delta_h^1 f(x) = \Delta_h f(x) = f(x+h) - f(x) = (T_h - \mathrm{id}) f(x)$$

be the *first order difference operator*; the higher order differences are defined inductively by

$$\Delta_h^r f(x) = \Delta_h(\Delta_h^{r-1} f)(x), \qquad r = 1, 2, \ldots.$$

Observe that $T_h^k f(x) = T_{kh} f(x) = f(x+kh)$. Plainly,

$$\Delta_h^r f(x) = (T_h - \mathrm{id})^r f(x) = \sum_{k=0}^{r} \binom{r}{k} (-1)^{r-k} T_{hk} f(x).$$

Let now $f \in L_{p,\mathrm{loc}}(\mathbb{R}^N)$ for some $p \in [1, \infty]$. Then we define the *r-th order modulus of L_p-continuity* as

$$\omega_p^r(f, t) = \sup_{|h| \le t} \|\Delta_h^r f\|_p.$$

The function $t \mapsto \omega_p^r(f, t)$ is non-decreasing, and satisfies $\omega_p^r(f, 2t) \le 2^r \|f\|_p$ and $\omega_p^r(f, 2t) \le 2^r \omega_p^r(f, t)$. The former estimate follows directly from the definition since $\Delta_h^r f(x) = (T_h - \mathrm{id})^r f(x)$ and the L_p norm of the right hand side is majorized by 2^r multiple of $\|f\|_p$. As to the latter estimate observe that $\Delta_{2h}^1 f(x) = f(x+2h) - f(x+h) + f(x+h) - f(x) = T_h[f(x+h) - f(x)] + [f(x+h) - f(x)] = (T_h + \mathrm{id})\Delta_h f(x)$, hence $\Delta_{2h}^r f(x) = (T_h + \mathrm{id})^r \Delta_h^r f(x)$ for $r \in \mathbb{N}$.

Now we are in a position to define the Besov spaces both with positive and negative smoothness. We shall do it in the whole $\mathbb{R}^N$ and the symbol $\mathbb{R}^N$ will be usually omitted.

2.4.2 Definition. Let $s > 0$ and let r be any positive integer, $r > s$. Let $1 \le p, q \le \infty$. The *Besov space* $B_{pq}^s = B_{pq}^s(\mathbb{R}^N)$ is defined as the linear set of all $f \in L_p(\mathbb{R}^N)$ with the finite norm

$$\|f\|_{B_{pq}^s} = \begin{cases} \|f\|_p + \left(\int_0^{+\infty} [t^{-s} \omega_p^r(f, t)]^q \frac{dt}{t} \right)^{1/q}, & \text{if } q < \infty \\ \|f\|_p + \sup_{t > 0} t^{-s} \omega_p^r(f, t), & \text{if } q = \infty. \end{cases} \tag{2.4.1}$$

The spaces $B_{pp}^s = B_{pp}^s(\mathbb{R}^N)$ will be called the *Sobolev-Slobodetskii spaces*.

If $0 < s < 1$ and $p = q = \infty$, the result are the *Hölder spaces* of order s.

If $s < 0$, and $1 < p, q < \infty$, then we put $B^s_{p\,q} = B^{-s}_{p'\,q'}$, where $p' = p/(p-1)$ and $q' = q/(q-1)$.

The reader should now compare the above definition with the abstract formula for the norm (2.3.8).

2.4.3 Remark. Since $\omega^r_p(f, 2t) \leq 2^r \omega^r_p(f, t)$ it is clear that after replacing $\sup_{t>0}$ by, say, $\sup_{0<t<1}$, we get an equivalent norm.

Second, it is not difficult to replace the moduli of continuity in the above definition by appropriate differences.

Third, one has to justify the freedom of the choice $r > s$. This is not straightforward and being of importance for the very definition we will pay some attention to it.

Last, but not least, note that the restriction $p, q > 1$ in the above definition is not necessary. One can start with arbitrary positive p and q, thus obtaining generally quasinorms in (2.4.1)

After introducing the Besov spaces we shall turn our attention to their real interpolation. We will present a formula for the K-functional between L_p and the Sobolev space W^k_p and we show that Besov spaces are results of a corresponding real interpolation between L_p and W^k_p. After getting familiar with the potential Sobolev spaces in the next section we shall explain connections between various function spaces appearing here.

Let us observe that Besov spaces can be introduced in many equivalent ways and we shall in fact briefly touch upon only some of them. The prominent Fourier analysis approach, including the case of vector-valued functions will be dealt with later; we will at least give a proof of the equivalence of the classical definition above and the definition using the decompositions and the Fourier analysis technique. The existing literature is rather vast. We refer to the standard monographs [26], [22], [148] also for a detailed account of references in the field.

Inequalities of Marchaud type

Our immediate program is to point out the equivalence of definitions of a Besov space, using an arbitrary integer r bigger than the smoothness parameter s. This can be based on an inequality of Marchaud type, which finds many useful applications in analysis and it is thus of independent interest (see [105] and [22], Chapter 5).

2.4.4 Proposition. *Let $0 < k < r$ be integers. Then*

$$2^{k-r}\omega_p^r(f,t) \le \omega_p^k(f,t) \le ct^k \int_t^{+\infty} \frac{\omega_p^r(f,\tau)}{\tau^k}\frac{d\tau}{\tau} \tag{2.4.2}$$

for all $t > 0$.

2.4.5 Corollary. *Let $s > 0$, $1 \le p, q \le \infty$, and $r_j > s$, $j = 1, 2$, be positive integers. Then the norms in (2.4.1) for $r = r_1$ and $r = r_2$ are equivalent.*

Proof. Let, for instance, $r_1 < r_2$ and denote here $\|f\|_{s,p,q,r_j}$ the norm in $B^s_{p\,q}$ for $r = r_j$, $j = 1, 2$. Then the first of the inequalities in (2.4.2) yields immediately $\|f\|_{s,p,q,r_1} \le c\|f\|_{s,p,q,r_2}$ with c independent of f. The converse estimate follows from the second inequality in (2.4.2) combined with Hardy's inequality (see Proposition 2.2.1). This can be left to the reader. □

Further more detailed look at the Besov spaces gives a basic ordering with respect to the parameters. We omit the proof; it is not difficult and it can be found in all the standard references quoted above.

2.4.6 Theorem. *Let $1 \le p \le \infty$. Further, let either $0 < s_1 < s_2 \le \infty$ and $1 \le q_1, q_2 \le \infty$, or $0 < s_1 = s_2 \le \infty$ and $q_1 \le q_2$. Then $B^{s_2}_{p\,q_1} \hookrightarrow B^{s_1}_{p\,q_2}$. In particular $W^{s_2}_p \hookrightarrow W^{s_1}_p$ for non-integer positive s_1 and s_2, $s_1 < s_2$.*

Our next step in our guided Besov spaces tour will be a theorem on an estimate for the K-functional between L_p and W^k_p, which is the key to real interpolation of Sobolev spaces. We state it without proof; it can be found e.g. in [22], Chapter 5.

2.4.7 Theorem. *Let $1 \le p \le \infty$ and let k be a positive integer. Then there exist positive constants c_1 and c_2 such that*

$$c_1K(t^k, f, L_p, W^k_p) \le \min(1, t^k)\|f\|_p + \omega^k_p(f,t) \le c_2K(t^k, f, L_p, W^k_p). \tag{2.4.3}$$

Immediately we get

2.4.8 Corollary. *Let $1 \le p, q \le \infty$, $0 < \theta < 1$, and suppose that k is a positive integer. Then*

$$(L_p, W^k_p)_{\theta,q} = B^{\theta k}_{p\,q}. \tag{2.4.4}$$

In particular $(L_p, W^k_p)_{\theta,p} = W^{\theta k}_p$ for non-integer θk.

Hence the Besov space $B^s_{p\,q}$ is of class $\theta = s/k$ (for $k > s$). Invoking the stability of the real interpolation method (see Theorem 2.3.13) we get

2.4.9 Theorem. *Let $0 < \theta < 1$, $1 \le p, q, q_1, q_2 \le \infty$, $s, s_1, s_2 > 0$, and let k, k_1, k_2 be positive integers, $k_1 \ne k_2$. Then*

(i) $(W_p^{k_1}, W_p^{k_2})_{\theta,q} = B_{p\,q}^s$ *if* $s = (1-\theta)k_1 + \theta k_2$,

(ii) $(W_p^k, B_{p\,q_2}^s)_{\theta,q} = B_{p\,q}^s$ *if* $s = (1-\theta)k + \theta s$ *and* $k \ne s$,

(iii) $(B_{p\,q_1}^{s_1}, B_{p\,q_2}^{s_2})_{\theta,q} = B_{p\,q}^s$ *if* $s = (1-\theta)s_1 + \theta s_2$ *and* $s_1 \ne s_2$.

The corresponding special formulae hold for interpolation with parameters $p = q$ on the left hand side and with Sobolev-Slobodetskii spaces W_p^s replacing Besov spaces on the right hand side of the above equalities provided s is not an integer.

For a detailed account of interpolation properties of Besov spaces, even within a broader range of parameters involved we refer to [23], Chapter 6, [148], Chapter 2, [149], Chapter 2.

Following this "more classical" method of handling the Besov spaces we will mention further basic imbedding properties. One of the alternative ways how to prove them is the use of Marchaud type inequalities, linking L_p-norms of derivatives of order k, where k is an integer, with moduli of continuity of order $s < k$. We shall state explicitly these estimates since they throw some more light on the integrability and differentiability properties of functions in Besov and Sobolev spaces and are therefore of independent interest. We omit the proofs; they can be found e.g. in [22], Chapter 5.

2.4.10 Theorem. *Let $0 < k < s$ be integers.*

(i) *There exist positive constants c_1 and c_2 such that*

$$c_1 \sup_{t>0} \left(t^{-k} \omega_p^k(f,t)\right) \le \sum_{|\alpha|=k} \|D^\alpha f\|_p \le c_2 \int_0^{+\infty} \frac{\omega_p^s(f,t)}{t^k} \frac{dt}{t}. \tag{2.4.5}$$

(ii) *If, for integers $0 < k < s$,*

$$\int_0^1 t^{-k} \omega_p^s(f,t) \frac{dt}{t} < +\infty, \tag{2.4.6}$$

then $f \in W_p^k$ and

$$\sum_{|\alpha|=k} \omega_p^s(D^\alpha f, t) \le \int_0^t \frac{\omega_p^s(f,\tau)}{\tau^k} \frac{d\tau}{\tau}, \qquad \tau > 0. \tag{2.4.7}$$

The reader will easily verify that Theorem 2.4.10 together with Hardy's inequality implies that given an integer k strictly between 0 and r, then a function f belongs to $B^s_{p\,q}$ if and only if all the derivatives $D^\alpha f$ of order k belong to $B^{r-k}_{p\,q}$.

In the first section of this chapter we have introduced the non-increasing rearrangement f^* of a function f and the integral average f^{**}. Both f^* and f^{**} are suitable for description of integrability properties of f. Hence the next theorem, which we also state without the proof (see e.g. [22], Chapter 5), can be directly used to establish imbeddings of Besov spaces into Lorentz spaces L_{pq} (Corollary 2.4.12), which in the particular case $p = q$ yields imbeddings into Lebesgue spaces. The latter, however, gives only a piece of information for general Besov spaces, a certain "screening on the diagonal", which is enough only in the case of Sobolev-Slobodetskii spaces.

2.4.11 Theorem. *Let $1 \le p \le \infty$. Then*

$$f^{**}(t) \le c \int_{t^{1/N}}^{+\infty} \frac{\omega_p^N(f,\tau)}{\tau^{N/p}} \frac{d\tau}{\tau}, \qquad \tau > 0$$

for all $f \in L_p$, with c independent of f.

2.4.12 Corollary. *The following imbeddings hold:*

(i) *Let $0 < s = N(1/p_1 - 1/p_2)$, where $1 \le p_1 < p_2 < \infty$, and suppose $1 \le q \le \infty$. Then $B^s_{p_1\,q} \hookrightarrow L_{p_2 q}$. In particular $B^s_{p_1\,p_2} \hookrightarrow L_{p_2}$, and if s is a non-integer, then $W^s_{p_1} \hookrightarrow L_{p_2 p_1} \hookrightarrow L_{p_2}$.*

(ii) *If $s = N/p$, $1 \le p < \infty$, then $B^s_{p\,1} \hookrightarrow L_\infty$. In particular $W^s_1 \hookrightarrow L_\infty$ if s is a non-integer.*

Let us observe that a natural target for imbeddings of Besov spaces are actually spaces of the same type. The last Corollary in turn implies another Marchaud type inequality and the latter yields an imbedding theorem in the scale of Besov spaces.

2.4.13 Theorem. *Let $1 \le p_1 < p_2 \le \infty$. Let $s \ge N$ be a positive integer and suppose that $s = N(1/p_1 - 1/p_2) > 0$. Then*

$$\omega^s_{p_2}(f,t) \le c \int_0^t \tau^{-s} \omega^s_{p_1}(f,\tau) \frac{ds}{s}, \qquad 0 < t < 1. \tag{2.4.8}$$

2.4.14 Corollary. *Let $s_2 - N/p_1 = s_1 - N/p_1$, $s_1, s_2 > 0$, $1 \le p_1 < p_2 \le \infty$, and $1 \le q \le \infty$. Then $B^{s_2}_{p_2\,q} \hookrightarrow B^{s_1}_{p_1\,q}$.*

Of particular importance are imbeddings describing relations between Sobolev and Besov spaces of integer order.

2.4.15 Theorem. *Let k be a positive integer. Then the following imbeddings are true:*

$$B^k_{p\,p} \hookrightarrow W^k_p \hookrightarrow B^k_{p\,2} \qquad \text{if } 1 < p \leq 2, \tag{2.4.9}$$

$$B^k_{p\,2} \hookrightarrow W^k_p \hookrightarrow B^k_{p\,p} \qquad \text{if } 2 \leq p < \infty. \tag{2.4.10}$$

In particular, $W^k_2 = B^k_{2\,2}$.

2.4.16 Remark. For a more general comparison theorem in the scale of Besov, Lizorkin-Triebel, and potential Sobolev spaces see for instance [119], [23], Chapter 3, [148], Chapter 2, [149], Chapter 2.

2.4.17 Remark. All the above imbedding properties have their dual counterpart (cf. Theorem 2.3.14) for negative values of the parameter s. Since the interpolation formulae for all the parameters s, p, q hold with the same value of θ as for $-s$, p', and q', respectively, the only difference in the corresponding theorems is only the assumption $s < 0$ and it is certainly not necessary to formulate them explicitly.

Let us note that some further imbedding theorems for Besov spaces have been established in [137].

2.4.2 The Fourier analysis approach

Let g be a non-negative C^∞ function in $\mathbb{R}^N$ supported in $\{\xi \in \mathbb{R}^N;\ |\xi| \leq 2\}$ and $g(\xi) = 1$ if $|\xi| \leq 1$. We define

$$g_0(\xi) = g(\xi), \qquad \xi \in \mathbb{R}^N,$$

and the telescoping sequence

$$g_j(\xi) = g(2^{-j}\xi) - g(2^{-j+1}\xi), \qquad j = 1, 2, \ldots.$$

Then

$$\operatorname{supp} g_j \subset \{\xi \in \mathbb{R}^N;\ 2^{j-1} \leq |\xi| \leq 2^{j+1}\}$$

and

$$\sum_{j=0}^{+\infty} g_j(\xi) = 1.$$

The sequence $\{g_j\}_{j=0}^{+\infty}$ is called a *smooth (dyadic) decomposition of unity*. Observe that in the above sum there are at most two non-vanishing terms at each $\xi \in \mathbb{R}^N$. Now define functions $\varphi_k \in \mathscr{S}(\mathbb{R}^N)$ in terms of their Fourier transforms by

$$\mathscr{F}(\varphi_j;\xi) = g_j(\xi) \qquad j = 0, 1, 2, \dots.$$

Observe that

$$\operatorname{supp} \mathscr{F}(\varphi_k) = \{\xi;\, 2^{k-1} \leq |\xi| \leq 2^{k+1}\}, \qquad k = 1, 2, \dots,$$

and

$$\operatorname{supp} \mathscr{F}(\varphi_0) = \{\xi;\, |\xi| \leq 2\}.$$

In other terms, if we put $h(\xi) = g(\xi) - g(2\xi)$, $\xi \in \mathbb{R}^N$, we have

$$\mathscr{F}(\varphi_k;\xi) = h(2^{-k}\xi), \qquad \xi \in \mathbb{R}^N,\ k = 1, 2, \dots, \tag{2.4.11}$$

h is non-negative, belongs to $\mathscr{D}(\mathbb{R}^N)$, it is supported in $\{\xi;\, 2^{-1} \leq |\xi| \leq 2\}$, and it is positive in the interior of $\operatorname{supp} h$. Conversely, if we start with h possessing these properties and take φ_k as in (2.4.11), and if we add a function g, then after a normalization we get a decomposition of unity, too. This is an alternative way of defining the decomposition that the reader can often meet in the literature.

2.4.18 Remark. Systems with similar properties as the above decomposition play a key role in the modern theory of function spaces. We shall restrict ourselves to considering the above special system $\{\varphi_k\}$, which will be quite sufficient for our purposes. At the same time we shall define Besov and Lizorkin-Triebel spaces in the following not for all possible values of parameters involved. We refer to [119], [148], [149], [23]. For reader's better orientation in the literature we present here the celebrated inequality due to Nikol'skii, which gives an idea of importance of functions whose Fourier transform is supported in a bounded set. In particular, $\varphi_k * f$ from the definition of the Besov space is a function (see the definition below) which enjoys this property.

The following theorem is one of the basic mighty tools in the function spaces theory in the Fourier analysis spirit.

2.4.19 Theorem (Nikol'skii's inequality). *Let $b > 0$, $1 \leq p \leq q \leq \infty$. Then there exists $c > 0$ such that*

$$\|D^\alpha g\|_q \leq c b^{|\alpha| + N(1/p - 1/q)} \|g\|_p \tag{2.4.12}$$

for all $g \in \mathscr{S}(\mathbb{R}^N)$ such that $\operatorname{supp} \mathscr{F}(g) \subset \{\xi;\, |\xi| \leq b\}$.

2.4.20 Definition. (i) Let $1 \leq p \leq \infty$, $1 \leq q \leq \infty$ and $s \in \mathbb{R}^1$. Then the *Besov space* $B^s_{p\,q} = B^s_{p\,q}(\mathbb{R}^N)$ is the space of all $f \in \mathscr{S}'(\mathbb{R}^N)$ with the finite norm

$$\|f\|_{B^s_{p\,q}} = \left(\sum_{k=0}^{+\infty} 2^{ksq}\|\varphi_k * f\|^q_{L_p}\right)^{1/q} \tag{2.4.13}$$

if $q < \infty$ and with the finite norm

$$\|f\|_{B^s_{p\,\infty}} = \sup_k 2^{ks}\|\varphi_k * f\|_{L_p} \tag{2.4.14}$$

if $q = \infty$.

(ii) Let $1 \leq p < \infty$, $1 \leq q \leq \infty$ and $s \in \mathbb{R}^1$. Then the *Lizorkin-Triebel space* $F^s_{p\,q} = F^s_{p\,q}(\mathbb{R}^N)$ is the space of all $f \in \mathscr{S}'(\mathbb{R}^N)$ with the finite norm

$$\|f\|_{F^s_{p\,q}} = \left\| \left(\sum_{k=0}^{+\infty} 2^{ksq}|(\varphi_k * f)(x)|^q\right)^{1/q} \right\|_{L_p} \tag{2.4.15}$$

with an analogous modification for $q = \infty$.

2.4.21 Remark. In the literature one often encounters the notation $\ell^s_q(L_p)$ and $L_q(\ell^s_p)$ for the mixed norm spaces consisting of sequences of functions $a_k = a_k(x)$; for instance $\|a_k\|_{\ell^s_q(L_p)} = \|\, 2^{sk}\|a_k(x)\|_{L_p}\,\|_{\ell_q}$. Used for the sequence $\{\varphi_k * f\}$, we have, for instance, $\|f\|_{B^s_{p\,q}} = \|\varphi_k * f\|_{\ell^s_q(L_p)}$ and so on.

2.4.3 Imbeddings and interpolation

We survey the basic properties of Besov and Lizorkin-Triebel spaces without proof (see e.g. [119], [23], Chapter 6, [148], Chapter 2, [149], Chapter 2).

The prominent feature of spaces of Besov, Lizorkin-Triebel (and Sobolev) type, widely used in applications, are the *imbedding theorems.* They are expressed in forms of norm inequalities, stating the fact, roughly speaking, that a function differentiable in such a sense belongs to a "better" space, either of the same type or even to other scales. The word "better" means improved integrability or certain differentiability in a pointwise sense (up to sets of measure zero).

Before stating the imbedding theorems we will recall the concept of Hölder spaces. Given an integer $k > 0$, then $C^k(\mathbb{R}^N)$ is the closure of $\mathscr{S}(\mathbb{R}^N)$ in the norm $\|f\|_{C^k} = \max_{|\alpha \leq k|} \sup_{x\in\mathbb{R}^N} |D^\alpha f(x)|$. If $\varkappa > 0$ is not an integer let us write $\varkappa = [\varkappa] + \{\varkappa\}$, where $[\varkappa]$ is the integer part of $\varkappa$

(hence $0 < \{\varkappa\} < 1$). Then the *Hölder space* $C^{\varkappa}(\mathbb{R}^N)$is defined as the linear set of all $f \in C^{[\varkappa]}(\mathbb{R}^N)$ with the finite norm

$$\|f\|_{C^{\varkappa}} = \|f\|_{C^{[\varkappa]}} + \sum_{|\alpha|=[\varkappa]} \sup_{x \neq y} \frac{|D^\alpha f(x) - D^\alpha f(y)|}{|x-y|^{\{\varkappa\}}}.$$

It is well known and easy to prove that $C^{\varkappa}(\mathbb{R}^N)$ is a Banach space.

Note also that for non-integer $\varkappa$ the space $C^{\varkappa}(\mathbb{R}^N)$ equals to the *Zygmund space* $\mathscr{C}^{\varkappa}$, which is defined as the linear set of all $f \in C^{[\varkappa]^-}(\mathbb{R}^N)$, where $\varkappa = [\varkappa]^- + \{\varkappa\}^+$ with $[\varkappa]^-$ an integer and $\{\varkappa\}^+ \in (0, 1]$, which has the finite norm

$$\|f\|_{\mathscr{C}^{\varkappa}} = \|f\|_{C^{[\varkappa]^-}} \sum_{|\alpha|=[\varkappa]^-} \sup_{x \neq y} \frac{|D^\alpha f(x) - 2D^\alpha((x+y)/2) + D^\alpha f(y)|}{|x-y|^{\{\varkappa\}^+}}.$$

2.4.22 Theorem. *The following imbedding statements are true:*

(i) *If* $1 < p_1 \leq p_2 < \infty$, $1 \leq q \leq \infty$, *and* $r, s \in \mathbb{R}^1$ *are such that*

$$s - \frac{N}{p_1} = r - \frac{N}{p_2}, \tag{2.4.16}$$

then

$$B^{s_2}_{p_2\,q}(\mathbb{R}^N) \hookrightarrow B^{s_2}_{p_1\,q}(\mathbb{R}^N).$$

(ii) *If* $1 < p_1 \leq p_2 < \infty$, $1 < q < \infty$, *and* $r, s \in \mathbb{R}^1$ *satisfy* (2.4.16), *then*

$$F^{s_2}_{p_2\,q}(\mathbb{R}^N) \hookrightarrow F^{s_2}_{p_1\,q}(\mathbb{R}^N).$$

(iii) *Let* $1 < p < \infty$ *and* $\varkappa \geq 0$. *Then*

$$B^{\varkappa+N/p}_{p\,1}(\mathbb{R}^N) \hookrightarrow C^{\varkappa}(\mathbb{R}^N) \quad \text{and} \quad B^{\varkappa+N/p}_{p\,1}(\mathbb{R}^N) \hookrightarrow \mathscr{C}^{\varkappa}(\mathbb{R}^N).$$

(iv) *Let* $1 < p, q < \infty$ *and let* $\varkappa$ *be non-integer. Then*

$$F^{\varkappa+N/p}_{p\,q}(\mathbb{R}^N) \hookrightarrow C^{\varkappa}(\mathbb{R}^N).$$

2.4.23 Theorem. *The following assertions hold:*

(i) *The spaces* $B^s_{p\,q}(\mathbb{R}^N)$ *and* $F^s_{p\,q}(\mathbb{R}^N)$ *are Banach spaces for all* $s \in \mathbb{R}^1$, $1 \leq p, q \leq \infty$.

(ii) *If* $s_1 < s_2$ *and* $1 \leq p, q \leq \infty$, *then* $B^{s_2}_{p\,q}(\mathbb{R}^N) \hookrightarrow B^{s_1}_{p\,q}(\mathbb{R}^N)$.

(iii) *If* $s \in \mathbb{R}^1$, $1 < p < \infty$, *and* $1 \leq q < \infty$, *then* $(B^s_{p\,q}(\mathbb{R}^N))' = B^{-s}_{p'\,q'}(\mathbb{R}^N)$.

(iv) *If* $s \in \mathbb{R}^1$, $1 \leq p \leq \infty$, *and* $1 \leq q_1 < q_2 \leq \infty$, *then* $B^s_{p\,q_1}(\mathbb{R}^N) \hookrightarrow B^s_{p\,q_2}(\mathbb{R}^N)$.

(v) *If* $k = 0, 1, 2, \ldots$, *then* $B^k_{2\,2}(\mathbb{R}^N) = W^k_2(\mathbb{R}^N)$. *This equality holds only for this special choice of parameters* k, p, *and* q.

(vi) *If* $k = 0, 1, 2, \ldots$, *and* $1 < p < \infty$, *then* $F^k_{p\,2}(\mathbb{R}^N) = W^k_p(\mathbb{R}^N)$.

(vii) *If* $s \in \mathbb{R}^1$ *and* $1 < p < \infty$, *then* $B^s_{p\,p}(\mathbb{R}^N) = F^s_{p\,p}(\mathbb{R}^N)$. *In particular, if* $s > 0$ *is a non-integer, then* $F^s_{p\,p}(\mathbb{R}^N)$ *is the Sobolev-Slobodetskii space* $W^s_p(\mathbb{R}^N)$.

2.4.24 Theorem. *Let* $s \in \mathbb{R}^1$ *and* $1 < p < \infty$. *Then*

$$B^s_{p\,q}(\mathbb{R}^N) \hookrightarrow F^s_{p\,q}(\mathbb{R}^N) \hookrightarrow B^s_{p\,p}(\mathbb{R}^N) \qquad \text{for all } 1 < q \leq p.$$

Without proof (see e.g. [148], Chapter 2) we present basic interpolation properties of Besov and Lizorkin-Triebel spaces. Below we formally put $1/\infty = 0$.

2.4.25 Theorem. *Let* $0 < \theta < 1$. *Then*

(i) $(B^{s_1}_{p\,q_1}(\mathbb{R}^N), B^{s_2}_{p\,q_2}(\mathbb{R}^N))_{\theta,q} = B^s_{p\,q}(\mathbb{R}^N)$
for all $s_1, s_2 \in \mathbb{R}^1$, $s_1 \neq s_2$, $1 < p \leq \infty$, *and* $1 \leq q_1, q_2 \leq \infty$, *where* $s = (1-\theta)s_1 + \theta s_2$,

(ii) $(B^{s_1}_{p_1\,q_1}(\mathbb{R}^N), B^{s_2}_{p_2\,q_2}(\mathbb{R}^N))_{\theta,q} = B^s_{p\,q}(\mathbb{R}^N)$
for all $s_1, s_2 \in \mathbb{R}^1$, $1 \leq q_1, q_2 < \infty$, $1 < p_1, p_2 < \infty$, $p_1 \neq p_2$, *and* $1 \leq q_1, q_2 < \infty$, *provided* $1/p = (1-\theta)/p_1 + \theta/p_2 = 1/q = (1-\theta)/q_1 + \theta/q_2$,

(iii) $[B^{s_1}_{p_1\,q_1}(\mathbb{R}^N), B^{s_2}_{p_2\,q_2}(\mathbb{R}^N)]_\theta = B^s_{p\,q}(\mathbb{R}^N)$
for all $s_1, s_2 \in \mathbb{R}^1$, $1 \leq q_1, q_2 \leq \infty$, $1 \leq p_1, p_2 \leq \infty$, *where* p *is as in* (ii) *and* $1/q = (1-\theta)/q_1 + \theta/q_2$.

2.4.26 Theorem. *Let* $0 < \theta < 1$. *Then*

(i) $(F^{s_1}_{p_1\,q_1}(\mathbb{R}^N), F^{s_2}_{p_2\,q_2}(\mathbb{R}^N))_{\theta,p} = F^s_{p\,p}(\mathbb{R}^N) = B^s_{p\,p}(\mathbb{R}^N)$
for all $s_1, s_2 \in \mathbb{R}^1$, $s_1 \neq s_2$, $1 < p_1, p_2, q_1, q_2 < \infty$, *where* $s = (1-\theta)s_1 + \theta s_2$, *and* $1/p = (1-\theta)/p_1 + \theta/p_2$,

(ii) $(F^s_{p_1\,q_1}(\mathbb{R}^N), F^s_{p_2\,q_2}(\mathbb{R}^N))_{\theta,p} = F^s_{p\,q}(\mathbb{R}^N)$
for all $s_1, s_2 \in \mathbb{R}^1$, $1 < q_1, q_2 < \infty$, $1 < p_1, p_2 < \infty$, $p_1 \neq p_2$, *provided* $1/p = (1-\theta)/p_1 + \theta/p_2 = 1/q = (1-\theta)/q_1 + \theta/q_2$,

(iii) $(F^s_{p_1\, q}(\mathbb{R}^N), F^s_{p_2\, q}(\mathbb{R}^N))_{\theta,p} = F^s_{p\, q}(\mathbb{R}^N)$
for all $s \in \mathbb{R}^1$, $1 < q < \infty$, *and* $1 < p_1, p_2 < \infty$, $p_1 \neq p_2$, *where* $1/p = (1-\theta)/p_1 + \theta/p_2$,

(iv) $[F^{s_1}_{p_1\, q_1}(\mathbb{R}^N), F^{s_2}_{p_2\, q_2}(\mathbb{R}^N)]_\theta = F^s_{p\, q}(\mathbb{R}^N)$
for all $s_1, s_2 \in \mathbb{R}^1$, $1 < p_1, p_2, q_1, q_2 < \infty$, *where* $s = (1-\theta)s_1 + \theta s_2$, $1/p = (1-\theta)/p_1 + \theta/p_2$, *and* $1/q = (1-\theta)/q_1 + \theta/q_2$.

2.4.27 Remark. A reader has certainly wondered why not all possible combinations of parameters have been considered for the real interpolation method. The reason is clear after looking at the very definition of Besov and Lizorkin-Triebel spaces. The respective norms are mixed norms and the scales of spaces in question are not closed with respect to the real interpolation. Refinement to Lorentz spaces and/or Lorentz sequence spaces will do; one obtains spaces based on Lorentz norms rather than on L_p- and/or ℓ_q-norms. Nevertheless, this is far beyond our needs here. See [148], Chapter 2 for details.

Considering equal parameters p and q and non-integer positive smoothness in theorems in this subsection, we arrive at particular statements for Sobolev-Slobodetskii spaces similarly as we explicitly did it when surveying the classical approach to Besov spaces. We leave it now to the reader.

2.4.28 Theorem. (i) *Let* $s \in \mathbb{R}^1$, $1 < p < \infty$, $1 \le q < \infty$, $p' = p/(p-1)$, *and* $q' = q/(q-1)$ ($q' = \infty$ *if* $q = 1$). *Then* $(B^s_{p\, q}(\mathbb{R}^N))' = B^{-s}_{p'\, q'}(\mathbb{R}^N)$.
(iii) *Let* $s \in \mathbb{R}^1$, $1 < p < \infty$, $p' = p/(p-1)$, *and let* $\widetilde{B}^s_{p\, \infty}(\mathbb{R}^N)$ *denote the closure of* $\mathscr{D}(\mathbb{R}^N)$ *in* $B^s_{p\, \infty}(\mathbb{R}^N)$. *Then* $(\widetilde{B}^s_{p\, \infty}(\mathbb{R}^N))' = B^{-s}_{p'\, 1}(\mathbb{R}^N)$.
(iii) *Let* $s \in \mathbb{R}^1$, $1 < p < \infty$, $1 \le q < \infty$, $p' = p/(p-1)$, *and* $q' = q/(q-1)$. *Then* $(F^s_{p\, q}(\mathbb{R}^N))' = F^{-s}_{p'\, q'}(\mathbb{R}^N)$.

For further properties of Besov spaces, particularly for their relationship to potential Sobolev spaces, see the next section.

2.4.29 Remark. The last two items in Theorem 2.4.23 are particularly worthy of a comment. Namely, considering $k = 0$, then W^k_2 is the Lebesgue space L_2 and therefore

$$\|f\|_2 \sim \left(\sum_{k=0}^{+\infty} \int_{\mathbb{R}^N} ((\varphi_k * f)(x))^2 \, dx \right)^{1/2}.$$

Furthermore, for any $1 < p < \infty$,

$$\|f\|_p \sim \left(\int_{\mathbb{R}^N} \left(\sum_{k=0}^{+\infty} ((\varphi_k * f)(\xi))^2 \right)^{p/2} d\xi \right)^{1/p}.$$

Last two equivalencies are usually called *inequalities of the Littlewood-Pólya type.*

The classical exposition of this topic can be found in Stein's monograph [143]. The reader should now return to the formula (2.4.15) on p. 108 to realize that $L_p(\mathbb{R}^N) = F^0_{p\,2}$.

2.4.4 Equivalence of definitions

In the following we shall tackle the coincidence of the above definition with the classical definition of Besov spaces. For these considerations it will be useful to express differences of a function as convolutions. To this end, note that for $g \in \mathscr{S}$,

$$
\begin{aligned}
(2\pi)^{N/2}\mathscr{F}(\Delta_h g;\xi) &= \int_{\mathbb{R}^N} e^{-i\xi\cdot x}\Delta_h g(x)\,dx \\
&= \int_{\mathbb{R}^N} (e^{-i\xi\cdot x} - e^{-i\xi\cdot(x-h)})g(x)\,dx \\
&= (1-e^{i\xi\cdot h})\int_{\mathbb{R}^N} e^{-i\xi\cdot x}g(x)\,dx \\
&= (1-e^{i\xi\cdot h})(2\pi)^{N/2}\mathscr{F}(g;\xi).
\end{aligned}
$$

Applying $\mathscr{F}^{-1}$ we get for $h^* : x \mapsto h\cdot x$:

$$\Delta_h g(x) = \mathscr{F}^{-1}((1-e^{ih^*})) * g(x).$$

Let $m = 1, 2, \ldots$, and $h \in \mathbb{R}^N$. Denote ϱ_h the function in $\mathscr{S}'(\mathbb{R}^N)$ whose Fourier image is $\left(1-e^{ih\cdot\xi}\right)^m$. Let ϱ_h^m be the m-times iterated convolution, $\varrho_h^m = \varrho_h * \cdots * \varrho_h$. It is

$$\mathscr{F}(\varrho_h^m;\xi) = \left(1-e^{ih\cdot\xi}\right)^m. \tag{2.4.17}$$

Plainly,

$$\varrho_h^m * g(x) = \Delta_h^m g(x) = \sum_{k=0}^{m}(-1)^k\binom{m}{k}g(x+kh).$$

Our next goal is a theorem on equivalence of the "classical" and the decomposition definition of Besov spaces. We shall need some auxiliary assertions. The notation ϱ_h^m in the following will be used for the function from (2.4.17) and $D_j^r f$ will stand for the "pure" derivative $\partial^r f/\partial x_j^r$.

2.4.30 Lemma. *For $m = 1, 2, \ldots,$ let $\varrho_{jk}^m(x) = \varrho_{2^{-k}e_j}^m(x)$, $x \in \mathbb{R}^N$, $k = 0, 1, \ldots,$ $j = 1, \ldots, N$, where e_j is the j-th unit coordinate vector. Let*

$\{\varphi_k\}_{k\geq 0}$ *be the standard dyadic decomposition of unity in* $\mathbb{R}^N$, *and* $1 \leq p \leq \infty$. *Then for every non-negative integer* r *there exists* $c > 0$ *such that*

$$\|\varphi_k * f\|_p \leq c2^{-rk} \sum_{j=1}^{N} \left\|\varrho_{jk}^m * D_j^r f\right\|_p, \qquad k = 1, 2, \ldots, \tag{2.4.18}$$

for every $f \in \mathscr{S}'$.

Proof. Let us fix k and let h be the basic "hat function" from the definition of the decomposition of unity, that is, $h \in \mathscr{D}(\mathbb{R}^N)$, $h \geq 0$, $\operatorname{supp} h = \{2^{-1} < |\xi| < 2\}$, $h > 0$ in the interior of its support (see p. 107). Let $v \in \mathscr{D}(\mathbb{R}^1)$, positive in the interior of $\operatorname{supp} v = (1/(2N), 3)$, and let $w \in \mathscr{D}(\mathbb{R}^{N-1})$, positive on the interior of $\operatorname{supp} w = \{\xi;\ |\xi| \leq 3\}$. Define functions z_j, $j = 1, \ldots, N$ in terms of their Fourier transform as

$$\begin{aligned}\mathscr{F}(z_j;\xi) =& v(\xi_j)w(\xi_1, \ldots, \xi_{j-1}, \xi_{j+1}, \ldots, \xi_N) \\ & \left(\sum_{j=1}^{N} v(\xi_j)w(\xi_1, \ldots, \xi_{j-1}, \xi_{j+1}, \ldots, \xi_N)\right)^{-1}.\end{aligned}$$

Then $\sum_{j=1}^N \mathscr{F} z_j \equiv 1$ on $\operatorname{supp} h$, thus $\sum_{j=1}^N \mathscr{F} z_j(2^{-k}\xi) \equiv 1$ on $\operatorname{supp} \varphi_k$. We have

$$\begin{aligned}\|\varphi_k * f\|_p &= \|\mathscr{F}^{-1}[h(2^{-k}\mathscr{F}\varphi)]\|_p \\ &\leq \sum_{j=1}^{N} \|\mathscr{F}^{-1}[\mathscr{F} z_j(2^{-k}\cdot)h(2^{-k}\cdot)\mathscr{F} f]\|_p \\ &= \sum_{j=1}^{N} \|\mathscr{F}^{-1}[\mathscr{F} z_j(2^{-k}\cdot)h(2^{-k}\cdot)\xi_j^r(1 - e^{i<2^{-k}e_j,\cdot>})^{-m} \\ &\qquad\qquad (1 - e^{i<2^{-k}e_j,\cdot>})^m \xi_j^r \mathscr{F} f]\|_p \\ &= \sum_{j=1}^{N} \left\|\mathscr{F}^{-1}[\mathscr{F} z_j(2^{-k}\cdot)h(2^{-k}\cdot)2^{-kr}(2^{-k}\xi_j)^{-r}\right. \\ &\qquad\qquad \left.(\mathscr{F}\varrho_{jk}^m(\cdot))^{-1}\mathscr{F}\varrho_{jk}^m(\cdot)\mathscr{F}(D_j^r f)]\right\|_p \\ &= 2^{-kr}\sum_{j=1}^{N} \left\|\mathscr{F}^{-1}[\mathscr{F} z_j(2^{-k}\cdot)h(2^{-k}\cdot)h(2^{-k}\cdot)(2^{-k}\xi_j)^{-r}\right. \\ &\qquad\qquad \left.\mathscr{F}(\varrho_{jk}^m(\cdot))^{-1}] * (\varrho_{jk}^m * D_j^r f)\right\|_p.\end{aligned}$$

The function $\xi \mapsto \mathscr{F} z_j(2^{-k}\xi)h(2^{-k}\xi)(2^{-k}\xi_j)^{-r}(\mathscr{F}\varrho_{jk}^m(\xi))^{-1}$ has the same M_p-norm as the function

$$\xi \mapsto \mathscr{F} z_j(\xi)h(\xi)(\xi_j)^{-r}(\mathscr{F}\varrho_{j0}^m(\xi))^{-1} = \mathscr{F} z_j(\xi)h(\xi)\xi_j^{-r}(1 - e^{i\xi_j})^{-m}$$

(see Lemma 2.3.8). A routine calculation shows that the assumptions of Lemma 2.3.5 are satisfied (the function in question is smoothly differentiable and the origin does not belong to its support). Hence we can conclude that (2.4.18) holds. □

2.4.31 Remark. Note that (2.4.18) implies

$$\|\varphi_k * f\|_p \le c2^{-rk} \sum_{j=1}^{N} \|\Delta^m_{2^{-k}e_j} f\|_p,$$

where e_j is the j-th unit coordinate vector.

Now we can prove the first half of the desired equivalence:

2.4.32 Proposition. *Let $s \in \mathbb{R}^1$, let m and r be integers, $m + r > s$, $0 \le r < s$, $1 \le p, q \le \infty$. Then*

$$\|f\|_{B^s_{p\,q}} \le c \left[\|f\|_p + \sum_{j=1}^{N} \left(\int_0^{+\infty} \left[t^{r-s} \omega^m_p \left(t, D^r_j f\right)\right]^q \frac{dt}{t} \right)^{1/q} \right] \tag{2.4.19}$$

for all $f \in \mathscr{S}(\mathbb{R}^N)$, with a constant c independent of f.

Proof. All the difficult job has been done in the previous lemma. In addition to that, note that $h \in M_1$. This follows at once from the multiplier criterion in Theorem 2.3.10. Thus invoking Lemma 2.4.30 and using Minkowski's inequality, we get

$$\begin{aligned}\|f\|_{B^s_{p\,q}} &\le c\|f\|_p + c\left(\sum_{k=1}^{+\infty} \left(2^{k(s-r)} \sum_{j=1}^{N} \left\| \varrho^m_{jk} * D^r_j f \right\|_p \right)^q \right)^{1/q} \\ &\le c\|f\|_p + c\left(\sum_{k=1}^{+\infty} \left(2^{k(s-r)} \Delta^m_p \left(2^{-k} e_j, D^r_j f\right) \right)^q \right)^{1/q}.\end{aligned}$$

Let us observe that the integral $\int_{2^k}^{2^{k+1}} dt/t$ is a constant independent of k. Hence monotonicity of ω^m_p yields (2.4.19). □

2.4.33 Remark. A byproduct of the above estimate (cf. also Remark 2.4.31) is

$$\|f\|_{B^s_{p\,q}} \le c\|f\|_p + c\sum_{j=1}^{N} \left(\int_{\mathbb{R}^N} \left[|h|^{r-s} \Delta^m_p \left(h, D^r_j f\right)\right]^q \frac{dh}{|h|^n} \right)^{1/q}.$$

Now we derive the converse estimate. Again we split the proof in several steps.

2.4.34 Lemma. *Let $s > 0$, m, r be integers, $m + r > s$, $0 \le r < s$, $1 \le p, q \le \infty$. Let $\{\varphi_k\}$ be the standard decomposition of unity and let ϱ_y^m be as in (2.4.17). Then there exists $c > 0$ such that*

$$\left\|\varrho_y^m * \varphi_k * D_j^r f\right\|_p \le c \min\left(1, |y|^m 2^{mk}\right) 2^{rk} \|\varphi_k * f\|_p, \qquad k = 1, 2, \ldots. \tag{2.4.20}$$

Proof. According to Lemma 2.3.9 the function $\xi \mapsto (1 - e^{ih\cdot\xi})^m$, $\xi \in \mathbb{R}^N$, where $h \in \mathbb{R}^N$ is fixed, has the same M_p-norm as $t \mapsto (1 - e^{it})^m$, $t \in \mathbb{R}^1$. Furthermore, in view of Lemma 2.3.8 this norm is the same for all functions $t \mapsto (1 - e^{at})^m$, $t \in \mathbb{R}^1$, where a is an arbitrary non-zero (complex) number. It suffices to choose $a < 0$ or $a = i$ and to apply Theorem 2.3.10 to see that there is $c > 0$ such that

$$\|\mathscr{F}(\varrho_y^m)\|_{M_1} \le c \qquad \text{for all } y \in \mathbb{R}^N.$$

(Note that, plainly, $\mathscr{F}(\varrho_y^m)$ belongs to M_p since L_p-norms of differences of any order of a function in L_p are estimated by a suitable multiple of the L_p-norm of this function.) An analogous argument shows that with the notation of the scalar product in $\mathbb{R}^N$ by $(\cdot,\cdot)_N$ we have

$$\left\|\mathscr{F}\left(\varrho_y^m (y, \cdot)_N^{-m}\right)\right\|_{M_1} \le c$$

and

$$\|(y/|y|, \cdot)_N^m \varphi(\cdot)\|_{M_1} \le c$$

with some c independent of y. Passing from ξ to 2^{-k} in the role of the variable we get

$$\|(y, \cdot)_N^m \varphi(2^{-k}\cdot)\|_{M_1} \le c|y|^m 2^{mk},$$

which implies that

$$\begin{aligned}\left\|\varrho_y^m * \varphi_k * D_j^r f\right\| &\le \min(c, c|y|^m 2^{mk}) \left\|\varphi_k * D_j^r f\right\|_p \\ &\le c \min(1, |y|^m 2^{mk}) \left\|\varphi_k * D_j^r f\right\|_p.\end{aligned}$$

It remains to estimate $\|\varphi_k * D_j^r f\|_p)\|$. If $k \ge 1$, then

$$\begin{aligned}\varphi_k * D_j^r f = \mathscr{F}^{-1}(\mathscr{F}\varphi_k \xi_j^r \mathscr{F} f(\xi)) &= (\varphi_k * \mathscr{F}^{-1})(\xi_j^r \mathscr{F} f(\xi)) \\ = \mathscr{F}^{-1}(\mathscr{F}\varphi_k \cdot \xi_j^r) * f.\end{aligned}$$

Hence it will be enough to find an appropriate estimate for the M_p norm of the function $\xi \mapsto \mathscr{F}\varphi_k(\xi)\xi_j^r$. However,

$$\begin{aligned}
\varphi_k * D_j^r f(x) &= \mathscr{F}^{-1}(\mathscr{F}\varphi_k(\xi)\xi_j^r) * f(x) \\
&= \mathscr{F}^{-1}(\mathscr{F}\varphi(2^{-k}\xi)(2^{-k}\xi_j)^r 2^{kr}) * f(x) \\
&= 2^{kr}\mathscr{F}^{-1}(\mathscr{F}\varphi(2^{-k}\xi)(2^{-k}\xi_j)^r \cdot \chi_{\operatorname{supp}\varphi}(2^{-k}\xi)) * f(x) \\
&= 2^{kr}\mathscr{F}^{-1}(\chi_{\operatorname{supp}\varphi}(2^{-k}\xi)(2^{-k}\xi_j^r)) * (\varphi_k * f), \qquad x \in \mathbb{R}^N.
\end{aligned}$$

The M_p-norm of the function $\xi \mapsto \chi_{\operatorname{supp}\varphi}(2^{-k}\xi)(2^{-k}\xi_j)$ is independent of k (cf. Lemma 2.3.8) and equals to a constant. Hence

$$\|\varphi_k * D_j^r f\|_p \le c 2^{rk}\|\varphi_k * f\|_p.$$

The same argument applies to φ_0. We have proved (2.4.20). □

We are in a position to prove the estimate converse to (2.4.19).

2.4.35 Proposition. *Let $s \in \mathbb{R}^1$, let m and r be integers, $m + r > s$, $1 \le p, q \le \infty$. Then*

$$\|f\|_p + \sum_{j=1}^{N}\left(\int_0^{+\infty}\left[t^{r-s}\omega_p^m(t, D_j^r f)\right]^q \frac{dt}{t}\right)^{1/q} \le c\|f\|_{B_{pq}^s} \tag{2.4.21}$$

for all $f \in \mathscr{S}(\mathbb{R}^N)$, with a constant c independent of f.

Proof. Again we use monotonicity of functions involved to see the equivalence

$$\left(\int_0^{+\infty}\left[t^{r-s}\omega_p^m(t, D_j^r f)\right]^q \frac{dt}{t}\right)^{1/q} \sim \left(\sum_{i=-\infty}^{+\infty}\left(2^{i(s-r)}\omega_p^m(2^{-i}, D_j^r f)\right)^q\right)^{1/q}.$$

Let $y \in \mathbb{R}^N$, $|y| \le 2^{-i}$. Then (δ denotes the Dirac distribution)

$$\begin{aligned}
2^{i(s-r)}\Delta_p^m(y, D_j^r f) &= 2^{i(s-r)}\|\varrho_y^m * D_j^r f\|_p \\
&= 2^{i(s-r)}\|\varrho_y^m * \delta * D_j^r f\|_p \\
&\le 2^{i(s-r)}\sum_{k=0}^{+\infty}\|\varrho_y^m * \varphi_k * D_j^r f\|_p \\
&\le c\sum_{k=0}^{+\infty} 2^{i(s-r)}\min\left(1, |y|^m 2^{mk}\right)2^{rk}\|\varphi_k * f\|_p
\end{aligned}$$

$$\le c \sum_{k=0}^{+\infty} 2^{i(s-r)} \min\left(1, 2^{-(i-k)m}\right) 2^{rk} \|\varphi_k * f\|_p.$$

Hence

$$2^{i(s-r)} \omega_p^m(2^{-i}, D_j^r f) \le \sum_{k=0}^{+\infty} 2^{(i-k)(s-r)} \min\left(1, 2^{-(i-k)m}\right) 2^{sk} \|\varphi_k * f\|_p.$$

The last expression can be viewed as a multiple of convolution of the sequences

$$\alpha_k = 2^{k(s-r)} \min(1, 2^{-km}), \quad k \in \mathbb{Z},$$

$$\beta_k = \begin{cases} 2^{sk} \|\varphi_k * f\|_p, & k \ge 0, \\ 0, & k < 0. \end{cases}$$

Because $\{\alpha_k\} \in \ell^1$ and $\{\beta_k\} \in \ell^q$, we have

$$\begin{aligned} \|\{\alpha_k\} * \{\beta_k\}\|_{\ell^q} &\le \|\{\alpha_k\}\|_{\ell^1} \cdot \|\{\beta_k\}\|_{\ell^q} \\ &\le c\|f\|_{B^s_{p\,q}}. \end{aligned}$$

□

2.4.36 Remark. Similarly as in Remark 2.4.33 we have also proved that

$$\|f\|_p + c \sum_{j=1}^{N} \left(\int_{\mathbb{R}^N} \left[|h|^{r-s} \Delta_p^m \left(h, D_j^r f\right) \right]^q \frac{dh}{|h|^n} \right)^{1/q} \le c\|f\|_{B^s_{p\,q}}.$$

Hence the expression on the left hand side is an equivalent norm in $B^s_{p\,q}$.

This implies also the "usual" formula for the norm in the Sobolev-Slobodetskii spaces $B^s_{p\,p}$. If $s > 0$ is not an integer, let $s = [s] + \{s\}$, where $[s]$ is the integer part of s (hence $0 < \{s\} < 1$). An application of Fubini's theorem yields then

$$\|f\|_{B^s_{p\,p}} \sim \|f\|_p + \sum_{|\alpha|=[s]} \left(\int_{\mathbb{R}^N} \int_{\mathbb{R}^N} \frac{|D^\alpha f(x) - D^\alpha f(y)|^p}{|x-y|^{N+\{s\}p}} \, dx dy \right)^{1/p}.$$

Summation over all $|\alpha| \le [s]$ leads to another equivalent norm. Let us also observe that in all integral expressions for the norm in $B^s_{p\,q}$ one can replace the integration with respect to t and h by integration over some interval $(0, \delta)$ and $(0, \delta)^N$, respectively. This follows at once since the terms with the L_p-norms of differences and with the L_p-modulus of continuity play a role only for small values t and $|h|$, respectively; near infinity they are trivially majorized by a multiple of $\|f\|_p$. We refer to [148], Chapter 2, for a detailed account of these and still other equivalent norms in Besov spaces.

We shall present several basic assertions mostly without proofs. We refer to [119], [23], Chapter 6, [148], Chapter 2. We shall, however, prove the next lemma, actually as an archetypical example, in order to throw some light on how the decomposition idea works.

2.4.37 Definition. Let $s \in \mathbb{R}^N$. Then the *Bessel potential of order s* is the mapping defined for $f \in \mathscr{S}(\mathbb{R}^N)$ as

$$J_s f = \mathscr{F}^{-1}((1 + |\xi|^2)^{s/2} \mathscr{F}(f)). \tag{2.4.22}$$

Note that in literature one often meets the "symbolic" definition of powers of $(I - \Delta)$, namely, $\mathscr{F}((I - \Delta)^{s/2} f) = (1 + |\xi|)^s \mathscr{F}(f)$. In the Hilbert case this is justified by the Plancherel formula.

2.4.38 Lemma. *Let $1 \le p \le \infty$, $f \in \mathscr{S}'(\mathbb{R}^N)$, and $\|\varphi_k * f\|_p < \infty$. Then for every $s \in \mathbb{R}^1$,*

$$\|J_s \varphi_k * f\|_p \le c 2^{sk} \|\varphi_k * f\|_p \tag{2.4.23}$$

with a constant c independent of p and k.

Proof. We shall prove (2.4.23) for $k \ge 1$. The proof of (2.4.23) for $k = 0$ is quite similar and it is left to the reader. Note that the function $\varphi_{k-1} + \varphi_k + \varphi_{k+1}$ equals 1 on $\operatorname{supp} \varphi_k$, that is, supports of all the remaining functions of the decompositions have an empty intersection with $\operatorname{supp} \varphi_k$. Hence it is sufficient to show that

$$\|\mathscr{F}(J_s \varphi_i)\|_p \le c 2^{ks}, \tag{2.4.24}$$

where $i = k - 1, k, k + 1$. Since $\varphi_i(\xi) = g(2^{-i}\xi) - g(2^{-i+1}\xi) = h(2^{-i}\xi)$ (in notation of (2.4.11)) we have

$$\mathscr{F}(J_s \varphi_i; \xi) = (1 + |\xi|^2)^{s/2} \mathscr{F}(f_i; \xi) = (1 + |\xi|^2)^{s/2} h(2^{-i}\xi). \tag{2.4.25}$$

According to Lemma 2.3.8 the M_p-norm of the right hand side of (2.4.25) is the same as the M_p-norm of the function

$$\varrho_i(\xi) = 2^{is} \left(2^{-2i} + |\xi|\right)^{s/2} h(\xi).$$

The L_2-norms of the last function and its derivatives are plainly finite and an elementary estimate gives $\|D^\alpha \varrho_i\| \le c 2^{ks}$ for $i = k - 1, k, k + 1$, hence Lemma 2.3.10 yields (2.4.24). □

2.4.39 Corollary. *The Bessel potential J_s is an isomorphism between the spaces $B^{s_0}_{pq}$ and $B^{s_0 - s}_{pq}$ for all real s, s_0 and all $1 \le p, q \le \infty$.*

2.5 The potential spaces

To fix the notation we shall use the symbol $W_p^k(G)$ and $W_p^k(G, X)$, $1 \leq p < \infty$, $k = 0, 1, \ldots$, for the Sobolev space of real valued and vector-valued functions, resp.

There are more equivalent ways of introducing various clones of Sobolev spaces of non-integer order. Since many of the spaces used in the following have a Hilbert structure we shall follow the Fourier analysis approach, which makes it possible at the same time to introduce non-integer derivatives in a rather transparent way, e.g. without digging deeply into the spectral theory and powers of general positive operators (see [148], Chapter 1); we shall restrict ourselves just to powers of $I - \Delta$. The underlying "classical" analysis of Bessel potentials is not trivial and we shall mostly omit the hard technical details.

If $0 < \alpha < N$, then the function $\xi \mapsto |\xi|^{\alpha - N}$, $\xi \in \mathbb{R}^N$, is locally integrable and it is therefore a regular tempered distribution. For these values of α and $f \in \mathscr{S}(\mathbb{R}^N)$ we define the *Riesz potential of order* α by

$$\mathscr{I}_\alpha f(x) = \mathscr{F}^{-1}\big((2\pi)^{N/2-\alpha/2}|\xi|^{-\alpha}\mathscr{F}(f;\xi)\big)(x), \qquad x \in \mathbb{R}^N. \tag{2.5.1}$$

Of course, the meaning of (2.5.1) is to be understood in the duality sense, that is,

$$\int_{\mathbb{R}^N} \mathscr{I}_\alpha f(x)\overline{\mathscr{F}(\varphi;x)}\,dx = \int_{\mathbb{R}^N} ((2\pi)^{N/2-\alpha/2}|x|^{-\alpha}\mathscr{F}(f;\xi)\overline{\varphi(x)}\,dx$$

for our f and all $\varphi \in \mathscr{S}(\mathbb{R}^N)$. Note that the last identity follows from the well-known formula for the Fourier transform of $|x|^{\alpha-N}$,

$$(\mathscr{F}(|x|^{\alpha-N};\xi) = \frac{2^{\alpha/2}\pi^{N/2-\alpha/2}\Gamma(\alpha/2)}{|\xi|^\alpha\Gamma(N/2-\alpha/2)}.$$

(See e.g. [143], Chapter 5 for the proof. Let us observe that the multiplicative constants here and the other formulae concerning the Fourier transform might differ according to the definition of the Fourier transform. In particular, the classical monograph [143] uses $\int_{\mathbb{R}^N} \exp(2\pi x \cdot \xi) f(x)\,dx$.) Further calculation (see again e.g. [143]) leads to the formula, which expresses the Riesz potential in terms of a convolution,

$$\mathscr{I}_\alpha f(x) = \frac{\Gamma(N/2-\alpha/2)}{\pi^{N/2}2^\alpha\Gamma(\alpha/2)}\int_{\mathbb{R}^N}\frac{f(y)}{|x-y|^{N-\alpha}}\,dy, \qquad x \in \mathbb{R}^N. \tag{2.5.2}$$

The kernel

$$I_\alpha(x) = \frac{\Gamma(N/2-\alpha/2)|x|^{\alpha-N}}{\pi^{N/2}2^\alpha\Gamma(\alpha/2)}, \qquad x \in \mathbb{R}^N,$$

is called the *Riesz kernel of order* α.

Let $f \in \mathscr{S}(\mathbb{R}^N)$. Then plainly $(-\Delta)f(x) = \mathscr{F}^{-1}\big(|\xi|^2\mathscr{F}(f;\xi);x\big)$, $x \in \mathbb{R}^N$, and this suggests to define positive powers of the Laplacean: For $\beta > 0$ we put

$$(-\Delta)^{\beta/2}f(x) = \mathscr{F}^{-1}\big(|\xi|^\beta\mathscr{F}(f;\xi);x\big), \qquad x \in \mathbb{R}^N. \tag{2.5.3}$$

The definition (2.5.3) has sense, however, for *any* f such that $|\xi|^\beta\mathscr{F}(f;\xi)$ is in $\mathscr{S}'(\mathbb{R}^N)$. Hence if we consider the function $\xi \mapsto |\xi|^\alpha$ with $0 < \alpha < N$ (which is locally integrable in $\mathbb{R}^N$), we can define the *negative power of the Laplacean* as

$$(-\Delta)^{-\alpha/2}f(x) = \mathscr{F}^{-1}\big(|\xi|^{-\alpha}\mathscr{F}(f;\xi);x\big), \qquad x \in \mathbb{R}^N. \tag{2.5.4}$$

A simple calculation yields $(\mathscr{I}_\alpha f)(x) = (2\pi)^{\alpha/2-N/2}(-\Delta)^{-\alpha/2}f(x)$. Putting $g = (2\pi)^{\alpha/2-N/2}(-\Delta)^{\alpha/2}f$ and invoking (2.5.2), we arrive at

$$f(x) = \mathscr{I}_\alpha g(x) = \frac{\pi^{N/2}2^\alpha\Gamma(\alpha/2)}{\Gamma(N/2-\alpha/2)}\int_{\mathbb{R}^N}\frac{g(y)}{|x-y|^{N-\alpha}}\,dy, \qquad x \in \mathbb{R}^N$$

(compare with the classical harmonic analysis formula if $\alpha = 2$).

If we define on $\mathscr{S}(\mathbb{R}^N)$ the *Riesz transforms*,

$$R_jf(x) = \mathscr{F}^{-1}\left(-\frac{i\xi_j}{|\xi|}\mathscr{F}(f);\,x\right)$$

(a special sort of a singular integral operator with a Calderón-Zygmund kernel, see e.g. [143], [126]), then by the well-known rules for calculation with the Fourier transform we have $R_j(D_jf)(x) = \mathscr{F}^{-1}\big(\xi_j^2|\xi|^{-1}\mathscr{F}(f);x\big)$ and $\mathscr{F}\left(\sum_{j=1}^N R_j(D_jf);\,\xi\right) = |\xi|\mathscr{F}(f;\xi)$. Consequently

$$f(x) = \mathscr{F}^{-1}\left(|\xi|^{-1}\mathscr{F}\left(\sum_{j=1}^N R_j(D_jf)\right);\,x\right), \qquad x \in \mathbb{R}^N,$$

and we obtain the representation formula

$$f(x) \sim \mathscr{I}_1\left(\sum_{j=1}^N R_j(D_jf)\right)(x), \qquad x \in \mathbb{R}^N.$$

Note that this is one of the possible keys to imbedding theorems for Sobolev spaces. The operators R_j take $L_p(\mathbb{R}^N)$ into the same space for $1 < p < \infty$ (this is a deep fact from the theory of singular integral operators) and (as we shall see below) $\mathscr{I}_1$ maps $L_p(\mathbb{R}^N)$ into $L_q(\mathbb{R}^N)$, where $1/q = 1/p - 1/N$ (the Sobolev exponent). Nevertheless, $\mathscr{I}_1$ does not map $L_p(\mathbb{R}^N)$ into itself. This can be seen from its behaviour near infinity—it is not integrable there; hence the mapping $f \mapsto I_1 * f$ for $f \in L_p(\mathbb{R}^N)$ is not bounded in $L_p(\mathbb{R}^N)$. This disadvantage can be removed by considering the Bessel kernels, which have the same behaviour at the origin, but nice properties at infinity.

2.5.1 Theorem. *Let* $0 < \alpha < N$, $1 \le p < \infty$, $\alpha p < N$, *and* $1/q = 1/p - \alpha/N$. *Then* I_α *is of weak type* (p, q).

A straightforward application of the Marcinkiewicz theorem yields

2.5.2 Corollary. *Let* $0 < \alpha < N$, $1 < p < \infty$, $\alpha p < N$, *and* $1/q = 1/p - \alpha/N$. *Then* I_α *is of type* (p, q).

Before we prove Theorem 2.5.1 let us observe that Corollary 2.5.2 contains the classical Sobolev imbedding theorem. Indeed, considering the Sobolev space of the first order and employing the well-known representation formula

$$f(x) = \frac{1}{|S_{N-1}|} \sum_{j=1}^{N} \int_{\mathbb{R}^N} \frac{\partial f}{\partial x_j}(x - y) \frac{y_j}{|y|^{N+1}} \, dy, \tag{2.5.5}$$

where $|S_{N-1}|$ denotes the area of the unit sphere in $\mathbb{R}^N$, we see that

$$|f(x)| \le c \sum_{j=1}^{N} \mathscr{I}_1 \left(\left| \frac{\partial f}{\partial x_j} \right| \right) (x),$$

and we are done. Imbeddings for higher order spaces (with integer derivatives) follow by simple iteration. Observe also that the identity (2.5.5) is a sum of convolutions of partial derivatives with (singular) kernels $y_i/|y|^{N+1}$, which are special cases of the so called Calderón-Zygmund singular operators (see e.g. [143]) and their L_p boundedness $(1 < p < \infty)$ follows from the general theory of singular integrals.

Proof of Theorem 2.5.1. It is sufficient to consider the convolution with the kernel $K(x) = |x|^{\alpha - n}$. Fix $\sigma > 0$, and put $K_1(x) = K(x)$ on the ball about the origin and of the radius σ and $K_2 = K - K_1$. It is easy to check that $K_1 \in L_1$ and $K_2 \in L_{p/(p-1)}$. Hence the integrals $K_1 * f(x)$ and $K_2 * f$ with $f \in L_p$ exist for a.e. $x \in \mathbb{R}^N$. We want to prove the inequality of weak type

$$\lambda^q \operatorname{mes}\{x \in \mathbb{R}^N;\ |K * f(x)| > \lambda\} \le c \|f\|_p^q, \qquad f \in L_p(\mathbb{R}^N). \tag{2.5.6}$$

Plainly

$$
\begin{aligned}
\operatorname{mes}\{x \in \mathbb{R}^N; |K * f(x)| > \lambda\} \\
\leq \operatorname{mes}\{x \in \mathbb{R}^N; |K_1 * f(x)| > \lambda/2\} \\
+ \operatorname{mes}\{x \in \mathbb{R}^N; |K_2 * f(x)| > \lambda/2\}.
\end{aligned}
\tag{2.5.7}
$$

Further, suppose that $\|f\|_p = 1$. We have

$$
\begin{aligned}
\operatorname{mes}\{x \in \mathbb{R}^N; |K_1 * f(x)| > \lambda/2\} &\leq \frac{2^p \|K_1 * f\|_p^p}{\lambda^p} \\
&\leq \frac{2^p \|K_1\|_1^p \|f\|_p^p}{\lambda^p} \\
&= \frac{2^p \|K_1\|_1^p}{\lambda^p}.
\end{aligned}
\tag{2.5.8}
$$

The norm $\|K_1\|_1$ can be computed directly; it equals $c_1\sigma^\alpha$. By Hölder's inequality,

$$
\|K_2 * f\|_\infty \leq \|K_2\|_{p/(p-1)} \|f\|_p \leq \|K_2\|_{p/(p-1)},
$$

and this gives $\|K_2 * f\|_\infty \leq c_2\sigma^{-N/q}$ with a suitable constant c_2. Hence $\|K_2\|_{p/(p-1)} = \lambda/2$ if $2c_2\sigma^{-N/q} = \lambda/2$. Fix such σ. Then $\|K_2 * f\|_\infty \leq \lambda/2$ and the second summand on the right hand side of (2.5.7) is zero. Combining (2.5.7) and (2.5.8), we get

$$
\operatorname{mes}\{x \in \mathbb{R}^N; |K * f(x)| > \lambda\} \leq c_3\sigma^{\alpha p}\lambda^{-p} = c_4\lambda^{-q} = c_4\left(\frac{\|f\|_p}{\lambda}\right)^q.
$$

□

A detailed discussion of convolution inequalities and imbedding theorems of Sobolev type can be found e.g. in [101] and [157]. In particular, much attention in the theory has been devoted to refined imbeddings of Sobolev spaces into Lorentz spaces and there are also some interesting applications to the PDE's theory.

Now we shall collect basic facts about the *Bessel potentials* defined in Subsection 2.4.2. Note that, symbolically, J_α corresponds to $(I - \Delta)^{-\alpha/2}$, where Δ is the Laplace operator. Their crucial role in the theory of Sobolev spaces is the lifting property: the (potential) Sobolev spaces on $\mathbb{R}^N$ are exactly isomorphic copies of $L_p(\mathbb{R}^N)$ and the isomorphism is just the Bessel potential of corresponding order. We shall see that the local behaviour of the Bessel potentials near the origin is the same as in the case of the

Riesz potentials; nevertheless, they have more suitable properties near the infinity. We refer to [143], Chapter V, for more details.

Let $\alpha > 0$. Denote

$$g_\alpha(x) = \frac{1}{(4\pi)^{N/2}\Gamma(\alpha/2)} \int_0^{+\infty} \exp(-2\pi^2|x|^2/t)\exp(-t)t^{(-N+\alpha)/2}\,\frac{dt}{t}. \tag{2.5.9}$$

Then (see e.g. [143], [126] for the calculation) $\mathscr{F}(g_\alpha)(\xi) = (1+|\xi|)^{-\alpha/2}$. Inspecting more closely the formula for g_α, it is possible to show that for $0 < \alpha < N$ (see again e.g. [143]),

$$g_\alpha(x) \sim |x|^{-N+\alpha} + o(|x|^{-N+\alpha}) \quad \text{near the origin.} \tag{2.5.10}$$

The kernel g_α has an exponential decay at the infinity and it is not difficult to show directly that

$$g_\alpha(x) = O(e^{-c|x|}) \qquad \text{as } |x| \to +\infty. \tag{2.5.11}$$

The last two estimates are the key to convolution estimates in L_p and in even finer scales of spaces; this is one of the ways of proving imbedding theorems for Sobolev and more general spaces. We saw the core in the proof of Theorem 2.5.1, where we considered the Riesz kernel $|x|^{-N+\alpha}$ (up to a multiplicative constant).

We shall formulate this as a separate statement (see e.g. [143], [126] for the proof).

2.5.3 Proposition. *Let $0 < \alpha$ and g_α be given by (2.5.9). Then $\mathscr{F}g_\alpha(\xi) = (1+|\xi|^2)^{-\alpha/2}$, $\xi \in \mathbb{R}^N$, and the asymptotic behaviour of g_α near the origin and near infinity is given by (2.5.10) and (2.5.11), respectively.*

2.5.4 Definition. Let $\alpha \in \mathbb{R}^1$, $1 < p < \infty$. Then we define the *potential isotropic Sobolev space* $H_p^\alpha = H_p^\alpha(\mathbb{R}^N)$ as $J_\alpha(L_p(\mathbb{R}^N))$, that is, as the space of all $f \in \mathscr{S}(\mathbb{R}^N)$ for which there exists $g \in L_p(\mathbb{R}^n)$ such that $f = J_\alpha g = \mathscr{F}^{-1}((1+|\xi|^2)^{\alpha/2}\mathscr{F})$, with the norm $\|f\|_{H_p^\alpha(\mathbb{R}^N)} = \|g\|_{L_p(\mathbb{R}^N)}$.

The space H_p^α is often called the *Bessel potential space*. Historical reasons lead to frequent notation H^α instead of H_2^α.

For brevity, we shall omit the symbol for the domain $\mathbb{R}^N$ if no confusion can occur. Observe that the definition has a good sense since J_α is plainly one-to-one mapping from $\mathscr{S}(\mathbb{R}^N)$ onto itself. Indeed, if $J_\alpha g_1 = J_\alpha g_2$, then we have $\int_{\mathbb{R}^N}(J_\alpha g_1 - J_\alpha g_2)\varphi(x)\,dx = \int_{\mathbb{R}^N\times\mathbb{R}^N} g_\alpha(x-y)(g_1 - g_2)(y)\varphi(x)\,dxdy = \int_{\mathbb{R}^N}(g_1 - g_2)J_\alpha\varphi\,dx$, and, consequently, $g_1 = g_2$ a.e. in

$\mathbb{R}^N$. The spaces $H_p^k(\mathbb{R}^N)$ are isomorphic copies of L_p. In particular they are reflexive separable Banach spaces and H_2^α is a Hilbert space. In accordance with the common usage we shall use the brief notation H^α or $H^\alpha(\mathbb{R}^N)$ if $p = 2$ in the next chapters.

2.5.5 Remark. If $\alpha = 0$, then trivially $H_p^\alpha = L_p$. If $\alpha = k$ is a positive integer, then H_p^α coincides with the usual Sobolev space W_p^k, consisting of all functions in L_p possessing distributional derivatives up to the order k. This is easy to see in the prominent Hilbert case H_2^k; it is just a consequence of the Plancherel formula. The general case is much harder and the coincidence of the "classical" and potential spaces with integer derivatives and the equivalence of the norms can be proved e.g. with the help of multiplier theorems of Mikhlin type. For completeness we state it as a separate theorem and we give the proof. Let us observe that for $1 < p < \infty$ we obtain an equivalence of the potential norm $\|f\|_{H_p^k}$ with the norm $\|f\|_p + \sum_{j=1}^N \|\partial^k f/\partial x_j^k\|_p$.

It is well known from the theory of Sobolev spaces that the norms

$$\|f\|^{(1)} = \sum_{|\alpha| \le m} \|D^\alpha f\|_p, \qquad \|f\|^{(2)} = \sum_{j=1}^N \|D_j^m f\|_p,$$

$$\|f\|^{(3)} = \sum_{k=1}^m \sum_{j=1}^N \|D_j^k f\|_p, \qquad \|f\|^{(4)} = \|f\|_p + \sum_{|\alpha|=m} \|D^\alpha f\|_p,$$

and

$$\|f\|^{(5)} = \|f\|_p + \sum_{j=1}^N \|D_j^m f\|_p, \tag{2.5.12}$$

and their variants where we neglect $\|f\|_p$ are equivalent. There are also various other quantitative inequalities resulting from estimates of this type we have seen in Subsection 2.7.2. Let us point out that $\|f\|_p$ must be present if we consider Sobolev spaces on domains with finite measure. Mikhlin's multiplier theorem gives an efficient tool to handle similar relations, at least for $1 < p < \infty$. In next theorem we prove an equivalence of the potential norm with the norm in (2.5.12).

2.5.6 Theorem. *Let m be a positive integer and $1 < p < \infty$. Then $H_p^m(\mathbb{R}^N) = W_p^m(\mathbb{R}^N)$.*

Proof. It is not difficult to verify that the functions $\xi \mapsto \xi_j^m(1 + |\xi|^2)^{-m/2}$, $\xi \in \mathbb{R}^N$, $j = 1, \ldots, N$, satisfy the assumption of the Mikhlin multiplier theorem. For $f \in \mathscr{S}$ denote $D_j^m f = \partial^m f/\partial x_j^m$. Then

$$\begin{aligned}
\|D_j^m f\|_p &= \|\mathscr{F}^{-1}[\xi_j^m((1+|\xi|^2)^{-m/2})((1+|\xi|^2)^{m/2})\mathscr{F}(f)]\|_p \\
&= \|\mathscr{F}^{-1}[\xi_j^m((1+|\xi|^2)^{-m/2})\mathscr{F}\mathscr{F}^{-1}((1+|\xi|^2)^{m/2})\mathscr{F}(f)]\|_p \\
&= c\|\mathscr{F}^{-1}[\xi_j^m((1+|\xi|^2)^{-m/2})\mathscr{F}(J_m\mathscr{F}(f))]\|_p \\
&\le c\|f\|_{H_p^m}.
\end{aligned}$$

Hence if $f \in H_p^m$, then $D_j^m f$ is a function in L_p whose norm is estimated by a multiple of the H_p^m-norm of f.

For the converse estimate let us write

$$\begin{aligned}
\|J_m f\|_p &= \left\|\mathscr{F}^{-1}((1+|\xi|^2)^{m/2}\mathscr{F}(f))\right\|_p \\
&= \left\|\mathscr{F}^{-1}\left((1+|\xi|^2)^{m/2}\left(1+\sum_{j=1}^N \eta(\xi_j)|\xi_j|^m\right)\right.\right. \\
&\qquad\qquad \left.\left.\left(1+\sum_{j=1}^N \eta(\xi_j)|\xi_j|^m\right)^{-1}\mathscr{F}(f)\right)\right\|_p,
\end{aligned}$$

where η is a non-negative C^∞ function on $\mathbb{R}^N$, vanishing in the unit ball and equal to 1 on the set $\{\xi;\ |\xi| \ge 2\}$. An elementary calculation yields that for any multiindex α,

$$\left|\frac{\partial^{|\alpha|}}{\partial\xi_1^{\alpha_1}\dots\partial\xi_N^{\alpha_N}}\left((1+|\xi|^2)^{m/2}\left(1+\sum_{j=1}^N \eta(\xi_j)|\xi_j|^m\right)^{-1}\right)\right| \le \frac{c_\alpha}{|\xi|^\alpha}, \qquad \xi \ne 0,$$

hence we can invoke Mikhlin's multiplier theorem (see Theorem 2.3.7) to get

$$\begin{aligned}
\|J_m f\|_p &\le c\left\|\mathscr{F}^{-1}\left(\left(1+\sum_{j=1}^N \eta(\xi_j)|\xi_j|^m\right)\mathscr{F}(f)\right)\right\|_p \\
&\le c\|f\|_p + c\sum_{j=1}^N \left\|\mathscr{F}^{-1}\left(\eta(\xi_j)|\xi_j|^m\xi_j^{-m}\mathscr{F}(D_j^m f)\right)\right\|_p.
\end{aligned}$$

Recalling Mikhlin's theorem once more it is not difficult to show that the function $\xi \mapsto \eta_j(\xi_j)|\xi_j|^m\xi_j^{-m}$ is a Fourier multiplier in any L_p, $1 < p < \infty$. Hence

$$\|J_m f\|_p \le c\|f\|_p + c\sum_{j=1}^N \|D_j^m f\|_p.$$

The proof is complete. □

2.5.7 Remark. Mikhlin's multiplier theorem provides a powerful tool for handling the problem of intermediate derivatives and equivalent norms even in "classical" Sobolev spaces. The preceding theorem states the equivalence of the H_p^m-norm with the sum of $\|f\|_p$ and the "pure" derivatives of the highest order. One can, however, consider also other equivalent norms on H_p^m. Essentially the question can be reduced to an estimate of the L_p-norm of a mixed derivative $D^{(\alpha_1,\dots,\alpha_N)f}$ by a multiple of the sum of the L_p-norms of $D_j^{\alpha_j}f$, and then to an estimate of the latter norms by the L_p-norms of the corresponding "pure" highest order derivatives and of the function itself. To avoid huge and clumsy formulae let us consider—without loss of generality—the case $N = 2$. If $D_1^2 f, D_2^2 f \in L_p$, then

$$\begin{aligned}\left\| \frac{\partial^2 f}{\partial x_1 \partial x_2} \right\|_p &= \left\| \mathscr{F}^{-1}\left(\frac{\xi_1\xi_2}{\xi_1^2+\xi_2^2}(\xi_1^2+\xi_2^2)\mathscr{F}(f) \right) \right\|_p \\ &\le \left\| \mathscr{F}^{-1}\left(\frac{\xi_1\xi_2}{\xi_1^2+\xi_2^2} \right) * (D_1^2 f + D_2^2 f) \right\|_p .\end{aligned} \tag{2.5.13}$$

It is easy to verify that the function $\xi \mapsto \xi_1\xi_2/(\xi_1^2+\xi_2^2)$ satisfies the assumptions of Mikhlin's multiplier theorem. Hence the right hand side of (2.5.13) is estimated by a multiple of $\|D_1^2 f\|_p + \|D_2^2 f\|_p$. As to the estimate of the "pure" derivatives, fix an integer $0 < k < m$ and consider, for instance, differentiation with respect to the first variable. We have

$$\|D_1^1 f\|_p = \|\mathscr{F}^{-1}(\xi_1(1+\xi_1^2)^{-1}) * (f + D_1^1 f)\|_p.$$

By an analogous multiplier argument as before we arrive at the estimate $\|D_1^k f\|_p \le c(\|f\|_p + \|D_1^m f\|_p)$. The general case consists of combinations of the above steps. Let us point out that estimates for mixed derivatives by the non-mixed ("pure") derivatives do not hold generally for $p = 1$ (see e.g. [148]).

A more general case occurs when one has to cope with a similar problem in anisotropic spaces. The answer is particularly aesthetically pleasing if we consider an arbitrary set of N-dimensional multiindices, say, $\mathcal{M}$ and assume that $D^\alpha f \in L_p$ for all $\alpha \in \mathcal{M}$. Then $D^\beta f \in L_p$ for all β, where β are the lattice points with integer coordinates that belong to the convex hull of $\mathcal{M}$ (see e.g. [26]).

The estimates for the intermediate derivatives imply multiplicative estimates analogous to that from Subsection 2.7.2. Indeed, it suffices to replace the function $x \mapsto f(x)$ by $x \mapsto f(\varepsilon x)$ and to use the simple argument in Subsection 2.7.2.

2.5.8 Remark. Passing to spaces on domains follows the well-known scheme. A potential space on a domain can be defined in a natural way as a factorspace modulo equality a.e. on the domain in question. If this domain has the extension property with respect to spaces considered we arrive even to the desired "classical" Sobolev (or Sobolev-Slobodetskii in the non-integer case) spaces. A standard sufficient condition for W_p^k is for instance the boundary of the class $\mathcal{C}^{k-1,1}$, that is, when the domain locally lies on one side of an $n-1$-dimensional manifold, which can be described by local maps with Lipschitz continuous derivatives of the $(k-1)$th order plus either some "uniformity" of corresponding Lipschitz constants or assumption about existence of finitely many local maps covering the boundary. Extension theorems for Sobolev spaces can be found in many standard monographs, recall [143], [26], [1], [148]. An attention to various smoothness properties of a boundary is paid e.g. in [1]. Recently, some authors (see [49], [50]) considered rather sophisticated potential type operators, working directly on smooth domains. We give some more information in Section 2.8.

2.5.9 Remark. It turns out that the quality of the target spaces for imbeddings of Sobolev spaces on domains heavily depends on the geometric properties of the domain in question. Sobolev [140] proved his celebrated theorem for domains with the cone property. There is a very large literature on imbedding theorems of Sobolev type in domains of various type. Let us recall at least Maz'ya [109] with classes J_α and $I_{p,\alpha}$ (capacity type assumptions), Reshetnyak [122], [60], dealing with the John domains, and Besov in [26], where the so called domains with the horn property are considered (for further generalizations see e.g. the second (Russian) edition of [26]). In particular, the situation becomes still more complicated if one considers anisotropic spaces as one can see in [26], Chapter 5. We return to extensions in more details in Section 2.7.

2.5.1 Interpolation in potential spaces

Our aim in the following will be interpolation theorems for the potential Sobolev spaces. Essential for us are theorems providing interpolation estimates both for derivatives and integrability exponents. As pointed out above this can be relatively easy if the spaces in question are Hilbert spaces since then one can invoke the Plancherel theorem.

Let $\{\mathscr{F}(\varphi_k)\}_{k\geq 0}$ be a standard dyadic decomposition of unity in $\mathbb{R}^N$. The construction was described in Subsection 2.4.2. Let us define the mappings $\mathcal{J}$ and $\mathcal{P}$ on $\mathscr{S}'$ by

$$(\mathcal{J}f)_0 = \varphi_0 * f,$$
$$(\mathcal{J}f)_k = \varphi_k * f, \qquad k = 1, 2, \dots,$$

and

$$\mathcal{P}a = \sum_{j=0}^{+\infty} \widetilde{\varphi}_j * a_j,$$

where $a = \{a_j\}_{j\geq 0} \subset \mathscr{S}$, $\widetilde{\varphi}_0(\xi) = \varphi_0(\xi) + \varphi_1(\xi)$, and $\widetilde{\varphi}_j(\xi) = \varphi_j(2\xi) + \varphi_j(\xi) + \varphi_j(\xi/2)$, $j = 1, 2, \ldots$, for sequences $\{a_j\}_{j\geq 0}$ such that the series for $\mathcal{P}a$ is convergent in $\mathscr{S}'$. Plainly $\mathcal{P}\mathcal{J}$ is identity on $\mathscr{S}'$.

Recall the notation ℓ_2^s from p. 92 and $L_p(\ell_2^s)$ from Remark 2.4.21. We have

2.5.10 Theorem. *Let $1 < p < \infty$ and $s \in \mathbb{R}^1$. Then*

$$\begin{aligned} \mathcal{J} :& H_p^s(\mathbb{R}^N) \to L_p(\ell_2^s), \\ \mathcal{P} :& L_p(\ell_2^s) \to H_p^s(\mathbb{R}^N), \end{aligned}$$

and the superposition $\mathcal{P}\mathcal{J}$ is identity on $H_p^s(\mathbb{R}^N)$.

Proof. Let $\rho \in \mathscr{S}'(\mathbb{R}^N, \ell_2^s)$ be given by

$$(\rho(\xi))_k = (1 + |\xi|^2)^{-s/2} \mathscr{F}\varphi_k(\xi), \qquad j = 0, 1, 2, \ldots.$$

Then

$$|\xi|^\alpha \|D^\alpha \rho(\xi)\|_{L(\mathbb{R}^N, \ell_2^s)} \leq |\xi|^\alpha \left(\sum_{k=0}^{+\infty} (2^{ks} |D^\alpha (\rho(\xi))_k|)^2 \right)^{1/2} \leq c^\alpha,$$

where here and in the sequel the symbol $L(X, Y)$ stands for the space of bounded linear operators from X to Y. By Mikhlin's multiplier theorem we get $\mathcal{J} : H_p^s(\mathbb{R}^N) \to L_p(\ell_2^s)$.

Next we show that $\mathcal{P} : L_p(\ell_2^s) \to H_p^s(\mathbb{R}^N)$. In view of the isometry of $H_p^s(\mathbb{R}^N)$ and $L_p(\mathbb{R}^N)$ this is equivalent to $J_s P : L_p(\ell_2^s) \to L_p(\mathbb{R}^N)$. We can write $J_s P(\alpha) = \mathscr{F}^{-1} h * \alpha^{(s)}$, where $\alpha = \{\alpha_j\}_{j\geq 0}$ and $\alpha^{(s)} = \{2^{js}\beta_k\}_{k\geq 0}$, and

$$h(\xi)\beta = \sum_{k=0}^{+\infty} 2^{-js} (1 + |\xi|^2)^{s/2} \widetilde{\varphi}_k(\xi) a_k.$$

We conclude that h an element of $L(\mathscr{S}(\mathbb{R}^N, \ell_2), \mathbb{C})$ and, moreover,

$$|\xi|^\alpha \|D^\alpha h(\xi)\|_{L(\mathscr{S}(\mathbb{R}^N, \ell_2), \mathbb{C})} \leq |\xi|^\alpha \left(\sum_{k=0}^{+\infty} |D^\alpha (1 + |\xi|^2)^{s/2} |\widetilde{\varphi}_k(\xi)_k(\xi))^2 \right)^{1/2},$$

which is bounded. Now it suffices to invoke Mikhlin's multiplier theorem. □

2.5.11 Remark. A theorem on the interpolation estimate for the Hilbert case is rather easy. We leave the proof to the reader.

2.5.12 Corollary. *Let* $0 < \theta < 1$, $1 < p_1, p_2 < \infty$ *and* $s_1, s_2 \in \mathbb{R}^1$. *Then* $[H^{s_1}_{p_1}, H^{s_2}_{p_2}]_\theta = H^s_p$, *where* $s = (1-\theta)s_1 + \theta s_2$ *and* $1/p = (1-\theta)p_1 + \theta p_2$.

The *proof* follows at once from Theorem 2.3.5, Proposition 2.3.12, and Theorem 2.5.10.

In an analogous manner we can prove a vector-valued version of the last theorem (see Theorem 2.6.10 on p. 136).

Next two theorems give information about ordering of potential spaces with respect to the Besov spaces and interpolation links between Besov and Sobolev potential spaces. A general reference is [23] and [148].

Observe that the result of real interpolation between the latter spaces falls into the same class only for a special choice of the parameters involved.

2.5.13 Theorem. *The following imbeddings hold:*

(i) $B^s_{p\,1}(\mathbb{R}^N) \hookrightarrow H^s_p(\mathbb{R}^N) \hookrightarrow B^s_{p\,\infty}(\mathbb{R}^N)$ *for all* $s \in \mathbb{R}^1$ *and all* $1 \le p \le \infty$

(ii) $B^s_{p\,p}(\mathbb{R}^N) \hookrightarrow H^s_p(\mathbb{R}^N) \hookrightarrow B^s_{p\,2}(\mathbb{R}^N)$ *for all* $s \in \mathbb{R}^1$ *and all* $p \in (1, 2]$

(iii) $B^s_{p\,2}(\mathbb{R}^N) \hookrightarrow H^s_p(\mathbb{R}^N) \hookrightarrow B^s_{p\,p}(\mathbb{R}^N)$ *for all* $s \in \mathbb{R}^1$ *and all* $p \in [2, \infty]$.

2.5.14 Theorem. *Let* $0 < \theta < 1$. *Then*

(i) $(H^{s_1}_p(\mathbb{R}^N), H^{s_2}_p(\mathbb{R}^N))_{\theta,q} = B^s_{p\,q}(\mathbb{R}^N)$ *for all* $s_1, s_2 \in \mathbb{R}^1$, $s_1 \neq s_2$, $1 \le p, q \le \infty$, *where* $s = (1-\theta)s_1 + \theta s_2$

(ii) $(H^s_{p_1}(\mathbb{R}^N), H^s_{p_2}(\mathbb{R}^N))_{\theta,p} = H^s_p(\mathbb{R}^N)$ *for all* $s \in \mathbb{R}^1$, $1 \le p_1, p_2 \le \infty$, *where* $1/p = (1-\theta)/p_1 + \theta/p_2$.

2.5.15 Remark. Using standard localization techniques we get the interpolation theorem for potential spaces of the type $H^s_p(I, H^\sigma_q)$. Formal proofs would require additional text and we feel that this is not necessary.

2.5.16 Remark. Later we shall need interpolation estimates in the scale consisting of Sobolev-Slobodetskii spaces $B^s_{p\,p}(\mathbb{R}^N)$ for s non-integer and of the Sobolev spaces $W^s_p(\mathbb{R}^N)$ for integer s. Since for an integer we have $B^s_{p\,p}(\mathbb{R}^N) = W^s_p(\mathbb{R}^N)$ if and only if $p = 2$ the above interpolation theorem does not provide desired estimates for this scale. A remedy is easy: One can use the reiteration in order to interpolate between the Sobolev spaces of integer order and Sobolev-Slobodetskii spaces, representing both of them as a result of interpolation of Sobolev spaces of integer order. The procedure is straightforward, using Theorem 2.3.13, and we give a proof for reader's convenience in a special case needed later.

2.5.17 Theorem. *Let* $0 < s < 1$, $1 < p < \infty$. *Then*

$$(W_p^s(\mathbb{R}^N), W_p^1(\mathbb{R}^N))_{\eta,p} = B_{p\,p}^{s_1}(\mathbb{R}^N)$$

for $s_1 = (1-\eta)s + \eta$.

Proof. According to Theorem 2.5.6 we have $W_p^k(\mathbb{R}^N) = H_p^k(\mathbb{R}^N)$ for integer k. Further, in view of Theorem 2.5.14, $B_{p\,p}^s(\mathbb{R}^N) = (L_p(\mathbb{R}^N), H_p^2(\mathbb{R}^N))_{s/2,p}$ and $W_p^1(\mathbb{R}^N) = H_p^1(\mathbb{R}^N) = (L_p(\mathbb{R}^N), H_p^2(\mathbb{R}^N))_{1/2,p}$. Hence invoking the theorem on stability of the real interpolation method (Theorem 2.3.13), we get

$$\begin{aligned}(W_p^s(\mathbb{R}^N), &W_p^1(\mathbb{R}^N))_{\eta,p} \\ &= ((L_p(\mathbb{R}^N), H_p^2(\mathbb{R}^N))_{s/2,p}, (L_p(\mathbb{R}^N), H_p^2(\mathbb{R}^N))_{1/2,p})_{\eta,p} \\ &= (H_p^0(\mathbb{R}^N), H_p^2(\mathbb{R}^N))_{\theta,p}\end{aligned}$$

provided $\theta = (1-\eta)s/2 + \eta/2$ and $s_1 = (1-\theta).0 + 2\theta = (1-\eta)s + \eta$. □

2.6 Vector-valued Sobolev and Besov spaces

Sobolev spaces

We are now in a position to define more complicated Sobolev and Besov type spaces. We shall use the notation $\overline{p} = (p_1, p_2, \ldots, p_N)$ for a multiindex with N components; if $\overline{r} = (r_1, \ldots, r_N)$ and $\overline{s} = (s_1, \ldots, s_N)$ are such multiindices, then $\overline{r} \le \overline{s}$ if $r_i \le s_i$ for all $1 \le i \le N$, and $\overline{r}/\overline{s}$ is the multiindex with components r_i/s_i, $1 \le i \le N$. Further, if $a \in \mathbb{R}^N$, then $\overline{a}$ is the N-tuple with all coordinates equal to a. By the symbol $L_{\overline{p}} = L_{\overline{p}}(\mathbb{R}^N)$ we denote the *Lebesgue space with mixed norm*, endowed with the norm

$$f \mapsto \| \ldots \| \|f\|_{L_{p_N}} \|_{L_{p_{n-1}}} \ldots \|_{L_{p_1}},$$

where the L_{p_j} norm is taken with respect to x_j, $1 \le j \le N$. If $n = 2$, this is a special case of the vector-valued Lebesgue space $L_{p_1}(\mathbb{R}^N, L_{p_2})$.

2.6.1 Definition. Let $\overline{m} = (m_1, \ldots, m_N)$ be a multiindex of non-negative integers and $\overline{1} \le \overline{p} < \overline{\infty}$. Then the *classical anisotropic Sobolev space with dominating mixed smoothness* $S_{\overline{p}}^{\overline{m}}W = S_{\overline{p}}^{\overline{m}}W(\mathbb{R}^N)$ is defined as the space of all $f \in L_{\overline{p}}$ with the finite norm

$$f \mapsto \|f\|_{S_{\overline{p}}^{\overline{m}}W} = \sum_{\substack{0 \le \alpha_j \le m_j \\ 1 \le j \le N}} \|D^{(\alpha_1,\ldots,\alpha_N)} f\|_{L_{\overline{p}}}.$$

Let $-\overline{\infty} < \overline{r} < \overline{\infty}$ and $\overline{1} < \overline{p} < \overline{\infty}$. Then we define the *potential Sobolev space with dominating mixed smoothness* $S_{\overline{p}}^{\overline{r}}H = S_{\overline{p}}^{\overline{r}}H(\mathbb{R}^N)$ as the space of all $f \in \mathscr{S}'(\mathbb{R}^N)$ with the finite norm

$$f \mapsto \|f\|_{S_{\overline{p}}^{\overline{r}}H} = \left\| \mathscr{F}^{-1} \left(\prod_{j=1}^{N} (1 + |\xi_j|^2) \right)^{r_j/2} \mathscr{F} f \right\|_{L_{\overline{p}}}.$$

Let $\overline{m}$ be a multiindex of non-negative integers and $\overline{1} \leq \overline{p} < \overline{\infty}$. Then we define the (*classical*) *anisotropic Sobolev space* (*with non-mixed derivatives*) $W_{\overline{p}}^{\overline{m}} = W_{\overline{p}}^{\overline{m}}(\mathbb{R}^N)$ as the space of all $f \in L_{\overline{p}}(\mathbb{R}^N)$ with the finite norm

$$f \mapsto \|f\|_{L_{\overline{p}}} + \sum_{\substack{\alpha_j \leq m_j \\ j=1,\dots,N}} \|D^{(0,\dots,\alpha_j,\dots,0)} f\|_{L_{\overline{p}}}.$$

First two kinds of spaces are of primary interest for us. Let us point out that the above definitions yield the same space provided $1 < \overline{p} < \overline{\infty}$. We shall omit the proof—a corresponding multiplier theorem for mixed L_p spaces can be used similarly as in the isotropic case (Theorem 2.5.6); see [135] for details. For convenience, we shall state it separately. Observe that an analogous theorem can be proved for the classical anisotropic Sobolev spaces.

2.6.2 Theorem. *Let* $1 < \overline{p} < \overline{\infty}$ *and let* $\overline{m}$ *be an* N*-tuple of non-negative integers. Then* $S_{\overline{p}}^{\overline{m}}W = S_{\overline{p}}^{\overline{m}}H$.

2.6.3 Remark. These definitions have their vector-valued counterpart. As was already observed relevant properties of scalar-valued spaces carry over to the vector case. Nevertheless, the analog is not complete. The first problem is a restriction as to duality. It is not generally true that $(L_p(\mathbb{R}^N, X))' = L_{p'}(\mathbb{R}^N, X')$ for every Banach space X and $1 < p < \infty$. The necessary and sufficient condition for it is the Radon-Nikodým property of X. (See [39] for details.) Second, there is a question about coincidence of Sobolev potential spaces (defined via Bessel potentials) with the "classical" Sobolev spaces with distributional derivatives, and the density of smooth functions in them. This leads to the so called UMD property of the space X (the property of unconditional martingale difference sequences). We refer to [6], [7], [8], [133], [134] for a discussion of these topics and further references.

Sobolev spaces with dominating mixed smoothness of integer order have been investigated by Nikol'skii, Amanov, Grisvard, Sparr, and further authors (see references in [133]). General spaces of this type in $\mathbb{R}^N$ have been introduced and studied in the framework of the Fourier analysis approach by Schmeisser; see e.g. the monograph by Schmeisser and Triebel [135].

Let us make several important observations.

If $\overline{p} = (p, \ldots, p)$, and the smoothness parameters are integers (or the same integer), then Fubini's theorem gives a relatively simple expression for the norm. If $\overline{p} = \overline{2}$, then all the spaces are Hilbert spaces and a suitable tool for handling them is just the properties of the Fourier transform (transformation of derivatives plus the Plancherel formula).

Spaces on domains in $\mathbb{R}^N$ can be defined in a standard way as factorspaces modulo equality a.e. in this domain. Of course, this definition has a good sense only if the domain in question has the extension property with respect to spaces considered. We do not know if there are extension theorems supporting the full generality of the above definitions. The most important case in the following is spaces of type $H^{\alpha}(I, H^{\beta}(\Omega))$, where I is a one-dimensional interval and Ω is a domain in $\mathbb{R}^N$ with a "nice" boundary. This case is sufficiently dealt with in the existing literature (see e.g. [102]).

Let us have a look at some more special cases and let us illustrate the difference between various concepts of the anisotropy here. Let $N = 2$, $p_1 = p_2$, $\alpha_1 = \alpha_2 = 1$. Then the corresponding anisotropic space consists of functions with

$$\left(\int_{\mathbb{R}^2} \left| \frac{\partial f(x)}{\partial x_1} \right|^p + \left| \frac{\partial f(x)}{\partial x_2} \right|^p \right)^{1/p} dx < +\infty.$$

Only the "pure" derivatives appear here and generally it is not true that $\partial^2 f(x)/\partial x_1 \partial x_2 \in L_p$. Still, one can get information about the L_p-integrability of derivatives (cf. e.g. [26], Chapter 3): It is possible to show it for $D^{(\gamma_1, \gamma_2)} f$, where $\gamma_1 + \gamma_2 \leq 1$. Generally, if $f \in W_p^{\overline{m}}$, then $D^{(\gamma_1, \ldots, \gamma_N)} f$ belongs to L_p, where $\gamma_1/m_1 + \cdots + \gamma_N/m_N \leq 1$.

On the other hand, the norm in $S_p^{(1,1)} W$ is given by

$$\left(\int_{\mathbb{R}^2} \left| \frac{\partial f(x)}{\partial x_1} \right|^p + \left| \frac{\partial^2 f(x)}{\partial x_1 \partial x_2} \right|^p + \left| \frac{\partial f(x)}{\partial x_2} \right|^p \right)^{1/p} dx.$$

This gives a justification to the terminology; clearly, the mixed derivatives dominate—they are of the highest order.

Besov spaces

We come to the vector-valued analog of scalar-valued Besov spaces. At the beginning we shall stick to real analysis terms. For $f \in L_p(X)$, where X is a Banach space, let us define the j-th difference of the m-th order, where

m is a positive integer and $j = 1, \ldots, N$, by

$$\Delta^m_{t,j} f(x) = \Delta^m_{te_j} f(x),$$

where e_j is the unit vector in the direction of the j-th coordinate axis. Further, let $s \in \mathbb{R}^1$ and put $s = [s]^- + \{s\}$, where $[s]^-$ is an integer and $0 < \{s\} \leq 1$.

For the moment it will be convenient to introduce "pure anisotropic" Sobolev spaces. In Section 2.3 we recalled the concept of an equibounded strongly continuous semigroup. Let X be a Banach space. For $m = 0, 1, \ldots,$ and $1 \leq p \leq \infty$ denote by $\mathcal{W}^m_p(X)$ the Sobolev space consisting of all functions f whose "pure" distributional derivatives $D^k_j f$ belong to $L_p(X)$ for all $k = 0, 1, \ldots$ and $j = 1, \ldots, N$. Consider the operators defined in (2.3.10). Let Λ_j, $j = 1, \ldots, N$ be the infinitesimal generators of $\{G_j(t)\}_{t \geq 0}$ and consider the space $\bigcap_{1 \leq j \leq N} D(\Lambda_j)$ and K^m, $m = 0, 1, 2 \ldots,$ from Section 2.3. Then $K^m = W^m_p(X)$ and $\bigcap_{1 \leq j \leq N} D(\Lambda_j) = \mathcal{W}^m_p(X)$. (See [133] for the proof.) In particular, $\Lambda_1^{\alpha_1} \cdots \Lambda_N^{\alpha_N} f = D^\alpha f$ for all $f \in K^m$ and $|\alpha| \leq m$, and $\Lambda^m_j f = D^m_j f$ for all $f \in \bigcap_{1 \leq j \leq N} D(\Lambda_j)$.

2.6.4 Definition. Let X be a Banach space, $1 \leq p, q \leq \infty$ and $0 < s < \infty$. Then we define the Besov space $B^s_{p\,q}(X) = B^s_{p\,q}(\mathbb{R}^N, X)$ as the space of all $f \in \mathcal{W}^{[s]^-}_p$ with the finite norm

$$\|f\|_{B^s_{p\,q}(X)} = \|f\|_{L_p(X)} + \sum_{j=1}^N \left(\int_0^{+\infty} t^{-\{s\}q} \|\Delta^2_{t,j} D_j^{[s]^-} f\|^q_{L_p(X)} \frac{dt}{t} \right)^{1/q}.$$

Interpolation relations between Sobolev and Besov spaces in the vector case are analogous to those known for scalar spaces.

2.6.5 Theorem. *Let X be a Banach space, $1 \leq p < \infty$, $1 \leq q \leq \infty$, and $0 < s < \infty$. If $m > s$ is an integer, then*

$$B^s_{p\,q}(X) = (L_p(X), W^m_p(X))_{s/m,q} = (L_p(X), \mathcal{W}^m_p(X))_{s/m,q}.$$

If s_i, $i = 1, 2$, are non-negative integers, $0 < \theta < 1$, and $s = (1-\theta)s_1 + \theta s_2$, then

$$B^s_{p\,q}(X) = (W^{s_1}_p(X), W^{s_2}_p(X))_{\theta,q} = (W^{s_1}_p(X), \mathcal{W}^{s_2}_p(X))_{\theta,q}.$$

For the proof see [133].

To give a general definition of a vector-valued Besov space $B^s_{p\,q}(X) = B^s_{p\,q}(\mathbb{R}^N, X)$ and that of a Lizorkin-Triebel vector-valued space $F^s_{p\,q}(X) =$

$F^s_{p\,q}(\mathbb{R}^N, X)$ in the framework of the Fourier analysis approach one needs first to look at the vector-valued counterpart of the Fourier transform and its inverse. The formulae are the same; one also gets the well-known relations for Fourier images of derivatives and convolutions. Similarly as in the scalar case the Fourier transform and its inverse are bounded one-to-one mappings from $\mathscr{S}'(\mathbb{R}^N, X)$ into itself. We shall denote the vector Fourier transform and its inverse by the same symbol $\mathscr{F}$ and $\mathscr{F}^{-1}$, respectively.

Nevertheless, not everything carries over to the vector-valued situation; particularly Plancherel's formula does not hold in general. (See e.g. [118].)

The definition of vector-valued Besov and Lizorkin-Triebel spaces is formally the same as in Definition 2.4.20 and we shall omit it. Note that the basic tool to handle these spaces include Mikhlin multiplier theorem for vector-valued functions, vector-valued variants of further multiplier criterions. Basic properties of $B^s_{p\,q}(X)$ are summarized in the following theorems (see Schmeisser [133]).

2.6.6 Theorem. *Let X be a Banach space. For $1 \le p, q \le \infty$ and $s \in \mathbb{R}^1$ the spaces $B^s_{p\,q}(X)$ are Banach spaces continuously imbedded into $\mathscr{S}'(\mathbb{R}^N, X)$.*

2.6.7 Theorem. *Let $s, s_1, s_2 \in \mathbb{R}^1$, $1 \le p, p_1, p_2 < \infty$, $1 \le q, q_1, q_2 \le \infty$. Then*

$$\begin{aligned}
&B^s_{p\,q_1}(X) \hookrightarrow B^s_{p\,q_2}(X), && \text{if } q_1 \le q_2,\\
&B^{s+\varepsilon}_{p\,q_1}(X) \hookrightarrow B^s_{p\,q_2}(X) && \text{for any } \varepsilon > 0,\\
&B^{s_1}_{p_1\,q}(X) \hookrightarrow B^{s_2}_{p_2\,q}(X) && \text{whenever } s_1 - \frac{N}{p_1} = s_2 - \frac{N}{p_2},\\
&(B^{s_1}_{p\,q_1}(X), B^{s_2}_{p\,q_2}(X))_{\theta,q} = B^s_{p\,q}(X) && \\
& && \text{where } s = (1-\theta)s_1 + \theta s_2 \text{ for any } \theta \in (0,1).
\end{aligned}$$

For the proof see [133].

Particularly important is the duality theorem. We state it as a separate statement.

2.6.8 Theorem. *Let X be a reflexive Banach space, $s > 0$, $1 < p < \infty$, $1 \le q < \infty$. Then*

$$(B^s_{p\,q}(X))' = B^{-s}_{p'\,q'}(X'),$$

where p' and q' are conjugate exponents to p and q, respectively.

The proof goes along the lines of the scalar case.

In an analogous manner, in accordance with the corresponding definition of the Bessel potential spaces, one can define their vector-valued variants $H_p^\alpha(X) = H_p^\alpha(\mathbb{R}^N, X)$; absolute values are replaced by norms and one has to work with vector valued tempered distributions and the Fourier transform for them.

Iterated Sobolev and Besov spaces

Let $\overline{p} = (p_1, p_2)$ and $N = N_1 + N_2$ in the following, that is, we split $x_1, \ldots, x_N$ into two groups, $x_1, \ldots, x_{N_1}$ and $x_{N_1+1}, \ldots, x_{N_1+N_2}$ and consider the mixed norms with the power p_1 and p_2 with respect to the first and the second group of variables, respectively. Then the mixed norm Lebesgue space $L_{\overline{p}} = L_{\overline{p}}(\mathbb{R}^N)$ can be simply identified with the "iterated" Lebesgue space $L_{p_1}(L_{p_2}) = L_{p_1}(\mathbb{R}^{N_1}, L_{p_2}(\mathbb{R}^{N_2}))$, that is, with the space of L_{p_1}-integrable functions defined on $\mathbb{R}^{N_1}$ whose values are elements of $L_{p_2} = L_{p_2}(\mathbb{R}^{N_2})$. The same holds for the iterated Sobolev and Besov spaces (cf. [133]). We shall state this property as a separate statement after we introduce an additional notation. These spaces are usually called Sobolev (Besov, etc.) spaces with dominating mixed smoothness. Sobolev spaces of this type have been introduced by Nikol'skii (cf. [103], [62], [142]). Nevertheless, this scale naturally falls into a more general scale of anisotropic Sobolev (Besov, etc.) spaces with dominating mixed smoothness and they allow the powerful and unified Fourier analysis approach. This has been done by Schmeisser in a series of papers, see e.g. in [131], [132], [133], and spaces of this type are subject of the monograph [135]. Note that there are considered spaces, where all the multiindices have two components; this is, however, only a formal restriction to simplify the notation.

2.6.9 Theorem. *Let $\overline{p} = (p_1, p_2)$, $1 \leq p_1, p_2 \leq \infty$, $\mathbb{R}^N = \mathbb{R}^{N_1} \times \mathbb{R}^{N_2}$. Let $\overline{k} = (k_1, k_2)$, where k_1 and k_2 are non-negative integers. Then*

$$L_{\overline{p}}(\mathbb{R}^{N_1} \times \mathbb{R}^{N_2}) = L_{p_1}(\mathbb{R}^{N_1}, L_{p_2}(\mathbb{R}^{N_2}).$$

Similarly,

$$SW_{\overline{p}}^{\overline{k}}(\mathbb{R}^{N_1} \times \mathbb{R}^{N_2}) = W_{p_1}^{k_1}(\mathbb{R}^{N_1}, W_{p_2}^{k_2}(\mathbb{R}^{N_2})).$$

The proof is not very difficult and we omit it. Hence $W_{\overline{p}}^{\overline{k}}(\mathbb{R}^{N_1} \times \mathbb{R}^{N_2})$ (see Definition 2.6.1) can be alternatively defined in terms of vector-valued Sobolev spaces $W_{p_1}^{k_1}(\mathbb{R}^{N_1}, X)$, where $X = W_{p_2}^{k_2}(\mathbb{R}^{N_2})$. (See [133] for details and further references.)

A consequence of Theorem 2.6.7 is

2.6.10 Theorem. *Let* $1 < p_1, p_2 < \infty$, $s, s_1, s_2 \in \mathbb{R}^1$, $0 < \theta < 1$, $r, r_1, r_2 \in \mathbb{R}^1$, $0 < \theta < 1$, $s = (1-\theta)s_1 + \theta s_2$, $r = (1-\theta)r_1 + \theta r_2$, $1/p = (1-\theta)/p_1 + \theta/p_2$. *Then*

$$[W^{s_1}_{p_1}(\mathbb{R}^1, W^{r_1}_{p_1}(\mathbb{R}^N)), W^{s_2}_{p_2}(\mathbb{R}^1, W^{r_2}_{p_2}(\mathbb{R}^N))]_\theta = W^s_p(\mathbb{R}^1, W^r_p(\mathbb{R}^N)).$$

2.6.11 Remark. There are a number of other types of anisotropic spaces considered in the literature. In classical terms, the general situation can be as follows: Consider a set M of multiindices and an associated set of exponents P, consisting of (possibly different) vectors $\overline{p^\alpha} = (p^\alpha_1, \ldots, p^\alpha_N)$, where $1 \leq p^\alpha_i \leq \infty$ for $\alpha \in M$ and $i = 1, \ldots, N$. Then one can define the anisotropic space $W^M_P(\mathbb{R}^N)$ as the space of all functions f with the finite norm

$$\|f\|_{W^M_P} = \sum_{\alpha \in M} \|D^\alpha\|_{L_{\overline{p^\alpha}}}.$$

2.6.1 Some special types of anisotropic spaces

Special cases, investigated by various authors, include the Sobolev spaces, where

$$M = \{\alpha;\ |\alpha| \leq m\} \quad \text{and } p^\alpha = \overline{p} = (p_1, \ldots, p_N) \text{ for all } \alpha \in M,$$

and one considers the condition $D^\alpha \in L_{\overline{p}}$. See [26] and references there for a detailed account.

If there is a scalar p^α for each $|\alpha| \leq m$, that is, every derivative has its own (non-mixed norm) L_p space, we get another type of anisotropic space, investigated by Troisi [151], [152] in the 1960s and used later in a number of papers on PDEs regularity (see e.g. [21] and references therein).

Particular attention has been paid to the so called reduced Sobolev spaces studied by Adams [2] and, more generally, by Amanov [4] and others. The basic case is when for $k \leq N$ one considers only those multiindices $|\alpha| \leq k$, for which $\alpha_i \leq 1$. A very surprising property of these spaces is no loss of imbedding properties; if $kp < N$, then they are imbedded into L_{p*}, where $p* = Np/(N-p)$ is the Sobolev conjugate exponent, exactly as under the "classical" assumption $D^\alpha f \in L_p$ for all $|\alpha| \leq k$, $kp < N$. The fine difference concerns behaviour of the imbedding norm when the product pk is close to the critical value N. Some of these spaces fall into the scale of potential Sobolev spaces with dominating smoothness (see [131], [132] for the theory of these spaces, and [97]).

We shall now briefly describe some special cases of iterated Besov spaces, that is, we will consider vector-valued Besov spaces, where the role of the space X is played by another Besov space. For the sake of simple notation let us consider the case $N = 2$. It turns out that we need the *mixed differences*. Let f be a function in $\mathbb{R}^2$ and $h_i \in \mathbb{R}^1$, $i = 1, 2$. Then we put

$$\Delta_{h_1,1} f(x) = \Delta_{h_1}^1 f(x) = f(x_1, x_2) - f(x_1 + h_1, x_2),$$
$$\Delta_{h_2,2} f(x) = \Delta_{h_1}^2 f(x) = f(x_1, x_2) - f(x_1, x_2 + h_2)$$

and

$$\Delta_{h_i,i}^m = \Delta_{h_i,i}(\Delta_{h_i,i}^{m-1}), \qquad i = 1, 2.$$

Now if $\overline{m} = (m_1, m_2)$ is a couple of positive integers and if $\overline{h} = (h_1, h_2) \in \mathbb{R}^1 \times \mathbb{R}^1$, we define the mixed difference of order $\overline{m}$ as

$$\Delta_{\overline{h}}^{\overline{m}} f(x) = \Delta_{h_2,2}^{m_2}(\Delta_{h_1,1}^{m_1} f(x)).$$

2.6.12 Definition. Let $N = 2$, $\overline{1} \leq \overline{p} = (p_1, p_2) \leq \overline{\infty}$, $\overline{1} \leq \overline{q} = (q_1, q_2) \leq \overline{\infty}$, and $\overline{0} < \overline{s} < \overline{\infty}$. Let $m_i > s_i$ be integers, $i = 1, 2$. Then we define the *Besov space* $SB_{\overline{p}\,\overline{q}}^{\overline{s}}$ *with dominating mixed smoothness* as the linear set of all functions f with the finite norm

$$\begin{aligned}
&\|f\|_{SB_{\overline{p}\,\overline{q}}^{\overline{s}}} \\
&= \|f\|_{L_{\overline{p}}} + \left\| \left(\int_0^{+\infty} h_1^{-s_1 q_1} \|\Delta_{h_1,1}^{m_1} f(x_1, x_2)\|_{L_{p_1}}^{q_1} \frac{dh_1}{h_1} \right)^{1/q_1} \right\|_{L_{p_2}} \\
&\quad + \left(\int_0^{+\infty} h_2^{-s_2 q_2} \|\Delta_{h_2,2}^{m_2} f(x_1, x_2)\|_{L_{\overline{p}}}^{q_2} \frac{dh_2}{h_2} \right)^{1/q_2} \\
&\quad + \Bigg(\int_0^{+\infty} h_2^{-s_2 q_2} \\
&\qquad \left\| \left(\int_0^{+\infty} h_1^{-s_1 q_1} \|\Delta_{\overline{h}}^{\overline{m}} f(x_1, x_2)\|_{L_{p_1}}^{q_1} \frac{dh_1}{h_1} \right)^{1/q_1} \right\|_{L_{p_2}}^{q_2} \frac{dh_2}{h_2} \Bigg)^{1/q_2}
\end{aligned}$$

(with an appropriate change for infinite values of p_i and q_i).

The definition is formally rather complicated. Nevertheless, one can prove the following iteration theorem, similar to Theorem 2.6.9. (See again [133], [135] for details.)

2.6.13 Theorem. *Under the notation and assumption on the parameters from the previous theorem we have*

$$SB_{\overline{p}\,\overline{q}}^{\overline{s}} = B_{p_2\,q_2}^{s_2}(\mathbb{R}^1, B_{p_1\,q_1}^{s_1}).$$

2.6.14 Remark. The above spaces fall into the Fourier analytic scheme after the proper concepts of an anisotropic resolution of unity is introduced. Spaces of Lizorkin-Triebel type can be introduced, too, and their coincidence with the Sobolev type spaces can be proved, analogous to what we already know from the scalar isotropic case. We will not pursue this topic here and we refer to [133], [135].

2.6.15 Remark. In the next chapters we shall mainly have to cope with the anisotropy resulting from splitting the variables into two groups, corresponding to the time and space variables, and in addition restricted to bounded domains. Whereas this is not a real problem in one dimension, one has to be careful about the quality of the boundary of the domain for the space variables as we have seen already—specifically, an anisotropy in the space variables leads to strong assumptions on the domain to guarantee e.g. validity of imbedding theorems (the horn property, see [26], Chapter 2). Hence our considerations could be based on more general function spaces but the price for doing this would be rather high. An assumption about isotropic behaviour of the space variables will make our life not easy but manageable.

It is exactly the point, where relevant applications create a challenge for the theory of function spaces, and on the other hand it raises natural questions about the descriptions of real processes by mathematical means.

The spaces which we will consider in the sequel are the special mixed norm Sobolev (Sobolev-Slobodetskii) spaces, which can be denoted—in the language of the iterated spaces—by $W^{s_1}_{p_1}(I, W^{s_2}_{p_2}(\Omega))$, where I is an interval in $\mathbb{R}^1$ and Ω is a domain in $\mathbb{R}^N$. The adaption of the last definition to the situation $\mathbb{R}^1 \times \mathbb{R}^N$ is clear. Extension theorems for spaces on intervals are well known (see e.g. [102]) and if Ω has the extension property with respect to (isotropic) Besov spaces, we can use the full power of the theory in $\mathbb{R}^N$ as developed in [151], [152], [26], [131], [132], [133], [135].

2.7 Extensions and traces

2.7.1 Geometrical properties of boundaries of domains

Let Ω be a domain in $\mathbb{R}^N$. Several times in the previous text we came across geometrical properties of $\partial\Omega$, the boundary of Ω. Since they play a very important role both in the theory and in the applications we shall give several definitions frequently met in the literature. In the following we say that a family $\mathscr{G}$ of open sets in $\mathbb{R}^N$ is *locally finite* if every compact set meets only a finite number of sets from $\mathscr{G}$.

Let $B \subset \mathbb{R}^N$ be a closed ball and let $x \in \mathbb{R}^N \setminus B$. A *bounded cone* with the vertex x is the set $\{z \in \mathbb{R}^N;\ z = (1-\theta)x + \theta y,\ \theta \in [0,1],\ y \in B\}$.

2.7.1 Definition. Let $\Omega \subset \mathbb{R}^N$ be a domain.

(i) The domain Ω is said to have the *segment property* if there exists a finite open covering $\{G_\kappa\}_{\kappa\in K}$ of $\partial\Omega$ and a family $\{y_\kappa\}_{\kappa\in K}$, $y_\kappa \in \mathbb{R}^N \setminus \{0\}$ such that if $x \in \overline{\Omega} \cap G_\kappa$ for some κ, then $x + ty_\kappa \in \Omega$ for all $t \in (0,1)$.

(ii) The domain Ω is said to have the *cone property* if there exists a bounded cone C in $\mathbb{R}^N$ such that $x + C_x \subset \Omega$ for every $x \in \Omega$ and some cone C_x congruent to C.

(iii) The domain Ω is said to have the *uniform cone property* if there exists a locally finite open covering $\{G_\kappa\}_{\kappa\in K}$ of $\partial\Omega$ and a family $\{C_\kappa\}_{\kappa\in K}$ of bounded cones congruent to some fixed cone $C \subset \mathbb{R}^N$ such that the following properties are satisfied:

it is $\sup_{\kappa\in K} \operatorname{diam} G_\kappa < \infty$ and there exists a constant $\varepsilon > 0$ such that $\{x \in \Omega;\ \operatorname{dist}(x, \partial\Omega) < \varepsilon\} \subset \bigcup_\kappa G_\kappa$;

it is $\bigcup_{x\in G_\kappa\cap\Omega}(x + C_\kappa) = H_\kappa \subset \Omega$;

there is a positive integer L such that the intersection of every subfamily of $\{H_\kappa\}_{\kappa\in K}$ consisting of more than L sets is empty.

(iv) The domain Ω is said to have the *strong local Lipschitz property* if there exist $\varepsilon > 0$, a locally finite open covering $\{G_\kappa\}_{\kappa\in K}$ of $\partial\Omega$, and a family of real functions $\{f_\kappa\}_{\kappa\in K}$ defined in some neighbourhood of zero in $\mathbb{R}^{N-1}$, possessing the following properties:

there is a positive integer L such that the intersection of any L sets from $\{G_\kappa\}_{\kappa\in K}$ is empty;

if $x_j \in \Omega$ are such that $\operatorname{dist}(x_j, \partial\Omega) < \varepsilon$, $j = 1, 2$, and $|x_1 - x_2| < \varepsilon$, then there is $\kappa_0 \in K$ such that $\operatorname{dist}(x_j, \partial G_{\kappa_0}) > \varepsilon$;

the family $\{f_\kappa\}_{\kappa\in K}$ is uniformly Lipschitz continuous;

for every $\kappa \in K$ there is a local Cartesian coordinate system, say, $(z_1, \dots, z_N)$ such that $\Omega \cap G_\kappa = \{(z_1, \dots, z_N);\ z_N > f_\kappa(z_1, \dots, z_{N-1})\}$.

(v) Let $\ell = (\ell_1, \dots, \ell_N)$ be a N-tuple of positive real numbers, $h > 0$, $\varepsilon > 0$, and $a_i \neq 0$, $i = 1, \dots, N$. The set

$$V(\ell, h) = \bigcup_{0<v<h} \left\{ x \in \mathbb{R}^N;\ \frac{x_i}{a_i} > 0,\ v < \left(\frac{x_i}{a_i}\right)^{\ell_i} < (1+\varepsilon)v,\ 1 \le i \le N \right\}$$

is called an ℓ-*horn* (with radius h and aperture ε). An open set $\Omega \subset \mathbb{R}^N$ is said to have the *weak horn property* if $\Omega = \bigcup_{k=1}^{M} \Omega_k$, where M is a finite integer, Ω_k are open subsets of $\mathbb{R}^N$, and for every $k = 1, \dots, M$ there exist an ℓ-horn $V_k(\ell^k, h^k)$ such that

$$\Omega = \bigcup_{k=1}^{M} (\Omega_k + V_k(\ell^k, h^k)).$$

(vi) The domain Ω is said to have the *uniform C^m-property* if there exist a locally finite open covering $\{G_k\}_{k\in\mathrm{N}}$ of $\partial\Omega$ and a family of one-to-one transformations $\{T_k\}_{k\in\mathrm{N}}$ of the class C^m, with bounded partial derivatives up to the order m, mapping G_k onto the unit ball in $\mathbb{R}^N$ and possessing the following properties:

there is $\varepsilon > 0$ such that $\{x \in \Omega;\ \mathrm{dist}(x, \partial\Omega) < \varepsilon\} \subset \bigcup_{k\in\mathrm{N}} T_k^{-1}\{y \in \mathbb{R}^N;\ |y| < 2^{-1}\}$;

there is a positive integer L such that every L sets from $\{G_k\}_{k\in\mathrm{N}}$ have empty intersection;

$T_j(G_j \cap \Omega) = \{y \in B;\ y_N > 0\}$, where B is the unit ball in $\mathbb{R}^N$.

2.7.2 Remark. Plainly all the families $\{G_\kappa\}_{\kappa\in K}$ can be countable and if Ω is bounded, then these families can be assumed to consist of finite number of sets. The item (iv) of the above definition turns then into the usual definition of a Lipschitz continuous boundary, namely, when $\partial\Omega$ is assumed to be locally a graph of a Lipschitz continuous function and Ω is (locally) situated on one side of the respective graph.

Sometimes it is relevant to consider the cone property in the sense of (ii) and (iii) for the complement of Ω. Then one usually talks about the *outer cone property* and the *outer uniform cone property*, respectively, and if one has to distinguish, then (ii) and (iii) above get the specific names the *inner cone property* and the *inner uniform cone property*, respectively. The usage varies, however. Closely related is the concept of domains with a cusp (see e.g. [1]).

The concept of the ℓ-horn has been introduced in connection with anisotropic spaces (see [26], Chapter 2). The reader will easily verify that if ℓ has equal components, then the weak horn property coincides with the cone property (hence consequently called by some authors the weak cone property).

It is not difficult to check that for any domain in $\mathbb{R}^N$ the strong local Lipschitz property implies the uniform cone property, which in turn gives the segment property. The cone property is missing here; nevertheless, the following theorem is true for bounded domains (see [57], [1] for the proof):

2.7.3 Theorem (Gagliardo). *Given a bounded domain $\Omega \subset \mathbb{R}^N$ with the cone property and $\varepsilon > 0$, there exist finitely many open subsets $\Omega_1, \dots, \Omega_K$ such that $\Omega = \bigcup_{k=1}^{K} \Omega_k$ and to every $k = 1, \dots, K$ there exist $H_j \subset \overline{\Omega_k}$ with diameter smaller than ε and an open parallepiped P_k with one vertex at the origin such that $\Omega_k = \bigcup_{x \in H_k} (x + P_k)$. Furthermore, if ε is small enough, then every Ω_k, $k = 1, \dots, K$, has the strong local Lipschitz property.*

Let us note that a byproduct of the proof of the last theorem is the fact that every domain with the cone property can be written as a union of translations of finitely many parallelepipeds. If Ω is bounded, then interiors of fixed dilatations of these translated parallelepipeds (with respect to some fixed interior points) yield an open covering of $\overline{\Omega}$ and one can choose a countable subcovering. Using the corresponding decomposition of unity subordinated to the latter countable covering, one can pass to Lipschitz domains.

There are many other concepts useful for description of geometrical properties of domains. We refer to [1], Chapter 4 and to [26], Chapter 2. The reader may draw some illustrative pictures. In particular, it is a bit surprising that even nice domains fail to have the horn property for general values of the parameters of a horn. For instance, it is not difficult to check that a disk in $\mathbb{R}^2$ enjoys the horn property only if $2^{-1}\ell_1 \le \ell_2 \le 2\ell_1$ (see [26] for a number of examples). Since the horn property plays a major role e.g. in imbedding theorems for anisotropic Sobolev spaces this is a rather unpleasant fact, leading to essential restrictions. Let us point out that our anisotropic spaces in applications are assumed to be isotropic in the space variables and we can thus bypass such complications. The reader should have in mind, however, that further consideration, taking into account more general anisotropic spaces, should be carried out very carefully since standard imbedding and trace theorems do not hold in general anisotropic spaces (even if the boundary is infinitely differentiable, for instance).

2.7.2 Estimates of Gagliardo-Nirenberg type

We have seen in Section 2.4 that the scale of Sobolev spaces is not closed with respect to the real interpolation. Hence interpolation estimates from this section are sometimes insufficient for estimates of intermediate derivatives. If $1 < p < \infty$, then the desired estimates in the scale of potential Sobolev spaces follow by complex interpolation. Nevertheless, this approach fails when $p = 1$. To fill the gap we shall prove now an interpolation type estimate for L_p-norms, including $p = 1$. Let us start with $f \in \mathscr{D}(\mathbb{R}^1)$. Let $x \in \mathbb{R}^1$. Then by virtue of the mean value theorem there is $\xi \in (x, x+1)$ such that $f(x+1) - f(x) = f'(x) + \int_x^{x+\xi} f''(\tau)\, d\tau$. Hence $|f'(x)| \le |f(x+1)| + |f(x)| + \int_x^{x+\xi} |f''(\tau)|\, d\tau$. This yields $|f'(x)|^p \le c\left(|f(x+1)|^p + |f(x)|^p + \int_x^{x+1} |f''(\tau)|^p\, d\tau\right)$ for some constant $c > 0$ independent of f and integration with respect to x implies that $\|f'\|_p \le c(\|f\|_p + \|f''\|_p)$. If $1 < \ell < m$, then by iteration argument we get $\|f^{(\ell)}\|_p \le c(\|f\|_p + \|f^{(m)}\|_p)$. If $f \in \mathscr{D}(\mathbb{R}^N)$ and if $0 < |\beta| < m$, then iterating once more, this time with respect to variables $x_1, \ldots, x_n$, gives

$$\|D^\beta f\|_p \le c\Big(\|f\|_p + \sum_{|\alpha|=m} \|D^\alpha f\|_p\Big). \tag{2.7.1}$$

Now $\|D^\gamma f(\varepsilon x)\|_p = \varepsilon^{-N/p+|\gamma|/p}\|D^\gamma f(x)\|_p$ for any $\varepsilon > 0$ and any multiindex γ. Plugging this into (2.7.1) we arrive at

$$\|D^\beta f\|_p \le c\varepsilon^{-|\beta|}\Big(\|f\|_p + \varepsilon^m \sum_{|\alpha|=m} \|D^\alpha f\|_p\Big). \tag{2.7.2}$$

Moreover, if the last term on the right hand side of (2.7.2) is positive, then the particular choice

$$\varepsilon = \|f\|_p^{1/m}\Big(\sum_{|\alpha|=m} \|D^\alpha f\|_p\Big)^{-1/m}$$

finally yields

$$\|D^\beta f\|_p \le c\|f\|_p^{1-|\beta|/m}\Big(\sum_{|\alpha|=m} \|D^\alpha f\|_p\Big)^{|\beta|/m}. \tag{2.7.3}$$

That (2.7.3) holds for all $f \in W_p^m$ follows now by standard density argument. Since similar estimates will play an important role in the following, we shall formulate it as a separate theorem. Note that we hit on these questions in Remark 2.5.7.

Note also that a refined estimate can be proved (see Gagliardo [57]): Given $f \in W_p^m(a,b)$, $m \ge 2$ an integer, $1 \le p < \infty$, $(a,b) = I \subset \mathbb{R}^1$, then for every $0 < t < \operatorname{mes} I$,

$$\sum_{i=1}^{m-1} t^i \|f^{(i)}\|_{L_p(I)} \le t^m \|f^{(m)}\|_{L_p(I)} + c_m \|f\|_{L_p(I)}$$

with c_m independent of f. Of course, from this the above estimates in $\mathbb{R}^N$ follow similarly. Let us observe that the case $p = \infty$ goes along the same lines as $p < \infty$. We have:

2.7.4 Theorem. *Let $f \in C^\infty(\mathbb{R}^N)$, $1 \le p \le \infty$. Let $1 \le i < m$ be integers and denote $D_j^i = \partial^i / \partial x_j$, $j = 1, \ldots, N$. Then*

$$\|D_j^i\|_p \le c_m \|f\|_p^{1-i/m} \|D_j^i f\|_p^{i/m}, \qquad j = 1, \ldots, N, \tag{2.7.4}$$

and

$$\|D^\beta f\|_p \le c_m \|f\|_p^{1-i/m} \Big(\sum_{|\alpha|=k} \|D^\alpha\|_p \Big)^{i/m} . \tag{2.7.5}$$

Furthermore, (2.7.2) *and* (2.7.3) *hold for every $\varepsilon > 0$.*

Note that by density these estimates extend to functions in W_p^k.

Sometimes it is useful to have analogous estimates for mixed derivatives. Observe, however, that in anisotropic spaces there is a substantial restriction $1 < p < \infty$. We state a typical estimate of this sort and refer to [26], Chapter 3, for further generalizations and counterexamples.

2.7.5 Theorem. *Let $m_i \ge 1$, $i = 1, \ldots, N$ be integers and $1 < p < \infty$. Then there is $c = c(m_1, \ldots, m_N)$ such that*

$$\|D^\alpha f\|_p \le c \sum_{i=1}^N \|D_i^{m_i}\|_p \tag{2.7.6}$$

for every $f \in C^\infty(\mathbb{R}^N)$ and all $\alpha = (\alpha_1, \ldots, \alpha_N)$ with $\sum_{i=1}^N \alpha_i / m_i \le 1$.

Again, the estimate (2.7.6) extends by density argument to the appropriate (anisotropic) Sobolev space.

Let us still formulate a corresponding isotropic theorem for domains. We refer e.g. to [1], Chapter IV for the proof.

2.7.6 Theorem. *Let $\Omega \subset \mathbb{R}^N$ be a domain with the uniform cone property. Then the conclusion of Theorem 2.7.4 holds with the respective norms on Ω replacing $\mathbb{R}^N$.*

Furthermore, the same is true for bounded Ω with the cone property.

We refer to [26], Chapter III, for a general theorem of this sort in anisotropic spaces on domains and to [134] for spaces on $\mathbb{R}^N$.

2.7.3 Extension operators

As mentioned already several times earlier (e.g. p. 126), the extension theorem makes it possible to pass from theorems on $\mathbb{R}^N$ to corresponding claims on domains. The extension theorems are often based (apart from a clever idea) on quite elaborate and tedious calculations so that we touch upon only the major points of the theory. Naturally, as in all considerations near the boundary the geometric properties of the boundary are important.

Next, we employ the Besov spaces for a complete description of traces of functions in Sobolev spaces in question. Namely, after appropriate localization and introduction of Besov spaces on the boundary of a sufficiently smooth domain, this scale includes spaces of functions that are traces of functions in Sobolev spaces on the domain in question. The local considerations are well known and well described in the existing literature (cf. e.g. [1], [99], [114]) and we shall omit them.

Our ultimate goal in this part is an exposition of very recent interpolation results for Lipschitz domains, which are consequences of theorems in $\mathbb{R}^N$ and of the corresponding general extension theorem. In the literature one meets often several less general standard extension methods so that we start with them.

2.7.7 Theorem. *Let $\Omega \subset \mathbb{R}^N$ be either a halfspace or a domain in $\mathbb{R}^N$ with a bounded boundary, possessing the uniform C^m-regularity property. Then there exists an operator $E : \bigcup_{p,\ k\leq m} W_p^k(\Omega) \to \bigcup_{p,\ k\leq m} W_p^k(\mathbb{R}^N)$ such that $W_p^m(\Omega) \to W_p^m(\mathbb{R}^N)$ for every $k \in \mathbb{N}$ and every $p \in [1,\infty)$ and there is a constant $A(k,p)$ such that*

$$\|Ef\|_{W_p^k(\mathbb{R}^N)} \leq A(k,p)\|f\|_{W_p^k(\Omega)} \tag{2.7.7}$$

for every $1 \leq k \leq m$.

Note that an operator E with the above properties is sometimes called a *strong k-extension operator*. The proof of Theorem 2.7.7 is based on suitable reflections near the boundary (Hestenes [69] and further authors). Sometimes also the concept of a *simple (k,p)-extension operator* is used, when the construction of E is specific for an individual couple (k,p) and the estimate (2.7.7) need not necessarily hold for other values of the smoothness and integrability parameters.

Seeley [136] has modified the reflection method to get a *total extension operator*, that is, a "universal" extension operator E, whose restriction to any $W_p^k(\Omega)$ is a strong k-extension operator:

2.7.8 Theorem. *Let $\Omega \subset \mathbb{R}^N$ be either a halfspace or a domain in $\mathbb{R}^N$ with a bounded boundary, possessing the uniform C^m-regularity property for ev-*

ery $m \in \mathbb{N}$. *Then, there exists a total extension operator* $E\colon \bigcup_{\substack{k\in\mathbb{N}\\1\le p<\infty}} W_p^k(\Omega) \to \bigcup_{\substack{k\in\mathbb{N}\\1\le p<\infty}} W_p^k(\mathbb{R}^N)$.

We shall not give proofs of two preceding theorems (they can be found e.g. in [1], Chapter IV). The proof of Theorem 2.7.7 for the halfspace is, however, easy, once a suitable explicit formula for the extension is written down: If $u \in C^m(\overline{\mathbb{R}_+^{N+1}})$ (one can suppose this because $C^m(\overline{\mathbb{R}_+^{N+1}})$ is dense in $W_p^m(\mathbb{R}_+^{N+1})$), we put, for $x = (x', x_N) \in \mathbb{R}_+^{N+1}$,

$$Eu(x) = \begin{cases} u(x) & \text{if } x_N > 0, \\ \sum_{j=1}^{m+1} \lambda_j u(x', -jx_N) & \text{if } x_N \le 0, \end{cases}$$

and

$$E_\alpha u(x) = \begin{cases} u(x) & \text{if } x_N > 0, \\ \sum_{j=1}^{m+1} (-1)^{\alpha_N} \lambda_j u(x', -jx_N) & \text{if } x_N \le 0, \end{cases}$$

where λ_j satisfy $\sum_{j=1}^{m+1} (-1)^k \lambda_j = 1$ for all $k = 0, 1, \ldots, m$. One can verify without much effort that $Eu \in C^m(\mathbb{R}^N)$ and that $D^\alpha u(x) = E_\alpha D^\alpha u(x)$ for all $x \in \mathbb{R}^N$ and all α with $|\alpha| \le m$. The claim on the boundedness of E is immediate.

The proof of Theorem 2.7.8 is more technically complicated.

Calderón [31] established existence of a simple (m, p)-extension operator for domains with a modified uniform cone property (especially for bounded domains with a Lipschitz boundary) for $1 < p < \infty$. The proof is based on the Calderón-Zygmund theory of singular integral operators:

2.7.9 Theorem (Calderón). *Let* $1 < p < \infty$, $m \in \mathbb{N}$, *and let* $\Omega \subset \mathbb{R}^N$ *be a domain satisfying the condition of Definition 2.7.1 (iii), where the covering is supposed to be finite and diameters of the sets in the covering need not be uniformly bounded. Then there exists a simple* (m, p)*-extension operator for* Ω.

The restrictions for p in Calderón's theorem can be removed to obtain an extension theorem for Lipschitz domains (roughly speaking) and all p with $1 \le p \le \infty$. The general theorem is due to Stein and can be found e.g. in his monograph [143]. We shall not go into details; the basic case is the epigraph of a Lipschitz function and then one can consider domains with the so called *minimal smoothness property*: it is assumed that there exists a sequence of open sets $\mathscr{G} = \{G_k\}_{1\le k<\infty}$, an integer k_0, $\varepsilon > 0$, and $M > 0$ such that every point in $\mathbb{R}^N$ is contained in at most k_0 sets from $\mathscr{G}$, every point $x \in \partial\Omega$ is contained in some G_i together with the ball centered at x and with the diameter ε, and, finally, it is assumed that $G_i \cap \Omega = G_i \cap \Omega_i$, where Ω_i is an epigraph of a Lipschitz function in $\mathbb{R}^{N-1}$ (after a suitable rotation and translation of the coordinates) with a Lipschitz constant not exceeding M. This of course includes bounded domains with Lipschitz boundary we are mostly concerned with. We have

2.7.10 Theorem (Stein). *Let $1 \le p \le \infty$ and let the domain $\Omega \subset \mathbb{R}^N$ possess the minimal smoothness property. Then there exists a total extension operator for Ω. In particular this claim holds true for bounded domains with a Lipschitz boundary.*

The reader will find further references on extension theorems e.g. in [26].

We also refer to Jones' paper [87] on the so called ε-δ-property. For $p = 2$ and domains in $\mathbb{R}^2$, this is even a necessary and sufficient condition for a domain to possess the extension property. Nevertheless, the general case $N > 2$ is a difficult problem, not solved yet completely in this sense.

Important progress in the theory, namely, an existence of an universal extension operator independent of the smoothness parameter, appeared recently due to Rychkov [125]. We postpone this theorem and its interpolation consequences to Subsection 2.8.2 (see page 153).

2.8 Spaces on domains

2.8.1 Traces

Now we turn our attention to traces. We present a theorem for Besov and Sobolev spaces in the halfspace.

2.8.1 Theorem. *Let $1 < p < \infty$, $1 \le q \le \infty$, and $s > 1/p$. Then the trace operator maps $B^s_{pq}(\mathbb{R}^N_+)$ onto $B^{s-1/p}_{pq}(\mathbb{R}^{N-1})$. If k is a positive integer, then the trace operator maps $W^k_p(\mathbb{R}^N)$ onto $B^{k-1/p}_{pp}(\mathbb{R}^{N-1})$.*

In particular, for non-integer $s > 1/p$, the trace operator maps the Sobolev-Slobodetskii space $W^s_p(\mathbb{R}^N)$ onto $W^{s-1/p}_p(\mathbb{R}^{N-1})$.

Sketch of the proof. The basic step is to show that Tr maps $W^k_p(\mathbb{R}^N) = H^k_p(\mathbb{R}^N)$ into $B^{k-1/p}_{pp}(\mathbb{R}^{N-1})$ when k is a positive integer. The rest will follow by interpolation.

Since Sobolev spaces of positive integer order on $\mathbb{R}^N_+$ can be extended to the corresponding Sobolev spaces on the whole $\mathbb{R}^N$ and their restriction from $\mathbb{R}^N$ to $\mathbb{R}^N_+$ is plainly a continuous operator we can consider spaces on $\mathbb{R}^N_+$.

If we put $X_1 = W_p^k(\mathbb{R}^{N-1})$ and $X_1 = L_p(\mathbb{R}^{N-1})$, then $B_{pp}^{k-1/p}(\mathbb{R}^{N-1}) = (X_1, X_2)_{1/kp,p}$ according to Theorems 2.3.15 and 2.4.25. For $f \in \mathscr{S}(\mathbb{R}^N)$ define $u(t) = f(x,t)$, $x \in \mathbb{R}^{N-1}$, $t > 0$. Then $\lim_{t\to 0} u(t) = \operatorname{Tr} f$ in $L_p(\mathbb{R}^{N-1})$ and

$$\begin{aligned} &\|\operatorname{Tr} f\|_{(X_1,X_0)_{1/kp,p}} \\ &\quad \le \text{const.} \max\left(\|t^{1/p}u(t)\|_{L_p(X_1,dt/t)}, \|t^{1/p}u^m(t)\|_{L_p(X_1,dt/t)}\right). \end{aligned}$$

An inspection of the right hand side shows that it is equivalent to the norm in $W_p^k(\mathbb{R}_+^{N-1})$. Further, interpolation of the Sobolev spaces yields the trace result for all $s \in (1,\infty)$.

Let $1/p < s \le 1$. Since $B_{pp}^{1/p}(\mathbb{R}^1)$ is imbedded into $L_\infty(\mathbb{R}^1)$ we have $|\operatorname{Tr} f(x,t)| \le c\|f(x,t)\|_{B_{pp}^{1/p}(\mathbb{R}^1)}$. Raising this to the p-th power and integrating with respect to $x \in \mathbb{R}^{N-1}$ we arrive at

$$\|Trf\|_{L_p(\mathbb{R}^{N-1})} \le c\|f\|_{B_{p1}^{1/p}(\mathbb{R}^N)}.$$

Recalling Theorem 2.4.25 we get the theorem for $B_{pq}^s(\mathbb{R}^N)$, where $s > 1/p$.

In view of Theorem 2.5.13 (ii), we have $H_p^s(\mathbb{R}^N) \subset B_{pp}^s(\mathbb{R}^N)$ for $2 \le p < \infty$. Hence for these values of p the case of $H_p^s(\mathbb{R}^N)$ follows.

We have

$$\begin{aligned} &\operatorname{Tr}: H_2^\sigma(\mathbb{R}^N) \to B_{22}^{\sigma-1/2}(\mathbb{R}^{N-1}) && \text{provided } \sigma > 1/2, \\ &\operatorname{Tr}: H_{p_1}^1(\mathbb{R}^N) \to B_{p_1 p_1}^{1-1/p_1}(\mathbb{R}^{N-1}) && \text{for all } 1 < p_1 < \infty. \end{aligned}$$

Given $p \in (1,2)$ and $s > 1/p$, one can choose $p_1 \in (1,p)$ in such a way that

$$\frac{1}{p} = \frac{1-\theta}{2} + \frac{\theta}{p_1}, \qquad s = (1-\theta)\sigma + \theta,$$

with $\sigma > 1/2$.

We refer to [148], Chapter 2 for the construction of the extension operator. □

Let us return to the basic classical situation in Sobolev spaces of k-th order since this case can be treated with "more classical" means. Let us assume that there exist local transformations of a part of the boundary such that we can consider spaces on $\mathbb{R}^N$ and traces on $\mathbb{R}^{N-1}$, where the latter set is naturally identified with $\{x = (x_1, \ldots, x_N) \in \mathbb{R}^N;\, x_N = 0\}$. For such local transformations of $W_p^k(\Omega)$ it is enough that the boundary $\partial\Omega$ of the original domain is of class $C^{k-1,1}$, that is, one requires that $\partial\Omega$ is described with help of finitely many local coordinate systems as a graph of a one-to-one function, which, together with its inverse, possesses Lipschitz continuous derivatives of $(k-1)$-st order and Ω is locally located at one side of these graphs.

Let us write points $x \in \mathbb{R}^N$ as (x', x_N), where $x' = (x_1, \ldots, x_{N-1}) \in \mathbb{R}^{N-1}$.

2.8.2 Lemma. *Let $f \in W_p^1(\mathbb{R}^N) \cap C^\infty(\overline{\mathbb{R}^N})$ with some $1 < p < \infty$. Then the function* $\operatorname{Tr} f(x')$, $x' \in \mathbb{R}^{N-1}$, *belongs to $W_p^{1-1/p}(\mathbb{R}^{N-1})$ and*

$$\| \operatorname{Tr} f \|_{W_p^{1-1/p}(\mathbb{R}^{N1})} \leq c \| f \|_{W_p^1(\mathbb{R}^N)}, \tag{2.8.1}$$

where c is independent of f.

Furthermore, Tr *maps $W_1^1(\mathbb{R}^N)$ continuously into $L_1(\mathbb{R}^{N-1})$.*

The proof for $p = 1$ is easy. If $1 < p < \infty$, then the usual technique for proving the estimate (2.8.1) is use of Hardy's inequality, going back to Gagliardo [56] (see also [114], [99]).

The trace spaces above are, moreover, the "best possible". We refer to [56], [114] and [99], Chapter 6, for construction of extension operators from the boundary. After localizing the problem with help of a local coordinate system describing the boundary and transferring the situation to $\mathbb{R}_+^N$ one can start locally with a function f in $C^\infty(\mathbb{R}^{N-1})$ and supported in $(-1,1)^{N-1}$, for instance. Then we can write an extension explicitly in a form of a convolution

$$Ef(x', x_N) = \frac{1}{x_N^{N-1}} \int_{|x'-y'|<x_N} f(y') \varphi\left(\frac{x'-y'}{x_N}\right) dy',$$

$$x' \in (-1,1)^{N-1}, \ x_N \in (0,1),$$

where $\varphi \in \mathcal{D}(\mathbb{R}^{N-1})$, $\operatorname{supp} \varphi \subset (-1,1)^{N-1}$, $\varphi \geq 0$, and $\|\varphi\|_{L_1(\mathbb{R}^{N-1})} = 1$ (see [56], [99], Chapter 6). This works for $1 < p < \infty$.

The situation for $p = 1$ is very much different. As it was already observed it is simple to show that the traces of functions in $W_1^m(\Omega)$ (for Ω appropriately smooth) belong to $W_1^{m-1}(\partial\Omega)$. As to the inverse problem Gagliardo [56] has given a construction of an extension of an L^1 function on the boundary to a function in $W_1^1(\Omega)$. A more general question about existence of a bounded extension operator for $p = 1$ has been considered by Peetre [120]. A certain modification of the trace problem for $p = 1$ was considered by Souček [141].

Let us present the following general theorem for Sobolev spaces:

2.8.3 Theorem. *Let $1 < p < \infty$, l be an integer, and suppose that $\Omega \in \mathcal{C}^{\ell-1,1}$. Then there exists a unique linear mapping*

$$\mathrm{Tr}_n : W_p^\ell(\Omega) \to \prod_{k=0}^{\ell-1} B_{pp}^{\ell-k-1/p}(\partial\Omega)$$

such that $\mathrm{Tr}_n f = (f, \partial f/\partial n, \dots, \partial^{\ell-1} f/\partial n^{\ell-1})$ *for all* $f \in C^\infty(\overline{\Omega})$, *where*

$$\frac{\partial^j f}{\partial n^j}(x) = \sum_{|\alpha|=j} \frac{j!}{\alpha!}(\partial^\alpha f(x)/\partial x^\alpha) n^\alpha, \qquad x \in \partial\Omega,$$

is the j-th order derivative of f with respect to the outer normal at $x \in \partial\Omega$.

2.8.2 The problem of norms and interpolation on domains

In preceding sections we could see and appreciate the power of the $\mathbb{R}^n$ theory. In applications, however, one has to work with Sobolev type spaces defined on domains with boundaries of various smoothness quality. Plainly, if Ω is a domain in $\mathbb{R}^N$ and the smoothness parameter s is non-negative, we can naturally define the Sobolev (Besov, Lizorkin-Triebel) space on Ω as a space of restrictions in the corresponding space on the whole $\mathbb{R}^N$. Given such a function f on Ω, we can consider all functions g in the appropriate space on $\mathbb{R}^N$ that equal to f on Ω and to define the norm of f as the lowest upper bound of norms over all such g. We get a Banach space (it is a factorspace) and it is easy to realize that now one can immediately formulate consequences of all the $\mathbb{R}^N$ claims (especially imbedding theorems). If, on the other hand, we start with a space of functions on Ω and if Ω has the extension property with respect to the spaces considered, that is, if there exists a bounded extension operator, then we can proceed along the same lines. Hence it turns out that extension theorems are the main unlocking device here. This can be naturally done similarly if elements of the spaces above are tempered distributions and not functions.

The first basic question is about an intrinsic norm in spaces on domains. It is not clear for instance if the norm in the space $W_p^k(\Omega)$ ($k > 0$, integer) defined in this way is equivalent to $\sum_{|\alpha| \le k} \|(\partial^\alpha/\partial x^\alpha) f\|_{L_p(\Omega)}$ as one naturally wishes. The situation with Besov spaces is still more complicated at first glance—one has to be careful about the difference operator since the translated variable should stay in Ω.

We state the following theorems without proofs, which would require a lot of space. The general strategy is clear: one has to consider local coordinates to situate the problem to the basic setting of $\mathbb{R}^N_+$ (for domains with the cone property this is possible by Gagliardo's Theorem 2.7.3) and then to use extension theorems. We refer to [148], Chapters 2 and 4 for all the details.

For the sake of a simple formulation of the following theorems let a script letter as a symbol for a space denote the space resulting from restrictions from $\mathbb{R}^N$ to Ω, that is, $\mathscr{W}_p^k(\Omega)$ denotes the space of all restrictions of $f \in W_p^k(\mathbb{R}^N)$ to Ω, equipped with the corresponding factornorm, and so on.

Let us also recall the symbol $\{s\}^+$, coming out of the equation $s = [s]^- + \{s\}^+$, where $[s]^-$ is an integer and $0 < \{s\}^+ \le 1$ (i.e. $[s]^-$ is the biggest integer among those smaller than s).

Since the cone property belongs to those frequently met in assumptions and it is also well documented in the literature let us look at the situation here:

2.8.4 Theorem. *Let $\Omega \subset \mathbb{R}^N$ be a bounded domain with the cone property. Let $k > 0$ be an integer and $1 < p < \infty$. Then all the formulae*

$$\Big(\sum_{|\alpha|\le k} \|D^\alpha f\|^p_{L_p(\Omega)}\Big)^{1/p}, \qquad \|f\|^p_{L_p(\Omega)} + \Big(\sum_{|\alpha|=k} \|D^\alpha f\|^p_{L_p(\Omega)}\Big)^{1/p},$$

$$\text{and} \qquad \|f\|^p_{L_p(\Omega)} + \Big(\sum_{j=1}^{n} \|(\partial^k/\partial x_j^k) f\|^p_{L_p(\Omega)}\Big)^{1/p}$$

are equivalent norms in $\mathscr{W}_p^k(\Omega)$.

Let $0 < s < \infty$, $1 < p < \infty$ and $1 \le q \le \infty$. Let k and ℓ be integers, $0 \le k < \ell$, $\ell > s - k$. For $h \in \mathbb{R}^N$ denote $\Omega_{h,\ell} = \bigcap_{j=0}^{\ell}\{x;\, x + jh \in \Omega\}$. Then

$$\|f\|_{L_p(\Omega)} + \sum_{j=1}^{n}\Big(\int_{\mathbb{R}^N} |h|^{-(s-k)q} \Big\|\Delta_\ell^h(\partial^k/\partial x_j^k) f\Big\|^q_{L_p(\Omega_{h,\ell})} \frac{dh}{|h|^n}\Big)^{1/q},$$

$$\|f\|_{L_p(\Omega)} + \sum_{|\alpha|\le k}\Big(\int_{\mathbb{R}^N} |h|^{-(s-k)q} \Big\|\Delta_\ell^h D^\alpha f\Big\|^q_{L_p(\Omega_{h,\ell})} \frac{dh}{|h|^n}\Big)^{1/q}$$

are equivalent norms in $\mathscr{B}^s_{p\,q}(\Omega)$.

Especially the norm in the Sobolev-Slobodetskii space $\mathscr{W}_p^s(\Omega) = \mathscr{B}^s_{p\,p}(\Omega)$ is equivalent to

$$\|f\|_{L_p(\Omega)} + \sum_{|\alpha|=[s]} \int_\Omega\int_\Omega \frac{|D^\alpha f(x) - D^\alpha f(y)|^p}{|x-y|^{N+\{s\}p}}\, dx dy$$

(*one can also replace the summation over* $|\alpha| = [s]$ *by* $|\alpha| \leq [s]$ *or use the "pure" derivatives* $\partial^k/\partial x_j^k$, $1 \leq j \leq N$).

Interpolation and duality on domains

The next step in building the theory on domains in the Fourier analysis spirit is to establish their interpolation properties. As in the preceding subsection all depends on the extension properties of the domain in question. Therefore, one expects that domains with the cone property, in particular domains with a Lipschitz boundary, inherit relevant interpolation properties—at least for spaces whose elements are regular distributions. Let us formulate the corresponding interpolation theorems. We refer to [148], Chapter 4 for proofs.

The general principle is contained in the next abstract theorem on interpolation spaces. Given two linear spaces X and Y, we say that a linear mapping $R : X \to Y$ is a *retraction* if there exists a linear mapping $S : Y \to X$ such that $SR = \mathrm{id}$ on X. The mapping S is called a *coretraction* (with respect to R).

If X is a linear space, then an operator $P : X \to X$ is called a *projection* (of P into P) if $P^2 = P$. A subspace Z of X is called a *complementary subspace* if $Z = PX$ for some projection P of X into itself.

The reader can think about extensions to $\mathbb{R}^N$ of functions defined on a domain Ω (as of S) and then about their restrictions to Ω (the operator R); this is exactly how the next theorem applies (see [148], Chapter 1 for the proof). The theorem seems to be a bit complicated at first glance but this is not the case (draw a picture!). One can alternatively proceed directly as in the proof of Corollary 2.8.9 (see p. 154).

2.8.5 Theorem. *Let* X_i, Y_i, $i = 1, 2$ *be compatible couples of Banach spaces and suppose that a linear operator* $R : X_1 + X_2 \to Y_1 + Y_2$ *takes continuously* X_i *into* Y_i, $i = 1, 2$. *Let* $S : Y_1 + Y_2 \to X_1 + X_2$ *be a linear operator such that the restriction of* S *to* Y_j *is a coretraction with respect to the restriction of* R *to* X_j, $j = 1, 2$. *Let* $\mathscr{T}$ *be an interpolation functor on the family of all compatible couples of Banach spaces. Then* S *is an isomorphic mapping of* $\mathscr{T}(Y_1, Y_2)$ *onto a complementary subspace of* $\mathscr{T}(X_1, X_2)$ *and this subspace coincides with the range of the restriction of* SR *to* $\mathscr{T}(X_1, X_2)$. *Moreover, the restriction of* SR *to* $\mathscr{T}(X_1, X_2)$ *is a projector in* $\mathscr{T}(X_1, X_2)$.

Hence invoking the extension theorem for spaces on domains with the cone property, we almost immediately get the following theorem. Let us point out that $s \geq 0$.

2.8.6 Theorem. *Let $\Omega \subset \mathbb{R}^N$ be a domain with the cone property. Then the following statements hold true:*

(i) *Let us suppose that $0 < \theta < 1$, $1 < p_1, p_2 < \infty$, $1/p = (1-\theta)/p_1 + \theta/p_2$, $s \geq 0$. Then $(H^s_{p_1}(\Omega), H^s_{p_2}(\Omega))_{\theta,p} = H^s_p(\Omega)$.*

(ii) *With the same parameters as in* (i), *assume additionally that $s = (1-\theta)s_1 + \theta s_2$ for some $s_1, s_2 \geq 0$. Then $[H^{s_1}_{p_1}(\Omega), H^{s_2}_{p_2}(\Omega)]_\theta = H^s_p(\Omega)$.*

(iii) *Let $1 < p < \infty$, $1 \leq q_i, \leq \infty$, $i = 1, 2$, $0 < \theta < 1$, $1/q = (1-\theta)/q_1 + \theta/q_2$, $0 < s_1, s_2 < \infty$, $s_1 \neq s_2$, $s = (1-\theta)s_1 + \theta s_2$. Then we have $(B^{s_1}_{p\,q_1}(\Omega), B^{s_2}_{p\,q_2}(\Omega))_{\theta,q} = B^s_{p\,q}(\Omega)$. In particular, if $p = q_1 = q_2$, we get the corresponding interpolation formula for the Sobolev-Slobodetskii spaces with positive smoothness.*

(iv) *Let $1 < p < \infty$, $1 \leq q_i, \leq \infty$, $i = 1, 2$, $0 < \theta < 1$, $1/q = (1-\theta)/q_1 + \theta/q_2$, $0 < s_1, s_2 < \infty$, $s = (1-\theta)s_1 + \theta s_2$. Then we have $(B^{s_1}_{p\,q_1}(\Omega), B^{s_2}_{p\,q_2}(\Omega))_{\theta,q} = B^s_{p\,q}(\Omega)$. In particular, if $p = q_1 = q_2$, we get the corresponding interpolation formula for the Sobolev-Slobodetskii spaces with positive smoothness.*

(v) *Let $0 < s_1 < \infty$, $0 \leq s_2 < \infty$, $0 < \theta < 1$, $s = (1-\theta)s_1 + \theta s_2$, $1 < p < \infty$, $1 \leq q_1, q \leq \infty$. Then $(B^{s_1}_{p\,q_1}(\Omega), H^{s_2}_p(\Omega))_{\theta,q} = (H^{s_1}_p(\Omega), H^{s_2}_p(\Omega))_{\theta,q} = B^s_{p\,q}(\Omega)$.*

(vi) *We have $(B^{s_1}_{p\,q_1}(\Omega), W^{s_2}_p(\Omega))_{\theta,q} = (H^{s_1}_p(\Omega), W^{s_2}_p(\Omega))_{\theta,q} = B^s_{p\,q}(\Omega)$ with the same parameters as in* (v).

In particular, if $p = q_1$, we get for the Sobolev-Slobodetskii spaces with non-negative smoothness that $(W^{s_1}_p(\Omega), W^{s_2}_p(\Omega))_{\theta,q} = B^s_{p\,q}(\Omega)$. If further $p = q$, then $(W^{s_1}_p(\Omega), W^{s_2}_p(\Omega))_{\theta,p} = W^s_p(\Omega)$.

Important in applications are Sobolev-Slobodetskii spaces with negative smoothness on domains. The situation is not straightforward at all. To go on in the same spirit as before and to preserve our strategy let us stick to the earlier definition of spaces on domains as restrictions of spaces on $\mathbb{R}^N$. We saw in the preceding subsection that under fairly general assumptions on the smoothness of Ω (the cone property) we get the right spaces, well-known from the "classical" theory of Sobolev and Besov spaces, at least for non-negative smoothness parameters.

Let $s \in \mathbb{R}^1$, $1 < p < \infty$, $1 \leq q \leq \infty$ and let Ω be an arbitrary domain in $\mathbb{R}^N$. Let us denote by $\widetilde{B}^s_{p\,q}(\Omega)$ and by $\widetilde{H}^s_p(\Omega)$ the space of all the elements in $f \in B^s_{p\,q}(\mathbb{R}^N)$ and in $H^s_p(\mathbb{R}^N)$, respectively, whose support is contained in $\overline{\Omega}$. Then we have (see e.g. [148], Chapter 4):

2.8.7 Theorem. *Let* $s \in \mathbb{R}^1$, $1 < p < \infty$, $1 \leq q < \infty$ *and let* Ω *be an arbitrary domain in* $\mathbb{R}^N$. *Let* p' *and* q' *be the conjugate exponents to* p *and* q', *respectively. Then* $(\widetilde{B}^s_{p\,q}(\Omega))' = B^{-s}_{p'\,q'}(\Omega)$ *and* $(\widetilde{H}^s_p(\Omega))' = W^{-s}_{p'}(\Omega)$.

Note that, for $s > 0$, the "more traditional" way of introducing the Sobolev spaces $H^{-s}_p(\Omega)$ and Sobolev-Slobodetskii spaces $W^{-s}_p(\Omega)$ is to define them as duals of appropriate spaces of functions with zero traces, i.e. taking the closure of $C^\infty_0(\Omega)$ in $H^s_{p'}(\Omega)$ and $W^s_{p'}(\Omega)$, obtaining thus the spaces $\mathring{H}^s_{p'}(\Omega)$ and $\mathring{W}^s_{p'}(\Omega)$, respectively, and to pass to their duals (see [102], for instance). But as is well known for some values of s the concept of the trace lacks its sense, in particular, $C^\infty_0(\Omega)$ is dense both in $H^s_{p'}(\Omega)$ and $W^s_{p'}(\Omega)$ provided $s \leq 1/p'$ and Ω is sufficiently regular (see e.g. [148], Chapter 4). Second, there is a problem of a suitable extension operator for tempered distributions to guarantee interpolation properties analogous to those valid in $\mathbb{R}^N$. There is a large number of papers, above all in the Russian school of function spaces, dealing with extension properties of various kinds of Sobolev spaces (in particular general anisotropic spaces in the sense of Nikol'skii, etc.) We list at least an extensive treatment by Burenkov [28], containing also a survey of various extension methods and a lot of further references, Fain [52], with the problem of extension from domains with less regular boundaries, discussing the price paid in the form of weighted target spaces or worse integrability of extended functions, further Vodop'yanov, Gol'dstein and Latfullin [153] on the geometric properties of unbounded domains in the plane, necessary and sufficient for the existence of an extension operator on W^1_2 spaces, and eventually Kalyabin [89], studying extension operators on Lizorkin-Triebel spaces (of positive integer order).

The interpolation theory for spaces on Lipschitz domains has not been dealt with in the existing literature at all until recently. Despite a large number of papers, considering extension properties of Sobolev spaces of various types, it was only in 1999, when Rychkov [125] succeeded in establishing a universal extension operator for Besov and Lizorkin-Triebel spaces for the whole range of the smoothness parameter s. The construction hangs on a highly non-trivial generalization of the Calderón extension operator and we refer to the original paper for details.[1]

Let us recall our agreement that the spaces on domains are defined as restrictions of spaces in the whole $\mathbb{R}^N$. Then the universal extension theorem can be formulated as follows.

2.8.8 Theorem. *Let* $s \in \mathbb{R}^1$, $0 < p \leq \infty$, $0 < q \leq \infty$ ($p < \infty$ *in the case of the Lizorkin-Triebel spaces*) *and let* $\Omega \subset \mathbb{R}^N$ *be a bounded domain with*

---

[1]We are grateful to Prof. Hans Triebel for valuable discussion on the subject of interpolation on Lipschitz domains, resulting in his recent paper [150], filling an important gap in the theory.

Lipschitz boundary. Then there exists a universal extension from $B^s_{pq}(\Omega)$ and $F^s_{pq}(\Omega)$ into $B^s_{pq}(\mathbb{R}^N)$ and $F^s_{pq}(\mathbb{R}^N)$, respectively.

Before stating an interpolation theorem for spaces in domains let us observe that the formulae in Theorem 2.8.6 hold also for $p, q \in (0,1)$, that is, also in the case when the spaces in question are quasi-Banach spaces. There are no essential problems in extending the Peetre's K-method to this case; nevertheless, the Calderón complex method works only in the category of Banach spaces. This restriction can be removed on the basis of a recent generalization of the complex method in [110], covering all the Besov and Lizorkin-Triebel spaces with $p \in (0,1)$ or $q \in (0,1)$. We note this just for the sake of completeness—and to be able to state the next theorem in its most general form.

2.8.9 Corollary. *Let Ω be a bounded domain in $\mathbb{R}^N$ with a Lipschitz boundary. Then all the interpolation formulae from Theorem 2.8.6 holds for all real s_1 and s_2, $0 < p \leq \infty$, $0 < q \leq \infty$ ($p < \infty$ in the case of F^S_{pq}).*

Proof. Let $s_1, s_2 \in \mathbb{R}^1$, and let p_k, q_k, $k = 1,2$ satisfy the above assumptions. To simplify the notation denote by $U^k(\Omega)$ and $U^k(\mathbb{R}^N)$, $k = 1,2$, the corresponding couples of either Besov or Lizorkin-Triebel spaces in Ω and in $\mathbb{R}^N$, respectively. Let $0 < \theta < 1$ and let $RU^\theta(\Omega)$ be the restriction of $U^\theta(\mathbb{R}^N)$ to Ω, where the latter space is the result of interpolation (real or complex) of the couple $(U^1(\mathbb{R}^N), U^2(\mathbb{R}^N))$ with the parameter θ and some p, q, satisfying our assumptions. Further, let $IU^\theta(\Omega)$ be the result of the corresponding interpolation of the couple $(U^1(\Omega), U^2(\Omega))$. Let Rest be the operator of restriction from $\mathbb{R}^N$ to Ω and let Ext stands for the Rychkov extension operator.

Since the operator Rest is bounded from $U^k(\mathbb{R}^N)$ to $RU^k(\Omega)$ it is also bounded from $U^\theta(\mathbb{R}^N)$ to $IU^\theta(\Omega)$, that is,

$$\|f\|_{IU^\theta(\Omega)} \leq c\|\operatorname{Ext} f\|_{U^\theta(\mathbb{R}^N)}. \tag{2.8.2}$$

By interpolation argument, the operator Ext is bounded from $IU^\theta(\Omega)$ into $U^\theta(\mathbb{R}^N)$. Thus recalling (2.8.2) we obtain

$$\|f\|_{IU^\theta(\Omega)} \leq c \leq \|f\|_{RU^\theta(\Omega)}. \tag{2.8.3}$$

By the definition of spaces in Ω (restrictions from R^N and taking the quotient norm) we have

$$\|f\|_{RU^\theta(\Omega)} \leq \|\operatorname{Ext} f\|_{U^\theta(\mathbb{R}^N)}. \tag{2.8.4}$$

Combining (2.8.3) and (2.8.4) we see that the spaces in Ω defined by the restriction and those resulting by interpolation of spaces in Ω are the same, with equivalent norms. □

2.8.10 Remark. What was said about the isotropic spaces above can be immediately carried out for the iterated Sobolev and Besov spaces (the spaces with dominating mixed smoothness, isotropic in space variables). The reader will be able to formulate the anisotropic variants of Theorems 2.8.6 and 2.8.9.

Function with zero traces on Lipschitz domains

Let us now discuss the concept of zero traces. If we work with Bessel potential spaces involving those with negative smoothness it is natural to work with the definition of zero traces of functions in $H_p^s(\Omega)$ in the sense of the $\widetilde{H}_p^s(\Omega)$ spaces introduced in Triebel [148] for C^∞ domains (see the preceding subsection).

Let Ω be from now on a Lipschitz domain in $\mathbb{R}^N$ and let us consider the spaces

1. $\mathring{H}_p^s(\Omega)$ as the completion of $\mathscr{D}(\Omega)$ in $H_p^s(\Omega)$, for $1 < p < \infty$, $s \in \mathbb{R}^1$;
2. $\widetilde{H}_p^s(\Omega)$ as the linear set of all $f \in \mathscr{D}'(\Omega)$ such that there exists $g \in H_p^s(\mathbb{R}^N)$ such that $g|_\Omega = f$ in $\mathscr{D}'(\Omega)$ and

$$\operatorname{supp} g \subset \overline{\Omega}, \tag{2.8.5}$$

endowed with the factornorm

$$\|f\|_{\widetilde{H}_p^s(\Omega)} = \inf \|g\|_{H_p^s(\mathbb{R}^N)},$$

where the infimum is taken over all $g \in H_p^s(\mathbb{R}^N)$ satisfying (2.8.5). Then we have

2.8.11 Theorem. *Let $1 < p < \infty$ and s be a real number. Then*

$$\mathring{H}_p^s(\Omega) = H_p^s(\Omega) = \widetilde{H}_p^s(\Omega) \qquad \textit{for all } \frac{1}{p} - 1 < s < \frac{1}{p}.$$

If we define

$$\overline{H}_p^s(\Omega) = H_p^s(\Omega) \qquad \textit{for } s < \frac{1}{p}$$

and

$$\overline{H}_p^s(\Omega) = \widetilde{H}_p^s(\Omega) \qquad \textit{for } s > \frac{1}{p} - 1,$$

then the standard interpolation formulae hold for couples of $\overline{H}_p^s(\Omega)$ spaces.

The proof of this theorem can be found in Triebel [150].

Hence it seems that the $\overline{H}_p^s(\Omega)$ are the good choice for the definition of spaces with zero traces on Lipschitz domains, at least from the point of view of the interpolation theory.

2.8.3 Compact imbeddings

Let us start with an observation that theorems on compact imbeddings follow directly from the abstract theorems from Subsection 2.3.4 and from the corresponding imbedding theorems on domains (which are sharp in given scales of function spaces). It is sufficient to start from the well-known compact imbedding theorems for the Sobolev spaces of integer order (or their anisotropic variants) and to consider the interpolation scale between the starting and the target space for $\theta \in (0,1)$. We also refer to [26], Chapter 8 for general compact imbeddings of Sobolev and Besov type spaces. In [1] one can find an approach, which does not explicitly use the abstract theorem on interpolation of compact mappings.

Nevertheless, this cannot be done in the whole of $\mathbb{R}^N$. *A priori*, any Ω with infinite Lebesgue measure is excluded:

2.8.12 Theorem. *Let $\Omega \subset \mathbb{R}^N$ be a domain, $1 \le p < \infty$, and suppose that there exists a compact imbedding $W_p^k(\Omega) \hookrightarrow L_q(\Omega)$ for some $q \ge p$ and some $k = 1, 2, \ldots$. Then* $\operatorname{mes}(\Omega) < \infty$.

Furthermore, if Ω satisfies the condition $\limsup_{r \to +\infty} \operatorname{mes}(\{x \in \Omega;\, r \le |x| < r+1\}) > 0$, *then none of the imbeddings $W_p^k(\Omega) \hookrightarrow L_q(\Omega)$ with $q > p$ is compact.*

Further refinements of the above theorems and compact imbeddings of Sobolev spaces on special unbounded domains can be found e.g. in [1], Chapter VI. In the following we restrict our attention on bounded domains. Let us start with the model case. We have (see [1], Chapter VI or [99], Chapter V)

2.8.13 Theorem (Kondrashov). *Let $\Omega \subset \mathbb{R}^N$ be a bounded domain with the cone property, and $1 \le p < N$. Then the imbedding $W_p^1(\Omega) \hookrightarrow L_p(\Omega)$ is compact.*

The reader will be able to prove the theorem easily, recalling the Riesz criterion for relative compactness of subsets of $L_p(\Omega)$. The L_p-mean equicontinuity of a bounded set in $W_p^1(\Omega)$ follows by a density argument and then by an estimate of $f(x+h) - f(x)$ for smooth f by means of the mean value theorem (the L_p-norms of derivatives are equibounded) and the rest is just a technical problem: one can use the mean value theorem only on sets, where the derivatives of (a smooth) f exist. But thanks to the cone property the set, on which such an estimate is impossible, can be done sufficiently small.

The space $W_p^1(\Omega)$ is imbedded into $L_q(\Omega)$ with $q = Np/(N-p)$. Hence invoking Theorem 2.3.20 and Corollary 2.3.21 we arrive at once at

2.8.14 Corollary. *Let Ω be as in the preceding theorem, $1 \leq p < N$. Then $W_p^1(\Omega)$ is compactly imbedded into $L_r(\Omega)$ for every $1 \leq r < Np/(N-p)$. Moreover, if $0 < s_1 < s_2 \leq 1$, then $W_p^{s_1}(\Omega)$ is compactly imbedded into $W_p^{s_2}(\Omega)$.*

Of course, the second claim in the preceding Corollary holds in the scale of Besov spaces, resulting from real interpolation between $L_p(\Omega)$ and $W_p^1(\Omega)$.

Let us observe that quite analogous theorems hold also for imbeddings of Sobolev (Besov) spaces into the spaces with the smoothness parameter smaller than that of the corresponding trace space. The reader will easily formulate and justify them, relying on the fact that the imbeddings of $W_p^1(\Omega)$ into $L_p(\partial\Omega)$ are compact.

Following the above scheme one obtains corollaries of the imbedding theorems met earlier. We shall just formulate the basic theorem (sometimes called the Rellich-Kondrashov theorem) for compact imbeddings of Sobolev spaces and leave further details to the reader.

We present the basic compact imbedding theorem for Sobolev spaces.

2.8.15 Theorem. *Let $\Omega \subset \mathbb{R}^N$ be a bounded domain with the cone property. Let $1 \leq p < \infty$ and k be an integer, $k \geq 1$. If $kp < N$, then the imbedding $W_p^k(\Omega) \hookrightarrow L_r(\Omega)$ is compact for every r with $1 \leq r < Np/(N-kp)$.*

If $kp = N$, then the imbedding $W_p^k(\Omega) \hookrightarrow L_r(\Omega)$ is compact for every finite r.

If $kp > N$, then the imbeddings $W_p^{j+k}(\Omega) \hookrightarrow C^j(\Omega)$ are compact for every $j = 0, 1, \ldots$.

If the assumption about the regularity of $\partial\Omega$ is omitted, then the imbeddings above are compact provided $W_p^k(\Omega)$ is replaced by the closure of $\mathscr{S}(\Omega)$ in $W_p^k(\Omega)$.

2.8.16 Remark. A particular consequence of the last theorem is the following assertion: If Ω is a domain with the cone property, $1 < p < \infty$, and $0 < s_1 < s_s \leq 1$, then the imbedding $W_p^{s_2}(\Omega) \hookrightarrow W_p^{s_1}(\Omega)$ is compact.

The general picture does not change much when passing to compact imbeddings for spaces with mixed norms. The machinery is the same, only this time one has to interpolate between mixed norm L_p spaces. We recall the classical theorem on interpolation of mixed norm L_p spaces due to Benedek and Panzone [20]. As a result we obtain compact imbeddings of the anisotropic Sobolev spaces into the mixed norm L_p spaces, in every case, when all the components of the vector integrability parameter are strictly smaller than its "best" value, for which an imbedding exists.

2.8.17 Theorem. *Let* $\overline{p}^j = (p_1^j, \ldots, p_N^j)$, $j = 1, 2$ *be multiindices,* $\overline{1} \leq \overline{p}^j \leq \overline{\infty}$ *and* $1/q_i = (1-\theta)/p_i^1 + \theta/p_i^2$, $i = 1, \ldots, N$. *For any measurable set* $A \subset \mathbb{R}^N$ *the space* $L_{\overline{q}}(A)$ *is an exact interpolation space with respect to the couple* $(L_{\overline{p}^1}(A), L_{\overline{p}^2}(A))$, *and it is a result of real interpolation of this couple of spaces with parameters* θ *and* q *and a result of complex interpolation of this couple with parameter* θ.

The reader will be able now to formulate the corresponding theorem on compact imbeddings of iterated Sobolev and Sobolev-Slobodetskii spaces.

Chapter 3

Static and quasistatic contact problems

In this chapter we revisit the static contact problem by first giving an alternative existence proof using the penalty method. The penalty method is closer to physical intuition and needs weaker requirements on the smoothness of the boundary than the fixed point approach employed in the introduction. We assume that the coefficient of friction is solution dependent. This may be useful in order to model a possible difference of the coefficients of friction of stick and of friction of slip.

The second section of this chapter contains an extension of these results to the semicoercive contact problem. There, no displacement boundary condition is prescribed on any part of the boundary; therefore the bilinear form of elastic energy is no longer equivalent to the scalar product in $\boldsymbol{H}^1(\Omega)$. The problem is solved under the assumption of an additional condition called *pressure condition* that was introduced in the context of contact problems by G. Fichera. This condition requires, roughly speaking, the outer forces to press the body onto the obstacle and not to draw it away from the obstacle. If this condition is valid, some kind of coercivity of the total problem is established via the combination of boundary forces and contact forces.

In the third section the results of the previous two sections are extended to the contact of two elastic bodies. The last section contains a result of L.E. Andersson [12] on quasistatic contact problems, obtained by a suitable time discretization.

3.1 Coercive static case

Let us recall the contact problem of an elastic body with a rigid foundation. If Ω denotes a bounded domain occupied by the body and Γ is its boundary consisting of the mutually disjoint, open (with respect to the topology of Γ) parts $\Gamma_{\boldsymbol{U}}$, $\Gamma_{\boldsymbol{\tau}}$ and Γ_C, then the static contact problem is given by the variational inequality

Find a function $\boldsymbol{u} \in \mathscr{C}$ *such that for all* $\boldsymbol{v} \in \mathscr{C}$ *there holds*

$$A(\boldsymbol{u}, \boldsymbol{v}-\boldsymbol{u}) + \int_{\Gamma_C} \mathfrak{F}(\boldsymbol{u})|\sigma_n(\boldsymbol{u})|\big(|\boldsymbol{v_t}-\boldsymbol{w_t}| - |\boldsymbol{u_t}-\boldsymbol{w_t}|\big)\, ds_{\boldsymbol{x}} \geq \mathscr{L}(\boldsymbol{v}-\boldsymbol{u}). \quad (3.1.1)$$

Here the convex cone

$$\mathscr{C} = \big\{\boldsymbol{v} \in \boldsymbol{H}^1(\Omega);\ \boldsymbol{v} = \boldsymbol{U} \text{ on } \Gamma_{\boldsymbol{U}} \text{ and } v_n \leq g_n \text{ on } \Gamma_C\big\},$$

the bilinear form

$$A(\boldsymbol{u}, \boldsymbol{v}) = \int_\Omega \big(d\,\boldsymbol{u}\cdot\boldsymbol{v} + a_{ijk\ell}e_{ij}(\boldsymbol{u})e_{k\ell}(\boldsymbol{v})\big)\, d\boldsymbol{x}$$

and the linear functional

$$\mathscr{L}(\boldsymbol{v}) = \int_\Omega \boldsymbol{f}\cdot\boldsymbol{v}\, d\boldsymbol{x} + \int_{\Gamma_{\boldsymbol{\tau}}} \boldsymbol{\tau}\cdot\boldsymbol{v}\, ds_{\boldsymbol{x}}$$

are used. The coefficient functions d, $a_{ijk\ell}$ and the given functions $\boldsymbol{f}$, $\boldsymbol{U}$, $\boldsymbol{\tau}$, g_n are supposed to satisfy the conditions of Assumption 1.5.1. The function $\boldsymbol{w} \in \boldsymbol{H}^{1/2}(\Gamma_C)$ may arise from a time discretization as described in Section 1.5, there $\boldsymbol{w} = \boldsymbol{u}^{(\ell-1)}$.

3.1.1 Approximate problems and limit procedures

Instead of the fixed-point approach used in the introduction we apply here the penalty method. The inequality constraint $v_n \leq g_n$ in the cone $\mathscr{C}$ of admissible functions is replaced by adding the penalty functional

$$\int_{\Gamma_C} \tfrac{1}{\varepsilon}\left[u_n - g_n\right]_+ (v_n - u_n)\, ds_{\boldsymbol{x}}$$

with $[\,\cdot\,]_+ \equiv \max\{0, \cdot\}$ to the left hand side of (3.1.1). This is equivalent to prescribe the normal component of boundary stress $\sigma_n(\boldsymbol{u}) = -\frac{1}{\varepsilon}\left[u_n - g_n\right]_+$ in a weak sense and leads to the variational inequality

Find $\boldsymbol{u} \in \boldsymbol{U} + \boldsymbol{H}_0^1(\Omega)$ *with* $\boldsymbol{H}_0^1(\Omega) \equiv \big\{\boldsymbol{v} \in \boldsymbol{H}^1(\Omega);\ \boldsymbol{v} = 0 \text{ on } \Gamma_{\boldsymbol{U}}\big\}$, *such that for all* $\boldsymbol{v} \in \boldsymbol{U} + \boldsymbol{H}_0^1(\Omega)$

$$\begin{aligned} A(\boldsymbol{u}, \boldsymbol{v}-\boldsymbol{u}) &+ \int_{\Gamma_C} \tfrac{1}{\varepsilon}\left[u_n - g_n\right]_+ (v_n - u_n)\, ds_{\boldsymbol{x}} \\ &+ \int_{\Gamma_C} \mathfrak{F}(\boldsymbol{u})\tfrac{1}{\varepsilon}\left[u_n - g_n\right]_+ \big(|\boldsymbol{v_t}-\boldsymbol{w_t}| - |\boldsymbol{u_t}-\boldsymbol{w_t}|\big)\, ds_{\boldsymbol{x}} \\ &\geq \mathscr{L}(\boldsymbol{v}-\boldsymbol{u}). \end{aligned} \quad (3.1.2)$$

Here we assume the function $\boldsymbol{U}$ to be defined on the whole domain Ω.

In a next approximation step, the norm $|\cdot|$ in the friction functional is replaced by a smooth approximation $\Phi_\eta(\cdot)$. Here, $\Phi_\eta : \mathbb{R}^N \to [0, +\infty)$ is a convex C^1-function which has its minimum in $\boldsymbol{x} = 0$ and satisfies the approximation property

$$\big|\Phi_\eta(\boldsymbol{x}) - |\boldsymbol{x}|\big| \le \eta.$$

Such a function is e.g. given by

$$\Phi_\eta : \boldsymbol{x} \mapsto \begin{cases} |\boldsymbol{x}|, & |\boldsymbol{x}| \ge \eta, \\ -\dfrac{|\boldsymbol{x}|^4}{8\eta^3} + \dfrac{3|\boldsymbol{x}|^2}{4\eta} + \dfrac{3}{8}\eta, & |\boldsymbol{x}| < \eta, \end{cases}$$

with the gradient

$$\nabla\Phi_\eta(\boldsymbol{x}) = \beta(\boldsymbol{x})\frac{\boldsymbol{x}}{|\boldsymbol{x}|}, \quad \beta(\boldsymbol{x}) = \begin{cases} 1, & |\boldsymbol{x}| \ge \eta, \\ -\dfrac{|\boldsymbol{x}|^3}{2\eta^3} + \dfrac{3|\boldsymbol{x}|}{2\eta}, & |\boldsymbol{x}| < \eta. \end{cases}$$

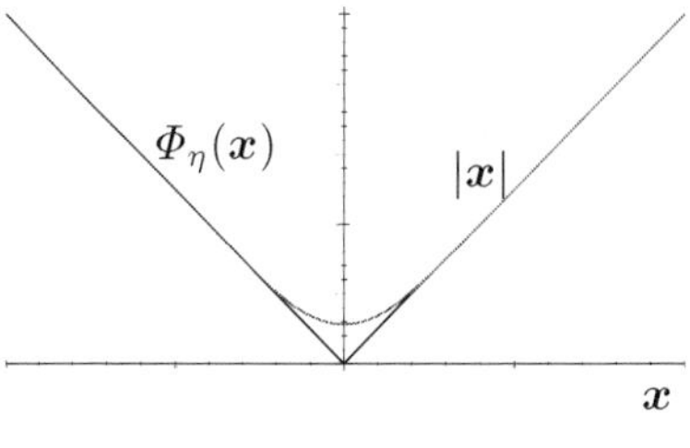

The resulting variational inequality is equivalent to the following variational equation

Find $\boldsymbol{u} \in \boldsymbol{U} + \boldsymbol{H}_0^1(\Omega)$ such that for all $\boldsymbol{v} \in \boldsymbol{H}_0^1(\Omega)$

$$\begin{aligned} A(\boldsymbol{u}, \boldsymbol{v}) &+ \int_{\Gamma_C} \tfrac{1}{\varepsilon}\left[u_n - g_n\right]_+ v_n \, ds_{\boldsymbol{x}} \\ &+ \int_{\Gamma_C} \mathfrak{F}(\boldsymbol{u}) \tfrac{1}{\varepsilon}\left[u_n - g_n\right]_+ \nabla\Phi_\eta(\boldsymbol{u_t} - \boldsymbol{w_t}) \cdot \boldsymbol{v_t} \, ds_{\boldsymbol{x}} = \mathscr{L}(\boldsymbol{v}). \end{aligned} \tag{3.1.3}$$

In fact, inserting $\boldsymbol{v} = \boldsymbol{u} + \lambda\widetilde{\boldsymbol{v}}$ into the smoothed version of (3.1.2), dividing the result by λ and passing to the limit $\lambda \to 0$ yields (3.1.3). On the other hand, choosing the test function $\boldsymbol{v} = \widetilde{\boldsymbol{v}} - \boldsymbol{u}$ in (3.1.3) and employing the relation

$$\nabla\Phi_\eta(\boldsymbol{u_t} - \boldsymbol{w_t}) \cdot \left(\widetilde{\boldsymbol{v}}_{\boldsymbol{t}} - \boldsymbol{u_t}\right) \leq \Phi_\eta(\widetilde{\boldsymbol{v}}_{\boldsymbol{t}} - \boldsymbol{w_t}) - \Phi_\eta(\boldsymbol{u_t} - \boldsymbol{w_t})$$

which is valid because Φ_η is convex, we obtain the smoothed version of (3.1.2).

The existence of solutions to this problem will be proved under the following set of assumptions:

3.1.1 Assumption. *Let Ω be a bounded domain with a Lipschitz boundary Γ. Let* either $\operatorname{mes}_\Gamma \Gamma_{\boldsymbol{U}} > 0$ *and* $d \geq 0$ or $\Gamma_{\boldsymbol{U}} = \emptyset$ *and* $d \geq d_0 > 0$ *on a set of positive measure. The coefficients $a_{ijk\ell}$ are symmetric, bounded and elliptic in the sense described in* (1.2.2)–(1.2.4). *The given data satisfy* $\boldsymbol{U} \in \boldsymbol{H}^1(\Omega)$, $\boldsymbol{f} \in \mathring{\boldsymbol{H}}^{-1}(\Omega)$, $\boldsymbol{\tau} \in \mathring{\boldsymbol{H}}^{-1/2}(\Gamma_{\boldsymbol{\tau}})$, $g_n \in H^{1/2}(\Gamma_C)$, $g_n \geq 0$ *a.e. in* Γ_C, *and* $\boldsymbol{w} \in \boldsymbol{H}^{1/2}(\Gamma_C)$. *The coefficient of friction* $\mathfrak{F} \equiv \mathfrak{F}(\boldsymbol{x}, \boldsymbol{u})$ *shall be globally bounded, non-negative and continuous in the sense of the Carathéodory property (i.e.* $\mathfrak{F}(\cdot, \boldsymbol{z})$ *is measurable for all* $\boldsymbol{z} \in \mathbb{R}^N$, *and* $\mathfrak{F}(\boldsymbol{x}, \cdot)$ *is continuous for a.e.* $\boldsymbol{x} \in \Gamma$*). The compatibility conditions* $U_n \leq g_n$ *and* $\boldsymbol{U_t} = 0$ *a.e. on* Γ_C *shall be valid.*

Then the following result holds:

3.1.2 Lemma. *Let Assumption 3.1.1 be valid. Then to each* $\varepsilon > 0$, $\eta > 0$ *there exists a solution* $\boldsymbol{u} = \boldsymbol{u}_{\varepsilon,\eta}$ *of problem* (3.1.3). *This solution satisfies the* a priori *estimate*

$$\|\boldsymbol{u}_{\varepsilon,\eta}\|_{\boldsymbol{H}^1(\Omega)} + \left\|\tfrac{1}{\varepsilon}\left[(u_{\varepsilon,\eta})_n - g_n\right]_+\right\|_{\mathring{H}^{-1/2}(\Gamma_C)} \leq c_1 \tag{3.1.4}$$

with a constant c_1 *independent of* ε *and* η. *For every fixed* $\varepsilon > 0$ *there exists a sequence* $\eta_k \to 0$ *and a corresponding sequence* $\boldsymbol{u}_{\varepsilon,\eta_k}$ *of solutions to the smoothed problem* (3.1.3) *that converges strongly in* $\boldsymbol{H}^1(\Omega)$ *to a solution* $\boldsymbol{u}_\varepsilon$ *of* (3.1.2).

Proof. The existence of solutions is proved by a fixed point approach. Let $\boldsymbol{z} \in \boldsymbol{U} + \boldsymbol{H}_0^1(\Omega)$ be an admissible displacement field. Then we define the problem

$$\begin{aligned} A(\boldsymbol{u}, \boldsymbol{v}) &+ \int_{\Gamma_C} \mathfrak{F}(\boldsymbol{z})\tfrac{1}{\varepsilon}\left[z_n - g_n\right]_+ \nabla\Phi_\eta(\boldsymbol{u_t} - \boldsymbol{w_t}) \cdot \boldsymbol{v_t}\, ds_{\boldsymbol{x}} \\ &+ \int_{\Gamma_C} \tfrac{1}{\varepsilon}\left[u_n - g_n\right]_+ v_n\, ds_{\boldsymbol{x}} = \mathscr{L}(\boldsymbol{v}). \end{aligned} \tag{3.1.5}$$

This problem can be interpreted as a (smoothed and penalized) contact problem with given friction; the term $\mathfrak{F}(\boldsymbol{z})\frac{1}{\varepsilon}[z_n - g_n]_+$ represents the given friction force. The problem is equivalent to the optimization problem

$$\boldsymbol{u} = \underset{\boldsymbol{v} \in \boldsymbol{U} + \boldsymbol{H}_0^1(\Omega)}{\operatorname{Argmin}} \left(J_0(\boldsymbol{v}) + j_{\boldsymbol{z}}(\boldsymbol{v}) \right)$$

with

$$J_0(\boldsymbol{v}) = \frac{1}{2} A(\boldsymbol{v}, \boldsymbol{v}) - \mathscr{L}(\boldsymbol{v})$$

and

$$j_{\boldsymbol{z}}(\boldsymbol{v}) = \int_{\Gamma_C} \left(\frac{1}{2\varepsilon} [v_n - g_n]_+^2 + \mathfrak{F}(\boldsymbol{z}) \frac{1}{\varepsilon} [z_n - g_n]_+ \Phi_\eta(\boldsymbol{v_t} - \boldsymbol{w_t}) \right) ds_{\boldsymbol{x}}.$$

As in the corresponding part of Section 1.5 the functional $J = J_0 + j_{\boldsymbol{z}} : \boldsymbol{U} + \boldsymbol{H}_0^1(\Omega) \to \mathbb{R}$ is strictly convex and continuous, hence it is weakly lower semicontinuous. It is coercive, since J_0 is coercive and $j_{\boldsymbol{z}}(\boldsymbol{v}) \geq 0$ for all $\boldsymbol{v} \in \boldsymbol{U} + \boldsymbol{H}_0^1(\Omega)$. According to Theorem 1.5.3, problem (3.1.5) has a unique solution $\boldsymbol{u}$. Inserting the test function $\boldsymbol{v} = \boldsymbol{u} - \boldsymbol{W}$ with a $\boldsymbol{W} \in \boldsymbol{H}^1(\Omega)$ that satisfies $\boldsymbol{W} = \boldsymbol{U}$ on Γ_U and $\boldsymbol{W_t} = 0$, $W_n \leq g_n$ on Γ_C yields the *a priori* estimate

$$\|\boldsymbol{u}\|_{\boldsymbol{H}^1(\Omega)} \leq c \tag{3.1.6}$$

with constant c independent of $\boldsymbol{z}$, ε and η. Due to the existence and uniqueness result for problem (3.1.5), the operator $\mathfrak{I}$ mapping any argument $\boldsymbol{z} \in \boldsymbol{L}_2(\Gamma)$ to the solution $\boldsymbol{u}$ of (3.1.5) is well defined. A solution of problem (3.1.2) is given by a fixed point of this operator. The existence of a fixed point follows from the fixed point theorem of Schauder, applied to the space $\mathscr{X} \equiv \boldsymbol{L}_2(\Gamma)$ and the subset

$$\mathscr{C} \equiv \left\{ \boldsymbol{v} \in \mathscr{X};\ \|\boldsymbol{v}\|_{\mathscr{X}} \leq \tilde{c} \right\}.$$

Here, the constant $\tilde{c}$ is taken from the estimate

$$\|\boldsymbol{u}\|_{\boldsymbol{H}^{1/2}(\Gamma)} \leq \tilde{c}$$

valid for the solution of (3.1.5); this is true due to the *a priori* estimate (3.1.6) and the trace theorem. In order to verify the complete continuity of the operator $\mathfrak{I} : \mathscr{X} \to \mathscr{X}$, let us consider a sequence $\boldsymbol{z}_k$ having the limit $\boldsymbol{z}$ in $\mathscr{X}$. Then the strong convergence $\frac{1}{\varepsilon}[(z_k)_n - g_n]_+ \to \frac{1}{\varepsilon}[z_n - g_n]_+$ in $L_2(\Gamma_C)$ holds. The Lebesgue dominated convergence theorem yields $\mathfrak{F}(\boldsymbol{z}_k) \to \mathfrak{F}(\boldsymbol{z})$

in $L_q(\Gamma_C)$ for any $q < \infty$. Let $\boldsymbol{u}_k$ be the solution of problem (3.1.5) to the given friction defined by $\boldsymbol{z}_k$. Due to the *a priori estimate* there exists a subsequence—denoted by $\boldsymbol{u}_k$ again—which converges weakly in $\boldsymbol{H}^1(\Omega)$ and, due to the trace theorem and Theorem 2.8.14, strongly in $\boldsymbol{L}_p(\Gamma_C)$ for any $p \in \left[1, 2 + 2/(N-2)\right)$ in the case $N \geq 3$ and $p \in [1, +\infty)$ in the case $N = 2$ to a function $\boldsymbol{u}$. Performing the limit $k \to \infty$ in problem (3.1.5) with given friction defined by $\boldsymbol{z}_k$ and with solution $\boldsymbol{u}_k$ it is verified that the limit $\boldsymbol{u}$ solves the problem with given friction defined by $\boldsymbol{z}$. Since this solution is unique, the whole sequence $\boldsymbol{u}_k$ converges to this limit. Adding the term $A(\boldsymbol{u}, \boldsymbol{u}_k - \boldsymbol{u})$ to both sides of (3.1.5) with solution $\boldsymbol{u}_k$ and test function $\boldsymbol{v} = \boldsymbol{u} - \boldsymbol{u}_k$ and passing to the limit $k \to \infty$ it is proved that the convergence $\boldsymbol{u}_k \to \boldsymbol{u}$ is in fact strong in $\boldsymbol{H}^1(\Omega)$. The trace theorem and the compact imbedding $\boldsymbol{H}^{1/2}(\Gamma) \hookrightarrow \boldsymbol{L}_2(\Gamma)$ yield the complete continuity of the operator $\mathfrak{J}$ and the assumptions of the fixed point theorem of Schauder are satisfied. Hence the existence of a solution to problem (3.1.3) and also the *a priori* estimate (3.1.4) are verified.

By a similar procedure we verify the convergence of a sequence $\boldsymbol{u}_{\varepsilon,\eta_k}$ of solutions of (3.1.3) with $\eta_k \to 0$ for fixed ε. In fact, for suitable sequences η_k and $\boldsymbol{u}_k = \boldsymbol{u}_{\varepsilon,\eta_k}$ the convergences $\boldsymbol{u}_k \rightharpoonup \boldsymbol{u}$ in $\boldsymbol{H}^1(\Omega)$ and $\boldsymbol{u}_k \to \boldsymbol{u}$ in $\boldsymbol{L}_p(\Gamma)$ with p defined above are valid. Passing to the limit in the smoothed variational inequality

$$
\begin{aligned}
&A(\boldsymbol{u}_k, \boldsymbol{v} - \boldsymbol{u}_k) + \int_{\Gamma_C} \tfrac{1}{\varepsilon}\left[(u_k)_n - g_n\right]_+ \left(v_n - (u_k)_n\right) ds_{\boldsymbol{x}} \\
&\quad + \int_{\Gamma_C} \mathfrak{F}(\boldsymbol{u}_k) \tfrac{1}{\varepsilon}\left[(u_k)_n - g_n\right]_+ \cdot \\
&\quad \cdot \left(\Phi_{\eta_k}(\boldsymbol{v_t} - \boldsymbol{w_t}) - \Phi_{\eta_k}\left((\boldsymbol{u}_k)_{\boldsymbol{t}} - \boldsymbol{w_t}\right)\right) ds_{\boldsymbol{x}} \geq \mathscr{L}(\boldsymbol{v} - \boldsymbol{u}_k)
\end{aligned}
\tag{3.1.7}
$$

and using the lower semicontinuity of positive (semi-)definite bilinear forms,

$$
\liminf_{k \to \infty} A(\boldsymbol{u}_k, \boldsymbol{u}_k) \geq A(\boldsymbol{u}, \boldsymbol{u})
$$

and the approximation property of Φ_{η_k} yields that $\boldsymbol{u}$ is a solution of (3.1.2). Then the usual trick of adding $A(\boldsymbol{u}, \boldsymbol{u}_k - \boldsymbol{u})$ to both sides of (3.1.7) with test function $\boldsymbol{v} = \boldsymbol{u}$ and passing to the limit again proves the strong convergence $\boldsymbol{u}_k \to \boldsymbol{u}$ in $\boldsymbol{H}^1(\Omega)$. □

In the proof of Lemma 3.1.2 the *completely continuous* character of the friction functional was essential. It enables overcoming the difficulties with the non-monotone friction term by the usage of *strong* convergences. However, in the limit $\varepsilon \to 0$ this complete continuity disappears. With the *a priori* estimates available from (3.1.4) it is therefore not possible to prove the convergence of an appropriate sequence $\boldsymbol{u}_{\varepsilon_k}$ of solutions to (3.1.2) for some sequence $\varepsilon_k \to 0$ to a solution of the original contact problem directly. Fortunately, with the techniques of the preceding chapters it is possible to restore the required strong convergence in the friction term by proving some additional regularity of the solution on the contact part of the boundary. The required additional assumptions are given in

3.1.3 Assumption. *In addition to Assumptions 3.1.1, let* $\Gamma_C \in C^{1+\beta}$ *with* $\beta > \frac{1}{2} + \alpha$ *for a parameter* $0 < \alpha \leq \frac{1}{2}$, $g_n \in H^{1/2+\alpha}(\Gamma_C)$, $\boldsymbol{w} \in H^{1/2+\mu}(\Gamma_C)$ *with* $\mu > \alpha$ *in the case* $\alpha < \frac{1}{2}$ *or* $\mu = \frac{1}{2}$ *in the case* $\alpha = \frac{1}{2}$ *and* $\boldsymbol{f} \in \mathring{\boldsymbol{H}}^{-1+\alpha}(\Omega_C)$ *with a subdomain* $\Omega_C \subset \Omega$ *satisfying* $\Gamma_C \subset \partial\Omega_C$. *The coefficients of the bilinear form* A *belong to* $C^{\beta-1/2}(\Omega)$. *For all* $\boldsymbol{u} \in \mathbb{R}^N$ *the support of* $\mathfrak{F}(\cdot, \boldsymbol{u})$ *is contained in a set* $\Gamma_{\mathfrak{F}} \subset \Gamma_C$ *having a positive distance to* $\Gamma \setminus \Gamma_C$. *The coefficient of friction is bounded by*

$$\|\mathfrak{F}\|_{L_\infty(\mathbb{R}^N \times \Gamma_C)} \leq C_{\mathfrak{F}}$$

with the constant $C_{\mathfrak{F}}$ *defined in* (3.1.24) *below.*

Then the following assertion holds

3.1.4 Theorem. *Under Assumption 3.1.3 every solution* $\boldsymbol{u} = \boldsymbol{u}_\varepsilon$ *of problem* (3.1.2) *satisfies the* a priori *estimate*

$$\|\boldsymbol{u}_\varepsilon\|_{\boldsymbol{H}^{1/2+\alpha}(\Gamma_{\mathfrak{F}})} + \left\| \tfrac{1}{\varepsilon} \left[(u_\varepsilon)_n - g_n \right]_+ \right\|_{H^{-1/2+\alpha}(\Gamma_{\mathfrak{F}})} \leq c_1$$

with a constant depending on the geometry of the domain and on the given data, but not on the penalty parameter ε.

Proof. The proof of this theorem is performed with the help of a local rectification of the boundary as described in Section 1.6 or Section 1.7.3. For simplicity of the presentation we employ the direct rectification from Section 1.6 only; the modifications required for the geometrical rectification mentioned at the end of Section 1.7.3 are described there. In order to get an optimal result it is important to apply the shift technique to the penalized but non-smoothed problem (3.1.2) in the case $\alpha < \frac{1}{2}$ and to the smoothed problem (3.1.3) in the case $\alpha = \frac{1}{2}$; the reason will be explained later. Both cases can be done in a uniform notation for the variational inequality

$$\begin{aligned} A(\boldsymbol{u}, \boldsymbol{v} - \boldsymbol{u}) &+ \int_{\Gamma_C} P_\varepsilon(\boldsymbol{u}) \, (v_n - u_n) \, ds_{\boldsymbol{x}} \\ &+ \int_{\Gamma_C} \mathfrak{G}(\boldsymbol{u}) \left(\mathfrak{T}(\boldsymbol{v} - \boldsymbol{w}) - \mathfrak{T}(\boldsymbol{u} - \boldsymbol{w}) \right) ds_{\boldsymbol{x}} \geq \mathscr{L}(\boldsymbol{v} - \boldsymbol{u}) \end{aligned} \tag{3.1.8}$$

with $P_\varepsilon(\boldsymbol{u}) = \frac{1}{\varepsilon}[u_n - g_n]_+$,

$$\mathfrak{G}(\boldsymbol{u}) = \begin{cases} \mathfrak{F}(\boldsymbol{u})P_\varepsilon(\boldsymbol{u}) & \text{for } \alpha < \frac{1}{2}, \\ \mathfrak{F}(\boldsymbol{u})P_\varepsilon(\boldsymbol{u})\nabla\Phi_\eta(\boldsymbol{u_t} - \boldsymbol{w_t}) & \text{for } \alpha = \frac{1}{2} \end{cases}$$

and

$$\mathfrak{T}(\boldsymbol{v}) = \begin{cases} |\boldsymbol{v_t}| & \text{for } \alpha < \frac{1}{2}, \\ \boldsymbol{v_t} & \text{for } \alpha = \frac{1}{2}. \end{cases}$$

In the case $\alpha = \frac{1}{2}$ inequality (3.1.8) follows from variational equation (3.1.3) for test function $\boldsymbol{v} - \boldsymbol{u}$.

Let us recall the notations of Section 1.6 for the direct rectification, namely the finite covering of the contact part of the boundary $\mathscr{U}_\delta \equiv \{U_i; i \in \mathscr{I}_\delta\}$, the appropriate local rectification maps $\mathfrak{R}_i = \Psi_i \circ \mathscr{O}_i : U_i \to O_i$ with $\operatorname{diam} O_i \leq \delta$, their inverse maps $\mathfrak{Q}_i = \mathfrak{R}_i^{-1}$, $i \in \mathscr{I}(\delta)$ and the partition of unity $\mathfrak{P}_\delta \equiv \{\varrho_i; i \in \mathscr{I}_\delta\}$ subordinate to the covering $\mathscr{U}_\delta$. The sets $O_\delta = B_{N-1}(\delta) \times (0, \delta)$ and $\Omega_{\varrho i} = \mathfrak{Q}_i(O_\delta)$ are introduced, and for a sufficiently small δ there holds $\Omega_{\varrho_i} \subset \Omega$. The localization and rectification for a fixed $i \in \mathscr{I}_\delta$ is done by taking the test function $\boldsymbol{u} + \varrho(\boldsymbol{v} - \varrho\boldsymbol{u})$ in relation (3.1.8), then shifting the cut-off function ϱ from the right hand side of each form to its left hand side and performing the transform of variables $\boldsymbol{y} = \mathfrak{R}(\boldsymbol{x})$. For simplicity of the notation we omit the localization index i. The result then reads

$$\begin{aligned} &\int_{O_\delta} \big(a(\varrho\boldsymbol{u}, \boldsymbol{v} - \varrho\boldsymbol{u}) + b(\boldsymbol{u}, \boldsymbol{v} - \varrho\boldsymbol{u}) + b_0(\varrho\boldsymbol{u}, \boldsymbol{v} - \varrho\boldsymbol{u})\big)\, d\boldsymbol{y} \\ &\qquad + \int_{S_\delta} \varrho P_\varepsilon(\boldsymbol{u})\big(v_n - (\varrho\boldsymbol{u})_n\big) J_\Gamma\, ds_{\boldsymbol{y}} \\ &\qquad + \int_{S_\delta} \varrho\mathfrak{G}(\boldsymbol{u})\left(\mathfrak{T}(\boldsymbol{v} - \varrho\boldsymbol{w}) - \mathfrak{T}(\varrho\boldsymbol{u} - \varrho\boldsymbol{w})\right) J_\Gamma\, ds_{\boldsymbol{y}} \\ &\quad \geq \int_{O_\delta} (\varrho\boldsymbol{f} - d\varrho\boldsymbol{u}) \cdot (\boldsymbol{v} - \varrho\boldsymbol{u})\, d\boldsymbol{y}. \end{aligned} \tag{3.1.9}$$

Here the notation $S_\delta \equiv B_{N-1}(\delta) \times \{0\}$ is used. The determinant of the Jacobian of this transformation is equal to 1, and the density of surface measure is equal to $J_\Gamma \equiv \sqrt{1 + |\nabla\psi|^2}$ with the function ψ from the representation $\Psi(x', x_N) = (x', x_N - \psi(x'))^\top$. In the friction term the estimate

$$\mathfrak{T}(\boldsymbol{u} + \varrho(\boldsymbol{v} - \varrho\boldsymbol{u}) - \boldsymbol{w}) - \mathfrak{T}(\boldsymbol{u} - \boldsymbol{w}) \leq \varrho\mathfrak{T}(\boldsymbol{v} - \varrho\boldsymbol{w}) - \varrho\mathfrak{T}(\varrho\boldsymbol{u} - \varrho\boldsymbol{w})$$

is used; it is valid for both definitions of $\mathfrak{T}$. For any function $\varphi : \Omega_\varrho \to \mathbb{R}$ we denote the transformed function $\varphi \circ \mathfrak{Q}$ again by φ. The forms a, b and

b_0 are defined on O_δ by the formulae (1.7.26) and (1.7.28). In the case of anisotropic material, the transformed bilinear form a has the same constants of ellipticity a_0 and of boundedness A_0; for homogeneous isotropic material it remains the same as the original one. The bilinear form $b : \boldsymbol{u}, \boldsymbol{v} \mapsto a(\boldsymbol{u}, \varrho\boldsymbol{v}) - a(\varrho\boldsymbol{u}, \boldsymbol{v})$ is a compact perturbation satisfying

$$|b(\boldsymbol{u}, \boldsymbol{v})| \leq c_1|\boldsymbol{u}||\nabla\boldsymbol{v}| + c_2|\nabla\boldsymbol{u}||\boldsymbol{v}|.$$

The form b_0 arises from the local rectification of Γ_C; it depends linearly on the components of $\nabla\psi$. Due to the special local coordinates it holds $|\nabla\psi(\boldsymbol{y}')| \leq c_8\delta^{\widetilde{\beta}}$, $\boldsymbol{y}' \in B_{N-1}(0,\delta)$, with $\widetilde{\beta} = \min\{\beta, 1\}$. Hence it follows for any suitable vector fields $\boldsymbol{v}, \boldsymbol{w}$

$$|b_0(\boldsymbol{v}, \boldsymbol{w})| \leq c\delta^{\widetilde{\beta}}|\nabla\boldsymbol{v}|\,|\nabla\boldsymbol{w}|. \tag{3.1.10}$$

In the anisotropic case we formally extend the coefficients of the bilinear form a from the domain of definition O_δ onto $O \equiv \mathbb{R}^{N-1} \times \mathbb{R}_+$ in such a way that they remain Lipschitz, bounded and elliptic with the same constants a_0 and A_0 as the original coefficients. In the homogeneous, isotropic case the appropriate extension of constant coefficients is clear. The coefficients of b vanish at the intersection of the boundary ∂O_δ of O_δ and O and can be extended by 0. The coefficients of b_0 and the density of surface measure J_Γ are extended from O_δ and $B_{N-1}(0,\delta)$ onto O and $\partial O \equiv \mathbb{R}^{N-1} \times \{0\}$, respectively, such that the extended functions are still Hölder continuous with the original exponent and bounded. The functions $\varrho\boldsymbol{u}$, ϱg_n, $\varrho\boldsymbol{w}$ and the modified volume force $\varrho\boldsymbol{f} - \varrho d\boldsymbol{u}$ are extended by 0 onto O and ∂O, respectively. Let us denote the extensions by the corresponding gothic letters $\mathfrak{u}$, $\mathfrak{g}_n$, $\mathfrak{w}$ and $\mathfrak{f}$. The argument $\boldsymbol{u}$ in the form $b(\boldsymbol{u}, \cdot)$ is extended to $\boldsymbol{H}^1(O)$ such that $\|\boldsymbol{u}\|_{\boldsymbol{H}^1(O)} \leq c\|\boldsymbol{u}\|_{H^1(O_\delta)}$ with a c independent of $\boldsymbol{u}$. Let us also formally extend the original outer normal vector $\boldsymbol{n}$ to the whole ∂O in such a way that $\boldsymbol{n}$ remains Hölder continuous with the original exponent and $|\boldsymbol{n} - (0, \ldots, 0, -1)| \leq c_1\delta^{\widetilde{\beta}}$. The density of surface measure is included in the penalty term,

$$\mathscr{S}(\mathfrak{u}) \equiv P_\varepsilon(\mathfrak{u})J_\Gamma,$$

or in the function $\mathfrak{G}$,

$$\mathfrak{G}_{\text{new}} = \mathfrak{G}_{\text{old}}J_\Gamma$$

and we forget the index "new". With this notation the following variational inequality is valid:

$$
\begin{aligned}
&\int_O \big(a(\mathfrak{u}, \boldsymbol{v}-\mathfrak{u}) + b(\boldsymbol{u}, \boldsymbol{v}-\mathfrak{u}) + b_0(\mathfrak{u}, \boldsymbol{v}-\mathfrak{u})\big)\, d\boldsymbol{x} \\
&\quad + \int_{\partial O} \mathscr{S}(\mathfrak{u})(v_{\widetilde{n}} - \mathfrak{u}_{\widetilde{n}})\, ds_{\boldsymbol{x}} \\
&\quad + \int_{\partial O} \mathfrak{G}(\mathfrak{u})\,(\mathfrak{T}(\boldsymbol{v}-\mathfrak{w}) - \mathfrak{T}(\mathfrak{u}-\mathfrak{w}))\, ds_{\boldsymbol{x}} \\
&\geq \int_O \mathfrak{f}\cdot(\boldsymbol{v}-\mathfrak{u})\, d\boldsymbol{x}.
\end{aligned}
\tag{3.1.11}
$$

The tildes at the normal indicate that they are calculated with respect of the original normal vector $\boldsymbol{n}$.

We study the regularity of the solution $\mathfrak{u}$ of this inequality with the shift technique as described in Subsection 1.7.3. For a function $z : O \to \mathbb{R}$ and for $\boldsymbol{h} \equiv (\boldsymbol{h}', 0) \in \mathbb{R}^N$ let

$$
z_{-\boldsymbol{h}}(\boldsymbol{x}, \boldsymbol{y}) = S_{\boldsymbol{h}} z(\boldsymbol{x}, \boldsymbol{y}) = z(\boldsymbol{x}+\boldsymbol{h}, \boldsymbol{y}) \text{ and } \Delta_{\boldsymbol{h}} z \equiv z_{-\boldsymbol{h}} - z
$$

denote the shifted function and the corresponding difference operator, respectively, cf. (1.7.4) and (1.7.5). We insert the test function $\boldsymbol{v} \equiv \mathfrak{u}_{-\boldsymbol{h}}$ into variational inequality (3.1.11). Then we shift the whole inequality (3.1.11) into the direction $\boldsymbol{h}$ and insert the test function $\boldsymbol{v}_{-\boldsymbol{h}} \equiv \mathfrak{u}$ into the shifted inequality. We add both inequalities, multiply the result by $|\boldsymbol{h}'|^{-N+1-2\alpha}$ and integrate the product with respect to $\boldsymbol{h}' \in \mathbb{R}^{N-1}$. Then most expressions can be estimated with the help of the given data. The coefficients c_i, $i \in \mathbb{N}$, may depend on the data

$$
\begin{aligned}
&a_0, A_0, \|\boldsymbol{f}\|_{\mathring{\boldsymbol{H}}^{-1+\alpha}(\Omega_C)}, \|g_n\|_{H^{1/2+\alpha}(\Gamma_C)}, \|\psi\|_{C^{1+\beta}(\mathbb{R}^{N-1})}, \\
&\qquad \|\varrho\|_{C^{1+\beta}(\Omega)}, \max_{\substack{i,j,k,\ell\\=1,\dots,N}} \|a_{ijk\ell}\|_{C^\beta(\Omega)}
\end{aligned}
\tag{3.1.12}
$$

and on the constants from the *a priori* estimate (3.1.4). For simplicity of the notation, let us omit the argument $\mathfrak{u}$ in $\mathscr{S} = \mathscr{S}(\mathfrak{u})$ and $\mathfrak{G} = \mathfrak{G}(\mathfrak{u})$ and let

$$
\mathscr{A}^2(\boldsymbol{u}) \equiv \int_{\mathbb{R}^{N-1}} |\boldsymbol{h}'|^{-N+1-2\alpha} A(\Delta_{\boldsymbol{h}} \boldsymbol{u}, \Delta_{\boldsymbol{h}} \boldsymbol{u})\, d\boldsymbol{h}'.
$$

The penalty term can be estimated with the decomposition

$$
\begin{aligned}
&\mathscr{S}_{-\boldsymbol{h}}\big(\mathfrak{u}_{-\boldsymbol{h}}\cdot\widetilde{\boldsymbol{n}}_{-\boldsymbol{h}} - \mathfrak{u}\cdot\widetilde{\boldsymbol{n}}_{-\boldsymbol{h}}\big) + \mathscr{S}\big(\mathfrak{u}\cdot\widetilde{\boldsymbol{n}} - \mathfrak{u}_{-\boldsymbol{h}}\cdot\widetilde{\boldsymbol{n}}\big) \\
&\quad = \Delta_{\boldsymbol{h}}\mathscr{S}\,\Delta_{\boldsymbol{h}}(\mathfrak{u}\cdot\widetilde{\boldsymbol{n}} - g_{\widetilde{n}}) + \Delta_{\boldsymbol{h}}\mathscr{S}\,\Delta_{\boldsymbol{h}} g_{\widetilde{n}} - \Delta_{\boldsymbol{h}}\mathscr{S}\,\mathfrak{u}\cdot\Delta_{\boldsymbol{h}}\widetilde{\boldsymbol{n}}
\end{aligned}
$$

$$+ \mathscr{S}\Delta_h \mathfrak{u}\Delta_h \tilde{n},$$

the monotonicity of the penalty functional and the C^β continuity of the normal vector $\boldsymbol{n}$ by

$$\begin{aligned}
&\int_{\mathbb{R}^{N-1}}\int_{\partial O} \frac{\mathscr{S}_{-h}\big(\mathfrak{u}_{-h}\cdot\tilde{n}_{-h} - \mathfrak{u}\cdot\tilde{n}_{-h}\big) + \mathscr{S}\big(\mathfrak{u}\cdot\tilde{n} - \mathfrak{u}_{-h}\cdot\tilde{n}\big)}{|h'|^{N-1+2\alpha}}\, dx\, dh' \\
&\geq -\left(\int_{\mathbb{R}^{N-1}} \frac{\|\Delta_h \mathscr{S}\|^2_{H^{-1/2}(\partial O)}}{|h'|^{N-1+2\alpha}}\, dh'\right)^{1/2} \cdot \\
&\qquad \cdot \Big(c_1\|g_{\tilde{n}}\|_{H^{1/2+\alpha}(\partial O)} + c_2\|\mathfrak{u}\|_{H^{1/2}(\partial O)}\Big) \\
&\quad - c_3\|\mathscr{S}\|_{H^{-1/2}(\partial O)}\left(\int_{\mathbb{R}^{N-1}} \frac{\|\Delta_h \mathfrak{u}\|^2_{H^{1/2}(\partial O)}}{|h'|^{N-1+2\alpha}}\, dh'\right)^{1/2}.
\end{aligned}$$

The friction term is estimated starting from the decomposition

$$\begin{aligned}
&\mathfrak{G}_{-h}(\mathfrak{T}_{-h}(\mathfrak{u}_{-h} - \mathfrak{w}_{-h}) - \mathfrak{T}_{-h}(\mathfrak{u} - \mathfrak{w}_{-h})) \\
&\quad + \mathfrak{G}(\mathfrak{T}(\mathfrak{u} - \mathfrak{w}) - \mathfrak{T}(\mathfrak{u}_{-h} - \mathfrak{w})) \\
&= \Delta_h \mathfrak{G}\Delta_h(\mathfrak{T}(\mathfrak{u} - \mathfrak{w})) + \mathfrak{G}_{-h}(\mathfrak{T}(\mathfrak{u} - \mathfrak{w}) - \mathfrak{T}_{-h}(\mathfrak{u} - \mathfrak{w}_{-h})) \\
&\quad + \mathfrak{G}(\mathfrak{T}_{-h}(\mathfrak{u}_{-h} - \mathfrak{w}_{-h}) - \mathfrak{T}(\mathfrak{u}_{-h} - \mathfrak{w})).
\end{aligned} \tag{3.1.13}$$

In the case $\alpha < \frac{1}{2}$ we have $\mathfrak{T}(v) = |v_{\tilde{t}}|$. The second term on the right hand side of (3.1.13) is bounded by

$$\begin{aligned}
&\big|\mathfrak{G}_{-h}(\mathfrak{T}(\mathfrak{u} - \mathfrak{w}) - \mathfrak{T}_{-h}(\mathfrak{u} - \mathfrak{w}_{-h}))\big| \\
&= \mathfrak{G}_{-h}\Big|(\mathfrak{T}(\mathfrak{u} - \mathfrak{w}) - \mathfrak{T}_{-h}(\mathfrak{u} - \mathfrak{w})) \\
&\qquad\qquad + (\mathfrak{T}_{-h}(\mathfrak{u} - \mathfrak{w}) - \mathfrak{T}_{-h}(\mathfrak{u} - \mathfrak{w}_{-h}))\Big| \\
&\leq \mathfrak{G}_{-h}\Big(c_1|h'|^\beta|\mathfrak{u} - \mathfrak{w}| + c_2|\mathfrak{w} - \mathfrak{w}_{-h}|\Big);
\end{aligned}$$

this follows from $\mathfrak{G} \geq 0$ and $\tilde{n} \in C^\beta$. Since $\beta > 2\alpha$ and $\boldsymbol{w} \in H^{1/2+\mu}(\partial O)$ holds with $\mu > \alpha$, the estimate

$$\begin{aligned}
&\left|\int_{\mathbb{R}^{N-1}}\int_{\partial O} \frac{\mathfrak{G}_{-h}(\mathfrak{T}(\mathfrak{u} - \mathfrak{w}) - \mathfrak{T}_{-h}(\mathfrak{u} - \mathfrak{w}_{-h}))}{|h'|^{N-1+2\alpha}}\, ds_{\boldsymbol{x}}\, dh'\right| \\
&\leq c_1(\beta)\|\mathfrak{G}\|_{H^{-1/2}(\partial O)}\|\mathfrak{u} - \mathfrak{w}\|_{H^{1/2}(\partial O)} \\
&\quad + c_2(\alpha,\mu)\|\mathfrak{G}\|_{H^{-1/2+\alpha}(\partial O)}\|\boldsymbol{w}\|_{H^{1/2+\mu}(\partial O)}
\end{aligned}$$

is valid. This relation follows with the help of

$$\int_{\mathbb{R}^{N-1}}\int_{\partial O} \frac{\mathfrak{G}_{-h}|\mathfrak{w} - \mathfrak{w}_{-h}|}{|h'|^{N-1+2\alpha}}\, ds_{\boldsymbol{x}}\, dh'$$

$$\leq \left(\int_{\mathbb{R}^{N-1}} \frac{\|\mathfrak{G}_{-h}\|^2_{H^{-1/2+\alpha}(\partial O)}}{|h'|^{N-1+2\alpha} \min\{|h'|,1\}^{-2\mu}} \, dh' \right)^{1/2} \cdot$$
$$\cdot \left(\int_{\mathbb{R}^{N-1}} \frac{\|\mathfrak{w}_{-h} - \mathfrak{w}\|^2_{H^{1/2-\alpha}(\partial O)}}{|h'|^{N-1+2\alpha} \min\{|h'|,1\}^{2\mu}} \, dh' \right)^{1/2}$$
$$\leq c_2(\alpha,\mu)\|\mathfrak{G}\|_{H^{-1/2+\alpha}(\partial O)}\|\mathfrak{w}\|_{H^{1/2+\mu}(\partial O)}.$$

Analogously, the last term in (3.1.13) is bounded by

$$\left|\mathfrak{G}(\mathfrak{T}_{-h}(\mathfrak{u}_{-h} - \mathfrak{w}_{-h}) - \mathfrak{T}(\mathfrak{u}_{-h} - \mathfrak{w}))\right|$$
$$\leq \mathfrak{G}\Big(c_1|h'|^{\beta}|\mathfrak{u}_{-h} - \mathfrak{w}_{-h}| + c_2|\mathfrak{w}_{-h} - \mathfrak{w}|\Big),$$

and the corresponding integral can be estimated as above. As a consequence, the friction term is bounded by

$$\left| \int_{\mathbb{R}^{N-1}} \int_{\partial O} \frac{\Delta_h \mathfrak{G} \Delta_h (\mathfrak{T}(\mathfrak{u} - \mathfrak{w}))}{|h'|^{N-1+2\alpha}} \, ds_{\boldsymbol{x}} \, dh' \right|$$
$$+ c_1(\beta)\|\mathfrak{G}\|_{H^{-1/2}(\partial O)}\big(\|\mathfrak{u}\|_{H^{1/2}(\partial O)} + \|\mathfrak{w}\|_{H^{1/2}(\partial O)}\big)$$
$$+ c_2\|\mathfrak{G}\|_{H^{-1/2+\alpha}(\partial O)}\|w\|_{H^{1/2+\mu}(\partial O)}.$$

Here the first term is further estimated by

$$c_{N-1}(\alpha)\|\mathfrak{F}\|_{L_\infty(\mathbb{R}^N \times \partial O)}\|P_\varepsilon(\mathfrak{u})\|_{H^{-(1/2-\alpha)}(\partial O)} \cdot$$
$$\cdot \big((1+\vartheta)d_{N-1}(\alpha)\|\mathfrak{T}(\mathfrak{u} - \mathfrak{w})\|_{\boldsymbol{H}^{1/2+\alpha}(\partial O)} + c\|\mathfrak{T}(u - \mathfrak{w})\|_{\boldsymbol{H}^{1/2}(\partial O)}\big)$$

with

$$d_{N-1}(\alpha) = \sqrt{\frac{c_{N-1}(\frac{1}{2} - \alpha)}{c_{N-1}(\frac{1}{2} + \alpha)}} \tag{3.1.14}$$

and $\vartheta = 0$ for $\alpha \leq \frac{1}{4}$, ϑ arbitrarily small at the expense of the constant $c = c(\vartheta)$ for $\frac{1}{4} < \alpha < \frac{1}{2}$ (c.f. (1.7.13)) and the estimate is done for the first case $\alpha < \frac{1}{2}$.

In the second case $\alpha = \frac{1}{2}$ the map $\mathfrak{T}$ is linear. As a consequence, the term $\mathfrak{w}$ in (3.1.13) cancels. We employ the decomposition

$$\begin{aligned} &\mathfrak{G}_{-h}(\mathfrak{T}_{-h}\mathfrak{u}_{-h} - \mathfrak{T}_{-h}\mathfrak{u}) + \mathfrak{G}(\mathfrak{T}\mathfrak{u} - \mathfrak{T}\mathfrak{u}_{-h}) \\ &= \Delta_h \mathfrak{G} \Delta_h (\mathfrak{T}\mathfrak{u}) - \Delta_h \mathfrak{G}(\Delta_h \mathfrak{T})\mathfrak{u} + \mathfrak{G}\Delta_h \mathfrak{T} \Delta_h \mathfrak{u}. \end{aligned} \tag{3.1.15}$$

Using the relation

$$\|\mathfrak{G}\|_{\boldsymbol{L}_2(\partial O)} = \|\mathfrak{F}P_\varepsilon(\mathfrak{u})\nabla\Phi_\eta(\mathfrak{u} - \mathfrak{w})J_\Gamma\|_{\boldsymbol{L}_2(\partial O)}$$
$$\leq C_1\|\mathfrak{F}\|_{L_\infty(\partial O \times \mathbb{R}^N)}\|P_\varepsilon(\mathfrak{u})\|_{L_2(\partial O)}$$

which is valid due to $|\nabla \Phi_\eta(v)| \le 1$, the integral over the first term on the right hand side of (3.1.15) is bounded by

$$\left| \int_{\mathbb{R}^{N-1}} \int_{\partial O} |h'|^{-N} \Delta_h \mathfrak{G} \Delta_h (\mathfrak{T}\mathbf{u})\, dx\, dh' \right| \\ \le c_{N-1}\big(\tfrac{1}{2}\big) \|\mathfrak{F}\|_{L_\infty(\partial O \times \mathbb{R}^N)} \|P_\varepsilon(\mathbf{u})\|_{L_2(\partial O)} \|\mathfrak{T}\mathbf{u}\|_{\boldsymbol{H}^1(\partial O)}.$$

The integral over the last term is estimated with the help of

$$|\mathfrak{G} \Delta_h \mathfrak{T} \Delta_h \mathbf{u}| \le c \min\{|h'|, 1\}\, |\mathfrak{G}|\, |\Delta_h \mathbf{u}|$$

by

$$\left| \int_{\mathbb{R}^{N-1}} \int_{\partial O} |h'|^{-N} \mathfrak{G} \Delta_h \mathfrak{T} \Delta_h \mathbf{u}\, dx\, dh' \right| \le c \|\mathfrak{G}\|_{\boldsymbol{L}_2(\partial O)} \|\mathbf{u}\|_{\boldsymbol{H}^{1/2}(\partial O)}.$$

For the remaining expression in (3.1.15) we use relation

$$\|(\Delta_h \mathfrak{T})\mathbf{u}\|_{\boldsymbol{H}^{1/2}(\partial O)} \le \|\Delta_h \mathfrak{T}\|_{L_\infty(\partial O)} \|\mathbf{u}\|_{\boldsymbol{H}^{1/2}(\partial O)} \\ + c(\beta) \|\Delta_h \mathfrak{T}\|_{W_\infty^{\beta/2}(\partial O)} \|\mathbf{u}\|_{L_2(\partial O)}.$$

This is further estimated with the obvious relation $\|\Delta_h \mathfrak{T}\|_{L_\infty(\partial O)} \le c|h'|$ and with

$$\|\Delta_h \mathfrak{T}\|_{W_\infty^{\beta/2}(\partial O)} \le c(\beta)|h'|^{\beta/2}.$$

The latter inequality is a consequence of the relation

$$|\mathfrak{T}(x + h + q) - \mathfrak{T}(x + h) - \mathfrak{T}(x + q) + \mathfrak{T}(x)| \le c|h'|^{\beta/2}|q|^{\beta/2}$$

valid for $h = (h', 0) \in \mathbb{R}^N$ and $q = (q', 0) \in \mathbb{R}^N$. This relation follows from the estimate

$$|\mathfrak{T}(x + q + h) - \mathfrak{T}(x + h) - \mathfrak{T}(x + q) + \mathfrak{T}(x)| \\ = \left| \int_0^1 \big(\nabla \mathfrak{T}(x + \vartheta h + q) - \nabla \mathfrak{T}(x + \vartheta h)\big) h\, d\vartheta \right| \\ \le c\|\mathfrak{T}\|_{C^\beta(\partial O)} |h'||q'|^{\beta - 1}$$

for $1 \le \beta \le 2$ and the similar relation

$$|\mathfrak{T}(x + q + h) - \mathfrak{T}(x + h) - \mathfrak{T}(x + q) + \mathfrak{T}(x)| \\ \le c\|\mathfrak{T}\|_{C^\beta(\partial O)} |q'||h'|^{\beta - 1}$$

derived by exchanging the roles of $\boldsymbol{q}$ and $\boldsymbol{h}$. Employing these auxiliary estimates we get

$$\left| \int_{\mathbb{R}^{N-1}} \int_{\partial O} \frac{\Delta_{\boldsymbol{h}} \mathfrak{G} \, (\Delta_{\boldsymbol{h}} \mathfrak{T}) \, \mathbf{u}}{|\boldsymbol{h}'|^N} \, dh' \, dx \right| \leq c \|\mathfrak{G}\|_{L_2(\partial O)} \|\mathbf{u}\|_{\boldsymbol{H}^{1/2}(\partial O)}.$$

The estimate of the total friction term in the case $\alpha = \frac{1}{2}$ is then given by

$$\begin{aligned} c_{N-1}\big(\tfrac{1}{2}\big) \|\mathfrak{F}\|_{L_\infty(\partial O \times \mathbb{R}^N)} \|P_\varepsilon(\mathbf{u})\|_{L_2(\partial O)} \|\mathfrak{T}\mathbf{u}\|_{\boldsymbol{H}^1(\partial O)} \\ + c_1 \|P_\varepsilon(\mathbf{u})\|_{L_2(\partial O)} \|\mathbf{u}\|_{\boldsymbol{H}^{1/2}(\partial O)}. \end{aligned}$$

Let us briefly explain why the different approaches for $\alpha < \frac{1}{2}$ and $\alpha = \frac{1}{2}$ are necessary. In the case $\alpha < \frac{1}{2}$ we can use the estimate $\|\Delta_{\boldsymbol{h}} \widetilde{\mathfrak{T}}\|_{L_\infty(\partial O)} \leq c|\boldsymbol{h}'|^\beta$ with $\beta > 2\alpha$ in order to cancel the factor $|\boldsymbol{h}'|^{-(N-1+2\alpha)}$ up to a locally integrable remainder $|\boldsymbol{h}'|^{-(N-1+2\alpha-\beta)}$. This is no longer possible for $\alpha = \frac{1}{2}$ any more. Therefore, we have to use "products of differences" in order to control the singular term $|\boldsymbol{h}'|^{-N}$. However, it is not possible to get those products using the non-linear norms. This is why the smoothing of the friction is necessary for $\alpha = \frac{1}{2}$. The smoothing cannot be applied for $\alpha < \frac{1}{2}$, because the estimate

$$\|\nabla \Phi_\eta P_\varepsilon\|_{L_2(\partial O)} \leq \|P_\varepsilon\|_{L_2(\partial O)}$$

is no longer valid, if the L_2-norm is replaced by a $H^{-(1/2-\alpha)}$-norm. Then some fractional derivatives of $\nabla \Phi_\eta$ will appear which are no longer uniformly bounded with respect to η.

Let us come back to the shift technique. All the other not yet mentioned terms are estimated as in Section 1.6. In fact, the estimates are easier as those given there, because the penalty method doesn't require the rotation of the normal vector as in the shift technique done for the fixed point approach. The result is the inequality

$$\begin{aligned} \mathscr{A}^2(\mathbf{u}) \leq \big(1 + c_1 \delta^{\widetilde{\beta}} + c_2 \vartheta\big) c_{N-1}(\alpha) d_{N-1}(\alpha) \|\mathfrak{F}\|_{L_\infty(\partial O \times \mathbb{R}^N)} \cdot \\ \cdot \|P_\varepsilon(\mathbf{u})\|_{H^{-(1/2-\alpha)}(\partial O)} \, \|\mathbf{u}_{\widetilde{t}}\|_{\boldsymbol{H}^{1/2+\alpha}(\partial O)} \\ + c_1(\vartheta) \|P_\varepsilon(\mathbf{u})\|_{H^{-(1/2-\alpha)}(\partial O)} \\ + c_2(\vartheta) \, \|\mathbf{u}_{\widetilde{t}}\|_{\boldsymbol{H}^{1/2+\alpha}(\partial O)} + c_3(\vartheta) \end{aligned} \tag{3.1.16}$$

with $d_{N-1}(\alpha)$ defined in (3.1.14) for $\alpha < \frac{1}{2}$ and $d_{N-1}(\frac{1}{2}) = 1$. The constant ϑ can be chosen arbitrarily small. The constants c_1, c_2 and c_3 depend only on the data from (3.1.12) and on ϑ, but neither on the penalty parameter ε nor on the smoothing parameter η.

For the further estimate of the term $\|\mathbf{u}_{\widetilde{t}}\|_{H^{1/2+\alpha}(\partial O)}$ we employ the special trace theorem

$$\|\mathbf{u}_{t}\|^2_{\boldsymbol{H}^{1/2}(\partial O)} \leq C_0^{(1)} \|\mathbf{u}\|^2_A + c\|\mathbf{u}_{t}\|^2_{L_2(\partial O)}. \tag{3.1.17}$$

In this estimate, whose proof is performed in the following subsection, the tangential component is taken with respect to the straightened boundary. This is admissible, since

$$\|\mathbf{u}_{\widetilde{t}}\|_{\boldsymbol{H}^{1/2+\alpha}(\partial O)} \leq \left(1 + c_1\delta^{\widetilde{\beta}}\right)\|\mathbf{u}_{t}\|_{\boldsymbol{H}^{1/2+\alpha}(\partial O)} + c_2\|\mathbf{u}\|_{L_2(\partial O)}$$

is valid due to the C^{β}-continuity of the normal vector $\widetilde{\boldsymbol{n}}$ and due to the fact that $|\widetilde{\boldsymbol{n}} - \boldsymbol{n}| \leq c\delta^{\widetilde{\beta}}$ on ∂O. Here it is necessary to have $\beta > \frac{1}{2} + \alpha$. Using this and the standard renormation technique described in Sections 1.7.1, 1.7.2, the estimate

$$\begin{aligned}\|\mathbf{u}_{\widetilde{t}}\|^2_{\boldsymbol{H}^{1/2+\alpha}(\partial O)} \leq{}& \left(1 + c_1\delta^{\widetilde{\beta}}\right) d_{N-1}^{-1}\left(\alpha, \tfrac{1}{2}\right) C_0^{(1)} \mathscr{A}^2(\mathbf{u}) \\ &+ c_2\|\mathbf{u}\|^2_{\boldsymbol{H}^{1/2}(\partial O)}\end{aligned} \tag{3.1.18}$$

is derived with $d_{N-1}(\alpha, \beta) = c_{N-1}(\alpha)c_{N-1}(\beta)/c_{N-1}(\alpha + \beta)$.

In order to finish the proof it is necessary to estimate the norm of the penalty term $\mathscr{S}$. This is done starting from the relation (1.7.7) which yields

$$\begin{aligned}\|\mathscr{S}\|^2_{H^{-(1/2-\alpha)}(\partial O)} \leq{}& (1 + \vartheta)\left(d^*_{N-1}(\alpha, \tfrac{1}{2})\right)^{-1} \cdot \\ &\cdot \int_{\mathbb{R}^{N-1}} |\boldsymbol{h}'|^{-(N-1+2\alpha)}\|\Delta_{\boldsymbol{h}}\mathscr{S}\|^2_{H^{-1/2}(\partial O)}\, d\boldsymbol{h}' \\ &+ c(\vartheta)\|\mathscr{S}\|^2_{H^{-1/2}(\partial O)}\end{aligned} \tag{3.1.19}$$

with an arbitrarily small $\vartheta > 0$ and $d^*_{N-1}(\alpha, \frac{1}{2}) = c_{N-1}(\alpha)c_{N-1}(\frac{1}{2} - \alpha)/c_{N-1}(\frac{1}{2})$, cf. (1.7.8). We employ the Green formula

$$\int_{\partial O} \mathscr{S}\, v_{\widetilde{n}}\, ds_{\boldsymbol{x}} = -\int_O \left(a(\mathbf{u}, \boldsymbol{v}) + b(\mathbf{u}, \boldsymbol{v}) + b_0(\boldsymbol{u}, \boldsymbol{v}) - \mathfrak{f}\cdot\boldsymbol{v}\right) d\boldsymbol{x}, \tag{3.1.20}$$

valid for functions $\boldsymbol{v}$ satisfying $\boldsymbol{v}_{\widetilde{t}} = 0$. The Green formula is used with the test function $\boldsymbol{v} = z\widetilde{\boldsymbol{n}} \in \boldsymbol{H}^{1/2}(\partial O)$, where z is defined by

$$\mathscr{F}(z;\boldsymbol{\xi}) \equiv \mathscr{F}(\mathscr{S};\boldsymbol{\xi})\left(1 + c_{N-1}(\tfrac{1}{2})\,|\boldsymbol{\xi}|\right)^{-1}.$$

This function fulfils the relations

$$\|z\|^2_{H^{1/2}(\partial O)} = \|\mathscr{S}\|^2_{H^{-1/2}(\partial O)} = \langle\mathscr{S}, z\rangle_{\partial O}.$$

We use a linear extension operator

$$\mathscr{E} : H^{1/2}(\partial O) \to H^1(O)$$

satisfying the inverse trace relation

$$A(\mathscr{E}\boldsymbol{v}, \mathscr{E}\boldsymbol{v}) \le C_0^{(2)} \|\boldsymbol{v}\|^2_{H^{1/2}(\partial O)} + c_1 \|\boldsymbol{v}\|^2_{L_2(\partial O)} \tag{3.1.21}$$

for all $\boldsymbol{v}$ with $\boldsymbol{v}_{\tilde{\boldsymbol{t}}} = \boldsymbol{0}$ and a minimal parameter $C_0^{(2)}$. The precise form of this extension will be given in the next section for different types of the constitutive relation. In fact, there this inverse trace estimate will be proved for a function satisfying $\boldsymbol{v}_{\boldsymbol{t}} = 0$; in other words, for the normal component of the *rectified* domain instead of the original domain. This is sufficient, because $\|v(\tilde{\boldsymbol{n}} - \boldsymbol{n})\|_{\boldsymbol{H}^{1/2}(\partial O)} \le c_1 \delta^{\tilde{\beta}} \|v\|_{\boldsymbol{H}^{1/2}(\partial O)} + c_2 \|v\|_{L_2(\partial O)}$ for every scalar function $v \in H^{1/2}(\partial O)$. We insert $\boldsymbol{v} = \mathscr{E}((z - z_{-\boldsymbol{h}})\tilde{\boldsymbol{n}})$ into (3.1.20), shift the equation (3.1.20) into the direction of $\boldsymbol{h}$, insert $\boldsymbol{v}_{-\boldsymbol{h}} = \mathscr{E}((z_{-\boldsymbol{h}} - z)\tilde{\boldsymbol{n}}_{-\boldsymbol{h}})$ into the shifted equation, add both equations, multiply the sum by $|\boldsymbol{h}'|^{-(N-1+2\alpha)}$, integrate the product with respect to $\boldsymbol{h}' \in \mathbb{R}^{N-1}$ and carry out the usual estimates. The result is

$$\begin{aligned}
&\int_{\mathbb{R}^{N-1}} |\boldsymbol{h}'|^{-(N-1+2\alpha)} \|\Delta_{\boldsymbol{h}} \mathscr{S}\|^2_{H^{-1/2}(\partial O)} \, d\boldsymbol{h}' \\
&\quad \le (1 + c_1 \delta^{\tilde{\beta}}) \int_{\mathbb{R}^{N-1}} |\boldsymbol{h}'|^{-(N-1+2\alpha)} \big(A(\Delta_{\boldsymbol{h}} \mathbf{u}, \Delta_{\boldsymbol{h}} \mathbf{u}) \cdot \\
&\qquad\qquad \cdot A\big(\mathscr{E}(\Delta_{\boldsymbol{h}}(z\tilde{\boldsymbol{n}})), \mathscr{E}(\Delta_{\boldsymbol{h}}(z\tilde{\boldsymbol{n}}))\big)\big)^{1/2} \, d\boldsymbol{h}' \\
&\quad + c_1 \left(\int_{\mathbb{R}^{N-1}} |\boldsymbol{h}'|^{-(N-1+2\alpha)} \|\Delta_{\boldsymbol{h}} \mathscr{S}\|^2_{H^{-1/2}(\partial O)} \, d\boldsymbol{h}' \right)^{1/2} \\
&\quad + c_2 \mathscr{A}(\mathbf{u}) + c_3.
\end{aligned}$$

Here the linear structure of $\mathscr{E}$ has been used in the formula $\mathscr{E}(\Delta_{\boldsymbol{h}}(z\tilde{\boldsymbol{n}})) = \Delta_{\boldsymbol{h}} \mathscr{E}(z\tilde{\boldsymbol{n}})$. The estimate of the term $A\big(\mathscr{E}(\Delta_{\boldsymbol{h}}(z\tilde{\boldsymbol{n}})), \mathscr{E}(\Delta_{\boldsymbol{h}}(z\tilde{\boldsymbol{n}}))\big)$ by means of the special trace relation gives

$$\begin{aligned}
&A\big(\mathscr{E}(\Delta_{\boldsymbol{h}}(z\tilde{\boldsymbol{n}})), \mathscr{E}(\Delta_{\boldsymbol{h}}(z\tilde{\boldsymbol{n}}))\big) \\
&\quad \le (1 + c_1 \delta^{\tilde{\beta}}) C_0^{(2)} \|\Delta_{\boldsymbol{h}} z\|'^2_{H^{1/2}(\partial O)} + c_2 \|\Delta_{\boldsymbol{h}} z\|^2_{L_2(\partial O)} \\
&\quad \le (1 + c_1 \delta^{\tilde{\beta}}) C_0^{(2)} \|\Delta_{\boldsymbol{h}} \mathscr{S}\|^2_{H^{-1/2}(\partial O)} + c_3 \|\Delta_{\boldsymbol{h}} \mathscr{S}\|^2_{H^{-1}(\partial O)}.
\end{aligned}$$

Here, the factor $1 + c_1 \delta^{\tilde{\beta}}$ is introduced, because then this estimate is also valid with the constant $C_0^{(2)}$ calculated for the extension of a scalar function by means of the normal n to the rectified domain; see the remark above. The estimate for the penalty term now reads

$$\begin{aligned}\|\mathscr{S}\|^2_{H^{-(1/2-\alpha)}(\partial O)} \leq (1 + c_1\delta^{\tilde{\beta}} + \vartheta)\big(d^*_{N-1}(\alpha, \tfrac{1}{2})\big)^{-1} \cdot \\ \cdot C_0^{(2)}\mathscr{A}^2(\mathbf{u}) + c_2\mathscr{A}(\mathbf{u}) + c_3(\vartheta)\end{aligned} \tag{3.1.22}$$

with $\vartheta > 0$ arbitrarily small. Hence from (3.1.16) we get

$$\begin{aligned}\mathscr{A}^2(\mathbf{u}) \leq (1 + c_1\delta^{\tilde{\beta}} + \vartheta)\|\mathfrak{F}\|_{L_\infty(\partial O\times\mathbb{R}^N)}(C_{\mathfrak{F}})^{-1}\mathscr{A}^2(\mathbf{u}) \\ + c_2\mathscr{A}(\mathbf{u}) + c_3(\vartheta)\end{aligned} \tag{3.1.23}$$

with

$$C_{\mathfrak{F}} = \left(C_0^{(1)}C_0^{(2)}\right)^{-1/2}. \tag{3.1.24}$$

If

$$\|\mathfrak{F}\|_{L_\infty(\partial O\times\mathbb{R}^N)} < C_{\mathfrak{F}} \tag{3.1.25}$$

holds and ϑ, δ are chosen sufficiently small, then $\mathscr{A}(\mathbf{u})$ is bounded. As a consequence the terms $\|\mathbf{u}\|_{\boldsymbol{H}^{1/2+\alpha}(\partial O)}$ and $\|\mathscr{S}\|_{H^{-(1/2-\alpha)}(\partial O)}$ are bounded, too. This is valid for each cut-off function $\varrho \in \mathfrak{P}_\delta$, if δ is sufficiently small. Hence the theorem is proved. □

Due to the better regularity of the solution on the contact part of the boundary proved in the previous theorem there exists a sequence of penalty parameters $\varepsilon_k \to 0$ and a corresponding sequence of solutions $\boldsymbol{u}_k = \boldsymbol{u}_{\varepsilon_k}$ to the penalized problem such that the following convergences are valid:

$$\begin{aligned}\boldsymbol{u}_k &\rightharpoonup \boldsymbol{u} && \text{in } \boldsymbol{H}^1(\Omega) \text{ and in } \boldsymbol{H}^{1/2+\alpha}(\Gamma_{\mathfrak{F}}),\\ \boldsymbol{u}_k &\to \boldsymbol{u} && \text{in } \boldsymbol{H}^\gamma(\Gamma_{\mathfrak{F}}) \text{ for every } \gamma < \tfrac{1}{2} + \alpha,\\ \mathfrak{F}(\boldsymbol{u}_k) &\to \mathfrak{F}(\boldsymbol{u}) && \text{in } L_q(\Gamma_{\mathfrak{F}}) \text{ for every } q < \infty \text{ and}\\ \sigma_n(\boldsymbol{u}_k) &\rightharpoonup \sigma_n(\boldsymbol{u}) && \text{in } H^{-(1/2-\alpha)}(\Gamma_{\mathfrak{F}}).\end{aligned} \tag{3.1.26}$$

The mentioned convergence of $\mathfrak{F}(\boldsymbol{u}_k)$ follows from the Lebesgue dominated convergence theorem combined with the uniform boundedness of $\mathfrak{F}(\boldsymbol{u}_k)$ in $L_\infty(\Gamma_C)$ and the strong convergence of $\boldsymbol{u}_k$ in $\boldsymbol{L}_2(\Gamma_{\mathfrak{F}})$.

If the coefficient of friction depends on the solution it is necessary to ensure the convergence $\mathfrak{F}(\boldsymbol{u}_k)\sigma_n(\boldsymbol{u}_k) \to \mathfrak{F}(\boldsymbol{u})\sigma_n(\boldsymbol{u})$. For $\alpha = \frac{1}{2}$ the weak convergence $\sigma_n(\boldsymbol{u}_k) \rightharpoonup \sigma_n(\boldsymbol{u})$ in $L_2(\Gamma_{\mathfrak{F}})$ and the strong convergence $\mathfrak{F}(\boldsymbol{u}_k) \to \mathfrak{F}(\boldsymbol{u})$ in $L_p(\Gamma_{\mathfrak{F}})$ with arbitrarily big $p < +\infty$ are sufficient to prove the weak convergence $\mathfrak{F}(\boldsymbol{u}_k)\sigma_n(\boldsymbol{u}_k) \rightharpoonup \mathfrak{F}(\boldsymbol{u})\sigma_n(\boldsymbol{u})$ in $L_q(\Gamma_{\mathfrak{F}})$ with $q < 2$. In the case $\alpha < \frac{1}{2}$ the situation is more complicated, because the convergence of $\sigma_n(\boldsymbol{u}_k)$ is proved only in a dual Sobolev space $H^{-(1/2-\alpha)}(\Gamma_{\mathfrak{F}})$. From the relation

$$\|\mathfrak{F}(\boldsymbol{u}_k)\sigma_n(\boldsymbol{u}_k)\|_{H^{-(1/2-\alpha)}(\Gamma_{\mathfrak{F}})} \leq \|\mathfrak{F}(\boldsymbol{u}_k)\|_{L_\infty(\Gamma_{\mathfrak{F}})}\|\sigma_n(\boldsymbol{u}_k)\|_{H^{-(1/2-\alpha)}(\Gamma_{\mathfrak{F}})}$$

that is valid due to the non-negativity of both $-\sigma_n(\boldsymbol{u}_k)$ and $\mathfrak{F}(\boldsymbol{u}_k)$ there follows $\mathfrak{F}(\boldsymbol{u}_k)\sigma_n(\boldsymbol{u}_k) \rightharpoonup \mathfrak{H}$ for a suitable subsequence at least, but it is not clear that $\mathfrak{H} = \mathfrak{F}(\boldsymbol{u})\sigma_n(\boldsymbol{u})$ holds. Therefore it is necessary to prove $\mathfrak{F}(\boldsymbol{u}_k)\sigma_n(\boldsymbol{u}_k) \rightharpoonup \mathfrak{F}(\boldsymbol{u})\sigma_n(\boldsymbol{u})$ in a suitable weak sense, e.g. in the sense of distributions. For $v \in C_0^1(\Gamma_{\mathfrak{F}})$ it holds

$$\begin{aligned}\langle \mathfrak{F}(\boldsymbol{u}_k)\sigma_n(\boldsymbol{u}_k), v\rangle_{\Gamma_{\mathfrak{F}}} &= \langle \sigma_n(\boldsymbol{u}_k), \mathfrak{F}(\boldsymbol{u}_k)v\rangle_{\Gamma_{\mathfrak{F}}} \to \langle \sigma_n(\boldsymbol{u}), \mathfrak{F}(\boldsymbol{u})v\rangle_{\Gamma_{\mathfrak{F}}} \\ &= \langle \mathfrak{F}(\boldsymbol{u})\sigma_n(\boldsymbol{u}), v\rangle_{\Gamma_{\mathfrak{F}}}\end{aligned}$$

if $\mathfrak{F}(\boldsymbol{u}_k) \to \mathfrak{F}(\boldsymbol{u})$ in $H^{1/2-\alpha}(\Gamma_{\mathfrak{F}})$. This property must be concluded from the strong convergence $\boldsymbol{u}_k \to \boldsymbol{u}$ in $\boldsymbol{H}^\gamma(\Gamma_{\mathfrak{F}})$ with the $\gamma < \frac{1}{2}+\alpha$ mentioned above and some continuity property of $\mathfrak{F}$. We show that the Hölder continuities

$$|\mathfrak{F}(\boldsymbol{x}, \boldsymbol{u}) - \mathfrak{F}(\boldsymbol{y}, \boldsymbol{u})| \leq C|\boldsymbol{x} - \boldsymbol{y}|^\lambda \text{ for every fixed } \boldsymbol{u} \in \mathbb{R}^N \\ \text{with } \lambda > \frac{1}{2} - \alpha \tag{3.1.27}$$

and simultaneously

$$|\mathfrak{F}(\boldsymbol{x}, \boldsymbol{u}) - \mathfrak{F}(\boldsymbol{x}, \boldsymbol{v})| \leq C|\boldsymbol{u} - \boldsymbol{v}|^\mu \text{ for every fixed } \boldsymbol{x} \in \Gamma_{\mathfrak{F}} \\ \text{with } \mu > \frac{1-2\alpha}{1+2\alpha} \tag{3.1.28}$$

are in fact sufficient. This follows via the decomposition

$$\begin{aligned}|\mathfrak{F}(\boldsymbol{x}, \boldsymbol{u}(\boldsymbol{x})) - \mathfrak{F}(\boldsymbol{y}, \boldsymbol{u}(\boldsymbol{y}))| &\leq |\mathfrak{F}(\boldsymbol{x}, \boldsymbol{u}(\boldsymbol{x})) - \mathfrak{F}(\boldsymbol{y}, \boldsymbol{u}(\boldsymbol{x}))| \\ &\quad + |\mathfrak{F}(\boldsymbol{y}, \boldsymbol{u}(\boldsymbol{x})) - \mathfrak{F}(\boldsymbol{y}, \boldsymbol{u}(\boldsymbol{y}))|,\end{aligned}$$

the inequality

$$\begin{aligned}&\int_{\Gamma_{\mathfrak{F}}}\int_{\Gamma_{\mathfrak{F}}} \frac{|\boldsymbol{u}(\boldsymbol{x}) - \boldsymbol{u}(\boldsymbol{y})|^{2\mu}}{|\boldsymbol{x} - \boldsymbol{y}|^{N-1+(1-2\alpha)}}\, d\boldsymbol{x}\, d\boldsymbol{y} \\ &\leq \left(\int_{\Gamma_{\mathfrak{F}}}\int_{\Gamma_{\mathfrak{F}}} \frac{1}{|\boldsymbol{x} - \boldsymbol{y}|^{N-1-\eta/(1-\mu)}}\, d\boldsymbol{x}\, d\boldsymbol{y}\right)^{1-\mu} \\ &\quad \left(\int_{\Gamma_{\mathfrak{F}}}\int_{\Gamma_{\mathfrak{F}}} \frac{|\boldsymbol{u}(\boldsymbol{x}) - \boldsymbol{u}(\boldsymbol{y})|^2}{|\boldsymbol{x} - \boldsymbol{y}|^{N-1+(1-2\alpha+\eta)/\mu}}\, d\boldsymbol{x}\, d\boldsymbol{y}\right)^{\mu}\end{aligned}$$

valid for $\mu \in (0,1)$ and a suitable $\eta > 0$ and the compact imbedding theorem. Condition $(1-2\alpha+\eta)/\mu < 1+2\alpha$ yields the given requirement for μ.

Using the established convergences we pass to the limit $k \to +\infty$ in the penalized variational inequality (3.1.2) with penalty parameter ε_k, solution $\boldsymbol{u}_k$ and test function $\boldsymbol{v} \in \mathscr{C}$. From the weak lower semicontinuity of positive definite bilinear forms it follows

$$\liminf_{k\to\infty} A(\boldsymbol{u}_k, \boldsymbol{u}_k) \geq A(\boldsymbol{u}, \boldsymbol{u}),$$

and for the penalty term there holds

$$\int_{\Gamma_C} \tfrac{1}{\varepsilon}\big[(u_k)_n - g_n\big]_+ \big(v_n - (u_k)_n\big)\, ds_{\boldsymbol{x}} \leq 0$$

due to $v_n \leq g_n$ on Γ_C. All the other convergences are trivial. The condition $u_n \leq g_n$ follows from the limit in the *a priori* estimate (3.1.4). Thus it follows that the limit function $\boldsymbol{u}$ solves the original contact problem (3.1.1). Moreover, adding the term $A(\boldsymbol{u}, \boldsymbol{u}_k - \boldsymbol{u})$ to variational inequality (3.1.2) for $\varepsilon = \varepsilon_k$, solution $\boldsymbol{u}_k$ and test function $\boldsymbol{v} = \boldsymbol{u}$ and passing to the limit $k \to 0$ again proves that the convergence $\boldsymbol{u}_k \to \boldsymbol{u}$ is strong in $\boldsymbol{H}^1(\Omega)$.

Hence the following theorem is proved:

3.1.5 Theorem. *Let the Assumptions 3.1.1 and 3.1.3 be valid. In the case* $\alpha < \frac{1}{2}$ *let* $\mathfrak{F}$ *satsify the continuity conditions* (3.1.27) *and* (3.1.28). *Then there exists a sequence* $\varepsilon_k \to 0$ *of penalty parameters and a corresponding sequence* $\boldsymbol{u}_{\varepsilon_k}$ *of solutions to* (3.1.2) *which converges strongly in* $\boldsymbol{H}^1(\Omega)$ *to a solution of the contact problem with friction* (3.1.1).

3.1.6 Remarks. 1. It is hardly possible to obtain any higher regularity than $\sigma_n(\boldsymbol{u}) \in L_{2,\mathrm{loc}}(\Gamma_C)$ and $\boldsymbol{u} \in \boldsymbol{H}^1_{\mathrm{loc}}(\Gamma_C)$. This is due to the fact that the crucial estimate $\|\,|\boldsymbol{u}|\,\|_{\boldsymbol{H}^\alpha(\Gamma)} \leq \|\boldsymbol{u}\|_{\boldsymbol{H}^\alpha(\Gamma)}$ is not valid for $\alpha > 1$.

2. The condition $\operatorname{supp}\mathfrak{F}(\cdot, \boldsymbol{u}) \subset \Gamma_{\mathfrak{F}}$ is, at a first glance, restrictive. It is necessary, since we are able to prove the regularity of the solution in the interior of Γ_C (with respect to the surface topology) only. If this condition shall be dropped, it is necessary to prove the regularity at the points or lines where different boundary conditions meet. This can be perhaps done by using techniques from [62]. On the other hand, the condition is not restrictive, if the contact part Γ_C is separated from the other parts. Moreover, if the contact boundary touches only the part with Neumann conditions, we may extend the potential contact boundary in such a way that contact occurs only in the interior part; then the coefficient of friction can be changed outside in such a way that this requirement on its support is satisfied.

3. Theorem 3.1.5 does not treat the case of a difference between the coefficients of friction of slip and friction of stick. In our static context let us consider the simplest case when $\mathfrak{F}$ depends only on the tangential displacement increment, $\mathfrak{F} = \mathfrak{F}(\boldsymbol{u}_t - \boldsymbol{w}_t)$ and the limit $\mathfrak{F}_0(0) \equiv \lim_{\boldsymbol{v}_t \to 0} \mathfrak{F}(\boldsymbol{v}_t)$ exists almost everywhere on Γ_C. We assume $\mathfrak{F}(0) \geq \mathfrak{F}_0(0)$ for almost every $x \in \Gamma_C$ and extend the definition of $\mathfrak{F}_0$ by $\mathfrak{F}_0(\boldsymbol{v}_t) = \mathfrak{F}(\boldsymbol{v}_t)$ for $\boldsymbol{v}_t \neq 0$. If the new coefficient $\mathfrak{F}_0$ satisfies all assumptions of Theorem 3.1.5, then there exists a solution $\boldsymbol{u}$ of the contact problem with friction (3.1.1) for this new coefficient $\mathfrak{F}_0$. However, this solution $\boldsymbol{u}$ also solves the problem (3.1.1) with coefficient $\mathfrak{F}$, because $\mathfrak{F}(\boldsymbol{u}_t - \boldsymbol{w}_t)|\sigma_n(\boldsymbol{u})|\,|\boldsymbol{u}_t - \boldsymbol{w}_t| = \mathfrak{F}_0(\boldsymbol{u}_t - \boldsymbol{w}_t)|\sigma_n(\boldsymbol{u})|\,|\boldsymbol{u}_t - \boldsymbol{w}_t|$ and $\mathfrak{F}(\boldsymbol{u}_t - \boldsymbol{w}_t)|\sigma_n(\boldsymbol{u})|\,|\boldsymbol{v}_t - \boldsymbol{w}_t| \geq \mathfrak{F}_0(\boldsymbol{u}_t - \boldsymbol{w}_t)|\sigma_n(\boldsymbol{u})|\,|\boldsymbol{v}_t - \boldsymbol{w}_t|$ for all test functions $\boldsymbol{v}$. On the other hand, the uniqueness of the solution is neither proved nor expected in this case. There could be a solution of the problem with coefficient $\mathfrak{F}$ which does not solve the problem with coefficient $\mathfrak{F}_0$.

4. The smoothing of the norm in the friction term is necessary for the case $\alpha = \frac{1}{2}$ only, because in this case the shift technique must be applied to the variational equation. For $\alpha < \frac{1}{2}$ the smoothing can be avoided completely. The solvability of the penalized problem can be established directly via the corresponding fixed point approach and results from convex analysis; the non-differentiability of the Euclidean norm does not lead to any substantial difficulties.

5. The requirement $\boldsymbol{w} \in H^{1/2+\mu}(\Gamma_C)$ with $\mu > \alpha$ for the case $\alpha < \frac{1}{2}$ is rather restrictive. As described in Section 1.5, the function $\boldsymbol{w}$ arises from the previous time step in a time discretization of a dynamic or quasistatic problem, and the condition $\mu > \alpha$ leads to a loss of regularity in every time step. Therefore the result for $\alpha = \frac{1}{2}$ with no loss of regularity is more advantageous.

3.1.2 Calculation of the admissible coefficient of friction

The admissible coefficient of friction for the existence result of the previous section is given in dependence of the constants in the special trace estimates (1.7.14) and (1.7.21) (cf. (3.1.17) and (3.1.21)). While the existence of these constants is clear, it is particularly important for applications to know their precise value. Since these trace estimates are required for halfspace domains only, it is possible to calculate optimal lower bounds for these constants.

We consider first the elasticity equations for homogeneous, isotropic material (cf. (1.2.5)) on the halfspace $O = \mathbb{R}^{N-1} \times \mathbb{R}_+$ with boundary ∂O:

$$\frac{E}{2(1+\nu)(1-2\nu)} \partial_i \partial_j u_j + \frac{E}{2+2\nu} \partial_j \partial_j u_i = 0 \text{ on } O \tag{3.1.29}$$

for $i = 1, \ldots, N$ with boundary conditions

$$\boldsymbol{u} = \boldsymbol{w} \text{ on } \partial O. \tag{3.1.30}$$

Written in terms of the Lamé constants, the first coefficient in (3.1.29) equals $\lambda + \mu$, while the second one is equal to μ. The corresponding energy norm will be denoted by

$$\|\boldsymbol{u}\|_A^2 \equiv A(\boldsymbol{u}, \boldsymbol{u}) = \int_O \frac{E}{1+\nu} \left[\frac{\nu}{1-2\nu} |\partial_j u_j|^2 + e_{jk}(\boldsymbol{u}) e_{jk}(\boldsymbol{u}) \right] d\boldsymbol{x}.$$

Then the following result is valid:

3.1.7 Theorem. (i) *Let $\boldsymbol{u} \in \boldsymbol{H}^1(O)$ satisfy $\boldsymbol{u} = \boldsymbol{w}$ on ∂O. Then for the tangential component $\boldsymbol{w_t} = (w_1, \ldots, w_{N-1}, 0)^\top$ there holds the estimate*

$$\|\boldsymbol{w_t}\|'^2_{\boldsymbol{H}^{1/2}(\partial O)} \le C_0^{(1)} \|\boldsymbol{u}\|_A^2 \tag{3.1.31}$$

with

$$C_0^{(1)} = \begin{cases} c_1\left(\frac{1}{2}\right) \dfrac{2(1-\nu^2)}{E}, & N = 2, \\ c_{N-1}\left(\frac{1}{2}\right) \dfrac{2+2\nu}{E}, & N \ge 3. \end{cases}$$

(ii) *Let $w \in H^{1/2}(\partial O)$. Then there exists a function $\boldsymbol{u} \in \boldsymbol{H}^1(O)$ with $\boldsymbol{u} = (0, \ldots, 0, w)^\top$ on ∂O such that the estimate*

$$\|\boldsymbol{u}\|_A^2 \le C_0^{(2)} \|w\|'^2_{H^{1/2}(\partial O)} \tag{3.1.32}$$

holds for every $N \in \mathbb{N}$ with

$$C_0^{(2)} = \frac{1}{c_{N-1}\left(\frac{1}{2}\right)} \frac{(2-2\nu)E}{(1+\nu)(3-4\nu)}.$$

Proof. Relations (3.1.31) and (3.1.32) will be proved for the solution $\boldsymbol{u}$ of equations (3.1.29) with boundary condition (3.1.30). This is sufficient, because this solution is a minimizer of the energy norm for the given boundary data,

$$\boldsymbol{u} = \underset{\substack{\boldsymbol{v} \in \boldsymbol{H}^1(O) \\ \boldsymbol{v} = \boldsymbol{w} \text{ on } \partial O}}{\operatorname{Argmin}} \|\boldsymbol{v}\|_A^2.$$

Hence, if relation (3.1.31) holds for $\boldsymbol{u}$, then it holds for all $\boldsymbol{v}$ satisfying the boundary condition. In what follows, we frequently use the partial Fourier transform with respect to the tangential variables

$$\tilde{\boldsymbol{u}}(\boldsymbol{\xi}, y) \equiv \mathscr{F}(\boldsymbol{u}; \boldsymbol{\xi}, y) \equiv \frac{1}{(2\pi)^{(N-1)/2}} \int_{\mathbb{R}^{N-1}} \boldsymbol{u}(\boldsymbol{x}', y) e^{-i\boldsymbol{x}'\cdot\boldsymbol{\xi}} \, d\boldsymbol{x}'$$

with $\boldsymbol{x}' = (x_1, \dots, x_{N-1})^\top$. The Fourier transforms of functions on the subspace ∂O are denoted by hats. Let us first consider the two-dimensional case. Then system (3.1.29) can be written in the form

$$\begin{aligned} \tilde{u}_1''(\xi, y) &= \frac{2-2\nu}{1-2\nu} \xi^2 \tilde{u}_1(\xi, y) - \frac{1}{1-2\nu} i\xi \tilde{u}_2'(\xi, y), \\ \tilde{u}_2''(\xi, y) &= \frac{1-2\nu}{2-2\nu} \xi^2 \tilde{u}_2(\xi, y) - \frac{1}{2-2\nu} i\xi \tilde{u}_1'(\xi, y) \end{aligned}$$

with $\tilde{u}_j' \equiv \dfrac{\partial \tilde{u}_j}{\partial y}$. For each fixed ξ this is a system of ordinary differential equations in y, which can be solved by standard methods. The general solution of the system is given by

$$\begin{aligned} \boldsymbol{u}(\xi, y) = {} & c_1(\xi) e^{\xi y} \begin{pmatrix} 1 \\ -i \end{pmatrix} + c_2(\xi) e^{-\xi y} \begin{pmatrix} 1 \\ i \end{pmatrix} \\ & + c_3(\xi) e^{\xi y} \left(\frac{2\xi y}{3-4\nu} \begin{pmatrix} 1 \\ -i \end{pmatrix} + \begin{pmatrix} 1 \\ i \end{pmatrix} \right) \\ & + c_4(\xi) e^{-\xi y} \left(\frac{2\xi y}{3-4\nu} \begin{pmatrix} 1 \\ i \end{pmatrix} + \begin{pmatrix} -1 \\ i \end{pmatrix} \right). \end{aligned}$$

The coefficients $c_\iota(\xi)$ are determined by the boundary conditions. Here the condition $\boldsymbol{u} \in \boldsymbol{H}^1(\Omega)$ can be interpreted as a boundary condition at $y = +\infty$. In the case $\xi > 0$ there follows $c_1(\xi) = c_3(\xi) = 0$ and

$$c_2(\xi) = \frac{\widehat{w}_1(\xi) - i\widehat{w}_2(\xi)}{2}, \quad c_4(\xi) = -\frac{\widehat{w}_1(\xi) + i\widehat{w}_2(\xi)}{2}.$$

Hence the solution is given by

$$\tilde{\boldsymbol{u}}(\xi, y) = \left[\begin{pmatrix} \widehat{w}_1(\xi) \\ \widehat{w}_2(\xi) \end{pmatrix} - \frac{\xi y}{3-4\nu} (\widehat{w}_1(\xi) + i\widehat{w}_2(\xi)) \begin{pmatrix} 1 \\ i \end{pmatrix} \right] e^{-\xi y}.$$

For $\xi < 0$ there holds $c_2(\xi) = c_4(\xi) = 0$ and

$$c_1(\xi) = \frac{\widehat{w}_1(\xi) + i\widehat{w}_2(\xi)}{2}, \quad c_3(\xi) = \frac{\widehat{w}_1(\xi) - i\widehat{w}_2(\xi)}{2}.$$

The corresponding solution is

$$\widetilde{\boldsymbol{u}}(\xi, y) = \left[\begin{pmatrix} \widehat{w}_1(\xi) \\ \widehat{w}_2(\xi) \end{pmatrix} + \frac{\xi y}{3-4\nu} (\widehat{w}_1(\xi) - i\widehat{w}_2(\xi)) \begin{pmatrix} 1 \\ -i \end{pmatrix} \right] e^{\xi y}.$$

In a compact form this solution can be written for all $\xi \in \mathbb{R}$ as

$$\widetilde{\boldsymbol{u}}(\xi, y) = \left[\begin{pmatrix} \widehat{w}_1(\xi) \\ \widehat{w}_2(\xi) \end{pmatrix} - \frac{y}{3-4\nu} \begin{pmatrix} |\xi|\widehat{w}_1(\xi) + i\xi\widehat{w}_2(\xi) \\ i\xi\widehat{w}_1(\xi) - |\xi|\widehat{w}_2(\xi) \end{pmatrix} \right] e^{-|\xi| y}. \quad (3.1.33)$$

Relations (3.1.31), (3.1.32) are now proved by the calculation of $\|\boldsymbol{u}\|_A^2$ for this solution. Using the easy relation

$$e_{jk}(\boldsymbol{u})e_{jk}(\boldsymbol{u}) = \tfrac{1}{2}(\partial_1 u_2 + \partial_2 u_1)^2 + (\partial_j u_j)^2 - 2\partial_1 u_1 \partial_2 u_2$$

the square of the energy norm can be represented by

$$\begin{aligned} \|\boldsymbol{u}\|_A^2 &= \frac{E}{2+2\nu} \int_O \left[\frac{2-2\nu}{1-2\nu} |\partial_j u_j|^2 + |\partial_2 u_1 + \partial_1 u_2|^2 \right. \\ &\qquad \left. - 4\partial_1 u_1 \partial_2 u_2 \right] dy\, dx' \\ &= \frac{E}{2+2\nu} \int_{\mathbb{R}} \int_0^1 \left[\frac{2-2\nu}{1-2\nu} \left| i\xi\widetilde{u}_1 + \frac{\partial \widetilde{u}_2}{\partial y} \right|^2 \right. \\ &\qquad \left. + \left| \frac{\partial \widetilde{u}_1}{\partial y} + i\xi\widetilde{u}_2 \right|^2 - 4\,\mathrm{Re}\left(i\xi\widetilde{u}_1 \overline{\frac{\partial \widetilde{u}_2}{\partial y}} \right) \right] dy\, d\xi. \end{aligned} \quad (3.1.34)$$

A standard calculation yields

$$\begin{aligned} i\xi\widetilde{u}_1 + \frac{\partial \widetilde{u}_2}{\partial y} &= \frac{2-4\nu}{3-4\nu} (i\xi\widehat{w}_1(\xi) - |\xi|\widehat{w}_2(\xi)) e^{-|\xi| y}, \\ \frac{\partial \widetilde{u}_1}{\partial y} + i\xi\widetilde{u}_2 &= \frac{2}{3-4\nu} \big[|\xi|\widehat{w}_1(\xi)\big(|\xi| y - (2-2\nu)\big) \\ &\qquad + i\xi\widehat{w}_2(\xi)\big(|\xi| y + (1-2\nu)\big)\big] e^{-|\xi| y}. \end{aligned}$$

Evaluation of the inner integral in the representation of the bilinear form with the help of the easy formula $\int_0^{+\infty} e^{-2|\xi| y} y^n\, dy = \dfrac{n!}{2|\xi|^{n+1}}$ yields

$$\begin{aligned} \|\boldsymbol{u}\|_A^2 &= \frac{E}{(1+\nu)(3-4\nu)} \cdot \\ &\quad \cdot \int_{\mathbb{R}} \Big[\left(|\widehat{w}_1(\xi)|^2 + |\widehat{w}_2(\xi)|^2 \right) |\xi| (2-2\nu) \\ &\qquad -2\,\mathrm{Re}\left(i\xi\widehat{w}_1(\xi)\overline{\widehat{w}_2(\xi)} \right) (1-2\nu) \Big]\, d\xi. \end{aligned} \quad (3.1.35)$$

In order to obtain relation (3.1.31), the mixed term—whose sign is not known—is estimated by

$$2\left||\xi|(1-2\nu)\widehat{w}_1\overline{\widehat{w}_2}\right| \le \frac{(1-2\nu)^2}{2-2\nu}|\xi|\,|\widehat{w}_1|^2 + (2-2\nu)|\xi|\,|\widehat{w}_2|^2 .$$

This yields

$$\|\boldsymbol{u}\|_A^2 \ge \int_{\mathbb{R}} \frac{E}{2(1-\nu^2)}\,|\widehat{w}_1(\xi)|^2\,|\xi|\,d\xi$$

which proves (3.1.31). For a function $\boldsymbol{w}$ with vanishing tangential component $w_1 = 0$ there holds

$$\|\boldsymbol{u}\|_A^2 = \int_{\mathbb{R}} \frac{(2-2\nu)E}{(1+\nu)(3-4\nu)}\,|\widehat{w}_2(\xi)|^2\,|\xi|\,d\xi.$$

Hence (3.1.32) is also proved.

In the case of dimension $N > 2$ different methods are employed for the proof of the trace theorem and of the inverse trace theorem. The trace theorem will be verified by using the Korn inequality for a half space and the corresponding trace theorem for the Laplace operator. This method is much easier to perform, and the result obtained is the same as that proved in [44] by a direct calculation for the three-dimensional case. Of course, the result for dimension $N > 3$ cannot be better.

Neglecting the contribution of $\operatorname{div}\boldsymbol{u}$ and employing the Korn inequality the energy norm can be estimated by

$$\|\boldsymbol{u}\|_A^2 \ge \frac{E}{1+\nu}\int_O e_{jk}(\boldsymbol{u})e_{jk}(\boldsymbol{u})\,d\boldsymbol{x} \ge \frac{E}{2+2\nu}\|\boldsymbol{u}\|'^2_{\boldsymbol{H}^1(O)}.$$

The estimate of the H^1-seminorm is done for a function $\boldsymbol{u}_L$ satisfying the Laplace equation $(\Delta\boldsymbol{u}_L)_i = 0$, $i = 1,\dots,N$, on O and the given boundary data $\boldsymbol{w}$ on ∂O. This is sufficient, since this solution minimizes the H^1-seminorm among all function satisfying the given boundary condition. The solution is given in the Fourier transformed representation by

$$\widetilde{\boldsymbol{u}}_L(\boldsymbol{\xi}, y) = \widehat{\boldsymbol{w}}(\boldsymbol{\xi})e^{-|\boldsymbol{\xi}|y}.$$

Evaluation of the seminorm leads to

$$\begin{aligned}\|\boldsymbol{u}_L\|'^2_{\boldsymbol{H}^1(O)} &= \int_{\mathbb{R}^{N-1}}\int_{\mathbb{R}_+} |\widehat{\boldsymbol{w}}(\boldsymbol{\xi})|^2 2|\boldsymbol{\xi}|^2 e^{-2|\boldsymbol{\xi}|y}\,dy\,d\boldsymbol{\xi}\\ &= \int_{\mathbb{R}^{N-1}} |\widehat{\boldsymbol{w}}(\boldsymbol{\xi})|^2|\boldsymbol{\xi}|\,d\boldsymbol{\xi} = \frac{1}{c_{N-1}\left(\frac{1}{2}\right)}\|\boldsymbol{w}\|'^2_{\boldsymbol{H}^{1/2}(\partial O)}.\end{aligned}$$

Hence the trace estimate (3.1.31) is proved for the case $N > 2$.

The inverse trace estimate is verified by a direct calculation, using the solution of the N-dimensional elasticity equations. For the Fourier transform $\widetilde{\boldsymbol{u}}$ with respect to the tangential variables these equations are given by

$$\widetilde{u}_\ell'' = \frac{1}{1-2\nu}\left(\sum_{j=1}^{N-1} \xi_\ell \xi_j \widetilde{u}_j - i\xi_\ell \widetilde{u}_N'\right) + |\boldsymbol{\xi}|^2 \widetilde{u}_\ell, \quad \ell = 1, \dots, N-1,$$

$$\widetilde{u}_N'' = \frac{1-2\nu}{2-2\nu}|\boldsymbol{\xi}|^2 \widetilde{u}_N - \frac{1}{2-2\nu}\sum_{j=1}^{N-1} i\xi_j \widetilde{u}_j'.$$

The solution of this system with boundary data $\boldsymbol{w}$ is given by

$$\widetilde{\boldsymbol{u}}(\boldsymbol{\xi}, y) = \left[\widehat{\boldsymbol{w}}(\boldsymbol{\xi}) - \frac{\xi_j \widehat{w}_j(\boldsymbol{\xi}) + i|\boldsymbol{\xi}|\widehat{w}_N(\boldsymbol{\xi})}{(3-4\nu)|\boldsymbol{\xi}|} y \begin{pmatrix} \boldsymbol{\xi} \\ i|\boldsymbol{\xi}| \end{pmatrix}\right] e^{-|\boldsymbol{\xi}| y}. \tag{3.1.36}$$

The inverse trace theorem is proved for functions with vanishing tangential components only. The corresponding solution is therefore defined by

$$\widetilde{\boldsymbol{u}}(\boldsymbol{\xi}, y) = \left[\begin{pmatrix} 0 \\ \widehat{w}_N(\boldsymbol{\xi}) \end{pmatrix} + \frac{y \widehat{w}_N(\boldsymbol{\xi})}{(3-4\nu)} \begin{pmatrix} -i\boldsymbol{\xi} \\ |\boldsymbol{\xi}| \end{pmatrix}\right] e^{-|\boldsymbol{\xi}| y}. \tag{3.1.37}$$

This solution is inserted into the bilinear form

$$\|\boldsymbol{u}\|_A^2 = \int_O \frac{E}{1+\nu}\left[\frac{\nu}{1-2\nu}|\partial_j u_j|^2 + e_{jk}(\boldsymbol{u}) e_{jk}(\boldsymbol{u})\right] d\boldsymbol{x}. \tag{3.1.38}$$

In the Fourier transformed representation we have

$$\mathscr{F}(\operatorname{div} \boldsymbol{u}; \boldsymbol{\xi}, y) = -\frac{2(1-2\nu)}{3-4\nu}|\boldsymbol{\xi}|\widehat{w}_N(\boldsymbol{\xi}) e^{-|\boldsymbol{\xi}| y},$$

$$\begin{aligned}
\mathscr{F}(e_{jk}(\boldsymbol{u}); \boldsymbol{\xi}, y)\overline{\mathscr{F}(e_{jk}(\boldsymbol{u}); \boldsymbol{\xi}, y)} &= \frac{|\boldsymbol{\xi}|^2}{2}|\widetilde{\boldsymbol{u}}|^2 + \frac{1}{2}|\widetilde{\boldsymbol{u}}'|^2 \\
+ \frac{1}{2}\left|\sum_{j=1}^{N-1} \xi_j \widetilde{u}_j\right|^2 &+ \sum_{j=1}^{N-1} i\xi_j \widetilde{u}_N \overline{\widetilde{u}_j'} + \frac{1}{2}|\widetilde{u}_N'|^2 \\
= \frac{4|\boldsymbol{\xi}|^4 y^2 + (8-16\nu)|\boldsymbol{\xi}|^3 y + 6(1-2\nu)^2 |\boldsymbol{\xi}|^2}{(3-4\nu)^2}&|\widehat{w}_N(\boldsymbol{\xi})|^2 e^{-2|\boldsymbol{\xi}| y}.
\end{aligned}$$

With the formula $\int_0^\infty y^n e^{-2|\boldsymbol{\xi}| y}\, dy = n!/(2|\boldsymbol{\xi}|)^{n+1}$ the inner integral of the energy norm can be calculated. The result is

$$\|\boldsymbol{u}\|_A^2 = \int_{\mathbb{R}^{N-1}} \frac{E}{1+\nu}\frac{2-2\nu}{3-4\nu}|\boldsymbol{\xi}||\widehat{w}_N(\boldsymbol{\xi})|^2\, d\boldsymbol{\xi}. \tag{3.1.39}$$

From this the inverse trace estimate follows immediately. □

The result of Theorem 3.1.7 can be easily extended to the case of anisotropic bodies. By the conditions of ellipticity and boundedness, the energy norm $\|\boldsymbol{u}\|_A$ is estimated from above and below in terms of the expression

$$\int_O e_{jk}(\boldsymbol{u})e_{jk}(\boldsymbol{u})\,d\boldsymbol{x}.$$

This term represents the energy norm of the isotropic problem with Young modulus $E = 1$ and Poisson ratio $\nu = 0$. Hence the constants $C_0^{(1)}$ and $C_0^{(2)}$ for the anisotropic case are given by products of the constants for these material data with the upper and lower bounds from the conditions of boundedness and ellipticity.

3.1.8 Corollary. *In the case of anisotropic material with constant of ellipticity a_0 and upper bound A_0 the relations* (3.1.31) *and* (3.1.32) *are valid with constants*

$$C_0^{(1)} = c_{N-1}\left(\tfrac{1}{2}\right)\frac{2}{a_0} \quad \textit{and}$$

$$C_0^{(2)} = \frac{1}{c_{N-1}\left(\tfrac{1}{2}\right)}\frac{2}{3}A_0.$$

From the just calculated constants it is easy to derive the following bounds for the coefficient of friction:

- In the case of homogeneous, isotropic material with Young modulus E and Poisson ration ν we have

$$C_{\mathfrak{F}} = \begin{cases} \dfrac{\sqrt{3-4\nu}}{2-2\nu}, & N = 2, \\ \sqrt{\dfrac{3-4\nu}{4-4\nu}}, & N \geq 3. \end{cases} \tag{3.1.40}$$

- For general material with coefficient of ellipticity a_0 and upper bound A_0 of the Hooke tensor there holds

$$C_{\mathfrak{F}} = \sqrt{\frac{3}{4}\frac{a_0}{A_0}}. \tag{3.1.41}$$

3.1.9 Remarks. 1. In the first existence result for static contact problems by Nečas, Jarušek and Haslinger [113] for an infinite strip in two dimensions the admissible coefficient of friction was given as $C_{\mathfrak{F}} = \sqrt{a_0/(2A_0)}$. The result was extended by Jarušek [73] to the case of more general sufficiently smooth domains and more space dimensions. There the admissible coefficient of friction for isotropic materal is given as $C_{\mathfrak{F}} = \sqrt[4]{(1-2\nu)/(2-2\nu)}$. The result presented here is better than both of these values. This amelioration of the result is due to the usage of the appropriate special trace estimates (3.1.31) and (3.1.32).

2. The localization technique shows that the constants a_0, A_0 in the formula (3.1.41) need not be valid for the whole domain Ω. The constant a_0 can be replaced by $\liminf_{\delta\to 0} a_0^\delta$ and A_0 by $\limsup_{\delta\to 0} A_0^\delta$, where a_0^δ, A_0^δ are the bounds for the bilinear form associated to $\Omega \cap (\Gamma_{\mathfrak{F}} + B(0,\delta))$. Here $B(0,\delta)$ is the ball with radius δ and midpoint in the origin.

3. The result for the isotropic case can be extended to the case of non-homogeneous material with a C^β-dependence of E and ν on the space variable, where $\beta > 1/2$. In fact, the trace theorems are applied to a localized version of the equation only, defined on a neighbourhood of diameter δ of some point $x_0 \in \Gamma_{\mathfrak{F}}$. The diameter can be chosen arbitrarily small. In this neighbourhood the values of the constants E and ν at x_0 can be taken, and the differences $E(x) - E(x_0)$, $\nu(x) - \nu(x_0)$ are estimated on the right hand side. Due to the required Hölder-continuity of E, ν this perturbation lead to an additional term bounded by $c\delta\mathscr{A}(\boldsymbol{u})$ on the right hand side. Since δ can be chosen as small as necessary, this term does not change the upper bound $C_{\mathfrak{F}}$.

3.2 Semicoercive contact problem

If neither the conditions $\operatorname{mes}_\Gamma \Gamma_U > 0$ nor $d \geq d_0 > 0$ on some subdomain of positive measure is satisfied, then the bilinear form A in the variational inequality is no longer equivalent to the $\boldsymbol{H}^1(\Omega)$-scalar product. In this case the existence proof of the preceding subsection is not valid. In the present section we show that the existence of solutions can be still proved, if some additional *pressure condition* for the outer forces is valid. The problem is given by the following variational inequality:

Find a function $\boldsymbol{u} \in \mathscr{C} \equiv \{\boldsymbol{v} \in \boldsymbol{H}^1(\Omega);\ v_n \leq g_n$ *on* $\Gamma_C\}$ *such that for all* $\boldsymbol{v} \in \mathscr{C}$ *there holds*

$$A(\boldsymbol{u}, \boldsymbol{v}-\boldsymbol{u}) + \int_{\Gamma_C} \mathfrak{F}(\boldsymbol{u})\,|\sigma_n(\boldsymbol{u})|\,(|\boldsymbol{v_t} - \boldsymbol{w_t}| - |\boldsymbol{u_t} - \boldsymbol{w_t}|)\,ds_{\boldsymbol{x}} \geq \mathscr{L}(\boldsymbol{v}-\boldsymbol{u}). \tag{3.2.1}$$

The problem will be solved under the following set of assumptions:

3.2.1 Assumption. *Let Assumption 3.1.3 be valid with the exception of the assumption there concerning the measure of* Γ_U *and the coefficient* d.

Instead of these, let the pressure condition

$$\mathscr{L}(\boldsymbol{v}) < 0 \quad \forall \boldsymbol{v} \in (\mathscr{C} - \{\boldsymbol{g}\}) \cap \mathscr{R} \setminus \{0\} \tag{3.2.2}$$

with the set of rigid motions $\mathscr{R}$ (cf. (1.2.15)*) be valid. Here, $\boldsymbol{g} \in \boldsymbol{H}^1(\Omega)$ is a suitable extension of $g_n \boldsymbol{n}$ from Γ_C to the whole domain.*

The only new assumption here is the pressure condition (3.2.2). Its physical interpretation is the following: the total outer force composed of the sum of volume and boundary forces is directed towards the obstacle. We will see that this condition restores some kind of coercivity of the whole problem.

In order to prove the solvability of this problem we use two approximations. The first one is the usual penalty method; it leads to the following variational inequality:

Find a function $\boldsymbol{u}_\varepsilon \in \boldsymbol{H}^1(\Omega)$ such that for all $\boldsymbol{v} \in \boldsymbol{H}^1(\Omega)$ there holds

$$\begin{aligned} A(\boldsymbol{u}_\varepsilon, \boldsymbol{v} - \boldsymbol{u}_\varepsilon) &+ \int_{\Gamma_C} \tfrac{1}{\varepsilon}\big[(u_\varepsilon)_n - g_n\big]_+ \big(v_n - (u_\varepsilon)_n\big)\, ds_{\boldsymbol{x}} \\ &+ \int_{\Gamma_C} \mathfrak{F}(\boldsymbol{u}_\varepsilon) \tfrac{1}{\varepsilon}\big[(u_\varepsilon)_n - g_n\big]_+ \\ &\qquad \big(|\boldsymbol{v_t} - \boldsymbol{w_t}| - |(\boldsymbol{u}_\varepsilon)_{\boldsymbol{t}} - \boldsymbol{w_t}|\big)\, ds_{\boldsymbol{x}} \ge \mathscr{L}(\boldsymbol{v} - \boldsymbol{u}_\varepsilon). \end{aligned} \tag{3.2.3}$$

The second one is a *coercive approximation* of the non-coercive problem. Let $\pi_{\mathscr{R}}$ denote the orthogonal projection of $\boldsymbol{L}_2(\Omega)$ onto the set of rigid motions $\mathscr{R}$. Then the coercive approximation is defined by

Find a function $\boldsymbol{u}_{\varepsilon,\gamma} \in \boldsymbol{H}^1(\Omega)$ such that for all $\boldsymbol{v} \in \boldsymbol{H}^1(\Omega)$ there holds

$$\begin{aligned} &\gamma\langle \pi_{\mathscr{R}} \boldsymbol{u}_{\varepsilon,\gamma}, \pi_{\mathscr{R}}(\boldsymbol{v} - \boldsymbol{u}_{\varepsilon,\gamma})\rangle_\Omega + A(\boldsymbol{u}_{\varepsilon,\gamma}, \boldsymbol{v} - \boldsymbol{u}_{\varepsilon,\gamma}) \\ &\quad + \int_{\Gamma_C} \tfrac{1}{\varepsilon}\big[(u_{\varepsilon,\gamma})_n - g_n\big]_+ \big(v_n - (u_{\varepsilon,\gamma})_n\big)\, ds_{\boldsymbol{x}} \\ &\quad + \int_{\Gamma_C} \mathfrak{F}(\boldsymbol{u}_{\varepsilon,\gamma}) \tfrac{1}{\varepsilon}\big[(u_{\varepsilon,\gamma})_n - g_n\big]_+ \cdot \\ &\qquad \cdot \big(|\boldsymbol{v_t} - \boldsymbol{w_t}| - |(\boldsymbol{u}_{\varepsilon,\gamma})_{\boldsymbol{t}} - \boldsymbol{w_t}|\big)\, ds_{\boldsymbol{x}} \ge \mathscr{L}(\boldsymbol{v} - \boldsymbol{u}_{\varepsilon,\gamma}). \end{aligned} \tag{3.2.4}$$

If $\gamma > 0$ holds, then this problem satisfies Assumption 3.1.1. Therefore, due to Lemma 3.1.2 there exists a solution. This solution satisfies an *a priori* estimate of the type (3.1.4), but the constant there depends on the approximation parameter γ, because the coercivity constant of the bilinear form $(\boldsymbol{u}, \boldsymbol{v}) \mapsto \gamma\langle \pi_{\mathscr{R}} \boldsymbol{u}, \pi_{\mathscr{R}} \boldsymbol{v}\rangle_\Omega + A(\boldsymbol{u}, \boldsymbol{v})$ is γ-dependent. Hence it is necessary to derive a suitable γ-independent *a priori* estimate. This estimate is essentially based on the next lemma.

3.2.2 Lemma. *Let the pressure condition* (3.2.2) *be valid and let the coefficients* $a_{ijk\ell}$ *of the bilinear form* A *be symmetric, bounded and elliptic. Then there holds*

$$\liminf_{\|\boldsymbol{v}\|_{\boldsymbol{H}^1(\Omega)}\to+\infty} \frac{A(\boldsymbol{v},\boldsymbol{v}) + \|[v_n - g_n]_+\|^2_{L_2(\Gamma_C)} - \mathscr{L}(\boldsymbol{v})}{\|\boldsymbol{v}\|_{\boldsymbol{H}^1(\Omega)}} > 0. \tag{3.2.5}$$

Proof. The assertion is proved by contradiction. Let us assume (3.2.5) is not true. Then there exists a sequence $\{\boldsymbol{u}_k\}_{k=1}^{+\infty}$ with $\|\boldsymbol{u}_k\|_{\boldsymbol{H}^1(\Omega)} \to +\infty$ and

$$\lim_{k\to+\infty} \frac{A(\boldsymbol{u}_k,\boldsymbol{u}_k) + \|[(u_k)_n - g_n]_+\|^2_{L_2(\Gamma_C)} - \mathscr{L}(\boldsymbol{u}_k)}{\|\boldsymbol{u}_k\|_{\boldsymbol{H}^1(\Omega)}} \le 0. \tag{3.2.6}$$

Let $\mathscr{Q}$ denote the $\boldsymbol{L}_2(\Omega)$-orthogonal complement to $\mathscr{R}$ in $\boldsymbol{H}^1(\Omega)$. Using the decomposition $\boldsymbol{u}_k = \boldsymbol{v}_k + \boldsymbol{z}_k$ with $\boldsymbol{v}_k = \pi_{\mathscr{Q}}\boldsymbol{u}_k$ and $\boldsymbol{z}_k = \pi_{\mathscr{R}}\boldsymbol{u}_k$ we obtain

$$\lim_{k\to+\infty} \frac{A(\boldsymbol{v}_k,\boldsymbol{v}_k)}{\|\boldsymbol{u}_k\|_{\boldsymbol{H}^1(\Omega)}} \le \|\mathscr{L}\|_{\mathring{\boldsymbol{H}}^{-1}(\Omega)}.$$

The bilinear form A is strongly elliptic on the subspace $\mathscr{Q}$; hence from this inequality and the decomposition

$$\frac{A(\boldsymbol{v}_k,\boldsymbol{v}_k)}{\|\boldsymbol{u}_k\|_{\boldsymbol{H}^1(\Omega)}} = \frac{A(\boldsymbol{v}_k,\boldsymbol{v}_k)}{\|\boldsymbol{v}_k\|_{\boldsymbol{H}^1(\Omega)}} \frac{\|\boldsymbol{v}_k\|_{\boldsymbol{H}^1(\Omega)}}{\|\boldsymbol{u}_k\|_{\boldsymbol{H}^1(\Omega)}}$$

there follows

$$\frac{\|\boldsymbol{v}_k\|_{\boldsymbol{H}^1(\Omega)}}{\|\boldsymbol{u}_k\|_{\boldsymbol{H}^1(\Omega)}} \to 0,$$

possibly after extraction of a subsequence. In fact, if $\|\boldsymbol{v}_k\|_{\boldsymbol{H}^1(\Omega)}$ is bounded, then this assertion is clearly true, and if there is a subsequence such that $\|\boldsymbol{v}_k\|_{\boldsymbol{H}^1(\Omega)}$ tends to $+\infty$, then $A(\boldsymbol{v}_k,\boldsymbol{v}_k)/\|\boldsymbol{v}_k\|_{\boldsymbol{H}^1(\Omega)}$ tends to $+\infty$ too; consequently $\|\boldsymbol{v}_k\|_{\boldsymbol{H}^1(\Omega)}/\|\boldsymbol{u}_k\|_{\boldsymbol{H}^1(\Omega)}$ has to tend to 0 in order to compensate this growth. For

$$\boldsymbol{s}_k \equiv \frac{\boldsymbol{u}_k}{\|\boldsymbol{u}_k\|_{\boldsymbol{H}^1(\Omega)}},$$

the sequence $\{\boldsymbol{s}_k\}$ is bounded; hence there exists a subsequence of $\{\boldsymbol{u}_k\}_{k=1}^{+\infty}$, which will be denoted by $\{\boldsymbol{u}_k\}_{k=1}^{+\infty}$ again, such that the corresponding sequence $\{\boldsymbol{s}_k\}$ tends to an element $\boldsymbol{u}^*$ weakly in $\boldsymbol{H}^1(\Omega)$ and strongly in $\boldsymbol{L}_p(\Gamma_C)$ for all indices $p < p_0$ with

$$p_0 = \infty \text{ for } N = 2 \text{ and } p_0 = 2 + \frac{4}{N-2} \text{ for } N \ge 3. \tag{3.2.7}$$

Moreover, $\boldsymbol{v}_k/\|\boldsymbol{u}_k\|_{\boldsymbol{H}^1(\Omega)}$ tends to 0 and $\boldsymbol{z}_k/\|\boldsymbol{u}_k\|_{\boldsymbol{H}^1(\Omega)}$ tends to $\boldsymbol{u}^*$; the second convergence holds because $\mathscr{R}$ is finite-dimensional. Therefore $\boldsymbol{s}_k$ converges to $\boldsymbol{u}^*$ in $\boldsymbol{H}^1(\Omega)$ and the limit $\boldsymbol{u}^*$ is an element of $\mathscr{R}$ with $\|\boldsymbol{u}^*\|_{\boldsymbol{H}^1(\Omega)} = 1$. Due to this strong convergence there holds

$$\frac{1}{\|\boldsymbol{u}_k\|_{\boldsymbol{H}^1(\Omega)}} \left\| [(u_k)_n - g_n]_+ \right\|_{L_2(\Gamma_C)}$$
$$= \left\| \left[\frac{(u_k)_n}{\|\boldsymbol{u}_k\|_{\boldsymbol{H}^1(\Omega)}} - \frac{g_n}{\|\boldsymbol{u}_k\|_{\boldsymbol{H}^1(\Omega)}} \right]_+ \right\|_{L_2(\Gamma_C)} \to \|[(u^*)_n]_+\|_{L_2(\Gamma_C)}.$$

The validity of relation (3.2.6) implies that $\|[(u^*)_n]_+\|_{L_2(\Gamma_C)} = 0$, because otherwise the limit of $\|[(u_k)_n - g_n]_+\|^2_{L_2(\Gamma_C)}/\|\boldsymbol{u}_k\|_{\boldsymbol{H}^1(\Omega)}$ should be $+\infty$. This shows $\boldsymbol{u}^* \in \mathscr{R} \cap (\mathscr{C} - \{\boldsymbol{g}\}) \setminus \{0\}$. The expression in (3.2.6) can be estimated by

$$\lim_{k\to+\infty} \frac{A(\boldsymbol{u}_k, \boldsymbol{u}_k) + \|[(u_k)_n - g_n]_+\|^2_{L_2(\Gamma_C)} - \mathscr{L}(\boldsymbol{u}_k)}{\|\boldsymbol{u}_k\|_{\boldsymbol{H}^1(\Omega)}}$$
$$\geq - \lim_{k\to+\infty} \mathscr{L}\left(\frac{\boldsymbol{u}_k}{\|\boldsymbol{u}_k\|_{\boldsymbol{H}^1(\Omega)}} \right) = -\mathscr{L}(\boldsymbol{u}^*).$$

Due to (3.2.2) there holds $\mathscr{L}(\boldsymbol{u}^*) < 0$; hence this is a contradiction to the assumption (3.2.6) and assertion (3.2.5) is verified. □

Using the previous lemma it is now easy to derive an *a priori* estimate uniform with respect to γ and ε. Let us extend the function $\boldsymbol{w}$ onto the whole domain such that $\boldsymbol{w}_n = 0$ on Γ_C holds. Inserting the test function $\boldsymbol{v} \equiv \boldsymbol{w}$ into variational inequality (3.2.4) and employing the non-negativity of the penalty and friction functionals and of g_n we get

$$\tfrac{1}{2}A(\boldsymbol{u}_{\varepsilon,\gamma}, \boldsymbol{u}_{\varepsilon,\gamma}) + \|[(u_{\varepsilon,\gamma})_n - g_n]_+\|^2_{L_2(\Gamma_C)} - \mathscr{L}(\boldsymbol{u}_{\varepsilon,\gamma})$$
$$\leq c_1\|\boldsymbol{w}\|^2_{\boldsymbol{H}^1(\Omega)} + c_2\|\boldsymbol{w}\|_{\boldsymbol{H}^1(\Omega)}$$

for $0 < \varepsilon < 1$ and $0 < \gamma \leq 1$. This inequality and Lemma 3.2.2—formally applied to the bilinear form $\frac{1}{2}A$—prove that $\|\boldsymbol{u}_{\varepsilon,\gamma}\|_{\boldsymbol{H}^1(\Omega)}$ is bounded uniformly with respect to ε and γ.

As a consequence, to every $\varepsilon > 0$ there exists a sequence of approximation parameters $\gamma_k \to 0$ and a corresponding sequence of solutions $\boldsymbol{u}_{\varepsilon,\gamma_k}$ converging to some limit function $\boldsymbol{u}_\varepsilon$ weakly in $\boldsymbol{H}^1(\Omega)$ and strongly in $\boldsymbol{L}_p(\Gamma_C)$ for all $p < p_0$ with p_0 defined in (3.2.7). Passing to the limit $k \to +\infty$ in problem (3.2.4) with solutions $\boldsymbol{u}_{\varepsilon,\gamma_k}$ it is verified that the limit $\boldsymbol{u}_\varepsilon$ is a solution of the penalized contact problem (3.2.3). Moreover, adding $A(\boldsymbol{u}_\varepsilon, \boldsymbol{u}_{\varepsilon,\gamma_k} - \boldsymbol{u}_\varepsilon)$ to both sides of inequality (3.2.4) for test function $\boldsymbol{v} = \boldsymbol{u}_\varepsilon$ and passing to the limit $k \to +\infty$ shows that $\pi_{\mathscr{Q}}\boldsymbol{u}_{\varepsilon,\gamma_k}$ tends to $\pi_{\mathscr{Q}}\boldsymbol{u}_\varepsilon$ strongly in $\boldsymbol{H}^1(\Omega)$. Since $\mathscr{R}$ is finite dimensional, from the weak convergence $\pi_{\mathscr{R}}\boldsymbol{u}_{\varepsilon,\gamma_k} \rightharpoonup \pi_{\mathscr{R}}\boldsymbol{u}_\varepsilon$ there follows strong convergence; hence the whole sequence $\boldsymbol{u}_{\varepsilon,\gamma_k}$ converges strongly in $\boldsymbol{H}^1(\Omega)$. Of course, the limit $\boldsymbol{u}_\varepsilon$ is also uniformly bounded in $\boldsymbol{H}^1(\Omega)$. Therefore the following Lemma is proved:

3.2.3 Lemma. *Let Assumption 3.2.1 be valid. Then there exists a sequence $\boldsymbol{u}_{\varepsilon,\gamma_k}$ of solutions to problem* (3.2.4) *with parameters $\gamma_k \to 0$ which converges in $\boldsymbol{H}^1(\Omega)$ to a solution $\boldsymbol{u}_\varepsilon$ of variational inequality* (3.2.3). *This solution satisfies the* a priori *estimate*

$$\|\boldsymbol{u}_\varepsilon\|_{\boldsymbol{H}^1(\Omega)} \le c \tag{3.2.8}$$

with constant c independent of the penalty parameter ε.

In the limit procedure $\varepsilon \to 0$ the same problem as in the preceding section appears: due to the loss of complete continuity of the friction functional the *a priori* estimate of the previous Theorem is not sufficient. Again this problem is solved by proving a better regularity result for the solution on the contact part of the boundary.

3.2.4 Lemma. *Let Assumption 3.2.1 be valid. Then the solutions to problem* (3.2.3) *satisfy the* a priori *estimate*

$$\|\boldsymbol{u}_\varepsilon\|_{\boldsymbol{H}^{1/2+\alpha}(\Gamma_{\mathfrak{F}})} + \left\| \frac{1}{\varepsilon}\big[(u_\varepsilon)_n - g_n\big]_+ \right\|_{H^{-(1/2-\alpha)}(\Gamma_{\mathfrak{F}})} \le const. \tag{3.2.9}$$

with α from Assumption 3.1.3 and a constant that depends on the geometry of the domain and on the given data but not on the penalty parameter ε.

The proof is the same as that given in the previous section for the coercive case. In fact, only the coercivity of the bilinear form A in the sense of (1.2.13) is used there. The contribution of the additional L_2-norm of the solution can be estimated using the known *a priori* estimate (3.2.8). In the case $\alpha = \frac{1}{2}$ it is, of course, necessary to carry out the regularity proof for the smoothed problem, and to pass to the limit in the smoothing parameter.

The final limit procedure $\varepsilon \to 0$ is the same as that described in the previous section. Therefore the following existence result is valid:

3.2.5 Theorem. *Let the assumption of Lemma 3.2.4 be satisfied. Then there exists a sequence $\boldsymbol{u}_{\varepsilon_k}$ of solutions to problem* (3.2.3) *with suitable penalty parameters $\varepsilon_k \to 0$ such that $\boldsymbol{u}_{\varepsilon_k}$ converges strongly in $\boldsymbol{H}^1(\Omega)$ to a solution of the contact problem with friction* (3.2.1).

3.2.6 Remark. An alternative proof of this theorem via the fixed-point approach instead of the penalty method is again possible.

Modification of the pressure condition

The pressure condition (3.2.2) cannot be prescribed if the set $\mathscr{R}\cap(\mathscr{C}-\{\boldsymbol{g}\})$ contains a non-trivial subspace. This is e.g. the case if the contact boundary is flat or flat in one direction. Here we show how this limitation can be overcome. Let us denote

$$\mathscr{R}_0 \equiv \{\boldsymbol{v} \in (\mathscr{C} - \{\boldsymbol{g}\}) \cap \mathscr{R};\ -\boldsymbol{v} \in (\mathscr{C} - \{\boldsymbol{g}\}) \cap \mathscr{R}\}.$$

Obviously, $\mathscr{R}_0$ is the maximal subspace of $\mathscr{R} \cap (\mathscr{C} - \{\boldsymbol{g}\})$. Let $\mathscr{R}_1$ be its $L_2(\Omega)$-orthogonal complement in $\mathscr{R}$. The given forces are assumed to vanish on $\mathscr{R}_0$,

$$\mathscr{L}(\boldsymbol{v}) = 0 \text{ for all } \boldsymbol{v} \in \mathscr{R}_0, \tag{3.2.10}$$

and to satisfy the *generalized pressure condition* on $\mathscr{R}_1$,

$$\mathscr{L}(\boldsymbol{v}) < 0 \text{ for all } \boldsymbol{v} \in (\mathscr{C} - \{\boldsymbol{g}\}) \cap \mathscr{R}_1 \setminus \{0\}. \tag{3.2.11}$$

Furthermore, Assumption 3.2.1 shall be valid for index $\alpha = \frac{1}{2}$ and boundary regularity $\Gamma_C \in C^{1+\beta}$ with $\beta > 1$. The pressure condition (3.2.2) there is replaced by (3.2.10) and (3.2.11). The restriction to $\alpha = \frac{1}{2}$ is necessary, because we use the shift technique for the smoothed problem.

The *coercive approximation* will be done for the smoothed variational equation

$$\begin{aligned} A(\boldsymbol{u}_{\varepsilon,\eta}, \boldsymbol{v}) &+ \int_{\Gamma_C} \tfrac{1}{\varepsilon}\left[(u_{\varepsilon,\eta})_n - g_n\right]_+ v_n \, ds_{\boldsymbol{x}} \\ &+ \int_{\Gamma_C} \mathfrak{F}(\boldsymbol{u}_{\varepsilon,\eta}) \tfrac{1}{\varepsilon}\left[(u_{\varepsilon,\eta})_n - g_n\right]_+ \nabla\Phi_\varepsilon\big((\boldsymbol{u}_{\varepsilon,\eta})_{\boldsymbol{t}} - \boldsymbol{w}_{\boldsymbol{t}}\big) \cdot \boldsymbol{v}_{\boldsymbol{t}} \, ds_{\boldsymbol{x}} \\ &= \mathscr{L}(\boldsymbol{v}). \end{aligned} \tag{3.2.12}$$

It consists of two steps representing the two sets $\mathscr{R}_0$ and $\mathscr{R}_1$. Instead of $\gamma\langle\pi_{\mathscr{R}}\boldsymbol{u}, \pi_{\mathscr{R}}\boldsymbol{v}\rangle_\Omega$ we add the term

$$\gamma_0\big\langle\pi_{\mathscr{R}_0}\boldsymbol{u}, \pi_{\mathscr{R}_0}\boldsymbol{v}\big\rangle_\Omega + \gamma_1\big\langle\pi_{\mathscr{R}_1}\boldsymbol{u}, \pi_{\mathscr{R}_1}\boldsymbol{v}\big\rangle_\Omega$$

with two possibly different constants γ_0 and γ_1 to the bilinear form. The resulting approximate variational equation is

Find a function $\boldsymbol{u} \in \boldsymbol{H}^1(\Omega)$ such that for all $\boldsymbol{v} \in \boldsymbol{H}^1(\Omega)$ there holds

$$\begin{aligned} \gamma_0\big\langle\pi_{\mathscr{R}_0}\boldsymbol{u}_{\varepsilon,\eta,\gamma_0,\gamma_1}, \pi_{\mathscr{R}_0}\boldsymbol{v}\big\rangle_\Omega &+ \gamma_1\big\langle\pi_{\mathscr{R}_1}\boldsymbol{u}_{\varepsilon,\eta,\gamma_0,\gamma_1}, \pi_{\mathscr{R}_1}\boldsymbol{v}\big\rangle_\Omega \\ + A(\boldsymbol{u}_{\varepsilon,\eta,\gamma_0,\gamma_1}, \boldsymbol{v}) &+ \int_{\Gamma_C} \tfrac{1}{\varepsilon}\big[(u_{\varepsilon,\eta,\gamma_0,\gamma_1})_n - g_n\big]_+ v_n \, ds_{\boldsymbol{x}} \\ + \int_{\Gamma_C} &\mathfrak{F}(\boldsymbol{u}_{\varepsilon,\eta,\gamma_0,\gamma_1}) \tfrac{1}{\varepsilon}\big[(u_{\varepsilon,\eta,\gamma_0,\gamma_1})_n - g_n\big]_+ \cdot \\ &\cdot \nabla\Phi_\eta\big((\boldsymbol{u}_{\varepsilon,\eta,\gamma_0,\gamma_1})_{\boldsymbol{t}} - \boldsymbol{w}_{\boldsymbol{t}}\big) \cdot \boldsymbol{v}_{\boldsymbol{t}} \, ds_{\boldsymbol{x}} = \mathscr{L}(\boldsymbol{v}). \end{aligned} \tag{3.2.13}$$

Due to the validity of the modified pressure condition (3.2.11) the crucial coercivity relation (3.2.5) is valid for the bilinear form

$$\widetilde{A} : (\boldsymbol{u}, \boldsymbol{v}) \mapsto \gamma_0 \langle \pi_{\mathscr{R}_0} \boldsymbol{u}, \pi_{\mathscr{R}_0} \boldsymbol{v} \rangle_{\boldsymbol{L}_2(\Omega)} + A(\boldsymbol{u}, \boldsymbol{v})$$

with fixed positive γ_0:

$$\liminf_{\|\boldsymbol{v}\|_{\boldsymbol{H}^1(\Omega)} \to +\infty} \frac{\widetilde{A}(\boldsymbol{v}, \boldsymbol{v}) + \|[v_n - g_n]_+\|^2_{L_2(\Gamma_C)} - \mathscr{L}(\boldsymbol{v})}{\|\boldsymbol{v}\|_{\boldsymbol{H}^1(\Omega)}} > 0.$$

Due to this relation the solution $\boldsymbol{u}_{\varepsilon,\eta,\gamma_0,\gamma_1}$ of the variational inequality is bounded in $\boldsymbol{H}^1(\Omega)$ uniformly with respect to γ_1 for any *fixed* γ_0. This yields the existence of a sequence $\{\gamma_{1,k}\}$ of approximation parameters tending to 0 such that a corresponding sequence of solutions $\boldsymbol{u}_{\varepsilon,\eta,\gamma_0,\gamma_{1,k}}$ converges weakly in $\boldsymbol{H}^1(\Omega)$ to a solution $\boldsymbol{u}_{\varepsilon,\eta,\gamma_0}$ of the variational equation

$$\begin{aligned}
&\gamma_0 \langle \pi_{\mathscr{R}_0} \boldsymbol{u}_{\varepsilon,\eta,\gamma_0}, \pi_{\mathscr{R}_0} \boldsymbol{v} \rangle_{\Omega} + A(\boldsymbol{u}_{\varepsilon,\eta,\gamma_0}, \boldsymbol{v}) \\
&\quad + \int_{\Gamma_C} \tfrac{1}{\varepsilon} \big[(u_{\varepsilon,\eta,\gamma_0})_n - g_n\big]_+ v_n \, ds_{\boldsymbol{x}} \\
&\quad + \int_{\Gamma_C} \mathfrak{F}(u_{\varepsilon,\eta,\gamma_0}) \tfrac{1}{\varepsilon} \big[(u_{\varepsilon,\eta,\gamma_0})_n - g_n\big]_+ \cdot \\
&\qquad\qquad \cdot \nabla \Phi_\eta \big((\boldsymbol{u}_{\varepsilon,\eta,\gamma_0})_{\boldsymbol{t}} - \boldsymbol{w}_{\boldsymbol{t}}\big) \cdot \boldsymbol{v}_{\boldsymbol{t}} \, ds_{\boldsymbol{x}} = \mathscr{L}(\boldsymbol{v}).
\end{aligned} \tag{3.2.14}$$

The next step is the proof of *a priori* estimates uniform with respect to the three parameters γ_0, η and ε. Let us define $\mathscr{X} = \mathscr{Q} \oplus \mathscr{R}_1$. Remember that $\mathscr{Q}$ is the $\boldsymbol{L}_2(\Omega)$-orthogonal complement of $\mathscr{R}$ in $\boldsymbol{H}^1(\Omega)$. Inserting the test function $\boldsymbol{v} = \boldsymbol{u}_{\varepsilon,\eta,\gamma_0} - \boldsymbol{w}$ into the variational equation and using $(\pi_{\mathscr{R}_0} \boldsymbol{u}_{\varepsilon,\eta,\gamma_0})_n = 0$ (this is valid due to the definition of $\mathscr{R}_0$), the relation

$$\nabla \Phi_\eta(\boldsymbol{v}_{\boldsymbol{t}}) \cdot \boldsymbol{v}_{\boldsymbol{t}} \geq 0 \quad \text{for all } \boldsymbol{v}$$

and $\mathscr{L}(\pi_{\mathscr{R}_0} \boldsymbol{u}_{\varepsilon,\eta,\gamma_0}) = 0$ yields

$$\begin{aligned}
&\tfrac{1}{2} A\big(\pi_{\mathscr{Z}} \boldsymbol{u}_{\varepsilon,\eta,\gamma_0}, \pi_{\mathscr{Z}} \boldsymbol{u}_{\varepsilon,\eta,\gamma_0}\big) \\
&\quad + \int_{\Gamma_C} \tfrac{1}{\varepsilon} \big[\big(\pi_{\mathscr{Z}} u_{\varepsilon,\eta,\gamma_0}\big)_n - g_n\big]_+ \big(\pi_{\mathscr{Z}} u_{\varepsilon,\eta,\gamma_0}\big)_n \, ds_{\boldsymbol{x}} \\
&\leq \mathscr{L}\big(\pi_{\mathscr{Z}} \boldsymbol{u}_{\varepsilon,\eta,\gamma_0}\big) + c_1 \|\boldsymbol{w}\|^2_{\boldsymbol{H}^1(\Omega)} + c_2 \|\boldsymbol{w}\|_{\boldsymbol{H}^1(\Omega)}.
\end{aligned}$$

This relation is valid for $\gamma_0, \gamma_1 < 1$ and an extension of $\boldsymbol{w}$ to the whole domain Ω that satisfies $w_n = 0$ on Γ_C. Employing the modified coercivity condition gives

$$\|\pi_{\mathscr{Z}} \boldsymbol{u}_{\varepsilon,\eta,\gamma_0}\|_{\boldsymbol{H}^1(\Omega)} \leq c \tag{3.2.15}$$

with constant independent of ε, η and γ_0.

Then the shift technique will be performed in order to get the required regularity of the solution on the contact part of the boundary. Since the uniform *a priori* estimate (3.2.15) is valid for the projections $\pi_{\mathscr{Q}}\boldsymbol{u}_{\varepsilon,\eta,\gamma_0}$ only, the shift technique will give only some regularity of this expression. The remainder $\pi_{\mathscr{R}_0}\boldsymbol{u}$, denoted by $\boldsymbol{q}_{\varepsilon,\eta,\gamma_0}$ in the sequel, appears only in the fricition functional in the terms $\mathfrak{F}(\boldsymbol{u}_{\varepsilon,\eta,\gamma_0})$ and $\nabla\Phi_\eta\big((\boldsymbol{u}_{\varepsilon,\eta,\gamma_0})_{\boldsymbol{t}}-\boldsymbol{w}_{\boldsymbol{t}}\big)$. In the shift technique we only need an upper bound for their product in $L_\infty(\Gamma_C)$, and this bound is independent of $\boldsymbol{q}_{\varepsilon,\eta,\gamma_0}$. Hence, the uniform boundedness of $\pi_{\mathscr{Q}}\boldsymbol{u}_{\varepsilon,\eta,\gamma_0}$ in $\boldsymbol{H}^1(\Gamma_{\mathfrak{F}})$ and of $\sigma_n(\boldsymbol{u}_{\varepsilon,\eta,\gamma_0}) = -\frac{1}{\varepsilon}\big[(u_{\varepsilon,\eta,\gamma_0})_n - g_n\big]_+$ in $L_2(\Gamma_{\mathfrak{F}})$ can be proved.

Employing these two estimates it is possible to pass to the limits $\eta \to 0$ and $\varepsilon \to 0$ for a fixed γ_0. Here $\boldsymbol{u}_{\varepsilon,\eta,\gamma_0}$ is bounded in $\boldsymbol{H}^1(\Omega)$ uniformly with respect to η and ε (but in dependence of γ_0). The function $\boldsymbol{q}_{\varepsilon,\eta,\gamma_0} = \pi_{\mathscr{R}_0}\boldsymbol{u}_{\varepsilon,\eta,\gamma_0}$ is therefore bounded in $\boldsymbol{H}^1(\Omega)$, too. Since $\mathscr{R}_0$ is finite-dimensional and all elements of $\mathscr{R}_0$ are smooth, $\boldsymbol{q}_{\varepsilon,\eta,\gamma_0}$ is also bounded in $\boldsymbol{H}^1(\Gamma_{\mathfrak{F}})$. Hence $\boldsymbol{u}_{\varepsilon,\eta,\gamma_0}$ is bounded in this space, too. Consequently for every fixed penalty parameter it is possible to find a sequence $\boldsymbol{u}_{\varepsilon,\eta_k,\gamma_0}$ of solutions with smoothing parameter $\eta_k \to 0$ such that all the convergences mentioned in (3.1.26) hold and its limit $\boldsymbol{u}_{\varepsilon,\gamma_0}$ satisfies the variational inequality

$$
\begin{aligned}
&\gamma_0\big\langle \pi_{\mathscr{R}_0}\boldsymbol{u}_{\varepsilon,\gamma_0}, \pi_{\mathscr{R}_0}(\boldsymbol{v}-\boldsymbol{u}_{\varepsilon,\gamma_0})\big\rangle_{\boldsymbol{L}_2(\Omega)} \\
&\quad + A(\pi_{\mathscr{Q}}\boldsymbol{u}_{\varepsilon,\gamma_0}, \boldsymbol{v}-\pi_{\mathscr{Q}}\boldsymbol{u}_{\varepsilon,\gamma_0}) \\
&\quad + \int_{\Gamma_C} \tfrac{1}{\varepsilon}\big[(u_{\varepsilon,\gamma_0})_n - g_n\big]_+\big(v_n - (u_{\varepsilon,\gamma_0})_n\big)\, ds_{\boldsymbol{x}} \\
&\quad + \int_{\Gamma_C} \mathfrak{F}(\boldsymbol{u}_{\varepsilon,\gamma_0})\tfrac{1}{\varepsilon}\big[(u_{\varepsilon,\gamma_0})_n - g_n\big]_+ \cdot \\
&\qquad \cdot\big(|\boldsymbol{v}_{\boldsymbol{t}}-\boldsymbol{w}_{\boldsymbol{t}}| - |(\boldsymbol{u}_{\varepsilon,\gamma_0})_{\boldsymbol{t}}-\boldsymbol{w}_{\boldsymbol{t}}|\big)\, ds_{\boldsymbol{x}} \ge \mathscr{L}(\boldsymbol{v}-\boldsymbol{u}_{\varepsilon,\gamma_0})
\end{aligned}
\tag{3.2.16}
$$

for all $\boldsymbol{v} \in \boldsymbol{H}^1(\Omega)$. Then, there exists a sequence of penalty parameters $\varepsilon_k \to 0$ and a corresponding sequence of solutions $\boldsymbol{u}_{\varepsilon_k,\gamma_0}$ which converges to a solution of the problem

$$
\begin{aligned}
&\gamma_0\big\langle \pi_{\mathscr{R}_0}\boldsymbol{u}_{\gamma_0}, \pi_{\mathscr{R}_0}(\boldsymbol{v}-\boldsymbol{u}_{\gamma_0})\big\rangle_{\Omega} + A(\pi_{\mathscr{Q}}\boldsymbol{u}_{\gamma_0}, \boldsymbol{v}-\pi_{\mathscr{Q}}\boldsymbol{u}_{\gamma_0}) \\
&\quad + \int_{\Gamma_C} \mathfrak{F}(\boldsymbol{u}_{\gamma_0})|\sigma_n(\boldsymbol{u}_{\gamma_0})|\big(|\boldsymbol{v}_{\boldsymbol{t}}-\boldsymbol{w}_{\boldsymbol{t}}| - |(\boldsymbol{u}_{\gamma_0})_{\boldsymbol{t}}-\boldsymbol{w}_{\boldsymbol{t}}|\big)\, ds_{\boldsymbol{x}} \\
&\ge \mathscr{L}(\boldsymbol{v}-\boldsymbol{u}_{\gamma_0}).
\end{aligned}
\tag{3.2.17}
$$

In the final limit procedure $\gamma_0 \to 0$ there is no *a priori* estimate for $\boldsymbol{q}_{\gamma_0} = \pi_{\mathscr{R}_0}\boldsymbol{u}_{\gamma_0}$ available. Therefore it is necessary to distinguish two cases.

In the first case we assume the existence of a sequence $\gamma_{0,k} \to 0$ such that the corresponding sequence $\boldsymbol{q}_k = \boldsymbol{q}_{\gamma_0,k}$ is *bounded.* Then, after the possible extraction of a subsequence, $\boldsymbol{q}_k$ converges to some limit $\boldsymbol{q} \in \mathscr{R}_0$. Consequently, again extracting a subsequence, we may assume that $\boldsymbol{u}_{\gamma_{0,k}}$ converges to a limit $\boldsymbol{u}$ in the sense described in (3.1.26). Hence the convergence $\gamma_{0,k} \to 0$ in inequality (3.2.17) can be performed and the limit $\boldsymbol{u}$ is a solution of the contact problem (3.2.1).

In the second case there holds $\|\boldsymbol{q}_{\gamma_{0,k}}\|_{\boldsymbol{H}^1(\Omega)} \to +\infty$ for a sequence $\gamma_{0,k} \to 0$ such that $\pi_{\mathscr{Q}}\boldsymbol{u}_{\gamma_{0,k}}$ converges to some $\boldsymbol{u}$ in the sense described in (3.1.26). Taking the test function $\boldsymbol{v} = 0$ in inequality (3.2.17), multiplying the result by $\|\boldsymbol{q}_{\gamma_{0,k}}\|^{-1}_{\boldsymbol{H}^1(\Omega)}$, passing to the limit $k \to \infty$ and employing the fact that both the bilinear and the linear forms do not depend on $\boldsymbol{q}_{\gamma_0,k}$, we prove

$$\lim_{k\to+\infty} \int_{\Gamma_C} \mathfrak{F}(\boldsymbol{u}_{\gamma_{0,k}}) \left|\sigma_n(\boldsymbol{u}_{\gamma_0,k})\right| \left| \frac{(\boldsymbol{q}_{\gamma_{0,k}})_{\boldsymbol{t}}}{\|\boldsymbol{q}_{\gamma_{0,k}}\|_{\boldsymbol{H}^1(\Omega)}} \right| ds_{\boldsymbol{x}} \leq 0.$$

Due to the *a priori* estimates there follows $\mathfrak{F}(\boldsymbol{u}_{\gamma_{0,k}}) \left|\sigma_n(\boldsymbol{u}_{\gamma_0,k})\right| \rightharpoonup \mathfrak{F}_0$ in $L_2(\Gamma_{\mathfrak{F}})$ with $\mathfrak{F}_0 \geq 0$ and $\boldsymbol{q}_{\gamma_{0,k}}/\|\boldsymbol{q}_{\gamma_{0,k}}\|_{\boldsymbol{H}^1(\Omega)} \to \boldsymbol{q}_0 \in \mathscr{R}_0$ (possibly after the extraction of a subsequence). By contradiction we prove $\mathfrak{F}_0 = 0$. Let us assume $\mathfrak{F}_0 > 0$ in a subset $M \subset \Gamma_{\mathfrak{F}}$ of positive surface-measure. Then there holds $\boldsymbol{q}_0 = 0$ in M. Since $\boldsymbol{q}_0$ is an element of $\mathscr{R}$ there holds $\boldsymbol{q}_0(\boldsymbol{x}) = B\boldsymbol{x} + \boldsymbol{r}$ with an antisymmetric matrix $B = -B^\top \in \mathbb{R}^{N,N}$ and $\boldsymbol{r} \in \mathbb{R}^N$. Let us consider sequences of the type $\{\boldsymbol{x}_k\}$ with $\boldsymbol{x}_k \in M$ and $\boldsymbol{x}_k \to \boldsymbol{x}_0 \in M$ for $k \to +\infty$. Taking the limit $k \to +\infty$ in the equation $0 = \boldsymbol{q}_0(\boldsymbol{x}_k) - \boldsymbol{q}_0(\boldsymbol{x}_0) = B(\boldsymbol{x}_k - \boldsymbol{x}_0)$ there follows $B\boldsymbol{y} = 0$ for every $\boldsymbol{y} \in \mathbb{R}^N$ being orthogonal to $\boldsymbol{n}_0 \equiv \boldsymbol{n}(\boldsymbol{x}_0)$. Due to $B^\top = -B$ there holds $\boldsymbol{n}_0^\top B\boldsymbol{n}_0 = 0$, and for every vector $\boldsymbol{y}$ orthogonal to $\boldsymbol{n}_0$ we also have $\boldsymbol{n}_0^\top B\boldsymbol{y} = 0$. Hence $B = 0$ and $\boldsymbol{q}_0$ is a constant vector. However, then from $\boldsymbol{q}_0 = 0$ in $\boldsymbol{x}_0$ there follows $\boldsymbol{q}_0 = 0$ everywhere, and this is a contradiction, because $\boldsymbol{q}_0$ is the limit of a sequence that consists of functions having $\boldsymbol{H}^1(\Omega)$-norm 1. Hence $\mathfrak{F}_0 = 0$ is proved. As a consequence, if $\mathfrak{F}(\boldsymbol{x}, \boldsymbol{u}) \equiv \mathfrak{F}(\boldsymbol{x}, \pi_{\mathscr{Z}}\boldsymbol{u})$ (in particular if $\mathfrak{F}$ is solution independent), then every function $\boldsymbol{u} + \boldsymbol{z}$ with $\boldsymbol{z} \in \mathscr{R}_0$ satisfies the variational inequality (3.2.1) and has $\mathfrak{F}(\boldsymbol{u})\sigma_n(\boldsymbol{u}) = 0$.

3.2.7 Theorem. *Let the Assumption 3.1.3 be valid with the exception of the assumptions concerning* mes Γ_U *and the coefficient d which are replaced by condition* (3.2.10) *and the pressure condition* (3.2.11). *Then the contact problem* (3.2.1) *has at least one solution.*

3.3 Contact problems for two bodies

We consider the contact problem of two elastic bodies occupying the two disjoint domains $\Omega^{(\iota)}$, $\iota = 1, 2$. For simplicity of the presentation we assume a common contact part Γ_C of both bodies. Hence the boundary $\Gamma^{(\iota)}$ of $\Omega^{(\iota)}$ shall be composed of the three measurable and mutally disjoint parts $\Gamma_{\boldsymbol{U}}^{(\iota)}$, $\Gamma_{\boldsymbol{\tau}}^{(\iota)}$ and Γ_C. Both boundaries $\Gamma^{(1)}$ and $\Gamma^{(2)}$ as well as all parts $\Gamma_X^{(\iota)}$, $\iota = 1, 2$, $X = \boldsymbol{U}, \boldsymbol{\tau}$ are assumed to be Lipschitz; the contact boundary Γ_C shall be of the class $C^{1+\beta}$ with $\beta > \frac{1}{2} + \alpha$ for $\alpha \in (0, \frac{1}{2}]$. The displacement field of body ι will be denoted by $\boldsymbol{u}^{(\iota)}$. The displacement fields are assumed to satisfy a suitable time-discretized version of the equations of linear elasticity

$$d^{(\iota)} u_i^{(\iota)} - \partial_j \sigma_{ij}(\boldsymbol{u}^{(\iota)}) = f_i^{(\iota)} \quad \text{in } \Omega^{(\iota)} \tag{3.3.1}$$

with volume force $f^{(\iota)}$. The constitutive relation in domain $\Omega^{(\iota)}$ shall be given by a Hooke law of the type (1.2.5) with Young modulus $E^{(\iota)}$ and Poisson ratio $\nu^{(\iota)}$ in the case of isotropic material or of the type (1.2.1) with the constants $a_0^{(\iota)}$, $A_0^{(\iota)}$ of ellipticity and boundedness in the anisotropic case. On the parts $\Gamma_{\boldsymbol{U}}^{(\iota)}$ of the boundary the displacements $\boldsymbol{U}^{(\iota)}$ are prescribed, and on the parts $\Gamma_{\boldsymbol{\tau}}^{(\iota)}$ tractions $\boldsymbol{\tau}^{(\iota)}$ are given. For simplicity of the notation let us define $\Omega = \Omega^{(1)} \cup \Omega^{(2)}$, $\Gamma_{\boldsymbol{U}} = \Gamma_{\boldsymbol{U}}^{(1)} \cup \Gamma_{\boldsymbol{U}}^{(2)}$ and $\Gamma_{\boldsymbol{\tau}} = \Gamma_{\boldsymbol{\tau}}^{(1)} \cup \Gamma_{\boldsymbol{\tau}}^{(2)}$. For $x \in \overline{\Omega^{(1)} \cup \Omega^{(2)}} \setminus \Gamma_C$ let the common displacement field $\boldsymbol{u}$ be defined by

$$\boldsymbol{u}(\boldsymbol{x}) = \begin{cases} \boldsymbol{u}^{(1)}(\boldsymbol{x}), & \boldsymbol{x} \in \Omega^{(1)} \cup \Gamma_{\boldsymbol{U}}^{(1)} \cup \Gamma_{\boldsymbol{\tau}}^{(1)}, \\ \boldsymbol{u}^{(2)}(\boldsymbol{x}), & \boldsymbol{x} \in \Omega^{(2)} \cup \Gamma_{\boldsymbol{U}}^{(2)} \cup \Gamma_{\boldsymbol{\tau}}^{(2)}. \end{cases}$$

Analogous definitions hold for the given data f, $\boldsymbol{U}$, $\boldsymbol{\tau}$, the coefficients $a_{ijk\ell}$ of the constitutive relation, the components of the stress and strain tensors $\sigma_{ij}(\boldsymbol{u})$ and $e_{ij}(\boldsymbol{u})$ and the coefficient d. Of course, $\boldsymbol{\tau}$ is defined on $\Gamma_{\boldsymbol{\tau}}^{(1)} \cup \Gamma_{\boldsymbol{\tau}}^{(2)}$ only. Sometimes the function $\boldsymbol{u}$ defined on the whole set Ω will be given in the form $\boldsymbol{u} = (\boldsymbol{u}^{(1)}, \boldsymbol{u}^{(2)})$ with $\boldsymbol{u}^{(1)}$ defined on $\Omega^{(1)}$ and $\boldsymbol{u}^{(2)}$ defined on $\Omega^{(2)}$.

The contact and friction is modelled in terms of the *relative displacement*

$$\boldsymbol{u}^{\text{rel}} = \boldsymbol{u}^{(1)} - \boldsymbol{u}^{(2)}$$

on the common contact part Γ_C of the boundary. To be in harmony with the common contact boundary of both bodies we assume the initial gap g_n to be zero, but also a non-zero g_n would not lead to any substantial mathematical difficulties. Let $\boldsymbol{n}^{(\iota)}$ be the normal vector on Γ_C directed to the exterior of $\Omega^{(\iota)}$ (hence $\boldsymbol{n}^{(2)} = -\boldsymbol{n}^{(1)}$). Due to the equilibrium of forces the condition

$$\sigma_{ij}\big(\boldsymbol{u}^{(1)}\big)\boldsymbol{n}_j^{(1)} = -\sigma_{ij}\big(\boldsymbol{u}^{(2)}\big)\boldsymbol{n}_j^{(2)} \tag{3.3.2}$$

must be satisfied. In order to simplify the notation, let us denote $\boldsymbol{\sigma}_\Gamma(\boldsymbol{u}) = \sigma_{ij}\big(\boldsymbol{u}^{(1)}\big)n_j^{(1)}$ and $\boldsymbol{n} = \boldsymbol{n}^{(1)}$. The appropriate Signorini condition for two bodies is given by

$$u_n^{\mathrm{rel}} \le 0, \quad \sigma_n(\boldsymbol{u}) \le 0, \quad u_n^{\mathrm{rel}}\sigma_n(\boldsymbol{u}) = 0 \tag{3.3.3}$$

with $\sigma_n(\boldsymbol{u}) = \boldsymbol{\sigma}_\Gamma(\boldsymbol{u}) \cdot \boldsymbol{n}$. The two-body friction law has the form

$$\begin{aligned} \boldsymbol{u}_{\boldsymbol{t}}^{\mathrm{rel}} = \boldsymbol{w}_{\boldsymbol{t}}^{\mathrm{rel}} \quad &\Rightarrow \quad |\boldsymbol{\sigma}_{\boldsymbol{t}}(\boldsymbol{u})| \le \mathfrak{F}|\sigma_n(\boldsymbol{u})|, \\ \boldsymbol{u}_{\boldsymbol{t}}^{\mathrm{rel}} \ne \boldsymbol{w}_{\boldsymbol{t}}^{\mathrm{rel}} \quad &\Rightarrow \quad \boldsymbol{\sigma}_{\boldsymbol{t}}(\boldsymbol{u}) = -\mathfrak{F}|\sigma_n(\boldsymbol{u})|\frac{\boldsymbol{u}_{\boldsymbol{t}}^{\mathrm{rel}} - \boldsymbol{w}_{\boldsymbol{t}}^{\mathrm{rel}}}{|\boldsymbol{u}_{\boldsymbol{t}}^{\mathrm{rel}} - \boldsymbol{w}_{\boldsymbol{t}}^{\mathrm{rel}}|}. \end{aligned} \tag{3.3.4}$$

The weak formulation of the two-body contact problem is again a variational inequality. The set of admissible functions is the cone

$$\mathscr{C} \equiv \{\boldsymbol{v} \in \boldsymbol{H}^1(\Omega)\,;\ \boldsymbol{v} = \boldsymbol{U} \text{ on } \Gamma_U \text{ and } v_n^{\mathrm{rel}} \le 0 \text{ on } \Gamma_C\}.$$

The problem is given as follows: find a function $\boldsymbol{u} \in \mathscr{C}$ such that for all $\boldsymbol{v} \in \mathscr{C}$ there holds

$$\begin{aligned} A(\boldsymbol{u}, \boldsymbol{v} - \boldsymbol{u}) + \int_{\Gamma_C} \mathfrak{F}|\sigma_n(\boldsymbol{u})|\big(\big|\boldsymbol{v}_{\boldsymbol{t}}^{\mathrm{rel}} - \boldsymbol{w}_{\boldsymbol{t}}^{\mathrm{rel}}\big| - \big|\boldsymbol{u}_{\boldsymbol{t}}^{\mathrm{rel}} - \boldsymbol{w}_{\boldsymbol{t}}^{\mathrm{rel}}\big|\big)\, ds_{\boldsymbol{x}} \\ \ge \mathscr{L}(\boldsymbol{v} - \boldsymbol{u}) \end{aligned} \tag{3.3.5}$$

with the bilinear form of the whole domain

$$A(\boldsymbol{u}, \boldsymbol{v}) = \int_\Omega \big(d\,u_i v_i + \sigma_{ij}(\boldsymbol{u})e_{ij}(\boldsymbol{v})\big)\, d\boldsymbol{x}$$

and the linear functional

$$\mathscr{L}(\boldsymbol{v}) = \int_\Omega \boldsymbol{f} \cdot \boldsymbol{v}\, d\boldsymbol{x} + \int_{\Gamma_\tau} \boldsymbol{\tau} \cdot \boldsymbol{v}\, ds_{\boldsymbol{x}}.$$

The integrals over Ω are given as the sums of the two integrals over both domains $\Omega^{(1)}$ and $\Omega^{(2)}$. With this notation, problem (3.3.5) has almost the same form as the corresponding formulation (3.1.1) of the contact problem of a body and a rigid obstacle. The only difference is the appearance of the relative displacements $\boldsymbol{u}^{\mathrm{rel}}$, $\boldsymbol{v}^{\mathrm{rel}}$ in the friction term and in the definition of the cone $\mathscr{C}$.

The equivalence of the variational formulation (3.3.5) with the corresponding classical formulation of the problem is proved similarly as in the introduction. Let us consider an admissible test function $\boldsymbol{v} = (\boldsymbol{v}^{(1)}, \boldsymbol{v}^{(2)})$ belonging to $\mathscr{C}$. We multiply the differential equations (3.3.1) in $\Omega^{(\iota)}$ with $\boldsymbol{v}^{(\iota)} - \boldsymbol{u}^{(\iota)}$, integrate the divergence of the stress tensor by parts and use the easy formula

$$\sum_{\iota=1}^{2} \int_{\Gamma_C} \sigma_{ij}(\boldsymbol{u}^{(\iota)}) n_j^{(\iota)} \varphi_i^{(\iota)} \, ds_{\boldsymbol{x}} = \int_{\Gamma_C} \sigma_{ij}(\boldsymbol{u}) n_j \varphi_i^{\text{rel}} \, ds_{\boldsymbol{x}}$$

which is valid due to the equilibrium of forces (3.3.2). Then we employ the appropriately modified weak formulations of the contact and friction conditions (3.3.3) and (3.3.4) and obtain the variational inequality (3.3.5). On the other hand, if $\boldsymbol{u} = (\boldsymbol{u}^{(1)}, \boldsymbol{u}^{(2)})$ is a twice continuously differentiable solution of (3.3.5), the differential equation (3.3.1) can be proved by inserting $\boldsymbol{v} = \boldsymbol{u} + \boldsymbol{\varphi}$ with test functions $\boldsymbol{\varphi} = (\boldsymbol{\varphi}^{(1)}, \boldsymbol{\varphi}^{(2)})$ satisfying $\boldsymbol{\varphi}^{(\iota)} = 0$ on $\Gamma^{(\iota)}$ into (3.3.5) and using the Green formula. Then the boundary conditions on $\Gamma_{\boldsymbol{\tau}}^{(\iota)}$ follow with the same procedure for test functions satisfying $\boldsymbol{\varphi}^{(\iota)} = 0$ on $\Gamma_{\boldsymbol{U}}^{(\iota)} \cup \Gamma_C^{(\iota)}$. Using test functions with $\boldsymbol{\varphi}^{(1)} = \boldsymbol{\varphi}^{(2)}$ on Γ_C yields the equilibrium of forces (3.3.2). Finally, taking general test functions $\boldsymbol{\varphi} \in \mathscr{C}$ leads to the weak formulations of the contact and friction conditions (3.3.3) and (3.3.4).

3.3.1 Coercive case

Let us first study the coercive case. Since the set Ω is not connected, it is necessary to assume the coercivity for both domains $\Omega^{(\iota)}, \iota = 1, 2$, separately. The assumptions are summed up in

3.3.1 Assumption. *In addition to the assumptions concerning the geometry of the domain and the coefficients* $a_{ijk\ell}^{(\iota)}$ *let* $a_{ijk\ell}^{(\iota)}, d^{(\iota)} \in C^{\beta-1/2}(\Omega)$, *and* $\Gamma_C \in C^{1+\beta}$ *with* $\beta > \frac{1}{2} + \alpha$ *and* $\alpha \in \left(0, \frac{1}{2}\right]$; *let* either $\operatorname{mes}_\Gamma \Gamma_{\boldsymbol{U}}^{(\iota)} > 0$ *and* $d^{(\iota)} \geq 0$ or $\Gamma_{\boldsymbol{U}}^{(\iota)} = \emptyset$ *and* $d^{(\iota)} \geq d_0 > 0$ *on a set of positive measure for* $\iota = 1, 2$. *The given data fulfil* $\boldsymbol{U} \in \boldsymbol{H}^1(\Omega)$, $\boldsymbol{f} \in \mathring{\boldsymbol{H}}^{-1}(\Omega)$ *and* $\boldsymbol{f} \in \mathring{\boldsymbol{H}}^{-1+\alpha}(\Omega_C)$, *where* $\Omega_C \subset \Omega$ *is a open set such that* $\operatorname{dist}\left(\Gamma_C, \Omega \setminus \overline{\Omega_C}\right) > 0$, $\boldsymbol{\tau} \in \mathring{\boldsymbol{H}}^{-1/2}(\Gamma_{\boldsymbol{\tau}})$, $w \in H^{1/2+\alpha}(\Gamma_C)$. *The compatibility condition* $\boldsymbol{U}^{(\iota)} = 0$ *a.e. on* Γ_C *shall be valid. The coefficient of friction* $\mathfrak{F} \equiv \mathfrak{F}(\boldsymbol{x}, \boldsymbol{u}^{\text{rel}})$ *is non-negative and for every* $\boldsymbol{u}^{\text{rel}}$ *its support shall be contained in a set* $\Gamma_{\mathfrak{F}} \subset \Gamma_C$ *having a positive distance to* $\Gamma \setminus \Gamma_C$. *In the case* $\alpha < \frac{1}{2}$ *the smoothness requirements* (3.1.27) *and* (3.1.28) *are valid; for* $\alpha = \frac{1}{2}$ *the coefficient is continuous in the sense of Carathéodory. The* L_∞*-norm of* $\mathfrak{F}$ *is smaller than the constant* $C_{\mathfrak{F}}$ *given below.*

As in the one-body case, the problem will be solved with the penalty method and, in the case $\alpha = \frac{1}{2}$, a smoothing of the friction functional. The appropriate penalized problem is:

Find $\boldsymbol{u}_\varepsilon \in \boldsymbol{U} + \boldsymbol{H}^1_0(\Omega)$ *with* $\boldsymbol{H}^1_0(\Omega) \equiv \{\boldsymbol{v} \in \boldsymbol{H}^1(\Omega);\ \boldsymbol{v} = 0 \text{ on } \Gamma_U\}$, *such that for all* $\boldsymbol{v} \in \boldsymbol{U} + \boldsymbol{H}^1_0(\Omega)$ *there holds*

$$
\begin{aligned}
A(\boldsymbol{u}_\varepsilon, \boldsymbol{v} - \boldsymbol{u}_\varepsilon) &+ \int_{\Gamma_C} \tfrac{1}{\varepsilon} \left[\left(u_\varepsilon^{\mathrm{rel}}\right)_n\right]_+ \left(v_n^{\mathrm{rel}} - \left(u_\varepsilon^{\mathrm{rel}}\right)_n\right) ds_{\boldsymbol{x}} \\
&+ \int_{\Gamma_C} \tfrac{1}{\varepsilon} \left[\left(u_\varepsilon^{\mathrm{rel}}\right)_n\right]_+ \mathfrak{F}\left(\boldsymbol{u}_\varepsilon^{\mathrm{rel}}\right) \cdot \\
&\qquad \cdot \left(\left|\boldsymbol{v}_{\boldsymbol{t}}^{\mathrm{rel}} - \boldsymbol{w}_{\boldsymbol{t}}^{\mathrm{rel}}\right| - \left|\left(\boldsymbol{u}_\varepsilon^{\mathrm{rel}}\right)_{\boldsymbol{t}} - \boldsymbol{w}_{\boldsymbol{t}}^{\mathrm{rel}}\right|\right) ds_{\boldsymbol{x}} \geq \mathscr{L}(\boldsymbol{v} - \boldsymbol{u}_\varepsilon).
\end{aligned}
\tag{3.3.6}
$$

The smoothed problem reads

Find $\boldsymbol{u}_{\varepsilon,\eta} \in \boldsymbol{U} + \boldsymbol{H}^1_0(\Omega)$ *such that for all* $\boldsymbol{v} \in \boldsymbol{H}^1_0(\Omega)$ *there holds*

$$
\begin{aligned}
A(\boldsymbol{u}_{\varepsilon,\eta}, \boldsymbol{v}) &+ \int_{\Gamma_C} \tfrac{1}{\varepsilon} \left[\left(u_{\varepsilon,\eta}^{\mathrm{rel}}\right)_n\right]_+ v_n^{\mathrm{rel}}\, ds_{\boldsymbol{x}} \\
&+ \int_{\Gamma_C} \tfrac{1}{\varepsilon} \left[\left(u_{\varepsilon,\eta}^{\mathrm{rel}}\right)_n\right]_+ \mathfrak{F}\left(\boldsymbol{u}_{\varepsilon,\eta}^{\mathrm{rel}}\right) \nabla \Phi_\varepsilon \left(\left(\boldsymbol{u}_{\varepsilon,\eta}^{\mathrm{rel}}\right)_{\boldsymbol{t}} - \boldsymbol{w}_{\boldsymbol{t}}^{\mathrm{rel}}\right) \cdot \boldsymbol{v}_{\boldsymbol{t}}^{\mathrm{rel}}\, ds_{\boldsymbol{x}} \\
&= \mathscr{L}(\boldsymbol{v}).
\end{aligned}
\tag{3.3.7}
$$

The solvability of the penalized problem is proved as in the case of the one body contact problem via an appropriate fixed-point approach with the help of convex analysis. In fact, the only change in the formulae is the substitution of $\boldsymbol{u}_\varepsilon$ and $\boldsymbol{v}_\varepsilon$ in the contact- and friction functionals by the relative displacements $\boldsymbol{u}_\varepsilon^{\mathrm{rel}}$ and $\boldsymbol{v}_\varepsilon^{\mathrm{rel}}$. Then the limit procedure $\eta \to 0$ is carried out. This modification of the proof of Lemma 3.1.2 leads to the proof of the following theorem:

3.3.2 Theorem. *Let Assumption 3.3.1 be valid. Then problem* (3.3.7) *has for each* $\varepsilon > 0$, $\eta > 0$ *at least one solution. Every solution satisfies the* a priori *estimate*

$$
\|\boldsymbol{u}_{\varepsilon,\eta}\|_{\boldsymbol{H}^1(\Omega)} + \left\| \tfrac{1}{\varepsilon} \left[\left(u_{\varepsilon,\eta}^{\mathrm{rel}}\right)_n\right]_+ \right\|_{\mathring{H}^{-1/2}(\Gamma_C)} \leq c_1
\tag{3.3.8}
$$

with constant c_1 *depending on the geometry of the domains and on the given data but not on the penalty parameter* ε *and the smoothing parameter* η. *For every fixed* $\varepsilon > 0$ *there exists a sequence* $\eta_k \to 0$ *and a corresponding sequence* $\boldsymbol{u}_{\varepsilon,k} = \boldsymbol{u}_{\varepsilon,\eta_k}$ *of solutions to* (3.3.7) *that converges strongly in* $\boldsymbol{H}^1(\Omega)$ *to a solution* $\boldsymbol{u}_\varepsilon$ *of* (3.3.6). *This solution also satisfies the* a priori *estimate* (3.3.8).

As in the one-body case it is now necessary to prove some better regularity of the solution on the contact part of the boundary. This is done for the smoothed version (3.3.7) of the problem in the case $\alpha = \frac{1}{2}$ and for the non-smoothed inequality (3.3.6) for $\alpha < \frac{1}{2}$.

3.3.3 Theorem. *Let Assumption 3.3.1 be valid. Then every solution* $\boldsymbol{u}_\varepsilon$ *of the penalized problem* (3.3.6) *satisfies the* a priori *estimate*

$$\|\boldsymbol{u}_\varepsilon\|_{\boldsymbol{H}^{1/2+\alpha}(\Gamma_{\mathfrak{F}})} + \left\| \tfrac{1}{\varepsilon}\big[\big(u_\varepsilon^{\mathrm{rel}}\big)_n\big]_+ \right\|_{H^{\alpha-1/2}(\Gamma_{\mathfrak{F}})} \le c_1$$

with a constant c_1 *indepenent of* ε.

Proof. Here the changes compared to the one-body case are more substantial than in the preceding proof. The changes concern in particular the localization technique. It is modified to the two-body case as follows: for a point $\boldsymbol{x}_0 \in \Gamma_{\mathfrak{F}}$ an affine coordinate transformation $\mathscr{O}_{\boldsymbol{x}_0}$ is chosen which maps this point to the origin, its tangent hyperplane onto the hyperplane $\mathbb{R}^{N-1} \times \{0\}$ and the normal vector $-\boldsymbol{n}$ at $\boldsymbol{x}_0$, directed into $\Omega^{(1)}$, onto the N-th basis vector $\boldsymbol{e}_N$. Then the boundary is locally represented as a graph of a function $\psi : B_{N-1}(\delta) \to \mathbb{R}$,

$$\Gamma_\delta(\boldsymbol{x}_0) \equiv \{\mathscr{O}_{\boldsymbol{x}_0}^{-1}(\boldsymbol{y}', \psi(\boldsymbol{y}'));\ \boldsymbol{y}' \in B_{N-1}(0,\delta)\},$$

and the sets

$$\begin{aligned}
\Omega_\delta^{(1)}(\boldsymbol{x}_0) &\equiv \big\{\mathscr{O}_{\boldsymbol{x}_0}^{-1}(\boldsymbol{y}', \psi(\boldsymbol{y}') + y_N);\ \boldsymbol{y}' \in B_{N-1}(0,\delta),\ y_N \in (0,\delta)\big\},\\
\Omega_\delta^{(2)}(\boldsymbol{x}_0) &\equiv \big\{\mathscr{O}_{\boldsymbol{x}_0}^{-1}(\boldsymbol{y}', \psi(\boldsymbol{y}') + y_N);\ \boldsymbol{y}' \in B_{N-1}(0,\delta),\ y_N \in (-\delta,0)\big\}
\end{aligned}$$

are defined. If δ is sufficiently small, then $\Omega_\delta^{(\iota)}(\boldsymbol{x}_0)$ is the intersection of $\Omega^{(\iota)}$ with the set

$$\boldsymbol{U}_\delta(\boldsymbol{x}_0) \equiv \big\{\mathscr{O}_{\boldsymbol{x}_0}^{-1}(\boldsymbol{y}', \psi(\boldsymbol{y}') + y_N);\ \boldsymbol{y}' \in B_{N-1}(0,\delta),\ y_N \in (-\delta,\delta)\big\}.$$

There exists a finite covering $\mathscr{U}_\delta \equiv \{U_\delta(\boldsymbol{x}_r); r = 1, \dots, r_\delta\}$ of $\Gamma_{\mathfrak{F}}$ analogous to that defined in Section 1.6 and a smooth partition of unity $\mathfrak{P}_\delta \equiv \{\varrho_r : \mathbb{R}^N \to [0,1],\ r = 1, \dots, r_\delta\}$ on $\Gamma_{\mathfrak{F}}$ subordinate to this covering. Using the transformation of coordinates

$$\begin{aligned}
\boldsymbol{x} &= \mathfrak{Q}(\boldsymbol{y}) = \mathscr{O}_{\boldsymbol{x}_0}^{-1}(\boldsymbol{y}', \psi(\boldsymbol{y}') + y_N),\\
\boldsymbol{y} &= (\boldsymbol{y}', y_N) \in O_\delta \equiv B_{N-1}(0,\delta) \times (-\delta,\delta)
\end{aligned}$$

the local rectification of the boundary can be done as in the one-body case. Here $\Omega_\delta^{(1)}$ is parametrized by the "upper half" $O_\delta^{(1)} \equiv B_{N-1}(0,\delta) \times (0,\delta)$ and $\Omega_\delta^{(2)}$ by the "lower half" $O_\delta^{(2)} \equiv B_{N-1}(0,\delta) \times (-\delta,0)$.

The result of the localization technique is a variational inequality on the domain $O = O^{(1)} \cup O^{(2)}$ with $O^{(1)} = \mathbb{R}^{N-1} \times \mathbb{R}_+$ and $O^{(2)} = \mathbb{R}^{N-1} \times \mathbb{R}_-$ with common boundary $S = \mathbb{R}^{N-1} \times \{0\}$,

$$\begin{aligned}
&\int_O \left(a(\mathfrak{u}, \boldsymbol{v} - \mathfrak{u}) + b(\boldsymbol{u}, \boldsymbol{v} - \mathfrak{u}) + b_0(\mathfrak{u}, \boldsymbol{v} - \mathfrak{u})\right) dx \\
&\qquad + \int_S \mathscr{S}\left(\mathfrak{u}^{\mathrm{rel}}\right) \cdot \left(v_n^{\mathrm{rel}} - \mathfrak{u}_n^{\mathrm{rel}}\right) ds_{\boldsymbol{x}} \\
&\qquad + \int_S \mathfrak{G}\left(\mathfrak{u}^{\mathrm{rel}}\right) \left(\mathfrak{T}\left(\boldsymbol{v}^{\mathrm{rel}} - \boldsymbol{w}^{\mathrm{rel}}\right) - \mathfrak{T}\left(\mathfrak{u}^{\mathrm{rel}} - \boldsymbol{w}^{\mathrm{rel}}\right)\right) ds_{\boldsymbol{x}} \\
&\quad \geq \int_O \mathfrak{f} \cdot (\boldsymbol{v} - \mathfrak{u})\, dx.
\end{aligned} \tag{3.3.9}$$

Here $\mathfrak{u}$ and $\mathfrak{f}$ denote the corresponding localized and straightened versions of $\boldsymbol{u}$ and $\boldsymbol{f}$ and the forms a, b and $b_0(\cdot,\cdot)$ are defined by obvious extensions of the corresponding formulae in the proof of Theorem 3.1.4 to the two-body case. In particular we have $A(\boldsymbol{u}, \boldsymbol{v}) = A^{(1)}(\boldsymbol{u}, \boldsymbol{v}) + A^{(2)}(\boldsymbol{u}, \boldsymbol{v})$ with

$$\begin{aligned}
A^{(1)}(\boldsymbol{u}, \boldsymbol{v}) &= \int_{O^{(1)}} a_{ijk\ell}\, e_{ij}(\boldsymbol{u}) e_{k\ell}(\boldsymbol{v})\, dx \text{ and} \\
A^{(2)}(\boldsymbol{u}, \boldsymbol{v}) &= \int_{O^{(2)}} a_{ijk\ell}\, e_{ij}(\boldsymbol{v}) e_{k\ell}(\boldsymbol{w})\, dx.
\end{aligned}$$

The definition of $\mathfrak{u}^{\mathrm{rel}}$ and $\boldsymbol{v}^{\mathrm{rel}}$ is obvious. The term $\mathscr{S} = \mathscr{S}\left(\mathfrak{u}^{\mathrm{rel}}\right) \equiv \frac{1}{\varepsilon}\left[\mathfrak{u}_n^{\mathrm{rel}}\right]_+ J_\Gamma$ is the penalty term multiplied by the density of surface measure J_Γ. The terms

$$\mathfrak{G}(\mathfrak{u}^{\mathrm{rel}}) \equiv \begin{cases} \mathfrak{F}(\mathfrak{u}^{\mathrm{rel}})\mathscr{S}(\mathfrak{u}^{\mathrm{rel}}) & \text{for } \alpha < \frac{1}{2}, \\ \mathfrak{F}(\mathfrak{u}^{\mathrm{rel}})\mathscr{S}(\mathfrak{u}^{\mathrm{rel}})\nabla\Phi_\eta(\mathfrak{u}_{\boldsymbol{t}}^{\mathrm{rel}} - \mathfrak{w}_{\boldsymbol{t}}^{\mathrm{rel}}) & \text{for } \alpha = \frac{1}{2} \end{cases}$$

and

$$\mathfrak{T}(\mathfrak{u}^{\mathrm{rel}}) \equiv \begin{cases} |\mathfrak{u}_{\boldsymbol{t}}^{\mathrm{rel}}| & \text{for } \alpha < \frac{1}{2}, \\ \mathfrak{u}_{\boldsymbol{t}}^{\mathrm{rel}} & \text{for } \alpha = \frac{1}{2} \end{cases}$$

are introduced in order to have a uniform notation for the shift technique applied to both the non-smoothed and the smoothed problem.

Starting from (3.3.9), the regularity proof with the shift technique can be performed as in the case of the one-body contact problem. The first result corresponding to inequality (3.1.16) in the one-body case is

$$\begin{aligned}
&\mathscr{A}_1^2(\mathfrak{u}) + \mathscr{A}_2^2(\mathfrak{u}) \leq \left(1 + c_1\delta^{\widetilde{\beta}} + c_2\vartheta\right)\|\mathfrak{F}\|_{L_\infty(\mathbb{R}^N \times \Gamma_C)} \cdot \\
&\quad \cdot d_{N-1}(\alpha) c_{N-1}(\alpha) \|\mathscr{S}\|_{H^{-(1/2-\alpha)}(S)} \left\|\mathfrak{u}_{\boldsymbol{t}}^{\mathrm{rel}}\right\|_{\boldsymbol{H}^{1/2+\alpha}(S)} \\
&\quad + c_1(\vartheta)\|\mathscr{S}\|_{H^{-(1/2-\alpha)}(S)} + c_2(\vartheta)\left\|\mathfrak{u}_{\boldsymbol{t}}^{\mathrm{rel}}\right\|_{\boldsymbol{H}^{1/2+\alpha}(S)} + c_3(\vartheta)
\end{aligned} \tag{3.3.10}$$

with ϑ arbitrarily small, $\widetilde{\boldsymbol{t}}$ denoting the tangential component with respect to the old domain, $\widetilde{\beta} = \min\{\beta, 1\}$, $d_{N-1}(\alpha) = \sqrt{\dfrac{c_{N-1}\left(\frac{1}{2}-\alpha\right)}{c_{N-1}\left(\frac{1}{2}+\alpha\right)}}$ and

$$\mathscr{A}_\iota^2(\mathbf{u}) = \int_{\mathbb{R}^{N-1}} |\boldsymbol{h}'|^{-(N-1+2\alpha)} A^{(\iota)}(\Delta_{\boldsymbol{h}}\mathbf{u}, \Delta_{\boldsymbol{h}}\mathbf{u})\, d\boldsymbol{h}'.$$

The estimate of $\left\|\mathbf{u}_{\widetilde{\boldsymbol{t}}}^{\mathrm{rel}}\right\|_{\boldsymbol{H}^1(S)}$ in the two-body case is

$$\begin{aligned} \left\|\mathbf{u}_{\widetilde{\boldsymbol{t}}}^{\mathrm{rel}}\right\|_{\boldsymbol{H}^{1/2+\alpha}(S)} &\le \left\|\mathbf{u}_{\widetilde{\boldsymbol{t}}}^{(1)}\right\|_{\boldsymbol{H}^{1/2+\alpha}(S)} + \left\|\mathbf{u}_{\widetilde{\boldsymbol{t}}}^{(2)}\right\|_{\boldsymbol{H}^{1/2+\alpha}(S)} \\ &\le \sum_{\iota=1}^{2} \left(\left(1 + c_1\delta^{\widetilde{\beta}}\right) \frac{\sqrt{C^{(1)}_{0,O^{(\iota)}}}}{d_{N-1}^{1/2}(\alpha, \frac{1}{2})} \mathscr{A}_\iota(\mathbf{u}) + c_2 \left\|\mathbf{u}^{(\iota)}\right\|_{\boldsymbol{H}^{1/2}(S)} \right) \end{aligned} \tag{3.3.11}$$

with $d_{N-1}(\alpha, \frac{1}{2}) = \dfrac{c_{N-1}(\alpha)\, c_{N-1}(\frac{1}{2})}{c_{N-1}(\frac{1}{2}+\alpha)}$. For the estimate of the contact force $\|\mathscr{S}\|_{H^{-(1/2-\alpha)}(S)}$ we have the freedom to choose the Green formula for either the "upper" domain $O^{(1)}$ or the "lower" domain $O^{(2)}$. This is a consequence of the equilibrium of forces on the contact boundary. As in the one-body case we derive

$$\begin{aligned} \|\mathscr{S}\|^2_{H^{-(1/2-\alpha)}(S)} \le{}& \left(1 + c_1\delta^{\widetilde{\beta}} + \vartheta\right) \left(d^*_{N-1}(\alpha, \tfrac{1}{2})\right)^{-1} C^{(2)}_{0,O^{(\iota)}} \mathscr{A}_\iota^2(\mathbf{u}) \\ &+ c_2(\vartheta) \end{aligned}$$

for $\iota = 1$ and $\iota = 2$. The result for the admissible coefficient of friction is optimal if we take the minimum of the values for $\iota = 1, 2$. Therefore the inequality

$$\begin{aligned} \mathscr{A}_1^2(\mathbf{u}) + \mathscr{A}_2^2(\mathbf{u}) \le{}& \left(1 + c\delta^{\widetilde{\beta}} + \vartheta\right) \|\mathfrak{F}\|_{L_\infty(\mathbb{R}^N \times \Gamma_C)} \cdot \\ & \cdot \left(\sqrt{C^{(1)}_{0,O^{(1)}}} \mathscr{A}_1(\mathbf{u}) + \sqrt{C^{(1)}_{0,O^{(2)}}} \mathscr{A}_2(\mathbf{u}) \right) \\ & \cdot \min\left\{ \sqrt{C^{(2)}_{0,O^{(1)}}} \mathscr{A}_1(\mathbf{u}), \sqrt{C^{(2)}_{0,O^{(2)}}} \mathscr{A}_2(\mathbf{u}) \right\} \\ & + c_2(\vartheta)\mathscr{A}_1(\mathbf{u}) + c_3(\vartheta)\mathscr{A}_2(\mathbf{u}) + c_4(\vartheta) \end{aligned} \tag{3.3.12}$$

is obtained. In order to get an estimate for $\mathscr{A}_1^2(\mathbf{u}) + \mathscr{A}_2^2(\mathbf{u})$, the value $\|\mathfrak{F}\|_{L_\infty(\mathbb{R}^N \times \Gamma_C)}$ must be smaller than the upper bound

$$C_{\mathfrak{F}} = \min_{\mathscr{A}_1, \mathscr{A}_2 > 0} \mathscr{T}(\mathscr{A}_1, \mathscr{A}_2)$$

with $\mathscr{T}$ defined as

$$\frac{\mathscr{A}_1^2+\mathscr{A}_2^2}{\left(\sqrt{C^{(1)}_{0,O^{(1)}}}\mathscr{A}_1+\sqrt{C^{(1)}_{0,O^{(2)}}}\mathscr{A}_2\right)\cdot\min\left\{\sqrt{C^{(2)}_{0,O^{(1)}}}\mathscr{A}_1,\sqrt{C^{(2)}_{0,O^{(2)}}}\mathscr{A}_2\right\}}.$$

Here the argument $\mathbf{u}$ is omitted for the sake of simplicity. If $\mathscr{A}_1\neq 0$ holds, then we can insert $\mathscr{A}_2=\gamma\mathscr{A}_1$ with an unknown non-negative constant γ. With the abbreviations $\mathfrak{x}_\iota=\sqrt{C^{(1)}_{0,O^{(\iota)}}}$ and $\mathfrak{y}_\iota=\sqrt{C^{(2)}_{0,O^{(\iota)}}}$, $\iota=1,2$, the optimization problem can be reformulated as

$$\begin{aligned}C_{\mathfrak{F}}&=\min_{\gamma>0}\left(\frac{1+\gamma^2}{(\mathfrak{x}_1+\mathfrak{x}_2\gamma)\cdot\min\{\mathfrak{y}_1,\mathfrak{y}_2\gamma\}}\right)\\&=\min_{\gamma>0}\max\left\{\frac{1}{\mathfrak{x}_2\mathfrak{y}_1}\frac{1+\gamma^2}{\frac{\mathfrak{x}_1}{\mathfrak{x}_2}+\gamma},\frac{1}{\mathfrak{x}_2\mathfrak{y}_2}\frac{1+\gamma^2}{\frac{\mathfrak{x}_1}{\mathfrak{x}_2}\gamma+\gamma^2}\right\}.\end{aligned}$$

The solution of this problem is obtained after some calculation,

$$C_{\mathfrak{F}}=\begin{cases}\dfrac{\mathfrak{y}_1^2+\mathfrak{y}_2^2}{\mathfrak{y}_1\mathfrak{y}_2(\mathfrak{y}_1\mathfrak{x}_2+\mathfrak{y}_2\mathfrak{x}_1)}, & \text{if } \dfrac{\mathfrak{x}_1}{\mathfrak{x}_2}\geq\dfrac{\mathfrak{y}_2^2-\mathfrak{y}_1^2}{2\mathfrak{y}_1\mathfrak{y}_2},\\[2ex] \dfrac{2}{\mathfrak{y}_1\left(\sqrt{\mathfrak{x}_1^2+\mathfrak{x}_2^2}+\mathfrak{x}_1\right)}, & \text{if } \dfrac{\mathfrak{x}_1}{\mathfrak{x}_2}<\dfrac{\mathfrak{y}_2^2-\mathfrak{y}_1^2}{2\mathfrak{y}_1\mathfrak{y}_2},\end{cases}$$

if $\mathfrak{y}_1\leq\mathfrak{y}_2$. In the case $\mathfrak{y}_2>\mathfrak{y}_1$ the indices of the bodies must be interchanged.

In the isotropic case this yields the formula

$$C_{\mathfrak{F}}=\begin{cases}\dfrac{1}{2\sqrt{S_1S_2}}\dfrac{R_1S_1+R_2S_2}{R_1\sqrt{S_1}+R_2\sqrt{S_2}}, & 4R_2^2\geq\dfrac{(R_1S_1-R_2S_2)^2}{S_1S_2},\\[2ex] \sqrt{\dfrac{R_2}{S_1}}\dfrac{1}{\sqrt{R_1+R_2}+\sqrt{R_2}}, & 4R_2^2\leq\dfrac{(R_1S_1-R_2S_2)^2}{S_1S_2}\end{cases}\tag{3.3.13}$$

for $R_1S_1\leq R_2S_2$ with

$$R_\iota=\frac{E^{(\iota)}}{1-\left(\nu^{(\iota)}\right)^2}\text{ and }S_\iota=\frac{\left(1-\nu^{(\iota)}\right)^2}{3-4\nu^{(\iota)}},\quad\iota=1,2,$$

in the two-dimensional case and

$$R_\iota=\frac{E^{(\iota)}}{1+\nu^{(\iota)}}\text{ and }S^{(\iota)}=\frac{1-\nu^{(\iota)}}{3-4\nu^{(\iota)}},\quad\iota=1,2,$$

for dimension $N\geq 3$. For $R_2S_2<R_1S_1$ the indices 1 and 2 of the bodies must be exchanged. For anisotropic material the upper bound of the

admissible coefficient of friction is given by

$$C_{\mathfrak{F}} = \begin{cases} \sqrt{\dfrac{3a_0^{(1)}a_0^{(2)}}{4A_0^{(1)}A_0^{(2)}}} \dfrac{A_0^{(1)} + A_0^{(2)}}{\sqrt{a_0^{(1)}A_0^{(1)}} + \sqrt{a_0^{(2)}A_0^{(2)}}}, & \dfrac{a_0^{(2)}}{a_0^{(1)}} \geq \dfrac{\left(A_0^{(1)} - A_0^{(2)}\right)^2}{4A_0^{(1)}A_0^{(2)}}, \\ \sqrt{\dfrac{3a_0^{(1)}a_0^{(2)}}{A_0^{(1)}}} \dfrac{1}{\sqrt{a_0^{(1)} + a_0^{(2)}} + \sqrt{a_0^{(2)}}}, & \dfrac{a_0^{(2)}}{a_0^{(1)}} \leq \dfrac{\left(A_0^{(1)} - A_0^{(2)}\right)^2}{4A_0^{(1)}A_0^{(2)}}, \end{cases} \tag{3.3.14}$$

if $A_0^{(2)} \geq A_0^{(1)}$. For the case $A_0^{(2)} \leq A_0^{(1)}$ the corresponding formula with interchanged indices of the two bodies is valid. □

From this regularity result, the convergence procedure for the penalty parameter can be carried out as described for the one body case. The result is the following Theorem:

3.3.4 Theorem. *Let Assumption 3.3.1 with the bound $C_{\mathfrak{F}}$ given by formula* (3.3.13) *for homogeneous isotropic bodies or by* (3.3.14) *for general bodies be satisfied. Then there exists at least one solution of the two-body contact problem* (3.3.5).

3.3.5 Remark. After a suitable adaptation to the two-body contact problem all remarks about Theorem 3.1.5 remain valid. In particular, the problem with different coefficients of friction of slip and friction of stick can be treated in a similar way: Let $\mathfrak{F}$ have a discontinuity at $\boldsymbol{u}_t^{\mathrm{rel}} - \boldsymbol{w}_t^{\mathrm{rel}} = 0$ and let $\mathfrak{F}_0(0) = \lim_{\boldsymbol{v}_t^{\mathrm{rel}} \to 0} \mathfrak{F}(\boldsymbol{v}_t^{\mathrm{rel}})$ exist and satisfy $\mathfrak{F}(0) \geq \mathfrak{F}_0(0)$. Let $\mathfrak{F}_0$ be extended by $\mathfrak{F}_0(\boldsymbol{v_t}) = \mathfrak{F}(\boldsymbol{v_t})$ for $\boldsymbol{v_t} \neq 0$. If $\mathfrak{F}_0$ satisfies Assumption 3.3.1, then any solution $\boldsymbol{u}$ of problem (3.3.5) with the coefficient of friction $\mathfrak{F}_0$ solves also the problem with coefficient of friction $\mathfrak{F}$.

3.3.2 Semicoercive problems

For two-body contact problems, there are two different cases of semicoercivity. In the first case both bodies are free. Then the cone of admissible functions is defined by

$$\mathscr{C} = \left\{ \boldsymbol{v} \in \boldsymbol{H}^1(\Omega);\ v_n^{\mathrm{rel}} \leq 0 \text{ on } \Gamma_C \right\}$$

and the appropriate set of rigid motions is

$$\mathscr{R} = \left\{ \boldsymbol{v} \in \boldsymbol{H}^1(\Omega);\ e_{ij}(\boldsymbol{v}) = 0 \text{ on } \Omega \right\}.$$

If the standard identification of $\boldsymbol{H}^1(\Omega)$ with $\boldsymbol{H}^1(\Omega^{(1)}) \times \boldsymbol{H}^1(\Omega^{(2)})$ is used, then $\mathscr{R}$ is equivalent to $\mathscr{R}^{(1)} \times \mathscr{R}^{(2)}$ with the sets of rigid motion $\mathscr{R}^{(\iota)}$ on $\Omega^{(\iota)}$. In the second case of semicoercivity one of the bodies is free, but the other one is "fixed" by either a Dirichlet boundary condition on a part of boundary Γ_U with positive surface measure or by the condition $d \geq d_0$ with a positive d_0 on a subset of positive measure. If the "free" body is denoted by $\Omega^{(1)}$ and the "fixed" one by $\Omega^{(2)}$, then the set of admissible functions of the problem is given by

$$\mathscr{C} = \left\{ \boldsymbol{v} \in \boldsymbol{H}^1(\Omega);\ \boldsymbol{v} = \boldsymbol{U} \text{ on } \Gamma_U^{(2)} \text{ and } v_n^{\mathrm{rel}} \leq 0 \text{ on } \Gamma_C \right\}$$

and the appropriate set of rigid motions is

$$\mathscr{R} = \left\{ \boldsymbol{v} \in \boldsymbol{H}^1(\Omega);\ e_{ij}(\boldsymbol{v}) = 0 \text{ on } \Omega^{(1)} \text{ and } \boldsymbol{v} = 0 \text{ on } \Omega^{(2)} \right\}.$$

Using the above mentioned identification, the latter set is equal to $\mathscr{R}^{(1)} \times \{0\}$.

The semicoercive problem can be solved under Assumption (3.3.1) supplemented with an additional *pressure condition.* The pressure condition has the same form (3.2.2) as in the semicoercive one-body case, if the new definitions of $\mathscr{C}$, $\boldsymbol{H}^1(\Omega)$, $\mathscr{R}$ and $\mathscr{L}$ for the two-body cases are used there. The crucial Lemma 3.2.2 is valid in the appropriate modified form, where the term u_n on Γ_C is replaced by u_n^{rel}; the proof of the modified lemma is the same as in the one-body case. From the modified lemma the existence of solutions to the penalized problem and the *a priori* estimate of the type (3.2.9) can be verified as in the semicoercive one-body case. Then the regularity result of Theorem 3.3.4 is proved as in the coercive two-body case under the same additional assumptions. In particular, the admissible coefficient of friction $C_{\mathfrak{F}}$ is the same. The convergence of a sequence of solutions of the penalized problem to a solution of the original contact problem is proved as in the coercive case. The result is the following theorem:

3.3.6 Theorem. *In addition to Assumption 3.3.1 let the pressure condition*

$$\mathscr{L}(\boldsymbol{v}) < 0 \text{ for all } \boldsymbol{v} \in \mathscr{C} \cap \mathscr{R} \setminus \{0\}$$

be valid. Then the semicoercive two-body contact problem has at least one solution. This solution can be represented as a limit of solutions of the corresponding penalized problem.

3.4 Quasistatic contact problem

In the quasistatic contact problem the body is assumed to be deformed very slowly and the inertial forces in the dynamic formulation are therefore neglected. The classical formulation of the problem is given by the set of relations

$$-\partial_j \sigma_{ij}(\boldsymbol{u}) = f_i \qquad \text{in } I_T \times \Omega, \tag{3.4.1}$$

$$\boldsymbol{u} = \boldsymbol{U} \qquad \text{on } I_T \times \Gamma_{\boldsymbol{U}}, \tag{3.4.2}$$

$$\boldsymbol{\sigma}_\Gamma(\boldsymbol{u}) = \boldsymbol{\tau} \qquad \text{on } I_T \times \Gamma_{\boldsymbol{\tau}}, \tag{3.4.3}$$

$$u_n \le g_n, \quad \sigma_n \le 0, \quad \sigma_n(u_n - g_n) = 0 \text{ on } I_T \times \Gamma_C, \tag{3.4.4}$$

$$\begin{aligned} \dot{\boldsymbol{u}}_{\boldsymbol{t}} = 0 \quad &\Rightarrow \quad |\boldsymbol{\sigma}_{\boldsymbol{t}}| \le \mathfrak{F}|\sigma_n|, \\ \dot{\boldsymbol{u}}_{\boldsymbol{t}} \neq 0 \quad &\Rightarrow \quad \boldsymbol{\sigma}_{\boldsymbol{t}} = -\mathfrak{F}|\sigma_n|\frac{\dot{\boldsymbol{u}}_{\boldsymbol{t}}}{|\dot{\boldsymbol{u}}_{\boldsymbol{t}}|} \end{aligned} \qquad \text{on } I_T \times \Gamma_C, \tag{3.4.5}$$

$$\boldsymbol{u}(0, x) = \boldsymbol{u}_0 \qquad \text{for } x \in \Omega. \tag{3.4.6}$$

In this formulation, the time derivative $\dot{\boldsymbol{u}}$ appears only in the friction law. The initial condition is given for the displacements only.

Let us introduce the variational formulation of this problem. The contact condition given here is formulated with the displacements, but the friction condition is formulated in the displacement velocities. Therefore the variational formulation of the quasistatic problem is a little more complicated than in the static case. Let us define the cone

$$\mathscr{C}(t) \equiv \left\{\boldsymbol{v} \in \boldsymbol{H}^1(\Omega);\ \boldsymbol{v} = \boldsymbol{U}(t, \cdot) \text{ on } \Gamma_{\boldsymbol{U}} \text{ and } v_n \le g_n \text{ on } \Gamma_C\right\}.$$

Then the quasistatic contact problem is given as follows:

Find a function $\boldsymbol{u} \in \boldsymbol{U} + H^1(I_T; \boldsymbol{H}_0^1(\Omega))$ with $\boldsymbol{u}(t, \cdot) \in \mathscr{C}(t)$ for almost every $t \in I_T$ such that for almost every $t \in I_T$ and every $\boldsymbol{v} \in \mathscr{C}(t)$ there holds

$$\begin{aligned} A(\boldsymbol{u}, \boldsymbol{v} - \boldsymbol{u}) + \int_{\Gamma_C} \mathfrak{F}|\sigma_n(\boldsymbol{u})|\big(|\boldsymbol{v}_{\boldsymbol{t}} - \boldsymbol{u}_{\boldsymbol{t}} + \dot{\boldsymbol{u}}_{\boldsymbol{t}}| - |\dot{\boldsymbol{u}}_{\boldsymbol{t}}|\big)\, ds_{\boldsymbol{x}} \\ \ge \mathscr{L}(\boldsymbol{v} - \boldsymbol{u}) \end{aligned} \tag{3.4.7}$$

with the bilinear form

$$A(\boldsymbol{u}, \boldsymbol{v}) = \int_\Omega \sigma_{ij}(\boldsymbol{u}) e_{ij}(\boldsymbol{v})\, d\boldsymbol{x}$$

and the linear functional

$$\mathscr{L}(\boldsymbol{v}) = \int_\Omega \boldsymbol{f} \cdot \boldsymbol{v}\, d\boldsymbol{x} + \int_{\Gamma_{\boldsymbol{\tau}}} \boldsymbol{\tau} \cdot \boldsymbol{v}\, ds_{\boldsymbol{x}}.$$

As before, the notation $\boldsymbol{H}_0^1(\Omega) = \left\{\boldsymbol{v} \in \boldsymbol{H}^1(\Omega);\ \boldsymbol{v} = 0 \text{ on } \Gamma_{\boldsymbol{U}}\right\}$ is employed. The weak formulation of the problem is derived from its classical version by the usual procedure: the differential equation (3.4.1) is multiplied by $v_i - u_i$, the sum over all indices i is taken and an integration by parts is performed. In the boundary integral we use the variational formulations (1.5.12) of the Signorini contact condition and

$$\boldsymbol{\sigma_t}\cdot(\boldsymbol{w_t}-\dot{\boldsymbol{u}}_t)+\mathfrak{F}|\sigma_n(\boldsymbol{u})|\big(|\boldsymbol{w_t}|-|\dot{\boldsymbol{u}}_t|\big) \geq 0 \text{ for all } \boldsymbol{w_t} \text{ orthogonal to } \boldsymbol{n} \tag{3.4.8}$$

of the Coulomb friction law (cf. also (1.5.13)). The latter is employed with the decomposition

$$\boldsymbol{v} - \boldsymbol{u} = (\boldsymbol{v} - \boldsymbol{u} + \dot{\boldsymbol{u}}) - \dot{\boldsymbol{u}} \equiv \boldsymbol{w} - \dot{\boldsymbol{u}}$$

which leads to the present form of the friction functional.

If the physical units are considered, then the variational formulation looks a little strange, because a displacement and a velocity appear in the same sum. In fact, the variational formulation can be modified in such a way that $\dot{\boldsymbol{u}}_t$ is replaced by $c\dot{\boldsymbol{u}}_t$ with an arbitrary parameter $c > 0$ that may be used to transform the physical units. This follows from the modified decomposition

$$\boldsymbol{v} - \boldsymbol{u} = (\boldsymbol{v} - \boldsymbol{u} + c\dot{\boldsymbol{u}}) - c\dot{\boldsymbol{u}} \equiv c\boldsymbol{w} - c\dot{\boldsymbol{u}},$$

taken in the product of (3.4.8) and c.

The existence proof to be presented here was published by L.E. Andersson in [12]. With the penalty method, the problem is approximated by the variational inequality

Find a function $\boldsymbol{u} \in \boldsymbol{U} + H^1\big(I_T; \boldsymbol{H}_0^1(\Omega)\big)$ *such that for almost every* $t \in I_T$ *and every* $\boldsymbol{v} \in \dot{\boldsymbol{U}}(t,\cdot) + \boldsymbol{H}_0^1(\Omega)$ *there holds*

$$\begin{aligned} A(\boldsymbol{u}, \boldsymbol{v} - \dot{\boldsymbol{u}}) &+ \int_{\Gamma_C} \tfrac{1}{\varepsilon}\big[u_n - g_n\big]_+ (v_n - \dot{u}_n)\, ds_{\boldsymbol{x}} \\ &+ \int_{\Gamma_C} \mathfrak{F}\tfrac{1}{\varepsilon}\big[u_n - g_n\big]_+ \big(|\boldsymbol{v_t}| - |\dot{\boldsymbol{u}}_t|\big)\, ds_{\boldsymbol{x}} \geq \mathscr{L}(\boldsymbol{v} - \dot{\boldsymbol{u}}). \end{aligned} \tag{3.4.9}$$

Here also a change of the test function is done. The solvability of this problem will be proved with a time discretization (Rothe method). We consider a partition of the time interval I_T with time steps of equal stepsize $\Delta t = T/L$. Let $t_\ell \equiv \ell\Delta t$, $\ell = 0, \ldots, L$, let $\boldsymbol{u}^{(\ell)}$ be an approximation for $\boldsymbol{u}(t_\ell)$ and $\Delta\boldsymbol{u}^{(\ell)} \equiv \boldsymbol{u}^{(\ell)} - \boldsymbol{u}^{(\ell-1)}$ be the time difference operator. The time-discretized problem is obtained from (3.4.9) by replacing $\boldsymbol{u}$ with $\boldsymbol{u}^{(\ell)}$ and $\dot{\boldsymbol{u}}^{(\ell)}$ with $\Delta\boldsymbol{u}^{(\ell)}/\Delta t$. If the result is multiplied by the time step Δt the following variational inequality is obtained:

Find a function $\boldsymbol{u}^{(\ell)} \in \boldsymbol{U}^{(\ell)} + \boldsymbol{H}_0^1(\Omega)$ *such that for all* $\boldsymbol{v} \in \Delta \boldsymbol{U}^{(\ell)} + \boldsymbol{H}_0^1(\Omega)$ *there holds*

$$\begin{aligned} &A\big(\boldsymbol{u}^{(\ell)}, \boldsymbol{v} - \Delta \boldsymbol{u}^{(\ell)}\big) + \int_{\Gamma_C} \tfrac{1}{\varepsilon}\big[u_n^{(\ell)} - g_n\big]_+ \big(v_n - \Delta u_n^{(\ell)}\big)\, ds_{\boldsymbol{x}} \\ &\quad + \int_{\Gamma_C} \mathfrak{F}\tfrac{1}{\varepsilon}\big[u_n^{(\ell)} - g_n\big]_+ \big(|\boldsymbol{v_t}| - |\Delta \boldsymbol{u_t}^{(\ell)}|\big)\, ds_{\boldsymbol{x}} \\ &\geq \mathscr{L}^{(\ell)}\big(\boldsymbol{v} - \Delta \boldsymbol{u}^{(\ell)}\big). \end{aligned} \tag{3.4.10}$$

Then, the smoothing of the norm in the friction functional is employed. The resulting variational inequality is

Find $\boldsymbol{u}^{(\ell)} \in \boldsymbol{U}^{(\ell)} + \boldsymbol{H}_0^1(\Omega)$ *such that for all* $\boldsymbol{v} \in \boldsymbol{H}_0^1(\Omega)$ *there holds*

$$\begin{aligned} &A\big(\boldsymbol{u}^{(\ell)}, \boldsymbol{v}\big) + \int_{\Gamma_C} \tfrac{1}{\varepsilon}\big[u_n^{(\ell)} - g_n\big]_+ v_n\, ds_{\boldsymbol{x}} \\ &\quad + \int_{\Gamma_C} \mathfrak{F}\tfrac{1}{\varepsilon}\big[u_n^{(\ell)} - g_n\big]_+ \nabla \Phi_\eta\big(\Delta \boldsymbol{u_t}^{(\ell)}\big) \cdot \boldsymbol{v_t}\, ds_{\boldsymbol{x}} \ = \ \mathscr{L}^{(\ell)}(\boldsymbol{v}). \end{aligned} \tag{3.4.11}$$

Problems (3.4.10) and (3.4.11) have the same forms as the corresponding versions of the static contact problem with $\boldsymbol{w} = \boldsymbol{u}^{(\ell-1)}$.

For simplicity of the presentation we assume $\boldsymbol{U}^{(\ell)} \equiv 0$ in the following proofs. This restriction can be relaxed: if a non-zero $\boldsymbol{U}^{(\ell)}$ can be extended onto the whole domain Ω in such a way that $\boldsymbol{U}^{(\ell)} = 0$ holds in a neighbourhood of Γ_C, then the transformed function

$$\widetilde{\boldsymbol{u}}^{(\ell)} = \boldsymbol{u}^{(\ell)} - \boldsymbol{U}^{(\ell)}$$

solves the problem with Dirichlet data $\widetilde{\boldsymbol{U}}^{(\ell)} = 0$ and the modified linear functional

$$\widetilde{\mathscr{L}}^{(\ell)}(\boldsymbol{v}) = \mathscr{L}^{(\ell)}(\boldsymbol{v}) - A\big(\boldsymbol{U}^{(\ell)}, \boldsymbol{v}\big).$$

In the assumptions of the assertions we will state the required assumptions for $\boldsymbol{U}^{(\ell)}$; they can be derived easily from the corresponding assumptions on $\mathscr{L}^{(\ell)}$.

The existence of a solution to (3.4.11) is proved as in the static case via a fixed point approach. Then an *a priori* estimate for its solution $\boldsymbol{u}^{(\ell)}$ is derived by taking the test function $\boldsymbol{v} = \boldsymbol{u}^{(\ell)}$. In order to have this estimate independent of the solution $\boldsymbol{u}^{(\ell-1)}$ from the previous time step, the friction functional is treated in a different way than in the static problem. We have

$$\begin{aligned}
\int_{\Gamma_C} & \mathfrak{F}\tfrac{1}{\varepsilon}\big[u_n^{(\ell)} - g_n\big]_+ \nabla\Phi_\eta\big(\Delta u_t^{(\ell)}\big) \cdot u_t^{(\ell)}\, ds_x \\
&\le \|\mathfrak{F}\|_{L_\infty(\Gamma_C)} \big\|\sigma_n\big(u^{(\ell)}\big)\big\|_{\mathring{H}^{-1/2}(\Gamma_C)} \big\|u_t^{(\ell)}\big\|_{H^{1/2}(\Gamma_C)} \\
&\le \|\mathfrak{F}\|_{L_\infty(\Gamma_C)} \sqrt{C^{(1)}C^{(2)}} \big\|u^{(\ell)}\big\|_A^2 + c_1\big\|u^{(\ell)}\big\|_A + C_2
\end{aligned}$$

with the energy norm $\|u\|_A^2 = \sqrt{A(u,u)}$ and the constants $C^{(1)}$, $C^{(2)}$ from the generalized trace and inverse trace relations

$$\big\|\sigma_n\big(u^{(\ell)}\big)\big\|_{\mathring{H}^{-1/2}(\Gamma_C)}^2 \le C^{(1)}\big\|u^{(\ell)}\big\|_A^2 + c_1\big\|\mathscr{L}^{(\ell)}\big\|_{\mathring{H}^{-1}(\Omega)}^2, \tag{3.4.12}$$

$$\big\|u_t^{(\ell)}\big\|_{H^{1/2}(\Gamma_C)}^2 \le C^{(2)}\big\|u^{(\ell)}\big\|_A^2 + c_2\big\|\mathscr{L}^{(\ell)}\big\|_{\mathring{H}^{-1}(\Omega)}^2. \tag{3.4.13}$$

The validity of these relations follows from the Green theorem and the usual trace and inverse trace theorems for Sobolev spaces. If

$$\|\mathfrak{F}\|_{L_\infty(\Gamma_C)} < \big(C^{(1)}C^{(2)}\big)^{-1/2}$$

holds, then we arrive at an *a priori* estimate

$$\big\|u^{(\ell)}\big\|_{H^1(\Omega)} \le c.$$

If $\|\mathscr{L}^{(\ell)}\|_{\mathring{H}^{-1}(\Omega)}$ is bounded uniformly with respect to ℓ, then the constant c there is independent of ℓ.

3.4.1 Remark. The constants $C^{(1)}$ and $C^{(2)}$ have a similar role as the constants $C_0^{(1)}$ and $C_0^{(2)}$ needed for the computation of the admissible coefficient of friction for the static problem. However, here they depend on the domain Ω. It is not clear how to calculate these constants for a general domain from e.g. some smoothness requirements. In the static case the localization technique enables the use of known constants for the half-space domain. However, by the localization and straightening of the boundary we obtain additional perturbation terms depending on lower-order space derivatives of the solution. In the proof of the space regularity these terms can be estimated with the help of existing *a priori* estimates in some lower-order Sobolev norm. Here, such an approach requires an additional knowledge about the time regularity of the solution in a lower-order Sobolev norm. It is not clear how to get such an estimate. This turns out to be a major drawback for the practical applicability of the obtained result, because the admissible coefficient of friction can be computed in special situations only.

After establishing a uniform bound for $\boldsymbol{u}^{(\ell)}$ in $\boldsymbol{H}^1(\Omega)$, a better regularity of the solution can be proved by the localization and shift techniques as in the static case. Here, it is again important to have the estimate independent of $\boldsymbol{u}^{(\ell-1)}$. This can be achieved by applying the shift technique to the smoothed problem (3.4.11) with regularity gain $\alpha = \frac{1}{2}$. In this case, the function $\boldsymbol{u}^{(\ell-1)}$ is only present in

$$\mathfrak{F}(\Delta \boldsymbol{u}_t^{(\ell)}) \nabla \Phi_\eta (\Delta \boldsymbol{u}_t^{(\ell)}).$$

In the shift technique, this term is estimated in $L_\infty(\partial O)$ (with the boundary ∂O of the straightended domain), and this estimate is independent of the argument $\Delta \boldsymbol{u}_t^{(\ell)}$. If the shift technique would be applied to the penalized but non-smoothed problem, then the estimate of the difference

$$\left|(\Delta \boldsymbol{u}_t^{(\ell)})_{-\boldsymbol{h}}\right| - \left|\Delta \boldsymbol{u}_t^{(\ell)}\right|$$

cannot be done without any dependence of the result on $\boldsymbol{u}^{(\ell-1)}$.

The results of this procedure are given in the next Theorem that is valid under the following assumptions:

3.4.2 Assumption. *Let the domain Ω be bounded and connected, let its boundary Γ be Lipschitz and be composed of the closures of the mutually disjoint parts Γ_U, Γ_τ and Γ_C which are open with respect to the surface topology and let* $\operatorname{mes}_\Gamma \Gamma_U > 0$. *The coefficients $a_{ijk\ell}$ are symmetric, bounded, elliptic in the sense described in* (1.2.2)–(1.2.4) *and C^β-smooth with $\beta > 2$. The given data satisfy $\boldsymbol{U}^{(\ell)} \in \boldsymbol{H}^1(\Omega)$, $\boldsymbol{f}^{(\ell)} \in \mathring{\boldsymbol{H}}^{-1}(\Omega)$, $\boldsymbol{\tau}^{(\ell)} \in \mathring{\boldsymbol{H}}^{-1/2}(\Gamma_\tau)$, $g_n \in H^1(\Gamma_C)$, $g_n \geq 0$ a.e. in Γ_C. For a set $\Omega_C \subset \Omega$ satisfying $\Gamma_C \subset \partial\Omega_C$ there holds $\boldsymbol{f}^{(\ell)} \in H^{-1/2}(\Omega_C)$ and $\boldsymbol{U}^{(\ell)} = 0$ on Ω_C. The norms of these functions in the corresponding spaces are independent of ℓ. The contact boundary Γ_C fulfils $\Gamma_C \in C^\beta$. The coefficient of friction $\mathfrak{F} \equiv \mathfrak{F}(\boldsymbol{x}, \boldsymbol{u})$ shall be non-negative and continuous in the sense of Carathéodory. For every $\boldsymbol{u} \in \mathbb{R}^N$ its support* $\operatorname{supp} \mathfrak{F}(\cdot, \boldsymbol{u})$ *is contained in a set $\Gamma_{\mathfrak{F}} \subset \Gamma_C$ having a positive distance to $\Gamma \setminus \Gamma_C$. Moreover, $\mathfrak{F}$ shall be globally bounded by*

$$\|\mathfrak{F}\|_{L_\infty(\Gamma_{\mathfrak{F}} \times \mathbb{R}^N)} < \widetilde{C}_{\mathfrak{F}}$$

with admissible coefficient of friction

$$\widetilde{C}_{\mathfrak{F}} = \left(\max\left\{ \sqrt{C_0^{(1)} C_0^{(2)}}, \sqrt{C^{(1)} C^{(2)}} \right\} \right)^{-1},$$

where the constants $C_0^{(1)}$, $C_0^{(2)}$ are computed from the coefficients of the bilinear form as in the static case and the constants $C^{(1)}$, $C^{(2)}$ are those from (3.4.12) *and* (3.4.13).

3.4.3 Theorem. *Let Assumption 3.4.2 be valid. Then for every* $\ell \in \{1, \dots, L\}$, $\varepsilon > 0$, $\eta > 0$ *problem* (3.4.11) *has a solution* $\boldsymbol{u}^{(\ell)}$. *This solution satisfies*

$$\left\|\boldsymbol{u}^{(\ell)}\right\|_{\boldsymbol{H}^1(\Omega)} \leq c_1 \tag{3.4.14}$$

and

$$\left\|\boldsymbol{u}^{(\ell)}\right\|_{\boldsymbol{H}^1(\Gamma_{\mathfrak{F}})} + \left\|\sigma_n\left(\boldsymbol{u}^{(\ell)}\right)\right\|_{L_2(\Gamma_{\mathfrak{F}})} \leq c_2 \tag{3.4.15}$$

with constants c_1, c_2 *independent of* ℓ, ε *and* η. *For every fixed* ℓ, ε *there exists a sequence* $\eta_k \to 0$ *such that a corresponding sequence* $\boldsymbol{u}_k^{(\ell)}$ *of solutions to* (3.4.11) *converges in* $\boldsymbol{H}^1(\Omega)$ *to a solution* $\boldsymbol{u}^{(\ell)}$ *of the penalized problem* (3.4.10). *This solution also satisfies the* a priori *estimates* (3.4.14) *and* (3.4.15).

In order to pass to the limit $\Delta t \to 0$ of the time step it is necessary to have a certain time regularity for the solutions of the penalized problem. This time regularity follows from the estimate for the time differences presented in the next Theorem:

3.4.4 Theorem. *In addition to Asumption 3.4.2 let the coefficient of friction be solution independent and satisfy the condition* (3.4.20) *given below. Then the solution of the time-discretized problem* (3.4.10) *satisfies the estimate*

$$\left\|\Delta\boldsymbol{u}^{(\ell)}\right\|_{\boldsymbol{H}^1(\Omega)} \leq c_1\left\|\Delta\mathscr{L}^{(\ell)}\right\|_{\boldsymbol{H}_0^1(\Omega)^*} + c_2\left\|\Delta\boldsymbol{U}^{(\ell)}\right\|_{\boldsymbol{H}^1(\Omega)} \tag{3.4.16}$$

with constants c_1, c_2 *independent of* ℓ *and* ε.

Proof. It is sufficient to consider the case $\boldsymbol{U}^{(\ell)} = 0$. Inserting the test function $\boldsymbol{v} = 0$ into variational inequality (3.4.10) yields

$$\begin{aligned} &A\left(\boldsymbol{u}^{(\ell)}, \Delta\boldsymbol{u}^{(\ell)}\right) + \left\langle \tfrac{1}{\varepsilon}\left[u_n^{(\ell)} - g_n\right]_+, \Delta u_n^{(\ell)} \right\rangle_{\Gamma_C} \\ &\quad + \left\langle \mathfrak{F}\tfrac{1}{\varepsilon}\left[u_n^{(\ell)} - g_n\right]_+, \left|\Delta\boldsymbol{u}_{\boldsymbol{t}}^{(\ell)}\right| \right\rangle_{\Gamma_C} \leq \mathscr{L}^{(\ell)}\left(\Delta\boldsymbol{u}^{(\ell)}\right). \end{aligned}$$

Inserting $\boldsymbol{v} = \Delta\boldsymbol{u}^{(\ell)} + \Delta\boldsymbol{u}^{(\ell-1)} = \boldsymbol{u}^{(\ell)} - \boldsymbol{u}^{(\ell-2)}$ into the equation for $\ell - 1$ gives

$$\begin{aligned} &A\left(\boldsymbol{u}^{(\ell-1)}, \Delta\boldsymbol{u}^{(\ell)}\right) + \left\langle \tfrac{1}{\varepsilon}\left[u_n^{(\ell-1)} - g_n\right]_+, \Delta u_n^{(\ell)} \right\rangle_{\Gamma_C} \\ &\quad + \left\langle \mathfrak{F}\tfrac{1}{\varepsilon}\left[u_n^{(\ell-1)} - g_n\right]_+, \left(\left|\Delta\boldsymbol{u}_{\boldsymbol{t}}^{(\ell)} + \Delta\boldsymbol{u}_{\boldsymbol{t}}^{(\ell-1)}\right| - \left|\Delta\boldsymbol{u}_{\boldsymbol{t}}^{(\ell-1)}\right|\right) \right\rangle_{\Gamma_C} \\ &\geq \mathscr{L}^{(\ell-1)}\left(\Delta\boldsymbol{u}^{(\ell)}\right). \end{aligned}$$

Taking the difference of both inequalities and using the estimate $\left|\Delta \boldsymbol{u}_{\boldsymbol{t}}^{(\ell)} + \Delta \boldsymbol{u}_{\boldsymbol{t}}^{(\ell-1)}\right| - \left|\Delta \boldsymbol{u}_{\boldsymbol{t}}^{(\ell-1)}\right| \leq \left|\Delta \boldsymbol{u}_{\boldsymbol{t}}^{(\ell)}\right|$ yields

$$
\begin{aligned}
A\big(\Delta \boldsymbol{u}^{(\ell)}, \Delta \boldsymbol{u}^{(\ell)}\big) &+ \left\langle \tfrac{1}{\varepsilon}\Delta\big(\big[u_n^{(\ell)} - g_n\big]_+\big), \Delta u_n^{(\ell)}\right\rangle_{\Gamma_C} \\
&+ \left\langle \tfrac{1}{\varepsilon}\Delta\big(\mathfrak{F}\big[u_n^{(\ell)} - g_n\big]_+\big), \left|\Delta \boldsymbol{u}_{\boldsymbol{t}}^{(\ell)}\right|\right\rangle_{\Gamma_C} \leq \Delta \mathscr{L}^{(\ell)}\big(\Delta \boldsymbol{u}^{(\ell)}\big).
\end{aligned} \tag{3.4.17}
$$

The contribution of the penalty term is bounded from below by 0. The friction term is estimated with the help of the trace theorem by

$$
\begin{aligned}
&\left\langle \tfrac{1}{\varepsilon}\Delta\big(\mathfrak{F}\big[u_n^{(\ell)} - g_n\big]_+\big), \left|\Delta \boldsymbol{u}_{\boldsymbol{t}}^{(\ell)}\right|\right\rangle_{\Gamma_C} \\
&\quad = -\left\langle \mathfrak{F}\sigma_n\big(\Delta \boldsymbol{u}^{(\ell)}\big), \left|\Delta \boldsymbol{u}_{\boldsymbol{t}}^{(\ell)}\right|\right\rangle_{\Gamma_C} \\
&\quad \geq -\|\mathfrak{F}\|_{H^{-1/2}(\Gamma_{\mathfrak{F}})\to H^{-1/2}(\Gamma_{\mathfrak{F}})} \left\|\sigma_n\big(\Delta \boldsymbol{u}^{(\ell)}\big)\right\|_{H^{-1/2}(\Gamma_{\mathfrak{F}})} \cdot \\
&\qquad \cdot \left\|\Delta \boldsymbol{u}_{\boldsymbol{t}}^{(\ell)}\right\|_{H^{1/2}(\Gamma_{\mathfrak{F}})}.
\end{aligned} \tag{3.4.18}
$$

Here, the norm of $\mathfrak{F}$

$$
\|\mathfrak{F}\|_{H^{-1/2}(\Gamma_{\mathfrak{F}})\to H^{-1/2}(\Gamma_{\mathfrak{F}})} \equiv \sup_{\substack{h \in H^{-1/2}(\Gamma_{\mathfrak{F}}) \\ \|h\|_{H^{-1/2}(\Gamma_{\mathfrak{F}})} \leq 1}} \|\mathfrak{F}h\|_{H^{-1/2}(\Gamma_{\mathfrak{F}})}
$$

is a *Sobolev multiplier norm.* The right-hand side of (3.4.18) can be estimated further with the help of the special trace and inverse trace estimates (3.4.12) and (3.4.13). The result of these estimates is the inequality

$$
\begin{aligned}
\left\|\Delta \boldsymbol{u}^{(\ell)}\right\|_A^2 \leq\ & \|\mathfrak{F}\|_{H^{-1/2}(\Gamma_{\mathfrak{F}})\to H^{-1/2}(\Gamma_{\mathfrak{F}})} \sqrt{C^{(1)}C^{(2)}} \left\|\Delta \boldsymbol{u}^{(\ell)}\right\|_A^2 \\
&+ c_1 \left\|\Delta \mathscr{L}^{(\ell)}\right\|_{\boldsymbol{H}_0^1(\Omega)^*} \left\|\Delta \boldsymbol{u}^{(\ell)}\right\|_A + c_2 \left\|\Delta \mathscr{L}^{(\ell)}\right\|_{\boldsymbol{H}_0^1(\Omega)^*}^2.
\end{aligned} \tag{3.4.19}
$$

If

$$
\|\mathfrak{F}\|_{H^{-1/2}(\Gamma_{\mathfrak{F}})\to H^{-1/2}(\Gamma_{\mathfrak{F}})} < \big(C^{(1)}C^{(2)}\big)^{-1/2} \tag{3.4.20}
$$

holds, then estimate (3.4.16) follows. □

3.4.5 Remark. From the duality

$$
\langle gf, v\rangle_M = \langle f, gv\rangle_M
$$

it is easily seen that $\|g\|_{H^{-1/2}(M)\to H^{-1/2}(M)}$ is equal to $\|g\|_{H^{1/2}(M)\to H^{1/2}(M)}$. The requirement that $\|\mathfrak{F}\|_{H^{1/2}(\Gamma_{\mathfrak{F}})\to H^{1/2}(\Gamma_{\mathfrak{F}})}$ is bounded by $\sqrt{C^{(1)}C^{(2)}}$ is rather restrictive. It not only needs some additional smoothness of $\mathfrak{F}$, but also prescribes an upper bound for the corresponding norm. In the static case we only have an upper bound for $\|\mathfrak{F}\|_{L_\infty(\Gamma_\mathfrak{F})}$, combined with some additional (moderate) smoothness of $\mathfrak{F}$ in the case $\alpha < \frac{1}{2}$ but without any additional conditions on the magnitude of the corresponding norm. From the estimate

$$\|g\, v\|_{H^{1/2}(M)} \leq \|g\|_{L_\infty(M)} \|v\|_{H^{1/2}(M)} \\ + c_1(\eta, p, M) \|g\|_{W_{2p}^{1/2+\eta}(M)} \|v\|_{L_{2p'}(M)}$$

valid for $\eta > 0$, $p, p' \geq 1$ with $1/p + 1/p' = 1$ and a certain constant $c_1(\eta, p, M)$ and from the Sobolev imbedding

$$\|v\|_{L_q(M)} \leq c_2(q) \|v\|_{H^{1/2}(M)}$$

valid for $q < +\infty$ in the case $\widetilde{N} = 1$ and $q \leq \frac{2\widetilde{N}}{\widetilde{N}-1}$ in the case $\widetilde{N} \geq 2$ with dimension $\widetilde{N}$ of M there follows an estimate of the type

$$\|g\|_{H^{1/2}(M) \to H^{1/2}(M)} \leq \|g\|_{L_\infty(M)} + c_4(\eta, p, M)) \|g\|_{W_{2p}^{1/2+\eta}(M)},$$

provided $p > \widetilde{N}$. If $\mathfrak{F}$ is solution-dependent, it is not possible to estimate $\|\mathfrak{F}(\dot{\boldsymbol{u}})\|_{W_{2p}^{1/2+\eta}(\Gamma_\mathfrak{F})}$ in terms of $\|\dot{\boldsymbol{u}}\|_{\boldsymbol{H}^{1/2}(\Gamma_\mathfrak{F})}$.

In the two previous theorems all the necessary *a priori* estimates required for the limit procedures $\Delta t \to 0$ and $\varepsilon \to 0$ are established. We consider first the limit of the time step in the penalized problem. In order to distinguish different levels of the time discretization let us denote $\Delta t_L \equiv T/L$, $t_\ell^{(L)} = \ell \Delta t_L$, and let $\boldsymbol{u}^{(L,\ell)}$ be the corresponding approximation of the solution at $t = t_\ell^{(L)}$. In order to relate the set of time-discrete equations (3.4.10) with the time continuous equation (3.4.9), it is necessary to extend the solutions of the time-discretized problems onto the whole time domain. This is done by

$$\boldsymbol{u}_L(t, \cdot) \equiv \boldsymbol{u}^{(L,\ell)}(\cdot) \text{ for } t \in \left[t_{\ell-1}^{(L)}, t_\ell^{(L)}\right), \quad \ell = 1, \ldots, L \tag{3.4.21}$$

for the displacement field and

$$\frac{\Delta \boldsymbol{u}_L}{\Delta t_L} \equiv \frac{\boldsymbol{u}^{(L,\ell)} - \boldsymbol{u}^{(L,\ell-1)}}{t_\ell^{(L)} - t_{\ell-1}^{(L)}} \text{ for } t \in \left[t_{\ell-1}^{(L)}, t_\ell^{(L)}\right), \quad \ell = 1, \ldots, L$$

for the time differences. Analogously for $\boldsymbol{Y} = \boldsymbol{f}, \boldsymbol{U}$ and $\boldsymbol{\tau}$, $\boldsymbol{Y}_L$ denote the corresponding extensions of the time discretizations $\boldsymbol{Y}^{(L,\ell)} = \boldsymbol{Y}\left(t_\ell^{(L)}\right)$ and $(\Delta \boldsymbol{Y}_L)/(\Delta t_L)$ the corresponding extensions of the time differences.

With this notation, the set of time-discrete problems (3.4.10) is equivalent to

$$\begin{aligned}
&\int_{I_T} \Bigg(A\left(\boldsymbol{u}_L, \boldsymbol{v}_L - \frac{\Delta \boldsymbol{u}_L}{\Delta t_L}\right) \\
&\qquad + \left\langle \tfrac{1}{\varepsilon}\big[(u_L)_n - (g_L)_n\big]_+, (v_L)_n - \left(\frac{\Delta u_L}{\Delta t_L}\right)_n \right\rangle_{\Gamma_C} \\
&\qquad + \left\langle \mathfrak{F}\tfrac{1}{\varepsilon}\big[(u_L)_n - (g_L)_n\big]_+, \big|(\boldsymbol{v}_L)_{\boldsymbol{t}}\big| - \left|\left(\frac{\Delta \boldsymbol{u}_L}{\Delta t_L}\right)_{\boldsymbol{t}}\right| \right\rangle_{\Gamma_C} \Bigg)\, dt \\
&\geq \int_{I_T} \mathscr{L}_L\left(\boldsymbol{v}_L - \frac{\Delta \boldsymbol{u}_L}{\Delta t_L}\right) dt.
\end{aligned} \tag{3.4.22}$$

This problem can be interpreted as a certain Galerkin discretization of the penalized quasistatic problem (3.4.9), if the given data $\boldsymbol{f}$, $\boldsymbol{U}$, $\boldsymbol{\tau}$ are also replaced by their time discretizations. For the limit procedure we need the following regularity assumptions on the given data:

3.4.6 Assumption. *Let* $\boldsymbol{f} \in H^1\big(I_T; \boldsymbol{H}_0^1(\Omega)^*\big)$, $\boldsymbol{\tau} \in H^1\big(I_T; \boldsymbol{H}^{1/2}(\Gamma_{\boldsymbol{\tau}})^*\big)$, $\boldsymbol{U} \in H^1\big(I_T; \boldsymbol{H}^1(\Omega)\big)$.

Due to these regularity properties there holds

$$\begin{aligned}
&\boldsymbol{f}_L(t,\cdot) \to \boldsymbol{f}(t,\cdot) \text{ in } \boldsymbol{H}_0^1(\Omega)^*, \\
&\boldsymbol{\tau}_L(t,\cdot) \to \boldsymbol{\tau}(t,\cdot) \text{ in } \boldsymbol{H}^{1/2}(\Gamma_{\boldsymbol{\tau}})^*, \\
&\boldsymbol{U}_L(t,\cdot) \to \boldsymbol{U}(t,\cdot) \text{ in } \boldsymbol{H}^1(\Omega)
\end{aligned} \tag{3.4.23}$$

for every $t \in I_T$, and

$$\begin{aligned}
&\frac{\Delta \boldsymbol{f}_L}{\Delta t_L} \to \dot{\boldsymbol{f}} \text{ in } L_2\big(I_T; \boldsymbol{H}_0^1(\Omega)^*\big), \\
&\frac{\Delta \boldsymbol{\tau}_L}{\Delta t_L} \to \dot{\boldsymbol{\tau}} \text{ in } L_2\big(I_T; \boldsymbol{H}^{1/2}(\Gamma_{\boldsymbol{\tau}})\big)^*, \\
&\frac{\Delta \boldsymbol{U}_L}{\Delta t_L} \to \dot{\boldsymbol{U}} \text{ in } L_2\big(I_T; \boldsymbol{H}^1(\Omega)\big).
\end{aligned} \tag{3.4.24}$$

The three last convergences follow from the representation

$$\frac{\Delta w_L}{\Delta t_L}(t) = \frac{1}{\Delta t}\int_{t_{\ell-1}^{(L)}}^{t_\ell^{(L)}} \dot{w}(s)\, ds \text{ for } t \in \left[t_{\ell-1}^{(L)}, t_\ell^{(L)}\right),$$

applied to $w \in \{\boldsymbol{f}, \boldsymbol{\tau}, \boldsymbol{U}\}$, and the fact that the mean value $\frac{1}{\Delta t}\int_t^{t+\Delta t} w(s)\, ds$ of a function $f \in L_2(\mathbb{R})$ converges to f strongly in $L_2(\mathbb{R})$ as $\Delta t \to 0$. Moreover, from the *a priori* estimates proved for $\boldsymbol{u}^{(\ell)}$ there follows the existence of a sequence $\Delta t_{L_k} \to 0$ of time steps and a corresponding sequence $\boldsymbol{u}_{L_k}$ of solutions of the time discrete equations such that the latter converges in the sense

$$\boldsymbol{u}_{L_k} \rightharpoonup \boldsymbol{u} \text{ in } L_2\big(I_T; \boldsymbol{H}^1(\Omega)\big), \tag{3.4.25}$$

$$\frac{\Delta \boldsymbol{u}_{L_k}}{\Delta t_{L_k}} \rightharpoonup \boldsymbol{w} \text{ in } L_2\big(I_T; \boldsymbol{H}^1(\Omega)\big). \tag{3.4.26}$$

In order to show $\boldsymbol{w} = \dot{\boldsymbol{u}}$ we consider a fixed test function $\boldsymbol{v} \in \mathring{C}^1\big(I_T; \boldsymbol{L}_2(\Omega)\big)$ and its sequence of time discretizations $\boldsymbol{v}_{L_k}$. Then the following discrete version of integration by parts is true:

$$\begin{aligned}
\int_{I_T} \frac{\Delta \boldsymbol{u}_{L_k}}{\Delta t_{L_k}} \cdot \boldsymbol{v}_{L_k}\, dt &= \sum_{\ell=1}^{L_k} \big(\boldsymbol{u}_{L_k}^{(\ell)} - \boldsymbol{u}_{L_k}^{(\ell-1)}\big)\boldsymbol{v}_{L_k}^{(\ell)} \\
&= \sum_{\ell=1}^{L_k-1} \boldsymbol{u}_{L_k}^{(\ell)}\big(\boldsymbol{v}_{L_k}^{(\ell)} - \boldsymbol{v}_{L_k}^{(\ell+1)}\big) + \boldsymbol{u}_{L_k}^{(L_k)} \cdot \boldsymbol{v}_{L_k}^{(L_k)} - \boldsymbol{u}_{L_k}^{(0)} \cdot \boldsymbol{v}_{L_k}^{(1)} \\
&= -\int_0^{T-\Delta t_{L_k}} \boldsymbol{u}_{L_k}(t) \cdot \frac{\Delta \boldsymbol{v}_{L_k}}{\Delta t_{L_k}}(t + \Delta t_{L_k})\, dt \\
&\quad + \boldsymbol{u}_{L_k}^{(L_k)} \cdot \boldsymbol{v}_{L_k}^{(L_k)} - \boldsymbol{u}_{L_k}^{(0)} \cdot \boldsymbol{v}_{L_k}^{(1)}.
\end{aligned}$$

Using the strong convergences $\boldsymbol{v}_{L_k} \to \boldsymbol{v}$ and $\dfrac{\Delta \boldsymbol{v}_{L_k}}{\Delta t_{L_k}} \to \dot{\boldsymbol{v}}$ in $L_2(I_T \times \Omega)$ together with (3.4.25) and (3.4.26) there follows

$$\int_{I_T \times \Omega} \boldsymbol{w} \cdot \boldsymbol{v}\, d\boldsymbol{x}\, dt = -\int_{I_T \times \Omega} \boldsymbol{u} \cdot \dot{\boldsymbol{v}}\, d\boldsymbol{x}\, dt$$

which proves $\boldsymbol{w} = \dot{\boldsymbol{u}}$.

In order to verify a strong convergence $\boldsymbol{u}_{L_k} \to \boldsymbol{u}$ in $L_2(I_T \times \Gamma)$ we first derive some time regularity of $\boldsymbol{u}_{L_k}$. For a test function $\boldsymbol{v} \in \mathring{C}(I_T; \boldsymbol{L}_2(\Omega))$ we have (in the distributional sense)

$$\begin{aligned}
\int_{I_T \times \Omega} \dot{\boldsymbol{u}}_{L_k} \cdot \boldsymbol{v}\, d\boldsymbol{x}\, dt &= \int_\Omega \sum_{\ell=1}^{L_k-1} \big(\boldsymbol{u}_{L_k}^{(\ell+1)} - \boldsymbol{u}_{L_k}^{(\ell)}\big) \cdot \boldsymbol{v}\big(t_{L_k}^{(\ell)}\big)\, d\boldsymbol{x} \\
&\le \left\| \frac{\Delta \boldsymbol{u}_{L_k}}{\Delta t_{L_k}} \right\|_{L_2(I_T \times \Omega)} \left(\sum_{\ell=1}^{L_k-1} \Delta t_{L_k} \big\|\boldsymbol{v}\big(t_{L_k}^{(\ell)}\big)\big\|_{L_2(\Omega)}^2 \right)^{1/2} \\
&\le c\|\boldsymbol{v}\|_{C(I_T; L_2(\Omega))}.
\end{aligned}$$

Due to the embedding $\mathring{H}^{1/2+\eta}(I_T; \boldsymbol{L}_2(\Omega)) \hookrightarrow \mathring{C}(I_T; \boldsymbol{L}_2(\Omega))$ this proves

$$\|\dot{\boldsymbol{u}}_{L_k}\|_{H^{-1/2-\eta}(I_T; \boldsymbol{L}_2(\Omega))} \le c$$

for every $\eta > 0$ with c independent of k. This implies that $\boldsymbol{u}_{L_k}$ is bounded in $H^{1/2-\eta,1}(I_T \times \Omega)$ uniformly with respect to k. By compact embedding we obtain $\boldsymbol{u}_{L_k} \to \boldsymbol{u}$ strongly in $L_2(I_T \times \Omega)$ and in $L_2(I_T \times \Gamma)$.

These convergences are sufficient to pass to the limit in the time-discretized version of problem (3.4.22). Indeed, all convergences are trivial with the exception of the following ones

$$A\left(\boldsymbol{u}_{L_k}, \frac{\Delta \boldsymbol{u}_{L_k}}{\Delta t_{L_k}}\right) \text{ and } \left\langle \mathfrak{F}\frac{1}{\varepsilon}\left[(u_{L_k})_n - (g_{L_k})_n\right]_+, \left|\left(\frac{\Delta \boldsymbol{u}_{L_k}}{\Delta t_{L_k}}\right)_{\boldsymbol{t}}\right|\right\rangle_{\Gamma_C}.$$

The convergence of the first term is proved by a discrete version of the integration by parts:

$$\begin{aligned}
\int_{I_T} A\left(\boldsymbol{u}_{L_k}, \frac{\Delta \boldsymbol{u}_{L_k}}{\Delta t_{L_k}}\right) dt &= \sum_{\ell=1}^{L_k} A\big(\boldsymbol{u}_{L_k}^{(\ell)}, \boldsymbol{u}_{L_k}^{(\ell)} - \boldsymbol{u}_{L_k}^{(\ell-1)}\big)\, dt \\
= \frac{1}{2}\sum_{\ell=1}^{L_k} \Big(A\big(\boldsymbol{u}_{L_k}^{(\ell)}, \boldsymbol{u}_{L_k}^{(\ell)}\big) &- A\big(\boldsymbol{u}_{L_k}^{(\ell-1)}, \boldsymbol{u}_{L_k}^{(\ell-1)}\big) \\
&+ A\big(\boldsymbol{u}_{L_k}^{(\ell)} - \boldsymbol{u}_{L_k}^{(\ell-1)}, \boldsymbol{u}_{L_k}^{(\ell)} - \boldsymbol{u}_{L_k}^{(\ell-1)}\big)\Big) \\
= \frac{1}{2}\Big(A\big(\boldsymbol{u}_{L_k}^{(L_k)}, \boldsymbol{u}_{L_k}^{(L_k)}\big) &- A(\boldsymbol{u}_0, \boldsymbol{u}_0) + \sum_{\ell=1}^{L_k} A\big(\Delta \boldsymbol{u}_{L_k}^{(\ell)}, \Delta \boldsymbol{u}_{L_k}^{(\ell)}\big)\Big).
\end{aligned} \tag{3.4.27}$$

The sum on the right-hand side converges to zero for $\Delta t_{L_k} \to 0$, because $\dfrac{\Delta \boldsymbol{u}_{L_k}}{\Delta t_{L_k}}$ is bounded in $L_2\big(I_T; \boldsymbol{H}^1(\Omega)\big)$, and therefore

$$\begin{aligned}
\sum_{\ell=1}^{L_k} A\big(\Delta \boldsymbol{u}_{L_k}^{(\ell)}, \Delta \boldsymbol{u}_{L_k}^{(\ell)}\big) &\le A_0 \sum_{\ell=1}^{L_k} \big\|\Delta \boldsymbol{u}_{L_k}^{(\ell)}\big\|_{\boldsymbol{H}^1(\Omega)}^2 \\
&= A_0 \int_0^T \Delta t_{L_k} \left\| \frac{\Delta \boldsymbol{u}_{L_k}^{(\ell)}}{\Delta t_{L_k}}(t) \right\|_{\boldsymbol{H}^1(\Omega)}^2 dt \le C \Delta t_{L_k}.
\end{aligned}$$

With the weak lower semicontinuity of positive semi-definite bilinear forms and the weak convergence $\boldsymbol{u}_{L_k}^{(L_k)} \rightharpoonup \boldsymbol{u}(T)$ in $\boldsymbol{H}^1(\Omega)$ we conclude

$$\begin{aligned}
&\liminf_{k\to+\infty} \int_{I_T} A\left(\boldsymbol{u}_{L_k}, \frac{\Delta \boldsymbol{u}_{L_k}}{\Delta t_{L_k}}\right) dt \\
&\quad \ge \frac{1}{2}\big(A(\boldsymbol{u}(T,\cdot), \boldsymbol{u}(T,\cdot)) - A(\boldsymbol{u}_0, \boldsymbol{u}_0)\big) = \int_{I_T} A(\boldsymbol{u}, \dot{\boldsymbol{u}})\, dt.
\end{aligned}$$

For the friction term we have

$$\left\langle \mathfrak{F}\frac{1}{\varepsilon}\left[(u_{L_k})_n - (g_{L_k})_n\right]_+, \left|\left(\frac{\Delta \boldsymbol{u}_{L_k}}{\Delta t_{L_k}}\right)_{\boldsymbol{t}}\right|\right\rangle_{\Gamma_C} \\ \geq \left\langle \mathfrak{F}\frac{1}{\varepsilon}\left[(u_{L_k})_n - (g_{L_k})_n\right]_+ \chi(\dot{\boldsymbol{u}}), \left(\frac{\Delta \boldsymbol{u}_{L_k}}{\Delta t_{L_k}}\right)_{\boldsymbol{t}}\right\rangle_{\Gamma_C} \tag{3.4.28}$$

with $\chi : \boldsymbol{v} \mapsto \begin{cases} \boldsymbol{v}/|\boldsymbol{v}|, & \boldsymbol{v} \neq 0, \\ 0, & \boldsymbol{v} = 0 \end{cases}$. The right-hand side of the previous inequality converges to

$$\left\langle \mathfrak{F}\frac{1}{\varepsilon}[u_n - g_n]_+ \chi(\dot{\boldsymbol{u}}), \dot{\boldsymbol{u}}_{\boldsymbol{t}}\right\rangle_{\Gamma_C} = \left\langle \mathfrak{F}\frac{1}{\varepsilon}[u_n - g_n]_+, |\dot{\boldsymbol{u}}_{\boldsymbol{t}}|\right\rangle_{\Gamma_C}.$$

Here, the weak convergence $\left((\Delta \boldsymbol{u}_{L_k})/(\Delta t_{L_k})\right)_{\boldsymbol{t}} \to \dot{\boldsymbol{u}}_{\boldsymbol{t}}$ follows from the weak convergence $(\Delta \boldsymbol{u}_{L_k})/(\Delta t_{L_k}) \to \dot{\boldsymbol{u}}$ and the representation $\boldsymbol{v_t} = \mathfrak{T}\boldsymbol{v}$ of the tangential component with a bounded matrix $\mathfrak{T} = \mathfrak{T}(x)$, $x \in \Gamma$. Hence the convergence of $\boldsymbol{u}_{L_k}$ to a weak solution of the penalized quasistatic problem (3.4.9) is proved. This solution satisfies the variational inequality (3.4.9) for almost every time t. Due to (3.4.14), (3.4.15) and (3.4.16) it is bounded in $H^1(I_T; \boldsymbol{H}^1(\Omega))$ and in $L_\infty(I_T; \boldsymbol{H}^1(\Gamma_{\mathfrak{F}}))$, and the penalty term is bounded in $L_\infty(I_T; L_2(\Gamma_{\mathfrak{F}}))$. With the Green formula

$$\left\langle \frac{1}{\varepsilon}[u_n - g_n]_+, v_n \right\rangle_{I_T \times \Gamma_C} = \mathscr{L}(\boldsymbol{v}) - A(\boldsymbol{u}, \boldsymbol{v})$$

valid for all $\boldsymbol{v} \in \boldsymbol{H}_0^1(\Omega)$ with $\boldsymbol{v_t} = 0$ on Γ_C we also conclude that $\frac{1}{\varepsilon}[u_n - g_n]_+$ is bounded in $H^1(I_T; H_0^{1/2}(\Gamma_C)^*)$. Here the space $H_0^{1/2}(\Gamma_C) = \{v \in H^{1/2}(\Gamma);\, v = 0 \text{ on } \Gamma_U\}$ is the space of traces of functions from $\boldsymbol{H}_0^1(\Omega)$. Hence the following theorem is verified:

3.4.7 Theorem. *Let the Assumptions 3.4.2 and 3.4.6 be valid and let the coefficient of friction be solution independent and satisfy the condition (3.4.20). Then there exists a sequence of time steps $\Delta t_{L_k} \to 0$ and a corresponding sequence of solutions $\boldsymbol{u}_{L_k}$ of the time-discretized problem (3.4.10) such that their extensions $\boldsymbol{u}_{L_k}$ defined by (3.4.21) converge strongly in $L_2(I_T; \boldsymbol{H}^1(\Omega))$ to a solution of the penalized quasistatic contact problem (3.4.9). This solution satisfies the* a priori *estimate*

$$\|\boldsymbol{u}\|_{H^1(I_T;\boldsymbol{H}^1(\Omega))} + \|\boldsymbol{u}\|_{L_\infty(I_T;\boldsymbol{H}^1(\Gamma_{\mathfrak{F}}))} \\ + \|P_\varepsilon(\boldsymbol{u})\|_{L_\infty(I_T;L_2(\Gamma_{\mathfrak{F}}))} + \|P_\varepsilon(\boldsymbol{u})\|_{H^1(I_T;H_0^{1/2}(\Gamma_C)^*)} \leq c \tag{3.4.29}$$

with a constant c independent of the penalty parameter ε and $P_\varepsilon : \boldsymbol{v} \mapsto \frac{1}{\varepsilon}[v_n - g_n]_+$.

The existence result for the quasistatic problem is now proved by passing to the limit $\varepsilon \to 0$ in the weak formulation of the penalized problem (3.4.9). Due to the *a priori* estimates proved just before, there exist a sequence $\varepsilon_k \to 0$ of penalty parameters and a corresponding sequence $\boldsymbol{u}_k$ of solutions to (3.4.9) with $\varepsilon = \varepsilon_k$ such that

$$\boldsymbol{u}_k \rightharpoonup \boldsymbol{u} \qquad \text{in } H^1\big(I_T; \boldsymbol{H}^1(\Omega)\big) \text{ and in } L_2\big(I_T; \boldsymbol{H}^1(\Gamma_{\mathfrak{F}})\big), \tag{3.4.30}$$

$$\boldsymbol{u}_k \to \boldsymbol{u} \qquad \text{in } L_2\big(I_T; \boldsymbol{H}^{1-\eta}(\Gamma_{\mathfrak{F}})\big) \text{ for any } \eta \in (0,1), \tag{3.4.31}$$

$$P_{\varepsilon_k}(\boldsymbol{u}_k) \rightharpoonup -\sigma_n(\boldsymbol{u}) \text{ in } H^1\big(I_T; H_0^{1/2}(\Gamma_C)^*\big) \cap L_2(I_T \times \Gamma_{\mathfrak{F}}), \tag{3.4.32}$$

$$P_{\varepsilon_k}(\boldsymbol{u}_k) \to -\sigma_n(\boldsymbol{u}) \text{ in } L_2\big(I_T; H^{-\eta}(\Gamma_{\mathfrak{F}})\big) \text{ and in } H^{1-\eta}\big(I_T; H^{-1/2}(\Gamma_{\mathfrak{F}})\big). \tag{3.4.33}$$

The last convergence here follows from compact imbedding theorems for Sobolev spaces. Moreover the convergence

$$\langle P_{\varepsilon_k}(\boldsymbol{u}_k), (u_k)_n \rangle_{I_T \times \Gamma_C} \to H_0$$

holds with a number $H_0 \in \mathbb{R}$. From the limit $k \to +\infty$ in the estimate

$$\frac{1}{\varepsilon_k} \Big\| \big[(u_k)_n - g_n\big]_+ \Big\|_{L_\infty(I_T; H^{1/2}(\Gamma_C)^*)} \le c$$

there follows $u_n \le g_n$ on Γ_C. For a test function $\boldsymbol{v} \in \boldsymbol{U} + L_2\big(I_T; \boldsymbol{H}_0^1(\Omega)\big)$ satisfying $v_n \le g_n$ there follows

$$\begin{aligned} &\left\langle \frac{1}{\varepsilon_k}\big[(u_k)_n - g_n\big]_+, v_n - (u_k)_n \right\rangle_{I_T \times \Gamma_C} \\ &\quad \le \left\langle \frac{1}{\varepsilon_k}\big[(u_k)_n - g_n\big]_+, g_n - (u_k)_n \right\rangle_{I_T \times \Gamma_C} \le 0. \end{aligned} \tag{3.4.34}$$

Moreover, due to $u_n \le g_n$ there holds

$$\left\langle \frac{1}{\varepsilon_k}\big[(u_k)_n - g_n\big]_+, (u - u_k)_n \right\rangle_{I_T \times \Gamma_C} \le 0. \tag{3.4.35}$$

The limit $k \to +\infty$ in this relation proves

$$-H_0 \le \langle \sigma_n(\boldsymbol{u}), u_n \rangle_{I_T \times \Gamma_C}. \tag{3.4.36}$$

In order to prove $-H_0 \ge \langle \sigma_n(\boldsymbol{u}), u_n \rangle_{I_T \times \Gamma_C}$ we take the test function $\boldsymbol{v} =$

$\dot{\boldsymbol{u}}_k - \boldsymbol{u}_k + \boldsymbol{u}$ in (3.4.9) and obtain

$$\left\langle \frac{1}{\varepsilon_k}\left[(u_k)_n - g_n\right]_+, (u - u_k)_n \right\rangle_{I_T \times \Gamma_C}$$
$$\geq \int_{I_T} \left(A(\boldsymbol{u}_k, \boldsymbol{u}_k - \boldsymbol{u}) + \mathscr{L}(\boldsymbol{u} - \boldsymbol{u}_k)\right) dt$$
$$+ \left\langle \mathfrak{F} P_{\varepsilon_k}(\boldsymbol{u}_k), \left|(\dot{\boldsymbol{u}}_k)_{\boldsymbol{t}}\right| - \left|(\dot{\boldsymbol{u}}_k - \boldsymbol{u}_k + \boldsymbol{u})_{\boldsymbol{t}}\right| \right\rangle_{I_T \times \Gamma_C}.$$

Here we pass to the limit $k \to +\infty$. Due to the weak lower semicontinuity of positive semi-definite bilinear forms, the relation

$$\mathfrak{F} P_{\varepsilon_k}(\boldsymbol{u}_k)\left(\left|(\dot{\boldsymbol{u}}_k)_{\boldsymbol{t}}\right| - \left|(\dot{\boldsymbol{u}}_k - \boldsymbol{u}_k + \boldsymbol{u})_{\boldsymbol{t}}\right|\right) \geq -\mathfrak{F} P_{\varepsilon_k}(\boldsymbol{u}_k)\left|(\boldsymbol{u}_k - \boldsymbol{u})_{\boldsymbol{t}}\right|$$

and the convergence $\mathfrak{F} P_{\varepsilon_k}(\boldsymbol{u}_k)\left|(\boldsymbol{u}_k - \boldsymbol{u})_{\boldsymbol{t}}\right| \to 0$ in $L_1(I_T \times \Gamma_C)$ there follows

$$-H_0 - \langle \sigma_n(\boldsymbol{u}), u_n \rangle_{I_T \times \Gamma_C} \geq 0.$$

In combination with (3.4.36) this proves

$$H_0 = -\langle \sigma_n(\boldsymbol{u}), u_n \rangle_{I_T \times \Gamma_C}.$$

Then the limit in (3.4.34) shows

$$\langle \sigma_n(\boldsymbol{u}), v_n - u_n \rangle_{I_T \times \Gamma_C} \geq 0 \tag{3.4.37}$$

for all $\boldsymbol{v} \in L_2\left(I_T; \boldsymbol{H}_0^{1/2}(\Gamma_C)\right)$ with $v_n \leq g_n$. Taking here $\boldsymbol{v}(t,\cdot) = \boldsymbol{u}(t \pm h, \cdot)$ and passing to the limit $h \to 0$ shows

$$\langle \sigma_n(\boldsymbol{u}), \dot{u}_n \rangle_{I_T \times \Gamma_C} = 0.$$

After these preparations we are able to pass to the limit $k \to +\infty$ in relation (3.4.9). Let us first consider the friction term. For any function $\chi \in \boldsymbol{L}_\infty\left(I_T \times \Gamma_{\mathfrak{F}}; \mathbb{R}^N\right)$ with $|\chi| \leq 1$ on $I_T \times \Gamma_{\mathfrak{F}}$ we have

$$\mathfrak{F} P_{\varepsilon_k}(\boldsymbol{u}_k)\left|(\dot{\boldsymbol{u}}_k)_{\boldsymbol{t}}\right| \geq \mathfrak{F} P_{\varepsilon_k}(\boldsymbol{u}_k)\chi \cdot (\dot{\boldsymbol{u}}_k)_{\boldsymbol{t}}.$$

Due to the convergence properties (3.4.30) and (3.4.33)

$$\langle \mathfrak{F} P_{\varepsilon_k}(\boldsymbol{u}_k)\chi, (\dot{\boldsymbol{u}}_k)_{\boldsymbol{t}} \rangle_{I_T \times \Gamma_C} \to \langle -\mathfrak{F}\sigma_n(\boldsymbol{u})\chi, \dot{\boldsymbol{u}}_{\boldsymbol{t}} \rangle_{I_T \times \Gamma_C}.$$

Using the function $\chi = \begin{cases} \dot{\boldsymbol{u}}_{\boldsymbol{t}}/|\dot{\boldsymbol{u}}_{\boldsymbol{t}}|, & \dot{\boldsymbol{u}}_{\boldsymbol{t}} \neq 0, \\ 0, & \dot{\boldsymbol{u}}_{\boldsymbol{t}} = 0, \end{cases}$ we obtain

$$\liminf_{k\to\infty} \left\langle \mathfrak{F} P_{\varepsilon_k}(\boldsymbol{u}_k), \left|(\dot{\boldsymbol{u}}_k)_{\boldsymbol{t}}\right| \right\rangle_{I_T \times \Gamma_{\mathfrak{F}}} \geq \left\langle \mathfrak{F}|\sigma_n(\boldsymbol{u})|, \left|\dot{\boldsymbol{u}}_{\boldsymbol{t}}\right| \right\rangle_{I_T \times \Gamma_{\mathfrak{F}}}.$$

With the weak lower semicontinuity of positive definite quadratic forms there follows after time integration

$$\liminf_{k\to\infty} \int_{I_T} A\big(\boldsymbol{u}_k, \dot{\boldsymbol{u}}_k\big)\, dt \ge \int_{I_T} A(\boldsymbol{u}, \dot{\boldsymbol{u}})\, dt.$$

The penalty term can be estimated with the help of the relation

$$\big[(u_k)_n - g_n\big]_+ (\dot{u}_k)_n = \tfrac{1}{2}\partial_t \big[(u_k)_n - g_n\big]_+^2$$

and of the assumption $(u_0)_n \le g_n$ by

$$\int_{I_T}\int_{\Gamma_C} \tfrac{1}{\varepsilon}\big[(u_k)_n - g_n\big]_+ (\dot{u}_k)_n \, ds_{\boldsymbol{x}}\, dt$$
$$= \frac{1}{2}\int_{\Gamma_C} \frac{1}{\varepsilon}\Big(\big[(u_k(T,\cdot))_n - g_n\big]_+^2 - \big[(u_0)_n - g_n\big]_+^2\Big)\, ds_{\boldsymbol{x}} \ \ge\ 0.$$

Consequently

$$\liminf_{k\to\infty} \int_{I_T}\int_{\Gamma_C} \tfrac{1}{\varepsilon}\big[(u_k)_n - g_n\big]_+ \big((\dot{u}_k)_n - v_n\big)\, ds_{\boldsymbol{x}}\, dt$$
$$\ge \int_{I_T} \langle \sigma_n(\boldsymbol{u}), v_n\rangle_{\Gamma_C}\, dt.$$

Hence the limit function $\boldsymbol{u}$ satisfies the variational inequality

$$\begin{aligned}\int_{I_T} \Big(A(\boldsymbol{u}, \boldsymbol{v} - \dot{\boldsymbol{u}}) - \langle \sigma_n(\boldsymbol{u}), v_n - \dot{u}_n\rangle_{\Gamma_C} \\ + \langle \mathfrak{F}|\sigma_n(\boldsymbol{u})|, |\boldsymbol{v_t}| - |\dot{\boldsymbol{u}}_{\boldsymbol{t}}|\rangle_{\Gamma_C}\Big)\, dt \ge \int_{I_T} \mathscr{L}(\boldsymbol{v} - \dot{\boldsymbol{u}})\, dt\end{aligned} \tag{3.4.38}$$

for all test functions $\boldsymbol{v} \in \dot{\boldsymbol{U}} + L_2(I_T; \boldsymbol{H}_0^1(\Omega))$. Here the term $\langle \sigma_n(\boldsymbol{u}), \dot{u}_n\rangle_{\Gamma_C}$ has been added. Choosing the new test function $\boldsymbol{v} = \widetilde{\boldsymbol{v}} + \dot{\boldsymbol{u}} - \boldsymbol{u}$ and employing relation (3.4.37) for the normal stresses proves that $\boldsymbol{u}$ indeed solves the weak formulation (3.4.7) of the quasistatic contact problem.

Moreover, with the usual procedure of adding $A(\boldsymbol{u}, \boldsymbol{u}_k - \boldsymbol{u})$ to both sides of (3.4.9) with solution $\boldsymbol{u}_k$ and passing to the limit it is verified that the convergence $\boldsymbol{u}_k \to 0$ is strong in $L_2\big(I_T; \boldsymbol{H}^1(\Omega)\big)$. As a consequence, the convergence $\boldsymbol{u}_k(t,\cdot) \to \boldsymbol{u}(t,\cdot)$ is also strong for almost every $t \in I_T$. Hence the following theorem is proved:

3.4.8 Theorem. *Let the assumptions of Theorem 3.4.7 be valid. Then there exists a sequence $\boldsymbol{u}_k$ of solutions to the penalized quasistatic problem (3.4.9) with penalty parameters $\varepsilon_k \to 0$ such that $\boldsymbol{u}_k$ converges strongly in $L_2\big(I_T; \boldsymbol{H}^1(\Omega)\big)$ to a solution of the quasistatic contact problem with friction (3.4.7).*

Chapter 4

Dynamic contact problems

In this chapter we survey relevant available results for dynamic contact problems with a physically well posed contact condition in displacements. We confine ourselves to the contact of a body with a rigid undeformable obstacle. The reason is the lack of a satisfactory model for the contact of two bodies with dimension higher than one respecting the non-penetrability of both bodies, in particular the time dependence of the appropriate couples of boundary points which should describe this condition. In this respect the dynamic problem differs from the static one, where the non-penetrability can be modelled on the basis of the original configuration and from the case of strings, where the geometry is essentially simplified (cf. e.g. [33]).

The main difficulty of the dynamic contact problems with friction arises from the hyperbolic nature of the equilibrium of inner forces. In variational inequalities for hyperbolic problems it is much more comfortable to have the *time derivatives* of the solution in the test function instead of the solution itself. With the velocity as a test function, the signs at the velocity and at the space gradient will be the same after integration by parts in time and space. On the other hand, if the displacement is used as a test function, the space and time terms have opposite signs with many unpleasant consequences on estimates and convergence procedures. Hence, from a mathematical point of view the contact condition in velocities is much easier to handle. Before the first results on the original contact condition appeared there was a large literature on the solution to the contact condition in velocities, whose results were mostly mentioned in the monograph [42]. They included also the problem with a given time-independent friction force.

The main tool in proofs of the existence of solutions to dynamic contact problems is the penalization of the contact condition. This method may yield energy preserving solutions which is why we have focused our approach to results of this type. On the other hand, these results represent an essential part of the knowledge in that field. There are not many of them and even fewer of them are applicable to contacts of bodies. The results for purely elastic and therefore purely hyperbolic situations are limited to strings, to polyharmonic problems, or to special problems on halfspaces.

These results are surveyed in the first section.

Including the viscous behaviour of the material, which is physically well based, we can parabolize the problems. Such a parabolization allows solving wider classes of problems. There is a remarkable difference between domain and boundary obstacles. While the first require some direct estimate of the penalty term which does not give much information, the latter are much simpler to solve, because dual estimates of the acceleration can be employed and the penalty term is (up to the sign) the normal boundary traction, which can be controlled by the volume terms. We have studied two kinds of viscosity: the long and the short memory. The long time-singular memory yields slightly weaker results which do not enable us to consider the friction. For the short memory materials, satisfactory results for problems with given friction are derived. However, the original problem with Coulomb friction for the contact condition in displacements remains unsolved.

4.1 A short survey about results for elastic materials

A number of publications followed the result of Amerio and Prouse [5], where piecewise smooth solutions to the problem of a homogeneous string were constructed. Most results in this field were obtained in the second half of the seventies and in the eighties. The main drawback of the mathematical modelling of dynamic contact problems is the lack of any satisfactory description of what will happen after a collision.

Let us first consider the simplest case of a string parallel to a plane obstacle moving perpendicularly towards it with constant velocity $\dot{u} \equiv -v_0$, $v_0 > 0$. Before the collision the equilibrium of forces with zero volume force $\ddot{u} - \partial_x^2 u = 0$ shall be satisfied. At the instant of collision t_0 the equality $u(t_0, \cdot) = 0$ holds, the left derivative $\dot{u}_-(t_0, \cdot) = -v_0$, but $\liminf\limits_{t \downarrow t_0} \dfrac{u(t, \cdot)}{t - t_0} \geq 0$. Hence $\ddot{u}(t_0, \cdot) > 0$ in some sense and the equilibrium of forces must be supplemented by an unknown contact force acting towards the string. The problem is therefore given by the system

$$u \geq 0, \; \ddot{u} - \partial_x^2 u \geq 0 \text{ and } u(\ddot{u} - \partial_x^2 u) = 0 \text{ on } \mathbb{R}_+ \times \tilde{I}, \tag{4.1.1}$$

where $\tilde{I}$ is the interval describing the position of the string in an appropriate reference configuration. Any extension of u beyond the collision such that $u : [t, x] \mapsto c(t - t_0)$, $t \geq t_0$, $x \in I$ with $0 \leq c \leq v_0$ satisfies this system. The condition $c \leq v_0$ ensures that the energy cannot grow during the process. If $c = \mathfrak{u}$ the contact force $(\partial_x^2 u - \ddot{u})(t_0, \cdot)$ is completely employed for the backward motion; if $c \in [0, v_0)$ it is partially or fully exhausted by a dissipation of the energy at contact. Such dissipation may occur e.g. due to some kind of adhesion.

The results mentioned below show how close the penalty approach is to the concept of energy preserving solutions. Further results or references for problems with purely elastic material can be found e.g. in [30] and [32]. The more recent research is mostly concentrated on some space discretization, leading to systems of ordinary differential equations, which allows formulating efficiently the energy balance at the contact.

4.1.1 A domain constraint for strings

The contact problem of a string with a concave obstacle under the assumption of energy conservation is given by

$$\begin{aligned} &\ddot{u} - \partial_x^2 u \geq 0, \quad u \geq \varphi \\ &\operatorname{supp}\left(\ddot{u} - \partial_x^2 u\right) \subset \{(t,x) \in \mathbb{R}_+ \times I;\, u(t,x) = \varphi(x)\}, \end{aligned} \tag{4.1.2}$$

$$u(0,\cdot) = u_0, \quad \dot{u}(0,\cdot) = u_1 \text{ a.e. in } I, \tag{4.1.3}$$

$$\begin{aligned} &\partial_x(-2\dot{u}\,\partial_x u) + \partial_t\left(|\partial_x u|^2 + |\dot{u}|^2\right) = 0 \\ &\quad \text{in the sense of distributions on } \mathbb{R}_+ \times I. \end{aligned} \tag{4.1.4}$$

Here I is the domain occupied by the string and the expression $\ddot{u} - \partial_x^2 u$ is understood in the sense of distributions. Observe that the last condition in (4.1.2) is a weak formulation of the corresponding verison of the orthogonality in (4.1.1). We consider two cases: a bounded interval $I = (0, L)$ or $I = \mathbb{R}$. In the first case the Dirichlet boundary condition

$$u(\cdot, 0) = u(\cdot, L) = 0 \text{ on } \mathbb{R}_+ \tag{4.1.5}$$

is added.

Condition (4.1.4) is a *local conservation of the energy.* It is equivalent to the condition that the divergence (∂_x, ∂_t) applied to the vector field

$$S_u \equiv \left(-2\partial_x u\dot{u}, (\partial_x u)^2 + \dot{u}^2\right)^\top$$

vanishes. The first component of this vector field is an energy flux density, similar to the Poynting vector known in electromagnetism; the second component is the total energy density given by the sum of kinetic and potential energy.

The main result was obtained by Schatzman in [127]:

4.1.1 Theorem. *Let $\partial_x^2 \varphi \geq 0$ on I in the sense of distributions, let u_0 belong to $H^1_{\text{loc}}(\bar{I})$ such that $u_0 \geq \varphi$ on I and let $u_1 \in L_{2,\text{loc}}(I)$. Let*

$\varphi(0) < 0$, $\varphi(L) < 0$ be satisfied if I is bounded. Then there exists a unique solution u to problem (4.1.2)–(4.1.4) for $I = \mathbb{R}$ or to problem (4.1.2)–(4.1.5) for bounded I such that $\partial_x u$ and $\dot{u} \in L_{\infty,\mathrm{loc}}(\mathbb{R}_+ ; L_{2,\mathrm{loc}}(I))$.

The *proof* of this result is very technical and is essentially based on a certain direct calculation. We avoid reproducing it here in detail and confine ourselves to outline its main steps. The concavity of the obstacle ensures that $u - \varphi$ satisfies the first condition in (4.1.2).

Let us first consider the case $I = \mathbb{R}$. Let w be a solution of the equation

$$\ddot{w} - \partial_x^2 w = 0 \text{ on } \mathbb{R}_+ \times I \tag{4.1.6}$$

satisfying the initial condition (4.1.3). We define the sets

$$\mathfrak{C} := \{[t, x] \in \mathbb{R}_+ \times \mathbb{R};\ t \geq |x|\},$$
$$\mathfrak{D} := \overline{\{[t, x] \in \mathbb{R}_+ \times \mathbb{R};\ w \leq \varphi\}}.$$

The boundary of the set $\mathfrak{C} + \mathfrak{D}$ is called *line of influence* of the obstacle. It is the graph of a Lipschitz function $\varkappa : \mathbb{R} \to \mathbb{R}_+$ with Lipschitz constant 1. Moreover, it holds $|\varkappa'(x)| < 1 \Rightarrow [\varkappa(x), x] \in \mathfrak{D}$. A calculation shows that the contact force acting along the line of influence can be represented as a measure $\mu(w)$ defined by

$$\langle \mu(w), \psi \rangle = 2 \int_{\varkappa(x)>0} \left(1 - (\varkappa'(x))^2\right) \psi(x, \varkappa(x))\, \dot{w}(x, \varkappa(x))\, dx. \tag{4.1.7}$$

As a consequence, the unique solution of problem (4.1.2)–(4.1.4) is

$$u = w - \mathscr{E} * \mu(w) \tag{4.1.8}$$

with the elementary solution of the wave equation

$$\mathscr{E}(t, x) = \begin{cases} \frac{1}{2}, & t \geq |x| \\ 0, & \text{elsewhere} \end{cases}$$

and w solving (4.1.3), (4.1.6). This is the way how the theorem is proved for $I = \mathbb{R}$. In addition the following identity holds:

$$|\dot{u}_-(\varkappa(x), x)| = |\dot{u}_+(\varkappa(x), x)| \text{ a.e. in } \{x \in \mathbb{R};\ |\varkappa'(x)| < 1\}.$$

For the analogous problem on a finite interval I, the data and the obstacle are extended by periodicity and symmetry to the whole real line. Then the above construction can be performed and yields the first contact line. This process can be repeated, but it has to be shown that the contact lines do not accumulate: this comes from the form of the Dirichlet conditions preventing the extremities of the string to touch the obstacle and therefore the distance between the contact lines is bounded away from zero by a number dependent on the energy of the initial data, $\varphi(0)$ and $\varphi(L)$ only.

For an affine obstacle and the initial data u_i such that u_1 and $\partial_x u_0$ have locally bounded variation, it can be proved that the energy-preserving solution is the strong limit of the solutions to the penalized problem

$$\ddot{u}_\varepsilon - \partial_x^2 u_\varepsilon + \tfrac{1}{\varepsilon}[u_\varepsilon]_-^2 = 0$$

with initial condition (4.1.3) and the Dirichlet boundary value condition for a bounded I. For details of the proofs see [15], [127] and mainly [128].

4.1.2 Obstacle at a finite number of points

We again consider the contact problem of a string, but now the obstacle is the union of a finite number of points moving in time. These points are given by functions $\mathfrak{x}_i : \mathbb{R}_+ \to \mathbb{R}$, $i = 1, \dots, k$ such that $\mathfrak{x}_1 < \dots < \mathfrak{x}_k$, $\inf_{\mathbb{R}_+} \mathfrak{x}_{i+1} - \mathfrak{x}_i > 0$, $i = 1, \dots, k-1$, and for all $i \in \{1, \dots, k\}$ the velocity satisfies $|\dot{\mathfrak{x}}_i| < 1$ on $\mathbb{R}_+$. The latter condition ensures that the functions $t \mapsto t \pm \mathfrak{x}_i(t)$, $i = 1, \dots, k$, are strictly increasing or, physically interpreted, the velocity of the obstacles is subsonic. The obstacle φ is defined by

$$\varphi(t,x) = \begin{cases} h_i(t) & \text{if } x = \mathfrak{x}_i(t),\ i \in \{1,\dots,k\}, \\ -\infty & \text{elsewhere.} \end{cases}$$

The problem to be solved is

$$\left.\begin{aligned} &\ddot{u} - \partial_x^2 u \geq f, \quad u \geq \varphi, \\ &(\ddot{u} - \partial_x^2 u - f)(u - \varphi) = 0 \end{aligned}\right\} \text{ on } \mathbb{R}_+ \times \mathbb{R}, \tag{4.1.9}$$

$$u(0,\cdot) = u_0, \quad \dot{u}(0,\cdot) = u_1 \quad \text{a.e. in } \mathbb{R}. \tag{4.1.10}$$

Schatzman has proved in [129]:

4.1.2 Theorem. *Let h_i belong to $C^0\left(\overline{\mathbb{R}_+}\right)$, let $u_0 \in W^1_{1,\mathrm{loc}}(\mathbb{R})$ such that $u_0(\mathfrak{x}_i) \geq h_i(0)$, $u_1 \in L_{1,\mathrm{loc}}(\mathbb{R})$ and $f \in L_{1,\mathrm{loc}}(\mathbb{R}_+ \times \mathbb{R})$. Then problem* (4.1.9)–(4.1.10) *has a unique solution u. The solutions of the corresponding penalized problems*

$$\ddot{u}_\varepsilon - \partial_x^2 u_\varepsilon = f + \tfrac{1}{\varepsilon} \sum_{i=1}^{k} [u_\varepsilon(\cdot, \mathfrak{x}_i(\cdot)) - h_i]_-$$

with initial condition (4.1.10) *tend to the solution u uniformly on compact sets in $\mathbb{R}_+ \times \mathbb{R}$.*

In special cases it is possible to prove some regularity of the solutions. In the case of a one-point obstacle with $\mathfrak{x} = const.$, $h = 0$ on $\mathbb{R}_+$, $f \in L_{2,\mathrm{loc}}(\mathbb{R} \times \mathbb{R}_+)$, $u_0 \in H^1_{\mathrm{loc}}(\mathbb{R})$ and $u_1 \in L_{2,\mathrm{loc}}(\mathbb{R})$, the solution u belongs to $C^0(\mathbb{R}_+; H^1_{\mathrm{loc}}) \cap C^1(\mathbb{R}_+; L_{2,\mathrm{loc}})$ and satisfies the pointwise condition of the conservation of energy

$$\partial_x(-2\dot{u}\,\partial_x u) + \partial_t\left((\partial_x u)^2 + \dot{u}^2\right) = 2f\dot{u}. \tag{4.1.11}$$

The interpretation of this condition is as in the previous case. There are also explicit expressions available for $\dot{u}_\pm$ at the times of collision.

The generalization of the result to the finite string is here straightforward (for both Dirichlet and Signorini boundary value condition).

As in the previous case the *proof* of the existence result is not based on the variational approach employed extensively throughout this book. Its idea becomes transparent if the simplest situation for $k = 1$, $h \equiv h_1 \equiv 0$, $\mathfrak{x}_1 \equiv \mathfrak{x} \equiv 0$ is considered. Let us denote by $\bar{u}$ the unique solution of the equation $\ddot{\bar{u}} - \partial_x^2 \bar{u} = f$ with the same initial data as in (4.1.10). Then $v = u - \bar{u}$ must solve the problem

$$\begin{aligned} &\ddot{v} - \partial_x^2 v = \mu(t) \otimes \delta_0(x) && \text{on } \mathbb{R}_+ \times \mathbb{R}, \\ &v(0,x) = \dot{v}(0,x) = 0, && x \in \mathbb{R}, \end{aligned} \tag{4.1.12}$$

$$\left.\begin{aligned} &\mu \geq 0, \quad v(\cdot,0) + \bar{u}(\cdot,0) \geq 0 \\ &\langle \mu, v(\cdot,0) + \bar{u}(\cdot,0)\rangle = 0 \end{aligned}\right\} \text{ on } \mathbb{R}_+. \tag{4.1.13}$$

Here, *a priori*, μ is a measure and δ_0 is the Dirac measure concentrated at 0. Thanks to d'Alembert, the solution of (4.1.12) has the following explicit form

$$v(t,x) = \begin{cases} 0 & \text{if } |x| > t, \\ \frac{1}{2}\langle \chi_{[0,t-|x|]}, \mu\rangle & \text{if } |x| \geq t. \end{cases}$$

In particular,

$$w(t) \equiv v(t+0,0) = \tfrac{1}{2}\int_{[0,t]} d\mu_s,$$

which means that (4.1.12)–(4.1.13) can be reduced to

$$w \geq -\bar{u}(\cdot,0), \quad \dot{w} \geq 0, \quad \langle \dot{w}, w + \bar{u}(\cdot,0)\rangle = 0, \tag{4.1.14}$$

which is a classical problem of the type $\dot{w} + \gamma(w + \bar{u}(\cdot,0)) = 0$ with a maximal monotone operator γ defined on $\overline{\mathbb{R}_+}$ by $\gamma \equiv 0$ on $\mathbb{R}_+$ and $\gamma(0) = \overline{\mathbb{R}_-}$. The regularity assumptions on f, u_0, u_1 guarantee that $\bar{u}(\cdot,0)$ is absolutely continuous, which is sufficient for the solvability of (4.1.14).

In contrast to the case of a domain obstacle the solution is unique here without an assumption on an energy conservation. The result can be obtained also with a variational method, if the input data f, u_0 and u_1 are sufficiently regular. A counterexample shows that the problem with a domain obstacle cannot be approximated by some limiting procedure, where the approximating problems are posed by a growing number of point obstacles. For details see [129].

4.1.3 Boundary contact of a halfspace

The result published in [100] is the only one in the literature giving the existence, uniqueness and conservation of energy of solutions of a dynamic contact problem formulated in displacements for a space dimension bigger than one. We consider the halfspace $\Omega = \mathbb{R}^{N-1} \times \mathbb{R}_+$, the time interval $I_t = (0,t)$, the time-space domain $Q_t \equiv I_t \times \Omega$, $t \in I_T$, and its lateral boundary $S_T \equiv I_T \times \Gamma$ with $\Gamma = \mathbb{R}^{N-1} \times \{0\}$. The problem to be solved is

$$\ddot{u} - \Delta u = f \text{ on } Q_T, \tag{4.1.15}$$

$$u(0,\cdot) = u_0, \quad \dot{u}(0,\cdot) = u_1 \text{ on } \Omega, \tag{4.1.16}$$

$$u \geq 0, \quad \partial_N u \geq 0, \quad u\,\partial_N u = 0 \text{ on } S_T. \tag{4.1.17}$$

The physical motivation of this formulation is both the problem for a membrane covering the halfspace and for an infinite homogeneous isotropic body occupying the halfspace whose displacement depends only on the normal variable. In the latter case, the Lamé system is composed of $N-1$ equations and of the unilateral problem of the given type. The main result of [100] is the following

4.1.3 Theorem. *Let* $0 < T$, $u_0 \in \mathring{H}^1(\Omega)$, $u_1 \in L_2(\Omega)$ *and* $f \in L_2(Q_T)$. *Then problem* (4.1.15)–(4.1.17) *possesses a unique solution* u *belonging to* $L_\infty(I_T; H^1(\Omega)) \cap W^1_\infty(I_T; L_2(\Omega))$. *If, moreover,* $u_0 \in \mathring{H}^1(\Omega) \cap H^{3/2}(\Omega)$, $u_1 \in H^{1/2}(\Omega)$ *and* $f \in H^{3/2}(Q_T)$, *then* u *belongs to* $L_\infty(I_T;\, H^{3/2}(\Omega)) \cap W^1_\infty(I_T; H^{1/2}(\Omega))$ *and the global conservation of energy*

$$\begin{aligned} &\int_\Omega \tfrac{1}{2}\left(|\dot{u}|^2 + |\nabla u|^2\right)(t,\cdot)\,dx - \int_\Omega \tfrac{1}{2}\left(|u_1|^2 + |\nabla u_0|^2\right)\,dx \\ &\quad = \int_{Q_t} f\dot{u}\,ds\,dx \end{aligned} \tag{4.1.18}$$

holds for all $t \in I_T$.

The main idea of the *proof* of this result is similar to the one used for the proof of Theorem 4.1.2, but considerably more technical. As above, we employ the solution $\bar{u}$ of the wave equation $\ddot{\bar{u}} - \Delta\bar{u} = f$ with initial data u_0 and u_1 and homogeneous Dirichlet boundary condition. Then the function $v = u - \bar{u}$ solves the problem

$$\begin{aligned} \ddot{v} - \Delta v = 0 \text{ on } Q_T, \quad v(0,\cdot) = \dot{v}(0,\cdot) = 0 \text{ on } \Omega, \\ v + \bar{u} \geq 0,\ \partial_N(v+\bar{u}) \geq 0,\ \langle \partial_N(v+\bar{u}), v+\bar{u}\rangle = 0 \text{ on } S_T. \end{aligned} \tag{4.1.19}$$

If we extend v by 0 for $t < 0$, the solution of (4.1.19) can be found by a partial Fourier transform in time and the tangential space variables via

$$\left(|\boldsymbol{\xi}|^2 - \omega^2\right) \mathscr{F}v - \partial_N^2(\mathscr{F}v) = 0, \tag{4.1.20}$$

where $\mathscr{F}v \equiv \mathscr{F}(v;\cdot)$. This ordinary differential equation with respect to the normal variable is parametrized by $\boldsymbol{\xi}$ and ω. Let λ be the causal determination of $\sqrt{|\boldsymbol{\xi}|^2 + \omega^2}$, that is, the boundary value of the holomorphic extension to $\operatorname{Im}\omega < 0$, given by

$$\lambda(\boldsymbol{\xi},\omega) = \begin{cases} \sqrt{\omega^2 - |\boldsymbol{\xi}|^2} & \text{if } |\boldsymbol{\xi}| < \omega \\ i\sqrt{\omega^2 - |\boldsymbol{\xi}|^2} & \text{if } \omega \geq |\boldsymbol{\xi}|, \\ -i\sqrt{\omega^2 - |\boldsymbol{\xi}|^2} & \text{if } \omega \leq -|\boldsymbol{\xi}|. \end{cases} \tag{4.1.21}$$

Then the general solution of (4.1.20) is of the form

$$\mathscr{F}(v;\omega,\boldsymbol{\xi},x_N) = A(\omega,\boldsymbol{\xi})e^{-x_N\lambda(\omega,\boldsymbol{\xi})} + B(\omega,\boldsymbol{\xi})e^{x_N\lambda(\omega,\boldsymbol{\xi})}$$

and the second term must vanish thanks to the energy and causality considerations. Therefore the Dirichlet to Neumann operator for the wave equation in a halfspace defined by

$$\begin{aligned} \ddot{w} - \Delta w = 0 \text{ in } \mathbb{R}^N \times \mathbb{R}_+, \quad w(\cdot,\cdot,0) = g, \\ A_e : g \mapsto \partial_N w(\cdot,\cdot,0) \end{aligned} \tag{4.1.22}$$

can be explicitly calculated in Fourier variables,

$$\mathscr{F}(A_e w;\omega,\boldsymbol{\xi}) = \lambda(\omega,\boldsymbol{\xi})\mathscr{F}(w;\omega,\boldsymbol{\xi},0),$$

and it is a monotone operator as we can see from (4.1.21). This essential observation requires many technicalities to be performed correctly and it is a heart of the result of Lebeau and Schatzman. Now (4.1.19) is reduced to the standard variatonal inequality related with the problem

$$A_e v + \partial_N\bar{u} + \gamma(v) \ni 0 \tag{4.1.23}$$

which can be solved easily by a penalty method.

The extension of the result to the case $\Omega = \mathbb{R}^{N-1} \times I$ for a bounded interval I is possible despite the fact that for the crucial monotonicity of the operator A_e the problem must be reduced to a small time interval. However, the extension to more general domains as well as to non-homogeneous materials remains open.

4.1.4 A result based on compensated compactness

We study now the boundary contact problem

$$\ddot{u} - \Delta u = f \qquad \text{in } Q_T \equiv (0,T) \times \Omega, \tag{4.1.24}$$

$$u(0,\cdot) = u_0, \ \dot{u}(0,\cdot) = u_1 \qquad \text{in } \Omega, \tag{4.1.25}$$

$$u = 0 \qquad \text{on } S_U \equiv (0,T) \times \Gamma_U, \tag{4.1.26}$$

$$u \geq g, \ \partial_n u \geq 0 \text{ and } (u-g)\partial_n u = 0 \text{ on } S_C \equiv (0,T) \times \Gamma_C. \tag{4.1.27}$$

The weak solvability of this problem is proved in [91] for a bounded domain Ω with a C^2-smooth boundary Γ decomposed into two mutually disjoint open parts Γ_U and Γ_C whose common relative boundary is sufficiently regular. A weak solution of problem (4.1.24)–(4.1.27) is a function $u \in L_\infty(0,T;\mathscr{C})$ that satisfies $\dot{u} \in L_\infty(0,T;L_2(\Omega)) \cap C(0,T;H^{-1/2}(\Omega))$, the initial condition (4.1.25) and the inequality

$$\begin{aligned}\int_{Q_T} &\big(\nabla u \cdot \nabla(v-u) - \dot{u}(\dot{v}-\dot{u})\big)\,d\boldsymbol{x}\,dt \\ &+ \int_\Omega \big(\dot{u}(T,\cdot)(v-u)(T,\cdot) - u_1(v(0,\cdot)-u_0)\big)\,d\boldsymbol{x} \\ \geq &\int_{Q_T} f(v-u)\,d\boldsymbol{x}\,dt\end{aligned} \tag{4.1.28}$$

holds for every $v \in \mathscr{C} := \{w \in H^1(\Omega);\ w = 0 \text{ on } \Gamma_U \text{ and } w \geq g \text{ on } \Gamma_C\}$. The assumptions about the data are $u_0 \in \mathring{H}^1(\Omega)$, $u_1 \in L_2(\Omega)$, $f \in L_2(Q_T)$; the gap function g belongs to $C^1(\overline{\Omega})$ and is non-positive on Γ_C.

The problem is solved via the penalization of the contact condition. The penalized contact condition introduced by Kim [91] has the form

$$\partial_n u_\varepsilon = -\tfrac{1}{\varepsilon}[u_\varepsilon - g]_- - \varepsilon\dot{u}_\varepsilon \text{ on } \Gamma_C \text{ for almost every } t \in (0,T).$$

As usual we denote $H_0^1(\Omega) = \{w \in H^1(\Omega);\ w = 0 \text{ on } \Gamma_U\}$. The variational formulation of the penalized problem reads

Find u_ε such that $\dot{u}_\varepsilon \in L_\infty(0,T;H_0^1(\Omega))$, $\ddot{u}_\varepsilon \in L_\infty(0,T;L_2(\Omega))$, the initial conditions (4.1.25) *holds and the variational equation*

$$\begin{aligned}\int_\Omega &(\ddot{u}_\varepsilon w + \nabla u_\varepsilon \cdot \nabla w - fw)\,d\boldsymbol{x}\,dt \\ &+ \int_{S_C} \Big(\tfrac{1}{\varepsilon}[u_\varepsilon - g]_- + \varepsilon\dot{u}_\varepsilon\Big)\,ds_{\boldsymbol{x}}\,dt = 0\end{aligned} \tag{4.1.29}$$

is satisfied for almost all $t \in (0,T)$ and all $w \in H_0^1(\Omega)$.

The penalized problem is solved via the appropriate Galerkin approximations under special regularity assumptions to f, u_0 and u_1. For these technicalities which are almost standard due to the compact character of the penalty term for a given $\varepsilon > 0$ we refer to the original paper [91]. The following *a priori* estimates are derived:

$$\sup_{t\in(0,T)} \Big(\|\dot{u}_\varepsilon(t,\cdot)\|^2_{L_2(\Omega)} + \|\nabla u_\varepsilon(t,\cdot)\|^2_{\boldsymbol{L}_2(\Omega)} + \tfrac{1}{\varepsilon}\|u_\varepsilon(t,\cdot)\|^2_{L_2(\Omega)}\Big) \\ + \varepsilon\|\dot{u}_\varepsilon\|^2_{L_2(Q_T)} \leq const. \tag{4.1.30}$$

with the constant independent of the parameter ε.

The convergence procedure for a sequence ε_k tending to 0 is based on the following lemma

4.1.4 Lemma. *Let $\{f_k\}$ be a weakly convergent sequence in $L_2(Q_T)$ and let $\{u_k\}$ be a sequence of functions such that*

$$\|u_k\|_{L_2(0,T;H^1(\Omega))} + \|\dot{u}\|_{L_2(Q_T)} \leq const.$$
$$\text{and } u_k \rightharpoonup u \text{ in } L_2(0,T;H^1(\Omega))$$

with the constant independent of k. Let each u_k satisfy the equation

$$\ddot{u}_k - \Delta u_k = f_k.$$

Then as $k \to +\infty$

$$\dot{u}_k^2 - |\nabla u_k|^2 \to \dot{u}^2 - |\nabla u|^2. \tag{4.1.31}$$

The *proof* of this lemma is based on Corollary 4.3 of the monograph [38].

With some technicalities this leads to the existence of a weak solution to problem (4.1.24)–(4.1.27).

Lemma 4.1.4 belongs to the compensated compactness theory which allows to conclude the convergence of a product of two weakly convergent sequences in some particular situations. This theory was founded by Luc Tartar [145] and developed mostly in France. The hope that it could be the main tool for solving a wider class of contact problems has not yet been approved; in particular, possible extensions of the presented result, e.g. to non-homogenous materials, are still open.

4.1.5 A result applicable to polyharmonic problems

We present now a general existence result for a certain class of variational inequalities. This result is applicable to a variety of contact problems. It is due to Maruo [108]; we present it in a modified version with the proof and applications adapted for this book by Schatzman. It is formulated for a more general situation than what we previously described, but the reader should not be misled: there are many conditions to be satisfied. For the case of a vibrating string with an arbitrary obstacle this theorem implies only the existence of a solution satisfying a basic energy inequality without information about the possible conservation of energy. It does not give more information for general elastodynamics in higher space dimension. It allows solving the problem of a two-dimensional plate with an obstacle and the problem of a beam with an obstacle, yet in both cases with an energy inequality.

4.1.5.1 Existence of solutions

Assume that H is a Hilbert space and that A is a monotone selfadjoint linear operator in H with a domain $D(A)$ that is dense in H. Let ϕ be a convex lower semicontinuous function from H to $\mathbb{R} \cup \{+\infty\}$ and let f be a continuous function from $I_T \times H$ to H. Moreover the existence of a function $h \in L_1(I_T)$ is assumed such that for all $[t, x, y] \in I_T \times H^2$ the inequalities

$$\|f(t,x) - f(t,y)\|_H \leq h(t)\|x - y\|_H \text{ and}$$
$$\|\partial_t f(t,x)\|_H \leq h(t)\,(1 + \|x\|_H)$$

are valid. We consider the problem

$$\ddot{u} + Au + \tilde{\partial}\phi \ni f(\cdot, u) \text{ on } I_T \equiv (0,T), \tag{4.1.32}$$
$$u(0) = u_0, \quad \dot{u}(0) = u_1. \tag{4.1.33}$$

Here, $\tilde{\partial}$ denotes the standard subdifferential of a convex function. We restrict ourselves to the case, where ϕ is an *indicatrix* of a closed convex set K in H, i.e.

$$\phi \equiv \delta_K : x \mapsto \begin{cases} 0 & \text{if } x \in K, \\ +\infty & \text{elsewhere.} \end{cases} \tag{4.1.34}$$

The functional conditions given by Maruo are as follows: denote by V the domain of $A^{1/2}$ equipped with the graph norm $\|u\|_V = \sqrt{\|u\|^2 + \langle Au, u\rangle_H}$ and assume the existence of two separable Banach spaces X_1 and X_2 such that the continuous injections

$$V \hookrightarrow X_1 \hookrightarrow X_2^* \hookrightarrow H \hookrightarrow X_2 \tag{4.1.35}$$

are valid and, moreover, the injection $V \hookrightarrow X_1$ is compact, V is dense in H and H is dense in X_2. Let us denote by π_K the projection of H onto

K, i.e. $\|z - \pi_K(z)\|_H \equiv \min_{y \in K} \|x - y\|_H$ (it is well known that the mapping π_K is well defined on H and is Lipschitz with the constant 1 in the norm of H). The hypotheses for K are:

(i) For all $R > 0$ there exists $z \in V$ and constants $c_1 > 0$ and $c_2 \geq 0$ such that for all x satisfying $\|x\|_V \leq R$ we have the estimate

$$\|x - \pi_K x\|_{X_2} \leq c_1 \langle x - \pi_K x, x - z \rangle_H + c_2. \tag{4.1.36}$$

(ii) The closure of $K \cap V$ in H is equal to the closure of K in H.

Let us observe that condition (i) is ensured in applications e.g. if z belongs to the interior of K taken in the norm topology of X_2^*. It yields the crucial estimate of the penalty term.

It remains to give an accurate definition of the notion of solution: *a weak solution of problem* (4.1.32)–(4.1.33) *is a Lipschitz function* $u : I_T \to H$ *possessing the following properties*:

1. *For all* $t \in I_T$, $u(t)$ *belongs to* $K \cap V$.

2. *There exist weak left and right derivatives* $\dot{u}_\pm$ *on* $\overline{I_T}$ *with the appropriate modifications at the end points.*

3. *Let* $\mathscr{K}$ *be the set of continuous functions from* I_T *to* X_1 *which take their values in* K. *The following variational inequality holds for all* $v \in \mathscr{K} \cap L_1(I_T; V) \cap W^1_\infty(I_T; H)$ *with* $v(t, \cdot) \in K$ *for almost every* $t \in I_T$:

$$\begin{aligned}\int_{I_T} &\langle Au - f(u), v - u \rangle_H - \langle \dot{u}, \dot{v} - \dot{u} \rangle_H \, dt \\ &+ \langle \dot{u}_-(T), v(T) - u(T) \rangle_H - \langle u_1, v(0) - u_0 \rangle_H \geq 0.\end{aligned} \tag{4.1.37}$$

4. *The initial condition is satisfied in the sense*

$$u(0) = u_0, \quad \langle \dot{u}_+(0) - u_1, v - u_0 \rangle_H \geq 0 \quad \forall v \in K.$$

4.1.5 Theorem. *Let* $u_0 \in V \cap K$, $u_1 \in H$ *and let all above listed assumptions hold. Then there is at least one solution of problem* (4.1.32)–(4.1.33) *such that the estimate*

$$\sup_{t \in I_T} \left(\|\dot{u}_\pm(t)\|_H^2 + \|u(t)\|_V^2 \right) \leq const. \left(\|u_1\|_H^2 + \|u_0\|_V^2 + 2 \int_{I_T} f(s, u(s)) \, ds \right)$$

holds with a constant depending only on the data of the problem.

Proof. It is a classical fact that the operator defined in $\mathscr{V} = V \times H$ by

$$\mathscr{A} : \begin{pmatrix} u \\ v \end{pmatrix} \mapsto \begin{pmatrix} -v \\ Au \end{pmatrix}, \quad (u, v) \in (V \times H)$$

generates a strongly continuous group of operators $\mathscr{S}$ in $\mathscr{A}$. The penalty approximation

$$\ddot{u}_\varepsilon + Au_\varepsilon + \frac{u_\varepsilon - \pi_K u_\varepsilon}{\varepsilon} = f(\cdot, u_\varepsilon) \text{ on } I_T \tag{4.1.38}$$

can be solved by a fixed point argument if it is rewritten as

$$\left.\begin{aligned} \dot{u}_\varepsilon - v_\varepsilon &= 0 \\ \dot{v}_\varepsilon + Au_\varepsilon &= f(\cdot, u_\varepsilon) - \frac{u_\varepsilon - \pi_K u_\varepsilon}{\varepsilon} \end{aligned}\right\} \text{on } I_T.$$

We apply Picard's iterations to

$$\begin{aligned} \begin{pmatrix} u_\varepsilon(t) \\ v_\varepsilon(t) \end{pmatrix} &= \mathscr{S}(t) \begin{pmatrix} u_0 \\ u_1 \end{pmatrix} \\ &\quad + \int_0^t \mathscr{S}(t-s) \begin{pmatrix} 0 \\ f(s, u_\varepsilon(s)) - (u_\varepsilon - \pi_K u_\varepsilon)/\varepsilon \end{pmatrix} ds \end{aligned}$$

and therefore the fixed point approximation converges. By construction, u_ε belongs to V and $\dot{u}_\varepsilon$ belongs to H. We multiply scalarly (4.1.38) by $(I + \alpha A)^{-1}\dot{u}_\varepsilon$ for $\alpha > 0$ which is permissible since this is a continuous function which takes its values in $D(A)$. Moreover, we infer from (4.1.38) that $\ddot{u}_\varepsilon$ is continuous with values in V^*; hence $(I + \alpha A)^{-1}\ddot{u}_\varepsilon$ is continuous into V and we have the identities

$$\begin{aligned} \left\langle \ddot{u}_\varepsilon, (I + \alpha A)^{-1}\dot{u}_\varepsilon \right\rangle_H &= \frac{1}{2}\frac{d}{dt} \left\langle \dot{u}_\varepsilon, (I + \alpha A)^{-1}\dot{u}_\varepsilon \right\rangle_H \text{ and} \\ \left\langle Au_\varepsilon, (I + \alpha A)^{-1}\dot{u}_\varepsilon \right\rangle_H &= \frac{1}{2}\frac{d}{dt} \left\langle Au_\varepsilon, (I + \alpha A)^{-1}u_\varepsilon \right\rangle_H . \end{aligned}$$

In both cases we have heavily used the fact that A is selfadjoint and monotone. Therefore, by integration we find that

$$\begin{aligned} &\tfrac{1}{2} \left\langle Au_\varepsilon, (I + \alpha A)^{-1}u_\varepsilon \right\rangle_H + \tfrac{1}{2} \left\langle \dot{u}_\varepsilon, (I + \alpha A)^{-1}\dot{u}_\varepsilon \right\rangle_H \\ &\quad = \int_0^t \left\langle f(s, u_\varepsilon(s)) - \varepsilon^{-1}(u_\varepsilon(s) - \pi_K u_\varepsilon(s)), (I + \alpha A)^{-1}\dot{u}_\varepsilon \right\rangle_H ds. \end{aligned}$$

By the convergence $\alpha \searrow 0$ we obtain the energy identity

$$
\begin{aligned}
&\tfrac{1}{2}\|\dot{u}_\varepsilon(t,\cdot)\|_H^2 + \tfrac{1}{2}\|u_\varepsilon(t,\cdot)\|_V^2 + \tfrac{1}{2\varepsilon}\|u_\varepsilon(t) - \pi_K u_\varepsilon(t)\|_H^2 \\
&\quad = \tfrac{1}{2}\|u_1\|_H^2 + \|u_0\|_V^2 + \int_0^t \langle f(s,u_\varepsilon(s)), \dot{u}_\varepsilon(s)\rangle_H \, ds \\
&\qquad + \frac{1}{2\varepsilon}\int_0^t \left\langle \pi_k(u_\varepsilon(s)) - u_\varepsilon(s), \frac{d}{dt}(\pi_K u(s))\right\rangle_H ds.
\end{aligned}
$$

Observe that on any Hilbert space H the function $x \mapsto \|x - \pi_K(x)\|_H^2$ is continuously Fréchet differentiable on H with the differential $x \mapsto 2(x - \pi_K x)$ if K is a closed convex subset of H. Hence for any function $v \in H^1(I;H)$ with a bounded derivative it holds

$$
\frac{d}{dt}\frac{1}{2}\|v(t) - \pi_K v(t)\|_H^2 = \langle v(t) - \pi_K v(t), \dot{v}(t)\rangle_H \, .
$$

A straightforward application of the Gronwall lemma yields the estimate

$$
\begin{aligned}
\max_{t\in(0,T)} \|\dot{u}_\varepsilon(t,\cdot)\|_H + \|u_\varepsilon(t,\cdot)\|_V + \varepsilon^{-1/2}\|u_\varepsilon(t) - \pi_K u_\varepsilon(t)\|_H& \\
\le C \equiv C(u_0, u_1, f).&
\end{aligned} \tag{4.1.39}
$$

We infer from the compactness of the injection $V \hookrightarrow X_1$ and from the Arzelà-Ascoli theorem that it is possible to extract a subsequence $\varepsilon_k \searrow 0$ such that for $u_k \equiv u_{\varepsilon_k}$ the following convergences hold

$$
\begin{aligned}
u_k &\to u \text{ in } C^0(0,T;X_1), \\
\dot{u}_k &\rightharpoonup^* \dot{u} \text{ in } L_\infty(0,T;H), \\
u_k(t) &\rightharpoonup u(t) \text{ in } V \text{ for any } t \in [0,T].
\end{aligned} \tag{4.1.40}
$$

In particular, $u(t)$ belongs to K for all $t \in I_T$.

We employ now assumption (4.1.36) with $R = C$ (C from (4.1.39)). It implies together with (4.1.38) and (4.1.39) that

$$
\int_{I_T} \varepsilon^{-1}\|u_\varepsilon(t) - \pi_K u_\varepsilon(t)\|_{X_2} dt \le C' \tag{4.1.41}
$$

independently of ε. We define

$$
\mu_\varepsilon = \frac{u_\varepsilon(t) - \pi_K u_\varepsilon(t)}{\varepsilon} \tag{4.1.42}
$$

and for a Banach space Y we denote by (Y,w) the space Y equipped with its weak topology and by (Y^*,w^*) its dual equipped with its weak* topology.

We know that $X_2 \hookrightarrow X_2^*$ and we infer from the continuous injection $V \hookrightarrow X_2^*$ that $X_2^{**} \hookrightarrow V^*$; therefore the estimate of μ_ε in X_2 implies an analogous estimate of μ_ε independent of $\varepsilon > 0$,

$$\int_{I_T} \|\mu_\varepsilon\|_{V^*} ds \leq C''. \tag{4.1.43}$$

By the weak* convergence in $L_\infty(I_T; V)^*$ we see that the limit $\varepsilon_k \to 0$ yields the equation

$$\ddot{u} + Au + \mu = f(\cdot, u) \tag{4.1.44}$$

in the sense of distributions with values in V^*, since $Au \in C^0(0,T;(V,w))$ and thus $Au \in C^0(0,T;V^*)$ and $\dot{u} \in L_\infty(0,T;H)$. The terms $f(\cdot,u)$ and Au are measures with values in V^*; hence $\dot{u}$ has limits from the right and from the left at each point of $(0,T)$ with the necessary modifications at endpoints and all these limits are, *a priori*, limits in V^*. However, by compatibility of topologies and by virtue of (4.1.39) after passing to the limit, we can find that $\dot{u}$ has a limit in V from the right and from the left at each point of $(0,T)$ with appropriate modification at the boundary. The validity of (4.1.37) is straightforward. □

4.1.5.2 Applications

The easiest physically reasonable applications are the following:

1. Contact problems for strings: here we have $A = -\partial_x^2$, $H = L_2(0,1)$, $V = \overset{\circ}{H}{}^1(0,1)$, $X_2 = L_1(0,1)$, $X_1 = X_2^* = L_\infty(0,1)$, $K = \{u \in H; u \geq \varphi\}$ for a given $\varphi \in H^1(0,1)$ and $z = \varphi + 1$. There follows $\pi_K u = \max\{u,\varphi\}$, $u - \pi_K u = -[\varphi - u]_+$ and $\langle u - \pi_K u, u - z\rangle_H = \int_0^1 [\varphi - u]_+(1+\varphi-u) ds \geq \|[\varphi - u]_+\|_{L_1(0,1)}$, hence (4.1.36) is satisfied.

2. Domain contact problems for membranes. For a bounded domain $\Omega \subset \mathbb{R}^N$, $N = 1,2$, we have $H = L_2(\Omega)$, $A = (-\Delta)^2$, $V = \overset{\circ}{H}{}^2(\Omega)$, $X_2 = L_1(\Omega)$. The rest is similar to the previous case.

3. Contact problems for viscoelastic materials. The problem is given by

$$\ddot{u} - Au - B\dot{u} + \widetilde{\partial}\delta_K(u) \ni f(\cdot, u). \tag{4.1.45}$$

The semi-group $\mathscr{S}$ is defined by $V = D(A) = D(B)$, $\mathscr{V} = V \times H$, $D(\mathscr{A}) = D(A) \times D(A)$,

$$\mathscr{A}\begin{pmatrix} u \\ v \end{pmatrix} = \begin{pmatrix} -v \\ Au + Bv \end{pmatrix}$$

and the Hille-Yosida theorem applies. It is advantageous to consider the equation $\ddot{u} + Au + B\dot{u} = 0$ as an integral equation,

$$\begin{aligned} u(t) &= u_0 + B^{-1}(I - e^{-tB})u_1 - \int_0^t \int_0^s e^{-(s-r)B} Au(r)\, dr\, ds \\ &= u_0 + B^{-1}(I - e^{-tB})u_1 \\ &\quad - \int_0^t B^{-1}(I - e^{-(t-r)B}) Au(r)\, dr, \end{aligned} \tag{4.1.46}$$

where we have changed the order of integration. Even if the kernel of B is not reduced to 0, we can define

$$B^{-1}(I - e^{-tB})z = \int_{\mathbb{R}} \frac{1 - e^{-t\lambda}}{\lambda} dP(\lambda)z,$$

where $dP(\lambda)$ is the spectral measure associated to B. As the spectrum of B is included into $\mathbb{R}_+$, this definition plainly makes sense. Then the main assumption needed to have a solution is that $(I + B)^{-1}A$ is bounded. In this case

$$B^{-1}(I - e^{-tB})A = B^{-1}(I + B)(I - e^{-tB})(I + B)^{-1}A$$

and

$$B^{-1}(I + B)(I - e^{-tB}) = \int_{\mathbb{R}} \frac{1 - e^{-t\lambda}}{\lambda}(1 + \lambda)dP(\lambda)$$

are clearly bounded operators. This shows that (4.1.46) can be solved by Picard's iterations. To meet the assumption (4.1.36) we need $N = 1$ for harmonic operators and $N = 2$ or 3 for biharmonic operators. The remainder of the proof is completely straightforward.

4.1.6 Remark. The unilateral conditions at the boundary do *not* enter this frame. For example, if we take $V = H^1(\Omega)$, $K := \{u \in V; u(0) \geq 0\}$ then K is not closed in $H = L_2(0, 1)$.

4.2 Results for materials with singular memory

Viscous material with singular (long time) memory is one of the possible models which could enable us to overcome the limits of pure elasticity outlined in the previous section. The models employed here are physically correctly justified from the physical point of view; hence the upcoming results are fully applicable.

4.2.1 Domain contact

The main application of the following problem is the contact of membranes. However, we shall formulate it slightly more generally. Let Ω be a domain in $\mathbb{R}^N$ having a $C^{1,1}$-smooth boundary Γ. The problem has the classical formulation

$$\left.\begin{array}{l}\ddot{u} = \operatorname{div} \boldsymbol{\sigma}(u) + g + f \\ u \geq 0, \quad f \geq 0, \quad uf = 0\end{array}\right\} \text{ on } Q_T = I_T \times \Omega, \tag{4.2.1}$$

where g is a given force. The contact force f denotes some kind of "slack variable". Here the constitutive law

$$\begin{aligned}\boldsymbol{\sigma}(u) = \boldsymbol{\sigma}^I(u) + \boldsymbol{\sigma}^M(u) \quad \text{with} \quad \sigma_i^I(u) = \partial_{\omega_i} W(\cdot, \nabla u) \quad \text{and} \\ \sigma_i^M(u)(t,\cdot) = \int_0^t \mathfrak{K}(t-s)\partial_{\omega_i} V\big(\cdot, \nabla u(t,\cdot) - \nabla u(s,\cdot)\big)\, ds\end{aligned} \tag{4.2.2}$$

is used, where $\partial_{\boldsymbol{\omega}}$ denotes the derivatives with respect to the second variable (the gradient). Here $\boldsymbol{\sigma}^I$ is the inviscid part of the stress tensor while $\boldsymbol{\sigma}^M$ represents its memory part.

We assume the stored energy function $W : \mathbb{R}^{2N} \to \mathbb{R}$ to be of class C^2 and to have a strongly elliptic and bounded partial Hessian with respect to the "dependent variables", i.e. there are $\beta_0^W, \beta_1^W > 0$ such that

$$\begin{aligned}\beta_0^W \xi_i \xi_i \leq \partial_{\omega_i}\partial_{\omega_j} W(\boldsymbol{x}, \boldsymbol{\omega})\, \xi_i \xi_j \quad \text{and} \\ \partial_{\omega_i}\partial_{\omega_j} W(\boldsymbol{x}, \boldsymbol{\omega}) \xi_i \zeta_j \leq \beta_1^W |\boldsymbol{\xi}|\, |\boldsymbol{\zeta}|\end{aligned} \tag{4.2.3}$$

hold for all $\boldsymbol{\xi}, \boldsymbol{\zeta}, \boldsymbol{\omega} \in \mathbb{R}^N$ and all $\boldsymbol{x} \in \Omega$. We assume that the function V appearing in the memory term satisfies conditions analogous to (4.2.3) with constants denoted by β_0^V, β_1^V, and, moreover, we impose

$$\begin{aligned}V(\cdot,0) = W(\cdot,0) = 0 \text{ and } \partial_{\omega_i} V(\cdot,0) = \partial_{\omega_i} W(\cdot,0) = 0 \text{ on } \Omega, \\ i = 1, \ldots, N.\end{aligned} \tag{4.2.4}$$

The kernel $\mathfrak{K}$ of the singular memory is integrable over $\mathbb{R}_+$; it has the form

$$\begin{aligned}&\mathfrak{K}(t) = t^{-2\alpha} a(t) + r(t), \ t \in \mathbb{R}_+ \equiv [0, +\infty) \text{ with } \alpha \in \left(0, \tfrac{1}{2}\right), \\ &\mathfrak{K}(t) = 0, \ t \leq 0.\end{aligned} \tag{4.2.5}$$

Both a and r belong to $C^1\big(\overline{R_+}\big)$; they are non-negative and non-increasing functions. Moreover, $a(t) > 0$ for t in a right neighbourhood of the origin and $t \mapsto t^{-2\alpha} a'(t)$ is integrable over $\mathbb{R}_+$. The assumptions about W yield

$$\begin{aligned}W(\cdot, \nabla w) &= \int_0^1 (1-\theta)\partial_{\omega_i}\partial_{\omega_j} W(\cdot, \nabla(\theta w))\, \partial_i w\, \partial_j w\, d\theta \\ &\in \left(\tfrac{1}{2}\beta_0^W |\nabla w|^2, \tfrac{1}{2}\beta_1^W |\nabla w|^2\right),\end{aligned} \tag{4.2.6}$$

and

$$
\begin{aligned}
&(\partial_{\omega_i} W(\cdot, \nabla v) - \partial \omega_i W(\cdot, \nabla w))\, \partial_i (v-w) \\
&\quad = \int_0^1 \partial_{\omega_i} \partial_{\omega_j} W(\cdot, \nabla(w + \theta(v-w))) \partial_i (v-w)\, \partial_j (v-w)\, d\theta \\
&\quad \geq \beta_0^W |\nabla(v-w)|^2, \\
&(\partial_{\omega_i} W(\cdot, \nabla v) - \partial_{\omega_i} W(\cdot, \nabla w))\, \partial_i \widetilde{w} \\
&\quad = \int_0^1 \partial_{\omega_i} \partial_{\omega_j} W(\cdot, \nabla(w + \theta(v-w))) \partial_i \widetilde{w}\, \partial_j (v-w)\, d\theta \\
&\quad \leq \beta_1^W |\nabla(v-w)|\, |\nabla \widetilde{w}|,
\end{aligned}
\tag{4.2.7}
$$

for arbitrary displacements v, w and $\widetilde{w}$ on Q_T. The formula (4.2.6) is proved with the help of relation $f(1) = \int_0^1 (1-\theta) f''(\theta)\, d\theta$; it is valid for functions satisfying $f(0) = f'(0) = 0$ and we apply it to $f : \theta \mapsto W(\cdot, \theta \nabla w)$. The formulae (4.2.7) are derived by the standard integration for the function $\theta \mapsto W(\cdot, \nabla(w + \theta(v-w)))$. Analogous relations hold for V, too. For $\mathfrak{K}$ we assume

$$
\mathfrak{c} \equiv 2\beta_1^V \int_{\mathbb{R}_+} \mathfrak{K}(s)\, ds < \beta_0^W . \tag{4.2.8}
$$

This assumption means that the memory is "small enough" and it ensures the strong monotonicity of the operator $\nabla u \mapsto \boldsymbol{\sigma}(u)$, i.e.

$$
\langle \boldsymbol{\sigma}(v) - \boldsymbol{\sigma}(w), \nabla(v-w) \rangle_{Q_T} \geq (\beta_0^W - \mathfrak{c}) \|\nabla(v-w)\|^2_{\boldsymbol{L}_2(Q_T)} \tag{4.2.9}
$$

for all v, w such that these expressions are defined. This relation is proved estimating the memory term; the following inequality holds:

$$
\begin{aligned}
&\Big| \int_{Q_T} \int_0^t \mathfrak{K}(t-s) \big(\partial_{\omega_i} V(\cdot, \nabla v(t,\cdot) - \nabla v(s,\cdot)) \\
&\qquad - \partial_{\omega_i} V(\cdot, \nabla w(t,\cdot) - \nabla w(s,\cdot)) \big) \partial_i \big(v(t,\cdot) - w(t,\cdot) \big)\, ds\, d\boldsymbol{x}\, dt \Big| \\
&\leq \beta_1^V \int_{Q_T} \int_0^t \mathfrak{K}(t-s) \big(|\nabla(v-w)(t,\cdot)| + |\nabla(v-w)(s,t)| \big) \\
&\qquad \cdot |\nabla(v-w)(t,\cdot)|\, ds\, d\boldsymbol{x}\, dt.
\end{aligned}
$$

Writing $z = v - w$ the right hand side of the above inequality can be estimated further by

$$
\begin{aligned}
&\frac{\mathfrak{c}}{2}\|\nabla z\|^2_{\boldsymbol{L}_2(Q_T)} + \beta_1^V \left(\int_{Q_T} \int_0^t \mathfrak{K}(t-s)\,|\nabla z(t,\cdot)|^2\,ds\,d\boldsymbol{x}\,dt \right)^{1/2} \\
&\quad \cdot \left(\int_{Q_T} \int_0^t \mathfrak{K}(t-s)\,|\nabla z(s,\cdot)|^2\,ds\,d\boldsymbol{x}\,dt \right)^{1/2} \\
&\le \mathfrak{c}\|\nabla z\|^2_{L_2(Q_T)}.
\end{aligned}
$$

Problem (4.2.1) is completed by the initial condition

$$
u(0,\cdot) = u_0, \quad \dot{u}(0,\cdot) - u_1 \ge 0, \quad (\dot{u}(0,\cdot) - u_1)u_0 = 0 \text{ on } \Omega, \tag{4.2.10}
$$

and the boundary condition

$$
\boldsymbol{\sigma}(u)\cdot n \ge 0, \quad u \ge 0, \quad u\,(\boldsymbol{\sigma}(u)\cdot n) = 0 \text{ on } S_T = I_T \times \Gamma. \tag{4.2.11}
$$

Instead of this condition the Dirichlet boundary condition can be used,

$$
u = U \text{ on } S_T. \tag{4.2.12}
$$

Let us introduce the cone

$$
\mathscr{C} := \{ v \in L_2(I_T; H^1(\Omega));\ v \ge 0 \text{ a.e. in } Q_T \}. \tag{4.2.13}
$$

A function $u \in \mathscr{C} \cap B_0(I_T; L_2(\Omega))$ *with* $\dot{u} \in L_\infty(I_T; L_2(\Omega))$ *and* $\dot{u}(T,\cdot) \in L_2(\Omega)$ *is called a weak solution to problem* [(4.2.1), (4.2.10), (4.2.11)] *if for every* $v \in \mathscr{C} \cap H^1(Q_T)$,

$$
\begin{aligned}
&\int_{Q_T} \big(\boldsymbol{\sigma}(u)\cdot\nabla(v-u) - \dot{u}(\dot{v}-\dot{u}) - g(v-u)\big)\,d\boldsymbol{x}\,dt \\
&\qquad + \int_\Omega (\dot{u}(v-u))(T,\cdot) - u_1(v(0,\cdot) - u_0)\,d\boldsymbol{x} \ge 0.
\end{aligned} \tag{4.2.14}
$$

This variational inequality will be solved under the assumptions

$$
\begin{aligned}
&g \in L_2(Q_T) \cup H^{1-\alpha}(I_T; \mathring{H}^{-1}(\Omega)), \\
&u_0 \in H^1(\Omega),\ u_0 \ge 0 \text{ a.e. in } \Omega \text{ and } u_1 \in L_2(\Omega).
\end{aligned} \tag{4.2.15}
$$

The assumption on g is motivated by the necessity to put a velocity as a test function.

For the problem with the Dirichlet boundary value condition we assume

$$
\begin{aligned}
&U \in L_2(I_T; H^{1/2}(\Gamma)) \text{ with } 0 \ge \dot{U} \in L_2(I_T; H^{1/2}(\Gamma)), \\
&U(0,\cdot) = u_0|_\Gamma, \operatorname*{ess\,inf}_{S_T} U \ge U_0 \text{ and } \ddot{U} \in L_2(S_T),
\end{aligned} \tag{4.2.16}
$$

where U_0 is a positive constant. It is not difficult to extend U to a function on $H^1(Q_T)$ such that $\dot{U} \in H^1(Q_T)$ and all assumptions about their bounds in (4.2.16) are preserved. Throughout this section we shall consider U to be such an extension of the original boundary value. In the appropriate version of (4.2.14) we require, moreover, that u, v belong to $U + L_2\big(I_T; \overset{\circ}{H}{}^1(\Omega)\big)$ for both the solution u and the test functions v.

Problem (4.2.14) is solved by penalizing the contact condition:
We look for u_ε in $H^1(Q_T)$ such that $\dot{u}_\varepsilon$ belongs to $B_0(I_T; L_2(\Omega))$, $\nabla u_\varepsilon \in L_\infty(I_T; \boldsymbol{L}_2(\Omega)) \cap H^\alpha(I_T; \boldsymbol{L}_2(\Omega))$, $\dot{u}_\varepsilon(T, \cdot)$ belongs to $L_2(\Omega)$, the initial conditions $u_\varepsilon(0, \cdot) = u_0$, $\dot{u}_\varepsilon(0, \cdot) = u_1$ are satisfied (both in the sense of Green's theorem and weak $L_2(\Omega)$-limit from the right side), and the equation

$$\int_{Q_T} \Big(\boldsymbol{\sigma}(u_\varepsilon) \cdot \nabla v + \ddot{u}_\varepsilon v - g v + \tfrac{1}{\varepsilon}[u_\varepsilon]_- v \Big) \, d\boldsymbol{x} \, dt = 0 \tag{4.2.17}$$

holds for every $v \in H^1(\Omega)$.
This variational equation corresponds to problem (4.2.1) with $f = \frac{1}{\varepsilon}[u_\varepsilon]_-$.

We define the characteristic function

$$\begin{aligned} \chi_t : \mathbb{R} \to \mathbb{R} \text{ such that } \chi_t(s) \equiv 1, \;\; s \in [0, t] \equiv \overline{I_t} \\ \text{and } \chi_t \equiv 0 \text{ on } \mathbb{R} \setminus \overline{I_t}. \end{aligned} \tag{4.2.18}$$

Let us insert $v = \chi_t \dot{u}_\varepsilon$ into (4.2.17). Since

$$\begin{aligned} \partial_{\boldsymbol{\omega}} V\big(\cdot, \nabla(u_\varepsilon(s, \cdot) - u_\varepsilon(r, \cdot))\big) \cdot \nabla \dot{u}(s, \cdot) &= \partial_s V\big(\cdot, \nabla(u_\varepsilon(s, \cdot) - u_\varepsilon(r, \cdot))\big), \\ \text{and} \quad \partial_{\boldsymbol{\omega}} W\big(\cdot, \nabla u_\varepsilon(s, \cdot)\big) \cdot \nabla \dot{u}(s, \cdot) &= \partial_s W\big(\cdot, \nabla u_\varepsilon(s, \cdot)\big), \end{aligned}$$

we obtain after integrating by parts in time

$$\begin{aligned} &\int_\Omega \big(\tfrac{1}{2}|\dot{u}_\varepsilon|^2 + W(\cdot, \nabla u_\varepsilon)\big)(t, \cdot) \, d\boldsymbol{x} + \int_\Omega (2\varepsilon)^{-1} [u_\varepsilon]_-^2 (t, \cdot) \, d\boldsymbol{x} \\ &\quad - \int_{Q_t} \int_0^s \mathfrak{K}'(s - r) V\big(\cdot, \nabla u_\varepsilon(s, \cdot) - \nabla u_\varepsilon(r, \cdot)\big) \, dr \, ds \, d\boldsymbol{x} \\ &\quad + \int_0^t \int_\Omega \mathfrak{K}(t - r) V\big(\cdot, \nabla u_\varepsilon(t, \cdot) - \nabla u_\varepsilon(r, \cdot)\big) \, dr \, d\boldsymbol{x} \\ &= \int_\Omega \big(\tfrac{1}{2}|u_1|^2 + W(\cdot, \nabla u_0)\big) \, d\boldsymbol{x} + \int_{Q_t} g \dot{u}_\varepsilon \, d\boldsymbol{x} \, ds. \end{aligned} \tag{4.2.19}$$

Observe that $\mathfrak{K}'(s) = -2\alpha s^{-1-2\alpha} a(s) + \gamma(s)$, where $\gamma : s \mapsto s^{-2\alpha} a'(s) + r'(s)$ is a negative function. We invoke the relations (4.2.6) and we use the formal integration by parts of the volume force term in (4.2.19) for the second possible space for g in (4.2.15). Then it is not diifficult to show that (4.2.19) implies the *a priori* estimate

$$\begin{aligned} &\|\dot{u}_\varepsilon\|^2_{L_\infty(I_T;L_2(\Omega))} + \|\nabla u_\varepsilon\|^2_{L_\infty(I_T;\boldsymbol{L}_2(\Omega))} + \|u_\varepsilon\|^2_{L_2(Q_T)} \\ &\quad + \tfrac{1}{\varepsilon}\|[u_\varepsilon]_-\|^2_{L_\infty(I_T;L_2(\Omega))} + \|\nabla u_\varepsilon\|^2_{H^\alpha(I_T;\boldsymbol{L}_2(\Omega))} \\ &\le c_0 \equiv c_0(\mathscr{J}) \text{ with } \mathscr{J} \equiv \big(\|u_0\|_{H^1(\Omega)}, \|u_1\|_{L_2(\Omega)}, \|g\|_X\big)\,, \end{aligned} \tag{4.2.20}$$

where the space X is one of the spaces appearing in (4.2.15) and the constant c_0 is independent of the penalty parameter ε.

In order to prove the solvability of (4.2.17) we use the standard Galerkin procedure. Let $\mathscr{Z} \equiv \{z_m;\, m \in \mathbb{N}\}$ be an L_2-orthogonal basis of $H^1(\Omega)$. We search for a solution $u_{\varepsilon,m} = \sum_{i=1}^{m} c_i z_i$, $c_i \in H^2(I_T; L_2(\Omega))$ of the approximate Galerkin problems

$$\begin{aligned} &\int_\Omega \Big(\boldsymbol{\sigma}(u_{\varepsilon,m}) \cdot \nabla z_i + \ddot{u}_{\varepsilon,m} z_i + \Big(\tfrac{1}{\varepsilon}[u_{\varepsilon,m}]_- - g\Big)\, z_i\Big)\, d\boldsymbol{x} = 0 \\ &\qquad \text{for } i = 1, \dots, m,\ t \in I_T, \end{aligned} \tag{4.2.21}$$

satisfying the approximate initial conditions

$$u_{\varepsilon,m}(0,\cdot) = \pi_{Z_m} u_0 \quad \text{and} \quad \dot{u}_{\varepsilon,m}(0,\cdot) = \pi_{Z_m} u_1.$$

The existence of this solution is ensured by the theory of Carathéodory solutions of systems of ordinary differential equations, see e.g. [36]. Hence, integrating (4.2.21) in time over I_T, the validity of the following equation is verified for any function $w \in Z_m \equiv \Big\{\sum_{i=1}^{m} c_i z_i;\ c_i \in L_2(I_T), i = 1, \dots, m\Big\}$:

$$\begin{aligned} &\int_{Q_T} \Big(\boldsymbol{\sigma}(u_{\varepsilon,m}) \cdot \nabla w + \ddot{u}_{\varepsilon,m} w + \big(\tfrac{1}{\varepsilon}[u_{\varepsilon,m}]_- w - g\big) w\Big)\, d\boldsymbol{x}\, dt \\ &\qquad = 0. \end{aligned} \tag{4.2.22}$$

In the same way as for the variational equation corresponding to the original penalized problem we prove the *a priori* estimate (4.2.20) for the solutions $u_{\varepsilon,m}$. Then for fixed $\varepsilon > 0$, the uniform boundedness of $\ddot{u}_{\varepsilon,m}$ in $L_2\big(I_T; \mathring{H}^{-1}(\Omega)\big)$ is derived as follows: for test functions z from Z_m the uniform boundedness of $\int_{Q_T} (\ddot{u}_{\varepsilon,m} z)(t,\cdot)\, d\boldsymbol{x}\, dt$ is easily derived from (4.2.22). The orthogonality of the basis yields

$$\int_\Omega (\ddot{u}_{\varepsilon,m} z)(t,\cdot)\, d\boldsymbol{x} = \int_\Omega (\ddot{u}_{\varepsilon,m} \pi_{Z_m} z)(t,\cdot)\, d\boldsymbol{x} \tag{4.2.23}$$

for any test function z and any time $t \in I_T$. The uniform boundedness of the projections π_{Z_m} with respect to m completes the proof of the dual estimate. The interpolation of this result with $u_{\varepsilon,m} \in H^\alpha\big(I_T; H^1(\Omega)\big)$ gives the uniform boundedness of $u_{\varepsilon,m}$ in $H^{1+\alpha/2}(I_T; L_2(\Omega))$. Hence, there is an element w_ε and a suitable sequence $m_k \to +\infty$ such that

$$\begin{aligned}
u_{\varepsilon,m_k} &\rightharpoonup w_\varepsilon && \text{in } H^{1+\alpha/2,1}(Q_T),\\
\dot u_{\varepsilon,m_k} &\to \dot w_\varepsilon && \text{in } L_2(Q_T),\\
[u_{\varepsilon,m_k}]_- &\to [w_\varepsilon]_- && \text{in } L_2(Q_T),\\
\dot u_{\varepsilon,m_k}(T,\cdot) &\rightharpoonup \dot w_\varepsilon(T,\cdot) && \text{in } L_2(\Omega),\\
\ddot u_{\varepsilon,m_k} &\rightharpoonup \ddot w_\varepsilon && \text{in } L_2\big(I_T; \mathring{H}^{-1}(\Omega)\big),\\
\boldsymbol{\sigma}(u_{\varepsilon,m_k}) &\rightharpoonup \boldsymbol{\kappa}_\varepsilon && \text{in } \boldsymbol{L}_2(\Omega).
\end{aligned} \tag{4.2.24}$$

The derivatives in T are understood as the left ones. From these convergences it is easy to see that w_ε satisfies (4.2.17) if $\boldsymbol{\sigma}(w_\varepsilon)$ is replaced by $\boldsymbol{\kappa}_\varepsilon$. We insert $v = u_{\varepsilon,m_k} - \pi_{Z_{m_k}} w_\varepsilon$ into (4.2.22) and add $\int_{Q_T} \boldsymbol{\sigma}(\pi_{Z_{m_k}} w_\varepsilon) \cdot \nabla(\pi_{Z_{m_k}} w_\varepsilon - u_{\varepsilon,m_k})\, d\boldsymbol{x}\, dt$ to both sides of the equation. Since $\pi_{Z_{m_k}} w_\varepsilon \to w_\varepsilon$ in $L_2(I_T; H^1(\Omega))$ and (4.2.7) holds, $\boldsymbol{\sigma}(\pi_{Z_{m_k}} w_\varepsilon)$ tends to $\boldsymbol{\sigma}(w_\varepsilon)$ strongly in $\boldsymbol{L}_2(Q_T)$. From this and all just proved convergences we derive

$$\begin{aligned}
\int_{Q_T} \big(\boldsymbol{\sigma}(\pi_{Z_{m_k}} w_\varepsilon) - \boldsymbol{\sigma}(u_{\varepsilon,m_k})\big) \cdot \nabla(\pi_{Z_{m_k}} w_\varepsilon - u_{\varepsilon,m_k})\, d\boldsymbol{x}\, dt&\\
\to 0 \text{ as } k \to +\infty.&
\end{aligned} \tag{4.2.25}$$

Now, (4.2.7) yields $\pi_{Z_{m_k}} w_\varepsilon - u_{\varepsilon,m_k} \to 0$ in $L_2\big(I_T; H^1(\Omega)\big)$, hence $u_{\varepsilon,m_k} \to w_\varepsilon$ there. Again from (4.2.7) there follows $\boldsymbol{\sigma}(u_{\varepsilon,m_k}) \to \boldsymbol{\sigma}(w_\varepsilon)$ in $\boldsymbol{L}_2(Q_T)$, consequently $\boldsymbol{\kappa}_\varepsilon = \boldsymbol{\sigma}(w_\varepsilon)$ and w_ε satisfying (4.2.17) can be denoted by u_ε. Since u_{ε,m_k} converges strongly to u_ε in $C^0(I_T; L_2(\Omega))$, the identity $u_\varepsilon(0,\cdot) = u_0$ holds. In order to prove the weak convergence $\dot u_{\varepsilon,m_k}(t,\cdot) \rightharpoonup \dot u_\varepsilon(t,\cdot)$ $L_2(\Omega)$ for a.e. $t \in I_T$ we take the test function $w = \chi_t v$ for χ_t from (4.2.18), any fixed $v \in Z_{m_k}$ for any k and $t \in I_T$. We formally integrate the acceleration term in (4.2.22) by parts in the time variable; we use the boundedness of $\{\dot u_{\varepsilon,m}(t,\cdot);\, t \in I_T,\ m \in \mathbb{N},\ \varepsilon > 0\}$ in $L_2(\Omega)$, the weak convergence of the accelerations and the convergence of the initial values and determine their right values. Similarly we determine their left values with the help of the characteristic function of $[t,T]$. Moreover, for any $v \in H^1(\Omega)$ and any sequence of intervals $(a_n, b_n) \subset I_T$ with $a_n \to a \in \overline{I_T}$ and $b_n \to a$ we have $\langle \ddot u_\varepsilon, \chi_{[a_n,b_n]} v\rangle_Q = \langle \dot u_\varepsilon([b_n]_-), v\rangle_\Omega - \langle \dot u_\varepsilon([a_n]_-), v\rangle_\Omega \to 0$, where $\chi_{[a_n,b_n]}$, the characteristic function of the interval, is 1 on it and vanishes outside and the subscripts denote the appropriate left and right derivatives. However, for a.e. $t \in I_T$ the left and right derivatives are identical. Hence they must be identical everywhere on I_T and $\dot{u}_\varepsilon \in C^0(I_T; \mathring{H}^{-1}(\Omega)) \cap B_0(I_T; L_2(\Omega))$, where the continuity is understood in the weak sense. This and standard density arguments imply the weak continuity of $\dot{u}$ from $\overline{I_T}$ to $L_2(\Omega)$. Then $\dot{u}_\varepsilon(0,\cdot) = u_1$ (as the right derivative) in the sense both of Green's theorem and of the weak right limit of $\dot{u}(t,\cdot)$ at 0. Hence u_ε is a solution of (4.2.17).

The uniqueness of the solution can be proved only under some additional assumption. Let us consider two different solutions $u_\varepsilon^{(1)}$, $u_\varepsilon^{(2)}$ of (4.2.17), use $\dot{u}_\varepsilon^{(1)} - \dot{u}_\varepsilon^{(2)}$ as a test function and take the difference of both equations. In general it is not clear that

$$\left(\boldsymbol{\sigma}\left(u_\varepsilon^{(1)}\right) - \boldsymbol{\sigma}\left(u_\varepsilon^{(2)}\right)\right) \cdot \nabla\left(\dot{u}_\varepsilon^{(1)} - \dot{u}_\varepsilon^{(2)}\right) \geq 0.$$

However, e.g. for the linear constitutive law, where

$$W(x, \nabla u) = \tfrac{1}{2} a_{ij}^{(0)}(\boldsymbol{x})\, \partial_i u\, \partial_j u \quad \text{and}$$
$$V(x, \nabla u) = \tfrac{1}{2} a_{ij}^{(1)}(\boldsymbol{x})\, \partial_i u\, \partial_j u$$

with symmetric matrices $\left(a_{ij}^{(\iota)}\right)$, $\iota = 0, 1$ on Ω such that the relations (4.2.3) for W and V are preserved, it is possible to use the standard integration by parts and the Gronwall lemma to prove the uniqueness. With the abbreviation $z = u_\varepsilon^{(1)} - u_\varepsilon^{(2)}$, the memory term is treated as follows:

$$\begin{aligned}
&\int_{Q_T} \int_0^t \mathfrak{K}(t-s) a_{ij}^{(1)} \partial_i (z(t,\cdot) - z(s,\cdot)) \partial_t \partial_j (z(t,\cdot) - z(s,\cdot))\, ds\, dt\, d\boldsymbol{x} \\
&= \int_{Q_T} \tfrac{1}{2} \mathfrak{K}(T-s) a_{ij}^{(1)} \partial_i (z(T,\cdot) - z(s,\cdot)) \partial_j (z(T,\cdot) - z(s,\cdot))\, ds\, d\boldsymbol{x} \\
&\quad - \int_{Q_T} \tfrac{1}{2} \int_0^t \mathfrak{K}'(t-s) a_{ij}^{(1)} \partial_i (z(t,\cdot) - z(s,\cdot)) \\
&\qquad \cdot\ \partial_i (z(t,\cdot) - z(s,\cdot))\, ds\, dt\, d\boldsymbol{x}
\end{aligned}$$

and both terms on the right hand side of this equality are positive.

Hence we have proved

4.2.1 Lemma. *Problem* (4.2.17) *has a solution u_ε for every $\varepsilon > 0$. If the operator $u \mapsto \sigma(u)$ is linear, then the solution is unique.*

4.2.2 Remark. An analogous lemma holds also for the problem with Dirichlet boundary condition (4.2.12) for which we define the cone

$$\mathscr{C}_U := \{v \in H^1(Q_T);\ v \equiv U \text{ on } S_T,\ v \geq 0 \text{ on } Q_T\}.$$

For the proof, we employ the variational formulation (4.2.14) with arbitrary $v \in \mathscr{C}_U$, and the variational formulation of the appropriate penalized problem (4.2.17) holds for each $v \in L_2\big(I_T; \mathring{H}^1(\Omega)\big)$. The estimate

(4.2.20) is an easy consequence of putting $v = \dot{U} - \dot{u}_\varepsilon$ into (4.2.17) and of the non-positivity of $\dot{U}$. To derive it, the integration by parts in the term containing $\ddot{u}_{\varepsilon\dot{U}}$ (requiring at least $\ddot{U} \in L_2(I_T; \overset{\circ}{H}{}^{-1}(\Omega)))$ should be performed. Of course, in this case $c_0 \equiv c_0\Big(\mathscr{J}, \|\dot{U}\|_{H^1(Q_T)}\Big)$ and $c_1 \equiv c_1\Big(\mathscr{J}, \|\dot{U}\|_{H^1(Q_T)}, \|U\|_{H^1(Q_T)}, U_0\Big)$.

For the limit $\varepsilon \to 0$ we need a new dual estimate of the acceleration. As only the equation (4.2.17) is at our disposal, we must first obtain some ε-independent estimate of the penalty term. We get it by taking the constant $v = 1$ as a test function. Via an integration by parts of the acceleration term this leads to

$$\frac{1}{\varepsilon}\int_{Q_T} [u_\varepsilon]_- \, d\boldsymbol{x}\, dt \in [0, c) \text{ with } c \equiv c(\mathscr{J}) \tag{4.2.26}$$

and $\mathscr{J}$ from (4.2.20). For the problem with a Dirichlet boundary value condition, an ε-independent estimate can be analogously proved by means of the test function $v = U - u_\varepsilon$ exploiting assumption (4.2.16). Thus we get an estimate $U_0 \frac{1}{\varepsilon}\int_{Q_T} [u_\varepsilon]_- \, d\boldsymbol{x}\, dt \le \frac{1}{\varepsilon}\int_{Q_T} [u_\varepsilon]_- U \, d\boldsymbol{x}\, dt$ from above by the remaining terms which can be estimated easily by the terms in (4.2.20). Of course, in this case $c \equiv c_0\big(\mathscr{J}, \|\dot{U}\|_{H^1(Q_T)}, U_0\big)$.

This uniform $L_1(Q_T)$-estimate of the penalty term and the "dual" imbedding theorem

$$L_\infty(\Omega)^* \hookrightarrow H^{N/2+\eta}(\Omega)^*$$

valid for any $\eta > 0$ yields

$$\left\| \frac{1}{\varepsilon}[u_\varepsilon]_- \right\|_{L_1(I_T; H^{N/2+\eta}(\Omega)^*)} \le c(\mathscr{J}); \tag{4.2.27}$$

in particular the constant is ε-independent. Using this in combination with (4.2.20) in the equation (4.2.17), we can see that the estimate

$$\|\ddot{u}_\varepsilon\|_{L_1(I_T; (H^{N/2+\eta}(\Omega))^*)} \le c(\mathscr{J}) \tag{4.2.28}$$

holds with c independent of $\varepsilon > 0$. The interpolation of this estimate with $u_\varepsilon \in W_1^\alpha\big(I_T; H^1(\Omega)\big)$ shows that u_ε is bounded in $W_1^\sigma(I_T; L_2(\Omega))$ for every $\sigma < \dfrac{4}{N+2} + \alpha\dfrac{N}{N+2}$. Here the condition

$$\alpha > 1 - \frac{2}{N} \tag{4.2.29}$$

is imposed to obtain $\sigma > 1$. By means of Hölder's inequality we get

$$\|\dot{u}_\varepsilon\|_{H^\lambda(I_T; L_2(\Omega))} \le c\|\dot{u}_\varepsilon\|^{1/2}_{L_\infty(I_T; L_2(\Omega))} \|\dot{u}_\varepsilon\|^{1/2}_{W_1^{2\lambda}(I_T; L_2(\Omega))}$$

if $\lambda \in (0, 1/2)$. From the requirement $2\lambda < \sigma - 1$ we obtain

$$\begin{aligned}&\{\dot{u}_\varepsilon;\ \varepsilon \geq 0\} \text{ is bounded in } H^\lambda(I_T; L_2(\Omega)) \text{ with}\\ &\qquad \lambda \in (0, \alpha/4) \text{ for } N = 2,\\ &\qquad \lambda \in \left(0, \tfrac{3}{10}\left(\alpha - \tfrac{1}{3}\right)\right) \text{ for } N = 3.\end{aligned} \tag{4.2.30}$$

Interpolating this result with the uniform estimate of u_ε in $H^\alpha(I_T; H^1(\Omega))$ we obtain that

$$\{\dot{u}_\varepsilon;\ \varepsilon \geq 0\} \text{ is bounded in } L_2\big(I_T; H^{\lambda/(1+\lambda-\alpha)}(\Omega)\big). \tag{4.2.31}$$

Now, Theorem 2.8.14 yields that $\{\dot{u}_\varepsilon;\ \varepsilon > 0\}$ is relatively compact in $L_2(Q_T)$. From the proved *a priori* estimates we infer the existence of a sequence ε_k tending to 0 such that

$$\begin{aligned}&\dot{u}_{\varepsilon_k} \to \dot{u} \text{ in } L_2(Q_T), \text{ hence } u_{\varepsilon_k} \to u \text{ in } C^0(I_T; L_2(\Omega)),\\ &u_{\varepsilon_k} \rightharpoonup u \text{ in } L_2(I_T; H^1(\Omega)),\\ &\dot{u}_{\varepsilon_k}(0,\cdot) \rightharpoonup \dot{u}(0,\cdot) \text{ and } \dot{u}_{\varepsilon_k}(T,\cdot) \rightharpoonup \dot{u}(T,\cdot) \text{ in } L_2(\Omega)\end{aligned} \tag{4.2.32}$$

with

$$\begin{aligned}&u \in H^1(Q_T) \cap L_\infty\big(I_T; H^1(\Omega)\big) \text{ and}\\ &\dot{u} \in H^{\lambda,\lambda/(1+\lambda)}(Q_T) \cap L_\infty(I_T; L_2(\Omega)).\end{aligned}$$

The limit u is non-negative a.e. in Q_T and $u(0,\cdot) = u_0$. Due to (4.2.3) the set $\{\boldsymbol{\sigma}(u_\varepsilon);\ \varepsilon > 0\}$ is bounded e.g. in $\boldsymbol{L}_2(Q_T)$; hence we can assume that there is a $\boldsymbol{\kappa} \in \boldsymbol{L}_2(Q_T)$ such that $\boldsymbol{\sigma}(u_{\varepsilon_k}) \rightharpoonup \boldsymbol{\kappa}$ in $\boldsymbol{L}_2(Q_T)$. Let us take $u - u_{\varepsilon_k}$ as a test function in (4.2.17). After performing the integration by parts in time, we obtain

$$\begin{aligned}&\int_{Q_T} \big(\boldsymbol{\sigma}(u_{\varepsilon_k}) \cdot \nabla(u - u_{\varepsilon_k}) - \dot{u}_{\varepsilon_k}(\dot{u} - \dot{u}_{\varepsilon_k}) - g(u - u_{\varepsilon_k})\big)\,d\boldsymbol{x}\,dt\\ &+ \int_\Omega \big(\dot{u}_{\varepsilon_k}(T,\cdot)(u - u_{\varepsilon_k})(T,\cdot) - u_1(u(0,\cdot) - u_0)\big)\,d\boldsymbol{x} \geq 0,\end{aligned} \tag{4.2.33}$$

because $[u_{\varepsilon_k}]_-(u - u_{\varepsilon_k}) \leq 0$. Adding $\int_{Q_T} \boldsymbol{\sigma}(u) \cdot \nabla(u_{\varepsilon_k} - u)\,d\boldsymbol{x}\,dt$ to both sides of (4.2.33) and using all just mentioned convergences we obtain

$$\int_{Q_T} \big(\boldsymbol{\sigma}(u) - \boldsymbol{\sigma}(u_{\varepsilon_k})\big) \cdot \nabla(u - u_{\varepsilon_k})\,d\boldsymbol{x}\,dt \to 0 \text{ as } k \to +\infty. \tag{4.2.34}$$

Then the strong monotonicity of the constitutive relation (4.2.9) together with (4.2.32) yields $u_{\varepsilon_k} \to u$ in $L_2(I_T; \boldsymbol{H}^1(\Omega))$. This and the last relation in (4.2.7) yields $\boldsymbol{\sigma}(u_{\varepsilon_k}) \to \boldsymbol{\sigma}(u)$ in $\boldsymbol{L}_2(Q_T)$ and $\boldsymbol{\kappa} = \boldsymbol{\sigma}(u)$. We can perform the limit procedure in (4.2.17) for any fixed $v \in \mathscr{C}$ and show that u solves (4.2.14). We have proved

4.2.3 Theorem. *Let the assumptions on Ω and conditions* (4.2.2), (4.2.3) *for both* V *and* W, (4.2.4), (4.2.5), (4.2.8) *and* (4.2.15) *be valid. If a Dirichlet boundary condition is considered, let* (4.2.16) *be true. Then there exists a weak solution to problem* (4.2.14) *if* $N = 2$ *and* $\alpha \in \left(0, \frac{1}{2}\right)$ *or* $N = 3$ *and* $\alpha \in \left(\frac{1}{3}, \frac{1}{2}\right)$.

4.2.4 Remark. If Γ, f, u_0, u_1 and possibly U are sufficiently regular, certain regularity results of u can be proved. There holds e.g.

$$u \in H^{1+\lambda, 1+\lambda/(1+\lambda-\alpha)}(Q_T)$$

for all $\lambda \in (0, \delta)$ with $\delta = \alpha/4$ for $N = 2$ and $\delta = \frac{3}{10}\left(\alpha - \frac{1}{3}\right)$ for $N = 3$. The result for the time variable was mentioned in (4.2.30). The space regularity will be obtained from (4.2.31) if we employ the localization and the translation (shift) method in arguments in the same way as in the previous chapter and consider the velocity as a part of the right hand side.

4.2.2 Boundary contact

We consider a similar viscoelastic material law as in the previous section,

$$\begin{aligned} \sigma_{ij}(\boldsymbol{u}) &= \sigma^I_{ij}(\boldsymbol{u}) + \sigma^M_{ij}(\boldsymbol{u}) \quad \text{with} \\ \sigma^I_{ij}(\boldsymbol{u}) &= \partial_{e_{ij}} W(\cdot, \varepsilon(\boldsymbol{u})) \quad \text{and} \\ \sigma^M_{ij}(\boldsymbol{u})(t, \cdot) &= \int_0^t \mathfrak{K}(t-s) \partial_{e_{ij}} V\left(\cdot, \varepsilon(\boldsymbol{u}(t, \cdot) - \boldsymbol{u}(s, \cdot))\right) ds. \end{aligned} \tag{4.2.35}$$

The viscoelastic body under consideration is in contact with a rigid obstacle. The linearized strain tensor ε has the components $e_{ij}(\boldsymbol{u}) = \frac{1}{2}\left(\partial_j u_i + \partial_i u_j\right)$ in terms of the displacement $\boldsymbol{u}$ and the stored energy functions $V, W : \mathbb{R}^{N+N^2} \to \mathbb{R}$ are of the class C^2 and have strongly elliptic and bounded partial Hessians, i.e. there are $\beta_0^Y, \beta_1^Y > 0$ such that

$$\begin{aligned} \beta_0^Y \xi_{ij}\xi_{ij} &\le \frac{\partial^2 Y}{\partial e_{ij} \partial e_{k\ell}}(\boldsymbol{x}, \boldsymbol{\omega})\, \xi_{ij}\xi_{k\ell} \quad \text{and} \\ \frac{\partial^2 Y}{\partial e_{ij} \partial e_{k\ell}}(\boldsymbol{x}, \boldsymbol{\omega}) \xi_{ij}\zeta_{k\ell} &\le \beta_1^Y \sqrt{\xi_{ij}\xi_{ij}} \sqrt{\zeta_{k\ell}\zeta_{k\ell}}, \ Y = V, W \end{aligned} \tag{4.2.36}$$

for all symmetric $N \times N$ matrices $\boldsymbol{\xi}, \boldsymbol{\zeta}, \boldsymbol{\omega}$ and all $\boldsymbol{x} \in \Omega$. Moreover, we assume that

$$V(\cdot, 0) = W(\cdot, 0) = 0 \text{ and } \partial_{e_{ij}} V(\cdot, 0) = \partial_{e_{ij}} W(\cdot, 0) = 0. \tag{4.2.37}$$

Similarly to the previous subsection there holds for $Y = V,\ W$:

$$
\begin{aligned}
Y(\cdot,\varepsilon(w)) &= \int_0^1 (1-\theta)\partial_{e_{ij}}\partial_{e_{k\ell}} Y(\cdot,\varepsilon(\theta \boldsymbol{w}))\, e_{ij}(\boldsymbol{w})\, e_{k\ell}(\boldsymbol{w})\, d\theta \\
&\in \left(\tfrac{1}{2}\beta_0^Y e_{ij}(\boldsymbol{w})\, e_{ij}(\boldsymbol{w}), \tfrac{1}{2}\beta_1^Y e_{ij}(\boldsymbol{w})\, e_{ij}(\boldsymbol{w})\right),
\end{aligned}
\tag{4.2.38}
$$

and

$$
\begin{aligned}
&\big(\partial_{e_{ij}} Y(\cdot,\varepsilon(\boldsymbol{v})) - \partial_{e_{ij}} Y(\cdot,\varepsilon(\boldsymbol{w}))\big) e_{ij}(\boldsymbol{v}-\boldsymbol{w}) \\
&\quad = \int_0^1 \partial_{e_{ij}}\partial_{e_{k\ell}} Y\big(\cdot,\varepsilon(\boldsymbol{w}+\theta(\boldsymbol{v}-\boldsymbol{w}))\big) e_{ij}(\boldsymbol{v}-\boldsymbol{w})\, e_{k\ell}(\boldsymbol{v}-\boldsymbol{w})\, d\theta \\
&\quad \ge \beta_0^Y e_{ij}(\boldsymbol{v}-\boldsymbol{w})\, e_{ij}(\boldsymbol{v}-\boldsymbol{w}), \\
&\big(\partial_{e_{ij}} Y(\cdot,\varepsilon(\boldsymbol{v})) - \partial_{e_{ij}} Y(\cdot,\varepsilon(\boldsymbol{w}))\big) e_{ij}(\tilde{\boldsymbol{w}}) \\
&\quad = \int_0^1 \partial_{e_{ij}}\partial_{e_{k\ell}} Y(\cdot,\varepsilon(\boldsymbol{w}+\theta(\boldsymbol{v}-\boldsymbol{w})))e_{ij}(\tilde{\boldsymbol{w}})\, e_{k\ell}(\boldsymbol{v}-\boldsymbol{w})\, d\theta \\
&\quad \le \beta_1^Y \sqrt{e_{ij}(\boldsymbol{v}-\boldsymbol{w})\, e_{ij}(\boldsymbol{v}-\boldsymbol{w})}\sqrt{e_{ij}(\tilde{\boldsymbol{w}})\, e_{ij}(\tilde{\boldsymbol{w}})}
\end{aligned}
\tag{4.2.39}
$$

for arbitrary displacements $\boldsymbol{v}$, $\boldsymbol{w}$ and $\tilde{\boldsymbol{w}}$ on Q_T. The kernel $\mathfrak{K}$ is assumed to have the form (4.2.5), where the employed functions satisfy all conditions required there. Moreover, the smallness of the memory (4.2.8) is assumed to ensure the strong monotonicity of the operator $\boldsymbol{\varepsilon} \to \boldsymbol{\sigma}$ formulated in the tensor coordinates (cf. (4.2.9)). We solve the problem

$$
\ddot{u}_i - \partial_j \sigma_{ij}(\boldsymbol{u}) = f_i,\ i = 1,\dots,N \quad \text{on } Q_T \equiv I_T \times \Omega, \tag{4.2.40}
$$

$$
\left.\begin{aligned}
u_n \le 0,\ \sigma_n(\boldsymbol{u}) \le 0,\ \sigma_n(\boldsymbol{u})\, u_n = 0,\\
\boldsymbol{\sigma_t}(\boldsymbol{u}) = 0
\end{aligned}\right\} \quad \text{on } S_{CT} \equiv I_T \times \Gamma_C, \tag{4.2.41}
$$

$$
\boldsymbol{\sigma}_\Gamma(\boldsymbol{u}) = \boldsymbol{\tau} \quad \text{on } S_{\boldsymbol{\tau}T} \equiv I_T \times \Gamma_{\boldsymbol{\tau}}, \tag{4.2.42}
$$

$$
\boldsymbol{u}(0,\cdot) = \boldsymbol{u}_0,\ \ \dot{\boldsymbol{u}}(0,\cdot) = \boldsymbol{u}_1 \quad \text{on } \Omega. \tag{4.2.43}
$$

Here, $\Omega \subset \mathbb{R}^N$ is a bounded domain occupied by the body with a $C^{1,1}$-smooth boundary Γ divided into two measurable and disjoint parts: a contact part Γ_C, where no friction occurs, and the remaining part $\Gamma_{\boldsymbol{\tau}}$ where the stress is prescribed as described above.

Let us denote

$$
\mathscr{C} := \{\boldsymbol{v} \in \boldsymbol{H}^1(\Omega);\ v_n \le 0 \text{ a.e. in } \Gamma_C\}. \tag{4.2.44}
$$

We introduce the variational formulation of the problem:
A weak solution of (4.2.40)–(4.2.43) *is a function* $\boldsymbol{u} \in B_0\big(I_T; \boldsymbol{H}^1(\Omega)\big)$ *such that* $\boldsymbol{u}(t,\cdot) \in \mathscr{C}$ *for a.e.* $t \in I_T$, $\dot{\boldsymbol{u}} \in L_\infty(I_T; \boldsymbol{L}_2(\Omega))$, $\dot{\boldsymbol{u}}(T,\cdot) \in \boldsymbol{L}_2(\Omega)$, *and*

such that for all $\boldsymbol{v} \in \boldsymbol{H}^1(Q_T)$ *with* $\boldsymbol{v}(t,\cdot) \in \mathscr{C}$ *a.e. in* I_T *the following inequality holds:*

$$\begin{aligned}\int_{Q_T} \big(\sigma_{ij}(\boldsymbol{u})e_{ij}(\boldsymbol{v}-\boldsymbol{u}) - \dot{\boldsymbol{u}}\cdot(\dot{\boldsymbol{v}}-\dot{\boldsymbol{u}})\big)\,d\boldsymbol{x}\,dt \\ + \int_{\Omega} \big(\dot{\boldsymbol{u}}\cdot(\boldsymbol{v}-\boldsymbol{u})\big)(T,\cdot)\,d\boldsymbol{x} \\ \geq \int_{\Omega} \boldsymbol{u}_1\cdot(\boldsymbol{v}(0,\cdot)-\boldsymbol{u}_0)\,d\boldsymbol{x} + \mathscr{L}(\boldsymbol{v}-\boldsymbol{u})\end{aligned} \tag{4.2.45}$$

with

$$\mathscr{L} : \boldsymbol{v} \mapsto \int_{Q_T} \boldsymbol{f}\cdot\boldsymbol{v}\,d\boldsymbol{x}\,dt + \int_{S_{\tau T}} \boldsymbol{\tau}\cdot\boldsymbol{v}\,ds_{\boldsymbol{x}}\,dt.$$

Problem (4.2.45) is solved with the help of the penalty method: *A function* $\boldsymbol{u}_\varepsilon$ *is a weak solution of the penalized problem, if* $\boldsymbol{u}_\varepsilon$ *belongs to* $B_0\big(I_T;\boldsymbol{H}^1(\Omega)\big)$, $\dot{\boldsymbol{u}}_\varepsilon \in B_0(I_T;\boldsymbol{L}_2(\Omega))$, *and* $\ddot{\boldsymbol{u}}_\varepsilon \in L_2\big(I_T;\mathring{\boldsymbol{H}}^{-1}(\Omega)\big)$, *the initial conditions in* (4.2.43) *are satisfied (in the sense both of Green's theorem and the limit from the right side), and the equation*

$$\begin{aligned}\int_{Q_T} \big(\ddot{\boldsymbol{u}}_\varepsilon\cdot\boldsymbol{v} + \sigma_{ij}(\boldsymbol{u}_\varepsilon)e_{ij}(\boldsymbol{v})\big)\,d\boldsymbol{x}\,dt \\ + \int_{S_{CT}} \tfrac{1}{\varepsilon}\,[(u_\varepsilon)_n]_+\,v_n\,ds_{\boldsymbol{x}}\,dt = \mathscr{L}(\boldsymbol{v})\end{aligned} \tag{4.2.46}$$

holds for all $\boldsymbol{v} \in L_2\big(I_T;\boldsymbol{H}^1(\Omega)\big)$.
The penalized problem is derived by replacing the Signorini boundary value condition on S_{CT} in (4.2.41) by the condition

$$\sigma_n(\boldsymbol{u}_\varepsilon) = -\tfrac{1}{\varepsilon}[(u_\varepsilon)_n]_+\,. \tag{4.2.47}$$

We solve our problems under the assumption

$$\begin{aligned}\boldsymbol{u}_0 \in \mathscr{C},\ \boldsymbol{u}_1 \in \boldsymbol{L}_2(\Omega),\ \boldsymbol{f} \in \boldsymbol{L}_2(Q_T) \cup H^{1-\alpha}\big(I_T;\mathring{\boldsymbol{H}}^{-1}(\Omega)\big) \\ \text{and } \boldsymbol{\tau} \in H^{1-\alpha}\big(I_T;\mathring{\boldsymbol{H}}^{1/2}(\Gamma_\tau)^*\big).\end{aligned} \tag{4.2.48}$$

Our aim is the proof of

4.2.5 Lemma. *Let the assumptions on* Ω *and its boundary, the assumptions* (4.2.35), (4.2.36), (4.2.37), (4.2.5), (4.2.8), *and* (4.2.48), *hold. Then there exists a solution to the penalized problem* (4.2.46).

Proof. The existence of a solution of problem (4.2.46) can be proved again by the Galerkin method. Denoting $S_t = (0,t) \times \Gamma$, $S_{Yt} = (0,t) \times \Gamma_Y$ for $Y = C, T$ and putting $v = \chi_t \dot{\boldsymbol{u}}_\varepsilon$ with χ_t from (4.2.18), we obtain

$$\begin{aligned}
&\int_\Omega \left(\tfrac{1}{2}|\dot{\boldsymbol{u}}_\varepsilon|^2 + W(\cdot, \boldsymbol{\varepsilon}(\boldsymbol{u}_\varepsilon))\right)(t,\cdot)\, d\boldsymbol{x} + \int_{\Gamma_C} \tfrac{1}{\varepsilon}|[(u_\varepsilon)_n]_+|^2(t,\cdot)\, ds_{\boldsymbol{x}} \\
&\quad - \int_{Q_t}\int_0^s \mathfrak{K}'(s-r) V(\cdot, \boldsymbol{\varepsilon}(\boldsymbol{u}_\varepsilon(s,\cdot)) - \boldsymbol{\varepsilon}(\boldsymbol{u}_\varepsilon(r,\cdot)))\, dr\, ds\, d\boldsymbol{x} \\
&\quad + \int_0^t \int_\Omega \mathfrak{K}(t-s) V(\cdot, \boldsymbol{\varepsilon}(\boldsymbol{u}_\varepsilon(t,\cdot)) - \boldsymbol{\varepsilon}(\boldsymbol{u}_\varepsilon(s,\cdot)))\, ds\, d\boldsymbol{x} \\
&= \int_\Omega \left(\tfrac{1}{2}|\boldsymbol{u}_1|^2 + W(\cdot, \boldsymbol{\varepsilon}(\boldsymbol{u}_0)\right)\, d\boldsymbol{x} - \int_{Q_t} \dot{\boldsymbol{f}} \cdot \boldsymbol{u}_\varepsilon\, d\boldsymbol{x}\, dt \\
&\quad + \int_\Omega \left(\boldsymbol{f}(t,\cdot)\cdot \boldsymbol{u}_\varepsilon(t,\cdot) - \boldsymbol{f}(0,\cdot)\cdot \boldsymbol{u}_0\right) ds_{\boldsymbol{x}} \\
&\quad - \int_{S_{\mathcal{T}t}} \dot{\boldsymbol{\tau}} \cdot \boldsymbol{u}_\varepsilon\, ds_{\boldsymbol{x}}\, ds + \int_{\Gamma_{\mathcal{T}}} \left(\boldsymbol{\tau}(t,\cdot)\cdot \boldsymbol{u}_\varepsilon(t,\cdot) - \boldsymbol{\tau}(0,\cdot)\cdot \boldsymbol{u}_0\right) ds_{\boldsymbol{x}}.
\end{aligned} \tag{4.2.49}$$

Here, we have used an integration by parts in the time variable for the "memory term" and for the terms with boundary and volume forces. The last one is redundant for $\boldsymbol{f} \in \boldsymbol{L}_2(Q_T)$. Using the expression $\boldsymbol{u}(t,\cdot) = \boldsymbol{u}_0 + \int_0^t \dot{\boldsymbol{u}}(s,\cdot)\, ds$ we infer from (4.2.38), (4.2.48), and (4.2.49)

$$\begin{aligned}
&\Big(\|\boldsymbol{u}_\varepsilon\|^2_{L_\infty(I_T;\boldsymbol{L}_2(\Omega))} + \|\dot{\boldsymbol{u}}_\varepsilon\|^2_{L_\infty(I_T;\boldsymbol{L}_2(\Omega))} \\
&\quad + \big\| |\boldsymbol{\varepsilon}(\boldsymbol{u}_\varepsilon)|^2 \big\|_{L_\infty(I_T;L_1(\Omega))} + \tfrac{1}{\varepsilon}\|[(u_\varepsilon)_n]_+\|^2_{L_\infty(I_T;L_2(\Omega))}\Big) \\
&\quad + \int_{Q_T}\int_0^T |t-s|^{-1-2\alpha} \left|\boldsymbol{\varepsilon}\left(\boldsymbol{u}_\varepsilon(t,\cdot) - \boldsymbol{u}_\varepsilon(s,\cdot)\right)\right|^2 d\boldsymbol{x}\, ds\, dt \\
&\le c\left(\beta_1^W, \beta_0^W, \beta_1^V, \beta_0^V\right) \mathscr{J}_0 \quad \text{with } |\boldsymbol{\varepsilon}|^2 = e_{ij}e_{ij} \text{ and} \\
&\mathscr{J}_0 = \|\boldsymbol{u}_0\|_{\boldsymbol{H}^1(\Omega)} + \|\boldsymbol{u}_1\|_{\boldsymbol{L}_2(\Omega)} + \|\boldsymbol{f}\|_{H^{1-\alpha}(I_T;\mathring{\boldsymbol{H}}^{-1}(\Omega))} \\
&\quad + \|\boldsymbol{\tau}\|_{H^{1-\alpha}(I_T;\boldsymbol{H}^{1/2}(\Gamma_{\mathcal{T}})^*)},
\end{aligned} \tag{4.2.50}$$

because the second "memory term" in (4.2.49) is non-negative. For the last terms in (4.2.49) the estimate by

$$c\|\dot{\boldsymbol{\tau}}\|_{H^\alpha(I_T;\boldsymbol{H}^{1/2}(\Gamma_{\mathcal{T}}))^*}\|\boldsymbol{u}_\varepsilon\|_{H^\alpha(I_T;\boldsymbol{H}^1(\Omega))} \text{ and}$$
$$c\|\boldsymbol{\tau}\|_{C^0(I_T;\boldsymbol{H}^1(\Omega))}\|\boldsymbol{u}_\varepsilon\|_{B_0(I_T;\boldsymbol{H}^{1/2}(\Gamma_{\mathcal{T}})^*)}$$

is used; it is based on the trace theorem and on the imbedding

$$H^\alpha\left(I_T; \boldsymbol{H}^{1/2}(\Gamma_{\mathcal{T}})^*\right) \hookrightarrow C^0\left(I_T; \boldsymbol{H}^{1/2}(\Gamma_{\mathcal{T}})^*\right).$$

Via the coercivity of the strains (Theorem 1.2.3) and the obvious imbedding $H^1 \hookrightarrow H^\alpha$ for $\alpha < 1$ we get from (4.2.50) the *a priori* estimate

$$\begin{aligned} &\|\dot{\boldsymbol{u}}_\varepsilon\|^2_{L_\infty(I_T;\boldsymbol{L}_2(\Omega))} + \|\nabla \boldsymbol{u}_\varepsilon\|^2_{L_\infty(I_T;L_2(\Omega;\mathbb{R}^{N^2}))} \\ &\quad + \tfrac{1}{\varepsilon}\|[(u_\varepsilon)_n]_+\|^2_{L_\infty(I_T;L_2(\Gamma_C))} + \|\nabla \boldsymbol{u}_\varepsilon\|^2_{H^\alpha(I_T;L_2(\Omega;\mathbb{R}^{N^2}))} \\ &\le c_0 \equiv c_0(\mathscr{J}) \quad \text{with} \\ &\mathscr{J} \equiv \Big(\beta_0^W, \beta_1^W, \beta_0^V, \beta_1^V, \|\boldsymbol{u}_0\|_{\boldsymbol{H}^1(\Omega)}, \|\boldsymbol{u}_1\|_{\boldsymbol{L}_2(\Omega)}, \|\boldsymbol{f}\|_X, \\ &\quad \|\boldsymbol{\tau}\|_{H^{1-\alpha}(I_T;\boldsymbol{H}^{1/2}(\Gamma_\tau)^*)}\Big), \end{aligned} \tag{4.2.51}$$

where the appropriate $X \in \Big\{\boldsymbol{L}_2(Q_T), H^{1-\alpha}\big(I_T; \mathring{\boldsymbol{H}}^{-1}(\Omega)\big)\Big\}$ is taken. The estimate of the fractional time derivative of the gradients follows easily from the last term on the left hand side in (4.2.50) which corresponds to the first "memory term" on the left hand side of (4.2.49). Moreover, in dynamic problems the semicoercive situation is prevented by the acceleration term via (1.2.12).

We employ the Galerkin method with an $\boldsymbol{L}_2(\Omega)$-orthogonal basis $\mathscr{Z} \equiv \{\boldsymbol{z}^{(i)};\, i \in \mathbb{N}\}$ of $\boldsymbol{H}^1(\Omega)$. By $\boldsymbol{u}_{\varepsilon,m}$ we denote a solution of the appropriate approximate problem to (4.2.46) for test functions from the space $X_m \equiv \Big\{\sum_{i=1}^{m} \boldsymbol{q}_i(t)\cdot \boldsymbol{z}^{(i)};\; \boldsymbol{q}_i \in \boldsymbol{L}_2(I_T)\Big\}$. The existence of such a solution follows easily from the theory of ordinary differential equations as in the previous subsection. The solutions $\boldsymbol{u}_{\varepsilon,m}$ satisfy for any $\varepsilon > 0$ and $m \in \mathbb{N}$ again the *a priori* estimate (4.2.51). As in the previous subsection, the ε-dependent *dual* estimate for $\ddot{u}_{\varepsilon,m}$ in $L_2\big(I_T; \mathring{\boldsymbol{H}}^{-1}(\Omega)\big)$ is derived and the estimate does not depend on $m \in \mathbb{N}$.

The appropriate variant of the convergences properties (4.2.24) for the sequence $\{u_{\varepsilon,m_k};\, k \in \mathbb{N}\}$ can be proved in the same manner as in the previous subsection. Moreover, analogous considerations to those made for the domain contact lead to an analogue to (4.2.25), to the strong convergence of gradients and to the closedness of the graph of $\varepsilon \mapsto \boldsymbol{\sigma}$ on the sets of solutions. The validity of the initial condition is proved also in the same manner as for the domain constraint. □

As in the previous subsection the solution of (4.2.46) is unique provided the constitutive relation is linear.

The limit procedure $\varepsilon \to 0$ is easier here. Taking a test function $\boldsymbol{v} \in L_2\big(I_T; \mathring{\boldsymbol{H}}^1(\Omega)\big)$ in (4.2.46) and using the ellipticity condition (4.2.36) and the uniform *a priori* estimate (4.2.51), we immediately get the dual estimate

$$\|\ddot{\boldsymbol{u}}_\varepsilon\|_{L_2(I_T;\boldsymbol{H}^{-1}(\Omega))} \le c_1 \equiv c_1(\mathscr{J}), \tag{4.2.52}$$

which is remarkably stronger than (4.2.28). This is exactly the point where the boundary constraint is easier to handle than the domain constraint because no additional estimate of the penalty term is needed. Theorem 2.6.10 from Chapter 2 yields for any $\varepsilon > 0$,

$$\begin{aligned} &\|\dot{\boldsymbol{u}}_\varepsilon\|_{H^{\alpha/2}(I_T;\boldsymbol{L}_2(\Omega))} \\ &\qquad \le const. \left(\|\boldsymbol{u}_\varepsilon\|_{H^\alpha(I_T;\boldsymbol{H}^1(\Omega))} + \|\ddot{\boldsymbol{u}}_\varepsilon\|_{L_2(I_T;\boldsymbol{H}^{-1}(\Omega))}\right), \\ &\|\dot{\boldsymbol{u}}_\varepsilon\|_{L_2(I_T;\boldsymbol{H}^{\alpha/(2-\alpha)}(\Omega))} \\ &\qquad \le const. \left(\|\boldsymbol{u}_\varepsilon\|_{H^\alpha(I_T;\boldsymbol{H}^1(\Omega))} + \|\ddot{\boldsymbol{u}}_\varepsilon\|_{L_2(I_T;\boldsymbol{H}^{-1}(\Omega))}\right). \end{aligned} \tag{4.2.53}$$

This result and the compact imbedding theorem (Theorem 2.8.14) yield that there is a sequence $\varepsilon_k \to 0$ and an element $\boldsymbol{u}$ such that

$$\begin{aligned} \boldsymbol{u}_{\varepsilon_k} &\rightharpoonup \boldsymbol{u} && \text{in } \boldsymbol{H}^{1+\alpha/2,\,1}(Q_T), \\ \ddot{\boldsymbol{u}}_{\varepsilon_k} &\rightharpoonup \ddot{\boldsymbol{u}} && \text{in } L_2(I_T;\boldsymbol{H}^{-1}(\Omega)), \\ \dot{\boldsymbol{u}}_{\varepsilon_k} &\to \dot{\boldsymbol{u}} && \text{in } \boldsymbol{H}^{\delta/2,\,\delta/(2-\delta)}(Q_T) \text{ for every } \delta < \alpha, \\ \dot{\boldsymbol{u}}_{\varepsilon_k}(t,\cdot) &\rightharpoonup \dot{\boldsymbol{u}}(t,\cdot) && \text{in } \boldsymbol{L}_2(\Omega) \text{ for } t = 0, T. \end{aligned} \tag{4.2.54}$$

These convergences lead to a relation similar to (4.2.34), which together with an appropriate version of (4.2.9) yields the strong convergence of the strain tensors. This, the just proved strong convergence in (4.2.54) and Theorem 1.2.3 from Chapter 1 prove the strong L_2 convergence of the space gradients. The strong convergence of the stress tensors is then a consequence of (4.2.39). Thus we have proved

4.2.6 Theorem. *Under the assumptions of Lemma 4.2.5 there exists a weak solution* $\boldsymbol{u} \in \boldsymbol{H}^{1+\alpha/2,1}(Q_T)$ *to the contact problem* (4.2.40) *such that* $\dot{\boldsymbol{u}}$ *belongs to* $\boldsymbol{H}^{\alpha/2,\,\alpha/(2-\alpha)}(Q_T)$.

4.2.7 Remarks. 1. From the space regularity of the velocity it is possible to derive $\boldsymbol{u} \in \boldsymbol{H}^{1+\alpha/2,\,1+\alpha/(2-\alpha)}(I_T \times \widetilde{\Omega})$ for any domain $\widetilde{\Omega}$ having its closure inside Ω. Such a result can be proved with help of the shift method, if the acceleration is considered like a right hand side of the problem. If the boundary Γ is sufficiently regular, then in the interiors of the separate parts of the boundary this technique yields the regularity of $\boldsymbol{u}$ in the tangential directions.

2. We have not considered a possible measurable part of the boundary Γ_U with a prescribed Dirichlet data $\boldsymbol{U}$ for the sake of simplicity only. If e.g. $\boldsymbol{U} \in \boldsymbol{H}^2(Q_T)$ with $\boldsymbol{U} = 0$ on Γ_C such that the usual compatibility conditions with the initial data are satisfied (cf. (4.4.15) below), the generalization of Theorem 4.2.6 to this case is an easy exercise. Unlike in the

case of the static problem, the existence or non-existence of the part of boundary with a Dirichlet boundary value condition is not relevant for the existence of solutions.

3. The $L_\infty(I_T; \boldsymbol{L}_2(\Omega))$-estimate for $\dot{\boldsymbol{u}}$ in (4.2.51) together with (4.2.52) leads to the weak continuity of the solution proved in Theorem 4.2.6 from $\overline{I_T}$ to $\boldsymbol{L}_2(\Omega)$. The proof is almost identical to that performed for the convergence of Galerkin solution to the solution of the penalized problem. The only remarkable difference which is in its first step, where we prove the weak convergence into $\boldsymbol{H}^{-1}(\Omega)$, does not make any difficulty because of the density of $\mathring{\boldsymbol{H}}^1(\Omega)$ in $\boldsymbol{L}_2(\Omega)$. In particular, for this and all further treated boundary contact problems the velocity is defined for every $t \in I_T$ as a function in $\boldsymbol{L}_2(\Omega)$. This is an essential difference from the cases of domain contacts, where at some time the left and right derivatives may pretty much differ on a subdomain of Ω with a non-zero measure.

4. The assumption about $\boldsymbol{f}$ required in (4.2.48) can be weakened. The result (4.2.53) shows that e.g. $\boldsymbol{f} \in L_2\big(I_T; \mathring{\boldsymbol{H}}^{-\alpha/(2-\alpha)}(\Omega)\big)$ is sufficient.

5. In order to include friction e.g. as in Section 4.4, the traces of the velocities should be at least in $\boldsymbol{L}_2(S_T)$ to justify the sense of $|\dot{\boldsymbol{u}}_{\boldsymbol{t}}|$ in the friction term. However, then we should prove $\dot{\boldsymbol{u}} \in L_2\big(I_T; \boldsymbol{H}^\gamma(\Omega)\big)$ for some $\gamma > \frac{1}{2}$ and by (4.2.53) this requires $\alpha > \frac{2}{3}$. Unfortunately, assumption (4.2.8) restricts α to $\big(0, \frac{1}{2}\big)$. Therefore we are not able to solve the frictional contact problem in the framework of the presented model.

4.3 Viscoelastic membranes

In this section we study the domain contact of a membrane consisting of material with short memory. We have again a bounded domain $\Omega \subset \mathbb{R}^N$ with a Lipschitz boundary Γ and a bounded time interval I_T. For the sake of simplicity we assume the linear constitutive relation

$$\sigma_i(u) = a_{ij}^{(1)} \partial_j \dot{u} + a_{ij}^{(0)} \partial_j u, \ i = 1, \dots, N, \tag{4.3.1}$$

although some non-linear relation like that in the previous section is also admissible to obtain the existence theorem below. The coefficients $a_{ij}^{(\iota)}$ are assumed to be symmetric (with respect to the lower indices) and to satisfy the usual ellipticity and boundedness conditions

$$\left.\begin{aligned} |\boldsymbol{\xi}|^2 a_0^{(\iota)} &\le a_{ij}^{(\iota)} \xi_i \xi_j, \\ a_{ij}^{(\iota)} \xi_i \zeta_j &\le a_1^{(\iota)} |\boldsymbol{\xi}|\,|\boldsymbol{\zeta}| \end{aligned}\right\} \text{ for all } \boldsymbol{\xi}, \boldsymbol{\zeta} \in \mathbb{R}^N, \ \iota = 0, 1, \tag{4.3.2}$$

almost everywhere on Ω with some given positive, real constants $a_k^{(\iota)}$, $\iota, k = 0, 1$. Our problem is

$$\left.\begin{array}{l}\ddot{u} = \operatorname{div} \boldsymbol{\sigma}(u) + g + f \\ u \geq 0, \quad f \geq 0, \quad uf = 0\end{array}\right\} \text{ on } Q_T, \tag{4.3.3}$$

$$u(0,\cdot) = u_0, \ \ \dot{u}(0,\cdot) - u_1 \geq 0, \ \ (\dot{u}(0,\cdot) - u_1)u_0 = 0 \text{ on } \Omega, \tag{4.3.4}$$

$$\boldsymbol{\sigma}(u) \cdot \boldsymbol{n} \geq 0, \quad u \geq 0, \quad u(\boldsymbol{\sigma}(u) \cdot \boldsymbol{n}) = 0 \text{ on } S_T, \tag{4.3.5}$$

where Q_T and S_T have the same meaning as in the previous section. Here, f plays again the role of a slack variable. The alternative Dirichlet boundary condition to (4.3.5) is

$$u = U \text{ on } S_T. \tag{4.3.6}$$

For the data we assume

$$\begin{array}{l} g \in L_2\big(I_T; \mathring{H}^{-1}(\Omega)\big), \ u_0 \in H^1(\Omega), \ u_0 \geq 0 \text{ a.e. in } \Omega, \\ u_1 \in L_2(\Omega). \end{array} \tag{4.3.7}$$

A weak solution of problem (4.3.3)–(4.3.5) *is a function*

$$u \in B_0\big(I_T; H^1(\Omega)\big) \cap \mathscr{C} \quad \text{with } \dot{u} \in L_\infty(I_T; L_2(\Omega)) \cap L_2\big(I_T; H^1(\Omega)\big)$$

and $\dot{u}(T,\cdot) \in L_2(\Omega)$ *that satisfies for every* $v \in \mathscr{C} \cap H^1(Q_T)$ *the following inequality:*

$$\begin{aligned} &\int_{Q_T} \big(\boldsymbol{\sigma}(u) \cdot \nabla(v-u) - \dot{u}(\dot{v}-\dot{u}) - g(v-u)\big)\, d\boldsymbol{x}\, dt \\ &\qquad + \int_{\Omega} \big(\dot{u}(v-u)(T,\cdot) - u_1(v(0,\cdot) - u_0)\big)\, d\boldsymbol{x} \geq 0. \end{aligned} \tag{4.3.8}$$

The cone $\mathscr{C}$ is defined in (4.2.13).

As earlier we solve problem (4.3.8) by the penalization of the contact condition. The penalized problem has the form:
Find $u_\varepsilon \in H^1(Q_T) \cap B_0\big(I_T; H^1(\Omega)\big)$ *such that*

$$\dot{u}_\varepsilon \in B_0(I_T; L_2(\Omega)) \cap L_2(I_T; H^1(\Omega)), \ddot{u} \in L_2(I_T; \mathring{H}^{-1}(\Omega)),$$

the initial conditions $u_\varepsilon(0,\cdot) = u_0$, $\dot{u}_\varepsilon(0,\cdot) = u_1$ *are satisfied and the equation*

$$\int_{Q_T} \Big(\boldsymbol{\sigma}(u_\varepsilon) \cdot \nabla v + \ddot{u}_\varepsilon v - gv + \tfrac{1}{\varepsilon}[u_\varepsilon]_- v\Big)\, d\boldsymbol{x}\, dt = 0 \tag{4.3.9}$$

holds for every $v \in L_2\big(I_T; H^1(\Omega)\big)$.

We again choose the velocity $\dot{u}_\varepsilon$ as a test function in (4.3.9). Recalling Theorem 1.2.3 we obtain the *a priori* estimate

$$\sup_{t\in I_T}\Big(\|\dot{u}_\varepsilon(t,\cdot)\|^2_{L_2(\Omega)}+\|u_\varepsilon(t,\cdot)\|^2_{H^1(\Omega)}+\tfrac{1}{\varepsilon}\|[u_\varepsilon]_-(t,\cdot)\|^2_{L_2(\Omega)}\Big)$$
$$+\|\nabla\dot{u}_\varepsilon\|^2_{L_2(I_T;\boldsymbol{L}_2(\Omega))}\le c_0\equiv c_0(\mathscr{I})\quad\text{with}$$
$$\mathscr{I}\equiv\Big[\|u_0\|_{H^1(\Omega)},\|u_1\|_{L_2(\Omega)},\|g\|_{L_2(I_T;\mathring{H}^{-1}(\Omega))}\Big] \tag{4.3.10}$$

in the same manner as in Subsection 4.2.1. The only difference here is the viscous term $\|\nabla\dot{u}_\varepsilon\|^2_{L_2(I_T;\boldsymbol{L}_2(\Omega))}$, which provides more regularity than the corresponding fractional time derivative in (4.2.20).

The penalized problem is solved again via the Galerkin approximation with the approximate spaces Z_m. Here, however, we have a complete parabolic situation for the velocities; the inviscid term is of a lower order. As L_2-orthogonal basis in $H^1(\Omega)$ e.g. a suitable set of eigenvectors of the problem $\partial_i\big(a^{(1)}_{ij}\partial_j u\big)=0$ can be taken. The *proof* of the existence of Galerkin solutions $u_{\varepsilon,m}$, $\varepsilon>0$, $m\in\mathbb{N}$, is based on the same results for systems of ordinary differential equations as in Subsection 4.2.1. The solutions $u_{\varepsilon,m}$ satisfy the estimate (4.3.10) again; its verification does not make any difference, and for a fixed $\varepsilon>0$ the penalty terms are bounded in $L_2(\Omega)$. Therefore, from the approximate Galerkin equation we can derive the dual estimate

$$\|\ddot{u}_{\varepsilon,m}\|_{L_2(I_T;\mathring{H}^{-1}(\Omega))}\le c\equiv c(\varepsilon), \tag{4.3.11}$$

where the constant on the right-hand side is independent of m. As earlier in the analogous situations, we have exploited here the orthogonality of the basis. From both estimates we can prove the validity of the following convergences for a fixed $\varepsilon>0$ and a suitable sequence $m_k\to+\infty$,

$$\begin{aligned}
\dot{u}_{\varepsilon,m_k}&\rightharpoonup\dot{u}_\varepsilon &&\text{in } H^{1/2,1}(Q_T),\\
\dot{u}_{\varepsilon,m_k}&\to\dot{u}_\varepsilon &&\text{in } L_2(Q_T),\\
\dot{u}_{\varepsilon,m_k}(T,\cdot)&\to\dot{u}_\varepsilon(T,\cdot) &&\text{in } L_2(\Omega),\\
\ddot{u}_{\varepsilon,m_k}&\rightharpoonup\ddot{u}_\varepsilon &&\text{in } L_2(I_T;\mathring{H}^{-1}(\Omega)),\\
[u_{\varepsilon,m_k}]_-&\to[u_\varepsilon]_- &&\text{in } L_2(Q_T),\\
\boldsymbol{\sigma}(u_{\varepsilon,m_k})&\to\boldsymbol{\sigma}(u_\varepsilon) &&\text{in } \boldsymbol{L}_2(Q_T),
\end{aligned} \tag{4.3.12}$$

and we can show that the limit u_ε is the unique solution to (4.3.9). In fact, the first convergence follows from the interpolation of (4.3.11) with the estimate of the viscous term in (4.3.10); the second one is then an easy consequence of a compact imbedding theorem (Theorem 2.8.14). The third and the fifth ones follow immediately from (4.3.10) while the fourth one from (4.3.11). The last one is obtained putting $v = \pi_{Z_{m_k}} \dot{u}_\varepsilon - \dot{u}_{\varepsilon,m_k}$ into the Galerkin approximation of (4.3.9), adding

$$\int_{Q_T} \boldsymbol{\sigma}(u_\varepsilon)(\dot{u}_{\varepsilon,m_k} - \dot{u}_\varepsilon) + \boldsymbol{\sigma}(u_{\varepsilon,m_k})(\pi_{Z_{m_k}} \dot{u}_\varepsilon - \dot{u}_{\varepsilon,m_k})\, dx\, dt$$

to both sides and using (4.3.2), the strong convergence of the projections and all the before proved convergences. This yields the strong convergence of the stresses. The validity of the initial condition is proved in the same way as in the case of the singular memory. Hence, u_ε solves (4.3.9). The linearity of the model makes the proof of its uniqueness quite standard and obvious.

For the limit procedure $\varepsilon \to 0$, we need again the estimate (4.2.26) of the penalty term, which can be proved here as in the previous section without any changes. This L_1-estimate and the appropriate dual version of Theorem 2.8.14 yield the dual estimate

$$\|\ddot{u}_\varepsilon\|_{\mathring{H}^{-1/2-\eta}(I_T;\mathring{H}^{-N/2-\eta}(\Omega)^*)} \le c(\mathscr{J}, \eta) \quad \text{for every } \eta > 0 \tag{4.3.13}$$

which is similar to (4.2.28) in the appropriate problem for material with long memory. Here we employ the dual imbedding theorem also for the time variable,

$$L_1\big(I_T; \mathring{H}^{-N/2-\eta}(\Omega)\big) \hookrightarrow \mathring{H}^{-1/2-\eta}\big(I_T; \mathring{H}^{-N/2-\eta}(\Omega)\big).$$

Interpolating (4.3.13) with the *a priori* estimate (4.3.10) for the viscous term, we arrive at the uniform estimate of $\|\dot{u}_\varepsilon\|_{H^\delta(I_T;L_2(\Omega))}$ for any $\delta \in (0, \frac{1}{N+2})$. This result allows proving the crucial strong convergence of the velocities for any dimension $N \in \mathbb{N}$, while for the long memory material only $N = 1, 2, 3$ was admissible. Of course, the case $N = 2$ is of physical importance. With this result we can find a sequence $\varepsilon_k \to 0$ such that

$$\begin{aligned}
\dot{u}_{\varepsilon_k} &\to \dot{u} && \text{in } H^{\delta,1-\eta}(Q_T), \\
u_{\varepsilon_k} &\rightharpoonup u && \text{in } H^1(Q_T), \\
\dot{u}_{\varepsilon_k} &\rightharpoonup \dot{u} && \text{in } L_2(I_T; H^1(\Omega)), \\
u_{\varepsilon_k}(T,\cdot) &\rightharpoonup u(T,\cdot) && \text{in } H^1(\Omega),
\end{aligned} \tag{4.3.14}$$

where $\eta > 0$ can be arbitrarily small. Considering $u - u_{\varepsilon_k}$ as a test function in (4.3.9) and adding

$$\int_{Q_T} a_{ij}^{(0)} \partial_i u\, \partial_j (u_{\varepsilon_k} - u)\, dx\, dt + \int_{Q_T} a_{ij}^{(1)} \partial_i \dot{u}\, \partial_j (u_{\varepsilon_k} - u)\, dx\, dt$$

to both sides of the resulting equation, we prove that the convergences $\nabla u_{\varepsilon_k} \to \nabla u$ and $\nabla u_{\varepsilon_k}(T,\cdot) \to \nabla u(T,\cdot)$ are in fact strong in the corresponding L_2-spaces. As the time interval I_T in the above mentioned considerations can be replaced by I_t, $t \in I_T$, we can also prove $u_{\varepsilon_k}(t,\cdot) \to u(t,\cdot)$ in $H^1(\Omega)$ for any $t \in I_T$.

On the other hand, there is no idea how to prove any strong L_2-convergence $\nabla \dot{u}_{\varepsilon_k} \to \nabla \dot{u}$ in general (unlike the boundary contact, where in the case $N = 1$ it is a byproduct of the result in [121], cf. Subsection 4.4.1.2). Such a relation together with the strong L_2 convergence $\dot{u}_{\varepsilon_k}(t,\cdot) \to \dot{u}(t,\cdot)$ for any $t \in I_T$ is necessary and sufficient to prove that the limit u, which is a solution of (4.3.8), satisfies the *global conservation of energy*:

$$\begin{aligned}\int_\Omega \tfrac{1}{2}\left(a_{ij}^{(0)}\partial_i u(t,\cdot)\,\partial_j u(t,\cdot) + \dot{u}^2(t,\cdot)\right) d\boldsymbol{x} \\ + \int_{Q_t} a_{ij}^{(1)}\partial_i \dot{u}\,\partial_j \dot{u}\, d\boldsymbol{x}\, ds \\ = \int_\Omega \tfrac{1}{2}\left(a_{ij}^{(0)}\partial_i u_0\,\partial_j u_0 + u_1^2\right) d\boldsymbol{x} + \int_{Q_t} g\dot{u}\, d\boldsymbol{x}\, ds\end{aligned} \tag{4.3.15}$$

for any $t \in I_T$. However, we have proved

4.3.1 Theorem. *Under the assumptions on Ω and the assumptions* (4.3.1), (4.3.2), (4.3.7) *there is at least one solution to problem* (4.3.8).

4.3.2 Remarks. 1. The corresponding version of Theorem 4.3.1 for the problem with Dirichlet boundary value condition (4.3.6) is valid like in the previous section. The requirements for U from (4.2.16) are preserved, but we may assume here that $U_0 = 0$. Contrary to material with singular memory, for the presented viscoelastic constitutive law we need not assume that the membrane is clamped in a distance from the obstacle. Indeed, the advantageous sign of $\dot{U}$ makes it possible to derive (in the same way as in Subsection 4.2.1) a modified version of (4.3.10) which includes the term $\|\nabla \dot{u}_\varepsilon\|^2_{L_2(Q_T)}$ on the left hand side. This estimate yields that for the sets Ω_k and M_k, $k \in \mathbb{N}$, defined as $\Omega_k \equiv \{\boldsymbol{x} \in \Omega; \operatorname{dist}(\boldsymbol{x}, \mathbb{R} \setminus \Omega) \ge 1/k\}$ and $M_k \equiv \Omega \setminus \Omega_k$ the $L_2(I \times M_k)$ norms of the velocities tend to 0 as $k \to +\infty$ uniformly with respect to both ε and m. Hence, to derive the necessary additional estimate of the penalty term we can use a set of non-negative C^1-smooth functions $\{\varrho_k; k \in \mathbb{N}\}$ defined on $\mathbb{R}^N$, supported in the closure of Ω and being equal to 1 on Ω_k. With them as test functions we can prove the boundedness of $\frac{1}{\varepsilon}\varrho_k[u_{\varepsilon,m}]_-$ uniformly with respect to m and of $\frac{1}{\varepsilon}\varrho_k u_\varepsilon$ uniformly with respect to ε, although such a uniformity with respect to k does not hold. We show the appropriate *a priori* and dual estimates for $\varrho_k u_{\varepsilon,m}$ and $\varrho_k u_\varepsilon$. Then we can prove the strong $L_2(Q_T)$ convergence of $\varrho_k \dot{u}_{\varepsilon,m_r}$ to $\varrho_k \dot{u}_\varepsilon$ for a suitable subsequence $m_r \to +\infty$ for any fixed k, possibly after extracting such a subsequence with the help of the usual diagonal method. An appropriate combination of this with the above mentioned uniform limit property on M_k concludes the crucial strong convergence $\dot{u}_{\varepsilon,m_r} \to \dot{u}_\varepsilon$ in $L_2(Q_T)$. This procedure can be employed in the next limit procedure for $\varepsilon \to 0$, too.

2. For Ω with boundary of the class $C^{1,1}$, for $f \in L_2(Q_T)$, $a_{ij}^{(0)} \in C^1(\Omega)$,

$$a_{ij} \equiv a_{ij}^{(0)} = a_{ij}^{(1)},\ i,j = 1,\dots,N \text{ on } \Omega, \tag{4.3.16}$$

and $u_0 \in H^2(\Omega)$ a certain space regularity of the solution can be proved. In fact, taking the test function $v = a_{ij}\partial_i\partial_j u_\varepsilon$ in (4.3.9) it can be proved after some calculation employing Green's theorem, (4.3.2), and (4.3.10)

$$\int_{Q_t} \big(a_0|\nabla \dot{u}_\varepsilon|^2 + a_{ij}\partial_i\partial_j\dot{u}_\varepsilon\ a_{ij}\partial_i\partial_j u_\varepsilon + a_{ij}\partial_i\partial_j u_\varepsilon\ a_{ij}\partial_i\partial_j u_\varepsilon \\ + \varepsilon^{-1} a_0|\nabla[u_\varepsilon]_-|^2\big)\, ds\, d\boldsymbol{x} \le c$$

with c independent of ε and $a_0 \equiv a_0^{(0)} = a_0^{(1)}$ introduced in (4.3.2). Hence

$$a_{ij}\partial_i\partial_j u \in L_\infty(I_T; L_2(\Omega)). \tag{4.3.17}$$

The standard localization and the translation (shift) method proves that $\nabla u \in L_\infty\big(I_T;\ \boldsymbol{H}^{1-\eta}(\widetilde{\Omega})\big)$ for any $\widetilde{\Omega} \subset \overline{\widetilde{\Omega}} \subset \Omega$ and any $\eta \in (0,1)$ without the assumption (4.3.16). Along the boundary such a result is proved only for the tangential directions. With (4.3.16) this regularity can be proved also for the normal direction using (4.3.17).

3. A non-linearity in the inviscid part of the constitutive relation as in Subsection 4.2.1 is possible. On the other hand, due to the mentioned difficulties with the strong convergence of the gradients of velocities in the limit procedure $\varepsilon \to 0$, the existence of solutions to the problem with non-linearities in the viscous part is still an open problem.

4.4 Problems with given friction

Comparing the just presented contact problems with domain obstacles for material with long and short memory, we see that the latter model leads to remarkably stronger results. For boundary contact this model allows including also a given friction into the formulation of the problem. The resulting problem is investigated here. At the end of this section, we also mention the appropriate analogous problem for the contact condition formulated in velocities.

4.4.1 Contact condition in displacements

We assume again a bounded domain $\Omega \subset \mathbb{R}^N$ having a boundary Γ of the class $C^{1,1}$ which is composed of three measurable pairwise disjoint parts Γ_C (the contact part), $\Gamma_{\boldsymbol{\tau}}$, and $\Gamma_{\boldsymbol{U}}$. The investigated model is

$$\ddot{u}_i - \partial_j \sigma_{ij}(\boldsymbol{u}) = f_i \quad \text{on } Q_T,\ i = 1, \ldots, N, \tag{4.4.1}$$

$$\boldsymbol{u} = \boldsymbol{U} \quad \text{on } S_{\boldsymbol{U}T} = I_T \times \Gamma_{\boldsymbol{U}}, \tag{4.4.2}$$

$$\boldsymbol{\sigma}_\Gamma(\boldsymbol{u}) = \boldsymbol{\tau} \quad \text{on } S_{\boldsymbol{\tau}T} = I_T \times \Gamma_{\boldsymbol{\tau}}, \tag{4.4.3}$$

$$u_n \le 0, \quad \sigma_n \le 0, \quad \sigma_n u_n = 0 \text{ on } S_{CT} = I_T \times \Gamma_{CT}, \tag{4.4.4}$$

$$\left.\begin{aligned} \dot{\boldsymbol{u}}_{\boldsymbol{t}} = 0 \quad &\Rightarrow \quad |\boldsymbol{\sigma}_{\boldsymbol{t}}| \le G \\ \dot{\boldsymbol{u}}_{\boldsymbol{t}} \ne 0 \quad &\Rightarrow \quad \boldsymbol{\sigma}_{\boldsymbol{t}} = -G\frac{\dot{\boldsymbol{u}}_{\boldsymbol{t}}}{|\dot{\boldsymbol{u}}_{\boldsymbol{t}}|} \end{aligned}\right\} \text{ on } S_{CT}, \tag{4.4.5}$$

$$\boldsymbol{u}(0,\boldsymbol{x}) = \boldsymbol{u}_0(\boldsymbol{x}), \quad \dot{\boldsymbol{u}}(0,\boldsymbol{x}) = \boldsymbol{u}_1(\boldsymbol{x}) \text{ on } \Omega. \tag{4.4.6}$$

The stress tensor $\boldsymbol{\sigma} \equiv \{\sigma_{ij};\ i,j = 1, \ldots, N\}$ is given by

$$\begin{aligned} &\sigma_{ij}(\boldsymbol{u}) = \sigma^I_{ij}(\boldsymbol{u}) + \sigma^V_{ij}(\dot{\boldsymbol{u}}),\ i,j = 1, \ldots, N, \quad \text{where} \\ &\sigma^I_{ij}(\boldsymbol{u}) = \partial_{e_{ij}} W(\cdot, \varepsilon(\boldsymbol{u})),\ i,j = 1, \ldots, N, \quad \text{and} \\ &\boldsymbol{\sigma}^V \equiv \{\sigma^V_{ij}(\dot{u})\} = A\varepsilon(\dot{\boldsymbol{u}}). \end{aligned} \tag{4.4.7}$$

We consider the linear operator $A : \varepsilon(\boldsymbol{u}) \mapsto a_{ijk\ell} e_{k\ell}(\boldsymbol{u})$ with coefficients satisfying

$$\begin{aligned} &|\boldsymbol{\xi}|^{-2} a_{ijk\ell}(\boldsymbol{x}) \xi_{ij} \xi_{k\ell} \in (a_0, a_1) \text{ for all } \boldsymbol{x} \in \Omega \\ &\qquad \text{and all symmetric } \boldsymbol{\xi} \in \mathbb{R}^{N^2} \end{aligned} \tag{4.4.8}$$

with some positive constants a_i, $i = 0, 1$, independent of $\boldsymbol{x}$ and $\boldsymbol{\xi}$. The viscous Hooke tensor satisfies the usual symmetries $a_{ijk\ell} = a_{jik\ell} = a_{k\ell ij}$ for $i,j,k,\ell = 1, \ldots, N$ on Ω. The space-dependent stored energy function $W : \mathbb{R}^{N+N^2} \to \mathbb{R}$ is assumed to satisfy conditions (4.2.36) and (4.2.37). The function G is non-negative; $-G$ plays the role of a given friction force.

For $\boldsymbol{w} \in \boldsymbol{H}^{1/2}(\Gamma)$ let us define the sets

$$\mathscr{C}_{\boldsymbol{w}} := \left\{\boldsymbol{v} \in \boldsymbol{H}^1(\Omega);\ \boldsymbol{v} = \boldsymbol{w} \text{ on } \Gamma_{\boldsymbol{U}},\ v_n \le 0 \text{ a.e. in } \Gamma_C\right\}$$

and $\boldsymbol{H}^1_0(\Omega) = \left\{\boldsymbol{w} \in \boldsymbol{H}^1(\Omega);\ \boldsymbol{w} = 0 \text{ a.e. on } \Gamma_{\boldsymbol{U}}\right\}$.

We introduce the variational formulation of the problem:
A weak solution of (4.4.1)–(4.4.6) is a function $\boldsymbol{u} \in B_0(I_T;\boldsymbol{H}^1(\Omega))$ *with* $\boldsymbol{u}(t,\cdot) \in \mathscr{C}_{\boldsymbol{U}(t,\cdot)}$ *for a.e.* $t \in I_T$, $\dot{\boldsymbol{u}} \in L_\infty(I_T;\boldsymbol{L}_2(\Omega)) \cap L_2(I_T;\boldsymbol{H}^1(\Omega))$, $\dot{\boldsymbol{u}}(T,\cdot) \in \boldsymbol{L}_2(\Omega)$ *such that for all* $\boldsymbol{v} \in \boldsymbol{H}^1(Q_T)$ *with* $\boldsymbol{v}(t,\cdot) \in \mathscr{C}_{\boldsymbol{U}(t,\cdot)}$ *a.e. in* I_T *the following inequality holds:*

$$\begin{aligned}
&\int_{Q_T} \left(\sigma_{ij}(\boldsymbol{u})e_{ij}(\boldsymbol{v}-\boldsymbol{u}) - \dot{\boldsymbol{u}}\cdot(\dot{\boldsymbol{v}}-\dot{\boldsymbol{u}})\right) d\boldsymbol{x}\, dt \\
&\quad + \int_{S_{CT}} G\left(|\boldsymbol{v}_{\boldsymbol{t}} + \dot{\boldsymbol{u}}_{\boldsymbol{t}} - \boldsymbol{u}_{\boldsymbol{t}}| - |\dot{\boldsymbol{u}}_{\boldsymbol{t}}|\right) ds_{\boldsymbol{x}}\, dt \\
&\quad + \int_\Omega \left(\dot{\boldsymbol{u}}\cdot(\boldsymbol{v}-\boldsymbol{u})\right)(T,\cdot)\, d\boldsymbol{x} \\
&\geq \mathscr{L}(\boldsymbol{v}-\boldsymbol{u}) + \int_\Omega \boldsymbol{u}_1\cdot(\boldsymbol{v}(0,\cdot)-\boldsymbol{u}_0)\, d\boldsymbol{x}.
\end{aligned} \tag{4.4.9}$$

Here

$$\mathscr{L} : \boldsymbol{v} \mapsto \int_{Q_T} \boldsymbol{f}\cdot\boldsymbol{v}\, d\boldsymbol{x}\, dt + \int_{S_{\boldsymbol{\tau}T}} \boldsymbol{\tau}\cdot\boldsymbol{v}\, ds_{\boldsymbol{x}}\, dt \tag{4.4.10}$$

denotes the linear functional of the given forces. Inequality (4.4.9) follows from (4.4.1) by multiplying the equilibrium of forces by $\boldsymbol{v}-\boldsymbol{u}$, by integrating the result over Q_T, using Green's theorem both in the time and space variables and employing the boundary value conditions (4.4.2)–(4.4.5) and the initial conditions (4.4.6). For the treatment of the friction term cf. the previous chapter.

4.4.1.1 Existence of solutions

Due to the non-smooth friction term we need to both penalize the contact condition and smooth the friction term in order to obtain a variational equation. The penalization consists in replacing the Signorini boundary value condition (4.4.4) by the condition

$$\sigma_n(\boldsymbol{u}_\varepsilon) = -\tfrac{1}{\varepsilon}\left[(u_\varepsilon)_n\right]_+ . \tag{4.4.11}$$

The variational formulation of the penalized problem is:
Find $\boldsymbol{u}_\varepsilon \in \boldsymbol{U} + B_0(I_T;\boldsymbol{H}_0^1(\Omega))$ *with* $\dot{\boldsymbol{u}}_\varepsilon \in L_\infty(I_T;\boldsymbol{L}_2(\Omega)) \cap L_2(I_T;\boldsymbol{H}^1(\Omega))$ *and* $\ddot{\boldsymbol{u}}_\varepsilon \in \boldsymbol{L}_2(I_T;\boldsymbol{H}_0^1(\Omega))^*$ *such that for all* $\boldsymbol{v} \in \dot{\boldsymbol{U}} + L_2(I_T;\boldsymbol{H}^1(\Omega))$ *the*

following inequality holds:

$$\begin{aligned}
&\int_{Q_T} (\ddot{\boldsymbol{u}}_\varepsilon \cdot (\boldsymbol{v} - \dot{\boldsymbol{u}}_\varepsilon) + \sigma_{ij}(\boldsymbol{u}_\varepsilon)e_{ij}(\boldsymbol{v} - \dot{\boldsymbol{u}}_\varepsilon))\, d\boldsymbol{x}\, dt \\
&\quad + \int_{S_{CT}} \Big(G\,(|\boldsymbol{v_t}| - |\dot{\boldsymbol{u}}_{\boldsymbol{t}}|) + \tfrac{1}{\varepsilon}[(u_\varepsilon)_n]_+(v_n - (\dot{\boldsymbol{u}}_\varepsilon)_n)\Big)\, ds_{\boldsymbol{x}}\, dt \\
&\quad \geq \mathscr{L}(\boldsymbol{v} - \dot{\boldsymbol{u}}_\varepsilon).
\end{aligned} \tag{4.4.12}$$

We remark that the test function $\boldsymbol{v}$ here corresponds with the test function $\boldsymbol{v} + \dot{\boldsymbol{u}} - \boldsymbol{u}$ in (4.4.9).

The smoothing of the friction functional is done by replacing in (4.4.12) the non-differentiable norms $|\cdot| \equiv \Phi_0$ in the friction term by a smooth and convex approximation Φ_η introduced in Subsection 3.1.1. The resulting inequality will be referred to as (4.4.12′). It is equivalent to the following variational equation:
Find a function $\boldsymbol{u}_{\varepsilon,\eta} \in \boldsymbol{U} + B_0\big(I_T; \boldsymbol{H}_0^1(\Omega)\big)$ *with* $\dot{\boldsymbol{u}}_{\varepsilon,\eta} \in B_0(I_T; \boldsymbol{L}_2(\Omega)) \cap L_2\big(I_T; \boldsymbol{H}^1(\Omega)\big)$ *and* $\ddot{\boldsymbol{u}}_{\varepsilon,\eta} \in L_2\big(I_T; \boldsymbol{H}_0^1(\Omega)^*\big)$ *such that the initial condition* (4.4.6) *is satisfied and the equation*

$$\begin{aligned}
&\int_{Q_T} (\ddot{\boldsymbol{u}}_{\varepsilon,\eta} \cdot \boldsymbol{v} + \sigma_{ij}(\boldsymbol{u}_{\varepsilon,\eta})e_{ij}(\boldsymbol{v}))\, d\boldsymbol{x}\, dt \\
&\quad + \int_{S_{CT}} \big(G\,(\nabla\Phi_\eta)((\dot{\boldsymbol{u}}_{\varepsilon,\eta})_{\boldsymbol{t}}) \cdot \boldsymbol{v_t} + \tfrac{1}{\varepsilon}\,[(u_{\varepsilon,\eta})_n]_+\, v_n\big) ds_{\boldsymbol{x}}\, dt \\
&\quad = \mathscr{L}(\boldsymbol{v})
\end{aligned} \tag{4.4.13}$$

holds for any $\boldsymbol{v} \in L_2\big(I_T; \boldsymbol{H}_0^1(\Omega)\big)$.
Indeed, the mentioned equivalence can be proved inserting $\boldsymbol{v} = \boldsymbol{w} - \dot{\boldsymbol{u}}_{\varepsilon,\eta}$ with an arbitrary $\boldsymbol{w} \in \dot{\boldsymbol{U}} + L_2\big(I_T; \boldsymbol{H}_0^1(\Omega)\big)$ into (4.4.13) and on applying the inequality

$$\begin{aligned}
&\int_{S_{CT}} G\,(\nabla\Phi_\eta)\big((\dot{\boldsymbol{u}}_{\varepsilon,\eta})_{\boldsymbol{t}}\big) \cdot (\boldsymbol{w_t} - (\dot{\boldsymbol{u}}_{\varepsilon,\eta})_{\boldsymbol{t}})\, ds_{\boldsymbol{x}}\, dt \\
&\quad \leq \int_{S_{CT}} G\,(\Phi_\eta(\boldsymbol{w_t}) - \Phi_\eta((\dot{\boldsymbol{u}}_{\varepsilon,\eta})_{\boldsymbol{t}}))\, ds_{\boldsymbol{x}}\, dt.
\end{aligned} \tag{4.4.14}$$

This inequality holds due to the convexity of Φ_η and the non-negativity of G for any $\eta, \varepsilon > 0$, $\boldsymbol{w} \in L_2\big(I_T; \boldsymbol{H}^{1/2}(\Gamma)\big)$, provided $G \in L_2\big(I_T; H^{1/2}(\Gamma_C)^*\big)$. The transition from the inequality (4.4.12′) (the smooothed version of (4.4.12)) to the equation (4.4.13) is easily proved by putting $\boldsymbol{v} = \dot{\boldsymbol{u}}_\varepsilon \pm \lambda\boldsymbol{w}$ with an arbitrary $\boldsymbol{w} \in L_2\big(I_T; \boldsymbol{H}_0^1(\Omega)\big)$ and $\lambda \in \mathbb{R}_+$, by dividing the whole inequality by λ and passing to the limit $\lambda \to 0$.

The introduced problem is solved under the following set of assumptions:

$$\begin{aligned}&\boldsymbol{u}_0 \in \mathscr{C}_{\boldsymbol{U}(0,\cdot)},\ \boldsymbol{u}_1 \in \boldsymbol{H}^1(\Omega),\ \boldsymbol{U} \in \boldsymbol{H}^2(Q_T) \text{ such that}\\ &\boldsymbol{U}(0,\cdot)_{|\Gamma_U} = \boldsymbol{u}_{0|\Gamma_U}, \dot{\boldsymbol{U}}(0,\cdot)_{|\Gamma_U} = \boldsymbol{u}_{1|\Gamma_U} \text{ and}\\ &\boldsymbol{U} = 0 \text{ a.e. in } S_{CT},\ \boldsymbol{\tau} \in L_2\big(I_T; \boldsymbol{H}^{1/2}(\Gamma_{\boldsymbol{\tau}})^*\big),\\ &\boldsymbol{f} \in L_2\big(I_T; \mathring{\boldsymbol{H}}^{-1}(\Omega)\big), \text{ and } 0 \le G \in L_2\big(I_T; \boldsymbol{H}^{1/2}(\Gamma_C)^*\big).\end{aligned} \tag{4.4.15}$$

The sign of G is understood in the usual dual sense. The assumptions can be a little weakened like it is done in the next chapter for problems with Coulomb friction.

To solve problem (4.4.13) we use the standard Galerkin approximation. We employ the usual notation $Q_t \equiv I_t \times \Omega$ and $S_{Yt} \equiv I_t \times \Gamma_Y$ for $t \in I_T$ and $Y \in \{C, \boldsymbol{\tau}, \boldsymbol{U}\}$. Taking $\boldsymbol{v} = \chi_t(\dot{\boldsymbol{u}}_{\varepsilon,\eta} - \dot{\boldsymbol{U}})$ for χ_t from (4.2.18) and any $t \in I_T$ and exploiting (4.4.14) we obtain

$$\begin{aligned}&\int_\Omega \big(\tfrac{1}{2}|\dot{\boldsymbol{u}}_{\varepsilon,\eta}|^2 + W(\cdot, \boldsymbol{\varepsilon}(\boldsymbol{u}_{\varepsilon,\eta}))\big)(t,\cdot)\, d\boldsymbol{x}\\ &\quad + \int_{Q_t} A(\boldsymbol{\varepsilon}(\dot{\boldsymbol{u}}_{\varepsilon,\eta}))\boldsymbol{\varepsilon}(\dot{\boldsymbol{u}}_{\varepsilon,\eta})\, d\boldsymbol{x}\, ds\\ &\quad + \int_{\Gamma_C} \tfrac{1}{\varepsilon} \left|[(u_{\varepsilon,\eta})_n]_+\right|^2 (t,\cdot)\, ds_{\boldsymbol{x}}\\ &\le \int_\Omega \Big(\tfrac{1}{2}|\boldsymbol{u}_1|^2 + W(\cdot, \boldsymbol{\varepsilon}(\boldsymbol{u}_0)) - \boldsymbol{u}_1\dot{\boldsymbol{U}}(0,\cdot) + (\dot{\boldsymbol{u}}_{\varepsilon,\eta}\dot{\boldsymbol{U}})(t,\cdot)\Big)\, d\boldsymbol{x}\\ &\quad + \int_{Q_t} \Big(A(\boldsymbol{\varepsilon}(\dot{\boldsymbol{u}}_{\varepsilon,\eta}))\boldsymbol{\varepsilon}(\dot{\boldsymbol{U}}) + \sigma^I_{ij}(\boldsymbol{u})e_{ij}(\dot{\boldsymbol{U}})\\ &\qquad\qquad - \dot{\boldsymbol{u}}_{\varepsilon,\eta}\ddot{\boldsymbol{U}} + \boldsymbol{f}\cdot(\dot{\boldsymbol{u}}_{\varepsilon,\eta} - \dot{\boldsymbol{U}})\Big)\, d\boldsymbol{x}\, ds\\ &\quad + \int_{S_{C,t}} G\Phi_\eta(\dot{\boldsymbol{U}}_{\boldsymbol{t}})\, ds_{\boldsymbol{x}}\, ds + \int_{S_{\boldsymbol{\tau},t}} \boldsymbol{\tau}\cdot(\dot{\boldsymbol{u}}_{\varepsilon,\eta} - \dot{\boldsymbol{U}})\, ds_{\boldsymbol{x}}\, ds.\end{aligned} \tag{4.4.16}$$

Here the relations (4.4.14) with $\boldsymbol{w} = \dot{\boldsymbol{U}}$, $\int_{S_{C,t}} G\,\Phi_\eta((\dot{\boldsymbol{u}}_{\varepsilon,\eta})_{\boldsymbol{t}})\, ds_{\boldsymbol{x}}\, ds \ge 0$, and $\frac{\partial}{\partial t}W(\cdot, \boldsymbol{\varepsilon}(\boldsymbol{u}_{\varepsilon,\eta})) = \partial_{e_{ij}}W(\cdot, \boldsymbol{\varepsilon}(\boldsymbol{u}_{\varepsilon,\eta}))\boldsymbol{\varepsilon}(\dot{\boldsymbol{u}}_{\varepsilon,\eta})$ have been used. With the help of relations (4.2.39), (4.4.8), and Theorem 1.2.3 it is not difficult to derive from (4.4.16) the *a priori* estimate

$$\begin{aligned}&\sup_{t\in I_T} \Big(\|\dot{\boldsymbol{u}}_{\varepsilon,\eta}(t,\cdot)\|^2_{\boldsymbol{L}_2(\Omega)} + \tfrac{1}{\varepsilon}\|[(u_{\varepsilon,\eta})_n]_+(t,\cdot)\|^2_{L_2(\Gamma_C)}\Big)\\ &\qquad + \|\dot{\boldsymbol{u}}_{\varepsilon,\eta}\|^2_{L_2(Q_T;\boldsymbol{H}^1(\Omega))} \le c_0 \equiv c_0(\mathscr{J}) \quad \text{with}\\ &\mathscr{J} \equiv \big(\beta^W_0, \beta^W_1, a_0, a_1, \|\boldsymbol{u}_0\|_{\boldsymbol{H}^1(\Omega)}, \|\boldsymbol{u}_1\|_{\boldsymbol{L}_2(\Omega)}, \|\boldsymbol{U}\|_{\boldsymbol{H}^2(Q_T)},\\ &\|G\|_{L_2(I_T;H^{1/2}(\Gamma_C)^*)}, \|\boldsymbol{f}\|_{L_2(I_T;\mathring{\boldsymbol{H}}^{-1}(\Omega))}, \|\boldsymbol{\tau}\|_{L_2(I_T;\boldsymbol{H}^{1/2}(\Gamma)^*)}\big).\end{aligned} \tag{4.4.17}$$

In order to derive a suitable dual estimate consider an arbitrary function $\boldsymbol{w} \in L_2(I_T; \mathring{\boldsymbol{H}}^1(\Omega))$ in (4.4.13). After standard estimates this yields

$$\begin{aligned} \|\ddot{\boldsymbol{u}}_{\varepsilon,\eta}\|^2_{L_2(I_T;\boldsymbol{H}^{-1}(\Omega))} \leq c_1 \|\nabla \boldsymbol{u}_{\varepsilon,\eta}\|^2_{L_2(Q_T;\mathbb{R}^{N^2})} \\ + c_2 \|\nabla \dot{\boldsymbol{u}}_{\varepsilon,\eta}\|^2_{L_2(Q_T;\mathbb{R}^{N^2})} + c_3 \|\boldsymbol{f}\|^2_{L_2(I_T;\mathring{\boldsymbol{H}}^{-1}(\Omega))} \end{aligned} \tag{4.4.18}$$

with c_1, c_2, c_3 independent of both ε, η and any boundary data. Now, we apply the interpolation theorem (Theorem 2.6.10) to estimates (4.4.17) and (4.4.18). This leads to

$$\begin{aligned} \|\dot{\boldsymbol{u}}_{\varepsilon,\eta}\|^2_{H^{1/2}(I_T;\boldsymbol{L}_2(\Omega))} \leq c_4 \|\nabla \boldsymbol{u}_{\varepsilon,\eta}\|^2_{L_2(Q_T;\mathbb{R}^{N^2})} \\ + c_5 \|\nabla \dot{\boldsymbol{u}}_{\varepsilon,\eta}\|^2_{L_2(Q_T;\mathbb{R}^{N^2})} + c_6 \|\boldsymbol{f}\|^2_{L_2(I_T;\mathring{\boldsymbol{H}}^{-1}(\Omega))}, \end{aligned} \tag{4.4.19}$$

where c_4, c_5, c_6 satisfy the same assertion as c_1, c_2, c_3 above. Then Theorem 2.8.3 yields

$$\|\dot{\boldsymbol{u}}_{\varepsilon,\eta}\|_{\boldsymbol{H}^{1/4,1/2}(S_T)} \leq c_5 \|\dot{\boldsymbol{u}}_{\varepsilon,\eta}\|_{\boldsymbol{H}^{1/2,1}(Q_T)}. \tag{4.4.20}$$

Let $\mathscr{Z} \equiv \{\boldsymbol{z}^{(i)};\ i \in \mathbb{N}\}$ be a $\boldsymbol{L}_2(\Omega)$-orthogonal basis of the space $\boldsymbol{H}_0^1(\Omega)$, $\mathscr{Z}_m$ be the subspace generated by the first m elements of $\mathscr{Z}$ and

$$\boldsymbol{Z}_m \equiv \left\{ \sum_{i=1}^m q_i \cdot \boldsymbol{z}^{(i)};\ \boldsymbol{q} \in \boldsymbol{L}_2(I_T) \right\}.$$

An element $\boldsymbol{u}_{\varepsilon,\eta,m} \in \boldsymbol{U} + Z_m$ is a Galerkin solution to (4.4.13) if it satisfies the approximate version of the initial conditions $\int_\Omega (\boldsymbol{u}_{\varepsilon,\eta,m} - \boldsymbol{u}_0) \cdot \boldsymbol{z}\, d\boldsymbol{x} = 0$ and $\int_\Omega (\dot{\boldsymbol{u}}_{\varepsilon,\eta,m} - \boldsymbol{u}_1) \cdot \boldsymbol{z}\, d\boldsymbol{x} = 0$ for any $\boldsymbol{z} \in \mathscr{Z}_m$ and if for every $\boldsymbol{v} \in Z_m$ the variational equation

$$\begin{aligned} &\int_{Q_T} \big(\ddot{\boldsymbol{u}}_{\varepsilon,\eta,m} \cdot \boldsymbol{v} + \sigma_{ij}(\boldsymbol{u}_{\varepsilon,\eta,m}) e_{ij}(\boldsymbol{v})\big)\, d\boldsymbol{x}\, dt \\ &\quad + \int_{S_{CT}} \left(\tfrac{1}{\varepsilon}[(u_{\varepsilon,\eta,m})_n]_+ v_n + G\, \nabla \Phi_\eta((\dot{\boldsymbol{u}}_{\varepsilon,\eta,m})_{\boldsymbol{t}}) \cdot \boldsymbol{v}_{\boldsymbol{t}} \right) ds_{\boldsymbol{x}}\, dt \\ &= \mathscr{L}(\boldsymbol{v}) \end{aligned} \tag{4.4.21}$$

holds. The existence and uniqueness of such a $\boldsymbol{u}_{\varepsilon,\eta,m}$ for $\eta, \varepsilon > 0$ and $m \in \mathbb{N}$ is obvious as seen earlier from the theory of ordinary differential equations. The estimate (4.4.17) can be verified again for $\boldsymbol{u}_{\varepsilon,\eta,m}$ uniformly with respect to $\varepsilon, \eta > 0$ and $m \in \mathbb{N}$.

Since $\mathscr{Z}$ is a L_2-orthogonal basis we have

$$\langle \ddot{\boldsymbol{u}}_{\varepsilon,\eta,m}, \boldsymbol{v} \rangle_{Q_T} = \langle \ddot{\boldsymbol{u}}_{\varepsilon,\eta,m}, \pi_{Z_m} \boldsymbol{v} \rangle_{Q_T} + \langle \ddot{\boldsymbol{U}}, \boldsymbol{v} - \pi_{Z_m} \boldsymbol{v} \rangle_{Q_T}$$

for any $\boldsymbol{v} \in L_2(I_T; \boldsymbol{H}_0^1(\Omega))$. From (4.4.21), (4.4.15), (4.4.17), and the uniform boundedness of the projections π_{Z_m} in $\boldsymbol{H}_0^1(\Omega)$ there follows the boundedness of $\{\ddot{\boldsymbol{u}}_{\varepsilon,\eta,m};\ \varepsilon, \eta > 0,\ m \in \mathbb{N}\}$ in $L_2(I_T; \boldsymbol{H}_0^1(\Omega)^*)$.

Now we prove the convergence of the Galerkin approximate solutions for fixed ε and η. The validity of (4.4.17), (4.4.18) (4.4.19), and (4.4.20) is verified for $\{\boldsymbol{u}_{\varepsilon,\eta,m};\ \varepsilon,\ \eta > 0,\ m \in \mathbb{N}\}$ by the same arguments as for $\{\boldsymbol{u}_{\varepsilon,\eta};\ \varepsilon, \eta > 0\}$. Consequently, for every fixed $\varepsilon > 0$, $\eta > 0$ there is a subsequence $m_k \to +\infty$ such that the following convergences are valid:

$$\begin{aligned}
\nabla\dot{\boldsymbol{u}}_{\varepsilon,\eta,m_k} &\rightharpoonup \nabla\dot{\boldsymbol{u}}_{\varepsilon,\eta} && \text{in } L_2(Q_T; \mathbb{R}^{N,N}),\\
\ddot{\boldsymbol{u}}_{\varepsilon,\eta,m_k} &\rightharpoonup \ddot{\boldsymbol{u}}_{\varepsilon,\eta} && \text{in } L_2(I_T; \boldsymbol{H}_0^1(\Omega)^*),\\
\dot{\boldsymbol{u}}_{\varepsilon,\eta,m_k}(t,\cdot) &\rightharpoonup \dot{\boldsymbol{u}}_{\varepsilon,\eta}(t,\cdot) && \text{in } \boldsymbol{L}_2(\Omega) \text{ for } t \in \{0,T\},\\
\dot{\boldsymbol{u}}_{\varepsilon,\eta,m_k} &\to \dot{\boldsymbol{u}}_{\varepsilon,\eta} && \text{in } \boldsymbol{L}_2(Q_T),\\
\dot{\boldsymbol{u}}_{\varepsilon,\eta,m_k} &\rightharpoonup \dot{\boldsymbol{u}}_{\varepsilon,\eta} && \text{in } \boldsymbol{H}^{1/4,1/2}(S_{CT}),\\
\dot{\boldsymbol{u}}_{\varepsilon,\eta,m_k} &\to \dot{\boldsymbol{u}}_{\varepsilon,\eta} && \text{in } \boldsymbol{H}^{1/4-\alpha,1/2-\alpha}(S_{CT})\\
& && \text{for any } \alpha \in \left(0, \tfrac{1}{4}\right),\\
\left.\begin{aligned}\nabla\boldsymbol{u}_{\varepsilon,\eta,m_k} &\rightharpoonup \nabla\boldsymbol{u}_{\varepsilon,\eta},\\ \boldsymbol{\sigma}(\boldsymbol{u}_{\varepsilon,\eta,m_k}) &\rightharpoonup \boldsymbol{\sigma}(\boldsymbol{u}_{\varepsilon,\eta})\end{aligned}\right\} & && \text{in } L_2(Q_T; \mathbb{R}^{N,N}).
\end{aligned} \tag{4.4.22}$$

The first three convergences here are consequences of (4.4.17) and (4.4.18); the other convergences follow by appropriate imbedding and trace theorems.

In order to obtain a strong convergence of $\nabla\boldsymbol{u}_{\varepsilon,\eta,m_k}$ in $\boldsymbol{L}_2(Q_T)$, we take the test function $\boldsymbol{v} = \pi_{m_k}\boldsymbol{u}_{\varepsilon,\eta} - \boldsymbol{u}_{\varepsilon,\eta,m_k}$ with the projection $\pi_{m_k} \equiv \pi_{\boldsymbol{U}+Z_{m_k}}$ in (4.4.21) and add

$$\int_{Q_T} \sigma_{ij}(\boldsymbol{u}_{\varepsilon,\eta})e_{ij}(\boldsymbol{u}_{\varepsilon,\eta,m_k} - \boldsymbol{u}_{\varepsilon,\eta}) + \sigma_{ij}(\boldsymbol{u}_{\varepsilon,\eta})e_{ij}(\boldsymbol{u}_{\varepsilon,\eta,m_k} - \boldsymbol{u}_{\varepsilon,\eta})\, d\boldsymbol{x}\, dt$$

to both sides of the obtained equation. The convergence of

$$\int_{Q_T} (\sigma_{ij}(\boldsymbol{u}_{\varepsilon,\eta}) - \sigma_{ij}(\boldsymbol{u}_{\varepsilon,\eta,m_k}))\, e_{ij}(\boldsymbol{u}_{\varepsilon,\eta} - \boldsymbol{u}_{\varepsilon,\eta,m_k})\, d\boldsymbol{x}\, dt$$

then follows from the convergences of all other terms. These convergences can be established with the help of relation (4.4.14), an estimate of the friction term by

$$\int_{S_{CT}} G(\Phi_\eta((\pi_{m_k}\boldsymbol{u}_{\varepsilon,\eta})_{\boldsymbol{t}} - (\boldsymbol{u}_{\varepsilon,\eta,m_k})_{\boldsymbol{t}} + (\dot{\boldsymbol{u}}_{\varepsilon,\eta,m_k})_{\boldsymbol{t}}) - \Phi_\eta((\dot{\boldsymbol{u}}_{\varepsilon,\eta,m_k})_{\boldsymbol{t}})),$$

the weak convergence of $\Phi_\eta((\pi_{m_k}\boldsymbol{u}_{\varepsilon,\eta})_{\boldsymbol{t}} - (\boldsymbol{u}_{\varepsilon,\eta,m_k})_{\boldsymbol{t}} + (\dot{\boldsymbol{u}}_{\varepsilon,\eta,m_k})_{\boldsymbol{t}})$, and the convergence $\Phi_\eta((\dot{\boldsymbol{u}}_{\varepsilon,\eta,m_k})_{\boldsymbol{t}}) \rightharpoonup \Phi_\eta((\dot{\boldsymbol{u}}_{\varepsilon,\eta})_{\boldsymbol{t}})$ in $H^{1/4,1/2}(S_{CT})$. The strong convergence $\nabla \boldsymbol{u}_{\varepsilon,\eta,m_k} \to \nabla \boldsymbol{u}_{\varepsilon,\eta}$ in $\boldsymbol{L}_2(Q_T)$ follows then from the strong monotonicity of the operator A and from (4.2.39).

Using the mentioned convergences it is possible to pass to the limit $k \to +\infty$ in (4.4.21) and to prove that $\boldsymbol{u}_{\varepsilon,\eta}$ is a solution of (4.4.13). In fact, to avoid possible problems with the convergence of $(\nabla \Phi_\eta)\big((\dot{\boldsymbol{u}}_{\varepsilon,\eta,m_k})_{\boldsymbol{t}}\big)$, this convergence is proved for the equivalent formulation of the Galerkin equations as the variational inequality corresponding to (4.4.13). The fulfilment of the initial conditions is verified in the same sense as in the previous cases, because $\boldsymbol{H}_0^1(\Omega)$ is again dense in $\boldsymbol{L}_2(\Omega)$. We have proved

4.4.1 Proposition. *Let the above mentioned assumptions on Ω, its boundary, the constitutive law, the operator A, the function W hold true and let assumptions* (4.4.15) *be fulfilled. Then there exists a solution of problem* (4.4.13) *for every $\varepsilon > 0$ and $\eta > 0$.*

4.4.2 Remark. The assumption $\boldsymbol{u}_1 \in \boldsymbol{H}^1(\Omega)$ in (4.4.15) is imposed to enable the consideration of a more general friction force G (see the theorem below). If we restrict ourselves to G from (4.4.15), the usual assumption $\boldsymbol{u}_1 \in \boldsymbol{L}_2(\Omega)$ is sufficient for both Proposition 4.4.1 and the theorem below.

Now the aim is to prove

4.4.3 Theorem. *Let the assumptions of Proposition 4.4.1 be valid with the exception of the assumption for G that is replaced by $G \in H^{1/4,1/2}(S_{CT})^*$. Then there exists a weak solution of the contact problem* (4.4.1)–(4.4.6).

Proof. The proof is done by passing to the two limits $\eta \to 0$ and $\varepsilon \to 0$. Let us first tackle the limit procedure for $\eta \to 0$. Due to the *a priori estimates* (4.4.17), (4.4.19) and (4.4.20) there is a sequence $\eta_k \to 0$ for $k \to +\infty$ such that the weak convergences $\boldsymbol{u}_{\varepsilon,\eta_k} \rightharpoonup \boldsymbol{u}_\varepsilon$ and $\dot{\boldsymbol{u}}_{\varepsilon,\eta_k} \rightharpoonup \dot{\boldsymbol{u}}_\varepsilon$ in $L_2\big(I_T; \boldsymbol{H}^1(\Omega)\big)$ as well as $\ddot{\boldsymbol{u}}_{\varepsilon,\eta_k} \rightharpoonup \ddot{\boldsymbol{u}}_\varepsilon$ in $L_2\big(I_T; \boldsymbol{H}_0^1(\Omega)^*\big)$ are valid. Due to compact imbedding theorems the sequence η_k can be chosen such that the strong convergences $\dot{\boldsymbol{u}}_{\varepsilon,\eta_k} \to \dot{\boldsymbol{u}}_\varepsilon$ in $\boldsymbol{L}_2(Q_T)$ and in $\boldsymbol{H}^{1/4-\alpha,1/2-\alpha}(S_T)$ hold with a sufficiently small $\alpha > 0$. Via the approximation properties of Φ_η we obtain $\Phi_{\eta_k}((\dot{\boldsymbol{u}}_{\varepsilon,\eta_k})_{\boldsymbol{t}}) \to \Phi_0((\dot{\boldsymbol{u}}_\varepsilon)_{\boldsymbol{t}})$ in $L_2(S_{CT})$. On the other hand, $\Phi_{\eta_k}((\dot{\boldsymbol{u}}_{\varepsilon,\eta_k})_{\boldsymbol{t}})$ is bounded in $H^{1/4,1/2}(S_{CT})$ because of the uniform Lipschitz continuity of Φ_η. Hence there is a subsequence that converges weakly in this space. Due to the mentioned convergence in $L_2(S_{CT})$ this limit coincides with $\Phi_0((\dot{\boldsymbol{u}}_\varepsilon)_{\boldsymbol{t}})$ and the whole sequence converges. Analogously, $\Phi_{\eta_k}((\boldsymbol{u}_\varepsilon)_{\boldsymbol{t}} - (\boldsymbol{u}_{\varepsilon,\eta_k})_{\boldsymbol{t}} + (\dot{\boldsymbol{u}}_{\varepsilon,\eta_k})_{\boldsymbol{t}}) \rightharpoonup \Phi_0((\dot{\boldsymbol{u}}_\varepsilon)_{\boldsymbol{t}})$ in $H^{1/4,1/2}(S_{CT})$. From these convergences we can prove the strong convergence $\nabla \boldsymbol{u}_{\varepsilon,\eta_k} \to \nabla \boldsymbol{u}_\varepsilon$ by the usual procedure; that is, by adding $\int_{Q_T} \sigma_{ij}(\boldsymbol{u}_\varepsilon) e_{ij}(\boldsymbol{u}_{\varepsilon,\eta_k} - \boldsymbol{u}_\varepsilon)\, d\boldsymbol{x}\, dt$ to both sides of the inequality (4.4.12′) with test function $\boldsymbol{v} = \boldsymbol{u}_\varepsilon + \dot{\boldsymbol{u}}_{\varepsilon,\eta_k} - \boldsymbol{u}_{\varepsilon,\eta_k}$, integrating by parts in time for the viscous part of the constitutive relation and passing to the limit $k \to +\infty$. This also yields the strong convergence of the elastic stresses. Then, by passing to the limit $k \to +\infty$ in this variational inequality with test function $\boldsymbol{v} = \boldsymbol{w} + \dot{\boldsymbol{u}}_{\varepsilon,\eta_k} - \boldsymbol{u}_{\varepsilon,\eta_k}$ for $\boldsymbol{w} \in \boldsymbol{U} + L_2(I_T; \boldsymbol{H}_0^1(\Omega))$ we prove that $\boldsymbol{u}_\varepsilon$ is the solution of the variational inequality

$$\begin{aligned}
&\int_{Q_T} \left(\ddot{\boldsymbol{u}}_\varepsilon \cdot (\boldsymbol{w} - \boldsymbol{u}_\varepsilon) + \sigma_{ij}(\boldsymbol{u}_\varepsilon) e_{ij}(\boldsymbol{w} - \boldsymbol{u}_\varepsilon)\right) d\boldsymbol{x}\, dt \\
&\quad + \int_{S_{CT}} \Big(\tfrac{1}{\varepsilon}[(u_\varepsilon)_n]_+ (w - u_\varepsilon)_n \\
&\qquad\qquad + G\left(|\boldsymbol{w}_{\boldsymbol{t}} + (\dot{\boldsymbol{u}}_\varepsilon)_{\boldsymbol{t}} - (\boldsymbol{u}_\varepsilon)_{\boldsymbol{t}}| - |\dot{\boldsymbol{u}}_{\boldsymbol{t}}|\right) \Big)\, ds_{\boldsymbol{x}}\, dt \\
&\geq \mathscr{L}(\boldsymbol{w} - \boldsymbol{u}_\varepsilon).
\end{aligned} \tag{4.4.23}$$

This inequality is identical to (4.4.12) with $\boldsymbol{v} = \boldsymbol{w} + \dot{\boldsymbol{u}}_\varepsilon - \boldsymbol{u}_\varepsilon$.

The last step of the proof is the limit procedure $\varepsilon \to 0$. For a suitable sequence $\varepsilon_k \to 0$ the convergences $\boldsymbol{u}_{\varepsilon_k} \rightharpoonup \boldsymbol{u}$ and $\dot{\boldsymbol{u}}_{\varepsilon_k} \rightharpoonup \dot{\boldsymbol{u}}$ in $L_2(I_T; \boldsymbol{H}^1(\Omega))$ hold as before, and the convergence $\ddot{\boldsymbol{u}}_{\varepsilon_k} \rightharpoonup \ddot{\boldsymbol{u}}$ now holds in $L_2(I_T; \boldsymbol{H}^{-1}(\Omega))$. The proofs of the required strong convergences are based on the same ideas as in the previous limit procedures and then the weak convergence of the stresses in $L_2(Q_T)$ is obvious; in fact the elastic parts of the stresses converge strongly. Restricting the set of test functions to that defined for problem (4.4.9), we prove after the usual integration by parts in the acceleration term that the limit $\boldsymbol{u}$ satisfies (4.4.9). Thus the existence of a solution for each $G \in H^{1/4,1/2}(S_{CT})^*$ is established. □

4.4.4 Corollary. *In addition to the assumptions of Theorem 4.4.3, let $\Gamma_C \in C^{2+\varepsilon}$ with $\varepsilon > 0$, let the coefficients of A be Hölder continuous with exponent $\frac{1}{2} + \varepsilon$ on $\overline{\Omega}$, $W \in C^2(\overline{\Omega} \times \mathbb{R}^{N^2})$, $\boldsymbol{f} \in \boldsymbol{L}_2(Q_T)$, and $G \in H^{1/4,0}(S_{CT})^*$. Then the solution $\boldsymbol{u}$ of (4.4.9) is in $B_0(I_T; \boldsymbol{H}^{3/2,1}(\widetilde{\Omega}))$, where $\widetilde{\Omega} \subset \Omega$ is a domain along the contact part of the boundary. Here the first index in $\boldsymbol{H}^{3/2,1}(\widetilde{\Omega})$ denotes the tangential and the second one the normal regularity of $\boldsymbol{u}$. Moreover, the solution belongs to $\boldsymbol{H}^{3/2,3/2,1}(\widetilde{Q}_T)$ for $\widetilde{Q}_T \equiv I_T \times \widetilde{\Omega}$.*

The proof of the time regularity can be done without any additional assumption. Due to the strong monotonicity of A and of $\frac{\partial W}{\partial \varepsilon}$, the time regularity and the space regularity in the tangential direction is proved by the translation (shift) method after local rectification of the boundary. In the estimates of the fractional tangential space derivative, the velocity is treated as a part of the right hand side of the problem and its already established space regularity (cf. (4.4.17)) is exploited. For the use of the method, the smoothness of all given elements in the constitutive law as well as of the boundary is needed.

4.4.5 Remark. The regularity of the solution along the contact part of the boundary that follows from Corollary 4.4.4 is not sufficient to solve the contact problem with Coulomb friction e.g. by means of the fixed point approach as it was seen in Chapter 1. Since it is not possibe to use velocities in the shift technique, the possible gain of regularity is limited. For the contact condition in displacements, the translation (shift) technique seems to be inadequate to prove the existence of a solution to the original contact problem with Coulomb friction, and this problem has not yet been solved.

4.4.1.2 Conservation of energy in viscoelastic contact problems

For several models in Section 4.1 the *local* conservation of energy could be proved for a suitable class of solutions, and the solution turned out to be unique. For another model the solutions satisfy a *global* conservation of energy. For the viscoelastic materials just studied there is only one result concerning the conservation of energy: Schatzman and Petrov proved in [121] the existence of solutions satisfying a *global* conservation of energy as in (4.3.15), i.e. the condition that the total final energy plus the energy lost by viscosity equals the total original energy plus the energy lost to or gained from the outer force. The result is restricted to a semi-infinite string in one space dimension ($N = 1$, $\Omega = \mathbb{R}_-$, $\Gamma = \Gamma_C = \{0\}$) with $W(x, \cdot) \equiv I$, $a_{11} = const.$ on Ω and $G = 0$. The proof is rather technical and complex. It is based on a transformation of the problem to a problem on the lateral boundary and on the obvious fact that the energy conservation is equivalent to the orthogonality of the boundary stress and the trace of the displacement velocity. The proof of the orthogonality relies on the facts that the trace of the velocity vanishes on the support of the boundary stress except on a countable set and the stress itself is a non-atomic measure. Both these facts are highly non-trivial and require sophisticated approximation ideas. The uniqueness of these solutions is not proved.

4.4.2 Contact condition in velocities

We again consider the contact problem with given friction force, but now the contact condition is formulated in velocities. This problem can be understood again as an auxiliary problem to the contact problem with Coulomb friction, useful e.g. in a fixed point approach. However, in the next Chapter we solve the contact problem with Coulomb friction directly with the help of a penalization of the contact condition and the smoothing of the friction term. Hence the introduction of the problem with given friction is not necessary, and we report the results only briefly.

The studied problem is given again by (4.4.1)–(4.4.3) for a domain Ω and a time interval I_T, but the condition (4.4.4) is replaced by

$$\dot{u}_n \le 0, \quad \sigma_n \le 0, \quad \sigma_n \dot{u}_n = 0 \quad \text{on } S_{CT} = I_T \times \Gamma_{CT}, \tag{4.4.24}$$

while the friction law (4.4.5) and the initial condition (4.4.6) remain unchanged. We assume the constitutive relation in the form $\boldsymbol{\sigma}^I + \boldsymbol{\sigma}^V$, where $\boldsymbol{\sigma}^I$ has the form (4.4.7) with the stored energy function W satisfying all assumptions as in the previous case. The operator $\varepsilon \mapsto \boldsymbol{\sigma}^V$ is in general non-linear and symmetric and satisfies the conditions

$$\begin{aligned} &\big(\sigma^V_{ij}(\boldsymbol{x}, \varepsilon(\dot{\boldsymbol{v}})) - \sigma^V_{ij}(\boldsymbol{x}, \varepsilon(\dot{\boldsymbol{w}}))\big) e_{ij}(\dot{\boldsymbol{v}} - \dot{\boldsymbol{w}}) \\ &\qquad \ge \beta^V_0 e_{ij}(\dot{\boldsymbol{v}} - \dot{\boldsymbol{w}}) e_{ij}(\dot{\boldsymbol{v}} - \dot{\boldsymbol{w}}), \\ &\big(\sigma^V_{ij}(\boldsymbol{x}, \varepsilon(\dot{\boldsymbol{v}})) - \sigma^V_{ij}(\boldsymbol{x}, \varepsilon(\dot{\boldsymbol{w}}))\big) e_{ij}(\widetilde{\boldsymbol{w}}) \\ &\qquad \le \beta^V_1 e_{ij}(\dot{\boldsymbol{v}} - \dot{\boldsymbol{w}}) e_{ij}(\widetilde{\boldsymbol{w}}) \end{aligned} \tag{4.4.25}$$

for any $\boldsymbol{v}, \boldsymbol{w}, \widetilde{\boldsymbol{w}} \in \boldsymbol{H}^1(\Omega)$ a.e. in Ω. Moreover, we assume $\boldsymbol{\sigma}(\cdot, 0) = 0$ and $\partial_{e_{ij}} \boldsymbol{\sigma}(\cdot, 0) \equiv 0$ for $i, j = 1, \ldots, N$. We also impose assumptions (4.4.15) for $\boldsymbol{U}, \boldsymbol{u}_0, \boldsymbol{u}_1, \boldsymbol{f}$; the assumption on G is weakened to $G \in H^{1/4,1/2}(S_{CT})^*$.

The simultaneous formulation of the contact condition and the friction law in velocities simplifies the variational formulation of the problem as follows:

Find a function $\boldsymbol{u} \in B_0\big(I_T; \boldsymbol{H}^1(\Omega)\big)$ *with*

$$\dot{\boldsymbol{u}} \in B_0\big(I_T; \boldsymbol{L}_2(\Omega)\big) \cap H^{1/2}\big(I_T; \boldsymbol{L}_2(\Omega)\big),$$

$\dot{\boldsymbol{u}}(t, \cdot) \in \mathscr{C}_{\dot{\boldsymbol{U}}(t,\cdot)}$ *for all* $t \in I_T$ *and* $\ddot{\boldsymbol{u}} \in L_2\big(I_T; \boldsymbol{H}_0^{-1}(\Omega)\big) \cap H^{1/2}\big(I_T; \boldsymbol{L}_2(\Omega)\big)^*$ *such that* (4.4.6) *holds and for each* $\boldsymbol{v} \in H^{1/2}(I_T; \boldsymbol{L}_2(\Omega))$ *satisfying* $\boldsymbol{v}(t, \cdot) \in \mathscr{C}_{\dot{\boldsymbol{U}}(t,\cdot)}$ *for all* $t \in I_T$ *the following inequality is valid:*

$$\begin{aligned} &\int_{Q_T} \big(\sigma_{ij}(\boldsymbol{u}) e_{ij}(\boldsymbol{v} - \dot{\boldsymbol{u}}) + \ddot{\boldsymbol{u}} \cdot (\boldsymbol{v} - \dot{\boldsymbol{u}})\big)\, d\boldsymbol{x}\, dt \\ &\qquad + \int_{S_{CT}} G\left(|\boldsymbol{v_t}| - |\dot{\boldsymbol{u}}_{\boldsymbol{t}}|\right) ds_{\boldsymbol{x}}\, dt \;\ge\; \mathscr{L}(\boldsymbol{v} - \dot{\boldsymbol{u}}_\varepsilon). \end{aligned} \tag{4.4.26}$$

As earlier we have $\mathscr{L} : \boldsymbol{w} \mapsto \int_{Q_T} \boldsymbol{f} \cdot \boldsymbol{w}\, d\boldsymbol{x}\, dt + \int_{S_{\tau T}} \boldsymbol{\tau} \cdot \boldsymbol{w}\, ds_{\boldsymbol{x}}\, dt$.

We start again to solve the problem for $G \in L_2\big(I_T; H^{1/2}(\Gamma_C)^*\big)$ under the assumptions (4.4.15) for the remaining data. After the penalization of the contact condition and the smoothing of the friction law, the problem is again solved by the Galerkin method. In the *a priori* estimate the term $\int_{S_{CT}} \frac{1}{\varepsilon}[\dot{u}_{\varepsilon,\eta}]_+^2 \, ds_{\boldsymbol{x}} \, dt$ replaces the earlier penalty term and the rest of (4.4.17) remains unchanged. Hence all the convergences in (4.4.22) are valid for both an appropriate sequence $\boldsymbol{u}_{\varepsilon,\eta,m_k}$ of solutions of a suitable Galerkin approximation and a sequence $\boldsymbol{u}_{\varepsilon,\eta_k}$ of solutions of the smoothed and penalized problem. They obviously hold also if the time interval I_T is reduced to I_t for $t \in I_T$. The proof of the strong L_2 convergence of the space gradients and their time derivatives is a bit more complicated. It is necessary to take velocities instead of displacements as test functions in the corresponding variational inequalities. For the Galerkin approximation we first restrict ourselves to some smaller time interval I_t. Adding

$$\int_{Q_t} \sigma_{ij}(\boldsymbol{u}_{\varepsilon,\eta}) e_{ij}(\dot{\boldsymbol{u}}_{\varepsilon,\eta,m_k} - \pi_{m_k}\dot{\boldsymbol{u}}_{\varepsilon,\eta}) \, d\boldsymbol{x} \, ds$$

to both sides of the appropriate variational equation, invoking (4.2.39), (4.4.25), using all the available convergence properties and the expression $\boldsymbol{v}(s) = \int_0^s \dot{\boldsymbol{v}}(r) \, dr + \boldsymbol{v}(0)$ we derive

$$\begin{aligned} &\int_{Q_t} \left(\sigma_{ij}^I(\boldsymbol{u}_{\varepsilon,\eta,m_k}) - \sigma_{ij}^I(\boldsymbol{u}_{\varepsilon,\eta})\right) e_{ij}(\dot{\boldsymbol{u}}_{\varepsilon,\eta,m_k} - \dot{\boldsymbol{u}}_{\varepsilon,\eta}) \, d\boldsymbol{x} \, ds \\ &\quad \leq \sqrt{t}\,(\beta_1^W + \lambda) \\ &\qquad \cdot \int_{Q_t} e_{ij}(\dot{\boldsymbol{u}}_{\varepsilon,\eta,m_k} - \dot{\boldsymbol{u}}_{\varepsilon,\eta}) \, e_{ij}(\dot{\boldsymbol{u}}_{\varepsilon,\eta,m_k} - \dot{\boldsymbol{u}}_{\varepsilon,\eta}) \, d\boldsymbol{x} \, ds \\ &\quad + c(\lambda) \int_{\Omega} \left(\pi_{Z_{m_k}} u_0 - u_0\right)^2 d\boldsymbol{x}, \end{aligned} \tag{4.4.27}$$

for an arbitrarily small $\lambda > 0$. Then it is easy to prove

$$\|\varepsilon(\dot{\boldsymbol{u}}_{\varepsilon,\eta,m_k} - \dot{\boldsymbol{u}}_{\varepsilon,\eta})\|_{L_2(Q_t;\mathbb{R}^{N,N})} \to 0,$$

if $t < \left(\beta_0^V/\beta_1^W\right)^2$ due to the convergence of the other terms in the variational inequality. Starting from the new time t, we can successively repeat the procedure and prove the strong convergence on the whole of Q_T. The corresponding result for the procedure $\eta \to 0$ is proved analogously. From this the strong convergence of the elastic stresses is a direct consequence of the assumption (4.4.25) and of (4.2.7). In such a way we can prove the solvability of the penalized problem:

Find a function $\boldsymbol{u}_\varepsilon \in B_0\big(I_T; \boldsymbol{H}^1(\Omega)\big)$ *such that* $\dot{\boldsymbol{u}}_\varepsilon \in L_\infty(I_T; \boldsymbol{L}_2(\Omega)) \cap L_2\big(I_T; \boldsymbol{H}^1(\Omega)\big)$, $\ddot{\boldsymbol{u}}_\varepsilon \in L_2\big(I_T; \boldsymbol{H}_0^1(\Omega)^*\big)$, *and for all* $\boldsymbol{v} \in L_2\big(I_T; \boldsymbol{H}^1(\Omega)\big)$ *the* *following inequality holds:*

$$\begin{aligned}\int_{Q_T} & \left(\sigma_{ij}(\boldsymbol{u}_\varepsilon)e_{ij}(\boldsymbol{v}-\dot{\boldsymbol{u}}_\varepsilon)+\ddot{\boldsymbol{u}}_\varepsilon\cdot(\boldsymbol{v}-\dot{\boldsymbol{u}}_\varepsilon)\right)d\boldsymbol{x}\,dt\\ &+\int_{S_{CT}} G\left(|\boldsymbol{v}_{\boldsymbol{t}}|-|(\dot{\boldsymbol{u}}_\varepsilon)_{\boldsymbol{t}}|\right)ds_{\boldsymbol{x}}\,dt \qquad (4.4.28)\\ &+\frac{1}{\varepsilon}\int_{S_{CT}}[(\dot{\boldsymbol{u}}_\varepsilon)_n]_+(v_n-(\dot{u}_\varepsilon)_n)\,ds_{\boldsymbol{x}}\,dt\geq \mathscr{L}(\boldsymbol{v}-\dot{\boldsymbol{u}}_\varepsilon).\end{aligned}$$

The uniqueness of the solution can be also established. Take $\boldsymbol{v}=\dot{\boldsymbol{u}}^{(1)}+\chi_t\big(\dot{\boldsymbol{u}}^{(2)}-\dot{\boldsymbol{u}}^{(1)}\big)$ for χ_t from (4.2.18) and any $t\in I_T$ in (4.4.28) with solution $\boldsymbol{u}^{(1)}$ and repeat this procedure with exchanged roles of $\boldsymbol{u}^{(1)}$ and $\boldsymbol{u}^{(2)}$. This yields after the addition of both obtained inequalities and after the use of(4.4.25) and (4.2.39)

$$\begin{aligned}\int_\Omega \left|(\dot{\boldsymbol{u}}^{(1)}-\dot{\boldsymbol{u}}^{(2)})(t,\cdot)\right|^2 d\boldsymbol{x}+\int_{Q_t} e_{ij}\big(\dot{\boldsymbol{u}}^{(1)}-\dot{\boldsymbol{u}}^{(2)}\big)e_{ij}\big(\dot{\boldsymbol{u}}^{(1)}-\dot{\boldsymbol{u}}^{(2)}\big)\,d\boldsymbol{x}\,ds\\ \leq c\int_{Q_t} e_{ij}\big(\dot{\boldsymbol{u}}^{(1)}-\dot{\boldsymbol{u}}^{(2)}\big)\,e_{ij}\big(\boldsymbol{u}^{(1)}-\boldsymbol{u}^{(2)}\big)\,d\boldsymbol{x}\,ds.\end{aligned}$$

A Gronwall-lemma type argument similar to that used for (4.4.27) yields $\boldsymbol{u}^{(1)}=\boldsymbol{u}^{(2)}$. The convergence of a suitable sequence of solutions $\boldsymbol{u}_{\varepsilon_k}$ of the penalized problem with $\varepsilon_k\to 0$ is obtained as in the previous limit procedures. Thus we have proved

4.4.6 Theorem. *Let the assumptions of the previous theorem hold with the exception of the assumption concerning the viscous stress which is replaced by assumption* (4.4.25). *Then there exists a unique solution of problem* (4.4.26).

4.4.7 Remark. In contrast to the contact problem with contact condition in the displacements it is here possible to obtain a regularity of the solution along S_{CT} that is sufficient to solve the contact problem with Coulomb friction using the fixed point approach. Since we solve this problem directly via the penalty method in the next chapter, we omit details here.

Chapter 5

Dynamic contact problems with Coulomb friction

This chapter is devoted to the solvability of dynamic contact problems with friction modeled by the Coulomb law. As pointed out in the previous chapter, even for the frictionless dynamic contact problem a certain viscous damping of the material is necessary in order to obtain existence results for a general multi-dimensional domain. Throughout this chapter we consider viscosity with short memory; this leads to a system of parabolic differential equations of second order formulated in the displacement velocities. The approach for the existence proof is similar to that in the static and quasistatic cases: the original problem is approximated by a sequence of auxiliary problems, the solvability and some limited regularity of the solutions of the auxiliary problems is proved, and a transition to the original problem is done. Again the contact problem with given friction in combination with a suitable fixed-point approach and the penalty method are the only up to now known efficient variants of the auxiliary problem. The regularity of the solutions is derived again by the shift technique combined with the local rectification of the boundary. However, the Coulomb friction law is formulated in the displacement velocities; therefore the regularity result has to yield some compactness in the traces of the displacement velocities and the boundary tractions. If the contact condition is formulated in the displacements, then the regularity would be proved for the displacements only; and the possible gain of regularity is not sufficient in order to get the required regularity of the velocities. In order to overcome this situation, we use here the contact condition in velocities

$$\dot{u}_n \leq 0, \ \sigma_n \leq 0, \ \sigma_n \dot{u}_n = 0.$$

This relation can be interpreted as a first-order approximation with respect to the time variable, realistic for a short time interval and for a vanishing initial gap between the body's boundary and the obstacle. This obviously limits the applicability of results presented below, but the model remains realistic e.g. for the first — and usually strongest — wave of an earthquake.

The regularity result necessary for the limit of the penalty parameter is proved for a coefficient of friction bounded by a definite constant arising from special trace and inverse trace theorems for the corresponding parabolic problem on a halfspace. These constants are calculated in a different manner for the homogeneous isotropic *viscous* part of the constitutive material law and for general viscoelastic materials. The results are much poorer than those in the static case: for isotropic material the admissible coefficient of friction is at most close to 1/2; for anisotropic material we have around 1/4 in the optimal cases.

The first section contains a description of the general scheme of the existence proof. We derive the existence result for the penalized problem, show the regularity of its solution under the validity of definite trace and inverse trace theorems and perform the limit procedure for the penalty parameter. In the next sections the required trace and inverse trace estimates are proved separately for the different cases described above. In the last section the contact problem is coupled with a heat equation in order to study the thermal aspects of friction. Two different models of heat conduction are employed: a linearized one which requires a certain linear limitation of the friction as a thermal source and a non-linear one, where the quadratic growth of this source is compensated by a suitable growth of the heat conductivity.

5.1 Solvability of frictional contact problems

We consider a viscoelastic body that occupies at time $t = 0$ the domain $\Omega \subset \mathbb{R}^N$ and is in contact with a rigid obstacle. The boundary Γ of the body consists of three measurable parts, the part Γ_U with prescribed displacements, the part Γ_τ with prescribed boundary tractions and the part Γ_C which is in contact with the rigid obstacle. The displacement field is a solution of the linear viscoelastic system

$$\ddot{u}_i - \partial_j \sigma_{ij}(\boldsymbol{u}) = f_i \tag{5.1.1}$$

on $Q_T = I_T \times \Omega$ with the considered time interval $I_T = (0, T)$. As usual, the summation convention is employed. The constitutive material law is that for linear viscoelastic material with short memory,

$$\sigma_{ij}(\boldsymbol{u}) = a^{(0)}_{ijk\ell} e_{k\ell}(\boldsymbol{u}) + a^{(1)}_{ijk\ell} e_{k\ell}(\dot{\boldsymbol{u}}),\ i, j = 1, \ldots, N, \tag{5.1.2}$$

with the components $e_{ij}(\boldsymbol{u}) \equiv \frac{1}{2}(\partial_i u_j + \partial_j u_i)$ of the linearized strain tensor and the analogously defined components $e_{ij}(\dot{\boldsymbol{u}}) \equiv \frac{1}{2}(\partial_i \dot{u}_j + \partial_j \dot{u}_i)$ of the strain-velocity tensor. The material coefficients are assumed to be symmetric,

$$a^{(\iota)}_{ijk\ell} = a^{(\iota)}_{jik\ell} = a^{(\iota)}_{k\ell ij},\ i,j,k,\ell = 1,\dots,N,\ \iota = 0,1, \tag{5.1.3}$$

bounded and elliptic,

$$a^{(\iota)}_0 \xi_{ij}\xi_{ij} \le a^{(\iota)}_{ijk\ell}\xi_{ij}\xi_{k\ell} \le A^{(\iota)}_0 \xi_{ij}\xi_{ij} \tag{5.1.4}$$

for all symmetric tensors $\{\xi_{ij}\}_{i,j=1}^N$ with constants $0 < a_0^{(\iota)} \le A_0^{(\iota)} < +\infty$. As in the previous chapter we denote $S_{XT} = I_T \times \Gamma_X$ for $X \in \{C, \boldsymbol{\tau}, \boldsymbol{U}\}$. Then the boundary conditions including the contact condition of Signorini type and the Coulomb friction law are given by

$$\boldsymbol{u} = \boldsymbol{U} \text{ on } S_{UT}, \tag{5.1.5}$$

$$\boldsymbol{\sigma}_\Gamma(\boldsymbol{u}) = \boldsymbol{\tau} \text{ on } S_{\boldsymbol{\tau}T}, \tag{5.1.6}$$

$$\dot{u}_n \le 0,\ \sigma_n(\boldsymbol{u}) \le 0,\ \sigma_n(\boldsymbol{u})\dot{u}_n = 0 \text{ on } S_{CT}, \tag{5.1.7}$$

$$\left.\begin{array}{lll} \dot{\boldsymbol{u}}_t = 0 & \Rightarrow\ |\boldsymbol{\sigma}_t(\boldsymbol{u})| & \le\ \mathfrak{F}|\sigma_n(\boldsymbol{u})| \\ \dot{\boldsymbol{u}}_t \ne 0 & \Rightarrow\ \boldsymbol{\sigma}_t(\boldsymbol{u}) & =\ -\mathfrak{F}|\sigma_n(\boldsymbol{u})|\dfrac{\dot{\boldsymbol{u}}_t}{|\dot{\boldsymbol{u}}_t|} \end{array}\right\} \text{ on } S_{CT}. \tag{5.1.8}$$

The initial conditions are

$$\boldsymbol{u}(0,x) = \boldsymbol{u}_0(\boldsymbol{x}) \text{ and } \dot{\boldsymbol{u}}(0,x) = \boldsymbol{u}_1(\boldsymbol{x}) \text{ in } \Omega. \tag{5.1.9}$$

The cone of admissible functions for this problem is

$$\mathscr{C} := \big\{\boldsymbol{v} \in \boldsymbol{H}^{1/2,\,1}(Q_T)\,;\, \boldsymbol{v} = \dot{\boldsymbol{U}} \text{ on } S_{UT} \text{ and } v_n \le 0 \text{ on } S_{CT}\big\}.$$

Then the weak formulation of the problem is given by the following variational inequality:

Find a function $\boldsymbol{u}$ *belonging to* $B_0\big(I_T; \boldsymbol{H}^1(\Omega)\big)$ *with* $\dot{\boldsymbol{u}} \in \mathscr{C}$ *and* $\ddot{\boldsymbol{u}} \in L_2\big(I_T; \boldsymbol{H}^1(\Omega)^*\big)$ *such that the initial condition* (5.1.9) *is satisfied and for each* $\boldsymbol{v} \in \mathscr{C}$ *there holds*

$$\begin{aligned} &\langle \ddot{\boldsymbol{u}}, \boldsymbol{v} - \dot{\boldsymbol{u}}\rangle_{Q_T} + A^{(0)}(\boldsymbol{u}, \boldsymbol{v} - \dot{\boldsymbol{u}}) + A^{(1)}(\dot{\boldsymbol{u}}, \boldsymbol{v} - \dot{\boldsymbol{u}}) \\ &\quad + \langle \mathfrak{F}\,|\sigma_n(\boldsymbol{u})|\,, |\boldsymbol{v}_t| - |\dot{\boldsymbol{u}}_t|\rangle_{S_{CT}} \ \ge\ \mathscr{L}(\boldsymbol{v} - \dot{\boldsymbol{u}}). \end{aligned} \tag{5.1.10}$$

Here

$$A^{(\iota)}(\boldsymbol{u}, \boldsymbol{v}) = \int_{Q_T} a^{(\iota)}(\boldsymbol{u}, \boldsymbol{v})\, d\boldsymbol{x}\, dt \tag{5.1.11}$$

with

$$a^{(\iota)}(\boldsymbol{u}, \boldsymbol{v}) \equiv a^{(\iota)}_{ijk\ell} e_{k\ell}(\boldsymbol{u}) e_{ij}(\boldsymbol{v})$$

for $\iota = 1, 2$ describe the bilinear forms of elastic energy and of viscous energy dissipation and

$$\mathscr{L}(\boldsymbol{v}) \equiv \int_{Q_T} \boldsymbol{f} \cdot \boldsymbol{v} \, d\boldsymbol{x} \, dt + \int_{S_{\boldsymbol{\tau} T}} \boldsymbol{\tau} \cdot \boldsymbol{v} \, ds_{\boldsymbol{x}} \, dt \tag{5.1.12}$$

is the linear functional of the given forces. The equivalence of this weak formulation with (5.1.1), (5.1.5)–(5.1.9) is valid in the usual sense.

In the first step of the existence proof we employ the penalty method and the smoothing of the friction in order to obtain an approximate problem of simpler mathematical structure. The penalization of the Signorini contact condition leads to the variational inequality

Find a function $\boldsymbol{u} \in B_0(I_T; \boldsymbol{H}^1(\Omega))$ *with* $\dot{\boldsymbol{u}} \in \boldsymbol{H}^{1/2,1}_{\dot{\boldsymbol{U}}}(Q_T)$ *and* $\ddot{\boldsymbol{u}} \in L_2(I_T; \boldsymbol{H}^1(\Omega)^*)$ *such that the initial condition* (5.1.9) *is satisfied and for each* $\boldsymbol{v} \in \boldsymbol{H}^{1/2,1}_{\dot{\boldsymbol{U}}}(Q_T)$ *there holds*

$$\begin{aligned} &\langle \ddot{\boldsymbol{u}}, \boldsymbol{v} - \dot{\boldsymbol{u}} \rangle_{Q_T} + A^{(0)}(\boldsymbol{u}, \boldsymbol{v} - \dot{\boldsymbol{u}}) + A^{(1)}(\dot{\boldsymbol{u}}, \boldsymbol{v} - \dot{\boldsymbol{u}}) \\ &\qquad + \left\langle \tfrac{1}{\varepsilon}[\dot{u}_n]_+, v_n - \dot{u}_n \right\rangle_{S_{CT}} + \left\langle \mathfrak{F} \tfrac{1}{\varepsilon}[\dot{u}_n]_+, |\boldsymbol{v}_{\boldsymbol{t}}| - |\dot{\boldsymbol{u}}_{\boldsymbol{t}}| \right\rangle_{S_{CT}} \\ &\geq \mathscr{L}(\boldsymbol{v} - \dot{\boldsymbol{u}}). \end{aligned} \tag{5.1.13}$$

Here for any function $\boldsymbol{w} \in \boldsymbol{H}^{1/2,1}(Q_T)$

$$\boldsymbol{H}^{1/2,1}_{\boldsymbol{w}}(Q_T) := \{\boldsymbol{v} \in \boldsymbol{H}^{1/2,1}(Q_T); \boldsymbol{v} = \boldsymbol{w} \text{ on } S_{UT}\}. \tag{5.1.14}$$

In the next step we replace the norms in the friction term of the previous inequality by differentiable, convex approximations $\Phi_\eta(\cdot)$ satisfying

$$\big|\Phi_\eta(\cdot) - |\cdot|\big| \leq \eta \text{ and } \nabla\Phi_\eta(0) = 0$$

as already described in Section 3.1. The resulting variational inequality is

Find a function $\boldsymbol{u} \in B_0(I_T; \boldsymbol{H}^1(\Omega))$ *with* $\dot{\boldsymbol{u}} \in \boldsymbol{H}^{1/2,1}_{\dot{\boldsymbol{U}}}(Q_T)$ *and* $\ddot{\boldsymbol{u}} \in L_2(I_T; \boldsymbol{H}^1(\Omega)^*)$ *such that the initial condition* (5.1.9) *is satisfied and for each* $\boldsymbol{v} \in \boldsymbol{H}^{1/2,1}_{\dot{\boldsymbol{U}}}(Q_T)$ *there holds*

$$\begin{aligned} &\langle \ddot{\boldsymbol{u}}, \boldsymbol{v} - \dot{\boldsymbol{u}} \rangle_{Q_T} + A^{(0)}(\boldsymbol{u}, \boldsymbol{v} - \dot{\boldsymbol{u}}) + A^{(1)}(\dot{\boldsymbol{u}}, \boldsymbol{v} - \dot{\boldsymbol{u}}) \\ &\quad + \left\langle \tfrac{1}{\varepsilon}[\dot{u}_n]_+, v_n - \dot{u}_n \right\rangle_{S_{CT}} \\ &\quad + \left\langle \mathfrak{F} \tfrac{1}{\varepsilon}[\dot{u}_n]_+, \Phi_\eta(\boldsymbol{v}_{\boldsymbol{t}}) - \Phi_\eta(\dot{\boldsymbol{u}}_{\boldsymbol{t}}) \right\rangle_{S_{CT}} \geq \mathscr{L}(\boldsymbol{v} - \dot{\boldsymbol{u}}). \end{aligned} \tag{5.1.15}$$

As in the static case it is equivalent to the following variational equation (see page 161):

Find a function $\boldsymbol{u} \in B_0\big(I_T; \boldsymbol{H}^1(\Omega)\big)$ *with* $\dot{\boldsymbol{u}} \in \boldsymbol{H}^{1/2,1}_{\dot{\boldsymbol{U}}}(Q_T)$ *and* $\ddot{\boldsymbol{u}} \in L_2\big(I_T; \boldsymbol{H}^1(\Omega)^*\big)$ *such that the initial condition* (5.1.9) *is valid and for each* $\boldsymbol{v} \in \boldsymbol{H}^{1/2,1}_0(Q_T)$ *there holds*

$$\begin{aligned} \langle \ddot{\boldsymbol{u}}, \boldsymbol{v}\rangle_{Q_T} + A^{(0)}(\boldsymbol{u},\boldsymbol{v}) + A^{(1)}(\dot{\boldsymbol{u}},\boldsymbol{v}) + \Big\langle \tfrac{1}{\varepsilon}[\dot{u}_n]_+\,, v_n\Big\rangle_{S_{CT}} \\ + \Big\langle \mathfrak{F}\tfrac{1}{\varepsilon}[\dot{u}_n]_+ \nabla\Phi_\eta(\dot{\boldsymbol{u}}_{\boldsymbol{t}}), \boldsymbol{v}_{\boldsymbol{t}}\Big\rangle_{S_{CT}} = \mathscr{L}(\boldsymbol{v}). \end{aligned} \tag{5.1.16}$$

The existence of a solution to the penalized and smoothed problem is proved under the following assumptions:

5.1.1 Assumption. *Let* Ω *be a bounded Lipschitz domain with a boundary* Γ *consisting of the measurable, mutually disjoint parts* $\Gamma_{\boldsymbol{U}}$, $\Gamma_{\boldsymbol{\tau}}$ *and* Γ_C. *The Hooke tensors* $\big\{a^{(\iota)}_{ijk\ell}\big\}^N_{i,j,k,\ell=1}$, $\iota = 0,1$, *are assumed to be measurable, symmetric, bounded and elliptic in the sense of* (5.1.3), (5.1.4). *Let the given data satisfy* $\boldsymbol{f} \in \mathring{\boldsymbol{H}}^{-1/2+\varepsilon',-1}(Q_T)$, $\boldsymbol{\tau} \in \mathring{\boldsymbol{H}}^{-1/4+\varepsilon',-1/2}(S_{\boldsymbol{\tau}T})$, $\dot{\boldsymbol{U}} \in \boldsymbol{H}^{1/2+\varepsilon',1}(Q_T)$ *with* $\varepsilon' > 0$ *and* $\boldsymbol{u}_0 \in \boldsymbol{H}^1(\Omega)$, $\boldsymbol{u}_1 \in \boldsymbol{H}^{1/2}(\Omega)$. *The compatibility conditions* $\boldsymbol{U}(0,x) = \boldsymbol{u}_0(\boldsymbol{x})$, $\dot{\boldsymbol{U}}(0,x) = \boldsymbol{u}_1(\boldsymbol{x})$ *for* $\boldsymbol{x} \in \Omega$ *and* $\dot{U}_n = 0$ *on* S_{CT} *shall be valid. The coefficient of friction* $\mathfrak{F}(\boldsymbol{x}, \dot{\boldsymbol{u}})$ *shall be continuous in the sense of Carathéodory, non-negative and bounded.*

5.1.2 Theorem. *Let Assumption 5.1.1 be valid. Then variational equation* (5.1.16) *has for all approximation parameters* $\varepsilon, \eta > 0$ *at least one solution. The solution satisfies the* a priori *estimate*

$$\|\dot{\boldsymbol{u}}\|_{L_\infty(I_T;\boldsymbol{L}_2(\Omega))} + \|\dot{\boldsymbol{u}}\|_{\boldsymbol{H}^{1/2,1}(Q_T)} \le c \tag{5.1.17}$$

with a constant c *independent of* ε *and* η. *If the coefficient of friction is independent of the solution,* $\mathfrak{F} = \mathfrak{F}(\boldsymbol{x})$, *then the solution is unique.*

Proof. In order to verify the existence of solutions the Galerkin method is employed. Let us recall the definition $\boldsymbol{H}^1_0(\Omega) := \big\{\boldsymbol{v} \in \boldsymbol{H}^1(\Omega); \boldsymbol{v} = 0 \text{ on } \Gamma_{\boldsymbol{U}}\big\}$. If no displacement boundary condition is prescribed, $\operatorname{mes}\Gamma_{\boldsymbol{U}} = 0$, then $\boldsymbol{H}^1_0(\Omega) = \boldsymbol{H}^1(\Omega)$. Let $\mathscr{X} := \big\{\boldsymbol{v}_j; j \in \mathbb{N}\big\}$ be an orthonormal basis of $\boldsymbol{H}^1_0(\Omega)$ whose elements are orthogonal with respect to the $L_2(\Omega)$ scalar product; this is e.g. satisfied for eigenfunctions of the Laplacian. Finally, let $\mathscr{X}_m := \operatorname{span}\big\{\boldsymbol{v}_1, \ldots, \boldsymbol{v}_m\big\}$ be the Galerkin space of order m.

For simplicity of the presentation we first consider the case of vanishing boundary and initial displacements, $\boldsymbol{U} \equiv 0$ (which also implies $\boldsymbol{u}_0 = \boldsymbol{u}_1 = 0$) and a little bit more regular functional $\mathscr{L} \in L_2\big(I_T, \mathring{\boldsymbol{H}}^{-1}(\Omega)\big)$. A solution of the Galerkin approximation of the order m is a function $\boldsymbol{u}_m(t,x) = \sum_{i=1}^m q_i(t)v_i(\boldsymbol{x})$, $\boldsymbol{u}_m \in L_\infty(I_T;\mathscr{X}_m)$, which satisfies the initial condition (5.1.9) in the sense that $\langle \boldsymbol{u}_m(0,\cdot),\boldsymbol{v}\rangle_\Omega = \langle \dot{\boldsymbol{u}}_m(0,\cdot),\boldsymbol{v}\rangle_\Omega = 0$ for any $\boldsymbol{v} \in \mathscr{X}_m$ and the system of equations

$$\begin{aligned}\int_\Omega \Big(\ddot{\boldsymbol{u}}_m \cdot \boldsymbol{v} + a^{(0)}(\boldsymbol{u}_m,\boldsymbol{v}) + a^{(1)}(\dot{\boldsymbol{u}}_m,\boldsymbol{v})\Big)\, d\boldsymbol{x} \\ + \int_{\Gamma_C} \tfrac{1}{\varepsilon}\big[(\dot{\boldsymbol{u}}_m)_n\big]_+ v_n \, ds_{\boldsymbol{x}} \\ + \int_{\Gamma_C} \mathfrak{F}\tfrac{1}{\varepsilon}\big[(\dot{\boldsymbol{u}}_m)_n\big]_+ \nabla\Phi_\eta\big((\dot{\boldsymbol{u}}_m)_{\boldsymbol{t}}\big)\cdot \boldsymbol{v}_{\boldsymbol{t}}\, ds_{\boldsymbol{x}} \\ = \int_\Omega \boldsymbol{f}\cdot\boldsymbol{v}\, d\boldsymbol{x} + \int_{\Gamma_\tau} \boldsymbol{\tau}\cdot\boldsymbol{v}\, ds_{\boldsymbol{x}}\end{aligned} \tag{5.1.18}$$

for all functions $\boldsymbol{v} \in \mathscr{X}_m$ and all $t \in I_T$. This problem is equivalent to a system of m ordinary differential equations for the unknown coefficient functions $q_i(t)$ whose solvability follows from the usual theory of ordinary differential equations.

Let us take the test function $\boldsymbol{v} = \dot{\boldsymbol{u}}_m$ in equation (5.1.18) and integrate with respect to $t \in (0,t_0)$. Employing the ellipticity and boundedness of the material tensors $\left\{a^{(\iota)}_{ijk\ell}\right\}$, the relations $\big[(\dot{\boldsymbol{u}}_m)_n\big]_+(\dot{\boldsymbol{u}}_m)_n \ge 0$, $\nabla\Phi_\eta\big((\dot{\boldsymbol{u}}_m)_{\boldsymbol{t}}\big)\cdot(\dot{\boldsymbol{u}}_m)_{\boldsymbol{t}} \ge 0$ which is valid because Φ_η is convex and has a minimum in 0, Theorem 1.2.3 and

$$\int_{t=0}^{t_0}\int_\Omega a^{(0)}(\boldsymbol{u}_m,\dot{\boldsymbol{u}}_m)\, d\boldsymbol{x}\, dt \ge -c\|\boldsymbol{u}_0\|^2_{\boldsymbol{H}^1(\Omega)}$$

with some non-negative c, the following estimate can be derived

$$\begin{aligned}\|\dot{\boldsymbol{u}}_m(t_0)\|^2_{\boldsymbol{L}_2(\Omega)} + \|\dot{\boldsymbol{u}}_m\|^2_{L_2(I_{t_0};\boldsymbol{H}^1(\Omega))} \\ \le c_1\|\dot{\boldsymbol{u}}_m\|_{L_2(I_{t_0};\boldsymbol{H}^1(\Omega))} + c_2\|\dot{\boldsymbol{u}}_m\|^2_{\boldsymbol{L}_2(Q_{t_0})} + c_3.\end{aligned} \tag{5.1.19}$$

Here and in the remaining part of this proof the constants c_i depend on the geometry of the domain, on the coefficient functions $a^{(\iota)}_{ijk\ell}$, on the given data $\boldsymbol{f}$, $\boldsymbol{\tau}_0$ and on the length T of the time interval, but not on the approximation parameters ε, η and m unless this dependence is explicitly indicated. Application of the Gronwall lemma to relation (5.1.19) yields

$$\|\dot{\boldsymbol{u}}_m\|^2_{L_\infty(I_T;\boldsymbol{L}_2(\Omega))} + \|\dot{\boldsymbol{u}}_m\|^2_{L_2(I_T;\boldsymbol{H}^1(\Omega))} \le c_1. \tag{5.1.20}$$

In order to derive a *dual estimate*, we consider a test function $\boldsymbol{v} \in L_2(I_T; \mathring{\boldsymbol{H}}^1(\Omega))$ with vanishing boundary data. We use the test function $\pi_{\mathscr{X}_m} v(t,\cdot)$

in the Galerkin equations (5.1.18) and integrate with respect to t over I_T. Due to the $\boldsymbol{L}_2(\Omega)$-orthogonality of the basis functions $\boldsymbol{v}_j$ there holds

$$\langle \ddot{\boldsymbol{u}}_m, \boldsymbol{v}\rangle_{Q_T} = \langle \ddot{\boldsymbol{u}}_m, \pi_{\mathscr{X}_m} v\rangle_{Q_T}.$$

Using also the boundedness of the operator $\pi_{\mathscr{X}_m} : \boldsymbol{H}^1(\Omega) \to \mathscr{X}_m$ and of the tensors $\{a^{(\iota)}_{ijk\ell}\}$, and the just proved estimate (5.1.20) yields

$$\|\ddot{\boldsymbol{u}}_m\|_{L_2(I_T;\boldsymbol{H}^{-1}(\Omega))} \le c(\varepsilon).$$

The interpolation $H^1(I_T;\boldsymbol{H}^{-1}(\Omega)) \cap L_2(I_T;\boldsymbol{H}^1(\Omega)) \hookrightarrow H^{1/2}(I_T;\boldsymbol{L}_2(\Omega))$ yields

$$\|\dot{\boldsymbol{u}}_m\|_{H^{1/2}(I_T;\boldsymbol{L}_2(\Omega))} \le c.$$

Combining this with (5.1.20) gives

$$\|\dot{\boldsymbol{u}}_m\|^2_{L_\infty(I_T;\boldsymbol{L}_2(\Omega))} + \|\dot{\boldsymbol{u}}_m\|^2_{\boldsymbol{H}^{1/2,\,1}(Q_T)} \le c(\varepsilon). \tag{5.1.21}$$

This estimate is sufficient in order to pass to the limit $m \to +\infty$. Due to compact imbedding theorems and trace theorems for Sobolev spaces there exists a sequence $m_k \to 0$ of Galerkin parameters and a corresponding sequence $\boldsymbol{u}_k = \boldsymbol{u}_{m_k}$ of solutions with limit $\boldsymbol{u}$ such that the following convergence properties are valid:

$$\begin{aligned}
&\boldsymbol{u}_k \rightharpoonup \boldsymbol{u} && \text{in } \boldsymbol{H}^1(Q_T),\\
&\dot{\boldsymbol{u}}_k \rightharpoonup \dot{\boldsymbol{u}} && \text{in } \boldsymbol{H}^{1/2,\,1}(Q_T) \text{ and strongly in } \boldsymbol{L}_2(Q_T)\\
& && \text{and in } \boldsymbol{L}_p(S_T) \text{ for all } p < p_0,\\
&\ddot{\boldsymbol{u}}_k \rightharpoonup \ddot{\boldsymbol{u}} && \text{in } L_2(I_T;\boldsymbol{H}^1_0(\Omega)^*),\\
&\dot{\boldsymbol{u}}_k(T,\cdot) \rightharpoonup \dot{\boldsymbol{u}}(T,\cdot) \text{ in } \boldsymbol{L}_2(\Omega),\\
&\mathfrak{F}(\dot{\boldsymbol{u}}_k) \to \mathfrak{F}(\dot{\boldsymbol{u}}) \text{ and } \nabla\Phi_\eta((\dot{\boldsymbol{u}}_k)_{\boldsymbol{t}}) \to \Phi_\eta(\dot{\boldsymbol{u}}_{\boldsymbol{t}})\\
& && \text{strongly in } L_q(S_T) \text{ for all } q < +\infty.
\end{aligned} \tag{5.1.22}$$

Here the value p_0 is chosen such that the imbedding $\boldsymbol{H}^{1/4,\,1/2}(S_{CT}) \hookrightarrow \boldsymbol{L}_{p_0}(S_{CT})$ is valid, that is $p_0 = 2 + 2/N$. Remember that $\boldsymbol{H}^{1/4,1/2}(S_T)$ is the space of traces of $\boldsymbol{H}^{1/2,1}(Q_T)$; see also Lemma 5.2.1 below. The convergence of $\mathfrak{F}(\dot{\boldsymbol{u}}_k)$ and $\nabla\Phi_\eta((\dot{\boldsymbol{u}}_k)_{\boldsymbol{t}})$ follows by the Lebesgue dominated convergence theorem from the combination of the strong convergence of $\dot{\boldsymbol{u}}_k$ in $\boldsymbol{L}_2(S_T)$ with the boundedness of $\mathfrak{F}$ and $\nabla\Phi_\eta$. Let us integrate equation (5.1.18) for Galerkin parameter m_k with solution $\boldsymbol{u}_k$ and test function $\boldsymbol{v} \in H^1(I_T;\mathscr{X}_{m_k})$ with respect to $t \in I_T$. Passing to the limit $k \to \infty$ in the resulting equation shows that the limit function $\boldsymbol{u}$ satisfies equation (5.1.16) for all test functions from $\bigcup_{m=1}^{+\infty} H^1(I_T; \mathscr{X}_m)$. Here in the acceleration term $\langle \ddot{\boldsymbol{u}}_k, \boldsymbol{v} \rangle_{Q_T}$ an integration by parts with respect to the time variable is employed. Since the space $\bigcup_{m=1}^{+\infty} H^1(I_T; \mathscr{X}_m)$ is dense in $\boldsymbol{H}_0^{1/2,1}(Q_T)$, the limit function $\boldsymbol{u}$ solves the variational equation for all admissible test functions. The constant in the *a priori* estimate (5.1.21) is independent of m; hence this estimate is also valid for the limit function $\boldsymbol{u}$.

The convergences $\dot{\boldsymbol{u}}_k \to \dot{\boldsymbol{u}}$ in $\boldsymbol{H}^{1/2,1}(Q_T)$ and $\dot{\boldsymbol{u}}_k(T,\cdot) \to \dot{\boldsymbol{u}}(T,\cdot)$ in $\boldsymbol{L}_2(\Omega)$ are in fact strong. Let us consider the Galerkin equation (5.1.18) at time t with $m = m_k$, solution $\boldsymbol{u}_k$ and test function $\boldsymbol{v} = \dot{\boldsymbol{u}}_k - \pi_k \dot{\boldsymbol{u}}$, where $\pi_k = \pi_{\mathscr{X}_{m_k}}$. We integrate with respect to t and add the term

$$\int_{Q_T} \left[a^{(0)}(\pi_k \boldsymbol{u}, \pi_k \dot{\boldsymbol{u}} - \dot{\boldsymbol{u}}_k) + a^{(1)}(\pi_k \dot{\boldsymbol{u}}, \pi_k \dot{\boldsymbol{u}} - \dot{\boldsymbol{u}}_k) \right] d\boldsymbol{x}\, dt$$
$$+ \int_{Q_T} \pi_k \ddot{\boldsymbol{u}} \cdot \left(\pi_k \dot{\boldsymbol{u}} - \dot{\boldsymbol{u}}_k \right) d\boldsymbol{x}\, dt$$

to both sides of the obtained equation. After an integration of the type

$$\int_{I_T} \dot{w}(t) w(t)\, dt = \frac{1}{2} \left(|w(T)|^2 - |w(0)|^2 \right)$$

we obtain

$$\int_{Q_T} a^{(1)} \left(\pi_k \dot{\boldsymbol{u}} - \dot{\boldsymbol{u}}_k, \pi_k \dot{\boldsymbol{u}} - \dot{\boldsymbol{u}}_k \right) d\boldsymbol{x}\, dt$$
$$+ \frac{1}{2} \int_{\Omega} a^{(0)} \left(\pi_k \boldsymbol{u}(T,\cdot) - \boldsymbol{u}_k(T,\cdot), \pi_k \boldsymbol{u}(T,\cdot) - \boldsymbol{u}_k(T,\cdot) \right) d\boldsymbol{x}$$
$$+ \frac{1}{2} \int_{\Omega} \left| \pi_k \dot{\boldsymbol{u}}(T,\cdot) - \dot{\boldsymbol{u}}_k(T,\cdot) \right|^2 d\boldsymbol{x}$$
$$= \frac{1}{2} \int_{\Omega} a^{(0)} \left(\pi_k \boldsymbol{u}_0 - \boldsymbol{u}_k(0,\cdot), \pi_k \boldsymbol{u}_0 - \boldsymbol{u}_k(0,\cdot) \right) d\boldsymbol{x}$$
$$+ \frac{1}{2} \int_{\Omega} \left| \pi_k \boldsymbol{u}_1 - \dot{\boldsymbol{u}}_k(0,\cdot) \right|^2 d\boldsymbol{x} + \mathscr{L}_k \left(\pi_k \dot{\boldsymbol{u}} - \dot{\boldsymbol{u}}_k \right)$$

with $\mathscr{L}_k$ defined by

$$\mathscr{L}_k(\boldsymbol{v}) = \int_{S_{CT}} \Big[\tfrac{1}{\varepsilon} \left[(\dot{u}_k)_n \right]_+ v_n + \mathfrak{F}(\dot{\boldsymbol{u}}_k) \tfrac{1}{\varepsilon} \left[(\dot{u}_k)_n \right]_+ \cdot$$
$$\cdot \nabla \Phi_\eta \left((\dot{\boldsymbol{u}}_k)_{\boldsymbol{t}} \right) \cdot \boldsymbol{v}_{\boldsymbol{t}} \Big] ds_{\boldsymbol{x}} dt - \mathscr{L}(\boldsymbol{v}).$$

From the convergence properties for $\boldsymbol{u}_k$ we conclude $\mathscr{L}_k \to \mathscr{L}$ strongly in the space $L_2\left(I_T; \boldsymbol{H}_0^1(\Omega)^*\right)$. Passing to the limit $k \to +\infty$ in the previous relation shows, in combination with the strong convergence $\dot{\boldsymbol{u}}_k \to \dot{\boldsymbol{u}}$

in $\boldsymbol{L}_2(Q_T)$ and the coercivity property of Theorem 1.2.3, that $\dot{\boldsymbol{u}}_k \to \dot{\boldsymbol{u}}$ in $L_2\big(I_T;\boldsymbol{H}^1(\Omega)\big)$, $\dot{\boldsymbol{u}}_k(T,\cdot) \to \dot{\boldsymbol{u}}(T,\cdot)$ in $\boldsymbol{L}_2(\Omega)$ and $\boldsymbol{u}_k(T,\cdot) \to \boldsymbol{u}(T,\cdot)$ in $\boldsymbol{H}^1(\Omega)$. Interpolating this result with the weak convergence of the accelerations we obtain the desired assertion.

For volume force $\boldsymbol{f} \in \mathring{\boldsymbol{H}}^{-1/2+\varepsilon',-1}(Q_T)$ we use a sequence $\boldsymbol{f}_k \in L_2\big(I_T;\mathring{\boldsymbol{H}}^{-1}(\Omega)\big)$ that converges to $\boldsymbol{f}$ in $\mathring{\boldsymbol{H}}^{-1/2+\varepsilon',-1}(Q_T)$. The estimate

$$\begin{aligned}\sup_{t\in I_T} \|\dot{\boldsymbol{u}}_k(t,\cdot)\|^2_{\boldsymbol{L}_2(\Omega)} &+ \|\dot{\boldsymbol{u}}_k\|^2_{L_2(I_T;\boldsymbol{H}^1(\Omega))}\\ &\le c_1\|\boldsymbol{f}_k\|_{\mathring{\boldsymbol{H}}^{-1/2+\varepsilon',-1}(Q_T)}\|\dot{\boldsymbol{u}}_k\|_{\boldsymbol{H}^{1/2,1}(Q_T)}\\ &\quad + c_2\|\dot{\boldsymbol{u}}_k\|^2_{L_2(Q_T)} + c_3\end{aligned} \tag{5.1.23}$$

holds. Then the dual estimate

$$\begin{aligned}\|\ddot{\boldsymbol{u}}_k\|_{\boldsymbol{H}^{-1/2+\varepsilon',-1}(Q_T)} \le\ & c_1\|\dot{\boldsymbol{u}}_k\|_{L_2(I_T;\boldsymbol{H}^1(\Omega))}\\ &+ c_2\|\boldsymbol{f}_k\|_{\mathring{\boldsymbol{H}}^{-1/2+\varepsilon',-1}(Q_T)}\end{aligned} \tag{5.1.24}$$

is obtained with the help of test functions from $\boldsymbol{H}^{1/2-\varepsilon',1}(Q_T)$ that vanish on S_T. Employing the extension technique described in Chapters 1 and 2 we extend the functions and their derivatives to $\mathbb{R}^{N+1}$ and use the Fourier transform for these extensions. We use the estimate

$$\begin{aligned}\|\dot{\boldsymbol{u}}_k\|'^2_{\boldsymbol{H}^{1/2,0}(\mathbb{R}^{N+1})} &= \int_{\mathbb{R}^{N+1}} |\mathscr{F}(\dot{\boldsymbol{u}}_k)|^2\, c_1\big(\tfrac{1}{2}\big)\,|\vartheta|\,d\xi\,d\vartheta\\ &\le \sqrt{\int_{\mathbb{R}^{N+1}} |\mathscr{F}(\ddot{\boldsymbol{u}}_k)|^2 \frac{c_1^2(\frac{1}{2})|\vartheta|^2}{1+c_1\big(\frac{1}{2}-\varepsilon'\big)\,|\vartheta|^{1-2\varepsilon'}+|\xi|^2}\,d\xi\,d\vartheta}\\ &\quad\cdot\sqrt{\int_{\mathbb{R}^{N+1}} |\mathscr{F}(\dot{\boldsymbol{u}}_k)|^2\,\big(1+c_1\big(\tfrac{1}{2}-\varepsilon'\big)\,|\vartheta|^{1-2\varepsilon'}+|\xi|^2\big)\,d\xi\,d\vartheta}\\ &= c_1\big(\tfrac{1}{2}\big)\,\|\ddot{\boldsymbol{u}}_k\|_{\boldsymbol{H}^{-1/2+\varepsilon',-1}(\mathbb{R}^{N+1})}\,\|\dot{\boldsymbol{u}}_k\|_{\boldsymbol{H}^{1/2-\varepsilon',1}(\mathbb{R}^{N+1})}\,.\end{aligned} \tag{5.1.25}$$

Combined with (5.1.24), an interpolation with respect to the time variable and Hölder's inequality this leads to the estimate

$$\begin{aligned}\|\dot{\boldsymbol{u}}_k\|'^2_{\boldsymbol{H}^{1/2,0}(Q_T)} &\le\\ &\le \Big(c_1\|\dot{\boldsymbol{u}}_k\|_{L_2(I_T;\boldsymbol{H}^1(\Omega))} + c_2\|\boldsymbol{f}_k\|_{\mathring{\boldsymbol{H}}^{-1/2+\varepsilon',-1}(Q_T)}\Big)\\ &\quad\cdot\Big(\varepsilon_0\|\dot{\boldsymbol{u}}_k\|_{\boldsymbol{H}^{1/2,0}(Q_T)} + c_3(\varepsilon_0,\varepsilon')\|\dot{\boldsymbol{u}}_k\|_{L_2(I_T;\boldsymbol{H}^1(\Omega))}\Big)\end{aligned} \tag{5.1.26}$$

for $\varepsilon_0 > 0$ arbitrarily small. Summing up the estimates (5.1.23)–(5.1.26), we arrive with the help of the Gronwall lemma to the *a priori* estimate

(5.1.17) independent of $k \in \mathbb{N}$. As a consequence, for a subsequence k_m tending to infinity the same weak and strong convergences as in the limit procedure for the Galerkin parameter are valid and the limit $\boldsymbol{u}$ is a weak solution of the problem (5.1.16) with the original $\boldsymbol{f} \in \mathring{\boldsymbol{H}}^{-1/2+\varepsilon,-1}(Q_T)$. The parameter $\varepsilon' > 0$ in the index of the space is required in order to obtain the estimate of $\sup\limits_{t \in I_T} \|\dot{\boldsymbol{u}}_k(t,\cdot)\|^2_{\boldsymbol{L}_2(\Omega)}$ in (5.1.23).

In the case $\boldsymbol{U} \not\equiv 0$ we rewrite variational equation (5.1.16) for the new function $\widetilde{u} \equiv u - \boldsymbol{U}$. This yields the same problem with the new functional

$$\widetilde{\mathscr{L}}(\boldsymbol{v}) \equiv \mathscr{L}(\boldsymbol{v}) - \langle \ddot{\boldsymbol{U}}, \boldsymbol{v}\rangle_{Q_T} - A^{(0)}(\boldsymbol{U}, \boldsymbol{v}) - A^{(1)}(\dot{\boldsymbol{U}}, \boldsymbol{v}).$$

If $\boldsymbol{U} \in \boldsymbol{H}^1(Q_T)$ and $\dot{\boldsymbol{U}} \in \boldsymbol{H}^{1/2+\varepsilon',1}(Q_T)$ holds, then the new functional is still an element of $\mathring{\boldsymbol{H}}^{-1/2+\varepsilon',-1}(Q_T)$ and the existence of the solution $\widetilde{\boldsymbol{u}}$ is proved as above.

The uniqueness of the solution for solution-independent coefficient of friction is verified as follows: let us consider two solutions $\boldsymbol{u}^{(1)}$ and $\boldsymbol{u}^{(2)}$. In the difference of the variational equations with these two solutions the volume force cancels. Then the difference of the accelerations is equal to a functional from $L_2\big(I_T; \boldsymbol{H}^1_0(\Omega)^*\big)$. Hence we are allowed to use there the test function $\big(\boldsymbol{u}^{(1)} - \boldsymbol{u}^{(2)}\big)\chi_t$ with the characteristic function χ_t of the time interval $(0,t)$. For the friction term there holds

$$\begin{aligned}
&\Big(\big[\dot{u}_n^{(1)}\big]_+ \nabla\Phi_\eta\big(\dot{\boldsymbol{u}}_t^{(1)}\big) - \big[\dot{u}_n^{(2)}\big]_+ \nabla\Phi_\eta\big(\dot{\boldsymbol{u}}_t^{(2)}\big)\Big) \cdot \Big(\dot{\boldsymbol{u}}_t^{(1)} - \dot{\boldsymbol{u}}_t^{(2)}\Big) \\
&= \frac{1}{2}\Big(\big[\dot{u}_n^{(1)}\big]_+ + \big[\dot{u}_n^{(2)}\big]_+\Big)\Big(\nabla\Phi_\eta\big(\dot{\boldsymbol{u}}_t^{(1)}\big) - \nabla\Phi_\eta\big(\dot{\boldsymbol{u}}_t^{(2)}\big)\Big) \cdot \Big(\dot{\boldsymbol{u}}_t^{(1)} - \dot{\boldsymbol{u}}_t^{(2)}\Big) \\
&\quad + \frac{1}{2}\Big(\nabla\Phi_\eta\big(\dot{\boldsymbol{u}}_t^{(1)}\big) + \nabla\Phi_\eta\big(\dot{\boldsymbol{u}}_t^{(2)}\big)\Big)\Big(\big[\dot{u}_n^{(1)}\big]_+ - \big[\dot{u}_n^{(2)}\big]_+\Big) \cdot \\
&\qquad \cdot \Big(\dot{\boldsymbol{u}}_t^{(1)} - \dot{\boldsymbol{u}}_t^{(2)}\Big) \\
&\geq -c_1\big|\dot{\boldsymbol{u}}^{(1)} - \dot{\boldsymbol{u}}^{(2)}\big|^2.
\end{aligned}$$

Here the last inequality is true, because Φ_η is convex and therefore $\nabla\Phi_\eta$ is monotone, and $\nabla\Phi_\eta$ is bounded. This yields the estimate

$$\begin{aligned}
&\big\|\big(\dot{\boldsymbol{u}}^{(1)} - \dot{\boldsymbol{u}}^{(2)}\big)(t)\big\|^2_{\boldsymbol{L}_2(\Omega)} + \big\|\dot{\boldsymbol{u}}^{(1)} - \dot{\boldsymbol{u}}^{(2)}\big\|^2_{L_2(I_t;\boldsymbol{H}^1(\Omega))} \\
&\qquad \leq c_1(\varepsilon)\big\|\dot{\boldsymbol{u}}^{(1)} - \dot{\boldsymbol{u}}^{(2)}\big\|^2_{L_2(I_t;\boldsymbol{L}_2(\Gamma_C))} + c_2(\varepsilon)\big\|\dot{\boldsymbol{u}}^{(1)} - \dot{\boldsymbol{u}}^{(2)}\big\|^2_{\boldsymbol{L}_2(Q_t)}
\end{aligned}$$

with constants c_1 and c_2 depending on the penalty parameter. The $L_2(\Gamma_C)$ norm is estimated with the help of the trace theorem and interpolation theorems by

$$\|\boldsymbol{v}\|_{\boldsymbol{L}_2(\Gamma_C)} \le c_1(\lambda)\|\boldsymbol{v}\|_{\boldsymbol{H}^{1/2+\lambda}(\Omega)} \le c_2(\lambda)\|v\|^{1/2-\lambda}_{\boldsymbol{L}_2(\Omega)}\|\boldsymbol{v}\|^{1/2+\lambda}_{\boldsymbol{H}^1(\Omega)}$$
$$\le \lambda\|\boldsymbol{v}\|_{\boldsymbol{H}^1(\Omega)} + c_3(\lambda)\|\boldsymbol{v}\|_{\boldsymbol{L}_2(\Omega)}$$

with λ arbitrarily small. Consequently the inequality

$$\big\|(\dot{\boldsymbol{u}}^{(1)} - \dot{\boldsymbol{u}}^{(2)})(t)\big\|^2_{\boldsymbol{L}_2(\Omega)} \le c_2\big\|\dot{\boldsymbol{u}}^{(1)} - \dot{\boldsymbol{u}}^{(2)}\big\|^2_{\boldsymbol{L}_2(Q_t)} \quad \text{for } t \in I_T$$

follows. Since both functions satisfy the same initial condition, application of the Gronwall lemma proves $\dot{\boldsymbol{u}}^{(1)} = \dot{\boldsymbol{u}}^{(2)}$. □

With the *a priori* estimate (5.1.17) there is no problem to perform the limit procedure for the smoothing parameter $\eta \to 0$ employing the compact character of the penalty- and the friction term. However, this compact character is lost if the penalty parameter ε tends to 0. Then the estimate (5.1.17) is no longer sufficient to pass to the limit $\varepsilon \to 0$ in the non-monotone friction term. For its convergence we need *strong* convergences of all but one factors, and from (5.1.17) we obtain only *weak* convergence. The situation is similar as in the static and quasistatic problem. The remedy of this situation is similar, too: we prove a better regularity of the solution on the contact part of the boundary, then we use compact imbedding theorems for Sobolev spaces and get the required *strong* convergences. Our aim is to prove this important regularity for a solution-dependent coefficient of friction $\mathfrak{F}$ as stated in Assumption 5.1.1. Therefore it is necessary to prove $\sigma_n(\boldsymbol{u}) \in L_2(S_{\mathfrak{F}T})$ (cf. Lemma 1.5.4). A comparison of the analogous situation in Section 3.1 for the regularization parameter $\alpha = 1/2$ outlines the necessity to prove the regularity of the solution for the smoothed problem (5.1.16).

In the proof of the regularity of the solutions certain special trace estimates will be required. They are valid for solutions of the parabolic differential equation with Dirichlet boundary conditions

$$\dot{v}_i - \partial_j\sigma_{ij}(\boldsymbol{v}) = \boldsymbol{f} \text{ on } Q, \tag{5.1.27}$$
$$\boldsymbol{v} = \boldsymbol{w} \text{ on } S \tag{5.1.28}$$

defined on the halfspace $Q \equiv \mathbb{R}\times\mathbb{R}^{N-1}\times\mathbb{R}_+$ with boundary $S \equiv \mathbb{R}\times\mathbb{R}^{N-1}\times\{0\}$ and spatial domain $O = \mathbb{R}^{N-1}\times\mathbb{R}_+$. Here, $\sigma_{ij}(\boldsymbol{u}) \equiv a_{ijk\ell}e_{k\ell}(\boldsymbol{u})$ denotes a stress tensor with bounded, symmetric and elliptic coefficients $a_{ijk\ell}$. The corresponding bilinear form is

$$A(\boldsymbol{u},\boldsymbol{v}) = \int_Q a(\boldsymbol{u},\boldsymbol{v})\,d\boldsymbol{x}\,dt \quad \text{with} \quad a(\boldsymbol{u},\boldsymbol{v}) = a_{ijk\ell}e_{k\ell}(\boldsymbol{u})e_{ij}(\boldsymbol{v}),$$

and

$$\|\boldsymbol{u}\|_A = \sqrt{A(\boldsymbol{u}, \boldsymbol{u})}$$

denotes the energy norm. Since we work on halfspaces, the correct space of weak solutions is no longer $\boldsymbol{H}^{1/2,\,1}(Q)$, because we cannot expect $\boldsymbol{u} \in \boldsymbol{L}_2(Q)$ in general. The right space is

$$\boldsymbol{V}^{1/2,\,1}(Q) := \big\{\boldsymbol{v} \in \boldsymbol{H}^{1/2,\,1}_{\text{loc}}(Q_T)\,;\, \|\boldsymbol{v}\|'_{\boldsymbol{H}^{1/2,\,1}(Q)} < +\infty, \\ \boldsymbol{v}_{|S} \in \boldsymbol{H}^{1/4,\,1/2}(S)\big\}.$$

This space is supplied with the norm $\|\cdot\|'_{\boldsymbol{H}^{1/2,\,1}(Q)}$. Due to the condition $\boldsymbol{v}_{|S} \in \boldsymbol{H}^{1/4,\,1/2}(S)$ this is indeed a norm, because the boundary data of any non-trivial rigid motion $\boldsymbol{u} \in \mathscr{R} \setminus \{0\}$ are not in $\boldsymbol{L}_2(S)$. The dual space of $\boldsymbol{V}^{1/2,\,1}(Q)$ is denoted by $\mathring{\boldsymbol{V}}^{-1/2,-1}(Q)$. For the dual space to $\mathring{\boldsymbol{V}}^{1/2,\,1}(Q) := \big\{\boldsymbol{v} \in \boldsymbol{V}^{1/2,1}(Q)\,;\, \boldsymbol{v} = 0 \text{ on } S\big\}$ the symbol $\boldsymbol{V}^{-1/2,-1}(Q)$ is used. Due to $\boldsymbol{H}^{1/2,\,1}(Q) \subsetneqq \boldsymbol{V}^{1/2,\,1}(Q)$ there holds $\mathring{\boldsymbol{V}}^{-1/2,-1}(Q) \subsetneqq \mathring{\boldsymbol{H}}^{-1/2,-1}(Q)$, and analogously $\boldsymbol{V}^{-1/2,-1}(Q) \subsetneqq \boldsymbol{H}^{-1/2,-1}(Q)$. In our applications, the functions $\boldsymbol{u}$ and $\boldsymbol{f}$ always arise from a localization technique. Hence they originally have compact support contained in a given domain; and in this case, the difference between the V-spaces and the H-spaces disappears. Let us denote the set of (weak) solutions of the equation (5.1.27) for given $\boldsymbol{f} \in \mathring{\boldsymbol{V}}^{-1/2,-1}(Q)$ by

$$\mathfrak{L}(\boldsymbol{f}) \equiv \big\{\boldsymbol{v} \in \boldsymbol{V}^{1/2,\,1}(Q)\,;\, \boldsymbol{v} \text{ solves (5.1.27)}\big\}.$$

As in the static case we use the Fourier transforms

$$\mathscr{F}(\boldsymbol{u};\vartheta,\boldsymbol{\xi}) \equiv (2\pi)^{-(N+1)/2} \int_{\mathbb{R}} \int_{\mathbb{R}^N} \boldsymbol{u}(t,\boldsymbol{x}) e^{-i(t\vartheta+\boldsymbol{x}\cdot\boldsymbol{\xi})}\, d\boldsymbol{x}\, dt \tag{5.1.29}$$

with respect to all variables and

$$\mathscr{F}(\boldsymbol{u};\vartheta,\boldsymbol{\xi}',x_N) \equiv \\ = (2\pi)^{-N/2} \int_{\mathbb{R}} \int_{\mathbb{R}^{N-1}} \boldsymbol{u}(t,\boldsymbol{x}',x_N) e^{-i(t\vartheta+\boldsymbol{x}'\cdot\boldsymbol{\xi}')}\, d\boldsymbol{x}'\, dt \tag{5.1.30}$$

with respect to the time variable and the tangential space variables only. From the inverse Fourier transform there follows

$$\mathscr{F}(\boldsymbol{u};\vartheta,\boldsymbol{\xi}',0) = \frac{1}{\sqrt{2\pi}} \int_{\mathbb{R}} \mathscr{F}(\boldsymbol{u};\vartheta,\boldsymbol{\xi}) d\xi_N.$$

Let us first prove the inverse trace estimate for the solutions of this equation:

5.1.3 Lemma. *Let* $\boldsymbol{f} \in \boldsymbol{V}^{-1/2,-1}(Q)$ *and* $\boldsymbol{w} \in \boldsymbol{H}^{1/4,\,1/2}(S)$. *Then the solution* $\boldsymbol{u}$ *of problem* (5.1.27), (5.1.28) *satisfies*

$$\|\boldsymbol{u}\|_{\boldsymbol{V}^{1/2,\,1}(Q)} \le c_1 \|\boldsymbol{w}\|_{\boldsymbol{H}^{1/4,\,1/2}(S)} + c_2 \|\boldsymbol{f}\|_{\boldsymbol{V}^{-1/2,-1}(Q)}.$$

Proof. We split the solution into a solution $\boldsymbol{u}_I$ of the *inhomogeneous* differential equations with *homogeneous* boundary conditions and a solution $\boldsymbol{u}_H$ of the *homogeneous* differential equations with *inhomogeneous* boundary data. Observe that $\boldsymbol{u}_H$ and $\boldsymbol{u}_I$ need not belong to $\boldsymbol{L}_2(Q)$, even if $\boldsymbol{u}$ does. Due to this reason it is necessary to consider the modified spaces $\boldsymbol{V}^{1/2,1}(Q)$ and $\boldsymbol{V}^{-1/2,-1}(Q)$ introduced above. Of course, the Fourier transforms of elements from these spaces exist in the sense of Fourier transforms of distributions, but they need not be square-integrable, either.

The function $\boldsymbol{u}_I$ is a solution of the variational formulation

$$\langle \dot{\boldsymbol{u}}_I, \boldsymbol{v} \rangle_Q + A(\boldsymbol{u}_I, \boldsymbol{v}) = \langle \boldsymbol{f}, \boldsymbol{v} \rangle_Q \tag{5.1.31}$$

valid for all $\boldsymbol{v} \in \mathring{\boldsymbol{V}}^{1/2,\,1}(Q)$. Hence we obtain the *a priori* estimate

$$\|\varepsilon(\boldsymbol{u}_I)\|^2_{L_2(Q;\mathbb{R}^{N,N})} \le c \|\boldsymbol{f}\|_{\boldsymbol{V}^{-1/2,-1}(Q)} \, \|\boldsymbol{u}_I\|_{\boldsymbol{V}^{1/2,\,1}(Q)} \tag{5.1.32}$$

and the dual estimate

$$\|\dot{\boldsymbol{u}}_I\|_{\boldsymbol{V}^{-1/2,-1}(Q)} \le c_1 \|\boldsymbol{f}\|_{\boldsymbol{V}^{-1/2,-1}(Q)} + c_2 \, \|\varepsilon(\boldsymbol{u}_I)\|_{L_2(Q;\mathbb{R}^{N,N})}. \tag{5.1.33}$$

The solution $\boldsymbol{u}_I$ is extended from the halfspace Q onto the whole space $\mathbb{R} \times \mathbb{R}^N$ by

$$\boldsymbol{u}_I(t, \boldsymbol{x}', -x_N) = -\,\boldsymbol{u}_I(t, \boldsymbol{x}', x_N) \text{ for } x_N < 0$$
$$\text{with } \boldsymbol{x}' = (x_1, \ldots, x_{N-1})^\top.$$

Since $\boldsymbol{u}_I$ is in $\boldsymbol{V}^{1/2,\,1}(Q)$ with homogeneous Dirichlet boundary conditions, the extended function is in $\boldsymbol{V}^{1/2,\,1}(\mathbb{R} \times \mathbb{R}^N) := \{\boldsymbol{v} \in \boldsymbol{H}^{1/2,\,1}_{\text{loc}}(\mathbb{R} \times \mathbb{R}^N)\,;\, \|\boldsymbol{v}\|'_{\boldsymbol{H}^{1/2,\,1}(\mathbb{R}\times\mathbb{R}^N)} < +\infty\}$. For any $\boldsymbol{v} \in \boldsymbol{V}^{1/2,\,1}(\mathbb{R} \times \mathbb{R}^N)$ there holds

$$\begin{aligned}\langle \dot{\boldsymbol{u}}_I, \boldsymbol{v} \rangle_{\mathbb{R}\times\mathbb{R}^N} &= \langle \dot{\boldsymbol{u}}_I, \boldsymbol{v} \rangle_{\mathbb{R}\times\mathbb{R}^{N-1}\times\mathbb{R}_+} + \langle \dot{\boldsymbol{u}}_I, \boldsymbol{v} \rangle_{\mathbb{R}\times\mathbb{R}^{N-1}\times\mathbb{R}_-}\\ &= \langle \dot{\boldsymbol{u}}_I, \boldsymbol{v}_+ - \boldsymbol{v}_- \rangle_Q\end{aligned}$$

with $\boldsymbol{v}_+(t, \boldsymbol{x}', x_N) = \boldsymbol{v}(t, \boldsymbol{x}', x_N)$ and $\boldsymbol{v}_-(t, \boldsymbol{x}', x_N) = \boldsymbol{v}(t, \boldsymbol{x}', -x_N)$ for $x_N > 0$. The difference $\boldsymbol{v}_+ - \boldsymbol{v}_-$ has vanishing boundary data on $\mathbb{R} \times \mathbb{R}^{N-1} \times \{0\}$ and is therefore an admissible test function for variational equation (5.1.31). Moreover, $\|\boldsymbol{v}_+ - \boldsymbol{v}_-\|_{\boldsymbol{V}^{1/2,\,1}(Q)} \le c\|\boldsymbol{v}\|_{\boldsymbol{V}^{1/2,1}(\mathbb{R}\times\mathbb{R}^N)}$ is valid. Consequently there holds

$$\|\dot{\boldsymbol{u}}_I\|_{\boldsymbol{V}^{-1/2,-1}(\mathbb{R}\times\mathbb{R}^N)} \le c_1 \|\boldsymbol{f}\|_{\boldsymbol{V}^{-1/2,-1}(Q)} + c_2 \, \|\varepsilon(\boldsymbol{u}_I)\|_{L_2(Q)}. \tag{5.1.34}$$

Via the Fourier transform with respect to all variables the following estimate is derived:

$$\|\boldsymbol{u}_I\|'^2_{\boldsymbol{H}^{1/2,0}(\mathbb{R}\times\mathbb{R}^N)} = \int_{\mathbb{R}\times\mathbb{R}^N} c_1\left(\tfrac{1}{2}\right) |\mathscr{F}(\boldsymbol{u}_I;\vartheta,\boldsymbol{\xi})|^2 \, |\vartheta| \, d\boldsymbol{\xi}\, d\vartheta$$

$$\leq c_1\left(\tfrac{1}{2}\right) \sqrt{\int_{\mathbb{R}\times\mathbb{R}^N} |\mathscr{F}(\boldsymbol{u}_I;\vartheta,\boldsymbol{\xi})|^2 \frac{|\vartheta|^2}{c_1\left(\frac{1}{2}\right)|\vartheta| + |\boldsymbol{\xi}|^2} \, d\boldsymbol{\xi}\, d\vartheta}$$

$$\sqrt{\int_{\mathbb{R}\times\mathbb{R}^N} |\mathscr{F}(\boldsymbol{u}_I;\vartheta,\boldsymbol{\xi})|^2 \left(c_1\left(\tfrac{1}{2}\right)|\vartheta| + |\boldsymbol{\xi}|^2\right) d\boldsymbol{\xi}\, d\vartheta}$$

$$= c \, \|\dot{\boldsymbol{u}}_I\|_{\boldsymbol{V}^{-1/2,-1}(Q)} \, \|\boldsymbol{u}_I\|_{\boldsymbol{V}^{1/2,1}(Q)} \, .$$

Therefore we have

$$\|\boldsymbol{u}_I\|'^2_{\boldsymbol{H}^{1/2,0}(Q)} \leq c \, \|\dot{\boldsymbol{u}}_I\|_{\boldsymbol{V}^{-1/2,-1}(Q)} \, \|\boldsymbol{u}_I\|_{\boldsymbol{V}^{1/2,1}(Q)} \, .$$

Due to the Korn inequality (Corollary 1.2.2) for the halfspace the norms $\|\varepsilon(\cdot)\|_{L_2(O)}$ and $\|\nabla\cdot\|_{L_2(O)}$ are equivalent. From this, the previous inequality and the estimates (5.1.32), (5.1.34) there follows

$$\|\boldsymbol{u}_I\|_{\boldsymbol{V}^{1/2,1}(Q)} \leq c\|\boldsymbol{f}\|_{\boldsymbol{V}^{-1/2,-1}(Q)}.$$

For the inhomogeneous boundary conditions we employ an extension of the boundary data onto the whole domain, defined via the heat equation

$$\mu\, \dot{z} - \Delta z = 0. \tag{5.1.35}$$

The solution of this equation is given in terms of its (scalar) boundary data w on S by

$$\mathscr{F}(z;\vartheta,\boldsymbol{\xi}',x_N) = e^{-\sqrt{|\boldsymbol{\xi}'|^2 + i\mu\vartheta}\, x_N} \mathscr{F}(w;\vartheta,\boldsymbol{\xi}'). \tag{5.1.36}$$

Here, the root of the complex value $|\boldsymbol{\xi}'|^2 + i\mu\vartheta$ is given by

$$\mathfrak{a} = \sqrt{|\boldsymbol{\xi}'|^2 + i\mu\vartheta} = \mathfrak{a}_1 + i\mathfrak{a}_2 \tag{5.1.37}$$

with $\mathfrak{a}_1 = \sqrt{\dfrac{\sqrt{|\boldsymbol{\xi}'|^4 + \mu^2|\vartheta|^2} + |\boldsymbol{\xi}'|^2}{2}}$, $\mathfrak{a}_2 = \operatorname{sign}\vartheta \sqrt{\dfrac{\sqrt{|\boldsymbol{\xi}'|^4 + \mu^2|\vartheta|^2} - |\boldsymbol{\xi}'|^2}{2}}$.
The parameter $\mu > 0$ introduced here will be important in order to get optimal estimates in the following section; here we restrict ourselves to $\mu = 1$. The extension $\mathscr{E}w$ of our vector-valued function is then defined by $(\mathscr{E}w)_i = z$ with z being the solution of (5.1.35) for boundary value w_i with

parameter $\mu = 1$. Using the corresponding representation (5.1.36) and the formula $\mathfrak{a}_2^2 + |\boldsymbol{\xi}'|^2 = \mathfrak{a}_1^2$, it is possible to calculate the seminorm

$$\begin{aligned}\|\mathscr{E}\boldsymbol{w}\|'^2_{\boldsymbol{V}^{1/2,\,1}(Q)} &= \int_{\mathbb{R}^N} |\mathscr{F}(\boldsymbol{w};\vartheta,\boldsymbol{\xi}')|^2 \int_{\mathbb{R}_+} \left(c_1\left(\tfrac{1}{2}\right)|\vartheta| + |\boldsymbol{\xi}'|^2 + |\mathfrak{a}|^2\right) \\ &\qquad\qquad \left|e^{-2\mathfrak{a}x_N}\right| \, dx_N \, d\boldsymbol{\xi}' \, d\vartheta \\ &= \int_{\mathbb{R}^N} |\mathscr{F}(\boldsymbol{w};\vartheta,\boldsymbol{\xi}')|^2 \left(c_1\left(\tfrac{1}{2}\right)\frac{|\vartheta|}{2\mathfrak{a}_1} + \mathfrak{a}_1\right) d\vartheta \, d\boldsymbol{\xi}' \\ &\le c_1 \|\boldsymbol{w}\|_{\boldsymbol{H}^{1/4,\,1/2}(Q)}.\end{aligned}$$

Here the last inequality is valid because of the relations

$$\max\left\{|\boldsymbol{\xi}'|, \sqrt{|\vartheta|/2}\right\} \le \mathfrak{a}_1 \le |\boldsymbol{\xi}'| + \sqrt{|\vartheta|/2}.$$

The function $\boldsymbol{u}_H$ can be represented by $\boldsymbol{u}_H = \mathscr{E}\boldsymbol{w} + \widetilde{\boldsymbol{u}}_H$, where $\widetilde{\boldsymbol{u}}_H$ now has vanishing boundary data. The variational formulation of the differential equations gives

$$\langle \dot{\widetilde{\boldsymbol{u}}}_H, \boldsymbol{v}\rangle_Q + A(\widetilde{\boldsymbol{u}}_H, \boldsymbol{v}) = -\big\langle (\mathscr{E}\boldsymbol{w})\dot{}, \boldsymbol{v}\big\rangle_Q - A(\mathscr{E}\boldsymbol{w}, \boldsymbol{v})$$

valid for every $\boldsymbol{v} \in \mathring{\boldsymbol{V}}^{1/2,1}(Q)$. For the right hand side of the previous equation there holds

$$\left|\big\langle (\mathscr{E}\boldsymbol{w})\dot{}, \boldsymbol{v}\big\rangle_Q + A(\mathscr{E}\boldsymbol{w}, \boldsymbol{v})\right| \le c\|\mathscr{E}\boldsymbol{w}\|_{\boldsymbol{V}^{1/2,\,1}(Q)} \|\boldsymbol{v}\|_{\boldsymbol{V}^{1/2,\,1}(Q)}.$$

Hence we can apply the just proved result for the homogeneous boundary conditions with linear functional $\boldsymbol{f} : v \mapsto -\big\langle (\mathscr{E}\boldsymbol{w})\dot{}, \boldsymbol{v}\big\rangle_Q - A(\mathscr{E}\boldsymbol{w}, \boldsymbol{v})$ and obtain

$$\|\widetilde{\boldsymbol{u}}_I\|_{\boldsymbol{V}^{1/2,1}(Q)} \le c_1 \|\mathscr{E}\boldsymbol{w}\|_{\boldsymbol{V}^{1/2,1}(Q)} \le c_2 \|\boldsymbol{w}\|_{\boldsymbol{H}^{1/4,\,1/2}(Q)}.$$

Application of the triangle inequality proves the Proposition. □

The special estimates required for our purpose are given in the next proposition.

5.1.4 Proposition. *Let* $\boldsymbol{f} \in \boldsymbol{V}^{-1/2,-1}(Q)$.

(i) *There exist constants* D_0 *and* K *and to every* $\boldsymbol{w} \in \boldsymbol{H}^{1/4,\,1/2}(S)$ *with* $\boldsymbol{w_t} = 0$ *there exists an extension* $\boldsymbol{v} \in \boldsymbol{V}^{1/2,\,1}(Q)$ *such that for any function* $\boldsymbol{u} \in \mathfrak{L}(\boldsymbol{f})$

$$\begin{aligned}&\big\langle \dot{\boldsymbol{u}}, \boldsymbol{v}\big\rangle_Q + A(\boldsymbol{u}, \boldsymbol{v}) \\ &\quad \le \Big(D_0\|\boldsymbol{u}\|_A + K\|\boldsymbol{f}\|_{\boldsymbol{V}^{-1/2,-1}(Q)}\Big) \|\boldsymbol{w}\|_{\boldsymbol{H}^{1/4,\,1/2}(S)}.\end{aligned} \tag{5.1.38}$$

(ii) There exist constants D_1, D_2, and K such that for every $\boldsymbol{u} \in \mathfrak{L}(\boldsymbol{f})$ satisfying (5.1.28) *there holds*

$$\|\boldsymbol{w_t}\|'_{\boldsymbol{H}^{0,1/2}(S)} \leq D_1\|\boldsymbol{u}\|_A + K\|\boldsymbol{f}\|_{\boldsymbol{V}^{-1/2,-1}(Q)} \quad \text{and} \tag{5.1.39}$$

$$\|\boldsymbol{w_t}\|'_{\boldsymbol{H}^{1/4,0}(S)} \leq D_2\|\boldsymbol{u}\|_A + K\|\boldsymbol{f}\|_{\boldsymbol{V}^{-1/2,-1}(Q)}. \tag{5.1.40}$$

In the case $\boldsymbol{f} = 0$ relations (5.1.39) and (5.1.40) represent a special trace estimate for the space $\boldsymbol{V}^{1/2,1}(Q)$ and the corresponding set of traces $\boldsymbol{H}^{1/4,1/2}(S)$ (equipped with the norm $\|\cdot\|'_{\boldsymbol{H}^{1/4,1/2}(S)}$), restricted to solutions of equation (5.1.27). Estimate (5.1.38) follows from the corresponding inverse trace theorem for the function w and from the relation

$$\|\boldsymbol{u}\|'_{\boldsymbol{V}^{1/2,1}(Q)} \leq c\|\boldsymbol{u}\|_A.$$

This inequality is true for solutions of equation (5.1.27) with $f = 0$ only; it is proved by means of the *a priori* estimate and the dual estimate for this equation. The form of the estimates is adapted to the needs in the regularity proof to be performed below. For our purpose it is important to get precise and optimal (as small as possible) values of the constants D_j, $j = 0, 1, 2$. The proof of (5.1.38), (5.1.39) and (5.1.40) including the calculation of these optimal values for different relevant cases will be given in Sections 5.2 and Subsections 5.3.1, 5.3.2.

For non-vanishing $\boldsymbol{f}$ the term $K\|\boldsymbol{f}\|_{\boldsymbol{V}^{-1/2,-1}(Q)}$ on the right hand sides comes from the already mentioned additive split $\boldsymbol{u} = \boldsymbol{u}_H + \boldsymbol{u}_I$ into a solution $\boldsymbol{u}_H$ of the *homogeneous* differential equations with *inhomogeneous* boundary data $\boldsymbol{u}_H = \boldsymbol{w}$ on S and a solution $\boldsymbol{u}_I$ of the *inhomogeneous* differential equation with *homogeneous* boundary conditions, and the relation $\|\boldsymbol{u}_I\|_{\boldsymbol{V}^{1/2,1}(Q)} \leq c\|\boldsymbol{f}\|_{\boldsymbol{V}^{-1/2,-1}(Q)}$ proved in Proposition 5.1.3. Then relations (5.1.39) and (5.1.40) follow easily from the triangle inequality $\|\boldsymbol{u}_H\|_A \leq \|\boldsymbol{u}\|_A + \|\boldsymbol{u}_I\|_A$. For the proof of relation (5.1.38) we have

$$\begin{aligned}
&\langle \dot{\boldsymbol{u}}, \boldsymbol{v}\rangle_Q + A(\boldsymbol{u}, \boldsymbol{v}) \\
&\quad = \langle \dot{\boldsymbol{u}}_H, \boldsymbol{v}\rangle_Q + A(\boldsymbol{u}_H, \boldsymbol{v}) + \langle \dot{\boldsymbol{u}}_I, \boldsymbol{v}\rangle_Q + A(\boldsymbol{u}_I, \boldsymbol{v}) \\
&\quad \leq D_0\|\boldsymbol{u}_H\|_A\|\boldsymbol{w}\|_{\boldsymbol{H}^{1/4,1/2}(S)} + c_1\|\boldsymbol{u}_I\|_{\boldsymbol{V}^{1/2,1}(Q)}\|\boldsymbol{v}\|_{\boldsymbol{V}^{1/2,1}(Q)}
\end{aligned}$$

and the inverse trace estimate

$$\|\boldsymbol{v}\|_{\boldsymbol{V}^{1/2,1}(Q)} \leq c_2\|\boldsymbol{w}\|_{\boldsymbol{H}^{1/4,1/2}(S)}. \tag{5.1.41}$$

Here it is important that the extension of $\boldsymbol{w}$ performed for the different cases in the following sections actually satisfies this relation. There the extension is done by either the heat equation (5.1.35) with parameter $\mu > 0$ or by the equation $\dot{v}_i - \partial_j \sigma_{ij}(\boldsymbol{v}) = 0$. In both cases (5.1.41) will be valid.

In the trace estimates for halfspaces the requirement $\boldsymbol{f} \in \boldsymbol{V}^{-1/2,-1}(Q)$ was sufficient. This corresponds to the condition $\mathscr{L} \in \boldsymbol{H}^{-1/2,-1}(Q_T)$ in the case of a bounded time interval, a bounded domain and more general boundary conditions. Indeed it is possible to prove the existence of solutions for our parabolic problem in this case, too. For simplicity of the presentation this will be explained at the example of a linear problem only. The elastic part of the bilinear form will be neglected; its inclusion does not change the proof and only enlarges the formulae. The displacement velocity of the problem will be denoted by $\boldsymbol{u}$. The boundary data consist of given boundary displacements on a measurable part Γ_U of the boundary and Neumann conditions on the remaining boundary. The latter are included in the functional $\mathscr{L}$. The initial data $\boldsymbol{u}_1$ and the Dirichlet data $\boldsymbol{U}$ will be set to zero. This is no restriction, because we are only interested in the influence of the functional $\mathscr{L}$, and any problem with non-zero data $\boldsymbol{u}_1$ and $\boldsymbol{U}$ can be represented as a sum of a solution for the problem with functional $\mathscr{L}$ and zero initial and Dirichlet data, and a solution with zero functional and the given data $(\boldsymbol{u}_1, \boldsymbol{U})$. Hence the variational formulation of the problem to be studied is

Find $\boldsymbol{u} \in \boldsymbol{H}_0^{1/2,1}(Q_T)$ *with* $\boldsymbol{u}(0,\cdot) = 0$ *such that*

$$\langle \dot{\boldsymbol{u}}, \boldsymbol{v} \rangle_{Q_T} + \langle \sigma_{ij}(\boldsymbol{u}), e_{ij}(\boldsymbol{v}) \rangle_{Q_T} = \langle \mathscr{L}, \boldsymbol{v} \rangle_{Q_T} \quad \text{for all } \boldsymbol{v} \in \boldsymbol{H}_0^{1/2,1}(Q_T). \tag{5.1.42}$$

The main problem in studying this equation for $\mathscr{L} \in \boldsymbol{H}^{-1/2,-1}(Q_T)$ is the proper definition of initial and terminal data. The weak properties of $\mathscr{L}$ do not permit defining the solution on every time level. However, the problem can be easily extended onto a larger time interval, say $\widetilde{I}_T \equiv (-1, T+1)$, with functional $\widetilde{\mathscr{L}} \in \boldsymbol{H}^{-1/2,-1}(\widetilde{Q}_T)$ defined by $\langle \widetilde{\mathscr{L}}, \boldsymbol{v} \rangle_{\widetilde{Q}_T} = \langle \mathscr{L}, \boldsymbol{v} \rangle_{Q_T}$, where $\widetilde{Q}_T = \widetilde{I}_T \times \Omega$. Due to the zero initial and Dirichlet data the solution $\widetilde{\boldsymbol{u}}$ of the extended problem remains zero for $t \in (-1, 0)$. Its restriction to Q_T is *defined* to be the solution of the original problem. This definition makes sense, if the extended problem is uniquely solvable. In order to prove this we use an approximating sequence $\mathscr{L}_k \in L_2(I_T; \boldsymbol{H}^{-1}(\Omega))$ tending to $\mathscr{L}$ in $\boldsymbol{H}^{-1/2,-1}(Q_T)$. For the solutions $\boldsymbol{u}_k$ to the problem with $\mathscr{L}_k$, the L_2-type *a priori* estimate

$$\|\boldsymbol{u}_k(T,\cdot)\|^2_{\boldsymbol{L}_2(\Omega)} + \|\varepsilon(\boldsymbol{u}_k)\|^2_{L_2(Q_T;\mathbb{R}^{N,N})} \le c_1 \|\boldsymbol{u}_k\|_{\boldsymbol{H}^{1/2,1}(Q_T)} \|\mathscr{L}_k\|_{\boldsymbol{H}^{-1/2,-1}(Q_T)}$$

and the dual estimate

$$\|\dot{\boldsymbol{u}}_k\|_{\boldsymbol{H}^{-1/2,-1}(Q_T)} \le c_2 \|\varepsilon(\boldsymbol{u}_k)\|_{L_2(Q_T;\mathbb{R}^{N,N})} + c_3 \|\mathscr{L}_k\|_{\boldsymbol{H}^{-1/2,-1}(Q_T)}$$

are valid with constants c_1, c_2, c_3 independent of k. This relation holds for both the extended and the original problems with the same constants. The initial condition $\boldsymbol{u}(0) = 0$ is defined in the sense that the extension of $\boldsymbol{u}$ by zero for $t \in (-1, 0)$ remains in $H^{1/2}\big((-1,T); \boldsymbol{L}_2(\Omega)\big)$. From the definition of the Sobolev-Slobodetskii norm it is easily seen that this requirement implies

$$\int_0^T \frac{\|\boldsymbol{u}_k(s)\|^2_{\boldsymbol{L}_2(\Omega)}}{|s|}\, ds < c_1 \|\boldsymbol{u}_k\|'^2_{H^{1/2}(\tilde{I}_T; \boldsymbol{L}_2(\Omega))}.$$

The left hand side of this relation is an upper bound for $\|\boldsymbol{u}_k\|^2_{\boldsymbol{L}_2(Q_T)}$, if T is finite. Moreover, from Theorem 1.2.3 about coerciveness of stresses there follows that the $\boldsymbol{H}^1(\Omega)$ norm is equivalent to $\|\cdot\|_{\boldsymbol{L}_2(\Omega)} + \|\varepsilon(\cdot)\|_{\boldsymbol{L}_2(\Omega;\mathbb{R}^{N,N})}$. As a consequence, the norm in $\boldsymbol{H}^{1/2,\,1}(Q_T)$ is equivalent to

$$\|\varepsilon(\cdot)\|_{L_2(Q_T;\mathbb{R}^{N,N})} + \|\cdot\|'_{H^{1/2}(\tilde{I}_T; \boldsymbol{L}_2(\Omega))}.$$

The $H^{1/2}\big(\tilde{I}_T; \boldsymbol{L}_2(\Omega)\big)$ seminorm of $\boldsymbol{u}_k$ is estimated with the interpolation

$$\|\boldsymbol{u}_k\|'_{H^{1/2}(\tilde{I}_T; \boldsymbol{L}_2(\Omega))} \le c \|\dot{\boldsymbol{u}}_k\|^{1/2}_{\boldsymbol{H}^{-1/2,-1}(\tilde{Q}_T)} \|\boldsymbol{u}_k\|^{1/2}_{\boldsymbol{H}^{1/2,1}(\tilde{Q}_T)}.$$

This relation follows easily with the use of the Fourier transform after an extension of $\boldsymbol{u}_k$ onto the full space $\mathbb{R} \times \mathbb{R}^N$, in a way already described on page 282. The extension with respect to the time variable is done by $\boldsymbol{u}(t,\cdot) = 0$ for $t < -1$. For $t \in (T+1, +\infty)$ we extend the functional $\mathscr{L}$ by 0 and we employ a smooth time localization function $\phi : \mathbb{R} \to [0,1]$ such that $\phi_{|(-\infty,T+1)} = 1$ and $\phi_{|(T+2,+\infty)} = 0$. Then $\phi \boldsymbol{u}_k$ is a solution of a parabolic problem with $\mathscr{L}$ supported on $[0,T] \cup (T+1, T+2)$ such that $\mathscr{L}(v) = \langle (\phi \boldsymbol{u})\dot{}, \boldsymbol{v}\rangle + \langle \sigma_{ij}(\phi \boldsymbol{u}), e_{ij}(\boldsymbol{v})\rangle$ on $(T+1, T+2)$. The extension with respect to the space variables is done by means of the local rectification of the boundary and the localization technique like it is done e.g. in the proofs of the next theorems. The combination of all estimates shows

$$\begin{aligned}\|\boldsymbol{u}_k\|^2_{\boldsymbol{H}^{1/2,1}(Q_T)} &\le c_1 \|\mathscr{L}_k\|^{1/2}_{\boldsymbol{H}^{-1/2,-1}(Q_T)} \|\boldsymbol{u}_k\|^{3/2}_{\boldsymbol{H}^{1/2,\,1}(Q_T)} \\ &\quad + c_2 \|\mathscr{L}_k\|_{\boldsymbol{H}^{-1/2,-1}(Q_T)} \|\boldsymbol{u}_k\|_{\boldsymbol{H}^{1/2,1}(Q_T)}\end{aligned}$$

which proves the *a priori* estimate

$$\|\boldsymbol{u}_k(T,\cdot)\|_{\boldsymbol{L}_2(\Omega)} + \|\boldsymbol{u}_k\|_{\boldsymbol{H}^{1/2,1}(Q_T)} \le c \|\mathscr{L}_k\|_{\boldsymbol{H}^{-1/2,-1}(Q_T)}$$

uniform in k. Since the problem is linear, this also proves that the solution $\boldsymbol{u}_k$ is unique for every $k \in \mathbb{N}$, and that $\boldsymbol{u}_k$ is a Cauchy sequence in

$\boldsymbol{H}^{1/2,\,1}(Q_T)$ and $\boldsymbol{u}_k(T,\cdot)$ is a Cauchy sequence in $\boldsymbol{L}_2(\Omega)$. According to the above definition, the limit $\boldsymbol{u}$ of the sequence $\boldsymbol{u}_k$ is a solution of the original problem with volume force $\boldsymbol{f}$. Obviously this solution is unique. Hence the following Theorem is proved:

5.1.5 Theorem. *Let Ω be a bounded Lipschitz domain and $\mathscr{L}$ belong to $\boldsymbol{H}^{-1/2,-1}(Q_T)$. Then there exists a unique solution $\boldsymbol{u} \in \boldsymbol{H}^{1/2,1}(Q_T)$ of problem* (5.1.42).

The proof of a better regularity for the contact problem with friction is possible if the following assumption is valid:

5.1.6 Assumption. *In addition to Assumption 5.1.1 let $\Gamma_C \in C^\beta$ for some $\beta > 2$. With a subdomain $\Omega_C \subset \Omega$ that satisfies $\Gamma_C \subset \partial\Omega_C$ and with $Q_{CT} \equiv I_T \times \Omega_C$ the local regularity properties $a^{(\iota)}_{ijk\ell} \in C^{\beta'}(\Omega_C)$ with $\beta' > \frac{1}{2}$, $i,j,k,\ell = 1,\ldots,N$, $\iota = 1,2$, $\boldsymbol{f} \in \mathring{\boldsymbol{H}}^{-1/4,-1/2}(Q_{CT})$, and $\boldsymbol{u}_0, \boldsymbol{u}_1 \in \boldsymbol{H}^{3/2}(\Omega_C)$ shall be valid; and $\dot{U} = 0$ on Q_{CT}. The coefficient of friction $\mathfrak{F} = \mathfrak{F}(\boldsymbol{x}, \dot{\boldsymbol{u}})$ shall be bounded by the constant*

$$C_{\mathfrak{F}} = \left(D_0\sqrt{D_1^2 + D_2^2}\right)^{-1}, \tag{5.1.43}$$

$\|\mathfrak{F}\|_{L_\infty(\Gamma_C\times\mathbb{R}^N)} < C_{\mathfrak{F}}$. The support of $\mathfrak{F}$ shall be contained in $\Gamma_{\mathfrak{F}} \times \mathbb{R}^N$ with a $\Gamma_{\mathfrak{F}} \subset \Gamma_C$ satisfying $\operatorname{dist}(\boldsymbol{x}, \Gamma \setminus \Gamma_C) \geq \delta_0 > 0$ *for all $\boldsymbol{x} \in \Gamma_{\mathfrak{F}}$.*

5.1.7 Theorem. *Under the above mentioned assumptions the solution of the smoothed and penalized dynamic contact problem* (5.1.16) *satisfies the* a priori *estimate*

$$\|\dot{\boldsymbol{u}}\|_{\boldsymbol{H}^{1/2,\,1}(S_{\mathfrak{F}T})} + \left\|\tfrac{1}{\varepsilon}[\dot{u}_n]_+\right\|_{L_2(S_{\mathfrak{F}T})} \leq c \tag{5.1.44}$$

with $S_{\mathfrak{F}T} \equiv I_T \times \Gamma_{\mathfrak{F}}$ and constant c independent of the approximation parameters ε, η.

Proof. The estimate is proved with the local rectification of the boundary and the shift technique as described in Section 1.7. In order to prove the result for a solution-dependent coefficient of friction, we need the boundary stress to be in $L_2(S_{\mathfrak{F}T})$. This corresponds to a regularity gain of $\frac{1}{2}$ with respect to the space and $\frac{1}{4}$ with respect to the time variable. Let us recall the basic notations of Section 1.7. For a sufficiently small parameter $\delta > 0$ we have a finite covering $\mathscr{U}_\delta := \left\{U_i; i \in \mathscr{I}_\delta\right\}$ of some neighbourhood of $\Gamma_{\mathfrak{F}}$. Each element U_i is transformed by a local rectification $\mathfrak{R}_i$ to a subset O_i of the halfspace $\mathbb{R}^{N-1} \times \mathbb{R}_+$ with $\operatorname{diam} O_i \leq \delta$. The rectification map is defined by $\mathfrak{R}_i = \Psi_i \circ \mathscr{O}_i$, where $\mathscr{O}_i$ is composed of a rotation and a shift transforming the tangent plane of Γ at a point $x \in U_i$ to $\mathbb{R}^{N-1} \times \{0\}$, and $\Psi : \boldsymbol{x} \equiv (\boldsymbol{x}', x_N) \mapsto (\boldsymbol{x}', x_N - \psi(\boldsymbol{x}'))$ with the function $\psi : \mathbb{R}^{N-1} \to \mathbb{R}$ from the local representation of the boundary as a graph over its tangent plane. Due to the required smoothness of the boundary and the precise form of the rectification there holds $\psi_i \in C^\beta$ and

$$\|\psi_i\|_{C^1} \le c_1\delta. \tag{5.1.45}$$

Let $\{\varrho_i; i \in \mathscr{I}_\delta\}$ be a partition of unity in a neighbourhood of $\Gamma_{\mathfrak{F}}$ subordinate to the covering $\mathscr{U}_\delta$. For a parameter $L \in \mathbb{R}$ let us define a cutoff function $\phi_L(t) \equiv \phi_0(t-L)$ with a non-increasing function $\phi_0 \in C^2(\mathbb{R};[0,1])$ satisfying $\phi_0(t) = 1$ for $t < -1$ and $\phi_0(t) = 0$ for $t > 0$. Then, for fixed parameters $i \in \mathscr{I}_\delta$ and $L \in \mathbb{R}$ a cutoff function $\rho(t,x)$ for time- and space variable is defined by $\rho(t,\boldsymbol{x}) \equiv \phi_L(t)\varrho_i(\boldsymbol{x})$. For simplicity of the notation we omit the indices i and L in the sequel.

We use the test function $\rho \boldsymbol{v}$ in variational equation (5.1.16) and shift the cutoff function ρ from the right hand side of each form to its left hand side; this gives

$$\begin{aligned}
&\int_{Q_T} \Big((\rho\dot{\boldsymbol{u}})^{\cdot} \cdot \boldsymbol{v} + a^{(0)}(\rho\boldsymbol{u},\boldsymbol{v}) + a^{(1)}(\rho\dot{\boldsymbol{u}},\boldsymbol{v}) \\
&\qquad\qquad + b^{(0)}(\boldsymbol{u},\boldsymbol{v}) + b^{(1)}(\dot{\boldsymbol{u}},\boldsymbol{v}) \Big)\, d\boldsymbol{x}\, dt \\
&\quad + \int_{S_{CT}} \Big(\tfrac{1}{\varepsilon}[\rho\dot{u}_n]_+ v_n + \mathfrak{F}\tfrac{1}{\varepsilon}[\rho\dot{u}_n]_+ \nabla\Phi_\eta(\dot{\boldsymbol{u}}_{\boldsymbol{t}}) \cdot \boldsymbol{v}_{\boldsymbol{t}} \Big)\, ds_{\boldsymbol{x}}\, dt \\
&= \int_{Q_T} (\rho\boldsymbol{f} + \dot{\rho}\dot{\boldsymbol{u}}) \cdot \boldsymbol{v}\, d\boldsymbol{x}\, dt
\end{aligned}$$

with the forms $b^{(\iota)}(\boldsymbol{u},\boldsymbol{v}) \equiv a^{(\iota)}(\boldsymbol{u},\rho\boldsymbol{v}) - a^{(\iota)}(\rho\boldsymbol{u},\boldsymbol{v})$ for $\iota = 0,1$. These forms satisfy

$$|b^{(\iota)}(\boldsymbol{u},\boldsymbol{v})| \le c_1|\nabla\boldsymbol{u}||\boldsymbol{v}| + c_2|\boldsymbol{u}||\nabla\boldsymbol{v}| \tag{5.1.46}$$

and $b^{(\iota)}(\boldsymbol{u},\boldsymbol{v}) = 0$ for $\boldsymbol{x} \notin U$.

In the next step, the local rectification is done by the transform of variables $\boldsymbol{x} = \mathfrak{R}^{-1}(\boldsymbol{y}) \equiv \mathfrak{Q}(\boldsymbol{y})$. The determinant of the Jacobian of this transform equals to 1 and the density of surface measure is $J_\Gamma \equiv \sqrt{1+|\nabla\psi|^2}$. For a moment, let us denote the transformed version of a function $g : U \to \mathbb{R}$ by $\tilde{g} \equiv g \circ \mathfrak{Q}$. The bilinear forms $a^{(\iota)}(\cdot,\cdot)$ are transformed into the forms

$$a^{(\iota)}(\boldsymbol{u},\boldsymbol{v}) \equiv \tilde{a}^{(\iota)}(\tilde{\boldsymbol{u}},\tilde{\boldsymbol{v}}) + b_0^{(\iota)}(\tilde{\boldsymbol{u}},\tilde{\boldsymbol{v}}),$$

where $\widetilde{a}(\cdot,\cdot)$ is the bilinear form with the coefficients

$$\widetilde{a}_{ijk\ell} = \sum_{r,s=1}^{N} a_{irks}\mathscr{O}_{jr}\mathscr{O}_{\ell s};$$

see also relation (1.7.28) on page 48. Here $\{\mathscr{O}_{ij}\}_{i,j=1}^{N}$ denotes the (constant) gradient of the transformation $\mathscr{O}$. The remainders $b_0^{(\iota)}(\boldsymbol{u},\boldsymbol{v})$ satisfy

$$|b_0^{(\iota)}(\boldsymbol{u},\boldsymbol{v})| \le c\delta|\nabla\boldsymbol{u}||\nabla\boldsymbol{v}| \tag{5.1.47}$$

due to relation (5.1.45). The corresponding transformed forms

$$\widetilde{b}^{(\iota)}(\boldsymbol{y};\widetilde{\boldsymbol{u}},\widetilde{\boldsymbol{v}}) \equiv b^{(\iota)}(\mathfrak{Q}(\boldsymbol{y});\widetilde{\boldsymbol{u}}\circ\mathfrak{R},\widetilde{\boldsymbol{v}}\circ\mathfrak{R})$$

still satisfy relation (5.1.46) with modified constants c_1 and c_2. Then the rectified variational equation is given by

$$\begin{aligned}\int_{Q_{\delta T}} \Big((\widetilde{\rho\dot{\boldsymbol{u}}})^{\cdot}\cdot\widetilde{\boldsymbol{v}} + \widetilde{a}^{(0)}(\widetilde{\rho\boldsymbol{u}},\widetilde{\boldsymbol{v}}) + \widetilde{a}^{(1)}(\widetilde{\rho\dot{\boldsymbol{u}}},\widetilde{\boldsymbol{v}}) + \widetilde{b}^{(0)}(\widetilde{\boldsymbol{u}},\widetilde{\boldsymbol{v}}) \\ + \widetilde{b}^{(1)}(\widetilde{\dot{\boldsymbol{u}}},\widetilde{\boldsymbol{v}}) + b_0^{(0)}(\widetilde{\rho\boldsymbol{u}},\widetilde{\boldsymbol{v}}) + b_0^{(1)}(\widetilde{\rho\dot{\boldsymbol{u}}},\widetilde{\boldsymbol{v}})\Big)\,d\boldsymbol{y}\,dt \\ + \int_{S_{\delta T}} \Big(\tfrac{1}{\varepsilon}[\widetilde{\rho\dot{u}_n}]_+\widetilde{v}_n + \widetilde{\mathfrak{F}}\tfrac{1}{\varepsilon}[\widetilde{\rho\dot{u}_n}]_+\nabla\Phi_\eta(\widetilde{\dot{\boldsymbol{u}}}_{\boldsymbol{t}})\cdot\widetilde{\boldsymbol{v}}_{\boldsymbol{t}}\Big)\cdot \\ \cdot J_\Gamma\,ds_{\boldsymbol{y}}\,dt \ge \int_{Q_{\delta T}} \Big(\widetilde{\rho\boldsymbol{f}} + \widetilde{\dot{\rho}\dot{\boldsymbol{u}}}\Big)\cdot\widetilde{\boldsymbol{v}}\,d\boldsymbol{y}\,dt\end{aligned} \tag{5.1.48}$$

with $Q_{\delta T} = I_T\times B_{N-1}(\delta)\times(0,\delta)$ and $S_{\delta T} = I_T\times B_{N-1}(\delta)\times\{0\}$. In order to extend this inequality from the domain of definition $Q_{\delta T}$ onto the halfspace $Q\equiv\mathbb{R}\times\mathbb{R}^{N-1}\times\mathbb{R}_+$ it is necessary to extend the functions and the coefficients. The extended variational inequality will be formulated for the functions

$$\breve{\boldsymbol{u}} : (t,\boldsymbol{y}) \mapsto \begin{cases} \widetilde{\rho}(\widetilde{\dot{\boldsymbol{u}}}-\widetilde{\boldsymbol{u}}_1) & \text{for } (t,\boldsymbol{y})\in I_T\times B_{N-1}(\delta)\times(0,\delta), \\ 0 & \text{otherwise} \end{cases} \tag{5.1.49}$$

and

$$\mathring{\boldsymbol{u}} : (t,\boldsymbol{y}) \mapsto \begin{cases} \widetilde{\rho}(\widetilde{\boldsymbol{u}}-\widetilde{\boldsymbol{u}}_0) & \text{for } (t,\boldsymbol{y})\in I_T\times B_{N-1}(\delta)\times(0,\delta), \\ 0 & \text{otherwise.} \end{cases} \tag{5.1.50}$$

The coefficients $\widetilde{a}_{ijk\ell}^{(\iota)}$ are extended in such a way that the constants of ellipticity and boundedness remain unchanged in the case of inhomogeneous material, or by their respective constant values for homogeneous material. The coefficients of $\widetilde{b}^{(\iota)}$ can be extended first by 0 onto $\mathbb{R}_+ \times \mathbb{R}^{N-1} \times \mathbb{R}_+$, and then to $t < 0$ such that they remain Hölder continuous with an exponent greater than $\frac{1}{2}$ and the relation (5.1.46) remains true. The coefficients of $\widetilde{b}_0^{(\iota)}$ are extended such that they remain Hölder continuous with an exponent greater than $\frac{1}{2}$ with respect to space and time variables and that relation (5.1.47) is still true. Let us omit the tilde in all extended functions and functionals, and let us denote the new space variable by $\boldsymbol{x}$ again. Then the transformed variational inequality has the form

$$\begin{aligned} &\int_Q \Big(\mathring{\dot{\boldsymbol{u}}} \cdot \boldsymbol{v} + a^{(0)}(\mathring{\boldsymbol{u}}, \boldsymbol{v}) + a^{(1)}(\breve{\boldsymbol{u}}, \boldsymbol{v})\Big)\, d\boldsymbol{x}\, dt \\ &\qquad + \int_S \big(\mathscr{S} v_n + \mathfrak{F}\, \mathscr{S}\, \nabla \Phi_\eta(\dot{\boldsymbol{u}}_t) \cdot \boldsymbol{v}_t\big)\, ds_{\boldsymbol{x}}\, dt = \mathscr{L}(\boldsymbol{u}, \boldsymbol{v}) \end{aligned} \tag{5.1.51}$$

with

$$\mathscr{S} = \tfrac{1}{\varepsilon}\big[\breve{u}_n + \rho u_{1n}\big]_+ J_\Gamma$$

and a new functional $\mathscr{L}(\boldsymbol{u}, \cdot)$ defined by

$$\begin{aligned} \mathscr{L}(\boldsymbol{u}, \boldsymbol{v}) \equiv \int_Q \Big(&(\rho \boldsymbol{f} + \dot{\rho} \dot{\boldsymbol{u}}) \cdot \boldsymbol{v} - a^{(0)}(\rho \boldsymbol{u}_0, \boldsymbol{v}) - a^{(1)}(\rho \boldsymbol{u}_1, \boldsymbol{v}) \\ &- b^{(0)}(\boldsymbol{u}, \boldsymbol{v}) - b^{(1)}(\dot{\boldsymbol{u}}, \boldsymbol{v}) - b_0^{(0)}(\mathring{\boldsymbol{u}}, \boldsymbol{v}) - b_0^{(0)}(\rho \boldsymbol{u}_0, \boldsymbol{v}) \\ &- b_0^{(1)}(\breve{\boldsymbol{u}}, \boldsymbol{v}) - b_0^{(1)}(\rho \boldsymbol{u}_1, \boldsymbol{v})\Big)\, d\boldsymbol{x}\, dt. \end{aligned}$$

Here, in the definition of the functional several functions must be formally extended, too. This extension should be done in such a way that the support of $\mathscr{L}(\boldsymbol{u}, \cdot)$ is contained in the set $I_T \times B_{N-1}(\delta) \times (0, \delta)$. Therefore, $\rho \boldsymbol{f} + \dot{\rho} \dot{\boldsymbol{u}}$, $\varrho \boldsymbol{u}_0$ and $\varrho \boldsymbol{u}_1$ are extended by 0. The arguments $\boldsymbol{u}$ and $\dot{\boldsymbol{u}}$ of the forms $b^{(\iota)}$ are extended first to $Q_+ \equiv \mathbb{R}_+ \times \mathbb{R}^{N-1} \times \mathbb{R}_+$ such that they remain bounded in the space $H^{1/2,\,1}(Q_+)$ and then by 0 for $t < 0$. The coefficients of the forms $b^{(\iota)}$ vanish outside $\boldsymbol{x} \in B_{N-1}(\delta) \times (0, \delta)$. Of course, the extension by 0 of several functions for $t < 0$ may create a discontinuity of the extended function at line $t = 0$. However, if a function $\boldsymbol{u} \in H^\alpha(\mathbb{R}_+; B)$ with a Banach space B and index $\alpha < \frac{1}{2}$ is extended by 0 onto $\mathbb{R}$, then the extended function remains in $H^\alpha(\mathbb{R}; B)$. The proved time regularity in the shift technique does not exceed $\alpha = \frac{1}{4}$; hence this discontinuity does not create any problems.

Inequality (5.1.51) is the basis for the regularity proof by the shift technique. Since the regularity of the solution must be increased for both the time and the space variable, we employ the shift technique with respect to both these variables simultaneously. Let us first consider the space variables. To a function $g : Q \to \mathbb{R}$ and a shift parameter $\boldsymbol{h} = (\boldsymbol{h}', 0)$ with $\boldsymbol{h}' \in \mathbb{R}^{N-1}$ let

$$g_{-\boldsymbol{h}} : (t, \boldsymbol{x}) \mapsto g(t, \boldsymbol{x} + \boldsymbol{h})$$

be the shift with respect to the tangential variables and

$$\Delta_{\boldsymbol{h}} g \equiv g_{-\boldsymbol{h}} - g$$

be the corresponding difference. We use the test functions $\boldsymbol{v} = \breve{\boldsymbol{u}} - \breve{\boldsymbol{u}}_{-\boldsymbol{h}}$ in equation (5.1.51) and $\boldsymbol{v}_{-\boldsymbol{h}} \equiv \breve{\boldsymbol{u}}_{-\boldsymbol{h}} - \breve{\boldsymbol{u}}$ in the shifted equation $(5.1.51)_{-\boldsymbol{h}}$; this is

$$\begin{aligned}
&\int_Q \Big(\dot{\breve{\boldsymbol{u}}}_{-\boldsymbol{h}} \cdot \boldsymbol{v}_{-\boldsymbol{h}} + a^{(0)}_{-\boldsymbol{h}}(\mathring{\boldsymbol{u}}_{-\boldsymbol{h}}, \boldsymbol{v}_{-\boldsymbol{h}}) + a^{(1)}_{-\boldsymbol{h}}(\breve{\boldsymbol{u}}_{-\boldsymbol{h}}, \boldsymbol{v}_{-\boldsymbol{h}}) \Big) \, d\boldsymbol{x}\, dt \\
&\qquad + \int_S \big(\mathscr{S}_{-\boldsymbol{h}} (v_n)_{-\boldsymbol{h}} + \mathfrak{F}_{-\boldsymbol{h}} \mathscr{S}_{-\boldsymbol{h}} \nabla \Phi_\eta((\dot{\boldsymbol{u}}_t)_{-\boldsymbol{h}}) \cdot (\boldsymbol{v}_t)_{-\boldsymbol{h}} \big) \, ds_{\boldsymbol{x}}\, dt \\
&\geq \mathscr{L}_{-\boldsymbol{h}}(\boldsymbol{u}_{-\boldsymbol{h}}, \boldsymbol{v}_{-\boldsymbol{h}})
\end{aligned}$$

with an obvious definition of $a_{-\boldsymbol{h}}$ and $\mathscr{L}_{-\boldsymbol{h}}$. Both resulting inequalities are added, their sum is multiplied by $|\boldsymbol{h}'|^{-N}$ and the product is integrated with respect to $\boldsymbol{h}' \in \mathbb{R}^{N-1}$. Then an inequality valid for the term

$$\mathscr{A}_1(\breve{\boldsymbol{u}}) \equiv \int_{\mathbb{R}^{N-1}} \int_Q |\boldsymbol{h}'|^{-N} a^{(1)}(\Delta_{\boldsymbol{h}} \breve{\boldsymbol{u}}, \Delta_{\boldsymbol{h}} \breve{\boldsymbol{u}}) \, d\boldsymbol{x}\, dt\, d\boldsymbol{h}' \tag{5.1.52}$$

is established by estimating all remaining terms. The acceleration term yields

$$\int_Q \Delta_{\boldsymbol{h}} \dot{\breve{\boldsymbol{u}}} \cdot \Delta_{\boldsymbol{h}} \breve{\boldsymbol{u}} \, d\boldsymbol{x}\, dt = \int_Q \frac{1}{2} \frac{d}{dt} |\Delta_{\boldsymbol{h}} \breve{\boldsymbol{u}}|^2 \, d\boldsymbol{x}\, dt = 0,$$

because $\breve{\boldsymbol{u}} = 0$ for $t < 0$ and for t sufficiently large. The term $\Delta_{\boldsymbol{h}}(\dot{\rho} \boldsymbol{u}) \cdot \Delta_{\boldsymbol{h}} \breve{\boldsymbol{u}}$ arising in the linear functional $\mathscr{L}(\cdot, \cdot)$ is of a lower order and can be estimated with the help of (5.1.17). The bilinear form of elastic energy can be estimated with Hölder's inequality by

$$\begin{aligned}
&\int_Q \Big(-a^{(0)}(\mathring{\boldsymbol{u}}, \Delta_{\boldsymbol{h}} \breve{\boldsymbol{u}}) + a^{(0)}_{-\boldsymbol{h}}(\mathring{\boldsymbol{u}}_{-\boldsymbol{h}}, \Delta_{\boldsymbol{h}} \breve{\boldsymbol{u}}) \Big) \, d\boldsymbol{x}\, dt \\
&= \int_Q \Big(a^{(0)}(\Delta_{\boldsymbol{h}} \mathring{\boldsymbol{u}}, \Delta_{\boldsymbol{h}} \breve{\boldsymbol{u}}) + (a^{(0)}_{-\boldsymbol{h}} - a^{(0)})(\mathring{\boldsymbol{u}}_{-\boldsymbol{h}}, \Delta_{\boldsymbol{h}} \breve{\boldsymbol{u}}) \Big) \, d\boldsymbol{x}\, dt \\
&\geq -\kappa \int_Q a^{(1)}(\Delta_{\boldsymbol{h}} \breve{\boldsymbol{u}}, \Delta_{\boldsymbol{h}} \breve{\boldsymbol{u}}) \, d\boldsymbol{x}\, dt - c_1(\kappa) \int_Q a^{(1)}(\Delta_{\boldsymbol{h}} \mathring{\boldsymbol{u}}, \Delta_{\boldsymbol{h}} \mathring{\boldsymbol{u}}) \, d\boldsymbol{x}\, dt \\
&\quad - c_2(\kappa) \min\{1, |\boldsymbol{h}'|\} \|\mathring{\boldsymbol{u}}\|^2_{L_2(\mathbb{R}; \boldsymbol{H}^1(O))}
\end{aligned}$$

with $\kappa > 0$ arbitrarily small. The remaining terms are estimated by the same techniques as described for the static problem in Section 3.1; the only difference here is the additional time integral. In particular, the penalty and friction terms can be estimated by

$$\int_{\mathbb{R}^{N-1}}\int_S |\boldsymbol{h}'|^{-N} \Delta_{\boldsymbol{h}}(\mathfrak{F}\,\mathscr{S}\,\nabla\Phi_\eta(\dot{\boldsymbol{u}}_t))\cdot\Delta_{\boldsymbol{h}}\breve{\boldsymbol{u}}_t\,ds_{\boldsymbol{x}}\,dt\,d\boldsymbol{h}' + c_1\sqrt{\mathscr{M}_1} + c_2$$

with

$$\mathscr{M}_1 \equiv \int_{\mathbb{R}^{N-1}} |\boldsymbol{h}'|^{-N}\left(\|\Delta_{\boldsymbol{h}}\breve{\boldsymbol{u}}\|^2_{\boldsymbol{H}^{1/4,\,1/2}(S)} + \|\Delta_{\boldsymbol{h}}\mathscr{S}\|^2_{H^{-1/4,-1/2}(S)}\right)d\boldsymbol{h}'.$$

As a consequence, we obtain the inequality

$$\begin{aligned}\mathscr{A}_1(\breve{\boldsymbol{u}}) &\le (1+\kappa)\cdot\\ &\cdot\int_{\mathbb{R}^{N-1}}\int_S |\boldsymbol{h}'|^{-N}\Delta_{\boldsymbol{h}}(\mathfrak{F}\nabla\Phi_\eta(\dot{\boldsymbol{u}}_t)\mathscr{S})\cdot\Delta_{\boldsymbol{h}}\breve{\boldsymbol{u}}_t\,d\boldsymbol{x}\,dt\,d\boldsymbol{h}'\\ &+ c_1(\kappa)\sqrt{\mathscr{M}_1} + c_2(\kappa)\mathscr{A}_1(\mathring{\boldsymbol{u}}) + c_3(\kappa).\end{aligned} \tag{5.1.53}$$

The parameter κ is composed from small parameters in suitable Hölder estimates and from the contribution of the estimate (5.1.47); it can be made arbitrarily small by choosing the diameter δ of the covering sufficiently small. The notation κ will be used in the sequel for such a constant with possibly different values in different relations.

An analogous technique can be applied with respect to the time shifts. For a parameter $q \in \mathbb{R}$ let

$$v_{-q}(t,\boldsymbol{x}) \equiv v(t+q,\boldsymbol{x}) \text{ and } \Delta_q v \equiv v_{-q} - v$$

denote the time shifted function and the difference of shifted and non-shifted function. We take the test function $\boldsymbol{v} = \breve{\boldsymbol{u}} - \breve{\boldsymbol{u}}_{-q}$ in variational equation (5.1.51), then we shift the whole equation into the direction q and take $\boldsymbol{v}_{-q} \equiv \breve{\boldsymbol{u}}_{-q} - \breve{\boldsymbol{u}}$ in the shifted equation $(5.1.51)_{-q}$. We add both equations, multiply the result by $|q|^{-3/2}$ and integrate the resulting product with respect to q over $\mathbb{R}$. Then we carry out analogous calculations and estimations as in the case of the tangential shifts. Finally, we arrive at a time-shift variant of the estimate (5.1.53), where the difference operator $\Delta_{\boldsymbol{h}}$ is replaced by Δ_q, the integral $\int_{\mathbb{R}^{N-1}} d\boldsymbol{h}'$ by $\int_{\mathbb{R}} dq$ and the factor $|\boldsymbol{h}'|^{-N}$ by $|q|^{-3/2}$. Here, the expression resulting from the displacements can be estimated by

$$\int_{\mathbb{R}}\int_Q |q|^{-3/2} a^{(0)}(\Delta_q\mathring{\boldsymbol{u}},\Delta_q\mathring{\boldsymbol{u}})\,d\boldsymbol{x}\,dt\,dq \le c_1\|\dot{\mathring{\boldsymbol{u}}}\|^{1/2}_{\boldsymbol{H}^{0,1}(Q)}\|\mathring{\boldsymbol{u}}\|^{3/2}_{\boldsymbol{H}^{0,1}(Q)} \le c_2.$$

Adding up the results from the tangential space and time shifts and using the additional abbreviations

$$\mathscr{A}_2(\breve{\boldsymbol{u}}) \equiv \int_{\mathbb{R}} \int_Q |q|^{3/2} a^{(1)}(\Delta_q \breve{\boldsymbol{u}}, \Delta_q \breve{\boldsymbol{u}})\, d\boldsymbol{x}\, dt\, dq,$$

$$\mathscr{I} \equiv \Bigg| \int_{\mathbb{R}^{N-1}} \int_{\mathbb{R}^N} |\boldsymbol{h}'|^{-N} \Delta_{\boldsymbol{h}}(\mathfrak{F}\nabla\Phi_\eta(\dot{\boldsymbol{u}}_t)\mathscr{S}) \cdot \Delta_{\boldsymbol{h}}(\breve{\boldsymbol{u}}_t)\, d\boldsymbol{x}\, dt\, d\boldsymbol{h}'$$
$$+ \int_{\mathbb{R}} \int_{\mathbb{R}^N} |q|^{-3/2} \Delta_q(\mathfrak{F}\nabla\Phi_\eta(\dot{\boldsymbol{u}}_t)\mathscr{S}) \cdot \Delta_q(\breve{\boldsymbol{u}}_t)\, d\boldsymbol{x}\, dt\, dq \Bigg|,$$

$\mathscr{M} \equiv \mathscr{M}_1 + \mathscr{M}_2$ with

$$\mathscr{M}_2 \equiv \int_{\mathbb{R}} |q|^{-3/2} \left(\|\Delta_q \breve{\boldsymbol{u}}\|^2_{\boldsymbol{H}^{1/4,\,1/2}(S)} + \|\Delta_q \mathscr{S}\|^2_{H^{-1/4,-1/2}(S)} \right) dq,$$

we arrive at the inequality

$$\mathscr{A}_1(\breve{\boldsymbol{u}}) + \mathscr{A}_2(\breve{\boldsymbol{u}}) \leq (1+\kappa)\mathscr{I} + c_1(\kappa)\mathscr{A}_1(\mathring{\boldsymbol{u}}) + c_2(\kappa)\sqrt{\mathscr{M}} + c_3(\kappa). \quad (5.1.54)$$

Now it is necessary to estimate the friction term $\mathscr{I}$ by the left hand side of the previous relation. Employing the Fourier transformation we have

$$\begin{aligned} \mathscr{I} &= \Bigg| \int_{\mathbb{R}} \left(c_{N-1}\left(\tfrac{1}{2}\right)|\boldsymbol{\xi}| + c_1\left(\tfrac{1}{4}\right)|\vartheta|^{1/2} \right) \\ &\qquad \operatorname{Re}\left(\mathscr{F}(\mathfrak{F}\nabla\Phi_\eta(\dot{\boldsymbol{u}}_t)\mathscr{S})\overline{\mathscr{F}(\breve{\boldsymbol{u}}_t)} \right) d\boldsymbol{\xi}\, d\vartheta \Bigg| \\ &\leq \|\mathfrak{F}\|_{L_\infty(\Gamma_{\mathfrak{F}} \times \mathbb{R}^N)} \|\mathscr{S}\|_{L_2(S)} \mathfrak{U}^{1/2}, \end{aligned} \quad (5.1.55)$$

with

$$\mathfrak{U} \equiv \int_{\mathbb{R}^N} \left(c_{N-1}\left(\tfrac{1}{2}\right)|\xi| + c_1\left(\tfrac{1}{4}\right)|\vartheta|^{1/2} \right)^2 |\mathscr{F}(\breve{\boldsymbol{u}}_t)|^2\, d\boldsymbol{\xi}\, d\vartheta.$$

Expression $\mathfrak{U}$ is equivalent to the $\boldsymbol{H}^{1/2,\,1}(S)$ seminorm of $\breve{\boldsymbol{u}}$. In order to obtain the best possible estimate, the special trace inequalities (5.1.39) and (5.1.40) are employed. Using the inequality $(x+y)^2 \leq (1+p)x^2 + \left(1+\frac{1}{p}\right)y^2$ for $x = c_{N-1}\left(\frac{1}{2}\right)|\boldsymbol{\xi}|$ and $y = c_1\left(\frac{1}{4}\right)|\vartheta|^{1/2}$ we get

$$\mathfrak{U} \leq (1+p) \int_{\mathbb{R}^{N-1}} |\boldsymbol{h}'|^{-N} \|\Delta_{\boldsymbol{h}} \breve{\boldsymbol{u}}_t\|'^2_{\boldsymbol{H}^{0,\,\frac{1}{2}}(S)}\, d\boldsymbol{h}'$$
$$+ \left(1+\tfrac{1}{p}\right) \int_{\mathbb{R}} |q|^{-3/2} \|\Delta_q \breve{\boldsymbol{u}}_t\|'^2_{\boldsymbol{H}^{1/4,\,0}(S)}\, dq + c.$$

Application of the special trace estimates (5.1.39) and (5.1.40) yields

$$\mathfrak{U} \leq (1+p)D_1^2 \mathscr{A}_1(\breve{\boldsymbol{u}}) + \left(1+\tfrac{1}{p}\right) D_2^2 \mathscr{A}_2(\breve{\boldsymbol{u}})$$
$$+ \kappa\left(\mathscr{A}_1(\breve{\boldsymbol{u}}) + \mathscr{A}_2(\breve{\boldsymbol{u}})\right) + c_1(\kappa)\mathscr{A}_3(\mathring{\boldsymbol{u}}) + c_2(\kappa).$$

Here, κ is again a constant that can be made arbitrarily small. The additional terms come from the estimate of the functional $\mathscr{L}(\boldsymbol{u})$. Optimization with respect to p gives the result

$$\mathfrak{U} \le (1+\kappa)\left(D_1\sqrt{\mathscr{A}_1(\breve{\boldsymbol{u}})} + D_2\sqrt{\mathscr{A}_2(\breve{\boldsymbol{u}})}\right)^2 + c_1(\kappa)\mathscr{A}_3(\mathring{\boldsymbol{u}}) + c_2(\kappa). \quad (5.1.56)$$

In the next step of the proof, it is necessary to estimate $\|\mathscr{S}\|^2_{L_2(S)}$. This term has the representation

$$\begin{aligned} \|\mathscr{S}\|^2_{L_2(S)} &= \|\mathscr{S}\|^2_{H^{-1/4,-1/2}(S)} \\ &+ \int_{\mathbb{R}^{N-1}} |\boldsymbol{h}'|^{-N} \|\Delta_{\boldsymbol{h}}\mathscr{S}\|^2_{H^{-1/4,-1/2}(S)}\, d\boldsymbol{h}' \\ &+ \int_{\mathbb{R}} |q|^{-3/2} \|\Delta_q \mathscr{S}\|^2_{H^{-1/4,-1/2}(S)}\, dq. \end{aligned} \quad (5.1.57)$$

The term $\mathscr{S}$ satisfies the Green formula

$$\begin{aligned} \langle \mathscr{S}, w_n \rangle_S &= \mathscr{L}(\boldsymbol{u}, \boldsymbol{w}) \\ &- \int_Q \left[\dot{\breve{\boldsymbol{u}}} \cdot \boldsymbol{w} + a^{(0)}(\mathring{\boldsymbol{u}}, \boldsymbol{w}) + a^{(1)}(\breve{\boldsymbol{u}}, \boldsymbol{w})\right] d\boldsymbol{x}\, dt \end{aligned} \quad (5.1.58)$$

for all test functions $\boldsymbol{w} \in H^{1/2,\,1}(Q)$ with $\boldsymbol{w_t} = 0$ on S. With this equation and the inverse trace theorem, the term $\|\mathscr{S}\|_{H^{-1/4,-1/2}(S)}$ can be estimated by $c_1\|\breve{\boldsymbol{u}}\|_{H^{1/2,\,1}(Q)} + c_2\|\mathring{\boldsymbol{u}}\|_{H^{1/2,\,1}(Q)} + c_3$ which is bounded due to the *a priori* estimates valid for $\boldsymbol{u}$. In order to estimate the remaining terms, we use the particular scalar test function $w^{\boldsymbol{h}}$ defined by

$$\mathscr{F}(w^{\boldsymbol{h}}; \vartheta, \boldsymbol{\xi}) = \mathscr{F}(\Delta_{\boldsymbol{h}}\mathscr{S}; \vartheta, \boldsymbol{\xi})\left(1 + c_1\left(\tfrac{1}{4}\right)|\vartheta|^{1/2} + c_{N-1}\left(\tfrac{1}{2}\right)|\boldsymbol{\xi}|\right)^{-1}$$

on S. This function fulfils the relations

$$\|w^{\boldsymbol{h}}\|^2_{H^{1/4,\,1/2}(S)} = \|\Delta_{\boldsymbol{h}}\mathscr{S}\|^2_{H^{-1/4,-1/2}(S)} = \langle \Delta_{\boldsymbol{h}}\mathscr{S}, w^{\boldsymbol{h}} \rangle_S.$$

The function $w^{\boldsymbol{h}}\boldsymbol{e}_N$ is extended onto the whole domain Q by means of the parabolic problem (5.1.27) with the form $a = a^{(1)}$. The extended function therefore fulfils estimate (5.1.38). We take $w^{\boldsymbol{h}}\boldsymbol{n}$ in (5.1.58), then shift (5.1.58) into the direction of $\boldsymbol{h}$, take $-w^{\boldsymbol{h}}\boldsymbol{n}_{-\boldsymbol{h}}$ in the shifted equation $(5.1.58)_{-\boldsymbol{h}}$, add both equations, multiply the sum by $|\boldsymbol{h}'|^{-N}$ and integrate with respect to $\boldsymbol{h}'$ over $\mathbb{R}^{N-1}$. Using the decomposition $w^{\boldsymbol{h}}\boldsymbol{n} = w^{\boldsymbol{h}}(\boldsymbol{n} - \boldsymbol{e}_N) + w^{\boldsymbol{h}}\boldsymbol{e}_N$ and the estimate (1.7.38) for the difference $\boldsymbol{n} - \boldsymbol{e}_N$ it is possible to prove with the help of relation (5.1.38) the inequality

$$
\begin{aligned}
&\int_{\mathbb{R}^{N-1}} |\boldsymbol{h}'|^{-N} \|\Delta_{\boldsymbol{h}} \mathscr{S}\|^2_{H^{-1/4,-1/2}(S)}\, d\boldsymbol{h}' \\
&\quad \le (1+\kappa) \int_{\mathbb{R}^{N-1}} |\boldsymbol{h}'|^{-N} D_0 \|\Delta_{\boldsymbol{h}} \breve{\boldsymbol{u}}\|_A \|\Delta_{\boldsymbol{h}} \mathscr{S}\|_{H^{-1/4,-1/2}(S)}\, d\boldsymbol{h}' \\
&\qquad + \Big(c_1 \sqrt{\mathscr{A}_1(\mathring{\boldsymbol{u}})} + c_2(\kappa)\Big) \cdot \\
&\qquad\quad \cdot \Big(\int_{\mathbb{R}^{N-1}} |\boldsymbol{h}'|^{-N} \|\Delta_{\boldsymbol{h}} \mathscr{S}\|^2_{H^{-1/4,-1/2}(S)}\, d\boldsymbol{h}'\Big)^{1/2} + c(\kappa).
\end{aligned}
$$

From this we obtain

$$
\begin{aligned}
&\int_{\mathbb{R}^{N-1}} |\boldsymbol{h}'|^{-N} \|\Delta_{\boldsymbol{h}} \mathscr{S}\|^2_{H^{-1/4,-1/2}(S)}\, d\boldsymbol{h}' \\
&\quad \le (1+\kappa) D_0^2 \mathscr{A}_1(\breve{\boldsymbol{u}}) + c_1(\kappa) \mathscr{A}_1(\mathring{\boldsymbol{u}}) + c_2(\kappa).
\end{aligned}
\tag{5.1.59}
$$

Repeating the same procedure for the time shifts with the special test function defined by

$$
\mathscr{F}(w^q; \vartheta, \boldsymbol{\xi}) = \mathscr{F}(\Delta_q \mathscr{S}; \vartheta, \boldsymbol{\xi}) \Big(1 + c_1(\tfrac{1}{4}) |\vartheta|^{1/2} + c_{N-1}(\tfrac{1}{2}) |\boldsymbol{\xi}|\Big)^{-1},
$$

the analogous relation

$$
\int_{\mathbb{R}} |q|^{-3/2} \|\Delta_q \mathscr{S}\|^2_{H^{-1/4,-1/2}(S)}\, dq \;\le\; (1+\kappa) D_0^2 \mathscr{A}_2(\breve{\boldsymbol{u}}) + c(\kappa) \tag{5.1.60}
$$

is established. Hence $\|\mathscr{S}\|^2_{L_2(S)}$ is bounded by

$$
\begin{aligned}
\|\mathscr{S}\|^2_{L_2(S)} &\le (1+\kappa) D_0^2 \big(\mathscr{A}_1(\breve{\boldsymbol{u}}) + \mathscr{A}_2(\breve{\boldsymbol{u}})\big) + c_1(\kappa) \mathscr{A}_1(\mathring{\boldsymbol{u}}) \\
&\quad + c_2(\kappa).
\end{aligned}
\tag{5.1.61}
$$

As a consequence of the inequalities (5.1.55), (5.1.56) and (5.1.61) relation (5.1.54) can be modified to

$$
\begin{aligned}
\mathscr{A}_1(\breve{\boldsymbol{u}}) + \mathscr{A}_2(\breve{\boldsymbol{u}}) \;&\le (1+\kappa)\; \|\mathfrak{F}\|_{L_\infty(\Gamma_{\mathfrak{z}} \times \mathbb{R}^N)} \\
&\quad D_0 \sqrt{\mathscr{A}_1(\breve{\boldsymbol{u}}) + \mathscr{A}_2(\breve{\boldsymbol{u}})} \Big(D_1 \sqrt{\mathscr{A}_1(\breve{\boldsymbol{u}})} + D_2 \sqrt{\mathscr{A}_2(\breve{\boldsymbol{u}})}\Big) \\
&\quad + c_1(\kappa) \mathscr{A}_1(\mathring{\boldsymbol{u}}) + c_2(\kappa).
\end{aligned}
\tag{5.1.62}
$$

The term $\sqrt{\mathscr{A}_1(\breve{\boldsymbol{u}}) + \mathscr{A}_2(\breve{\boldsymbol{u}})} \Big(D_1 \sqrt{\mathscr{A}_1(\breve{\boldsymbol{u}})} + D_2 \sqrt{\mathscr{A}_2(\breve{\boldsymbol{u}})}\Big)$ is bounded by $\sqrt{D_1^2 + D_2^2} \big(\mathscr{A}_1(\breve{\boldsymbol{u}}) + \mathscr{A}_2(\breve{\boldsymbol{u}})\big)$ due to Hölder's inequality for $\mathbb{R}^2$, applied to the vectors $(D_1, D_2)^\top$ and $\left(\sqrt{\mathscr{A}_1(\breve{\boldsymbol{u}})}, \sqrt{\mathscr{A}_2(\breve{\boldsymbol{u}})}\,\right)^\top$. Hence the result of all our estimates is the relation

$$\begin{aligned}\mathscr{A}_1(\breve{\boldsymbol{u}}) + \mathscr{A}_2(\breve{\boldsymbol{u}}) &\leq (1+\kappa)\|\mathfrak{F}\|_{L_\infty(\Gamma_{\mathfrak{F}}\times\mathbb{R}^N)} C_{\mathfrak{F}}^{-1} \cdot \\ &\cdot \big(\mathscr{A}_1(\breve{\boldsymbol{u}}) + \mathscr{A}_2(\breve{\boldsymbol{u}})\big) + c_1(\kappa)\mathscr{A}_1(\mathring{\boldsymbol{u}}) + c_2(\kappa)\end{aligned} \tag{5.1.63}$$

with the constant $C_{\mathfrak{F}}$ defined in Assumption 5.1.6.

Now, at the end of the proof, it is necessary to find an estimate for $\mathscr{A}_1(\mathring{\boldsymbol{u}})$. Here, the special dependence of the cutoff function $\rho = \rho(L; t, x)$ on the time variable t and the parameter L is essential. This function is non-decreasing with respect to L, non-increasing with respect to t and satisfies $\rho(L; t, x) = 0$ for $t > L$ and $\partial_L \rho(L; t, \boldsymbol{x}) = -\partial_t \rho(L; t, \boldsymbol{x})$. The functions $\breve{\boldsymbol{u}}$ and $\mathring{\boldsymbol{u}}$ also depend on L; and due to the extensions of $\rho(\boldsymbol{u} - \boldsymbol{u}_0)$ and of $\rho(\dot{\boldsymbol{u}} - \boldsymbol{u}_1)$ by zero for $t < 0$ we have $\breve{\boldsymbol{u}}(L; t, x) = \mathring{\boldsymbol{u}}(L; t, x) = 0$ for $L < 0$. The parameter L is now interpreted as a time evolution variable. Using the properties mentioned above and a partial integration with respect to t we derive

$$\begin{aligned}\partial_L \mathscr{A}_1(\mathring{\boldsymbol{u}}; L) &= \int_{\mathbb{R}^{N-1}} \int_Q \frac{\partial \phi_0^2}{\partial L}(t-L)|\boldsymbol{h}'|^{-N} a^{(1)}\big(\Delta_{\boldsymbol{h}} \varrho_i(\boldsymbol{u}-\boldsymbol{u}_0), \\ &\qquad\qquad \Delta_{\boldsymbol{h}} \varrho_i(\boldsymbol{u}-\boldsymbol{u}_0)\big)\, d\boldsymbol{x}\, dt\, d\boldsymbol{h}' \\ &= -\int_{\mathbb{R}^{N-1}} \int_Q \frac{\partial \phi_0^2}{\partial t}(t-L)|\boldsymbol{h}'|^{-N} a^{(1)}\big(\Delta_{\boldsymbol{h}} \varrho_i(\boldsymbol{u}-\boldsymbol{u}_0), \\ &\qquad\qquad \Delta_{\boldsymbol{h}} \varrho_i(\boldsymbol{u}-\boldsymbol{u}_0)\big)\, d\boldsymbol{x}\, dt\, d\boldsymbol{h}' \\ &= \int_{\mathbb{R}^{N-1}} \int_Q 2\phi_0^2(t-L)|\boldsymbol{h}'|^{-N} a^{(1)}\big(\Delta_{\boldsymbol{h}}(\varrho_i(\dot{\boldsymbol{u}}-\boldsymbol{u}_1+\boldsymbol{u}_1)), \\ &\qquad\qquad \Delta_{\boldsymbol{h}}(\varrho_i(\boldsymbol{u}-\boldsymbol{u}_0))\big)\, d\boldsymbol{x}\, dt\, d\boldsymbol{h}' \\ &\leq 2\big(\sqrt{\mathscr{A}_1(\breve{\boldsymbol{u}}; L)} + \sqrt{L}\, c\big)\sqrt{\mathscr{A}_1(\mathring{\boldsymbol{u}}; L)},\end{aligned}$$

where $c \equiv c(\boldsymbol{u}_1)$. Integration with respect to L yields

$$\begin{aligned}\mathscr{A}_1(\mathring{\boldsymbol{u}}; L) &\leq \int_0^L 2\sqrt{\mathscr{A}_1(\breve{\boldsymbol{u}}; s)}\sqrt{\mathscr{A}_1(\mathring{\boldsymbol{u}}; s)}\, ds + 2c\sqrt{L}\int_0^L \sqrt{\mathscr{A}_1(\mathring{\boldsymbol{u}}; s)}\, ds \\ &\leq 2\left(\left(\int_0^L \mathscr{A}_1(\breve{\boldsymbol{u}}; s)\, ds\right)^{1/2} + cL\right)\left(\int_0^L \mathscr{A}_1(\mathring{\boldsymbol{u}}; s)\, ds\right)^{1/2} \\ &\leq 2\sqrt{L}\sqrt{\mathscr{A}_1(\mathring{\boldsymbol{u}}; L)}\left(\left(\int_0^L \mathscr{A}_1(\breve{\boldsymbol{u}}; s)\, ds\right)^{1/2} + cL\right).\end{aligned}$$

In the last estimate we have used the fact that $\mathscr{A}_1(\mathring{\boldsymbol{u}}; L)$ is non-decreasing with respect to L. This gives

$$\mathscr{A}_1(\mathring{\boldsymbol{u}}; L) \leq 8L\left(\int_0^L \mathscr{A}_1(\breve{\boldsymbol{u}}; s)\, ds + c^2 L^2\right) \tag{5.1.64}$$

for all $L < T$. Hence, if the crucial estimate

$$\|\mathfrak{F}\|_{L_\infty(\Gamma_{\mathfrak{F}} \times \mathbb{R}^N)} < C_{\mathfrak{F}}$$

for the coefficient of friction holds, then by (5.1.63) and (5.1.64) and by the Gronwall lemma for a sufficiently small value of the parameter κ we get the estimate

$$\mathscr{A}_1(\breve{\boldsymbol{u}}) + \mathscr{A}_2(\breve{\boldsymbol{u}}) \leq c.$$

The constant c depends on the $\mathring{\boldsymbol{H}}^{-1/4,-1/2}(I_T \times \Omega_C)$ norm of the given volume data $\boldsymbol{f}$, on the coefficients of the bilinear forms $a^{(0)}$ and $a^{(1)}$, on the C^2 norms of the cutoff functions ϱ_i and ϕ_0, on the C^β norms of the local maps from the rectification of the boundary and on the $\boldsymbol{H}^{1/2,\,1}(Q_T)$ norm of the solution $\boldsymbol{u}$. Due to Assumption 5.1.6 and the *a priori* estimate (5.1.17) for the solution $\boldsymbol{u}$ of the penalized problem, all these terms are bounded; hence c can be chosen independently of the solution $\boldsymbol{u}$ and in particular of the approximation parameters ε and η. As a consequence, the norms $\|\boldsymbol{u}\|_{\boldsymbol{H}^{1/2,\,1}(S)}$ and $\|\mathscr{S}\|_{L_2(S)}$ are bounded. This is valid for all cutoff functions ϱ_i. Hence the global norms $\|\dot{\boldsymbol{u}}\|_{\boldsymbol{H}^{1/2,\,1}(S_{\mathfrak{F}T})}$ and $\left\|\frac{1}{\varepsilon}[\dot{u}_n]_+\right\|_{L_2(S_{\mathfrak{F}T})}$ are also bounded and the Theorem is proved. □

Since the *a priori* estimates (5.1.17) and (5.1.44) are uniform both in η and in ε, it is easy to prove the existence of a solution to the original contact problem (5.1.10). In fact, there exist sequences of smoothing parameters η_k and penalty parameters ε_k that both converge to 0 and a corresponding sequence of solutions $\boldsymbol{u}_k$ such that

$$\begin{aligned}
&\dot{\boldsymbol{u}}_k \rightharpoonup \dot{\boldsymbol{u}} \text{ in } \boldsymbol{H}^{1/2,\,1}(S_{\mathfrak{F}T}) \text{ and in } \boldsymbol{H}^{1/2,\,1}(Q_T),\\
&\dot{\boldsymbol{u}}_k(t,\cdot) \rightharpoonup \dot{\boldsymbol{u}}(t,\cdot) \text{ in } \boldsymbol{L}_2(\Omega) \text{ for almost every } t \in I_T\\
&\qquad\qquad\qquad\qquad \text{and for } t = T,\\
&\sigma_n(\boldsymbol{u}_k) = -\frac{1}{\varepsilon_k}\big[(\dot{u}_k)_n\big]_+ \rightharpoonup \sigma_n(\boldsymbol{u}) \text{ in } L_2(I_T \times \Gamma_{\mathfrak{F}}),\\
&\dot{\boldsymbol{u}}_k \rightharpoonup \dot{\boldsymbol{u}} \text{ in } \boldsymbol{H}^{1/2,\,1}(I_T \times \Gamma_{\mathfrak{F}}) \text{ and strongly in}\\
&\qquad \boldsymbol{L}_q(I_T \times \Gamma_{\mathfrak{F}}) \text{ for all } q < 2 + \tfrac{4}{N-1} \text{ and}\\
&\mathfrak{F}(\dot{\boldsymbol{u}}_k) \to \mathfrak{F}(\dot{\boldsymbol{u}}) \text{ in } L_p(S_{\mathfrak{F}T}) \text{ for all } p \in [1, +\infty).
\end{aligned} \tag{5.1.65}$$

The last convergence follows from the Lebesgue dominated convergence theorem. The variational equation (5.1.16) is equivalent to variational inequality (5.1.15). We pass to the limit $k \to +\infty$ in this inequality for solution $\boldsymbol{u}_k$ with approximation parameters ε_k, η_k and test function $\boldsymbol{v} \in \mathscr{C}$. Using the approximation property and the uniform Lipschitz continuity of Φ_η as well as the relations

$$\liminf_{k\to\infty} \langle \ddot{\boldsymbol{u}}_k, \dot{\boldsymbol{u}}_k \rangle_{Q_T} \geq \langle \ddot{\boldsymbol{u}}, \dot{\boldsymbol{u}} \rangle_{Q_T} \quad \text{and}$$
$$\liminf_{k\to\infty} A^{(0)}(\boldsymbol{u}_k, \dot{\boldsymbol{u}}_k) \geq A^{(0)}(\boldsymbol{u}, \dot{\boldsymbol{u}})$$

it is proved that the limit $\boldsymbol{u}$ is a solution of problem (5.1.10).

As in the proof of Theorem 5.1.2 the convergences $\dot{\boldsymbol{u}}_k \to \boldsymbol{u}$ in the space $\boldsymbol{H}^{1/2,\,1}(Q_T)$ and $\dot{\boldsymbol{u}}_k(T,\cdot) \to \dot{\boldsymbol{u}}(T,\cdot)$ in $\boldsymbol{L}_2(\Omega)$ are strong: choosing the test function $\boldsymbol{v} = \dot{\boldsymbol{u}}$ in (5.1.15) with solution $\boldsymbol{u}_k$ and smoothing parameter η_k, adding

$$\int_{Q_T} \Big(\ddot{\boldsymbol{u}} \cdot \big(\dot{\boldsymbol{u}}_k - \dot{\boldsymbol{u}}\big) + a^{(0)}\big(\boldsymbol{u}, \dot{\boldsymbol{u}}_k - \boldsymbol{u}\big) + a^{(1)}\big(\dot{\boldsymbol{u}}, \dot{\boldsymbol{u}}_k - \dot{\boldsymbol{u}}\big) \Big)\, d\boldsymbol{x}\, dt$$

to both sides of the resulting inequality and passing to the limit $k \to 0$ we conclude $\dot{\boldsymbol{u}}_k \to \dot{\boldsymbol{u}}$ in $L_2\big(I_T; \boldsymbol{H}^1(\Omega)\big)$ and $\dot{\boldsymbol{u}}_k(T,\cdot) \to \dot{\boldsymbol{u}}(T,\cdot)$ in $\boldsymbol{L}_2(\Omega)$. The functions $\boldsymbol{u}$ and $\boldsymbol{u}_k$ have the same initial data; hence this also implies $\boldsymbol{u}_k \to \boldsymbol{u}$ in the same space. Employing the test function $\boldsymbol{v} = \dot{\boldsymbol{u}}_k + \boldsymbol{w}$ and $\boldsymbol{v} = \dot{\boldsymbol{u}} - \boldsymbol{w}$ in (5.1.15) and (5.1.10) with $\boldsymbol{w} \in L_2\big(I_T; \mathring{\boldsymbol{H}}^1(\Omega)\big)$ and adding the result yields

$$\|\ddot{\boldsymbol{u}} - \ddot{\boldsymbol{u}}_k\|_{L_2(I_T;\mathring{\boldsymbol{H}}^{-1}(\Omega))} \leq c_1 \|\dot{\boldsymbol{u}} - \dot{\boldsymbol{u}}_k\|_{L_2(I_T;\boldsymbol{H}^1(\Omega))} + c_2 \|\boldsymbol{u} - \boldsymbol{u}_k\|_{L_2(I_T;\boldsymbol{H}^1(\Omega))}.$$

Consequently there holds $\ddot{\boldsymbol{u}}_k \to \ddot{\boldsymbol{u}}$ in $L_2\big(I_T; \boldsymbol{H}^{-1}(\Omega)\big)$ and the usual interpolation with $\dot{\boldsymbol{u}}_k \to \dot{\boldsymbol{u}}$ in $L_2\big(I_T; \boldsymbol{H}^1(\Omega)\big)$ gives $\dot{\boldsymbol{u}}_k \to \dot{\boldsymbol{u}}$ in $\boldsymbol{H}^{1/2,1}(Q_T)$. Hence the following proposition is proved:

5.1.8 Theorem. *Let the Assumption 5.1.1 and 5.1.6 be valid. Then there exist sequences ε_k of penalty parameters and η_k of smoothing parameters converging both to* 0 *and a corresponding sequence $\boldsymbol{u}_k$ of solutions of the penalized and smoothed problem* (5.1.16) *which converges strongly in the space $\boldsymbol{H}^{1/2,\,1}(Q_T)$ to a solution of the original problem* (5.1.10). *This solution satisfies the* a priori *estimate* (5.1.17).

5.1.9 Remarks. 1. By a corresponding limit procedure the existence of solutions to the penalized problem (5.1.13) can be proved. Here, Assumption 5.1.6 is not necessary. If the coefficient of friction is independent of the solution, then the uniqueness of the solution can be proved similarly as for the Galerkin approximation in the proof of Theorem 5.1.2.

2. As in the static case it is possible to treat a difference between the coefficients of friction of slip and friction of stick. Let $\mathfrak{F} \equiv \mathfrak{F}(\boldsymbol{x}, \dot{\boldsymbol{u}}_t)$ be a possibly discontinuous function such that the limit $\lim_{\boldsymbol{v}_t \to 0} \mathfrak{F}(\cdot, \boldsymbol{v}_t)$ exists, $\mathfrak{F}(\cdot, 0) \geq \lim_{\boldsymbol{v}_t \to 0} \mathfrak{F}(\cdot, \boldsymbol{v}_t)$ holds on Γ_C and

$$\mathfrak{F}_0(\boldsymbol{x}, \boldsymbol{v}_t) \equiv \begin{cases} \mathfrak{F}(\boldsymbol{x}, \boldsymbol{v}_t), & \boldsymbol{v}_t \neq 0, \\ \lim\limits_{\boldsymbol{v}_t \to 0} \mathfrak{F}(\boldsymbol{x}, \boldsymbol{v}_t), & \boldsymbol{v}_t = 0 \end{cases}$$

satisfies all requirements of Assumptions 5.1.1 and 5.1.6. Then all solutions of (5.1.10) with the coefficient of friction $\mathfrak{F}_0$ solve also the problem with the coefficient of friction $\mathfrak{F}$.

5.2 Anisotropic material

The aim of this section is the proof of Proposition 5.1.4 for the case of general anisotropic material, and in particular the derivation of optimal values for the constants D_0, D_1 and D_2. These constants depend on the properties of the *viscous* part of the constitutive relation only. Therefore, throughout this section, let $a_{ijk\ell}$ denote the coefficients of the Hooke tensor for this viscous part, and let a_0, A_0 be its lower and upper bounds in the sense of (5.1.4). By $\sigma_{ij}(\boldsymbol{u})$ the viscous part of the constitutive relation is denoted,

$$\sigma_{ij}(\boldsymbol{u}) = a_{ijk\ell} e_{k\ell}(\boldsymbol{u}), \tag{5.2.1}$$

and $\boldsymbol{u}$ has the role of a velocity. The corresponding bilinear form on the halfspace $Q = \mathbb{R} \times \mathbb{R}^{N-1} \times \mathbb{R}_+$ is

$$A(\boldsymbol{u}, \boldsymbol{v}) = \int_Q a(\boldsymbol{u}, \boldsymbol{v})\, d\boldsymbol{x}\, dt \quad \text{with } a(\boldsymbol{u}, \boldsymbol{v}) = a_{ijk\ell} e_{k\ell}(\boldsymbol{u}) e_{ij}(\boldsymbol{v}),$$

and $\|\boldsymbol{u}\|_A = \sqrt{A(u,u)}$ is the energy norm. The boundary of Q is denoted by $S = \mathbb{R} \times \mathbb{R}^{N-1} \times \{0\}$.

In a first step of the proof, we derive a specific variant of the trace theorem for the space $H^{1/2,1}(Q)$.

5.2.1 Lemma. *For $u \in H^{1/2,1}(Q)$ there holds*

$$\|u\|'^2_{H^{1/4,0}(S)} \leq 2 \frac{c_1\left(\frac{1}{4}\right)}{\sqrt{c_1\left(\frac{1}{2}\right)}} \|u\|'_{H^{1/2,0}(Q)} \|\partial_N u\|_{L_2(Q)} \quad \textit{and} \tag{5.2.2}$$

$$\|u\|'^2_{H^{0,1/2}(S)} \leq c_{N-1}\left(\tfrac{1}{2}\right) \|\nabla u\|^2_{L_2(Q)}. \tag{5.2.3}$$

Proof. For the proof of estimate (5.2.2) we extend the function u from Q onto the whole space $\mathbb{R}^{N+1}$ by $u(t, \boldsymbol{x}', x_N) = u(t, \boldsymbol{x}', -x_N)$ for $t \in \mathbb{R}$, $\boldsymbol{x}' \in \mathbb{R}^{N-1}$ and $x_N \in \mathbb{R}_-$. We use the Fourier transforms $\mathscr{F}(u)$ with respect to all variables (see (5.1.29)) and $\widetilde{\mathscr{F}}(u)$ with respect to the time variable and the tangential space variables only (cf. (5.1.30)). There holds

$$\begin{aligned} \|u\|'^2_{H^{1/4,0}(S)} &= \int_{\mathbb{R}^N} |\widetilde{\mathscr{F}}(u;\vartheta,\boldsymbol{\xi}',0)|^2 c_1\left(\tfrac{1}{4}\right)|\vartheta|^{1/2}\,d\vartheta\,d\boldsymbol{\xi}' \\ &= \frac{1}{2\pi}\int_{\mathbb{R}^N}\left|\int_{\mathbb{R}}\mathscr{F}(u;\vartheta,\boldsymbol{\xi})\,d\xi_N\right|^2 c_1\left(\tfrac{1}{4}\right)|\vartheta|^{1/2}\,d\vartheta\,d\boldsymbol{\xi}' \\ &\le \frac{1}{2\pi}\int_{\mathbb{R}}\Bigg(\int_{\mathbb{R}^N}|\mathscr{F}(u;\vartheta,\boldsymbol{\xi})|^2\left(\gamma^2 c_1\left(\tfrac{1}{2}\right)|\vartheta| + |\xi_N|^2\right)d\boldsymbol{\xi} \\ &\qquad \int_{\mathbb{R}}\frac{c_1\left(\tfrac{1}{4}\right)|\vartheta|^{1/2}}{\gamma^2 c_1\left(\tfrac{1}{2}\right)|\vartheta| + |\xi_N|^2}\,d\xi_N\Bigg)\,d\vartheta \\ &= \frac{c_1\left(\tfrac{1}{4}\right)}{\sqrt{c_1\left(\tfrac{1}{2}\right)}}\left(\gamma\|u\|'^2_{H^{1/2,0}(Q)} + \frac{1}{\gamma}\|\partial_N u\|^2_{L_2(Q)}\right) \end{aligned} \tag{5.2.4}$$

where $\gamma > 0$ is an arbitrary number. Here the relation

$$\|u\|'_{H^{\alpha,\beta}(\mathbb{R}^N)} = 2\|u\|'_{H^{\alpha,\beta}(Q)}$$

valid for $\alpha \ge 0$ and $\beta \in \{0,1\}$ is used. Choosing the optimal parameter $\gamma = \|\partial_N u\|_{L_2(Q)}/\|u\|'_{H^{1/2,0}(Q)}$ we obtain (5.2.2).

Relation (5.2.3) follows from the inequality

$$\|\nabla u\|^2_{L_2(Q)} \ge \|\nabla v\|^2_{L_2(Q)}$$

valid for the solution v of the problem

$$\Delta v = 0 \text{ in } Q \text{ and } v = u \text{ on } S.$$

Using the partial Fourier transform $\widetilde{\mathscr{F}}$, the function v can be expressed by

$$\widetilde{\mathscr{F}}(v;\vartheta,\boldsymbol{\xi}',x_N) = \widetilde{\mathscr{F}}(u;\vartheta,\boldsymbol{\xi}',0)e^{-|\boldsymbol{\xi}'|x_N}.$$

The $L_2(Q)$-norm of its gradient is given by

$$\begin{aligned} &\int_{\mathbb{R}}\int_{\mathbb{R}^{N-1}}\int_0^{+\infty}|\widetilde{\mathscr{F}}(u;\vartheta,\boldsymbol{\xi}',0)|^2 2|\boldsymbol{\xi}'|^2 e^{-2|\boldsymbol{\xi}'|x_N}\,dx_N\,d\boldsymbol{\xi}'\,d\vartheta \\ &\quad = \int_{\mathbb{R}}\int_{\mathbb{R}^{N-1}}|\widetilde{\mathscr{F}}(u;\vartheta,\boldsymbol{\xi}',0)|^2|\boldsymbol{\xi}'|\,d\boldsymbol{\xi}'\,d\vartheta = \frac{1}{c_{N-1}\left(\tfrac{1}{2}\right)}\|u\|'^2_{\boldsymbol{H}^{0,1/2}(Q)} \end{aligned}$$

and the proof is finished. □

Now the following inverse trace theorem for $H^{1/4,\,1/2}(S)$ is established.

5.2.2 Lemma. *For every function $\boldsymbol{w}$ from $\boldsymbol{H}^{1/4,\,1/2}(S)$ that satisfies $\boldsymbol{w_t} = (w_1,\dots,w_{N-1},0)^\top = 0$ and for every positive parameter μ there exists an extension $\boldsymbol{u}$ onto Q such that*

$$\begin{aligned}\|\nabla \boldsymbol{u}\|^2_{\boldsymbol{L}_2(Q)} \leq & \frac{1}{c_{N-1}\left(\frac{1}{2}\right)}\|\boldsymbol{w}\|'^2_{\boldsymbol{H}^{0,1/2}(S)} \\ & + \frac{\sqrt{\mu}}{\sqrt{2}c_1\left(\frac{1}{4}\right)}\|\boldsymbol{w}\|'^2_{\boldsymbol{H}^{1/4,0}(S)},\end{aligned} \tag{5.2.5}$$

$$\|\boldsymbol{u}\|'^2_{\boldsymbol{H}^{1/2,0}(Q)} \leq \frac{c_1\left(\frac{1}{2}\right)}{\sqrt{2\mu}c_1\left(\frac{1}{4}\right)}\|\boldsymbol{w}\|'^2_{\boldsymbol{H}^{1/4,0}(S)}. \tag{5.2.6}$$

Proof. The function $\boldsymbol{w}$ is extended onto Q by means of the parabolic problem (5.1.35). The extended function is given by formula (5.1.36). From this representation we calculate the seminorms

$$\begin{aligned}\|\nabla \boldsymbol{u}\|^2_{\boldsymbol{L}_2(Q)} &= \\ &= \int_{\mathbb{R}^N} \left|\mathscr{F}(\boldsymbol{w};\vartheta,\boldsymbol{\xi}')\right|^2 \int_{\mathbb{R}_+} \left(|\boldsymbol{\xi}'|^2 + |\mathfrak{a}|^2\right)\left|e^{-2\mathfrak{a}x_N}\right| \, dx_N \, d\boldsymbol{\xi}' \, d\vartheta \\ &= \int_{\mathbb{R}^N} |\mathscr{F}(\boldsymbol{w};\vartheta,\boldsymbol{\xi}')|^2 \mathfrak{a}_1 \, d\vartheta \, d\boldsymbol{\xi}'\end{aligned}$$

with $\mathfrak{a}$ defined in (5.1.37) and

$$\|\boldsymbol{u}\|'^2_{H^{1/2}(\mathbb{R};\boldsymbol{L}_2(O))} = \int_{\mathbb{R}^N} |\mathscr{F}(\boldsymbol{w};\vartheta,\boldsymbol{\xi}')|^2 c_1\left(\tfrac{1}{2}\right)\frac{|\vartheta|}{2\mathfrak{a}_1}\, d\boldsymbol{\xi}' \, d\vartheta.$$

From the inequalities $\max\left\{|\boldsymbol{\xi}'|, \sqrt{\frac{\mu|\vartheta|}{2}}\right\} \leq \mathfrak{a}_1 \leq |\boldsymbol{\xi}'| + \sqrt{\frac{\mu|\vartheta|}{2}}$ we obtain the proposed relations. □

5.2.3 Proposition. *Let $\boldsymbol{u} \in \boldsymbol{H}^{1/2,\,1}(Q)$ be a solution of the equation*

$$\dot{u}_i - \partial_j \sigma_{ij}(\boldsymbol{u}) = 0 \tag{5.2.7}$$

with general constitutive relation (5.2.1). *Then there holds*

$$\|\boldsymbol{u}\|'_{\boldsymbol{H}^{1/4,\,0}(S)} \leq (1+\kappa)D_1\|\boldsymbol{u}\|_A + c_1\|\boldsymbol{u}\|_{\boldsymbol{L}_2(Q)} \text{ and} \tag{5.2.8}$$

$$\|\boldsymbol{u}\|'_{\boldsymbol{H}^{0,1/2}(S)} \leq D_2\|\boldsymbol{u}\|_A \tag{5.2.9}$$

with constants

$$D_1 = \sqrt{c_1\left(\tfrac{1}{4}\right)\frac{2}{a_1}\sqrt[4]{8a_1A_1}} \text{ and} \tag{5.2.10}$$

$$D_2 = \sqrt{c_{N-1}\left(\tfrac{1}{2}\right)\frac{2}{a_1}}, \tag{5.2.11}$$

an arbitrarily small $\kappa > 0$ *and* c_1 *depending only on* κ, a_1, A_1 *and the dimension* N.

Proof. The usual dual estimate, applied to equation (5.2.7), yields

$$\|\dot{\boldsymbol{u}}\|_{L_2(\mathbb{R};\boldsymbol{H}^{-1}(O))} \leq \sqrt{A_1}\|\boldsymbol{u}\|_A,$$

where $O = \mathbb{R}_+ \times \mathbb{R}^{N-1}$. With the help of the Fourier transform the interpolation result

$$\|\boldsymbol{u}\|'^2_{\boldsymbol{H}^{1/2,0}(Q)} \leq c_1\left(\tfrac{1}{2}\right)\|\dot{\boldsymbol{u}}\|_{L_2(\mathbb{R};\boldsymbol{H}^{-1}(O))}\|\boldsymbol{u}\|_{L_2(\mathbb{R};\boldsymbol{H}^1(O))}$$

is derived. Using the Korn inequality for halfspaces from Corollary 1.2.2 and the coercivity of the general elastic bilinear form we conclude

$$\|\nabla\boldsymbol{u}\|_{L_2(Q;\mathbb{R}^{N\times N})} \leq \sqrt{\frac{2}{a_1}}\|\boldsymbol{u}\|_A. \tag{5.2.12}$$

Consequently there holds

$$\|\boldsymbol{u}\|'^2_{\boldsymbol{H}^{1/2,0}(Q)} \leq c_1\left(\tfrac{1}{2}\right)\sqrt{A_1}\|\boldsymbol{u}\|_A\left(\sqrt{\frac{2}{a_1}}\|\boldsymbol{u}\|_A + c_1\|\boldsymbol{u}\|_{\boldsymbol{L}_2(Q)}\right). \tag{5.2.13}$$

Application of the estimates (5.2.2) and (5.2.3) proves (5.2.8) and (5.2.9). □

5.2.4 Proposition. *Let* $\boldsymbol{u} \in \boldsymbol{H}^{1/2,1}(Q)$ *be a solution of equation (5.2.7) and let* $\boldsymbol{v} \in \boldsymbol{H}^{1/4,1/2}(S)$. *Then there exists an extension of* $\boldsymbol{v}$ *onto* Q *such that*

$$\left|\left\langle\dot{\boldsymbol{u}},\boldsymbol{v}\right\rangle_Q + A(\boldsymbol{u},\boldsymbol{v})\right| \leq \left(D_0\|\boldsymbol{u}\|_A + c_1\|\boldsymbol{u}\|_{\boldsymbol{L}_2(Q)}\right)\|\boldsymbol{v}\|'_{\boldsymbol{H}^{1/4,1/2}(S)}$$

with constant D_0 *defined by*

$$D_0 = \sqrt{A_1}\begin{cases}\sqrt{\dfrac{z+1}{c_{N-1}\left(\frac{1}{2}\right)z}}, & z \geq 1,\\[2ex] \sqrt{\dfrac{2}{c_{N-1}\left(\frac{1}{2}\right)\sqrt{z}}}, & z \leq 1,\end{cases} \tag{5.2.14}$$

where $z = \pi \dfrac{8\sqrt{2a_1A_1}}{c_{N-1}^2\left(\frac{1}{2}\right)}$. *The constant* c_1 *depends on* a_1, A_1 *and the dimension* N.

Proof. Let us extend the function $\boldsymbol{v}$ by the parabolic differential equation (5.1.35) with parameter μ to be specified later. This extension satisfies estimates (5.2.5) and (5.2.6). Using the Fourier transform with respect to the time variable and the dual estimate (5.2.13) there follows

$$\begin{aligned}\left|\langle\dot{\boldsymbol{u}},\boldsymbol{v}\rangle_Q\right| &\le c_1\left(\tfrac{1}{2}\right)^{-1}\|\boldsymbol{u}\|'_{\boldsymbol{H}^{1/2,0}(Q)}\|\boldsymbol{v}\|'_{\boldsymbol{H}^{1/2,0}(Q)}\\ &\le c_1\left(\tfrac{1}{2}\right)^{-1/2}\sqrt[4]{\frac{2A_1}{a_1}}\left(\|\boldsymbol{u}\|_A + c_1\|\boldsymbol{u}\|_{\boldsymbol{L}_2(Q)}\right)\|\boldsymbol{v}\|'_{\boldsymbol{H}^{1/2,0}(Q)}.\end{aligned}$$

Hence, from relations (5.2.5) and (5.2.6) we conclude

$$\begin{aligned}|\langle\dot{\boldsymbol{u}},\boldsymbol{v}\rangle + A(\boldsymbol{u},\boldsymbol{v})| &\le \left(\|\boldsymbol{u}\|_A + c_1\|\boldsymbol{u}\|_{\boldsymbol{L}_2(Q)}\right)\sqrt{A_1}\\ &\quad\left(c_1\left(\tfrac{1}{2}\right)^{-1/2}\sqrt[4]{\frac{2}{a_1A_1}}\|\boldsymbol{v}\|'_{\boldsymbol{H}^{1/2,0}(Q)} + \|\nabla\boldsymbol{v}\|_{\boldsymbol{L}_2(Q)}\right)\\ &\le \left(\|\boldsymbol{u}\|_A + c_1\|\boldsymbol{u}\|_{\boldsymbol{L}_2(Q)}\right)\sqrt{A_1}\left(\sqrt[4]{\frac{1}{\mu a_1A_1c_1\left(\frac{1}{4}\right)^2}}\|\boldsymbol{v}\|'_{\boldsymbol{H}^{1/4,0}(S)}\right.\\ &\quad\left. + \sqrt{c_{N-1}\left(\tfrac{1}{2}\right)^{-1}\|\boldsymbol{v}\|'^2_{\boldsymbol{H}^{0,1/2}(S)} + \frac{\sqrt{\mu}}{\sqrt{2}c_1\left(\frac{1}{4}\right)}\|\boldsymbol{v}\|'^2_{\boldsymbol{H}^{1/4,0}(S)}}\right).\end{aligned}$$

Employing the inequality $(y_1+y_2)^2 \le (1+p)y_1^2 + \left(1+p^{-1}\right)y_2^2$ for

$$y_1 = \sqrt{\frac{\|\boldsymbol{v}\|'^2_{\boldsymbol{H}^{1/4,0}(S)}}{\sqrt[4]{\mu a_1A_1}\sqrt{c_1\left(\frac{1}{4}\right)}}} \quad \text{and}$$

$$y_2 = \sqrt{\frac{\sqrt{\mu}}{\sqrt{2}c_1\left(\frac{1}{4}\right)}\|\boldsymbol{v}\|'^2_{\boldsymbol{H}^{1/4,0}(S)} + c_{N-1}\left(\tfrac{1}{2}\right)^{-1}\|\boldsymbol{v}\|'^2_{\boldsymbol{H}^{0,1/2}(S)}}$$

with arbitrary $p > 0$ yields

$$\begin{aligned}|\langle\dot{\boldsymbol{u}},\boldsymbol{v}\rangle + A(\boldsymbol{u},\boldsymbol{v})| &\le \left(\|\boldsymbol{u}\|_A + c_1\|\boldsymbol{u}\|_{\boldsymbol{L}_2(Q)}\right)\sqrt{A_1}\\ &\quad\left(\left(\frac{1+p}{c_1\left(\frac{1}{4}\right)\sqrt{\mu a_1A_1}} + \frac{\left(1+p^{-1}\right)\sqrt{\mu}}{\sqrt{2}c_1\left(\frac{1}{4}\right)}\right)\|\boldsymbol{v}\|'^2_{\boldsymbol{H}^{1/4,0}(S)}\right.\\ &\quad\left. + \frac{1+p^{-1}}{c_{N-1}\left(\frac{1}{2}\right)}\|\boldsymbol{v}\|'^2_{\boldsymbol{H}^{0,1/2}(S)}\right)^{1/2}.\end{aligned} \tag{5.2.15}$$

Let us use the abbreviations

$$X \equiv \|\boldsymbol{v}\|'^2_{\boldsymbol{H}^{1/4,0}(S)}, \; Y \equiv \|\boldsymbol{v}\|'^2_{\boldsymbol{H}^{0,1/2}(S)}, \; \mathscr{E}_1 \equiv \frac{1}{c_1\left(\frac{1}{4}\right)\sqrt{a_1 A_1}},$$

$$\mathscr{E}_2 \equiv \frac{1}{\sqrt{2}c_1\left(\frac{1}{4}\right)}, \; \mathscr{E}_3 \equiv \frac{1}{c_{N-1}\left(\frac{1}{2}\right)}, \; \mathscr{B}_2(p) \equiv \left(1+\frac{1}{p}\right)\mathscr{E}_3 \text{ and}$$

$$\mathscr{B}_1(p,\mu) \equiv \frac{1+p}{\sqrt{\mu}}\mathscr{E}_1 + \left(1+p^{-1}\right)\sqrt{\mu}\mathscr{E}_2.$$

Then the right hand side of (5.2.15) has the representation

$$\left(\|\boldsymbol{u}\|_A + c_1\|\boldsymbol{u}\|_{\boldsymbol{L}_2(Q)}\right)\sqrt{A_1}\sqrt{\mathscr{B}_1(p,\mu)X + \mathscr{B}_2(p)Y}.$$

In order to get an optimal value of D_0 we solve the optimization problem

$$\inf_{\substack{p>0\\ \mu>0}} \sup_{X,Y>0} \frac{\mathscr{B}_1(p,\mu)X + \mathscr{B}_2(p)Y}{X+Y} = \inf_{\substack{p>0\\ \mu>0}} \max\{\mathscr{B}_1(p,\mu), \mathscr{B}_2(p)\}. \quad (5.2.16)$$

The function $f(x) = A/x + Bx$ has its optimum in $x = \sqrt{A/B}$ with value $2\sqrt{AB}$. Hence the optimization of $\mathscr{B}_1$ with respect to μ gives

$$\mathscr{B}_3(p) \equiv \inf_{\mu>0} \mathscr{B}_1(p,\mu) = \left(\sqrt{p}+\frac{1}{\sqrt{p}}\right)2\sqrt{\mathscr{E}_1\mathscr{E}_2}.$$

The minimum of $\mathscr{B}_3$ is taken at $p=1$ with value $4\sqrt{\mathscr{E}_1\mathscr{E}_2}$. The function $\mathscr{B}_2$ is decreasing; and its point of intersection with $\mathscr{B}_3$ is given by $p = z \equiv \dfrac{\mathscr{E}_3^2}{4\mathscr{E}_1\mathscr{E}_2}$. Consequently, for $z \geq 1$ the optimization problem (5.2.16) has the solution $p = z$ with optimum $(1+1/z)\,\mathscr{E}_3$, while for $z \leq 1$ the solution is taken at $p=1$ and has the value $\mathscr{B}_3(1) = 4\sqrt{\mathscr{E}_1\mathscr{E}_2}$. This gives

$$D_0 = \sqrt{A_1}\begin{cases} \sqrt{\dfrac{1+z}{z}\mathscr{E}_3}, & z \geq 1, \\ 2\sqrt[4]{\mathscr{E}_1\mathscr{E}_2}, & z \leq 1. \end{cases}$$

Using the definitions of $\mathscr{E}_1, \mathscr{E}_2, \mathscr{E}_3$ and the relations $\sqrt{\mathscr{E}_1\mathscr{E}_2} = \dfrac{\mathscr{E}_3}{2\sqrt{z}}$ and $c_1\left(\frac{1}{4}\right) = 4\sqrt{2\pi}$ we obtain $z = \pi\dfrac{8\sqrt{2a_1A_1}}{c_{N-1}^2\left(\frac{1}{2}\right)}$ and the expression for D_0 given in the proposition. □

As an easy consequence of the two preceding propositions we obtain the upper bound for the admissible coefficient of friction.

5.2.5 Corollary. *In the case of anisotropic material with lower and upper bounds a_0 and A_0 for the* viscous *part of the constitutive relation, the assertions of Theorems 5.1.7 and 5.1.8 are valid with upper bound*

$$C_{\mathfrak{F}} \equiv \sqrt{\frac{a_1}{A_1}} \cdot \begin{cases} \sqrt{\dfrac{z}{2(z+1)(1+\sqrt{8z})}} & \text{for } z \geq 1, \\ \sqrt{\dfrac{\sqrt{z}}{4(1+\sqrt{8z})}} & \text{for } z \leq 1, \end{cases} \qquad (5.2.17)$$

for the admissible coefficient of friction. The parameter z is defined by

$$z = \pi \frac{8\sqrt{2a_1 A_1}}{c_{N-1}^2\left(\frac{1}{2}\right)}.$$

This formula indeed follows from $C_{\mathfrak{F}}^{-1} = D_0\sqrt{D_1^2 + D_2^2}$ with the relation

$$D_1 = \sqrt{\frac{2}{a_1}}\sqrt{c_{N-1}\left(\tfrac{1}{2}\right)}\,\sqrt[4]{8z}.$$

5.2.6 Remark. The values a_1, A_1 can be changed by the choice of the time unit. The definition of a “new time variable” $t^{\text{new}} = \lambda t^{\text{old}}$ for any constant $\lambda > 0$ transforms the system (5.1.1) to

$$\ddot{u}_i^{\text{new}} - \partial_j\left(\lambda^{-2} a_{ijk\ell}^{(0)} e_{k\ell}(\boldsymbol{u}^{\text{new}}) + \lambda^{-1} a_{ijk\ell}^{(1)} e_{k\ell}(\dot{\boldsymbol{u}}^{\text{new}})\right) = \lambda^{-2} f_i, \qquad i = 1, \ldots, N,$$

with $\boldsymbol{u}^{\text{new}}(t^{\text{new}}, \cdot) = \boldsymbol{u}^{\text{old}}(t^{\text{old}}, \cdot)$, $a_1^{\text{new}} = \lambda^{-1} a_1^{\text{old}}$, $A_1^{\text{new}} = \lambda^{-1} A_1^{\text{old}}$, hence $z^{\text{new}} = \lambda^{-1} z^{\text{old}}$ in (5.2.17). Due to the scaling of this equation the stress tensor changes to $\sigma_{ij}^{\text{new}}(\boldsymbol{u}^{\text{new}}) = \lambda^{-2}\sigma_{ij}^{\text{old}}(\boldsymbol{u}^{\text{old}})$, but in particular the ratio between normal and tangential components of the boundary traction does not change. By a suitable time scaling it is therefore possible to choose an arbitrary parameter z in the result (5.2.17). The maximal value of $C_{\mathfrak{F}}$ is obtained for $z \geq 1$. Optimization of the corresponding expression leads to the formula $z^3 - z - 2^{-1/2} = 0$. The solution of this equation is

$$z_{\max} = \frac{\sqrt[3]{3\sqrt{3} + \sqrt{19}} + \sqrt[3]{3\sqrt{3} - \sqrt{19}}}{\sqrt{6}} \approx 1.2510786$$

and the optimal value for $C_{\mathfrak{F}}$ is

$$C_{\mathfrak{F}} = k_{\max}\sqrt{\frac{a_1}{A_1}} \text{ with } k_{\max} \approx 0.258342. \qquad (5.2.18)$$

Formula (5.2.18) shows that the bound $C_{\mathfrak{F}}$ is low in this general case. Hence it is worth calculating bounds for different special materials.

5.3 Isotropic material

In the case of homogeneous isotropic viscous part of the constitutive relation the crucial trace constants D_0, D_1 and D_2 can be computed via an exact solution of a corresponding parabolic differential equation containing only the viscous part of the constitutive relations. As we will see, the obtained admissible coefficient of friction will be essentially bigger in this case. The results are also applicable to non-homogeneous material if the variation of the material data is sufficiently smooth.

For homogeneous, isotropic material with modulus of viscosity E (the viscous counterpart to the Young modulus of elasticity) and viscous Poisson ratio ν the viscous part of the constitutive law is given by

$$\sigma_{ij}(\boldsymbol{u}) = \frac{E\nu}{(1+\nu)(1-2\nu)}\delta_{ij}\partial_k u_k + \frac{E}{2(1+\nu)}(\partial_i u_j + \partial_j u_i).$$

Here $\boldsymbol{u}$ plays again the role of a velocity. Since the problem with non-vanishing load $\boldsymbol{f}$ can be treated again with the decomposition used in the proof of Lemma 5.1.3, we confine ourselves to the case of $\boldsymbol{f} \equiv 0$. Using the abbreviations

$$C_1 = \frac{2+2\nu}{E}, \; C_2 = \frac{(1+\nu)(1-2\nu)}{E(1-\nu)}$$

and the relation $\dfrac{E}{2(1+\nu)(1-2\nu)} = \dfrac{1}{C_2} - \dfrac{1}{C_1}$, the differential equations

$$\dot{u}_i - \partial_j \sigma_{ij}(\boldsymbol{u}) = 0, \; i = 1, \ldots, N, \tag{5.3.1}$$

can be written as

$$\dot{u}_j = \frac{1}{C_2}\partial_j^2 u_j + \left(\frac{1}{C_2} - \frac{1}{C_1}\right) \sum_{\substack{\ell=1,\ldots,N \\ \ell \neq j}} \partial_j \partial_\ell u_\ell + \frac{1}{C_1} \sum_{\substack{\ell=1,\ldots,N \\ \ell \neq j}} \partial_\ell^2 u_j \tag{5.3.2}$$

for $j = 1, \ldots, N$. The above system is solved for the Dirichlet boundary condition

$$\boldsymbol{u}(t, \boldsymbol{x}', 0) = \boldsymbol{w}(t, \boldsymbol{x}') \text{ for } (t, \boldsymbol{x}') \in \mathbb{R} \times \mathbb{R}^{N-1} \tag{5.3.3}$$

with $\boldsymbol{x}' = (x_1, \ldots, x_{N-1})^\top$.

In order to solve this system we employ the partial Fourier transform with respect to the time and the tangential space variables. In order to simplify formulae it is advantageous to use the transform $\vartheta_{\text{new}} = C_1 \vartheta$ for the dual time variable. For simplicity of the notation we keep the old symbol ϑ for the transformed variable ϑ_{new}. This transformation of ϑ leads to an additional factor in integrals with respect to ϑ, arising from the transformation formula for integrals. Since this factor is the same in all norms, we can neglect it in calculations. Then the system is transformed to

$$
\begin{aligned}
\widetilde{u}_j'' &= \left(i\vartheta + \frac{1}{k}\xi_j^2 + \sum_{\substack{\ell=1,\dots,N-1 \\ \ell\neq j}} \xi_\ell^2 \right) \widetilde{u}_j \\
&\quad + \frac{1-k}{k} \left(\sum_{\substack{\ell=1,\dots,N-1 \\ \ell\neq j}} \xi_j \xi_\ell \widetilde{u}_\ell - i\xi_j \widetilde{u}_N' \right), \\
\widetilde{u}_N'' &= k \left(i\vartheta + \sum_{\ell=1}^{N-1} \xi_\ell^2 \right) \widetilde{u}_N + (k-1) \sum_{\ell=1}^{N-1} i\xi_\ell \widetilde{u}_\ell'.
\end{aligned}
\tag{5.3.4}
$$

Here the short notation $\widetilde{\boldsymbol{u}} = \widetilde{\mathscr{F}}(\boldsymbol{u};\cdot)$ and $k = C_2/C_1$ is used, and the prime $'$ denotes the derivative with respect to $x_N =: y$. This system will be solved for the cases $N = 2$, $N = 3$ and the case of a general dimension separately.

5.3.1 The case of two space dimensions

In two space dimensions the system (5.3.4) has the form

$$
\begin{aligned}
\widetilde{u}_1'' &= \left(i\vartheta + \frac{\xi^2}{k} \right) \widetilde{u}_1 + \frac{k-1}{k} i\xi \widetilde{u}_2', \\
\widetilde{u}_2'' &= k(i\vartheta + \xi^2)\widetilde{u}_2 + (k-1) i\xi \widetilde{u}_1'.
\end{aligned}
\tag{5.3.5}
$$

This system has four independent solutions

$$
\widetilde{\boldsymbol{u}}^{(1)} = \begin{pmatrix} \mathfrak{a} \\ -i\xi \end{pmatrix} e^{\mathfrak{a}y}, \quad \widetilde{\boldsymbol{u}}^{(2)} = \begin{pmatrix} \mathfrak{a} \\ i\xi \end{pmatrix} e^{-\mathfrak{a}y}, \quad \widetilde{\boldsymbol{u}}^{(3)} = \begin{pmatrix} i\xi \\ \mathfrak{b} \end{pmatrix} e^{\mathfrak{b}y},
$$
$$
\widetilde{\boldsymbol{u}}^{(4)} = \begin{pmatrix} i\xi \\ -\mathfrak{b} \end{pmatrix} e^{-\mathfrak{b}y}
$$

with the complex roots $\mathfrak{a} = \mathfrak{a}_1 + i\mathfrak{a}_2$, $\mathfrak{b} = \mathfrak{b}_1 + i\mathfrak{b}_2$ defined by

$$\begin{aligned} \mathfrak{a} &= \sqrt{\xi^2 + i\vartheta} \\ &= \sqrt{\frac{\sqrt{\xi^4+\vartheta^2}+\xi^2}{2}} + i\,\mathrm{sign}(\vartheta)\sqrt{\frac{\sqrt{\xi^4+\vartheta^2}-\xi^2}{2}}, \\ \mathfrak{b} &= \sqrt{\xi^2 + ik\vartheta} \\ &= \sqrt{\frac{\sqrt{\xi^4+k^2\vartheta^2}+\xi^2}{2}} + i\,\mathrm{sign}(\vartheta)\sqrt{\frac{\sqrt{\xi^4+k^2\vartheta^2}-\xi^2}{2}}. \end{aligned} \tag{5.3.6}$$

The solution of the boundary value problem is that linear combination of these independent functions which satisfies the boundary conditions

$$\widetilde{\boldsymbol{u}}(\vartheta,\xi,0) = \widehat{\boldsymbol{w}}(\vartheta,\xi) = \begin{pmatrix} \widehat{w}_1(\vartheta,\xi) \\ \widehat{w}_2(\vartheta,\xi) \end{pmatrix} \quad \text{and } \widetilde{u}(\vartheta,\xi,y) \to 0 \text{ for } y \to +\infty.$$

This function is uniquely determined by

$$\begin{aligned} \widetilde{\boldsymbol{u}}(\vartheta,\xi,y) = \mathscr{D}^{-1} \Bigg[&(\mathfrak{b}\widehat{w}_1 + i\xi\widehat{w}_2) \begin{pmatrix} \mathfrak{a} \\ i\xi \end{pmatrix} e^{-\mathfrak{a}y} \\ &+ (i\xi\widehat{w}_1 - \mathfrak{a}\widehat{w}_2) \begin{pmatrix} i\xi \\ -\mathfrak{b} \end{pmatrix} e^{-\mathfrak{b}y} \Bigg] \end{aligned} \tag{5.3.7}$$

with $\mathscr{D} = \mathfrak{a}\mathfrak{b} - \xi^2$. Motivated by the form of this solution, we perform the change of variables

$$\mathfrak{p}_1 = \frac{\mathfrak{b}\widehat{w}_1 + i\xi\widehat{w}_2}{\mathscr{D}}, \quad \mathfrak{p}_2 = \frac{i\xi\widehat{w}_1 - \mathfrak{a}\widehat{w}_2}{\mathscr{D}}. \tag{5.3.8}$$

Observe that this relation can be easily inverted,

$$\widehat{w}_1 = \mathfrak{a}\mathfrak{p}_1 + i\xi\mathfrak{p}_2, \quad \widehat{w}_2 = i\xi\mathfrak{p}_1 - \mathfrak{b}\mathfrak{p}_2.$$

In the new variables, the solution is

$$\widetilde{\boldsymbol{u}}(\vartheta,\xi,y) = \mathfrak{p}_1 \begin{pmatrix} \mathfrak{a} \\ i\xi \end{pmatrix} e^{-\mathfrak{a}y} + \mathfrak{p}_2 \begin{pmatrix} i\xi \\ -\mathfrak{b} \end{pmatrix} e^{-\mathfrak{b}y}. \tag{5.3.9}$$

In order to prove the trace estimates (5.1.39) and (5.1.40), we consider the energy norm $\|\boldsymbol{u}\|_A^2$. It can be written in Fourier-transformed variables as

$$\begin{aligned} \|\boldsymbol{u}\|_A^2 = \int_Q \sigma_{ij}(\boldsymbol{u})e_{ij}(\boldsymbol{u})\,d\boldsymbol{x}\,dt \;=\; M \int_{\mathbb{R}}\int_{\mathbb{R}\times\mathbb{R}_+} \Big[\nu|\widetilde{\mathrm{div}(\boldsymbol{u})}|^2 \\ + (1-2\nu)\left(|\widetilde{\partial_1 u_1}|^2 + |\widetilde{\partial_2 u_2}|^2 + 2|\widetilde{e_{12}(\boldsymbol{u})}|^2 \right) \Big]\, dy\,d\xi\,d\vartheta \end{aligned} \tag{5.3.10}$$

with $M = \dfrac{E}{(1+\nu)(1-2\nu)}$. The Fourier-transformed values in this integral are given by

$$
\begin{aligned}
\widetilde{\partial_1 u_1} &= i\xi\mathfrak{a}\mathfrak{p}_1 e^{-\mathfrak{a}y} - \xi^2\mathfrak{p}_2 e^{-\mathfrak{b}y},\\
\widetilde{\partial_2 u_2} &= -i\xi\mathfrak{a}\mathfrak{p}_1 e^{-\mathfrak{a}y} + \mathfrak{b}^2\mathfrak{p}_2 e^{-\mathfrak{b}y},\\
2\widetilde{e_{12}(\boldsymbol{u})} &= \partial_y\tilde{u}_1 + i\xi\tilde{u}_2 = -\left(\mathfrak{a}^2+\xi^2\right)\mathfrak{p}_1 e^{-\mathfrak{a}y} - 2i\xi\mathfrak{b}\mathfrak{p}_2 e^{-\mathfrak{b}y},\\
\widetilde{\operatorname{div}(\boldsymbol{u})} &= i\xi\tilde{u}_1 + \partial_y\tilde{u}_2 = \left(\mathfrak{b}^2-\xi^2\right)\mathfrak{p}_2 e^{-\mathfrak{b}y}.
\end{aligned}
$$

Using these formulae in (5.3.10) and performing the inner integration with respect to y yields after some calculation the representation

$$
\begin{aligned}
\|\boldsymbol{u}\|_A^2 = M\int_{\mathbb{R}}\int_{\mathbb{R}} \Bigg[&(1-2\nu)\frac{2|\mathfrak{a}|^2\xi^2 + \frac{1}{2}|\mathfrak{a}^2+\xi^2|^2}{2\mathfrak{a}_1}|\mathfrak{p}_1|^2\\
&+ \left(\nu\frac{|\mathfrak{b}^2-\xi^2|^2}{2\mathfrak{b}_1} + (1-2\nu)\frac{|\mathfrak{b}|^4 + 2\xi^2|\mathfrak{b}|^2+\xi^4}{2\mathfrak{b}_1}\right)|\mathfrak{p}_2|^2\\
&- 2\operatorname{Re}\big((1-2\nu)i\xi(\mathfrak{a}\bar{\mathfrak{b}}+\xi^2)\mathfrak{p}_1\bar{\mathfrak{p}}_2\big)\Bigg]\,d\xi\,d\vartheta.
\end{aligned}
$$

From the special form of the complex roots $\mathfrak{a} = \mathfrak{a}_1 + i\mathfrak{a}_2$ and $\mathfrak{b} = \mathfrak{b}_1 + i\mathfrak{b}_2$ it is easy to derive the formulae $2\mathfrak{a}_1\mathfrak{a}_2 = \operatorname{Im}(\mathfrak{a}^2) = \vartheta$, $2\mathfrak{b}_1\mathfrak{b}_2 = \operatorname{Im}(\mathfrak{b}^2) = k\vartheta$, $|\mathfrak{a}|^2 + \xi^2 = 2\mathfrak{a}_1^2$, $|\mathfrak{b}|^2 + \xi^2 = 2\mathfrak{b}_1^2$ and

$$
\begin{aligned}
\left|\mathfrak{a}^2+\xi^2\right|^2 &= 4\xi^4 + \vartheta^2,\\
|\mathfrak{b}|^4 + 2\xi^2|\mathfrak{b}|^2 + \xi^4 &= 4\mathfrak{b}_1^2\xi^2 + k^2\vartheta^2.
\end{aligned}
$$

With their application the energy norm is transformed to

$$
\begin{aligned}
\|\boldsymbol{u}\|_A^2 &= M\int_{\mathbb{R}}\int_{\mathbb{R}} \Bigg[|\mathfrak{p}_1|^2\frac{1-2\nu}{2}(\vartheta\mathfrak{a}_2 + 4\mathfrak{a}_1\xi^2)\\
&\quad - 2\operatorname{Re}\big((1-2\nu)i\xi(\mathfrak{a}\bar{\mathfrak{b}}+\xi^2)\mathfrak{p}_1\bar{\mathfrak{p}}_2\big)\\
&\quad + |\mathfrak{p}_2|^2\left((1-\nu)k\vartheta\mathfrak{b}_2 + 2(1-2\nu)\mathfrak{b}_1\xi^2\right)\Bigg]\,d\xi\,d\vartheta.
\end{aligned}
$$

The definitions of the values k, C_1, and C_2 yield $1-2\nu = 2(1-\nu)k$. Hence the result for the energy norm is

$$
\begin{aligned}
\|\boldsymbol{u}\|_A^2 = \frac{E}{1+\nu}\int_{\mathbb{R}}\int_{\mathbb{R}} \Bigg[&|\mathfrak{p}_1|^2\left(\frac{\vartheta\mathfrak{a}_2}{2} + 2\mathfrak{a}_1\xi^2\right)\\
+ |\mathfrak{p}_2|^2\left(\frac{\vartheta\mathfrak{b}_2}{2} + 2\mathfrak{b}_1\xi^2\right) &- 2\operatorname{Re}\left(i\xi(\mathfrak{a}\bar{\mathfrak{b}}+\xi^2)\mathfrak{p}_1\bar{\mathfrak{p}}_2\right)\Bigg]\,d\xi\,d\vartheta.
\end{aligned}
\tag{5.3.11}
$$

With this representation the formulae for the constants D_1 and D_2 shall be derived. The Sobolev-Slobodetskii seminorm with the new dual time variable $\vartheta = \vartheta_{\text{new}} = C_1\vartheta_{\text{old}}$ is

$$\|w_1\|'^2_{H^{\alpha,\beta}(S)} = \int_{\mathbb{R}}\int_{\mathbb{R}} |\widehat{w}_1(\vartheta,\xi)|^2 \left(c_1(\alpha)\left|C_1^{-1}\vartheta\right|^{2\alpha} + c_1(\beta)|\xi|^{2\beta}\right) d\xi\, d\vartheta$$

for $\alpha, \beta \in [0,1)$ with $c_1(0) = 0$, where we neglect the factor arising from the transformation formula for integrals. With the representation

$$\widehat{w}_1 = \mathfrak{a}\mathfrak{p}_1 + i\xi\mathfrak{p}_2$$

we see that the optimal constants D_1, D_2 are the solutions of the optimization problems

$$\begin{aligned} D_1^2 &= \frac{1+\nu}{E} \sup_{\substack{\mathfrak{p}_1,\mathfrak{p}_2\in\mathbb{C}\\ \xi,\vartheta\in\mathbb{R}}} \mathfrak{E}(\xi,\vartheta,\mathfrak{a},\mathfrak{b},\mathfrak{p}_1,\mathfrak{p}_2)\; c_1\left(\tfrac{1}{2}\right)|\xi| \\ D_2^2 &= \frac{1+\nu}{E} \sup_{\substack{\mathfrak{p}_1,\mathfrak{p}_2\in\mathbb{C}\\ \xi,\vartheta\in\mathbb{R}}} \mathfrak{E}(\xi,\vartheta,\mathfrak{a},\mathfrak{b},\mathfrak{p}_1,\mathfrak{p}_2)\; c_1\left(\tfrac{1}{4}\right)C_1^{-1/2}|\vartheta|^{1/2} \end{aligned} \tag{5.3.12}$$

where the expression $\mathfrak{E} \equiv \mathfrak{E}(\xi,\vartheta,\mathfrak{a},\mathfrak{b},\mathfrak{p}_1,\mathfrak{p}_2)$ is defined as

$$\frac{|\mathfrak{a}|^2|\mathfrak{p}_1|^2 + 2\,\mathrm{Im}(\mathfrak{a}\xi\mathfrak{p}_1\bar{\mathfrak{p}}_2) + \xi^2|\mathfrak{p}_2|^2}{|\mathfrak{p}_1|^2\left(\frac{1}{2}\vartheta\mathfrak{a}_2 + 2\mathfrak{a}_1\xi^2\right) + 2\xi\,\mathrm{Im}\big((\mathfrak{a}\bar{\mathfrak{b}} + \xi^2)\mathfrak{p}_1\bar{\mathfrak{p}}_2\big) + |\mathfrak{p}_2|^2\left(\frac{1}{2}\vartheta\mathfrak{b}_2 + 2\mathfrak{b}_1\xi^2\right)}\,.$$

The results of the optimization problems do not change, if we restrict the optimization to non-negative values of ϑ only. This can be seen as follows: Changing the sign of ϑ in $\mathfrak{a}$ and $\mathfrak{b}$ leads to the conjugate complex values $\bar{\mathfrak{a}}$ and $\bar{\mathfrak{b}}$. Hence the values $\mathfrak{a}_1$, $\mathfrak{b}_1$, $\vartheta\mathfrak{a}_2$ and $\vartheta\mathfrak{b}_2$ do not depend on the sign of ϑ. Using the formula $\mathrm{Im}(\bar{\mathfrak{c}}\mathfrak{p}_1\bar{\mathfrak{p}}_2) = \mathrm{Im}(\mathfrak{c}(-\overline{\mathfrak{p}_1})\mathfrak{p}_2)$ in the imaginary parts arising in the functions to be optimized it is seen that the ranges of these functions are preserved by the restriction to non-negative values of ϑ. Due to the continuity of the numerator and denominator of these functions it is possible to exclude the parameters $\xi = 0$, $\vartheta = 0$ and $\mathfrak{p}_2 = 0$ from the optimization without changing its result. Hence we are able to use the new variables $\mathfrak{p}_1 = z\mathfrak{p}_2$ with a complex parameter z and $\vartheta = \gamma^2\xi^2$ with a positive parameter γ. A close look at the optimization problems and at the definition of the complex roots $\mathfrak{a}$, $\mathfrak{b}$ shows that the variable ξ cancels completely. The old roots $\mathfrak{a}$, $\mathfrak{b}$ are thereby transformed to

$$\mathfrak{x} = \sqrt{1 + i\gamma^2} = \sqrt{\frac{\sqrt{1+\gamma^4}+1}{2}} + i\sqrt{\frac{\sqrt{1+\gamma^4}-1}{2}} = \mathfrak{x}_1 + i\mathfrak{x}_2,$$

$$\mathfrak{y} = \sqrt{1 + ik\gamma^2} = \sqrt{\frac{\sqrt{1+k^2\gamma^4}+1}{2}} + i\sqrt{\frac{\sqrt{1+k^2\gamma^4}-1}{2}}$$
$$= \mathfrak{y}_1 + i\mathfrak{y}_2.$$

The result of this transform of variables is the optimization problems

$$D_1^2 = \frac{1}{2}\sup_{\substack{z\in\mathbb{C}\\ \gamma>0}} C_1 c_1\left(\tfrac{1}{2}\right) G(\gamma, z) \text{ and}$$

$$D_2^2 = \frac{1}{2}\sup_{\substack{z\in\mathbb{C}\\ \gamma>0}} \sqrt{C_1} c_1\left(\tfrac{1}{4}\right) \gamma\, G(\gamma, z)$$

in the two variables $\gamma > 0$ and $z \in \mathbb{C}$ with

$$G(\gamma, z) = \frac{|\mathfrak{x}|^2|z|^2 + 2\operatorname{Im}(\mathfrak{x}z) + 1}{\left(\frac{1}{2}\gamma^2\mathfrak{x}_2 + 2\mathfrak{x}_1\right)|z|^2 + 2\operatorname{Im}((\mathfrak{x}\overline{\mathfrak{y}} + 1)z) + \left(\frac{1}{2}\gamma^2\mathfrak{y}_2 + 2\mathfrak{y}_1\right)}.$$

The optimization with respect to the complex parameter z can be performed analytically. The function G has the representation

$$G(\gamma, z) = \frac{|\mathfrak{x}|^2}{\frac{1}{2}\gamma^2\mathfrak{x}_2 + 2\mathfrak{x}_1} \cdot \frac{|z + a|^2}{|z|^2 + 2\operatorname{Re}(\overline{b}z) + |c|^2}$$

with the complex parameters

$$a = \frac{i}{\mathfrak{x}}, \quad \overline{b} = -i\frac{2(\mathfrak{x}\overline{\mathfrak{y}} + 1)}{\gamma^2\mathfrak{x}_2 + 4\mathfrak{x}_1}, \quad |c|^2 = \frac{\gamma^2\mathfrak{y}_2 + 4\mathfrak{y}_1}{\gamma^2\mathfrak{x}_2 + 4\mathfrak{x}_1}. \tag{5.3.13}$$

The optimization problem

$$\sup_{z\in\mathbb{C}} \frac{|z + a|^2}{|z|^2 + 2\operatorname{Re}(\overline{b}z) + |c|^2} \tag{5.3.14}$$

can be reformulated by an easy transform of the variable to

$$\sup_{z\in\mathbb{C}} \frac{|z + (a - b)|^2}{|z|^2 + |c|^2 - |b|^2} = \sup_{r>0} \frac{(r + |a - b|)^2}{r^2 + |c|^2 - |b|^2}.$$

With the help of the formulae $\gamma^2 = 2\mathfrak{x}_1\mathfrak{x}_2$, $\mathfrak{x}_1^2 = \mathfrak{x}_2^2 + 1$ and $\mathfrak{y}_1^2 = \mathfrak{y}_2^2 + 1$ there follows first

$$|\mathfrak{x}|^2|\mathfrak{y}|^2 = 2\mathfrak{x}_2^2\mathfrak{y}_2^2 + 2\mathfrak{x}_2^2\mathfrak{y}_1^2 + 2\mathfrak{y}_1^2 - 1$$

and then

$$|c|^2 - |b|^2 = \frac{\gamma^4 \mathfrak{x}_2 \mathfrak{y}_2 + 8\mathfrak{x}_2^2 \mathfrak{y}_2(\mathfrak{x}_2 - \mathfrak{y}_2) + 8\mathfrak{y}_1(\mathfrak{x}_1 - \mathfrak{y}_1) + 8\mathfrak{x}_2^2 \mathfrak{y}_1(\mathfrak{x}_1 - \mathfrak{y}_1)}{(\gamma^2 \mathfrak{x}_2 + 4\mathfrak{x}_1)^2}.$$

This term is strictly positive, provided $\gamma \in (0, +\infty)$. The function $f : r \mapsto \frac{(r+A)^2}{r^2+B^2}$ takes its maximum at $r = \frac{B^2}{A}$ with value $1 + \frac{A^2}{B^2}$. Therefore the optimization problem has a unique solution with the optimal value

$$1 + \frac{|a-b|^2}{|c|^2 - |b|^2}. \tag{5.3.15}$$

Using the abbreviations

$$\alpha_{\mathfrak{x}} = \frac{\gamma^2 \mathfrak{x}_2}{2} + 2\mathfrak{x}_1, \quad \alpha_{\mathfrak{y}} = \frac{\gamma^2 \mathfrak{y}_2}{2} + 2\mathfrak{y}_1, \quad \mathfrak{s} = \mathfrak{x}\bar{\mathfrak{y}} + 1$$

the values a, b and c have the representations

$$a = \frac{i}{\mathfrak{x}}, \quad b = \frac{i\,\mathfrak{s}}{\alpha_{\mathfrak{x}}} \text{ and } |c|^2 = \frac{\alpha_{\mathfrak{y}}}{\alpha_{\mathfrak{x}}}$$

and there follows

$$\begin{aligned} |c|^2 - |b|^2 + |a-b|^2 &= |c|^2 + |a|^2 - 2\operatorname{Re}(a\bar{b}) \\ &= \frac{\alpha_{\mathfrak{y}}}{\alpha_{\mathfrak{x}}} + \frac{1}{|\mathfrak{x}|^2} - 2\operatorname{Re}\left(\frac{\mathfrak{s}\bar{\mathfrak{x}}}{\alpha_{\mathfrak{x}}|\mathfrak{x}|^2}\right) \\ &= \frac{\alpha_{\mathfrak{y}}|\mathfrak{x}|^2 + \alpha_{\mathfrak{x}} - 2\operatorname{Re}(\mathfrak{s}\bar{\mathfrak{x}})}{\alpha_{\mathfrak{x}}|\mathfrak{x}|^2} = \frac{\gamma^2\left(|\mathfrak{x}|^2\mathfrak{y}_2 + \mathfrak{x}_2\right)}{\left(\gamma^2\mathfrak{x}_2 + 4\mathfrak{x}_1\right)|\mathfrak{x}|^2}. \end{aligned} \tag{5.3.16}$$

With the representation

$$|c|^2 - |b|^2 = \frac{\gamma^4 \mathfrak{x}_2 \mathfrak{y}_2 + 4\gamma^2 \operatorname{Im}(\mathfrak{x}\mathfrak{y}) - 4|\mathfrak{x}\mathfrak{y} - 1|^2}{\left(\gamma^2 \mathfrak{x}_2 + 4\mathfrak{x}_1\right)^2}$$

the result of the optimization (5.3.14) yields

$$\sup_{z \in \mathbb{C}} G(\gamma, z) = 2\frac{\gamma^2\left(|\mathfrak{x}|^2\mathfrak{y}_2 + \mathfrak{x}_2\right)}{\gamma^4 \mathfrak{x}_2 \mathfrak{y}_2 + 4\gamma^2 \operatorname{Im}(\mathfrak{x}\mathfrak{y}) - 4|\mathfrak{x}\mathfrak{y} - 1|^2}.$$

This leads to the formulae

$$D_1^2 = \sup_{\gamma > 0} \frac{q_1\gamma^2\left(|\mathfrak{x}|^2\mathfrak{y}_2 + \mathfrak{x}_2\right)}{\gamma^4 \mathfrak{x}_2 \mathfrak{y}_2 + 4\gamma^2 \operatorname{Im}(\mathfrak{x}\mathfrak{y}) - 4|\mathfrak{x}\mathfrak{y} - 1|^2}, \tag{5.3.17}$$

$$D_2^2 = \sup_{\gamma > 0} \frac{q_2\gamma^3\left(|\mathfrak{x}|^2\mathfrak{y}_2 + \mathfrak{x}_2\right)}{\gamma^4 \mathfrak{x}_2 \mathfrak{y}_2 + 4\gamma^2 \operatorname{Im}(\mathfrak{x}\mathfrak{y}) - 4|\mathfrak{x}\mathfrak{y} - 1|^2} \tag{5.3.18}$$

with $q_1 = C_1 c_1\left(\frac{1}{2}\right) = 2\pi c_1$ and $q_2 = \sqrt{C_1} c_1\left(\frac{1}{4}\right) = 4\sqrt{2\pi C_1}$. Due to the complicated structure of the complex roots $\mathfrak{x}$ and $\mathfrak{y}$ these optimization problems cannot be solved analytically. It is possible to prove that the functions to be optimized are continuous for $0 < \gamma < +\infty$ and that their limits for $\gamma \to 0$ and $\gamma \to +\infty$ exist and are bounded. Hence the optimization problems are solvable and the results can be calculated numerically with standard optimization techniques.

In order to calculate the constant D_0 of the trace estimate (5.1.38) we first evaluate the expression

$$\left\langle \dot{\boldsymbol{u}}, \boldsymbol{v} \right\rangle_Q + A(\boldsymbol{u}, \boldsymbol{v})$$

for two functions $\boldsymbol{u}, \boldsymbol{v} \in \mathscr{L}(0)$ with $A(\boldsymbol{u}, \boldsymbol{v}) \equiv \int_Q a(\boldsymbol{u}, \boldsymbol{v})\, d\boldsymbol{x}\, dt$. The tangential component v_1 of $\boldsymbol{v}$ is assumed to vanish on the boundary S. Consequently, the function $\boldsymbol{u}$ is defined by (5.3.9) in terms of its boundary data $\boldsymbol{w}$ via formula (5.3.8) and $\boldsymbol{v}$ is given by the same formula, if $\mathfrak{p}_1$ and $\mathfrak{p}_2$ are replaced by

$$\mathfrak{q}_1 = \frac{i\xi\widehat{v}_2}{\mathscr{D}} \text{ and } \mathfrak{q}_2 = -\frac{\mathfrak{a}\widehat{v}_2}{\mathscr{D}} \tag{5.3.19}$$

with the normal component of boundary data v_2. The bilinear form $A(\boldsymbol{u}, \boldsymbol{v})$ can be easily calculated from (5.3.11) with the help of the formula

$$A(\boldsymbol{u}, \boldsymbol{v}) = \frac{1}{2}\big(A(\boldsymbol{u}+\boldsymbol{v}, \boldsymbol{u}+\boldsymbol{v}) - A(\boldsymbol{u}, \boldsymbol{u}) - A(\boldsymbol{v}, \boldsymbol{v})\big).$$

The result is given by

$$\begin{aligned} A(\boldsymbol{u}, \boldsymbol{v}) &= \\ &= \frac{E}{1+\nu} \int_{\mathbb{R}} \int_{\mathbb{R}} \left[\frac{1}{2}\left(|\mathfrak{p}_1 + \mathfrak{q}_1|^2 - |\mathfrak{p}_1|^2 - |\mathfrak{q}_1|^2\right) \left(\frac{\vartheta \mathfrak{a}_2}{2} + 2\mathfrak{a}_1 \xi^2 \right) \right. \\ &\quad + \frac{1}{2}\left(|\mathfrak{p}_2 + \mathfrak{q}_2|^2 - |\mathfrak{p}_2|^2 - |\mathfrak{q}_2|^2\right) \left(\frac{\vartheta \mathfrak{b}_2}{2} + 2\mathfrak{b}_1 \xi^2 \right) \\ &\quad \left. - 2\operatorname{Re}\left(i\xi(\mathfrak{a}\overline{\mathfrak{b}} + \xi^2) \frac{1}{2}\big((\mathfrak{p}_1 + \mathfrak{q}_1)(\overline{\mathfrak{p}}_2 + \overline{\mathfrak{q}}_2) - \mathfrak{p}_1 \overline{\mathfrak{p}}_2 - \mathfrak{q}_1 \overline{\mathfrak{q}}_2\big) \right) \right] d\xi\, d\vartheta. \end{aligned}$$

This can be simplified to

$$\begin{aligned} A(\boldsymbol{u}, \boldsymbol{v}) &= \frac{E}{1+\nu} \int_{\mathbb{R}} \int_{\mathbb{R}} \left[\operatorname{Re}(\mathfrak{p}_1 \overline{\mathfrak{q}}_1) \left(\frac{\vartheta \mathfrak{a}_2}{2} + 2\mathfrak{a}_1 \xi^2 \right) + \operatorname{Re}(\mathfrak{p}_2 \overline{\mathfrak{q}}_2) \cdot \right. \\ &\quad \left. \cdot \left(\frac{\vartheta \mathfrak{b}_2}{2} + 2\mathfrak{b}_1 \xi^2 \right) - 2\operatorname{Re}\left(i\xi(\mathfrak{a}\overline{\mathfrak{b}} + \xi^2) \frac{\mathfrak{p}_1 \overline{\mathfrak{q}}_2 + \mathfrak{q}_1 \overline{\mathfrak{p}}_2}{2} \right) \right] d\xi\, d\vartheta. \end{aligned}$$

It remains to calculate the acceleration part $\langle \dot{\boldsymbol{u}}, \boldsymbol{v} \rangle_Q$. It is given in Fourier transformed functions by

$$\langle \dot{\boldsymbol{u}}, \boldsymbol{v} \rangle_Q = \mathrm{Re}\left(\int_{\mathbb{R}} \int_{\mathbb{R}\times\mathbb{R}_+} iC_1^{-1}\vartheta\, \widetilde{u}(\vartheta,\xi,y)\overline{\widetilde{v}(\vartheta,\xi,y)}\, dy\, d\xi\, d\vartheta \right).$$

From the representations of $\widetilde{\boldsymbol{u}}$ and $\widetilde{\boldsymbol{v}}$ there follows

$$\begin{aligned}\widetilde{\boldsymbol{u}}(\vartheta,\xi,y)\overline{\widetilde{\boldsymbol{v}}(\vartheta,\xi,y)} &= \left(|\mathfrak{a}|^2+\xi^2\right)\mathfrak{p}_1\overline{\mathfrak{q}_1}\, e^{-2\mathfrak{a}_1 y} + \left(|\mathfrak{b}|^2+\xi^2\right)\mathfrak{p}_2\overline{\mathfrak{q}_2}\, e^{-2\mathfrak{b}_1 y} \\ &\quad - i\xi\left(\mathfrak{a}+\overline{\mathfrak{b}}\right)\mathfrak{p}_1\overline{\mathfrak{q}_2}\, e^{-(\mathfrak{a}+\overline{\mathfrak{b}})y} + i\xi\left(\overline{\mathfrak{a}}+\mathfrak{b}\right)\mathfrak{p}_2\overline{\mathfrak{q}_1}\, e^{-(\overline{\mathfrak{a}}+\mathfrak{b})y}.\end{aligned}$$

Performing the inner integration with respect to y yields

$$\begin{aligned}\langle \dot{\boldsymbol{u}}, \boldsymbol{v} \rangle_Q = \mathrm{Re} \int_{\mathbb{R}} \int_{\mathbb{R}} iC_1^{-1}\vartheta \Bigg[\mathfrak{p}_1\overline{\mathfrak{q}}_1 \frac{|\mathfrak{a}|^2+\xi^2}{2\mathfrak{a}_1} + \mathfrak{p}_2\overline{\mathfrak{q}}_2 \frac{|\mathfrak{b}|^2+\xi^2}{2\mathfrak{b}_1} \\ + i\xi\left(\mathfrak{p}_2\overline{\mathfrak{q}}_1 - \mathfrak{p}_1\overline{\mathfrak{q}}_2\right)\Bigg]\, d\xi\, d\vartheta.\end{aligned}$$

With the formulae $|\mathfrak{a}|^2+\xi^2 = 2\mathfrak{a}_1^2$ and $|\mathfrak{b}|^2+\xi^2 = 2\mathfrak{b}_1^2$ we obtain

$$\langle \dot{\boldsymbol{u}}, \boldsymbol{v} \rangle_Q = \mathrm{Re} \int_{\mathbb{R}} \int_{\mathbb{R}} C_1^{-1}\left[i\vartheta\, \mathfrak{a}_1\mathfrak{p}_1\overline{\mathfrak{q}}_1 + i\vartheta\, \mathfrak{b}_1\mathfrak{p}_2\overline{\mathfrak{q}}_2 + \xi\vartheta\left(\mathfrak{p}_1\overline{\mathfrak{q}}_2 - \mathfrak{p}_2\overline{\mathfrak{q}}_1\right)\right] d\xi\, d\vartheta.$$

Adding the expressions for $A(\boldsymbol{u},\boldsymbol{v})$ and $\langle \dot{\boldsymbol{u}}, \boldsymbol{v} \rangle_Q$ and using the definition of $C_1 = \dfrac{2+2\nu}{E}$ the following result is derived:

$$\begin{aligned}&\langle \dot{\boldsymbol{u}}, \boldsymbol{v} \rangle_Q + A(\boldsymbol{u},\boldsymbol{v}) = \\ &= C_1^{-1}\,\mathrm{Re} \int_{\mathbb{R}} \int_{\mathbb{R}} \Big[\left(i\vartheta\,\overline{\mathfrak{a}} + 4\mathfrak{a}_1\xi^2\right)\mathfrak{p}_1\overline{\mathfrak{q}}_1 + \left(i\vartheta\,\overline{\mathfrak{b}} + 4\mathfrak{b}_1\xi^2\right)\mathfrak{p}_2\overline{\mathfrak{q}}_2 \\ &\quad - i\xi\left(i\vartheta + 2(\mathfrak{a}\overline{\mathfrak{b}}+\xi^2)\right)\mathfrak{p}_1\overline{\mathfrak{q}}_2 + i\xi\left(i\vartheta + 2(\overline{\mathfrak{a}}\mathfrak{b}+\xi^2)\right)\mathfrak{p}_2\overline{\mathfrak{q}}_1 \Big]\, d\xi\, d\vartheta.\end{aligned}$$

This can be simplified with the formula (5.3.19) for the expressions $\mathfrak{q}_1$ and $\mathfrak{q}_2$. After some calculation one obtains

$$\langle \dot{\boldsymbol{u}}, \boldsymbol{v} \rangle_Q + A(\boldsymbol{u},\boldsymbol{v}) = \frac{1}{C_1}\,\mathrm{Re} \int_{\mathbb{R}} \int_{\mathbb{R}} \left[2i\xi\,\mathfrak{a}\mathfrak{p}_1 - (2\xi^2+i\vartheta)\mathfrak{p}_2\right]\overline{\widehat{v}}_2\, d\xi\, d\vartheta.$$

Now a standard use of Hölder's inequality gives

$$\begin{aligned}&\langle \dot{\boldsymbol{u}}, \boldsymbol{v} \rangle_Q + A(\boldsymbol{u},\boldsymbol{v}) \le \\ &\le \|v_2\|'_{H^{1/4,1/2}(S)} \frac{1}{C_1} \left(\int_{\mathbb{R}^2} \frac{\left|2i\xi\,\mathfrak{a}\mathfrak{p}_1 - (2\xi^2+i\vartheta)\mathfrak{p}_2\right|^2}{c_1\left(\frac{1}{4}\right)C_1^{-1/2}|\vartheta|^{1/2} + c_1\left(\frac{1}{2}\right)|\xi|}\, d\xi\, d\vartheta \right)^{1/2}.\end{aligned}$$

The second factor on the right hand side of the previous formula shall be estimated by $D_0\|\boldsymbol{u}\|_A$. Due to the representation (5.3.11) of the energy norm $\|\boldsymbol{u}\|_A$ it can be seen that D_0^2 is the solution of the optimization problem

$$D_0^2 = \sup_{\substack{\mathfrak{p}_1,\mathfrak{p}_2\in\mathbb{C}\\ \xi,\vartheta\in\mathbb{R}}} \frac{1}{2C_1}\left(c_1\left(\tfrac{1}{4}\right)C_1^{-1/2}|\vartheta|^{1/2} + c_1\left(\tfrac{1}{2}\right)|\xi|\right)^{-1} \cdot \mathfrak{E}(\xi,\vartheta,\mathfrak{a},\mathfrak{b},\mathfrak{p}_1,\mathfrak{p}_2) \tag{5.3.20}$$

with the same $\mathfrak{E}$ as in (5.3.12). As in the optimization problems for the calculation of D_1, D_2 we use the transform of variables $\mathfrak{p}_1 = z\mathfrak{p}_2$ with $z\in\mathbb{C}$ and $\vartheta = \gamma^2\xi^2$. The resulting problem is

$$\sup_{\gamma>0}\left\{\frac{1}{2\left(\sqrt{C_1}c_1\left(\frac{1}{4}\right)\gamma + C_1c_1\left(\frac{1}{2}\right)\right)}\frac{4|\mathfrak{x}|^2}{\frac{1}{2}\gamma^2\mathfrak{x}_2 + 2\mathfrak{x}_1}\sup_{z\in\mathbb{C}}\frac{|z+a|^2}{|z|^2+2\,\mathrm{Re}(\bar{b}z)+|c|^2}\right\}$$

with the parameters

$$a = \frac{i}{\mathfrak{x}}\left(1 + i\frac{\gamma^2}{2}\right)$$

and b, c defined as in (5.3.13). The result of the optimization with respect to z is given by formula (5.3.15). Compared to expression (5.3.16), in the numerator $|c|^2 - |b|^2 + |a-b|^2$ there appear the additional terms

$$\frac{\gamma^4}{4|\mathfrak{x}|^2} - 2\,\mathrm{Re}\left(\frac{\gamma^2}{2}\frac{i\,\mathfrak{s}}{\alpha_{\mathfrak{x}}}\right).$$

Those terms arise from the different form of a. These additional terms can be modified to

$$\frac{\frac{1}{4}\gamma^4\alpha_{\mathfrak{x}} + 2\gamma^2\,\mathrm{Im}(\mathfrak{s}\bar{\mathfrak{x}})}{|\mathfrak{x}|^2\alpha_{\mathfrak{x}}} = \frac{\frac{1}{4}\gamma^2\left(\frac{1}{2}\gamma^2\mathfrak{x}_2 + 2\mathfrak{x}_1\right) - 2\gamma^2\left(|\mathfrak{x}|^2\mathfrak{y}_2 + \mathfrak{x}_2\right)}{|\mathfrak{x}|^2\left(\frac{1}{2}\gamma^2\mathfrak{x}_2 + 2\mathfrak{x}_1\right)}.$$

Hence we have

$$|c|^2 - |b|^2 + |a-b|^2 = \frac{1}{4}\frac{\gamma^4\left(\gamma^2\mathfrak{x}_2 + 4\mathfrak{x}_1\right) - \gamma^2\left(|\mathfrak{x}|^2\mathfrak{y}_2 + \mathfrak{x}_2\right)}{|\mathfrak{x}|^2\left(\frac{1}{2}\gamma^2\mathfrak{x}_2 + 2\mathfrak{x}_1\right)}$$

and the following one-parametric optimization problem for D_0 is obtained:

$$D_0^2 = \sup_{\gamma>0}\frac{\gamma^4(\gamma^2\mathfrak{x}_2 + 4\mathfrak{x}_1) - 4\gamma^2(|\mathfrak{x}|^2\mathfrak{y}_2 + \mathfrak{x}_2)}{(q_1 + q_2\gamma)(\gamma^4\mathfrak{x}_2\mathfrak{y}_2 + 4\gamma^2\,\mathrm{Im}(\mathfrak{x}\mathfrak{y}) - 4|\mathfrak{x}\mathfrak{y} - 1|^2)}. \tag{5.3.21}$$

As in the problems before, it is not possible to obtain an analytical solution. If the viscous Poisson ratio ν satisfies $-1 < \nu < \frac{1}{2}$ then the function to be

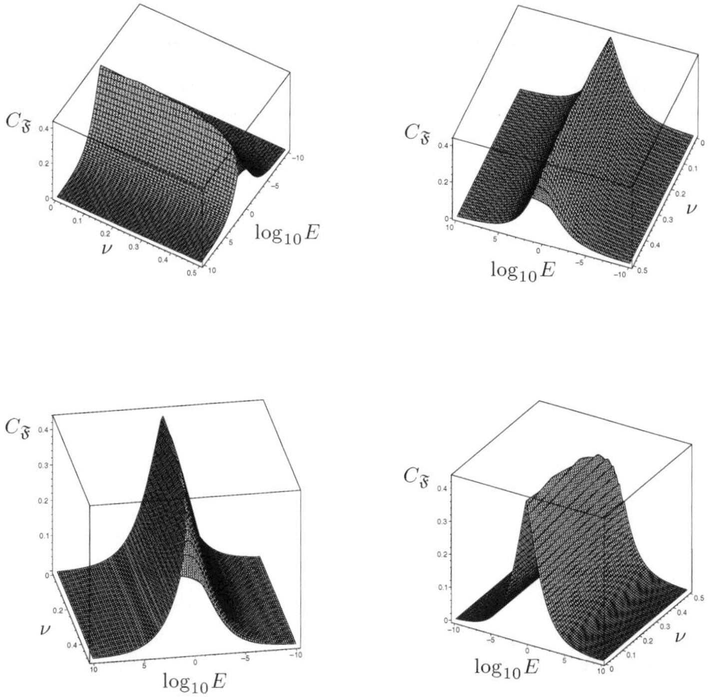

Figure 5.1: Upper bound for the coefficient of friction in dependence on the modulus of viscosity (E) and the viscous Poisson ratio (ν) for $0 \leq \nu < 0.5$.

optimized is non-negative, continuous and its limits for $\gamma \to 0$ and $\gamma \to +\infty$ exist and are bounded. Therefore the problem has a solution.

The optimization problems defining the constants D_0, D_1 and D_2 in the two-dimensional homogeneous isotropic case can be solved numerically. Since the functions to be optimized are rather complicated, we employ a simple optimization algorithm based on curve-fitting with a three-point pattern as described e.g. in [104], Section 7.3. After the solution of the three optimization problems for D_0, D_1 and D_2, the upper bound for the admissible coefficient of friction can be calculated. In Figures 5.1 and 5.2 the coefficient of friction is depicted in dependence of the modulus of viscosity E and the viscous Poisson ratio ν for positive and negative values of ν. The admissible coefficient of friction converges to 0 for both very small and very large values of E. Hence our existence result does not give

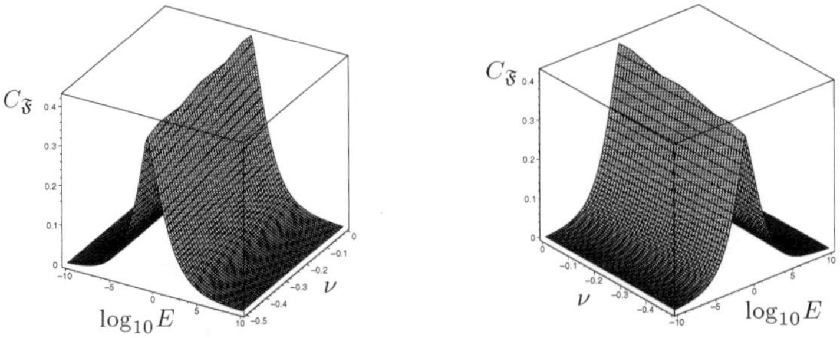

Figure 5.2: Upper bound for the coefficient of friction in dependence on the modulus of viscosity (E) and the viscous Poisson ratio (ν) for $-1 < \nu < 0$.

any information about the quasistatic contact problem and the purely elastic contact problem without any viscous damping. However, for moderate values of E and ν the magnitude of the admissible coefficient of friction varies between 0.4 and 0.5. This is sufficient for many cases of practical relevance where the coefficient of friction has values around 0.3; and it is in particular much better than that for the general case with a maximal possible value of around 0.25. The case of $\nu = 0.5$ is not included in our result, because then the representation (5.3.4) of the differential equations becomes singular. In the limit case $\nu \to 0.5$ the admissible coefficient of friction we have obtained here tends also to zero.

The rescaling of the time variable mentioned in the final remark of the previous section enables optimizing these formulae with respect to E and obtaining an upper bound that only depends on ν, $C_{\mathfrak{F}}(\nu) = \max\limits_{E>0} C_{\mathfrak{F}}(\nu, E)$. In the next section we compare $C_{\mathfrak{F}}(\nu)$ for the cases $N = 2$ and $N = 3$.

5.3.2 The case of three space dimensions

The result for $N = 2$ derived in the preceding subsection seems to be quite complex. In this subsection we show that its generalization for higher space dimension does not additionally complicate it.

In the case $N = 3$ the solutions are searched in the form $\widetilde{\boldsymbol{u}}(\vartheta, \xi_1, \xi_2, y) = e^{\lambda y}\widehat{\boldsymbol{v}}(\vartheta, \xi_1, \xi_2)$. This leads to the equation $\mathfrak{A}(\lambda)\boldsymbol{v} = 0$ with $\mathfrak{A}(\lambda)$ defined by

$$\begin{pmatrix} i\vartheta + \frac{1}{k}\xi_1^2 + \xi_2^2 - \lambda^2 & \frac{1-k}{k}\xi_1\xi_2 & \lambda\frac{k-1}{k}i\xi_1 \\ \frac{1-k}{k}\xi_1\xi_2 & i\vartheta + \xi_1^2 + \frac{1}{k}\xi_2^2 - \lambda^2 & \lambda\frac{k-1}{k}i\xi_2 \\ \lambda(k-1)i\xi_1 & \lambda(k-1)i\xi_2 & ki\vartheta + k(\xi_1^2 + \xi_2^2) - \lambda^2 \end{pmatrix}.$$

The determinant of this matrix is

$$\det \mathfrak{A}(\lambda) = -\big(\lambda^2 - (|\boldsymbol{\xi}|^2 + i\vartheta)\big)^2 \big(\lambda^2 - (|\boldsymbol{\xi}|^2 + ik\vartheta)\big).$$

Its roots

$$\lambda_{1,2} = \pm\sqrt{|\boldsymbol{\xi}|^2 + i\vartheta}, \quad \lambda_{3,4} = \pm\sqrt{|\boldsymbol{\xi}|^2 + ik\vartheta}$$

differ from the two-dimensional case only by the formula $|\boldsymbol{\xi}|^2 = \xi_1^2 + \xi_2^2$, the appropriate version of $\boldsymbol{\xi}^2$ there. We introduce the complex roots

$$\mathfrak{a} \equiv \sqrt{|\boldsymbol{\xi}|^2 + i\vartheta} \text{ and } \mathfrak{b} \equiv \sqrt{|\boldsymbol{\xi}|^2 + ik\vartheta}$$

in the same way as in (5.3.6) with the obvious change of ξ by $|\boldsymbol{\xi}|$. The solutions $\widetilde{u}(\vartheta, \boldsymbol{\xi}, y)$ with $\widetilde{u}(\vartheta, \boldsymbol{\xi}, y) \to 0$ for $y \to +\infty$ are characterized by the roots $-\mathfrak{a}$ and $-\mathfrak{b}$. Independent eigenvectors for $-\mathfrak{a}$ are $(\xi_2, -\xi_1, 0)^\top$ and $(\xi_2\mathfrak{a}, \xi_1\mathfrak{a}, 2i\xi_1\xi_2)^\top$ while for $-\mathfrak{b}$ we have the eigenvector $(\xi_1, \xi_2, \mathfrak{b})^\top$. Therefore the solution has the form

$$\widetilde{\boldsymbol{u}}(\vartheta, \xi_1, \xi_2, y) = c_1 e^{-\mathfrak{a}y} \begin{pmatrix} \xi_2 \\ -\xi_1 \\ 0 \end{pmatrix} + c_2 e^{-\mathfrak{a}y} \begin{pmatrix} \xi_2\mathfrak{a} \\ \xi_1\mathfrak{a} \\ 2i\xi_1\xi_2 \end{pmatrix} + c_3 e^{-\mathfrak{b}y} \begin{pmatrix} \xi_1 \\ \xi_2 \\ i\mathfrak{b} \end{pmatrix}.$$

We are searching for a solution with the boundary condition

$$\widetilde{\boldsymbol{u}}(\vartheta, \boldsymbol{\xi}, 0) = \widehat{\boldsymbol{w}}(\vartheta, \boldsymbol{\xi}).$$

This leads to the system of linear equations

$$\begin{pmatrix} \xi_2 & \xi_2\mathfrak{a} & \xi_1 \\ -\xi_1 & \xi_1\mathfrak{a} & \xi_2 \\ 0 & 2i\xi_1\xi_2 & i\mathfrak{b} \end{pmatrix} \begin{pmatrix} c_1 \\ c_2 \\ c_3 \end{pmatrix} = \widehat{\boldsymbol{w}}.$$

This system has the solution

$$c_3 = -\frac{z_1 + \mathfrak{a}z_3}{\mathscr{D}}, \qquad c_2 = \frac{\mathfrak{b}z_1 + |\boldsymbol{\xi}|^2 z_3}{2\xi_1\xi_2\mathscr{D}}, \qquad c_1 = \frac{\widehat{w}_1 - \xi_2\mathfrak{a}c_2 - \xi_1 c_3}{\xi_2}$$

with the abbreviations

$$z_1 = \xi_1\widehat{w}_1 + \xi_2\widehat{w}_2, \quad z_3 = i\widehat{w}_3, \quad \mathscr{D} = \mathfrak{a}\mathfrak{b} - |\boldsymbol{\xi}|^2.$$

After some calculation we finally arrive at the following expression of the solution

$$\begin{aligned}
\mathscr{D}\widetilde{u}_1 &= \left[(\mathscr{D} + \xi_1^2)\widehat{w}_1 + \xi_1\xi_2\widehat{w}_2 + i\xi_1\mathfrak{a}\widehat{w}_3\right] e^{-\mathfrak{a}y} + i\xi_1\mathfrak{p}_0 e^{-\mathfrak{b}y}, \\
\mathscr{D}\widetilde{u}_2 &= \left[\xi_1\xi_2\widehat{w}_1 + (\mathscr{D} + \xi_2^2)\widehat{w}_2 + i\xi_2\mathfrak{a}\widehat{w}_3\right] e^{-\mathfrak{a}y} + i\xi_2\mathfrak{p}_0 e^{-\mathfrak{b}y}, \qquad (5.3.22) \\
\mathscr{D}\widetilde{u}_3 &= \left[i\xi_1\mathfrak{b}\widehat{w}_1 + i\xi_2\mathfrak{b}\widehat{w}_2 + (\mathscr{D} - \mathfrak{a}\mathfrak{b})\widehat{w}_3\right] e^{-\mathfrak{a}y} - \mathfrak{b}\mathfrak{p}_0 e^{-\mathfrak{b}y}
\end{aligned}$$

with

$$\mathfrak{p}_0 = i\xi_1\widehat{w}_1 + i\xi_2\widehat{w}_2 - \mathfrak{a}\widehat{w}_3.$$

Denoting the coefficients at $e^{-\mathfrak{a}y}$ in the rows of (5.3.22) by $\mathscr{D}^{-1}\mathfrak{a}\mathfrak{p}_1$, $\mathscr{D}^{-1}\mathfrak{a}\mathfrak{p}_2$ and $i\mathscr{D}^{-1}(\xi_1\mathfrak{p}_1 + \xi_2\mathfrak{p}_2)$ and defining $\mathfrak{p}_3 = \mathscr{D}^{-1}\mathfrak{p}_0$ we get

$$\begin{aligned}
\widetilde{u}_1 &= \mathfrak{a}\mathfrak{p}_1 e^{-\mathfrak{a}y} + i\xi_1\mathfrak{p}_3 e^{-\mathfrak{b}y},\\
\widetilde{u}_2 &= \mathfrak{a}\mathfrak{p}_2 e^{-\mathfrak{a}y} + i\xi_2\mathfrak{p}_3 e^{-\mathfrak{b}y},\\
\widetilde{u}_3 &= (i\xi_1\mathfrak{p}_1 + i\xi_2\mathfrak{p}_2)e^{-\mathfrak{a}y} - \mathfrak{b}\mathfrak{p}_3 e^{-\mathfrak{b}y}.
\end{aligned} \tag{5.3.23}$$

Conversely, it holds

$$\widehat{w}_1 = \mathfrak{a}\mathfrak{p}_1 + i\xi_1\mathfrak{p}_3,\ \ \widehat{w}_2 = \mathfrak{a}\mathfrak{p}_2 + i\xi_2\mathfrak{p}_3 \text{ and } \widehat{w}_3 = i(\xi_1\mathfrak{p}_1 + \xi_2\mathfrak{p}_2) - \mathfrak{b}\mathfrak{p}_3.$$

Hence it is easy to see that

$$\begin{aligned}
|\widehat{w}_1|^2 + |\widehat{w}_2|^2 =& |\mathfrak{a}|^2\big(|\mathfrak{p}_1|^2 + |\mathfrak{p}_2|^2\big) + 2\,\mathrm{Im}\big(\mathfrak{a}(\xi_1\mathfrak{p}_1 + \xi_2\mathfrak{p}_2)\overline{\mathfrak{p}_3}\big)\\
&+ |\boldsymbol{\xi}|^2|\mathfrak{p}_3|^2.
\end{aligned} \tag{5.3.24}$$

After some extended calculation similar to that presented below for the general case we arrive at the expression

$$\begin{aligned}
\|\boldsymbol{u}\|_A^2 = \frac{E}{1+\nu}\int_{\mathbb{R}^3}&\Big[\big(\tfrac{1}{2}\mathfrak{b}_2\vartheta + 2\mathfrak{b}_1|\boldsymbol{\xi}|^2\big)|\mathfrak{p}_3|^2\\
&+ \tfrac{1}{2}\big(\mathfrak{a}_2\vartheta + \mathfrak{a}_1|\boldsymbol{\xi}|^2\big)\big(|\mathfrak{p}_1|^2 + |\mathfrak{p}_2|^2\big) + \tfrac{3}{2}\mathfrak{a}_1|\xi_1\mathfrak{p}_1 + \xi_2\mathfrak{p}_2|^2\\
&- 2\,\mathrm{Re}\,\big(i\big(|\boldsymbol{\xi}|^2 + \mathfrak{a}\overline{\mathfrak{b}}\big)(\xi_1\mathfrak{p}_1 + \xi_2\mathfrak{p}_2)\overline{\mathfrak{p}_3}\big)\Big]\,d\boldsymbol{\xi}\,d\vartheta
\end{aligned} \tag{5.3.25}$$

for the energy norm. For the bilinear form we get

$$\begin{aligned}
\langle\dot{\boldsymbol{u}},\boldsymbol{v}\rangle_Q + A(\boldsymbol{u},\boldsymbol{v}) = \frac{1}{C_1}\mathrm{Re}\int_{\mathbb{R}^3}&\Big[\big(i\vartheta\overline{\mathfrak{b}} + 4\mathfrak{b}_1|\boldsymbol{\xi}|^2\big)\,\mathfrak{p}_3\overline{\mathfrak{q}_3}\\
&+ \big(i\vartheta\overline{\mathfrak{a}} + \mathfrak{a}_1|\boldsymbol{\xi}|^2\big)\,(\mathfrak{p}_1\overline{\mathfrak{q}_1} + \mathfrak{p}_2\overline{\mathfrak{q}_2})\\
&+ 3\mathfrak{a}_1(\xi_1\mathfrak{p}_1 + \xi_2\mathfrak{p}_2)\,(\xi_1\overline{\mathfrak{q}_1} + \xi_2\overline{\mathfrak{q}_2})\\
&+ \tfrac{1}{\mathfrak{a}_1}(\xi_1\mathfrak{p}_2 - \xi_2\mathfrak{p}_1)\,(\xi_2\overline{\mathfrak{q}_1} - \xi_1\overline{\mathfrak{q}_2})\\
&+ \big(\vartheta - 2i\big(\mathfrak{a}\overline{\mathfrak{b}} + |\boldsymbol{\xi}|^2\big)\big)\,(\xi_1\mathfrak{p}_1 + \xi_2\mathfrak{p}_2)\overline{\mathfrak{q}_3}\\
&- \big(\vartheta - 2i\big(\overline{\mathfrak{a}}\mathfrak{b} + |\boldsymbol{\xi}|^2\big)\big)\,\mathfrak{p}_3\,(\xi_1\overline{\mathfrak{q}_1} + \xi_2\overline{\mathfrak{q}_2})\Big]\,d\boldsymbol{\xi}\,d\vartheta
\end{aligned} \tag{5.3.26}$$

with $\mathfrak{q} = \mathscr{D}^{-1}[i\xi_1, i\xi_2, -\mathfrak{a}]\widehat{v}_3$. This can be simplified to

$$\langle\dot{\boldsymbol{u}},\boldsymbol{v}\rangle_Q + A(\boldsymbol{u},\boldsymbol{v}) = \frac{1}{C_1}\mathrm{Re}\int_{\mathbb{R}^3}\big[2i\mathfrak{a}(\mathfrak{p}_1\xi_1 + \mathfrak{p}_2\xi_2) - \mathfrak{p}_3\big(i\vartheta + 2|\boldsymbol{\xi}|^2\big)\big]$$

$$\cdot \overline{\widehat{v}_3}\, d\boldsymbol{\xi}\, d\vartheta.$$

Let us define the notation

$$\begin{aligned} &\boldsymbol{q} \equiv [q_1, q_2],\ q_j = \frac{\mathfrak{p}_i}{\mathfrak{p}_3},\ j = 1, 2,\ \boldsymbol{\xi}_{\text{new}} \equiv \frac{\boldsymbol{\xi}}{\sqrt{|\vartheta|}}, \\ &\mathfrak{x} \equiv \sqrt{|\boldsymbol{\xi}_{\text{new}}|^2 + i} = \mathfrak{x}_1 + i\mathfrak{x}_2, \\ &\mathfrak{y} \equiv \sqrt{|\boldsymbol{\xi}_{\text{new}}|^2 + ik} = \mathfrak{y}_1 + i\mathfrak{y}_2 \text{ with} \\ &\mathfrak{x}_{1,2} = \sqrt{\frac{\sqrt{|\boldsymbol{\xi}_{\text{new}}|^4 + 1} \pm |\boldsymbol{\xi}_{\text{new}}|^2}{2}},\ \mathfrak{y}_{1,2} = \sqrt{\frac{\sqrt{|\boldsymbol{\xi}_{\text{new}}|^4 + k^2} \pm |\boldsymbol{\xi}_{\text{new}}|^2}{2}}, \\ &\mathfrak{J} \equiv \tfrac{1}{2}\left(|\mathfrak{x}_2| + \mathfrak{x}_1|\boldsymbol{\xi}_{\text{new}}|^2\right)|\boldsymbol{q}|^2 + \tfrac{3}{2}\mathfrak{x}_1|\boldsymbol{q}\cdot\boldsymbol{\xi}_{\text{new}}|^2 \\ &\qquad + 2\,\mathrm{Im}\big((\mathfrak{x}\overline{\mathfrak{y}} + |\boldsymbol{\xi}_{\text{new}}|^2)\boldsymbol{q}\cdot\boldsymbol{\xi}_{\text{new}}\big) + \left(\tfrac{1}{2}\mathfrak{y}_2 + 2\mathfrak{y}_1|\boldsymbol{\xi}_{\text{new}}|^2\right) \end{aligned} \tag{5.3.27}$$

and use the old symbol $\boldsymbol{\xi}$ instead of $\boldsymbol{\xi}_{\text{new}}$ in the sequel. Then the constants D_j, $j = 0, 1, 2$, are the solutions of the problems

$$D_1^2 = \frac{C_1}{2} \sup_{\substack{\boldsymbol{q}\in\mathbb{C}^2 \\ \boldsymbol{\xi}\in\mathbb{R}^2}} \frac{\left(|\mathfrak{x}|^2|\boldsymbol{q}|^2 - 2\,\mathrm{Re}(\mathfrak{x}\,\boldsymbol{q}\cdot i\boldsymbol{\xi}) + |\boldsymbol{\xi}|^2\right) c_2\left(\tfrac{1}{2}\right)|\boldsymbol{\xi}|}{\mathfrak{J}}, \tag{5.3.28}$$

$$D_2^2 = \frac{\sqrt{C_1}}{2} \sup_{\substack{\boldsymbol{q}\in\mathbb{C}^2 \\ \boldsymbol{\xi}\in\mathbb{R}^2}} \frac{\left(|\mathfrak{x}|^2|\boldsymbol{q}|^2 - 2\,\mathrm{Re}(\mathfrak{x}\,\boldsymbol{q}\cdot i\boldsymbol{\xi}) + |\boldsymbol{\xi}|^2\right) c_1\left(\tfrac{1}{4}\right)}{\mathfrak{J}}, \tag{5.3.29}$$

$$D_0^2 = \frac{1}{2} \sup_{\substack{\boldsymbol{q}\in\mathbb{C}^2 \\ \boldsymbol{\xi}\in\mathbb{R}^2}} \frac{\left|2\mathfrak{x}(q_1\xi_1 + q_2\xi_2) - 1 + 2i|\boldsymbol{\xi}|^2\right|^2}{\left(c_1\left(\tfrac{1}{4}\right)\sqrt{C_1} + c_2\left(\tfrac{1}{2}\right)|\boldsymbol{\xi}|C_1\right)\mathfrak{J}}. \tag{5.3.30}$$

Here we use the change of variables

$$\begin{aligned} &\begin{matrix} |\boldsymbol{\xi}|^{-1}(\xi_1 q_1 + \xi_2 q_2) \equiv s_1 \\ |\boldsymbol{\xi}|^{-1}(-\xi_2 q_1 + \xi_1 q_2) \equiv s_2 \end{matrix},\ \text{i.e.}\ \begin{matrix} q_1 = |\boldsymbol{\xi}|^{-1}(\xi_1 s_1 - \xi_2 s_2), \\ q_2 = |\boldsymbol{\xi}|^{-1}(\xi_2 s_1 + \xi_1 s_2), \end{matrix} \\ &\text{and}\ |q_1|^2 + |q_2|^2 = |s_1|^2 + |s_2|^2. \end{aligned} \tag{5.3.31}$$

Then both optimization problems (5.3.28) and (5.3.29) have the form

$$\sup_{\substack{\boldsymbol{\xi}\in\mathbb{R}^2 \\ \boldsymbol{s}\in\mathbb{C}^2}} \frac{\left(|\mathfrak{x}|^2\left(|s_1|^2 + |s_2|^2\right) - 2|\boldsymbol{\xi}|\,\mathrm{Re}(i\mathfrak{x}s_1) + |\boldsymbol{\xi}|^2\right)\mathfrak{f}(|\boldsymbol{\xi}|)}{\mathfrak{X}_{\mathfrak{x}\mathfrak{y}}(\boldsymbol{\xi}, s_1, s_2)}$$

with

$$\mathfrak{X}_{\mathfrak{x}\mathfrak{y}}(\boldsymbol{\xi}, s_1, s_2) = \tfrac{1}{2}\left(\mathfrak{x}_2 + \mathfrak{x}_1|\boldsymbol{\xi}|^2\right)\left(|s_1|^2 + |s_2|^2\right) + |\boldsymbol{\xi}|^2\tfrac{3}{2}\mathfrak{x}_1|s_1|^2$$

$$+ \operatorname{Im}\big((\mathfrak{x}\bar{\mathfrak{y}} + |\boldsymbol{\xi}|^2)s_1\big) + \big(\tfrac{1}{2}\mathfrak{y}_2 + 2\mathfrak{y}_1|\boldsymbol{\xi}|^2\big)$$

and $\mathfrak{f}(|\boldsymbol{\xi}|) = \frac{1}{2}C_1 c_2\big(\frac{1}{2}\big)|\boldsymbol{\xi}|$ or $f(|\boldsymbol{\xi}|) = \frac{1}{2}\sqrt{C_1}c_1\big(\frac{1}{4}\big)$ for D_1, D_2, respectively.

The optimization in s_2 is in fact performed in $|s_2|^2$. The optimum can be achieved only for $s_2 = 0$ or $|s_2| = +\infty$. For $|s_2| = +\infty$ the result is

$$S_2 \equiv \sup_{|\boldsymbol{\xi}| \in \mathbb{R}_+} \frac{|\mathfrak{x}|^2 \mathfrak{f}(|\boldsymbol{\xi}|)}{\frac{1}{2}\left(\mathfrak{x}_2 + \mathfrak{x}_1|\boldsymbol{\xi}|^2\right)} = \sup_{t \in \mathbb{R}_+} \frac{2\sqrt{2}\,\mathfrak{f}(\sqrt{t})}{\sqrt{t + \sqrt{1+t^2}}}$$
$$= \begin{cases} c_2\big(\frac{1}{2}\big)C_1 = 4\pi C_1 & \text{for } D_1, \\ c_1\big(\frac{1}{4}\big)\sqrt{2C_1} = 8\sqrt{\pi C_1} & \text{for } D_2. \end{cases}$$

For $s_2 = 0$ the optimum is achieved at

$$S_1 \equiv \sup_{\substack{\boldsymbol{\xi} \in \mathbb{R}^2 \\ s \in \mathbb{C}}} \frac{\left(|\mathfrak{x}|^2|s|^2 - 2|\boldsymbol{\xi}|\operatorname{Re}(i\mathfrak{x}\cdot s) + |\boldsymbol{\xi}|^2\right)\mathfrak{f}(|\boldsymbol{\xi}|)}{\mathfrak{Z}_{\mathfrak{x}\mathfrak{y}}(\boldsymbol{\xi}, s)},$$

where

$$\mathfrak{Z}_{\mathfrak{x}\mathfrak{y}}(\boldsymbol{\xi}, s) = \big(\tfrac{1}{2}\mathfrak{x}_2 + 2\mathfrak{x}_1|\boldsymbol{\xi}|^2\big)|s|^2 + 2\operatorname{Im}\big((\mathfrak{x}\bar{\mathfrak{y}} + |\boldsymbol{\xi}|^2)s\big) + \big(\tfrac{1}{2}\mathfrak{y}_2 + 2\mathfrak{y}_1|\boldsymbol{\xi}|^2\big). \tag{5.3.32}$$

The optimal value for the respective constant is $\max\{S_1, S_2\}$.

The change of variables (5.3.31) and the notation (5.3.32) lead to

$$D_0^2 = \frac{1}{2}\sup_{\substack{s \in \mathbb{C} \\ \boldsymbol{\xi} \in \mathbb{R}^2_+}} \frac{\left|2\mathfrak{x}s - 1 + 2i|\boldsymbol{\xi}|^2\right|^2}{\left(c_1\big(\frac{1}{4}\big)\sqrt{C_1} + c_2\big(\frac{1}{2}\big)C_1|\boldsymbol{\xi}|\right)\mathfrak{Z}_{\mathfrak{x}\mathfrak{y}}(\boldsymbol{\xi}, s)} \tag{5.3.33}$$

because the optimal s_2 is equal to 0. The calculations of S_1 or D_0 in (5.3.33) are the same as the corresponding calculations in the two-dimensional case, if $\gamma = |\boldsymbol{\xi}|^{-1}$ is taken and $c_1\big(\frac{1}{2}\big)$ is replaced by $c_2\big(\frac{1}{2}\big)$. Hence the only difference to the two-dimensional situation is the necessity to take $\max\{S_1, S_2\}$ in the calculation of D_1. In Figures 5.3 and 5.4 the computed dependence of $C_{\mathfrak{F}}$ on E and ν is depicted.

As in the two-dimensional case it is possible to optimize with respect to the Young modulus E by an appropriate rescaling of the time variable. The resulting bounds depend on ν only; they are compared to their values for the two-dimensional case in Figures 5.5 and 5.6.

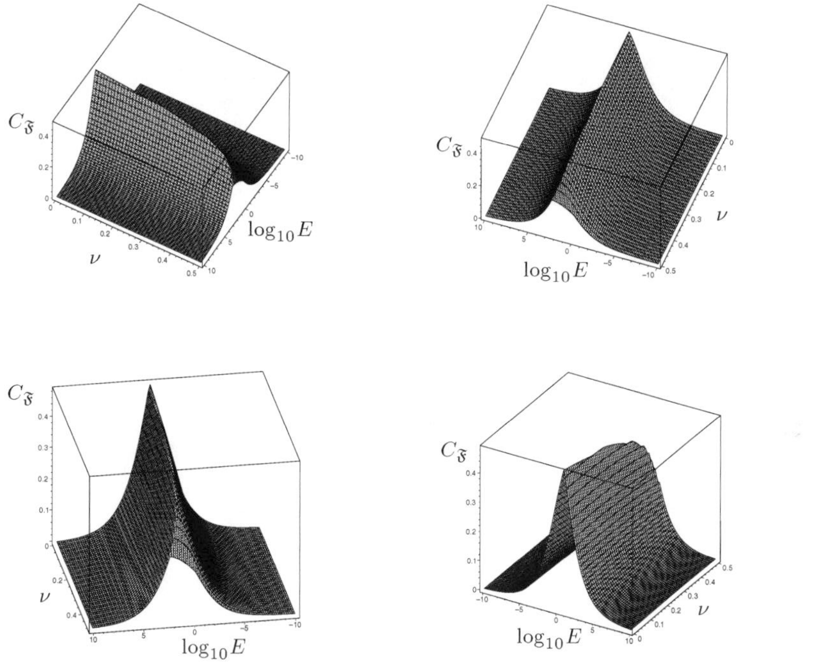

Figure 5.3: Upper bound for the coefficient of friction in dependence on the modulus of viscosity (E) and the viscous Poisson ratio (ν) for $0 \le \nu < 0.5$.

5.3.3 The case of a general dimension

Motivated by the form of the solution in three dimensions we assume that the solution for dimension $N \ge 4$ has the form

$$\begin{aligned} \widetilde{u}_j &= \mathfrak{a}\mathfrak{p}_j e^{-\mathfrak{a}y} + i\xi_j \mathfrak{p}_N e^{-\mathfrak{b}y}, \quad j = 1, \dots, N-1, \\ \widetilde{u}_N &= i\mathfrak{p}'_{N-1} \cdot \boldsymbol{\xi} e^{-\mathfrak{a}y} - \mathfrak{b}\mathfrak{p}_N e^{-\mathfrak{b}y} \end{aligned} \tag{5.3.34}$$

with $\mathfrak{p}'_{N-1} \equiv (\mathfrak{p}_1, \dots, \mathfrak{p}_{N-1})^\top$ and

$$\mathfrak{p}_j = (\mathscr{D}\mathfrak{a})^{-1} \left[\mathscr{D}\widehat{w}_j + \sum_{\ell=1}^{N-1} \xi_j \xi_\ell \widehat{w}_\ell + i\xi_j \mathfrak{a}\widehat{w}_N \right], \quad j = 1, \dots, N-1,$$

$$\mathfrak{p}_N = \mathscr{D}^{-1} \left[\sum_{\ell=1}^{N-1} i\, \xi_\ell \widehat{w}_\ell - \mathfrak{a}\widehat{w}_N \right].$$

Using this representation in the differential equations and the boundary conditions we prove that this is indeed the case. The boundary conditions

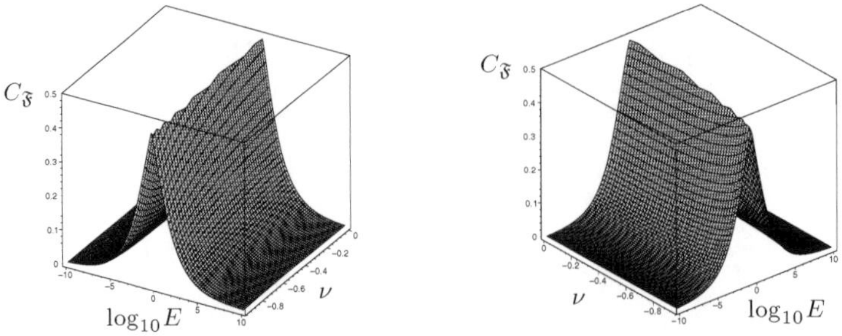

Figure 5.4: Upper bound for the coefficient of friction in dependence on the modulus of viscosity (E) and the viscous Poisson ratio (ν) for $-1 < \nu < 0$.

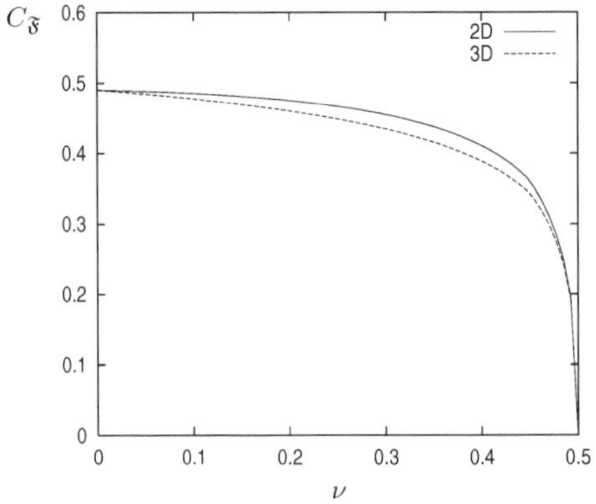

Figure 5.5: Optimal upper bound for the coefficient of friction in dependence on the viscous Poisson ratio $\nu \in \left[0, \frac{1}{2}\right)$.

are equivalent to

$$
\begin{aligned}
\widehat{w}_j &= \mathfrak{a}\mathfrak{p}_j + i\,\xi_j \mathfrak{p}_N, \ j = 1, \ldots, N-1, \\
\widehat{w}_N &= i\,\boldsymbol{\xi} \cdot \mathfrak{p}'_{N-1} - \mathfrak{b}\mathfrak{p}_N.
\end{aligned}
\tag{5.3.35}
$$

In order to calculate $\|\cdot\|_A$ we first compute

$$
\begin{aligned}
\widetilde{\operatorname{div}(\boldsymbol{u})} &= \left(\mathfrak{b}^2 - |\boldsymbol{\xi}|^2\right) \mathfrak{p}_N e^{-\mathfrak{b}y}, \\
\widetilde{e_{jj}(\boldsymbol{u})} &= i\mathfrak{a}\xi_j \mathfrak{p}_j e^{-\mathfrak{a}y} - \xi_j^2 \mathfrak{p}_N e^{-\mathfrak{b}y}, \qquad j \neq N,
\end{aligned}
$$

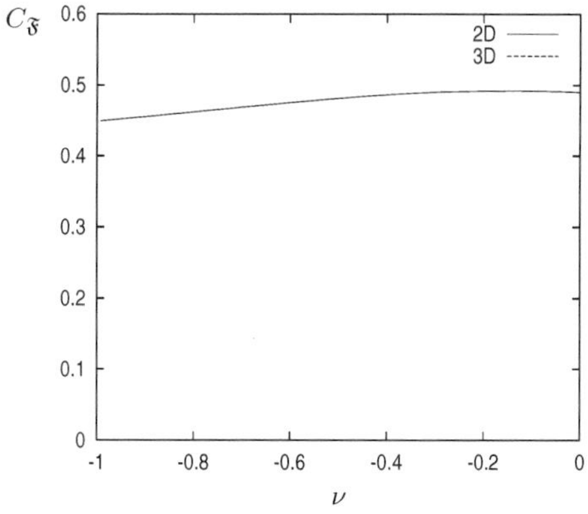

Figure 5.6: Optimal upper bound for the coefficient of friction in dependence on the viscous Poisson ratio $\nu \in (-1, 0]$.

$$\begin{aligned}
\widetilde{e_{NN}(\boldsymbol{u})} &= -i\mathfrak{a}(\mathfrak{p}'_{N-1} \cdot \boldsymbol{\xi})e^{-\mathfrak{a}y} + \mathfrak{b}^2\mathfrak{p}_N e^{-\mathfrak{b}y}, \\
2\widetilde{e_{j\ell}(\boldsymbol{u})} &= i\mathfrak{a}(\xi_j\mathfrak{p}_\ell + \xi_\ell\mathfrak{p}_j)e^{-\mathfrak{a}y} - 2\xi_j\xi_\ell\mathfrak{p}_N e^{-\mathfrak{b}y}, \ j,\ell \neq N, \\
2\widetilde{e_{jN}(\boldsymbol{u})} &= -\left(\xi_j\mathfrak{p}'_{N-1} \cdot \boldsymbol{\xi} + \mathfrak{a}^2\mathfrak{p}_j\right)e^{-\mathfrak{a}y} - 2i\mathfrak{b}\xi_j\mathfrak{p}_N e^{-\mathfrak{b}y}, \ j \neq N.
\end{aligned}$$

The integrals $I(D) \equiv \int_{\mathbb{R}_+} |D|^2 \, d\boldsymbol{x}_N$ are

$$\begin{aligned}
I\left(\widetilde{\operatorname{div}(\boldsymbol{u})}\right) &= \tfrac{1}{2}k\mathfrak{b}_2\vartheta|\mathfrak{p}_N|^2, \\
I\left(\widetilde{e_{jj}(\boldsymbol{u})}\right) &= \frac{\xi_j^2|\mathfrak{a}\mathfrak{p}_j|^2}{2\mathfrak{a}_1} + \frac{\xi_j^4|\mathfrak{p}_N|^2}{2\mathfrak{b}_1} - 2\operatorname{Re}\frac{i\mathfrak{a}\xi_j^3\mathfrak{p}_j\overline{\mathfrak{p}_N}}{\mathfrak{a}+\overline{\mathfrak{b}}}, \ j \neq N, \\
I\left(\widetilde{e_{NN}(\boldsymbol{u})}\right) &= \frac{|\mathfrak{a}|^2|\mathfrak{p}'_{N-1}\cdot\boldsymbol{\xi}|^2}{2\mathfrak{a}_1} + \frac{|\mathfrak{b}^2\mathfrak{p}_N|^2}{2\mathfrak{b}_1} - 2\operatorname{Re}\frac{i\mathfrak{a}(\mathfrak{p}'_{N-1}\cdot\boldsymbol{\xi})\overline{\mathfrak{b}^2\mathfrak{p}_N}}{\mathfrak{a}+\overline{\mathfrak{b}}}, \\
I\left(\widetilde{e_{j\ell}(\boldsymbol{u})}\right) &= |\mathfrak{a}|^2\frac{\xi_j^2|\mathfrak{p}_\ell|^2 + \xi_\ell^2|\mathfrak{p}_j|^2 + 2\xi_j\xi_\ell\operatorname{Re}(\mathfrak{p}_j\overline{\mathfrak{p}_\ell})}{8\mathfrak{a}_1} \\
&\quad + \frac{\xi_j^2\xi_\ell^2|\mathfrak{p}_N|^2}{2\mathfrak{b}_1} - \xi_j\xi_\ell\operatorname{Re}\left(i\mathfrak{a}\frac{(\xi_j\mathfrak{p}_\ell + \xi_\ell\mathfrak{p}_j)\overline{\mathfrak{p}_N}}{\mathfrak{a}+\overline{\mathfrak{b}}}\right), \ j,\ell \neq N, \\
I\left(\widetilde{e_{jN}(\boldsymbol{u})}\right) &= \frac{\xi_j^2|\boldsymbol{\xi}\cdot\mathfrak{p}'_{N-1}|^2 + |\mathfrak{a}^2\mathfrak{p}_j|^2 2\xi_j\operatorname{Re}\left(\mathfrak{a}^2\mathfrak{p}_j\boldsymbol{\xi}\cdot\overline{\mathfrak{p}'_{N-1}}\right)}{8\mathfrak{a}_1} \\
&\quad + \frac{\xi_j^2|\mathfrak{b}\mathfrak{p}_N|^2}{2\mathfrak{b}_1} - \xi_j\operatorname{Re}\left(\frac{i\left(\xi_j\mathfrak{p}'_{N-1}\cdot\boldsymbol{\xi} + \mathfrak{a}^2\mathfrak{p}_j\right)\overline{\mathfrak{b}\mathfrak{p}_N}}{\mathfrak{a}+\overline{\mathfrak{b}}}\right), \ j \neq N.
\end{aligned}$$

Hence

$$\sum_{j,\ell=1}^{N-1} I\Big(\widetilde{e_{j\ell}(\boldsymbol{u})}\Big) = \frac{|\mathfrak{a}|^2}{4\mathfrak{a}_1}|\boldsymbol{\xi}|^2\left[|\mathfrak{p}'_{N-1}|^2 + |\mathfrak{p}'_{N-1}\cdot\boldsymbol{\xi}|^2\right] + \frac{|\boldsymbol{\xi}|^4}{2\mathfrak{b}_1}|\mathfrak{p}_N|^2 - 2|\boldsymbol{\xi}|^2\operatorname{Re}\left(\frac{i\mathfrak{a}\mathfrak{p}'_{N-1}\cdot\boldsymbol{\xi}\overline{\mathfrak{p}_N}}{\mathfrak{a}+\overline{\mathfrak{b}}}\right),$$

$$\sum_{j=1}^{N-1} I\Big(\widetilde{e_{jN}(\boldsymbol{u})}\Big) = \frac{|\boldsymbol{\xi}|^2|\mathfrak{p}'_{N-1}\cdot\boldsymbol{\xi}|^2 + |\mathfrak{a}|^4|\mathfrak{p}'_{N-1}|^2 + 2\operatorname{Re}(\mathfrak{a}^2)|\mathfrak{p}'_{N-1}\cdot\boldsymbol{\xi}|^2}{8\mathfrak{a}_1} + \frac{|\boldsymbol{\xi}|^2|\mathfrak{b}\mathfrak{p}_N|^2}{2\mathfrak{b}_1} - \operatorname{Re}\left(\frac{i\left(|\boldsymbol{\xi}|^2\mathfrak{p}'_{N-1}\cdot\boldsymbol{\xi} + \mathfrak{a}^2\mathfrak{p}'_{N-1}\cdot\boldsymbol{\xi}\right)\overline{\mathfrak{b}\mathfrak{p}_N}}{\mathfrak{a}+\overline{\mathfrak{b}}}\right),$$

and therefore

$$\sum_{j,\ell=1}^{N} I\Big(\widetilde{e_{j\ell}(\boldsymbol{u})}\Big) = \tfrac{1}{2}\Big[\left(\mathfrak{a}_2\vartheta + \mathfrak{a}_1|\boldsymbol{\xi}|^2\right)|\mathfrak{p}'_{N-1}|^2 + 3\mathfrak{a}_1|\mathfrak{p}'_{N-1}\cdot\boldsymbol{\xi}|^2 + \left(\mathfrak{b}_2 k\vartheta + 4\mathfrak{b}_1|\boldsymbol{\xi}|^2\right)|\mathfrak{p}_N|^2\Big] - 2\operatorname{Re}\left[i\left(|\boldsymbol{\xi}|^2 + \mathfrak{a}\overline{\mathfrak{b}}\right)(\mathfrak{p}'_{N-1}\cdot\boldsymbol{\xi})\overline{\mathfrak{p}_N}\right].$$

Summing this with $I(\operatorname{div}(\boldsymbol{u}))$, we see that the norm $\|\boldsymbol{u}\|_A$ has the form (5.3.25) with the appropriate N-dimensional variant of $|\mathfrak{p}'_{N-1}|^2 \equiv |\mathfrak{p}_1|^2 + \ldots + |\mathfrak{p}_{N-1}|^2$ and $\mathfrak{p}'_{N-1}\cdot\boldsymbol{\xi} \equiv \mathfrak{p}_1\xi_1 + \ldots + \mathfrak{p}_{N-1}\xi_{N-1}$. Naturally, this holds also for the appropriate general version of the bilinear form (5.3.26). It is easy to generalize (5.3.24) to the expression

$$|\widehat{w}_1|^2 + \ldots + |\widehat{w}_{N-1}|^2 = |\mathfrak{a}|^2|\mathfrak{p}'_{N-1}|^2 + |\boldsymbol{\xi}|^2|\mathfrak{p}_N|^2 + 2\operatorname{Im}\big(\mathfrak{a}(\boldsymbol{\xi}\cdot\mathfrak{p}'_{N-1})\overline{\mathfrak{p}_N}\big). \tag{5.3.36}$$

For the solution $\boldsymbol{v}$ of (5.3.1) with the Dirichlet boundary condition $\boldsymbol{w} = (0,\ldots,0,v_N)$ on $\mathbb{R}^N\times\{0\}$ holds

$$\langle\dot{\boldsymbol{u}},\boldsymbol{v}\rangle_Q = \operatorname{Re}\left(\frac{1}{C_1\overline{\mathscr{D}}}\int_{\mathbb{R}^N}\vartheta\left(i\mathfrak{a}_2\mathfrak{p}'_{N-1}\cdot\boldsymbol{\xi} - i\overline{\mathfrak{a}}\mathfrak{b}_1\mathfrak{p}_N + i|\boldsymbol{\xi}|^2\mathfrak{p}_N\right)\overline{\widehat{v}_N}\,d\boldsymbol{\xi}\,d\vartheta\right),$$

$$A(\boldsymbol{u},\boldsymbol{v}) = -\operatorname{Re}\Bigg(\frac{1}{C_1\overline{\mathscr{D}}}\int_{\mathbb{R}^N}\Big(i\big(\mathfrak{a}_2\vartheta + 4\mathfrak{a}_1|\boldsymbol{\xi}|^2 - 2\overline{\mathfrak{a}}|\boldsymbol{\xi}|^2 - 2|\mathfrak{a}|^2\overline{\mathfrak{b}}\big)\mathfrak{p}'_{N-1}\cdot\boldsymbol{\xi} - 2\big(|\boldsymbol{\xi}|^2 + \overline{\mathfrak{a}}\mathfrak{b}\big)|\boldsymbol{\xi}|^2\mathfrak{p}_N + \big(\mathfrak{b}_2\vartheta + 4\mathfrak{b}_1|\boldsymbol{\xi}|^2\big)\overline{\mathfrak{a}}\mathfrak{p}_N\Big)\overline{\widehat{v}_N}\,d\boldsymbol{\xi}\,d\vartheta\Bigg).$$

This leads after some calculation to the expression

$$\langle \dot{\boldsymbol{u}}, \boldsymbol{v} \rangle_Q + A(\boldsymbol{u}, \boldsymbol{v}) =$$
$$= \frac{1}{C_1} \mathrm{Re} \left(\int_{\mathbb{R}^N} \left[2i\mathfrak{a}(\mathfrak{p}'_{N-1} \cdot \boldsymbol{\xi}) - \mathfrak{p}_N \left(i\vartheta + 2|\boldsymbol{\xi}|^2 \right) \right] \overline{\widehat{v}_N} \, d\boldsymbol{\xi} \, d\vartheta \right) . \tag{5.3.37}$$

As above we derive the optimization problems (5.3.28)-(5.3.30) with the corresponding N–dimensional generalizations of $\mathfrak{a}$, $\mathfrak{b}$ and $\boldsymbol{q} = \mathfrak{p}'_{N-1}/\mathfrak{p}_N \in \mathbb{C}^{N-1}$ (c.f. (5.3.27)). Let us take an orthonormal basis $\{\boldsymbol{e}_j;\ j = 1, \ldots N-1\}$ of $\mathbb{C}^{N-1}$ such that $\boldsymbol{e}_1 = \boldsymbol{\xi}/|\boldsymbol{\xi}|$ for a fixed $\boldsymbol{\xi} \neq 0$. Then $\boldsymbol{q} = \sum_{j=1}^{N-1} s_j \boldsymbol{e}_j$ with $s_j = (\boldsymbol{q}, \overline{\boldsymbol{e}_j})$, $j = 1, \ldots, N-1$, and for $\boldsymbol{s} \equiv [s_1, \ldots, s_{N-1}]$ the identity $|\boldsymbol{q}|^2 = |\boldsymbol{s}|^2$ holds. Since the expression $|s_2|^2 + \ldots + |s_{N-1}|^2$ can be handled as $|s_2|^2$ in the three-dimensional case, the optimization problems to calculate the constants D_j, $j = 0, 1, 2$, are the same as earlier and their values differ from their three-dimensional variants only via the dimension-dependent constant $c_{N-1}\left(\frac{1}{2}\right)$. Again it is possible to rescale the time variable and eliminate the dependence of $C_{\mathfrak{F}}$ on E.

5.3.4 The non-homogeneous case

As in the static case (see the final remark of Section 3.1), the results presented above can be applied also to the case of non-homogeneous isotropic material with a $C^{\beta'}$–smooth dependence of the viscous Young modulus E and the Poisson ratio ν on the space variables (and no dependence on the time variable). It is advantageous to use the rescaling of the time variable for the localized problems, because it is then possible to choose the optimal time scaling for every local problem separately. Since the upper bound for the optimal time scaling only depends on the viscous Poisson ration, $C_{\mathfrak{F}} = C_{\mathfrak{F}}(\nu)$, the global upper bound can be computed as the minimum

$$C_{\mathfrak{F}} = \inf_{x \in \Gamma_{\mathfrak{F}}} C_{\mathfrak{F}}(\nu(\boldsymbol{x}))$$

of the local bounds, depending on the corresponding Poisson ratios.

5.4 Thermo-viscoelastic problems

Friction is a dissipative term in the viscoelastic equation of motion. The dissipated mechanical energy is not lost. It is transformed into thermal energy. The existence theorems in dynamic contact problems are based on the presence of a viscosity which can be interpreted as inner friction. This term also dissipates mechanical energy into thermal energy. If these phenomena are included in the model, we must add a heat equation to the viscoelastic system and we must formulate the coupling between these two problems. The coupling terms describe the heat generated by boundary friction (the *frictional heat*), the heat generated by viscous inner friction (the *viscous heat*) and an additional term describing the mutual exchange of energy between heat equation and viscoelastic system. We will find therefore a *deformation heat* in the heat equation, and thermal stresses in the viscoelastic system.

Let us first present the basic principles of the physics of thermoelastic material. The model of thermo-viscoelasticity employed in this section is based on the equations of motion

$$\ddot{u}_i = \partial_j \sigma_{ij}(\boldsymbol{u}, \Theta)$$

and on the heat equation

$$\dot{\mathfrak{e}} = \partial_i (c_{ij} \partial_j \Theta) + g$$

with displacement field $\boldsymbol{u}$, temperature Θ, stress tensor $\sigma(\boldsymbol{u}, \Theta)$, internal energy $\mathfrak{e}$, components of the (possibly anisotropic) heat diffusivity c_{ij} and external heat source g. The constitutive relation is given by a linear thermo-viscoelastic law of the Kelvin-Voigt type

$$\sigma_{ij} \equiv \sigma_{ij}(\boldsymbol{u}, \Theta) = a^{(0)}_{ijk\ell} e_{k\ell}(\boldsymbol{u}) + a^{(1)}_{ijk\ell} e_{k\ell}(\dot{\boldsymbol{u}}) - b_{ij}\Theta, \quad i,j = 1, \ldots, N, \tag{5.4.1}$$

where the tensor b_{ij} describes the thermal stresses. Let us first study the case of purely elastic behaviour, $a^{(1)}_{ijk\ell} = 0$. If we consider the simplest case where the heat capacity is constant and coefficients $a^{(0)}_{ijk\ell}$ and b_{ij} do not depend on the temperature, the density of free energy associated to the system is

$$\mathfrak{f} = \mathfrak{f}(\boldsymbol{u}, \Theta) = \frac{1}{2} a^{(0)}_{ijk\ell} e_{k\ell}(\boldsymbol{u}) e_{ij}(\boldsymbol{u}) - b_{ij}\Theta e_{ij}(\boldsymbol{u}) - \Theta \ln(\Theta).$$

The corresponding entropy is

$$\mathfrak{s} = -\partial_\Theta \mathfrak{f} = b_{ij} e_{ij}(\boldsymbol{u}) + 1 + \ln(\Theta), \tag{5.4.2}$$

and the internal energy reads

$$\mathfrak{e} = \mathfrak{f} + \Theta \mathfrak{s} = \frac{1}{2} a^{(0)}_{ijk\ell} e_{k\ell}(\boldsymbol{u}) e_{ij}(\boldsymbol{u}) + \Theta,$$

it is the sum of elastic and thermal energy. The second law of thermodynamics can be written as $\dot{\mathfrak{s}} = \dot{\mathfrak{e}}/\Theta$ and therefore

$$\dot{\mathfrak{e}} = \Theta\dot{\mathfrak{s}} = \dot{\Theta} + \Theta b_{ij} e_{ij}(\dot{\boldsymbol{u}}),$$

which leads to the heat equation

$$\dot{\Theta} = \partial_i(c_{ij}\partial_j\Theta) - b_{ij}\Theta e_{ij}(\dot{\boldsymbol{u}}) + g. \tag{5.4.3}$$

In the case of the included viscosity, there is the additional viscous dissipation $a^{(1)}_{ijk\ell} e_{k\ell}(\dot{\boldsymbol{u}}) e_{ij}(\dot{\boldsymbol{u}})$. Due to its dissipative character, it is not added to one of the thermodynamical potentials, but it is included as a heat source in the heat equation, in order to satisfy the first law of thermodynamics. The heat equation of thermo-viscoelastic material then reads

$$\dot{\Theta} - \partial_i(c_{ij}\partial_j\Theta) = a^{(1)}_{ijk\ell}\, e_{ij}(\dot{\boldsymbol{u}}) e_{k\ell}(\dot{\boldsymbol{u}}) - b_{ij}\Theta\, e_{ij}(\dot{\boldsymbol{u}}). \tag{5.4.4}$$

We consider a thermo-viscoelastic contact problem with friction given on a bounded domain $\Omega \subset \mathbb{R}^N$, its boundary Γ is partitioned into the measurable and mutually disjoint subsets $\Gamma_{\boldsymbol{U}}$, $\Gamma_{\boldsymbol{\tau}}$ and Γ_C, and the time interval is $I_T \equiv [0,T]$. We use again the notation $Q_T = I_T \times \Omega$, $S_T = I_T \times \Gamma$ and its parts $S_{YT} = I_T \times \Gamma_Y$ for $Y = C, \boldsymbol{\tau}, \boldsymbol{U}$. Then the thermo-viscoelastic contact problem with Coulomb friction is the following:

Find a couple of displacement field and temperature $[u, \Theta]$ *satisfying the differential equations*

$$\ddot{u}_i - \partial_j\sigma_{ij}(\boldsymbol{u},\Theta) = f_i, \quad i = 1,\dots,N, \qquad \text{in } Q_T, \tag{5.4.5}$$

$$\dot{\Theta} - \partial_i(c_{ij}\partial_j\Theta) = a^{(1)}_{ijk\ell}\, e_{ij}(\dot{\boldsymbol{u}}) e_{k\ell}(\dot{\boldsymbol{u}}) - b_{ij}\Theta\, e_{ij}(\dot{\boldsymbol{u}}) \text{ in } Q_T, \tag{5.4.6}$$

the boundary value conditions

$$\boldsymbol{u} = \boldsymbol{U} \qquad \text{on } S_{\boldsymbol{U}T}, \tag{5.4.7}$$

$$\boldsymbol{\sigma}_\Gamma(\boldsymbol{u},\Theta) = \boldsymbol{\tau} \qquad \text{on } S_{\boldsymbol{\tau}T}, \tag{5.4.8}$$

$$c_{ij}\partial_j\Theta n_i = K(\Upsilon - \Theta) \text{ on } S_{\boldsymbol{U}T} \cup S_{\boldsymbol{\tau}T}, \tag{5.4.9}$$

$$\left.\begin{array}{l} \dot{u}_n \le 0, \quad \sigma_n \le 0, \quad \sigma_n\dot{u}_n = 0, \\ \dot{\boldsymbol{u}}_{\boldsymbol{t}} = \boldsymbol{0} \quad \Rightarrow \quad |\boldsymbol{\sigma}_{\boldsymbol{t}}| \le \mathfrak{F}|\sigma_n|, \\ \dot{\boldsymbol{u}}_{\boldsymbol{t}} \ne \boldsymbol{0} \quad \Rightarrow \quad \boldsymbol{\sigma}_{\boldsymbol{t}} = -\mathfrak{F}|\sigma_n|\dfrac{\dot{\boldsymbol{u}}_{\boldsymbol{t}}}{|\dot{\boldsymbol{u}}_{\boldsymbol{t}}|} \end{array}\right\} \text{ on } S_{CT}, \tag{5.4.10}$$

$$c_{ij}\partial_j\Theta n_i = \mathfrak{F}|\sigma_n||\dot{\boldsymbol{u}}_{\boldsymbol{t}}| + K(\Upsilon - \Theta) \text{ on } S_{CT}, \tag{5.4.11}$$

and the initial conditions

$$\boldsymbol{u}(0,x) = \boldsymbol{u}_0(\boldsymbol{x}),\ \ \dot{\boldsymbol{u}}(0,x) = \boldsymbol{u}_1(\boldsymbol{x}) \text{ and } \Theta(0,x) = \Theta_0(\boldsymbol{x}) \text{ for } x \in \Omega. \tag{5.4.12}$$

We assume that the coefficients $a^{(\iota)}_{ijk\ell}$ satisfy the same symmetries as in (5.1.3) and the same boundedness and ellipticity conditions as in (5.1.4). The tensor $\{b_{ij}\}$ of thermal expansion will be symmetric, Hölder continuous with respect to the space variable and globally bounded. The thermal conductivity tensor c_{ij} is also assumed to be symmetric and elliptic. Since different particular models treated in the sequel are mutually distinguished by further properties of this tensor, we do not get into more details here.

The "right-hand side" of equation (5.4.6) represents the respective heat source created by viscosity (the viscous heat) and deformation of the material (the deformation heat), as described earlier. The first term on the "right-hand side" of (5.4.11) is the source created by friction, the frictional heat.

Let us introduce the following sets of admissible functions

$$\mathscr{C} := \left\{ \boldsymbol{v} \in \boldsymbol{H}^{1/2,\,1}(Q_T);\; \boldsymbol{v}|_{S_U T} = \dot{\boldsymbol{U}}|_{S_U T} \text{ and } v_n|_{S_C T} \le 0 \right\} \tag{5.4.13}$$

and

$$\mathfrak{V} := H^{1/2,\,1}(Q_T). \tag{5.4.14}$$

A weak solution to the thermo-viscoelastic contact problem with friction is a pair $(\boldsymbol{u}, \Theta)$ such that $\dot{\boldsymbol{u}}$ belongs to $\mathscr{C}$, Θ to $\mathfrak{V}$, $\dot{\Theta}$ to $\mathfrak{V}^*$ and $\ddot{\boldsymbol{u}}$ to $\left\{ \boldsymbol{v} \in \boldsymbol{H}^{1/2,1}(Q_T);\; \boldsymbol{v} = 0 \text{ on } S_{UT} \right\}^*$, the initial conditions

$$\boldsymbol{u}(0,\cdot) = \boldsymbol{u}_0, \quad \dot{\boldsymbol{u}}(0,\cdot) = \boldsymbol{u}_1, \quad \text{and} \quad \Theta(0,\cdot) = \Theta_0$$

are satisfied and the variational formulation of the contact part and the variational equation of the heat transfer part of the problem simultaneously hold. The variational inequality of the contact part of the thermo-viscoelastic contact problem with friction is given by

$$\begin{aligned} \left\langle \ddot{\boldsymbol{u}}, \boldsymbol{v} - \dot{\boldsymbol{u}} \right\rangle_{Q_T} + \int_{Q_T} \Big(a^{(0)}(\boldsymbol{u}, \boldsymbol{v} - \dot{\boldsymbol{u}}) + a^{(1)}(\dot{\boldsymbol{u}}, \boldsymbol{v} - \dot{\boldsymbol{u}}) \\ - b_{ij}\Theta\, e_{ij}(\boldsymbol{v} - \dot{\boldsymbol{u}}) \Big)\, d\boldsymbol{x}\, dt \\ + \int_{S_T} \mathfrak{F} |\sigma_n(\boldsymbol{u}, \Theta)| \big(|\boldsymbol{v}_t| - |\dot{\boldsymbol{u}}_t| \big)\, ds_{\boldsymbol{x}}\, dt \ge \mathscr{L}(\boldsymbol{v} - \dot{\boldsymbol{u}}) \end{aligned} \tag{5.4.15}$$

for all $\boldsymbol{v} \in \mathscr{C}$. Here we have used the usual bilinear forms of elastic energy and viscous energy dissipation

$$a^{(\iota)}(\boldsymbol{u}, \boldsymbol{v}) = a^{(\iota)}_{ijk\ell} e_{k\ell}(\boldsymbol{u}) e_{ij}(\boldsymbol{v}), \; \iota = 0, 1,$$

and the linear functional

$$\mathscr{L}(\boldsymbol{v}) = \int_{Q_T} \boldsymbol{f} \cdot \boldsymbol{v}\, d\boldsymbol{x}\, dt + \int_{S_\tau T} \boldsymbol{\tau} \cdot \boldsymbol{v}\, ds_{\boldsymbol{x}}\, dt.$$

The variational equation of the heat transfer part of the problem is defined by

$$\begin{aligned}\langle\dot{\Theta},\varphi\rangle_{Q_T} &+ \int_{Q_T} \left(c_{ij}\partial_j\Theta\,\partial_i\varphi + b_{ij}\Theta\, e_{ij}(\dot{\boldsymbol{u}})\,\varphi\right) d\boldsymbol{x}\,dt \\ &+ \langle K(\Theta-\Upsilon),\varphi\rangle_{S_T} = \int_{Q_T} a^{(1)}_{ijk\ell}e_{k\ell}(\dot{\boldsymbol{u}})e_{ij}(\dot{\boldsymbol{u}})\,\varphi\,d\boldsymbol{x}\,dt \\ &+ \int_{S_{CT}} \mathfrak{F}|\sigma_n(\boldsymbol{u},\Theta)||\dot{\boldsymbol{u}}_{\boldsymbol{t}}|\,\varphi\,ds_{\boldsymbol{x}}\,dt\end{aligned} \tag{5.4.16}$$

for all $\varphi\in\mathfrak{V}$.

Let us briefly show that the employed model is thermodynamically consistent. For simplicity of the presentation we shall assume $\Gamma_U=\emptyset$, because in this case the contribution of boundary terms (transport of energy and entropy across the boundary) can be understood more easily. The conservation of energy (the first law of thermodynamics) follows by using the test functions $\boldsymbol{v}=\dot{\boldsymbol{u}}\chi_{(t_0,T)}$ and $\boldsymbol{v}=(1+\chi_{(0,t_0)})\dot{\boldsymbol{u}}$ in (5.4.15) and $\varphi=\chi_{(0,t_0)}$ in (5.4.16), where χ_M denotes the characteristic function of M. Combining both results we observe that the coupling terms (viscous heat, deformation heat and frictional heat) cancel and we obtain

$$\begin{aligned}\int_\Omega &\left(\frac{1}{2}|\dot{\boldsymbol{u}}(t_0)|^2+\frac{1}{2}a^{(0)}_{ijk\ell}e_{k\ell}(\boldsymbol{u}(t_0))e_{ij}(\boldsymbol{u}(t_0))+\Theta(t_0)\right)d\boldsymbol{x} \\ &= \int_\Omega\left(\frac{1}{2}|\boldsymbol{u}_1|^2+\frac{1}{2}a^{(0)}_{ijk\ell}e_{k\ell}(\boldsymbol{u}_0)e_{ij}(\boldsymbol{u}_0)+\Theta_0\right)d\boldsymbol{x} \\ &\quad+\int_{Q_{t_0}}\boldsymbol{f}\cdot\dot{\boldsymbol{u}}\,d\boldsymbol{x}\,dt+\int_{S_{t_0}}\left(\boldsymbol{\tau}\cdot\dot{\boldsymbol{u}}-K(\Theta-\Upsilon)\right)ds_{\boldsymbol{x}}\,dt.\end{aligned}$$

Hence the total energy of the system, consisting of the sum of kinetic energy $\int_\Omega\frac{1}{2}|\dot{\boldsymbol{u}}|^2\,d\boldsymbol{x}$, elastic energy $\int_\Omega\frac{1}{2}a^{(0)}_{ijk\ell}e_{k\ell}(\boldsymbol{u})e_{ij}(\boldsymbol{u})\,d\boldsymbol{x}$, and total heat $\int_\Omega\Theta\,d\boldsymbol{x}$, at time t_0 is equal to initial energy plus energy supplied to the system. The latter is given as a sum of outer forces multiplied with the displacement velocities, integrated over time (work = integral of power over time, power = force times velocity)

$$\int_{Q_{t_0}}\boldsymbol{f}\cdot\dot{\boldsymbol{u}}\,d\boldsymbol{x}\,dt+\int_{S_{t_0}}\boldsymbol{\tau}\cdot\dot{\boldsymbol{u}}\,ds_{\boldsymbol{x}}\,dt$$

and external heat flux into the domain,

$$\int_{S_{t_0}}K(\Upsilon-\Theta)\,ds_{\boldsymbol{x}}\,dt.$$

Here the frictional heat is not considered as an external energy source, because it is created by energy already present in the system. The second law of thermodynamics is obtained with the test function $\varphi = \chi_{(0,t_0)}\Theta^{-1}$ in the heat equation (5.4.16). The result is

$$\int_\Omega (\ln(\Theta(t_0)) + b_{ij}e_{ij}(\boldsymbol{u}(t_0)))\,d\boldsymbol{x} = \int_\Omega (\ln(\Theta_0) + b_{ij}e_{ij}(\boldsymbol{u}_0))\,d\boldsymbol{x}$$
$$+ \int_{Q_{t_0}} \left(\frac{c_{ij}\partial_j\Theta\partial_i\Theta}{\Theta^2} + \frac{a^{(1)}_{ijk\ell}e_{k\ell}(\dot{\boldsymbol{u}})e_{ij}(\dot{\boldsymbol{u}})}{\Theta} \right) d\boldsymbol{x}\,dt$$
$$+ \int_{S_{t_0}} \left(\frac{\mathfrak{F}|\sigma_n(\boldsymbol{u},\Theta)||\dot{\boldsymbol{u}}_{\boldsymbol{t}}|}{\Theta} + \frac{K(\Upsilon - \Theta)}{\Theta} \right) ds_{\boldsymbol{x}}\,dt.$$

The density of entropy is given by $\mathfrak{s} = \ln(\Theta) + b_{ij}e_{ij}(\boldsymbol{u})$; see (5.4.2) (the constant term 1 there can be neglected). Hence, according to the previous equation, the entropy at time t_0 is equal to the initial entropy plus the entropy produced by heat diffusion $\int_{Q_{t_0}} \Theta^{-2}c_{ij}\partial_j\Theta\partial_i\Theta\,d\boldsymbol{x}\,dt$ plus the entropy produced by viscous heat $\int_{Q_{t_0}} \Theta^{-1}a^{(1)}_{ijk\ell}e_{k\ell}(\dot{\boldsymbol{u}})e_{ij}(\dot{\boldsymbol{u}})\,d\boldsymbol{x}\,dt$ plus the entropy flux accross the boundary $\int_{S_{t_0}} \Theta^{-1}c_{ij}\partial_j\Theta n_i\,ds_{\boldsymbol{x}}\,dt$. Due to the physical requirement $\Theta > 0$, both entropy production terms in the second line of the previous formula are in fact non-negative.

In this section we study the solvability of thermo-viscoelastic contact problems. The main difficulty consists in the coupling terms (viscous heat, deformation heat and frictional heat) on the "right-hand side" of (5.4.6) which have a quadratic growth, while the "left-hand side" is only linear. The limited regularity of solutions to the dynamic contact problem with Coulomb friction studied in the previous sections does not allow solving such a system by means of the variational methods employed throughout this book in the original form. In order to proceed, some simplifications of the model will be necessary.

Basically there are two possible approaches to solve this kind of problem:

- *the fixed-point approach*, where a fictitious given temperature $\Theta^{(0)}$ replaces the real temperature Θ in (5.4.5). The mechanical and thermal problems are solved separately and we seek a fixed point of the composition of the corresponding operators;

- *the energetic approach*, where the whole system is treated; we derive some very weak *a priori* estimate coming from the energy conservation and we improve them in subsequent steps.

We shall employ these two approaches in the next subsections. In the second approach the compensation of deformation heat and sometimes also of viscous heat leads to weaker conditions on the growth of the heat diffusivity than the compensation of the frictional heat. In order to obtain optimal results we also introduce a non-linear dissipation of energy along the boundary, described by a function $R(\Theta)$. A possible physical source of this term could be a simplified model of heat radiation. The corresponding boundary condition then is

$$c_{ij}\partial_j\Theta n_i = \mathfrak{F}|\sigma_n||\dot{\boldsymbol{u}}_{\boldsymbol{t}}| + K(\Upsilon - \Theta) - R(\Theta) + R(\Upsilon).$$

Here R will be an increasing function with certain growth properties.

5.4.1 Fixed-point approach

We apply this approach to a linearized version of the above described system. The equation (5.4.5) is preserved in the original form while the heat equation is linearized around a given reference temperature $\widetilde{\Theta}$ and zero displacement velocities. The linearized equation then reads

$$\dot{\Theta} - \partial_i(c_{ij}\partial_j\Theta) = -d_{ij}\partial_j\dot{u}_i \text{ in } Q_T \tag{5.4.17}$$

with $d_{ij} = \widetilde{\Theta}\, b_{ij}$. The viscous heat cancels due to the linearization around $\dot{\boldsymbol{u}} = 0$. In the boundary conditions the frictional heat cannot be linearized, because it is not smooth. However, in order to enable the existence proof and to be in harmony with the linear structure of the other term, we linearize the growth of the frictional heat. This is done by replacing condition (5.4.11) with

$$c_{ij}\partial_j\Theta n_i = J(\cdot, \mathfrak{F}|\sigma_n|, |\dot{\boldsymbol{u}}_{\boldsymbol{t}}|) + K(\Upsilon - \Theta) \text{ on } S_{CT}, \tag{5.4.18}$$

where the functional $J \equiv J(\boldsymbol{x}, \mathfrak{F}|\sigma_n|, |\dot{\boldsymbol{u}}_{\boldsymbol{t}}|)$ is monotone in the second and third argument and satisfies the growth condition

$$J(\boldsymbol{x}, y, z) \leq c(1 + |y| + |z|) \tag{5.4.19}$$

with some constant $c \in \mathbb{R}$. In addition, the operator $\mathscr{J} : (f, g) \mapsto J(\cdot, f, g)$ is assumed to satisfy the continuity relation

$$\mathscr{J}(f_n, g_n) \rightharpoonup \mathscr{J}(f, g) \text{ in } L_\alpha(S_{CT}) \tag{5.4.20}$$

for some $\alpha > 2 - 2/(2 + N)$ if $f_n \rightharpoonup f$ in $L_2(S_{CT})$ and $g_n \to g$ in $L_2(S_{CT})$. Moreover, if both former convergences are strong, then the convergence in (5.4.20) is assumed to be strong, too. A function having such a continuity property is for example given by

$$J : (\boldsymbol{x}, f, g) \mapsto G(\boldsymbol{x}, g)f + H(\boldsymbol{x}, g)$$

where H and G satisfy the usual Carathéodory condition, G is uniformly bounded and $|H(\boldsymbol{x}, y)| \leq c_1|y| + c_2$, $c_j \in \mathbb{R}$, $j = 1, 2$.

The initial conditions (5.4.12) and the remaining boundary value conditions (5.4.7), (5.4.8) (5.4.9), and (5.4.10) remain unchanged. Since we need the uniqueness of the solution to the appropriate penalized version of the problem, we assume the coefficient of friction to be independent of the solution throughout this subsection. The heat diffusivity is assumed to be elliptic and bounded,

$$
\begin{aligned}
c_0 \xi_i \xi_i \leq c_{ij} \xi_i \xi_j \leq C_0 \xi_i \xi_i, \ \boldsymbol{\xi} \in \mathbb{R}^N, \ \boldsymbol{x} \in \Omega, \\
\text{with } 0 < c_0 \leq C_0 < +\infty.
\end{aligned} \tag{5.4.21}
$$

A weak solution of this linearized thermo-viscoelastic contact problem is a pair of functions $(\boldsymbol{u}, \Theta)$ with $\dot{\boldsymbol{u}} \in \mathscr{C}$, $\ddot{\boldsymbol{u}} \in \mathfrak{U}^*$ with

$$
\mathfrak{U} := \boldsymbol{H}^{1/2,\,1}(Q_T) \equiv \left\{ v \in \boldsymbol{H}^{1/2,\,1}(Q_T);\, \boldsymbol{v} = 0 \text{ on } S_{UT} \right\}, \tag{5.4.22}
$$

$\Theta \in \mathfrak{V}$, $\dot{\Theta} \in \mathfrak{V}^*$, which satisfy the initial conditions $\boldsymbol{u}(0, \cdot) = \boldsymbol{u}_0$, $\dot{\boldsymbol{u}}(0, \cdot) = \boldsymbol{u}_1$, and $\Theta(0, \cdot) = \Theta_0$ such that

$$
\begin{aligned}
&\left\langle \ddot{\boldsymbol{u}}, \boldsymbol{v} - \dot{\boldsymbol{u}} \right\rangle_{Q_T} + A^{(0)}(\boldsymbol{u}, \boldsymbol{v} - \dot{\boldsymbol{u}}) + A^{(1)}(\dot{\boldsymbol{u}}, \boldsymbol{v} - \dot{\boldsymbol{u}}) \\
&\quad - \langle b_{ij}\Theta, e_{ij}(\boldsymbol{v} - \dot{\boldsymbol{u}}) \rangle_{Q_T} + \langle \mathfrak{F}|\sigma_n(\boldsymbol{u}, \Theta)|, |\boldsymbol{v_t}| - |\dot{\boldsymbol{u}}_{\boldsymbol{t}}| \rangle_{S_{CT}} \\
&\geq \mathscr{L}(\boldsymbol{v} - \dot{\boldsymbol{u}})
\end{aligned} \tag{5.4.23}
$$

holds for all $\boldsymbol{v} \in \mathscr{C}$ and

$$
\begin{aligned}
&\left\langle \dot{\Theta}, \varphi \right\rangle_{Q_T} + \langle c_{ij}\partial_j \Theta, \partial_i \varphi \rangle_{Q_T} + \langle d_{ij} \partial_j \dot{u}_i, \varphi \rangle_{Q_T} \\
&\quad + \langle K(\Theta - \Upsilon), \varphi \rangle_{S_T} = \langle \mathscr{J}(\mathfrak{F}|\sigma_n(\boldsymbol{u}, \Theta)|, |\dot{\boldsymbol{u}}_{\boldsymbol{t}}|), \varphi \rangle_{S_{CT}}
\end{aligned} \tag{5.4.24}
$$

is satisfied for all $\varphi \in \mathfrak{V}$. The bilinear forms $A^{(0)}$ and $A^{(1)}$ are defined in (5.1.11). Of course, due to the extensive linearizations carried out here the presented model is no longer thermodynamically consistent.

5.4.1.1 Approximate problem

In order to solve this problem, we first penalize the Signorini contact condition in (5.4.10) by the condition $\sigma_n(\boldsymbol{u}) = -\frac{1}{\varepsilon}[\dot{u}_n]_+$ like in the preceding sections. The resulting penalized contact problem for a given temperature $\Theta^{(0)}$ is formulated by the following variational inequality:

Find a function $\boldsymbol{u}$ *with* $\boldsymbol{u}(0,\cdot) = \boldsymbol{u}_0$, $\dot{\boldsymbol{u}}(0,\cdot) = \boldsymbol{u}_1$, $\dot{\boldsymbol{u}} \in \dot{\boldsymbol{U}} + \mathfrak{U}$, *and* $\ddot{\boldsymbol{u}} \in \mathfrak{U}^*$, *such that for all* $\boldsymbol{v} \in \boldsymbol{U} + \mathfrak{U}$ *there holds*

$$\begin{aligned}
&\langle \ddot{\boldsymbol{u}}, \boldsymbol{v} - \dot{\boldsymbol{u}} \rangle_{Q_T} + A^{(0)}(\boldsymbol{u}, \boldsymbol{v} - \dot{\boldsymbol{u}}) + A^{(1)}(\dot{\boldsymbol{u}}, \boldsymbol{v} - \dot{\boldsymbol{u}}) \\
&\quad - \left\langle b_{ij}\Theta^{(0)}, e_{ij}(\boldsymbol{v} - \dot{\boldsymbol{u}}) \right\rangle_{Q_T} + \left\langle \tfrac{1}{\varepsilon}[\dot{u}_n]_+ , v_n - \dot{u}_n \right\rangle_{S_{CT}} \\
&\quad + \left\langle \mathfrak{F}\tfrac{1}{\varepsilon}[\dot{u}_n]_+, |\boldsymbol{v}_t| - |\dot{\boldsymbol{u}}_t| \right\rangle_{S_{CT}} \ge \mathscr{L}(\boldsymbol{v} - \dot{\boldsymbol{u}}).
\end{aligned} \tag{5.4.25}$$

In the heat equation the only change concerns the heat generated by friction. The penalized heat equation is:

Find a function $\Theta \in \mathfrak{V}$ *with* $\dot{\Theta} \in \mathfrak{V}^*$, $\Theta(0,\cdot) = \Theta_0$ *such that for all* $\varphi \in \mathfrak{V}$

$$\begin{aligned}
&\langle \dot{\Theta}, \varphi \rangle_{Q_T} + \langle c_{ij}\partial_j\Theta, \partial_i\varphi \rangle_{Q_T} + \langle d_{ij}\partial_j\dot{u}_i, \varphi \rangle_{Q_T} \\
&\quad + \langle K(\Theta - \Upsilon), \varphi \rangle_{S_T} = \left\langle \mathscr{I}\left(\mathfrak{F}\tfrac{1}{\varepsilon}[\dot{u}_n]_+, |\dot{\boldsymbol{u}}_t|\right), \varphi \right\rangle_{S_{CT}}.
\end{aligned} \tag{5.4.26}$$

A solution of the coupled problem consisting of the contact problem (5.4.25) and the modified heat transfer problem (5.4.26) will be constructed as a fixed point: Let $\Theta^{(0)} \in \mathfrak{V}$ be a fixed temperature field. Then the problem (5.4.25) with this $\Theta^{(0)}$ defines a displacement field $\boldsymbol{u} = \boldsymbol{u}(\Theta^{(0)}) \in \mathfrak{U}$. We denote the solution operator $\Theta^{(0)} \mapsto \boldsymbol{u}$ by Ξ_1. An other operator Ξ_2 is defined by the heat transfer problem: if the displacement field $\boldsymbol{u}$ is given, we obtain a function $\Theta = \Xi_2(\boldsymbol{u})$. If the problems (5.4.25) and (5.4.26) are uniquely solvable, i.e. Ξ_1 and Ξ_2 are single-valued, then by this procedure a single-valued operator

$$\Phi = \Xi_2 \circ \Xi_1 \tag{5.4.27}$$

is defined, and a solution of the approximate thermo-viscoelastic contact problem is given by a fixed point of this operator and the corresponding solution $\boldsymbol{u}$ of the contact problem. In order to prove the existence of such a fixed point, we apply Theorem 2.1.21 (Schauder's fixed-point theorem).

We start with the existence result for the approximate contact problem. Often in the sequel the dependence of constants arising in estimates on the data of the problems is important. If a constant c depends on the geometry of the domain and the coefficients $a^{(\iota)}_{ijk\ell}$, b_{ij}, d_{ij}, $\mathfrak{F}$ at most, then this dependence is indicated by $c = c(\mathscr{G})$. If a constant depends on the given data $\boldsymbol{u}_0$, $\boldsymbol{u}_1$, Θ_0, $\mathscr{L}$, Υ, then we write $c = c(\mathscr{I})$. The dependence on other quantities, in particular the penalty parameter ε, will be separately indicated.

5.4.1 Proposition. *Let Assumption 5.1.1 be satisfied and $b_{ij} \in L_\infty(\Omega)$ for $i, j = 1, \ldots, N$. Let $\Theta^{(0)} \in L_2(Q_T)$ be a fixed temperature field. Then the approximate contact problem* (5.4.25) *has a unique solution which satisfies the* a priori *estimates*

$$\begin{aligned} \|\dot{\boldsymbol{u}}\|^2_{L_\infty(I_T;\boldsymbol{L}_2(\Omega))} + \|\dot{\boldsymbol{u}}\|^2_{L_2(I_T;\boldsymbol{H}^1(\Omega))} &\le c_1(\mathscr{G})\|\Theta^{(0)}\|^2_{L_2(Q_T)} \\ &+ c_2(\mathscr{G})\left(\|\mathscr{L}\|^2_{L_2(I_T;\boldsymbol{H}^1(\Omega))^*} + \|\boldsymbol{u}_0\|^2_{\boldsymbol{H}^1(\Omega)} + \|\boldsymbol{u}_1\|^2_{\boldsymbol{L}_2(\Omega)}\right) \end{aligned} \tag{5.4.28}$$

and

$$\|\dot{\boldsymbol{u}}\|^2_{\boldsymbol{H}^{1/2,1}(Q_T)} \le c_3(\mathscr{I})\|\Theta^{(0)}\|_{L_2(Q_T)} + c_4(\mathscr{I}). \tag{5.4.29}$$

The solution $\boldsymbol{u}$ depends continuously on the temperature: if $\boldsymbol{u}^{(1)}$ and $\boldsymbol{u}^{(2)}$ are two solutions with corresponding temperature fields $\Theta_1^{(0)}$ and $\Theta_2^{(0)}$, then

$$\left\|\dot{\boldsymbol{u}}^{(1)} - \dot{\boldsymbol{u}}^{(2)}\right\|_{\boldsymbol{H}^{1/2,1}(Q_T)} \le c_4(\varepsilon,\mathscr{G})\left\|\Theta_1^{(0)} - \Theta_2^{(0)}\right\|_{L_2(Q_T)}. \tag{5.4.30}$$

Proof. The existence of the solution and the *a priori* estimate (5.4.29) is a result just proved as a combination of Theorem 5.1.2 and Remark 1 to Theorem 5.1.8. There the heat deformation $\boldsymbol{v} \mapsto \langle b_{ij}\Theta^{(0)}, e_{ij}(\boldsymbol{v})\rangle_{Q_T}$ is interpreted as part of the linear functional; its $L_2(I_T; H^1(\Omega)^*)$ norm is obviously bounded by $c(\mathscr{G})\|\Theta^{(0)}\|_{L_2(Q_T)}$. In order to prove the continuous dependence (5.4.30), we assume that $\boldsymbol{u}^{(1)}$ and $\boldsymbol{u}^{(2)}$ are two solutions with corresponding temperature fields $\Theta_1^{(0)}$ and $\Theta_2^{(0)}$. We fix a time $t_0 > 0$ and define the test functions $\boldsymbol{v}^{(r)}$, $r = 1, 2$, by $\boldsymbol{v}^{(1)} = \dot{\boldsymbol{u}}^{(2)}$ and $\boldsymbol{v}^{(2)} = \dot{\boldsymbol{u}}^{(1)}$ for $t \in I_{t_0} \equiv (0, t_0)$ and $\boldsymbol{v}^{(r)} = \dot{\boldsymbol{u}}^{(r)}$ for $t > t_0$. Employing the test function $\boldsymbol{v}^{(r)}$ in the inequality with solution $\boldsymbol{u}^{(r)}$ and temperature $\Theta_r^{(0)}$ and adding the results for $r = 1, 2$, we obtain

$$\begin{aligned} &\left\|\left(\dot{\boldsymbol{u}}^{(1)} - \dot{\boldsymbol{u}}^{(2)}\right)(t_0)\right\|^2_{\boldsymbol{L}_2(\Omega)} + \left\|\nabla\left(\dot{\boldsymbol{u}}^{(1)} - \dot{\boldsymbol{u}}^{(2)}\right)\right\|^2_{L_2(I_{t_0}\times\Omega)} \\ &\quad\le c(\mathscr{G})\left(\left\|\Theta_1^{(0)} - \Theta_2^{(0)}\right\|_{L_2(I_{t_0}\times\Omega)}\left\|\dot{\boldsymbol{u}}^{(1)} - \dot{\boldsymbol{u}}^{(2)}\right\|_{L_2((0,t_0);\boldsymbol{H}^1(\Omega))}\right. \\ &\qquad\left. + \|\mathfrak{F}\|_{L_\infty(\Gamma_C)}\frac{1}{\varepsilon}\left\|\dot{\boldsymbol{u}}^{(1)} - \dot{\boldsymbol{u}}^{(2)}\right\|^2_{\boldsymbol{L}_2(I_{t_0}\times\Gamma_C)}\right). \end{aligned}$$

In order to estimate the $L_2(I_{t_0} \times \Gamma_C)$ norm we use the trace theorem $H^{2/3}(\Omega) \to L_2(\Gamma_C)$. This, the interpolation inequality

$$\begin{aligned} \|v\|^2_{H^{2/3}(\Omega)} &\le c_1(\mathscr{G})\|v\|^{2/3}_{L_2(\Omega)}\|v\|^{4/3}_{H^1(\Omega)} \\ &\le c_2(\eta,\mathscr{G})\|v\|^2_{L_2(\Omega)} + \eta\|v\|^2_{H^1(\Omega)} \end{aligned}$$

with sufficiently small $\eta > 0$ and an obvious simplification yield

$$
\begin{aligned}
&\left\| \left(\dot{\boldsymbol{u}}^{(1)} - \dot{\boldsymbol{u}}^{(2)} \right)(t_0) \right\|^2_{\boldsymbol{L}_2(\Omega)} + \left\| \dot{\boldsymbol{u}}^{(1)} - \dot{\boldsymbol{u}}^{(2)} \right\|^2_{L_2(I_{t_0};\boldsymbol{H}^1(\Omega))} \\
&\qquad \leq c_1(\mathscr{G}) \left\| \Theta_1^{(0)} - \Theta_2^{(0)} \right\|^2_{L_2(I_{t_0} \times \Omega)} \\
&\qquad\quad + c_2(\varepsilon, \mathscr{G}) \int_0^{t_0} \left\| \left(\dot{\boldsymbol{u}}^{(1)} - \dot{\boldsymbol{u}}^{(2)} \right)(t) \right\|^2_{\boldsymbol{L}_2(\Omega)} dt.
\end{aligned}
\tag{5.4.31}
$$

Application of the Gronwall lemma leads to

$$
\left\| \left(\dot{\boldsymbol{u}}^{(1)} - \dot{\boldsymbol{u}}^{(2)} \right)(t_0) \right\|^2_{\boldsymbol{L}_2(\Omega)} \leq c_1 \left\| \Theta_1^{(0)} - \Theta_2^{(0)} \right\|^2_{L_2((0,t_0)\times\Omega)} e^{c_2(\varepsilon,\mathscr{G})t_0}.
$$

Combined with estimate (5.4.31) this yields

$$
\left\| \dot{\boldsymbol{u}}^{(1)} - \dot{\boldsymbol{u}}^{(2)} \right\|^2_{L_2(I_T;\boldsymbol{H}^1(\Omega))} \leq c(\varepsilon, \mathscr{G}) \left\| \Theta_1^{(0)} - \Theta_2^{(0)} \right\|^2_{L_2(Q_T)}. \tag{5.4.32}
$$

This inequality is completed by an appropriate dual estimate: As usual we choose a test function $\boldsymbol{w} \in L_2(I_T; \mathring{\boldsymbol{H}}^1(\Omega))$ with vanishing boundary values. Using the test functions $\boldsymbol{v} = \boldsymbol{w} + \dot{\boldsymbol{u}}^{(1)}$ in (5.4.25) with solution $\boldsymbol{u}^{(1)}$, $\boldsymbol{v} = -\boldsymbol{w} + \dot{\boldsymbol{u}}^{(2)}$ in (5.4.25) with solution $\boldsymbol{u}^{(2)}$ and adding the results yields

$$
\begin{aligned}
\left\| \ddot{\boldsymbol{u}}^{(1)} - \ddot{\boldsymbol{u}}^{(2)} \right\|_{L_2(I_T;\boldsymbol{H}^{-1}(\Omega))} &\leq c_1(\mathscr{G}) \left\| \dot{\boldsymbol{u}}^{(1)} - \dot{\boldsymbol{u}}^{(2)} \right\|_{L_2(I_T;\boldsymbol{H}^1(\Omega))} \\
&\quad + c_2(\mathscr{G}) \left\| \Theta_1^{(0)} - \Theta_2^{(0)} \right\|_{L_2(Q_T)},
\end{aligned}
$$

because the boundary terms vanish. We use again the standard interpolation and get

$$
\|\boldsymbol{v}\|^2_{H^{1/2}(I_T;\boldsymbol{L}_2(\Omega))} \leq c(\mathscr{G}) \|\boldsymbol{v}\|_{L_2(I_T;\boldsymbol{H}^1(\Omega))} \|\boldsymbol{v}\|_{H^1(I_T;\boldsymbol{H}^{-1}(\Omega))}.
$$

Hence from (5.4.32) we obtain also

$$
\left\| \dot{\boldsymbol{u}}^{(1)} - \dot{\boldsymbol{u}}^{(2)} \right\|^2_{H^{1/2}(I_T;\boldsymbol{L}_2(\Omega))} \leq c(\varepsilon, \mathscr{G}) \left\| \Theta_1^{(0)} - \Theta_2^{(0)} \right\|^2_{L_2(Q_T)}
$$

and the continuity relation (5.4.30) is proved. The uniqueness of the solution follows directly from this estimate with $\Theta_1^{(0)} = \Theta_2^{(0)}$.

□

For the heat equation the following existence and uniqueness result is valid:

5.4.2 Proposition. *Let* $\dot{\boldsymbol{u}} \in \boldsymbol{H}^{1/2,1}(Q_T)$ *be a fixed displacement velocity,* Ω *be a bounded Lipschitz domain, the coefficients* $c_{ij} \in L_\infty(\Omega)$, $i, j = 1, \dots, N$, *be symmetric, bounded and elliptic. Let the coefficients of thermal expansion* $d_{ij} \in L_\infty(\Omega)$ *be bounded and symmetric,* $\Upsilon \in L_2(I_T; H^{-1/2}(\Gamma))$ *and* $\Theta_0 \in L_2(\Omega)$. *Let, moreover, the coefficient of heat exchange* K *and the coefficient of friction be bounded and non-negative. Then the approximate heat transfer problem* (5.4.26) *has a unique solution which satisfies the* a priori *estimate*

$$\begin{aligned}
&\|\Theta\|^2_{L_\infty(I_T;L_2(\Omega))} + \|\Theta\|^2_{H^{1/2,\,1}(Q_T)} \le c_1(\mathscr{G})\mathfrak{J} + c_2(\mathscr{G}) \quad \text{with} \\
&\quad \mathfrak{J} \equiv \|\dot{\boldsymbol{u}}\|^2_{L_2(I_T;\boldsymbol{H}^1(\Omega))} + \left\| \mathscr{I}\left(\mathfrak{F}_{\frac{1}{\varepsilon}}[\dot{u}_n]_+\,, |\dot{\boldsymbol{u}}_{\boldsymbol{t}}|\right)\right\|^2_{L_2(S_{CT})} \\
&\qquad + \|\Theta_0\|^2_{L_2(\Omega)} + \|\Upsilon\|^2_{L_2(I_T;H^{-1/2}(\Gamma))}
\end{aligned} \tag{5.4.33}$$

with constants c_1, c_2 *independent of* Θ_0, Υ *and* ε. *The solution* Θ *depends continuously on the displacement velocity: if* $\dot{\boldsymbol{u}}^{(k)}$ *converges to* $\dot{\boldsymbol{u}}$ *in* $\boldsymbol{H}^{1/2,\,1}(Q_T)$ *and* $\Theta^{(k)}$, Θ *are the corresponding solutions of the heat transfer problem, then* $\Theta^{(k)}$ *converges to* Θ *in* $H^{1/2,\,1}(Q_T)$.

Proof. The existence of solutions is proved by the usual Galerkin method. Let $\{v_i\}$ be a basis of $H^1(\Omega)$ being $L_2(\Omega)$-orthogonal and

$$\mathfrak{V}_m \equiv \left\{ (t, \boldsymbol{x}) \mapsto \sum_{i=1}^{m} q_i(t) v_i(\boldsymbol{x})\,;\; q_i \in L_\infty(I_T),\; i = 1, \dots m,\; (t, \boldsymbol{x}) \in Q_T \right\}.$$

Then $\bigcup_{m=1}^{+\infty} \mathfrak{V}_m$ is dense in $L_2(I_T; H^1(\Omega))$. A Galerkin solution Θ_m of the approximate heat transfer problem is a function from $\mathfrak{V}_m$ which satisfies the Galerkin equations

$$\begin{aligned}
&\langle \dot{\Theta}_m, \varphi \rangle_\Omega + \langle c_{ij}\partial_j\Theta_m, \partial_i\varphi\rangle_\Omega + \langle d_{ij}\partial_j\dot{u}_i, \varphi\rangle_\Omega \\
&\quad + \langle K(\Theta_m - \Upsilon), \varphi\rangle_\Gamma = \left\langle \mathscr{I}\left(\mathfrak{F}_{\frac{1}{\varepsilon}}[\dot{u}_n]_+, |\dot{\boldsymbol{u}}_{\boldsymbol{t}}|\right), \varphi \right\rangle_{\Gamma_C}.
\end{aligned} \tag{5.4.34}$$

As usual, such a system of ordinary differential equations is solved by means of the standard results concerning its Carathéodory solutions. If we integrate the Galerkin equation (5.4.34) in time over I_T with the test function $\varphi(t) = \Theta_m(t)$ for $t \le t_0$ and $\varphi(t) = 0$ for $t > t_0$ and employ the ellipticity of the tensor c_{ij} as well as the trace theorem for $L_2(I_T; H^1(\Omega))$, we obtain the estimate

$$
\begin{aligned}
&\|\Theta_m(t_0)\|^2_{L_2(\Omega)} + \|\nabla\Theta_m\|^2_{L_2(Q_{t_0})} \\
&\quad \le c(\mathscr{G})\Big(\|\dot{\boldsymbol{u}}\|^2_{L_2(I_{t_0};\boldsymbol{H}^1(\Omega))} + \|\Theta_m\|^2_{L_2(Q_{t_0})} + \|\Theta_0\|^2_{L_2(\Omega)} \\
&\qquad + \|\Upsilon\|^2_{L_2(I_T;H^{-1/2}(\Gamma))} + \Big\| \mathscr{J}\big(\mathfrak{F}_{\frac{1}{\varepsilon}}[\dot{u}_n]_+, |\dot{\boldsymbol{u}}_{\boldsymbol{t}}|\big)\Big\|^2_{L_2(S_{CT})}\Big)
\end{aligned}
\tag{5.4.35}
$$

with constant $c(\mathscr{G})$ independent of ε, m, Θ_m and t_0. Application of the Gronwall lemma yields

$$
\begin{aligned}
&\|\Theta_m\|^2_{L_\infty(I_T;L_2(\Omega))} + \|\Theta_m\|^2_{L_2(I_T;H^1(\Omega))} \\
&\quad \le c(\mathscr{G})\Big(\|\dot{\boldsymbol{u}}\|^2_{L_2(I_T;\boldsymbol{H}^1(\Omega))} + \|\Theta_0\|^2_{L_2(\Omega)} \\
&\qquad + \|\Upsilon\|^2_{L_2(I_T;H^{-1/2}(\Gamma))} + \Big\| \mathscr{J}\big(\mathfrak{F}_{\frac{1}{\varepsilon}}[\dot{u}_n]_+\, , |\dot{\boldsymbol{u}}_{\boldsymbol{t}}|\big)\Big\|^2_{L_2(S_{CT})}\Big)
\end{aligned}
\tag{5.4.36}
$$

with $c(\mathscr{G})$ independent of ε, m and Θ_m. From the condition $J(\boldsymbol{x}, y, z) \le c\,(1 + |y| + |z|)$ follows

$$
\Big\| \mathscr{J}\big(\mathfrak{F}_{\frac{1}{\varepsilon}}[\dot{u}_n]_+, |\dot{\boldsymbol{u}}_{\boldsymbol{t}}|\big)\Big\|^2_{L_2(S_{CT})} \le c_1(\varepsilon, \mathscr{G})\|\dot{\boldsymbol{u}}\|^2_{L_2(I_T;\boldsymbol{H}^1(\Omega))} + c_2(\varepsilon, \mathscr{G}),
$$

where the constants $c_1(\varepsilon, \mathscr{G})$ and $c_2(\varepsilon, \mathscr{G})$ are independent of m, Θ_m, and $\dot{\boldsymbol{u}}$. Consequently, there exists a sequence $\{m_k\}$ converging to $+\infty$ and a corresponding sequence Θ_{m_k} of solutions to the Galerkin equations converging to a limit function Θ weakly in $L_2\big(I_T, H^1(\Omega)\big)$ such that $\Theta_{m_k}(T, \cdot) \rightharpoonup \Theta(T, \cdot)$ in $H^1(\Omega)$. Passing to the limit $m_k \to +\infty$ we prove in a standard way that Θ is a solution of (5.4.26). This solution also satisfies the *a priori* estimate (5.4.36). By taking a test function $\varphi \in L_2\big(I_T;\, \mathring{H}^1(\Omega)\big)$ in the variational equation the dual estimate

$$
\begin{aligned}
\|\dot{\Theta}\|_{L_2(I_T;H^{-1}(\Omega))} \le\; & c_1(\varepsilon, \mathscr{G})\|\dot{\boldsymbol{u}}\|_{L_2(I_T;\boldsymbol{H}^1(\Omega))} \\
&+ c_2(\varepsilon, \mathscr{G})\|\Theta\|_{L_2(I_T;H^1(\Omega))}
\end{aligned}
$$

is verified. Interpolation with $\Theta \in L_2\big(I_T; H^1(\Omega)\big)$ completes the proof of the *a priori* estimate (5.4.33).

In order to verify the uniqueness of the solution, we assume that $\Theta^{(1)}$ and $\Theta^{(2)}$ are two solutions. Employing the test function $\boldsymbol{v}$ defined by $\boldsymbol{v} = \Theta^{(1)} - \Theta^{(2)}$ for $t < t_0$ and $\boldsymbol{v} = 0$ for $t \ge t_0$ into the variational equation with solution $\Theta^{(1)}$, the test function $-v$ into the equation with solution $\Theta^{(2)}$ and adding the results yields $\big\| \big(\Theta^{(2)} - \Theta^{(1)}\big)\,(t_0)\big\|_{\boldsymbol{L}_2(\Omega)} = 0$ for all $t_0 \in I_T$.

It remains to show the continuous dependence of the solution on the displacement velocities. Let $\dot{\boldsymbol{u}}^{(k)}$ be a sequence converging to $\dot{\boldsymbol{u}}$ in $\boldsymbol{H}^{1/2,\,1}(Q_T)$ and let $\Theta^{(k)}$, Θ be the solutions of the heat transfer problem with velocities $\dot{\boldsymbol{u}}^{(k)}$, $\dot{\boldsymbol{u}}$. Then the inequality

$$\begin{aligned}
\left\|\Theta^{(k)}-\Theta\right\|^2_{L_2(I_T;H^1(\Omega))} &\leq c_1(\varepsilon,\mathscr{G})\left\|\dot{\boldsymbol{u}}^{(k)}-\dot{\boldsymbol{u}}\right\|^2_{L_2(I_T;\boldsymbol{H}^1(\Omega))}\\
&+c_2(\varepsilon,\mathscr{G})\left\|\varpi\left(\dot{\boldsymbol{u}}^{(k)}\right)-\varpi\left(\dot{\boldsymbol{u}}\right)\right\|_{L_{2-2/(N+2)}(S_{CT})}\cdot\\
&\cdot\left\|\Theta^{(k)}-\Theta\right\|_{L_{2+2/N}(S_{CT})}
\end{aligned}\tag{5.4.37}$$

with the abbreviation $\varpi(\boldsymbol{u})=\mathscr{J}\left(\mathfrak{F}\frac{1}{\varepsilon}[u_n]_+,|\boldsymbol{u_t}|\right)$ can be derived. Due to the continuity condition for the operator $\mathscr{J}$ and to the fact that the trace of a function from $H^{1/2,\,1}(Q_T)$ on S_T is an element of $L_{2+2/N}(S_T)$, the right hand side of this inequality converges to 0, hence $\Theta^{(k)}\to\Theta$ in $L_2\big(I_T;H^1(\Omega)\big)$. Moreover, for $k\in\mathbb{N}$ the dual estimate

$$\begin{aligned}
\left\|\dot{\Theta}^{(k)}-\dot{\Theta}\right\|_{L_2(I_T;H^{-1}(\Omega))} &\leq c_1(\mathscr{G})\left\|\dot{\boldsymbol{u}}^{(k)}-\dot{\boldsymbol{u}}\right\|_{L_2(I_T;\boldsymbol{H}^1(\Omega))}\\
&+c_2(\mathscr{G})\left\|\Theta^{(k)}-\Theta\right\|_{L_2(I_T;H^1(\Omega))}
\end{aligned}$$

is valid, hence there holds also $\dot{\Theta}^{(k)}\to\dot{\Theta}$ in the space $L_2\big(I_T;H^{-1}(\Omega)\big)$. The application of interpolation theorems as described above yields that $\Theta^{(k)}$ converges to Θ in $H^{1/2,\,1}(Q_T)$. □

With the help of the two preceding propositions the existence of solutions of the approximate coupled thermo-viscoelastic contact problem can be ensured.

5.4.3 Proposition. *Let the assumptions of Propositions 5.4.1 and 5.4.2 be valid. Then there exists a solution of problem* (5.4.25), (5.4.26). *This solution satisfies the* a priori *estimate*

$$\|\dot{\boldsymbol{u}}\|_{H^{1/2,\,1}(Q_T)}+\|\Theta\|_{H^{1/2,\,1}(Q_T)}\leq c(\varepsilon,\mathscr{I}).\tag{5.4.38}$$

Proof. It is sufficient to show that the mapping Φ defined above satisfies the assumptions of Theorem 2.1.21 (Schauder's fixed-point theorem). We prove first that Φ maps an appropriate convex bounded set $B_R^{(t_0)}\equiv\left\{v\in L_2(Q_{t_0});\|v\|_{L_2(Q_{t_0})}\leq R\right\}$ with $R>0$ into itself. Here we take a time interval $I_{t_0}\equiv[0,t_0]$ for an appropriate $t_0\in I_T$ chosen below and $Q_{t_0}\equiv I_{t_0}\times\Omega$. Let $\Theta^{(0)}$ be an initial temperature field, $\boldsymbol{u}$ be the solution of problem (5.4.25) with $\Theta = \Theta^{(0)}$ and $\Theta \equiv \Phi(\Theta^{(0)})$. Inserting the estimate (5.4.28) for the displacement velocities and time interval I_{t_0} into the *a priori* estimate (5.4.35) of the heat equation yields the inequality

$$\|\Theta(t)\|^2_{L_2(\Omega)} \le c(\mathscr{G})\Big(\|\Theta\|^2_{L_2(Q_t)} + \Big\|\Theta^{(0)}\Big\|^2_{L_2(Q_t)} + \|\mathscr{L}\|^2_{L_2(I_T;\boldsymbol{H}^1(\Omega))^*} \\ + \|\Theta_0\|^2_{L_2(\Omega)} + \|\Upsilon\|^2_{L_2(I_T;H^{-1/2}(\Gamma))} + \|\boldsymbol{u}_0\|^2_{\boldsymbol{H}^1(\Omega)} + \|\boldsymbol{u}_1\|^2_{\boldsymbol{L}_2(\Omega)}\Big)$$

for every $t \in I_{t_0}$. Here it is important that the constants remain bounded for $t_0 \to 0$. This is indeed the case, because the constants in both (5.4.28) and (5.4.35) do not depend on the length of the time interval, if this length is uniformly bounded. Then the integration of the previous equation with respect to time over I_{t_0} yields

$$\|\Theta\|^2_{L_2(Q_{t_0})} \le c(\varepsilon,\mathscr{G})\Big(\|\Theta\|^2_{L_2(Q_{t_0})} + \Big\|\Theta^{(0)}\Big\|^2_{L_2(Q_{t_0})} \\ + \|\mathscr{L}\|^2_{L_2(I_T;\boldsymbol{H}^1(\Omega))^*} + \|\Upsilon\|^2_{L_2(I_T;H^{-1/2}(\Gamma))} \\ + \|\Theta_0\|^2_{L_2(\Omega)} + \|\boldsymbol{u}_0\|^2_{\boldsymbol{H}^1(\Omega)} + \|\boldsymbol{u}_1\|^2_{\boldsymbol{L}_2(\Omega)}\Big) \cdot t_0$$

with c independent of t_0, of the initial data $\boldsymbol{u}_0$, $\boldsymbol{u}_1$, of the boundary data Υ, $\boldsymbol{\tau}$ and of the volume force $\boldsymbol{f}$. If t_0 is small enough, $t_0 < T_0 \equiv 1/(2c(\varepsilon,\mathscr{G}))$, then there exists a constant $R > 0$ such that for $\|\Theta^{(0)}\|_{L_2(Q_{t_0})} \le R$ there follows $\|\Theta\|_{L_2(Q_{t_0})} \le R$. The ball $B_R^{(t_0)}$ is a bounded, convex, closed subset of $L_2(Q_{t_0})$, and the mapping $\Phi : L_2(Q_{t_0}) \ni \Theta^{(0)} \mapsto \Theta \in L_2(Q_{t_0})$ is continuous. Moreover, from Propositions 5.4.1 and 5.4.2 the estimate

$$\|\Theta\|_{H^{1/2,1}(Q_{t_0})} \le c_1(\varepsilon,\mathscr{I})\|\Theta^{(0)}\|_{L_2(Q_{t_0})} + c_2(\varepsilon,\mathscr{I})$$

follows; this proves the complete continuity of Φ. Hence at least for a small time interval with length $t_0 < T_0$, the approximate thermo-viscoelastic contact problem (5.4.25), (5.4.26) has a solution. However, the supremum T_0 of admissible diameters of this time interval is independent of the initial data; hence this consideration can be repeated iteratively for all intervals $[\ell t_0, (\ell+1)t_0]$, $\ell = 1, 2, 3, \ldots$, with the new initial data $\boldsymbol{u}(\ell t_0)$, $\dot{\boldsymbol{u}}(\ell t_0)$ and $\Theta(\ell t_0)$ until the existence of the solution is proved for the whole time interval $[0, T]$. The *a priori* estimate (5.4.38) is a consequence of the uniform boundedness of Θ in $L_2(Q_T)$ and of the inequalities (5.4.29) and (5.4.33). □

5.4.1.2 Existence of the solution of the thermo-viscoelastic contact problem

The existence of solutions to the original contact problem is proved by passing to the limit $\varepsilon \to 0$ of the penalty parameter. Therefore it is necessary to have *a priori* estimates which are valid uniformly with respect to the penalty parameter. For the viscoelastic contact problem without consideration of the temperature field such an estimate has been derived in Theorem 5.1.7 under the validity of Assumption 5.1.6. The only difference of our problem (5.4.25) is the appearence of the additional term $\boldsymbol{v} \mapsto \langle b_{ij}\Theta, \partial_j v_i\rangle_{Q_T}$ in (5.4.25) as an additional volume force. The following theorem gives the corresponding regularity result for such a volume force.

5.4.4 Lemma. *In addition to the conditions of Theorem 5.1.7 let* $b_{ij} \in C^{\beta'}(\Omega)$ *with a* $\beta' > 1/2$ *and* $\Theta = \Theta^{(0)} \in L_2\big(I_T; H^{\beta'}(\Omega)\big)$ *be satisfied. Then the solution* $\boldsymbol{u}$ *of problem* (5.4.25) *satisfies the regularity estimate*

$$\begin{aligned} &\|\dot{\boldsymbol{u}}\|_{\boldsymbol{H}^{1/2,1}(Q_T)} + \|\dot{\boldsymbol{u}}\|_{\boldsymbol{H}^{1/2,1}(S_{\mathfrak{F}T})} + \left\| \tfrac{1}{\varepsilon}[\dot{u}_n]_+ \right\|_{L_2(S_{\mathfrak{F}T})} \\ &\qquad \le c_1\|\Theta\|_{L_2(I_T;H^{\alpha}(\Omega))} + c_2 \end{aligned} \tag{5.4.39}$$

with constants c_1, c_2 *independent of* ε, $\boldsymbol{u}$ *and* Θ.

Proof. It is sufficient to consider the terms that arise from the additional functional $\boldsymbol{v} \mapsto \langle b_{ij}\Theta, \partial_j v_i\rangle_{Q_T}$ in the proof of Theorem 5.1.7 by the shift technique. After the localization technique and the local rectification of the boundary we have terms of the type $\langle \Psi, \partial_j \boldsymbol{v}\rangle_Q$, where Ψ comes from $b_{ij}\Theta$ and $Q = \mathbb{R} \times \mathbb{R}^{N-1} \times \mathbb{R}^+$. We have to estimate terms of three different types.

From the shift technique with respect to the tangential space variables there arises the term

$$\int_{\mathbb{R}^{N-1}} \int_Q |\boldsymbol{h}|^{-N} \Delta_{\boldsymbol{h}}\Psi \Delta_{\boldsymbol{h}}\big(\partial_j \breve{u}_\ell\big)\, d\boldsymbol{x}\, dt\, d\boldsymbol{h}.$$

The estimate of this term is rather easy. Employing the Fourier transform with respect to the tangential space variables and the usual renormalization procedure we obtain

$$\begin{aligned} &\operatorname{Re}\left(\int_{\mathbb{R}} \int_O \mathscr{F}(\Psi; t, \boldsymbol{\xi}, x_N) \overline{i\xi_j \mathscr{F}(\partial_j \breve{u}_\ell; t, \boldsymbol{\xi}, x_N)} c_{N-1}\big(\tfrac{1}{2}\big) |\boldsymbol{\xi}|\, dx_N\, d\boldsymbol{\xi}\, dt \right) \\ &\qquad \le c\|\Psi\|_{L_2(\mathbb{R};H^{1/2}(\mathbb{R}^{N-1};L_2(\mathbb{R}_+)))} \|\breve{u}_\ell\|_{L_2(\mathbb{R};H^{3/2}(O))} \end{aligned}$$

with $O = \mathbb{R}^{N-1} \times \mathbb{R}_+$. Here the last factor can be bounded by the term $\mathscr{A}_1(\breve{\boldsymbol{u}})$ defined in (5.1.52).

In the shift technique with respect to the time variable we have to distinguish between the normal and tangential derivatives of $\breve{\boldsymbol{u}}_\ell$. For the tangential derivative, that is, $j \in \{1, \ldots, N-1\}$, the estimate is also rather easy:

$$\int_{\mathbb{R}} \int_Q |q|^{-3/2} \Delta_q \Psi \Delta_q (\partial_j \breve{u}_\ell) \, d\boldsymbol{x} \, dt \, d\boldsymbol{h}$$

$$= \mathrm{Re}\Big(\int_{\mathbb{R}} \int_O \mathscr{F}(\Psi; \vartheta, \boldsymbol{\xi}, x_N) \overline{i\xi_j \mathscr{F}(\breve{u}_\ell; \vartheta, \boldsymbol{\xi}, x_N)} \cdot$$

$$\cdot \, c_1\big(\tfrac{1}{4}\big) |\vartheta|^{1/2} \Big) \, dx_N \, d\boldsymbol{\xi} \, d\vartheta$$

$$\leq c \|\Psi\|_{L_2(\mathbb{R}; H^{1/2}(\mathbb{R}^{N-1}; L_2(\mathbb{R}_+)))} \|\breve{u}_\ell\|_{H^{1/2}(I_T; H^{1/2}(\mathbb{R}^{N-1}; L_2(\mathbb{R}_+)))}$$

and the last term here can be further estimated by interpolation

$$\|\breve{u}_\ell\|_{H^{1/2}(I_T; H^{1/2}(\mathbb{R}^{N-1}; L_2(\mathbb{R}_+)))}$$

$$\leq c \Big(\|\breve{u}_\ell\|_{H^{3/4}(I_T; L_2(O))} + \|\breve{u}_\ell\|_{L_2(I_T; H^{3/2}(\mathbb{R}^{N-1}; L_2(\mathbb{R}_+)))} \Big).$$

The estimate of the first term on the right hand side here is explained below.

The term associated to the time shifts and the normal derivative $\partial_N \boldsymbol{u}$ is

$$\int_{\mathbb{R}} \int_Q |q|^{-3/2} \Delta_q \Psi \, \Delta_q (\partial_N \breve{u}_\ell) \, d\boldsymbol{x} \, dt \, dq.$$

In order to get an estimate of the desired type we wish to provide both Ψ and $\partial_N \breve{u}_\ell$ with half of the space derivative, as we did before. In order to use the Fourier transform it is necessary to extend the functions onto the whole space $\mathbb{R}^N$ instead of O. The difficulty arises from the fact that $H^{1/2}(O)$ is exactly the border case for which such an extension is not clear. Therefore we first employ an integration by parts,

$$\int_Q \Psi \, \partial_N \breve{u}_\ell \, d\boldsymbol{x} \, dt = \int_S \Psi \, \breve{u}_\ell \, ds_{\boldsymbol{x}} \, dt - \int_Q \partial_N \Psi \, \breve{u}_\ell \, d\boldsymbol{x} \, dt,$$

where $S = \mathbb{R} \times \mathbb{R}^{N-1} \times \{0\}$. The boundary term then leads to

$$\int_{\mathbb{R}} \int_S |q|^{-3/2} \Delta_q \Psi \, \Delta_q \breve{u}_\ell \, ds_{\boldsymbol{x}} \, dt \, dq$$

$$= \mathrm{Re} \Big(\int_{\mathbb{R}^N} \mathscr{F}(\Psi; \vartheta, \boldsymbol{\xi}) \overline{\mathscr{F}(\breve{u}_\ell; \vartheta, \boldsymbol{\xi})} \, c_1\big(\tfrac{1}{4}\big) |\vartheta|^{1/2} \, d\boldsymbol{\xi} \, d\vartheta \Big)$$

$$\leq c \|\Psi\|_{L_2(S)} \|\breve{u}_\ell\|_{H^{1/2,0}(S)}.$$

The remaining volume integral can be estimated by

$$\int_{\mathbb{R}}\int_{Q} |q|^{-3/2} \Delta_q\left(\partial_N \Psi\right) \Delta_q \breve{u}_\ell \, d\boldsymbol{x}\, dt\, dq$$
$$= \operatorname{Re}\left(\int_{\mathbb{R}}\int_{O} \mathscr{F}(\partial_N \Psi; \vartheta, \boldsymbol{\xi})\overline{\mathscr{F}(\breve{u}_\ell; \vartheta, \boldsymbol{\xi})} c_1\left(\tfrac{1}{4}\right)|\vartheta|^{1/2}\, d\boldsymbol{\xi}\, d\vartheta\right)$$
$$\leq c\|\partial_N \Psi\|_{L_2(\mathbb{R};L_2(\mathbb{R}^{N-1};H^{\alpha-1}(\mathbb{R}_+)))}\|\breve{u}_\ell\|_{H^{1/2}(\mathbb{R};L_2(\mathbb{R}^{N-1};H^{1-\alpha}(\mathbb{R}_+)))}.$$

Due to $1/2 < \alpha < 1$ there is a bounded extension operator from both $H^{\alpha-1}(\mathbb{R}_+)$ to $H^{\alpha-1}(\mathbb{R})$ and from $H^{1-\alpha}(\mathbb{R}_+)$ to $H^{1-\alpha}(\mathbb{R})$. For the extended functions an easy calculation with the Fourier transform for the whole space shows

$$\|\partial_N \Psi\|_{L_2(\mathbb{R}^{N-1};H^{\alpha-1}(\mathbb{R}))} \leq c\|\Psi\|_{L_2(\mathbb{R}^{N-1};H^{\alpha}(\mathbb{R}))}.$$

In order to finish the proof it is necessary to estimate $\|\breve{\boldsymbol{u}}\|_{H^{3/4}(I_T;\boldsymbol{L}_2(O))}$. This is concluded from the variational equation

$$\langle \dot{\breve{\boldsymbol{u}}}, \boldsymbol{v}\rangle_Q + \int_Q \left(a^{(1)}(\breve{\boldsymbol{u}}, \boldsymbol{v}) + a^{(0)}(\mathring{\boldsymbol{u}}, \boldsymbol{v})\right) d\boldsymbol{x}\, dt$$
$$= \mathscr{L}(\boldsymbol{u}, \boldsymbol{v}) + \int_Q \boldsymbol{\Psi} \cdot \nabla \boldsymbol{v}\, d\boldsymbol{x}\, dt \tag{5.4.40}$$

that holds for $\boldsymbol{v} \in L_2\big(\mathbb{R}; \mathring{\boldsymbol{H}}^1(O)\big)$. In the distributional sense, $\dot{\breve{\boldsymbol{u}}}$ is composed of the three linear functionals

$$\boldsymbol{w}^{(1)} : \boldsymbol{v} \mapsto \int_Q \left(a^{(1)}(\breve{\boldsymbol{u}}, \boldsymbol{v}) + a^{(0)}(\mathring{\boldsymbol{u}}, \boldsymbol{v})\right) d\boldsymbol{x}\, dt,$$
$$\boldsymbol{w}^{(2)} : \boldsymbol{v} \mapsto \mathscr{L}(\boldsymbol{u}, \boldsymbol{v}) \quad \text{and}$$
$$\boldsymbol{w}^{(3)} : \boldsymbol{v} \mapsto \int_Q \boldsymbol{\Psi} \cdot \nabla \boldsymbol{v}\, d\boldsymbol{x}\, dt.$$

These functionals are bounded by

$$\|\boldsymbol{w}^{(1)}\|_{H^{1/4}(\mathbb{R};H^{-1}(O))} \leq c_1\|\breve{\boldsymbol{u}}\|_{H^{1/4}(\mathbb{R};H^1(O))} + c_2\|\mathring{\boldsymbol{u}}\|_{H^{1/4}(\mathbb{R};H^1(O))},$$
$$\|\boldsymbol{w}^{(2)}\|_{H^{-1/4,-1/2}(Q)} = \|\mathscr{L}(\boldsymbol{u})\|_{H^{-1/4,-1/2}(Q)} \quad \text{and}$$
$$\|\boldsymbol{w}^{(3)}\|_{L_2(\mathbb{R};H^{-1/2}(O))} \leq c_3\|\Psi\|_{L_2(\mathbb{R};H^{\alpha}(O))}.$$

Since we work with distributions, that means, since the test functions have vanishing boundary data, we can easily extend the functionals on $H^{-\alpha}(O)$ by symmetry onto $H^{-\alpha}(\mathbb{R}^N)$, $\langle \boldsymbol{w}, \boldsymbol{v}\rangle_{\mathbb{R}^N} = \langle \boldsymbol{w}, \boldsymbol{v}_+ - \boldsymbol{v}_-\rangle_O$ with

$\boldsymbol{v}_+(\boldsymbol{x}', x_N) = \boldsymbol{v}(\boldsymbol{x}', x_N)$ and $\boldsymbol{v}_-(\boldsymbol{x}', x_N) = \boldsymbol{v}(\boldsymbol{x}', -x_N)$. Then we can work with the Fourier transform for the extended functions and functionals on the full space, and the desired result is obtained by an easy computation. For the seminorm in $H^{3/4}(\mathbb{R}; L_2(\mathbb{R}^N))$ we have

$$
\begin{aligned}
&\|\breve{\boldsymbol{u}}\|'^2_{H^{3/4}(\mathbb{R};\boldsymbol{L}_2(\mathbb{R}^N))} = c_1\left(\tfrac{3}{4}\right) \int_{\mathbb{R}} \int_{\mathbb{R}^N} |\mathscr{F}(\breve{\boldsymbol{u}}; \vartheta, \boldsymbol{\xi})|^2 |\vartheta|^{3/2} \, d\boldsymbol{\xi} \, d\vartheta \\
&= c_1\left(\tfrac{3}{4}\right) \int_{\mathbb{R}} \int_{\mathbb{R}^N} |\mathscr{F}(\breve{\boldsymbol{u}}; \vartheta, \boldsymbol{\xi})| |\mathscr{F}(\dot{\breve{\boldsymbol{u}}}; \vartheta, \boldsymbol{\xi})| |\vartheta|^{1/2} \, d\boldsymbol{\xi} \, d\vartheta \\
&\le c_1\left(\tfrac{3}{4}\right) \int_{\mathbb{R}} \int_{\mathbb{R}^N} |\mathscr{F}(\breve{\boldsymbol{u}}; \vartheta, \boldsymbol{\xi})| \sum_{\ell=1}^{3} \left|\mathscr{F}\left(\boldsymbol{w}^{(\ell)}; \vartheta, \boldsymbol{\xi}\right)\right| |\vartheta|^{1/2} \, d\boldsymbol{\xi} \, d\vartheta \\
&\le c\Big(\left\|\boldsymbol{w}^{(1)}\right\|_{H^{1/4}(\mathbb{R};\boldsymbol{H}^{-1}(\mathbb{R}^N))} + \left\|\boldsymbol{w}^{(2)}\right\|_{\boldsymbol{H}^{-1/4,-1/2}(\mathbb{R}\times\mathbb{R}^N)} \\
&\quad + \left\|\boldsymbol{w}^{(3)}\right\|_{L_2(\mathbb{R};\boldsymbol{H}^{-1/2}(\mathbb{R}^N))}\Big) \cdot \Big(\int_{\mathbb{R}^{N+1}} |\mathscr{F}(\breve{\boldsymbol{u}}; \vartheta, \boldsymbol{\xi})|^2 \cdot \\
&\quad \cdot \left(|\vartheta|^{1/2}(1 + |\boldsymbol{\xi}|^2) + |\vartheta|(1 + |\vartheta|^{1/2} + |\boldsymbol{\xi}|) + |\vartheta||\boldsymbol{\xi}| \right) d\boldsymbol{\xi} \, d\vartheta \Big)^{1/2}.
\end{aligned}
$$

The weight in the last integral is obviously bounded by

$$
c\big(|\vartheta|^{1/2}|\boldsymbol{\xi}|^2 + |\vartheta|^{3/2} + 1\big);
$$

this shows the estimate

$$
\begin{aligned}
\|\breve{\boldsymbol{u}}\|^2_{H^{3/4}(\mathbb{R};\boldsymbol{L}_2(\mathbb{R}^N))} \le c\Big(&\|\breve{\boldsymbol{u}}\|_{H^{1/4}(\mathbb{R};\boldsymbol{H}^1(\mathbb{R}^N))} + \|\mathring{\boldsymbol{u}}\|_{H^{1/4}(\mathbb{R};\boldsymbol{H}^1(\mathbb{R}^N))} \\
&+ \|\mathscr{L}(\boldsymbol{u}, \cdot)\|_{H^{-1/4,-1/2}(\mathbb{R}\times\mathbb{R}^N)} + \|\Psi\|_{L_2(\mathbb{R};\boldsymbol{H}^{\alpha}(\mathbb{R}^N))}\Big)
\end{aligned}
$$

and the proof of $\breve{\boldsymbol{u}} \in H^{3/4}\big(\mathbb{R}; \boldsymbol{L}_2(O)\big)$ is done. □

5.4.5 Remark. The proof has also shown the time regularity

$$
\dot{\boldsymbol{u}} \in H^{3/4}(I_T; \boldsymbol{L}_2(\Omega_{\mathfrak{F}}))
$$

for a domain $\Omega_{\mathfrak{F}} \subset \Omega$ that has a positive distance to $\Gamma \setminus \Gamma_{\mathfrak{F}}$. This time regularity will be needed for further considerations.

In the sequel it is important that relation (5.4.39) holds for varying finite time T with constants independent of T. This can be verified as follows: for a time interval $I_{t_0} = (0, t_0) \subset I_T$ we consider the contact problem on the full time interval I_T with temperature field $\Theta\chi_{t_0}$, where χ_{t_0} is the indicator function of I_{t_0}, and denote its solution by $\widetilde{\boldsymbol{u}}$. Due to the unique solvability of the contact problem, $\widetilde{\boldsymbol{u}}$ coincides with $\boldsymbol{u}$ on I_{t_0}. Relation (5.4.39) holds for

$\tilde{\boldsymbol{u}}$ with Θ replaced by $\Theta\chi_{I_{t_0}}$ and the same contants. Obviously it remains valid, if the time interval I_T on the left hand side is replaced by I_{t_0}. Then the resulting inequality implies (5.4.39) for the original solution $\boldsymbol{u}$ with T replaced by t_0.

With this estimate it is not difficult to derive uniform estimates for the pair of solutions $(\boldsymbol{u}, \Theta)$ of the coupled thermo-viscoelastic contact problem.

5.4.6 Proposition. *Let the assumptions of Propositions 5.4.1, 5.4.2, Assumption 5.1.6 be valid and let* $b_{ij} \in C^{\beta}(\Omega)$ *with a* $\beta > 1/2$. *Then the solution* $(\boldsymbol{u}, \Theta)$ *of problem* (5.4.25), (5.4.26) *satisfies the inequality*

$$\begin{aligned} \|\dot{\boldsymbol{u}}\|_{\boldsymbol{H}^{1/2,1}(Q_T)} + \|\dot{\boldsymbol{u}}\|_{\boldsymbol{H}^{1/2,1}(S_{\mathfrak{F}T})} + \left\| \tfrac{1}{\varepsilon}[\dot{u}_n]_+ \right\|_{L_2(S_{\mathfrak{F}T})} \\ + \|\Theta\|_{H^{1/2,1}(Q_T)} \leq c, \end{aligned} \tag{5.4.41}$$

with c *independent of* ε, $\boldsymbol{u}$ *and* Θ.

Proof. The *a priori* estimate (5.4.35) of the heat equation, applied to the solution of the coupled problem, yields

$$\begin{aligned} &\|\Theta(t_0)\|^2_{L_2(\Omega)} + \|\nabla\Theta\|^2_{L_2(Q_{t_0})} \\ &\leq c_1 \left(\|\dot{\boldsymbol{u}}\|^2_{L_2(I_{t_0};H^1(\Omega))} + \left\| \mathscr{J}\left(\mathfrak{F}\tfrac{1}{\varepsilon}[\dot{u}_n]_+, |\dot{\boldsymbol{u}}_{\boldsymbol{t}}|\right) \right\|^2_{L_2(S_{\mathfrak{F}t_0})} \right) \\ &\quad + c_2. \end{aligned} \tag{5.4.42}$$

With the growth condition for the function $\mathscr{J}$ and inequality (5.4.39) applied for T replaced by t_0, the right hand side of (5.4.42) can be estimated by $c_1\|\Theta\|^2_{L_2(I_{t_0};H^{\alpha}(\Omega))} + c_2$ with $\alpha > 1/2$. Using interpolation theorems for Sobolev spaces and the inequality $x \cdot y \leq \eta x^2 + \eta^{-1} y^2$ for $x, y \in \mathbb{R}$, the norm $\|\Theta\|^2_{H^{\alpha}(\Omega)}$ can be estimated by $\eta\|\Theta\|^2_{H^1(\Omega)} + c(\eta)\|\Theta\|^2_{L_2(\Omega)}$ with arbitrarily small $\eta > 0$. Therefore inequality (5.4.42) can be simplified to

$$\|\Theta(t_0)\|^2_{L_2(\Omega)} + \|\nabla\Theta\|^2_{L_2(Q_{t_0})} \leq c_1\|\Theta\|^2_{L_2(Q_{t_0})} + c_2.$$

Application of the Gronwall lemma and of estimates (5.4.33), (5.4.39) shows (5.4.41). □

From the *a priori* estimate (5.4.41) there follows the existence of a sequence $\{\varepsilon_k\}$ and of a corresponding sequence of solutions $\left(\boldsymbol{u}^{(k)}, \Theta^{(k)}\right)$ such that the following convergence properties are valid:

$$\begin{gathered} \varepsilon_k \to 0, \quad \dot{\boldsymbol{u}}_k \rightharpoonup \dot{\boldsymbol{u}} \text{ in } \boldsymbol{H}^{1/2,1}(Q_T) \text{ and strongly in } \boldsymbol{L}_2(S_T), \\ -\tfrac{1}{\varepsilon_k}\left[\dot{u}_n^{(k)}\right]_+ \rightharpoonup \sigma_n(\boldsymbol{u}) \text{ in } L_2(S_{\mathfrak{F}T}), \quad \Theta_k \rightharpoonup \Theta \text{ in } H^{1/2,1}(Q_T) \text{ and} \\ \Theta_k \to \Theta \text{ (strongly) both in } L_2(Q_T) \text{ and in } L_2(S_{CT}). \end{gathered}$$

Performing the limit $k \to +\infty$ both in the variational inequality (5.4.25) and in the modified variational equation (5.4.26) shows that the limit $(\boldsymbol{u}, \Theta)$ is a solution of the original thermo-viscoelastic contact problem with Coulomb friction (5.4.23), (5.4.24). Let us sum up the assumptions as follows

5.4.7 Assumption. *Let Ω be a bounded Lipschitz domain with a boundary Γ consisting of the measurable pairwisely disjoint parts Γ_U, Γ_τ and Γ_C and $\Gamma_C \in C^\beta$ with a $\beta > 2$. Let $I_T \equiv [0,T]$ with $0 < T < +\infty$. The given data satisfy the regularity properties $\boldsymbol{u}_0 \in \boldsymbol{H}^1(\Omega)$, $\boldsymbol{u}_1 \in \boldsymbol{L}_2(\Omega)$, $\Theta_0 \in L_2(\Omega)$, $\boldsymbol{U} \in \boldsymbol{H}^{1/2+\eta,1}(Q_T)$ with $\eta > 0$ such that the compatibility conditions $\boldsymbol{U}(0,\cdot) = \boldsymbol{u}_0$, $\dot{\boldsymbol{U}}(0,\cdot) = \boldsymbol{u}_1$ are satisfied and $\boldsymbol{U} = 0$ on S_{CT}, $\boldsymbol{f} \in \mathring{\boldsymbol{H}}^{-1/2+\eta,-1}(Q_T)$, $\boldsymbol{\tau} \in \mathring{\boldsymbol{H}}^{-1/4+\eta,-1/2}(S_{\tau T})$ and $\Upsilon \in L_2(I_T; H^{-1/2}(\Gamma))$. The coefficients $a^{(\iota)}_{ijkl}$, $\iota = 0,1$, c_{ij}, b_{ij}, d_{ij} are symmetric in their appropriate manner and satisfy the ellipticity and boundedness conditions* (5.1.3), (5.1.4) *and* (5.4.21). *For a domain $\Omega_C \subset \Omega$ such that $\Gamma_C \subset \partial\Omega_C$ and $Q_{CT} \equiv I_T \times \Omega_C$ the local regularity requirements $\boldsymbol{u}_0, \boldsymbol{u}_1 \in \boldsymbol{H}^{3/2}(\Omega_C)$, $\boldsymbol{f} \in \mathring{\boldsymbol{H}}^{-1/4,-1/2}(Q_{CT})$, and $a^{(\iota)}_{ijkl}, c_{ij}, b_{ij} \in C^{\beta'}(\Omega_C)$ with $\beta' > \frac{1}{2}$ hold. The coefficient of friction $\mathfrak{F} : \Omega \to \mathbb{R}$ is non-negative, bounded by the constant $C_\mathfrak{F}$ from* (5.2.17) *or from the respective cases from Section 5.3 in the isotropic case and its support is contained in $I_T \times \Gamma_\mathfrak{F} \subset \Gamma_C$ with $\Gamma_\mathfrak{F}$ satisfying* $\operatorname{dist}(\boldsymbol{x}, \Gamma \setminus \Gamma_C) \geq \delta_0 > 0$ *for all $\boldsymbol{x} \in \Gamma_\mathfrak{F}$. The function J satisfies the growth condition* (5.4.19) *and the continuity condition* (5.4.20). *The coefficient K in the heat exchange condition is a non-negative, bounded, measurable function of the space variable.*

Then the following theorem is proved:

5.4.8 Theorem. *Under the above summed assumptions the thermo-visco-elastic contact problem with Coulomb friction* (5.4.23), (5.4.24) *has at least one solution.*

5.4.2 Energetic approach

The linearization employed above leads to a loss of physical consistency; the linearized model satisfies neither the first nor the second law of thermodynamics. In a physically consistent model the fixed point approach is no longer optimal, because the separate treatment of the mechanical and thermal parts leads to redundantly high requirements on the growth of the diffusion coefficients, cf. e.g. [81], where both approaches are compared.

We return to the original formulation of the thermo-viscoelastic problem (5.4.5)–(5.4.12) described in the introduction of this section. In order to derive *a priori* estimates for this model it is necessary to compensate the quadratic growth of the viscous heat, the frictional heat and the deformation heat. There is an estimate describing the energy conservation of the system, see (5.4.65) or (5.4.89) below, where the coupling terms cancel, but it is rather weak. For sufficiently regular solutions the result of the energy estimate can be used to compensate the growth of the quadratic coupling terms with the help of interpolation and imbedding theorems. However, the regularity of solutions to the contact problem is limited by the non-smooth Coulomb law of friction, and an additional compensation will be needed. It will be achieved with the help of a heat diffusivity depending on the solution. Two different models will be treated: in the first one the coefficients depend on the temperature itself while in the second one they depend on its gradient. None of the models violates a basic law of thermodynamics.

For the solvability of the models described below the following lemma is important:

5.4.9 Lemma. *Let Ω be a domain with a boundary $\Gamma \in C^{\beta}$ with $\beta > 2$. The whole boundary is assumed to be in possible contact with a rigid foundation, $\Gamma_C = \Gamma$. Let Assumptions 5.1.1 and 5.1.6 with $\Omega_C = \Omega$ be satisfied for the appropriate constant $C_{\mathfrak{F}}$ (with the exception of the redundant assumptions about $\boldsymbol{U}$ and $\boldsymbol{\tau}$) and let $\Theta \in L_2\big(I_T; H^{\beta'}(\Omega)\big)$ with $\beta' > 1/2$ and $b_{ij} \in C^{\beta'}(\Omega)$. Then there exists a solution to the penalized problem* (5.1.13) *satisfying the following estimate*

$$\begin{aligned} &\|\dot{\boldsymbol{u}}\|_{\boldsymbol{H}^{3/4,3/2}(Q_T)} + \left\| \tfrac{1}{\varepsilon}[\dot{u}_n]_+ \right\|_{L_2(S_T)} + \|\dot{\boldsymbol{u}}\|_{\boldsymbol{H}^{1/2,1}(S_T)} \\ &\quad \le c_1 \|f\|_{L_2(I_T;\boldsymbol{H}^{1/2}(\Omega)^*)} + c_2 \|\Theta\|_{L_2(I_T;H^{\alpha}(\Omega))} + c_3. \end{aligned} \tag{5.4.43}$$

The constants c_1, c_2 and c_3 depend on the coefficient matrices and on the geometry of the domain only.

Proof. The proof of the regularity result from Theorem 5.1.7 and Lemma 5.4.4 by the localization- and shift techniques has in fact shown the relation

$$\nabla \breve{\boldsymbol{u}} \in H^{1/4}\big(\mathbb{R}; L_2(O; \mathbb{R}^{N,N})\big) \cap L_2\big(\mathbb{R}; H^{1/2}(\mathbb{R}^{N-1}; L_2(\mathbb{R}_+; \mathbb{R}^{N,N}))\big)$$

for the gradient of a localized and rectified variant $\breve{\boldsymbol{u}}$ of $\dot{\boldsymbol{u}}$ and the time regularity $\breve{\boldsymbol{u}} \in H^{3/4}\big(\mathbb{R}; \boldsymbol{L}_2(O)\big)$ with upper bounds for the corresponding norms given by the right hand side of (5.4.43). It remains to prove the normal regularity $\partial_N \breve{\boldsymbol{u}} \in L_2\big(\mathbb{R}; \boldsymbol{H}^{1/2}(O)\big)$. The definition of $\breve{\boldsymbol{u}}$ and $\mathring{\boldsymbol{u}}$ is given in (5.1.49) and (5.1.50).

In order to prove the normal regularity we represent the derivatives of second order in normal direction via the differential equation. This is done for the function localized with respect to the space variables only, without the cutoff function ϕ_L for the time variable. For simplicity of the notation we use the symbol $\dot{\boldsymbol{u}}$ for the localized function, too. Due to the requirements on the coefficients there holds

$$\sum_{j=1}^{N} \alpha_{ij}^{(0)} \partial_N^2 \dot{u}_j = \ddot{u}_i - \sum_{j=1}^{N} \partial_N \alpha_{ij}^{(0)} \partial_N \dot{u}_j - \sum_{\substack{j,k,\ell=1 \\ (j,k)\neq(N,N)}}^{N} \partial_j \big(\beta_{ijk\ell}^{(1)} \partial_k \dot{u}_\ell\big)$$

$$- \sum_{j,k,\ell=1}^{N} \partial_j \big(\beta_{ijk\ell}^{(0)} \partial_k u_\ell\big) + \partial_j \big(b_{ij} \Theta\big) + \mathscr{L}_i(\boldsymbol{u})$$

with coefficients $\alpha_{ij}^{(0)}, \beta_{ijk\ell}^{(\iota)} \in C^{\beta'}(O)$, a positive definite matrix $\{\alpha_{ij}^{(0)}\}$, and $\mathscr{L}_i \in \mathring{H}^{-1/4,-1/2}(I_T \times O)$. The summation convention is not employed here. Due to the already proved facts $\ddot{\boldsymbol{u}}_i \in H^{-1/4}\big(I_T; \boldsymbol{L}_2(O)\big)$ and $\partial_j \partial_k \boldsymbol{u} \in L_2\big(I_T; \boldsymbol{H}^{-1/2}(O)\big)$ for $k \neq N$ there follows

$$\|\partial_N^2 \dot{\boldsymbol{u}} \chi_{t_0}\|_{\boldsymbol{H}^{-1/4,-1/2}(I_T \times O)} \leq c_1 \|\partial_N^2 \boldsymbol{u} \chi_{t_0}\|_{\boldsymbol{H}^{-1/4,-1/2}(I_T \times O)} + c_2 \quad (5.4.44)$$

with the characteristic function χ_{t_0} of the time-interval $(0, t_0)$ for a $t_0 \in I_T$ and constants c_1, c_2 depending on $\mathscr{L}$, Θ, and the $C^{\beta'}(O)$ norm of the coefficients. After the extension of $\boldsymbol{u}$ onto the whole time-space domain $\mathbb{R}^{N+1}$ the usage of the Fourier transform yields

$$\begin{aligned}
&\int_{\mathbb{R}^{N+1}} |\xi_N|^3 |\mathscr{F}(\dot{\boldsymbol{u}}; \vartheta, \boldsymbol{\xi})|^2 \, d\boldsymbol{\xi} \, d\vartheta \\
&\quad \leq c_1 \int_{\mathbb{R}^{N+1}} \frac{|\vartheta|^2 + |\boldsymbol{\xi}|^4}{1 + c_1(\frac{1}{4})|\vartheta|^{1/2} + |\boldsymbol{\xi}|} \mathscr{F}(\dot{\boldsymbol{u}}; \vartheta, \boldsymbol{\xi}) \, d\boldsymbol{\xi} \, d\vartheta \\
&\quad \leq c_2 \Big(\|\ddot{\boldsymbol{u}}\|^2_{\boldsymbol{H}^{-1/4,-1/2}(\mathbb{R}^{N+1})} + \|\nabla^2 \boldsymbol{u}\|^2_{\boldsymbol{H}^{-1/4,-1/2}(\mathbb{R}^{N+1})} \Big) .
\end{aligned} \quad (5.4.45)$$

Using this relation in (5.4.44) shows

$$\|\dot{\boldsymbol{u}}\|_{L_2(I_{t_0}; \boldsymbol{H}^{3/2}(O))} \leq c_1 \|\boldsymbol{u}\|_{L_2(I_{t_0}; \boldsymbol{H}^{3/2}(O))} + c_2. \quad (5.4.46)$$

Application of a Gronwall lemma argument shows $\dot{\boldsymbol{u}} \in L_2\big(I_T; \boldsymbol{H}^{3/2}(O)\big)$. □

5.4.10 Remark. Since the estimate is independent of the smoothing parameters ε and η, it is also true for the solution of the original problem (5.1.10). A similar estimate can be proved if $\Gamma = \Gamma_U$ or $\Gamma = \Gamma_\mathcal{T}$ with sufficiently regular data, or if $\Gamma = \Gamma_C \cup \Gamma_\mathcal{T} \cup \Gamma_U$ with positive distances between these three parts. However, in the general case there are relative boundaries of these parts with respect to the boundary topology. The behaviour of the solution around the relative boundary of Γ_C is not yet known for the dynamic problem even in the two-dimensional case. Its investigation requires special techniques which are far from the scope of this book. All results in the rest of this chapter are based on the estimate (5.4.43), typically used in the equivalent version

$$\begin{aligned} &\|\nabla\dot{\boldsymbol{u}}\|_{H^{1/4,1/2}(Q_T;\mathbb{R}^{N^2})} + \|\sigma_n(\boldsymbol{u},\Theta)\|_{L_2(S_T)} + \|\dot{\boldsymbol{u}}\|_{H^{1/2,1}(S_T)} \\ &\quad \le c_1\|\Theta\|_{L_2(I_T;H^\alpha(\Omega))} + c_2 \end{aligned} \tag{5.4.47}$$

for the respective penalized problem with $\alpha > 1/2$ and constants dependent on the input data only. The validity of inequality (5.4.47) will be specified as an additional assumption. According to Lemma 5.4.9 it is true for $\Gamma = \Gamma_C$ at least. By the consideration explained after Remark 5.4.5 it is seen that this estimate holds with constants independent of the length T of the time interval, if this length is bounded by a fixed constant.

5.4.2.1 A model with heat conductivity depending on the temperature

In the model studied now the heat conductivity is assumed to depend on the space variable and the temperature, $c_{ij} = c_{ij}(\boldsymbol{x},\Theta)$. It shall be continuous in the sense of Carathéodory; this means $\boldsymbol{x} \mapsto c_{ij}(\boldsymbol{x},\Theta)$ is measurable for every $\Theta \in \mathbb{R}$ and $\Theta \mapsto c_{ij}(\boldsymbol{x},\Theta)$ is continuous for almost every $\boldsymbol{x} \in \Omega$. In order to compensate the growth of the non-linear coupling terms, the following special growth condition is required,

$$c_0(1+|\Theta|^\gamma)\xi_i\xi_i \le c_{ij}(\Theta)\xi_i\xi_j \le C_0(1+|\Theta|^\gamma)\xi_i\xi_i \tag{5.4.48}$$

for all vectors $\{\xi_i\} \in \mathbb{R}^N$ and all $\boldsymbol{x} \in \Omega$ with $0 < c_0 \le C_0 < +\infty$, with parameter $\gamma > 0$ to be specified later. On the boundary of the domain we assume heat exchange with a non-negative coefficient K and a given outer temperature Υ, heat generation by friction, and the boundary dissipation term caused e.g. by heat radiation. The total boundary condition then reads

$$c_{ij}\partial_j\Theta n_i = \mathfrak{F}|\sigma_n(\boldsymbol{u},\Theta)||\dot{\boldsymbol{u}}_t| + K(\Upsilon - \Theta) - R(\Theta) \text{ on } S_T, \tag{5.4.49}$$

if $\mathfrak{F}$ is extended by $\mathfrak{F} = 0$ outside Γ_C. In such a way the condition (5.4.49) replaces (5.4.9) and (5.4.11) in the original formulation of the problem. The function R depends on the space variable and on the temperature. We study both cases with and without boundary dissipation simultaneously. In the second case we formally prescribe $R(\Theta) = 0$. In the first case R shall be continuous in the sense of Carathéodory, non-decreasing in the variable Θ and satisfy $R(\Theta) = 0$ for $\Theta \leq 0$ as well as the growth condition

$$c_1|\Theta|^{r_R} \leq R(\Theta) \leq c_2|\Theta|^{r_R} + c_3 \text{ for } \Theta \geq 0 \tag{5.4.50}$$

with constants $c_1, c_2, c_3 > 0$ and growth exponent $r_R > 1$. In order to treat both cases simultaneously, we define a constant c_R by

$$c_R = \begin{cases} 0 & \text{for } R \equiv 0, \\ 1 & \text{otherwise.} \end{cases} \tag{5.4.51}$$

For the variational formulation of the problem we preserve the definition (5.4.13) of the cone $\mathscr{C}$ while the space $\mathfrak{V}$ has the form

$$\mathfrak{V} := L_2\big(I_T; H^1(\Omega)\big). \tag{5.4.52}$$

A weak solution of this thermo-viscoelastic contact problem is a pair of functions $(\boldsymbol{u}, \Theta) : Q_T \to \mathbb{R}^N \times \mathbb{R}$ with $\dot{\boldsymbol{u}} \in \mathscr{C}$, $\ddot{\boldsymbol{u}} \in L_2\big(I_T; H^1(\Omega)^*\big)$, $\Theta \in \mathfrak{V}$, $\dot{\Theta} \in \mathfrak{V}^*$ which satisfy the initial conditions $\boldsymbol{u}(0, \cdot) = \boldsymbol{u}_0$, $\dot{\boldsymbol{u}}(0, \cdot) = \boldsymbol{u}_1$ and $\Theta(0, \cdot) = \Theta_0$ such that the following two variational relations are satisfied: the variational inequality of the viscoelastic contact problem

$$\begin{aligned} &\big\langle \ddot{\boldsymbol{u}}, \boldsymbol{v} - \dot{\boldsymbol{u}} \big\rangle_{Q_T} + A^{(0)}(\boldsymbol{u}, \boldsymbol{v} - \dot{\boldsymbol{u}}) + A^{(1)}(\dot{\boldsymbol{u}}, \boldsymbol{v} - \dot{\boldsymbol{u}}) \\ &\quad - \big\langle b_{ij}\Theta_+ , e_{ij}(\boldsymbol{v} - \dot{\boldsymbol{u}})\big\rangle_{Q_T} + \big\langle \mathfrak{F}|\sigma_n(\boldsymbol{u}, \Theta)|, |\boldsymbol{v}_t| - |\dot{\boldsymbol{u}}_t|\big\rangle_{S_{CT}} \\ &\geq \langle \mathscr{L}, \boldsymbol{v} - \dot{\boldsymbol{u}}\rangle_{Q_T} \end{aligned} \tag{5.4.53}$$

for all $\boldsymbol{v} \in \mathscr{C}$ and the variational equation of the heat transfer problem

$$\begin{aligned} &\Big\langle \dot{\Theta}, \varphi \Big\rangle_{Q_T} + \langle c_{ij}(\Theta)\partial_i\Theta, \partial_j\varphi\rangle_{Q_T} + \langle b_{ij}\Theta_+ e_{ij}(\dot{\boldsymbol{u}}), \varphi\rangle_{Q_T} \\ &\qquad + \langle K(\Theta - \Upsilon), \varphi\rangle_{S_T} + \big\langle R(\Theta), \varphi\big\rangle_{S_T} \\ &= \Big\langle a^{(1)}_{ijk\ell} e_{ij}(\dot{\boldsymbol{u}}) e_{k\ell}(\dot{\boldsymbol{u}}), \varphi \Big\rangle_{Q_T} + \big\langle \mathfrak{F}|\sigma_n(\boldsymbol{u}, \Theta)||\dot{\boldsymbol{u}}_t|, \varphi\big\rangle_{S_T} \end{aligned} \tag{5.4.54}$$

for all $\varphi \in W^1_\infty\big(I_T; H^\ell(\Omega)\big)$ with $\ell > N/2+1$. Here in the viscous heat term we have replaced the temperature by its positive part Θ_+. This replacement will be justified later by a proof that Θ is indeed non-negative.

The existence of solutions to the thermo-viscoelastic contact problem (5.4.53), (5.4.54) is proved under the following assumptions:

5.4.11 Assumption. *In addition to Assumption 5.1.1 and 5.1.6 with* $\Omega = \Omega_C$*, the assumption* (5.4.47) *and the above mentioned assumptions for the coefficient functions* b_{ij}*,* c_{ij} *and the function* R *let* $0 \leq \Upsilon \in L_2(S_T)$*,* $0 \leq$

$K \in L_\infty(\Gamma)$, $b_{ij} \in C^\alpha(\Omega)$ with $\alpha > 1/2$. The data in (5.4.53) *satisfy $\boldsymbol{\tau} \in \boldsymbol{L}_2(S_{\boldsymbol{\tau}T})$, and $\boldsymbol{U}, \dot{\boldsymbol{U}} \in \boldsymbol{H}^{3/4,3/2}(Q_T)$. The initial data satisfy $\Theta_0 \in L_2(\Omega)$ and $\Theta_0 \geq 0$. The growth parameters γ, r_R and the function R fulfil either $1 - 2/N < \gamma < 2 + 4/N$ and $r_R > 2 + 1/N$ or $1 - 1/N - 1/(N+1) < \gamma < 2 + 4/N$ and $R \equiv 0$.*

Then the following existence result is valid

5.4.12 Theorem. *Under Assumption 5.4.11 problem* (5.4.53), (5.4.54) *has at least one solution. The temperature field of this solution is non-negative. Every solution satisfies the* a priori *estimate*

$$\|\dot{\boldsymbol{u}}\|_{\boldsymbol{H}^{3/4,\,3/2}(Q_T)} + \|\Theta\|_{L_\infty(I_T, L_{2-\gamma}(\Omega))} + \|\Theta\|_{L_2(I_T;H^1(\Omega))} \\ + c_R\|\Theta\|_{L_{r_R+1-\gamma}(S_T)} \leq c. \tag{5.4.55}$$

First we briefly outline the sketch of the proof. Besides the usual approximations consisting both in penalizing the unilateral contact condition and in smoothing the norm in the friction functional an additional limiting of the growth of the viscous heat, the deformation heat and the frictional heat is employed. Thereby a variational equation is obtained and its solvability is verified with the Galerkin method. In order to perform the limit procedures for the approximation parameters, suitable *a priori* estimates independent of these parameters are necessary. In a first step the energy of the system is estimated; this yields a very weak estimate which is then improved via suitable test functions with the help of the imbedding and interpolation theorems from Chapter 2.

We use the usual penalty approximation of the contact condition and we apply the cutoff function

$$\mathsf{h}_M : [0, +\infty) \to \mathbb{R}, \quad \mathsf{h}_M : x \mapsto \begin{cases} x, & x \leq M, \\ M, & x > M \end{cases}$$

to all heat sources and to the heat-deformation term in the mechanical part of the system. The resulting approximate variational inequality is

Find a couple $(\boldsymbol{u}, \Theta) : Q_T \to \mathbb{R}^N$ with $\boldsymbol{u}(0,\cdot) = \boldsymbol{u}_0$, $\dot{\boldsymbol{u}}(0,\cdot) = \boldsymbol{u}_1$, $\dot{\boldsymbol{u}} \in \dot{\boldsymbol{U}} + \mathfrak{U}$ with $\mathfrak{U} = \boldsymbol{H}_0^{1/2,1}(Q_T)$, and $\ddot{\boldsymbol{u}} \in \mathfrak{U}^*$, $\Theta \in \mathfrak{V}$, $\dot{\Theta} \in \mathfrak{V}^*$ such that for all $\boldsymbol{v} \in \dot{\boldsymbol{U}} + \mathfrak{U}$ and all $\varphi \in L_2(I_T; H^\ell(\Omega))$ with $\ell > N/2 + 1$ there holds

$$\langle \ddot{\boldsymbol{u}}, \boldsymbol{v} - \dot{\boldsymbol{u}} \rangle_{Q_T} + A^{(0)}(\boldsymbol{u}, \boldsymbol{v} - \dot{\boldsymbol{u}}) + A^{(1)}(\dot{\boldsymbol{u}}, \boldsymbol{v} - \dot{\boldsymbol{u}}) \\ - \langle b_{ij}\mathsf{h}_M(\Theta_+), e_{ij}(\boldsymbol{v} - \dot{\boldsymbol{u}}) \rangle_{Q_T} + \langle \tfrac{1}{\varepsilon}[\dot{u}_n]_+, v_n - \dot{u}_n \rangle_{S_{CT}} \\ + \langle \mathfrak{F}\tfrac{1}{\varepsilon}[\dot{u}_n]_+, |\boldsymbol{v_t}| - |\dot{\boldsymbol{u}}_{\boldsymbol{t}}| \rangle_{S_{CT}} \geq \langle \mathscr{L}, \boldsymbol{v} - \dot{\boldsymbol{u}} \rangle_{Q_T} \tag{5.4.56}$$

and

$$
\begin{aligned}
&\langle\dot\Theta,\varphi\rangle_{Q_T}+\langle c_{ij}(\Theta)\partial_i\Theta,\partial_j\varphi\rangle_{Q_T}+\langle b_{ij}\mathsf{h}_M(\Theta_+)e_{ij}(\dot{\boldsymbol u}),\varphi\rangle_{Q_T}\\
&\qquad+\langle K(\Theta-\Upsilon),\varphi\rangle_{S_T}+\big\langle R(\Theta),\varphi\big\rangle_{S_T}\\
&\quad=\Big\langle\mathsf{h}_M\big(a^{(1)}_{ijk\ell}e_{ij}(\dot{\boldsymbol u})e_{k\ell}(\dot{\boldsymbol u})\big),\varphi\Big\rangle_{Q_T}\\
&\qquad+\big\langle\mathfrak{F}\mathsf{h}_M\big(\tfrac{1}{\varepsilon}[\dot u_n]_+|\dot{\boldsymbol u}_{\boldsymbol t}|\big),\varphi\big\rangle_{S_T}.
\end{aligned}\tag{5.4.57}
$$

Then the smoothing of the norm in the Coulomb law yields a variational inequality that is equivalent to the following variational equation:

Find a couple $(\boldsymbol u,\Theta):Q_T\to\mathbb{R}^N$ *with* $\boldsymbol u(0,\cdot)=\boldsymbol u_0$, $\dot{\boldsymbol u}(0,\cdot)=\boldsymbol u_1$, $\dot{\boldsymbol u}\in\dot{\boldsymbol U}+\mathfrak{U}$ *and* $\ddot{\boldsymbol u}\in\mathfrak{U}^*$, $\Theta\in\mathfrak{V}$, $\dot\Theta\in\mathfrak{V}^*$ *such that for all* $\boldsymbol v\in\mathfrak{U}$ *and all* $\varphi\in L_2\big(I_T;H^\ell(\Omega)\big)$ *with* $\ell>N/2+1$ *there holds*

$$
\begin{aligned}
&\big\langle\ddot{\boldsymbol u},\boldsymbol v\big\rangle_{Q_T}+A^{(0)}(\boldsymbol u,\boldsymbol v)+A^{(1)}(\dot{\boldsymbol u},\boldsymbol v)-\big\langle b_{ij}\mathsf{h}_M(\Theta_+),e_{ij}(\boldsymbol v)\big\rangle_{Q_T}\\
&\qquad+\Big\langle\tfrac{1}{\varepsilon}[\dot u_n]_+,v_n\Big\rangle_{S_{CT}}+\Big\langle\mathfrak{F}\tfrac{1}{\varepsilon}[\dot u_n]_+\nabla\Phi_\eta(\dot{\boldsymbol u}_{\boldsymbol t}),\boldsymbol v_{\boldsymbol t}\Big\rangle_{S_{CT}}\\
&\quad=\langle\mathscr{L},\boldsymbol v\rangle_{Q_T},
\end{aligned}\tag{5.4.58}
$$

$$
\begin{aligned}
&\big\langle\dot\Theta,\varphi\big\rangle_{Q_T}+\big\langle c_{ij}(\Theta)\partial_i\Theta,\partial_j\varphi\big\rangle_{Q_T}+\big\langle b_{ij}\mathsf{h}_M(\Theta_+)e_{ij}(\dot{\boldsymbol u}),\varphi\big\rangle_{Q_T}\\
&\qquad+\big\langle K(\Theta-\Upsilon),\varphi\big\rangle_{S_T}+\big\langle R(\Theta),\varphi\big\rangle_{S_T}\\
&\quad=\Big\langle\mathsf{h}_M\Big(a^{(1)}_{ijk\ell}e_{ij}(\dot{\boldsymbol u})e_{k\ell}(\dot{\boldsymbol u})\Big),\varphi\Big\rangle_{Q_T}\\
&\qquad+\Big\langle\mathfrak{F}\mathsf{h}_M\Big(\tfrac{1}{\varepsilon}[\dot u_n]_+\Phi_\eta(\dot{\boldsymbol u}_{\boldsymbol t})\Big),\varphi\Big\rangle_{S_{CT}}.
\end{aligned}\tag{5.4.59}
$$

For simplicity of the presentation this problem will be studied for vanishing Dirichlet data $\boldsymbol U\equiv 0$. This condition is, of course, also covered by the case of no Dirichlet data $\Gamma_U=0$. The case of non-vanishing Dirichlet data $\boldsymbol U\neq 0$ can be easily treated by the transform $\boldsymbol u=\boldsymbol U+\widetilde{\boldsymbol u}$ in the contact problem with an obvious modification of the right hand side, as already described in various other proofs.

The solvability of this problem is stated in the next Lemma.

5.4.13 Lemma. *Let Assumption 5.4.11 be valid. Then, for fixed approximation parameters* $\varepsilon,M,\eta>0$ *problem* (5.4.58), (5.4.59) *has a solution* $(\boldsymbol u,\Theta)$. *Every solution satisfies the* a priori *estimate*

$$
\begin{aligned}
&\|\Theta\|_{H^{s,1}(Q_T)}+\big\||\Theta|^{\gamma/2}\nabla\Theta\big\|_{L_2(Q_T)}+c_R\|\Theta\|_{L_{r_R+1}(S_T)}\\
&\qquad+\|\dot{\boldsymbol u}\|_{\boldsymbol H^{1/2,1}(Q_T)}\ \le\ c
\end{aligned}\tag{5.4.60}
$$

with $s<1/(N+4)$ *and constant* c *independent of* η *but dependent on* ε *and* M.

Proof. The existence of a solution is proved with the Galerkin approximation. Let $\{\boldsymbol{v}_i\,;\, i \in \mathbb{N}\}$ be a basis of $\{\boldsymbol{v} \in \boldsymbol{H}^1(\Omega)\,;\, v = 0 \text{ on } \Gamma_U\}$ and $\{\varphi_i\,;\, i \in \mathbb{N}\}$ be a basis of $H^1(\Omega)$. Let

$$\mathscr{U}_m := \operatorname{span}\{\boldsymbol{v}_1, \ldots, \boldsymbol{v}_m\} \text{ and}$$
$$\mathscr{V}_m := \operatorname{span}\{\varphi_1, \ldots, \varphi_m\}$$

be the corresponding finite dimensional Galerkin spaces, $\pi_{\mathscr{U}_m}$ denote the L_2-orthogonal projection of $\boldsymbol{L}_2(\Omega)$ onto $\mathscr{U}_m$ and $\pi_{\mathscr{V}_m}$ have the same meaning for the space $\mathscr{V}_m$. It is required later that $\pi_{\mathscr{U}_m}$ is bounded in $\boldsymbol{H}^1(\Omega)$ and $\pi_{\mathscr{V}_m}$ is bounded in $H^\ell(\Omega)$ for an integer ℓ larger than $1+N/2$, and that their bounds do not depend on m. This is a condition for the proper choice of the bases $\{\boldsymbol{v}_j\}$ and $\{\varphi_j\}$. Such bases exist. A possible choice is a basis $\{\boldsymbol{v}_j\}$ of eigenfunctions to the Laplacian and a basis $\{\varphi_j\}$ of eigenfunctions to the linear operator $A^\ell : H^\ell(\Omega) \to H^\ell(\Omega)^*$ defined by $A^\ell : (u, v) \mapsto (u, v)_{H^\ell(\Omega)}$ with the $H^\ell(\Omega)$ scalar product $(\cdot,\cdot)_{H^\ell(\Omega)}$. Since $\pi_{\mathscr{V}_m}$ is bounded in both $L_2(\Omega)$ and $H^\ell(\Omega)$, it is, by interpolation, also bounded in $H^\beta(\Omega)$ for all $\beta \in (0, \ell)$. The Galerkin approximation of problem (5.4.58), (5.4.59) is defined by:

Find functions

$$\boldsymbol{u}_m(t, x) \equiv \sum_{i=1}^{m} q_i(t)\boldsymbol{v}_i(\boldsymbol{x}) \in L_\infty(I_T, \mathscr{U}_m) \text{ and}$$
$$\Theta_m(t, x) \equiv \sum_{i=1}^{m} r_i(t)\varphi_i(\boldsymbol{x}) \in L_\infty(I_T, \mathscr{V}_m)$$

such that for all $\boldsymbol{v} \in \mathscr{U}_m$, all $\varphi \in \mathscr{V}_m$ and all $t \in I_T$ the following equations hold:

$$\begin{aligned}
&\langle \ddot{\boldsymbol{u}}_m, \boldsymbol{v}\rangle_\Omega + \langle a^{(0)}_{ijk\ell} e_{k\ell}(\boldsymbol{u}_m) + a^{(1)}_{ijk\ell} e_{k\ell}(\dot{\boldsymbol{u}}_m), e_{ij}(\boldsymbol{v})\rangle_\Omega \\
&\quad - \langle b_{ij} \mathsf{h}_M((\Theta_m)_+), e_{ij}(\boldsymbol{v})\rangle_\Omega + \left\langle \tfrac{1}{\varepsilon}\left[(\dot{u}_m)_n\right]_+, v_n \right\rangle_{\Gamma_C} \\
&\quad + \left\langle \mathfrak{F}\tfrac{1}{\varepsilon}\left[(\dot{u}_m)_n\right]_+ \nabla\Phi_\eta((\dot{\boldsymbol{u}}_m)_{\boldsymbol{t}}), \boldsymbol{v}_{\boldsymbol{t}} \right\rangle_{\Gamma_C} = \langle \boldsymbol{f}, \boldsymbol{v}\rangle_\Omega + \langle \boldsymbol{\tau}, \boldsymbol{v}\rangle_{\Gamma_\tau}
\end{aligned} \tag{5.4.61}$$

and

$$\begin{aligned}
&\langle \dot{\Theta}_m, \varphi\rangle_\Omega + \langle c_{ij}(\Theta_m)\partial_i\Theta_m, \partial_j\varphi\rangle_\Omega + \langle K(\Theta_m - \Upsilon), \varphi\rangle_\Gamma \\
&\quad + \langle R(\Theta_m), \varphi\rangle_\Gamma + \langle b_{ij}\mathsf{h}_M((\Theta_m)_+) e_{ij}(\dot{\boldsymbol{u}}_m), \varphi\rangle_\Omega \\
&= \left\langle \mathsf{h}_M(a^{(1)}_{ijk\ell} e_{ij}(\dot{\boldsymbol{u}}_m) e_{k\ell}(\dot{\boldsymbol{u}}_m)), \varphi \right\rangle_\Omega \\
&\quad + \left\langle \mathfrak{F}\mathsf{h}_M\left(\tfrac{1}{\varepsilon}\left[(\dot{u}_m)_n\right]_+ \Phi_\eta((\dot{\boldsymbol{u}}_m)_{\boldsymbol{t}})\right), \varphi \right\rangle_{\Gamma_C} .
\end{aligned} \tag{5.4.62}$$

The initial conditions shall be satisfied in the appropriate approximate sense,

$$\langle \boldsymbol{u}_m(0,\cdot), \boldsymbol{v} \rangle_\Omega = \langle \boldsymbol{u}_0, \boldsymbol{v} \rangle_\Omega \text{ and } \langle \dot{\boldsymbol{u}}_m(0,\cdot), \boldsymbol{v} \rangle_\Omega = \langle \boldsymbol{u}_1, \boldsymbol{v} \rangle_\Omega$$
$$\text{for all } \boldsymbol{v} \in \mathscr{U}_m \text{ and}$$
$$\langle \Theta_m(0,\cdot), \varphi \rangle_\Omega = \langle \Theta_0, \varphi \rangle_\Omega \text{ for all } \varphi \in \mathscr{V}_m.$$

The Galerkin equations are equivalent to an initial value problem for a system of ordinary differential equations. The solvability of such a system follows from the theory of ordinary differential equations.

Let us take the test function $\boldsymbol{v} = \dot{\boldsymbol{u}}_m$ in equation (5.4.61) and integrate the result with respect to $t \in (0, t_0)$ for a $t_0 \in I_T$. Using the usual coerciveness of stresses, the positivity of the friction term and the boundedness of $\mathsf{h}_M\big((\Theta_m)_+\big)$, the relation

$$\|\dot{\boldsymbol{u}}_m(t_0)\|^2_{\boldsymbol{L}_2(\Omega)} + \|\nabla \boldsymbol{u}_m(t_0)\|^2_{\boldsymbol{L}_2(\Omega)} + \|\nabla \dot{\boldsymbol{u}}_m\|^2_{\boldsymbol{L}_2(Q_{t_0})}$$
$$\leq c_1 \|\dot{\boldsymbol{u}}_m\|_{L_2(I_{t_0};\boldsymbol{H}^1(\Omega))} + c_2$$

is derived. The constants here depend on M and other data of the problem, but not on m, η, and ε. By the Gronwall lemma there follows

$$\|\dot{\boldsymbol{u}}_m\|^2_{L_\infty(I_T;\boldsymbol{L}_2(\Omega))} + \|\boldsymbol{u}_m\|^2_{L_\infty(I_T;\boldsymbol{H}^1(\Omega))} + \|\dot{\boldsymbol{u}}_m\|^2_{L_2(I_T;\boldsymbol{H}^1(\Omega))} \leq c.$$

The same procedure is repeated for the heat equation (5.4.62) with test function $\varphi = \Theta_m$. Here, the heat diffusion term is estimated with the help of the ellipticity condition for the coefficients c_{ij} by

$$\big\langle c_{ij}(\Theta_m)\partial_i\Theta_m, \partial_j\Theta_m \big\rangle_{Q_{t_0}}$$
$$\geq c_1 \left(\|\nabla\Theta_m\|^2_{L_2(Q_{t_0})} + \big\||\Theta_m|^{\gamma/2}\nabla\Theta_m\big\|^2_{L_2(Q_{t_0})} \right).$$

The heat dissipation term on the boundary is bounded from below by $\mathsf{c}\|(\Theta_m)_+\|^{r_R+1}_{L_{r_R+1}(S_{t_0})}$. The estimate of the frictional heat, of the deformation heat and of the viscous heat is trivial, since the growth of these terms is limited at value M. Hence the relation

$$\|\Theta_m\|^2_{L_\infty(I_T;L_2(\Omega))} + \big\||\Theta_m|^{\gamma/2}\nabla\Theta_m\big\|^2_{L_2(Q_T)} + \|\Theta_m\|^2_{L_2(I_T;H^1(\Omega))}$$
$$+ c_R\|\Theta_m\|^{r_R+1}_{L_{r_R+1}(S_T)} \leq c$$

is obtained with constant c independent of the approximation parameters ε, η and m.

In the limit procedure $m \to +\infty$ of the Galerkin parameter it is necessary to have strong convergences of the arguments of non-linear terms. This can be achieved with the help of compact imbedding theorems, if some additional small time regularity of the solution is available. The proof of this time regularity is done by a standard dual estimate. For a test function $\boldsymbol{v} \in L_2(I_T; \boldsymbol{H}^1(\Omega))$ the $L_2(\Omega)$ projection $\pi_{\mathscr{U}_m} \boldsymbol{v}$ is used in (5.4.61) and the result is integrated with respect to $t \in I_T$. Then, due to the just proved bounds for $\boldsymbol{u}_m$ the relation

$$\langle \ddot{\boldsymbol{u}}_m, \boldsymbol{v} \rangle_{Q_T} = \langle \ddot{\boldsymbol{u}}_m, \pi_{\mathscr{U}_m} \boldsymbol{v} \rangle_{Q_T} \le c_1 \|\boldsymbol{v}\|_{L_2(I_T; \boldsymbol{H}^1(\Omega))} + c_2$$

is proved. In the last step, the boundedness of the L_2 projection $\pi_{\mathscr{U}_m}$ in $\boldsymbol{H}^1(\Omega)$ is essential. The constants depend on ε and M, but not on η and on the Galerkin parameter m. This equation proves that $\ddot{\boldsymbol{u}}_m$ is bounded uniformly with respect to m in $L_2(I_T; \boldsymbol{H}^{-1}(\Omega))$. Interpolation of $\dot{\boldsymbol{u}}_m \in H^1(I_T; \boldsymbol{H}^{-1}(\Omega))$ with $\dot{\boldsymbol{u}}_m \in L_2(I_T; \boldsymbol{H}^1(\Omega))$ gives a bound in $H^{1/2}(I_T; \boldsymbol{L}_2(\Omega))$.

This procedure is then repeated for the heat equation with a test function $\pi_{\mathscr{V}_m} \varphi$, where $\varphi \in L_\infty(I_T; H^\beta(\Omega))$ with $1 + N/2 < \beta \le \ell$. Here, the diffusion term is estimated by

$$\int_{Q_T} |\Theta_m|^\gamma |\nabla \Theta_m| |\nabla \pi_{\mathscr{V}_m} \varphi| \, d\boldsymbol{x} \, dt \le c_1 \| |\Theta_m|^{\gamma/2} \nabla \Theta_m \|_{L_2(Q_T)} \cdot \| |\Theta_m|^{\gamma/2} \|_{L_2(Q_T)} \| \pi_{\mathscr{V}_m} \varphi \|_{L_\infty(I_T; W^1_\infty(\Omega))}.$$

In order to estimate the norm $\| |\Theta_m|^{\gamma/2} \|_{L_2(Q_T)}$ for possibly large values of γ we use the relation $|\Theta|^{\gamma/2} |\nabla \Theta| = c(\gamma) |\nabla (|\Theta|^{1+\gamma/2})|$ and the imbedding $H^1(\Omega) \hookrightarrow L_p(\Omega)$ for $p < \infty$ in the case $N = 2$ and $p = 2N/(N-2)$ for $N \ge 3$. Then

$$\|\Theta\|^{1+\gamma/2}_{L_{p(1+\gamma/2)}(\Omega)} \le c_1 \| |\Theta|^{1+\gamma/2} \|_{L_p(\Omega)} \le c_2(\gamma) \| |\Theta|^{1+\gamma/2} \|_{H^1(\Omega)},$$

and here the last term can be estimated by

$$c_1(\gamma) \| |\Theta|^{\gamma/2} \nabla \Theta \|_{L_2(\Omega)} + c_2(\gamma) \|\Theta\|^{2+\gamma}_{L_{1+\gamma/2}(\Omega)}.$$

This shows that Θ_m is bounded in $L_{2+\gamma}(I_T; L_{p(1+\gamma/2)}(\Omega))$ and $|\Theta_m|^{\gamma/2}$ is bounded in $L_2(Q_T)$ for arbitrarily large γ. Interpolation of

$$\Theta_m \in L_{2+\gamma}(I_T; L_{p(1+\gamma/2)}(\Omega)) \text{ with } \Theta_m \in L_\infty(I_T; L_2(\Omega))$$

shows that Θ_m is bounded in $L_{2+\gamma+4/N}(Q_T)$. For $1 + N/2 < \beta \le \ell$ the space $H^\beta(\Omega)$ is imbedded into $W^1_\infty(\Omega)$, and the projection $\pi_{\mathscr{V}_m}$ is bounded in $H^\beta(\Omega)$. Since, moreover, $H^\alpha(I_T;\ldots)$ is imbedded in $L_\infty(I_T;\ldots)$ for $\alpha > 1/2$, the diffusion term is bounded by

$$c\|\varphi\|_{H^\alpha(I_T;H^\beta(\Omega))}$$

with constant independent of m. All the other terms in the dual estimate are bounded by

$$c_1\|\varphi\|_{H^\alpha(I_T;H^\beta(\Omega))} + c_2.$$

As a consequence, the dual estimate gives a bound for $\dot{\Theta}_m$ in the dual space $H^\alpha\big(I_T;H^\beta(\Omega)\big)^*$. By interpolation with $\Theta \in L_2\big(I_T;H^1(\Omega)\big)$, a bound in $H^s\big(I_T;L_2(\Omega)\big)$ with $s = (1-\alpha)/(\beta+1)$ is obtained. The choices of α and β mentioned above allow any value of s smaller than $1/(4+N)$. Collecting all the estimates it is evident that the Galerkin solutions $\boldsymbol{u}_m, \Theta_m$ satisfy the *a priori* estimate

$$\begin{aligned}
&\|\Theta_m\|_{L_\infty(I_T;L_2(\Omega))} + \|\Theta_m\|_{H^{s,1}(Q_T)} + c_R\|\Theta_m\|_{L_{r_R+1}(S_T)} \\
&\quad + \|\Theta_m\|_{L_{2+\gamma+4/N}(Q_T)} + \|\dot{\boldsymbol{u}}_m\|_{L_\infty(I_T;\boldsymbol{L}_2(\Omega))} + \|\boldsymbol{u}_m\|_{L_\infty(I_T;\boldsymbol{H}^1(\Omega))} \\
&\quad + \|\dot{\boldsymbol{u}}_m\|_{\boldsymbol{H}^{1/2,1}(Q_T)} \le c
\end{aligned}$$

with constant independent of m, η and with s from above.

Due to compact imbedding theorems and trace theorems for Sobolev spaces, there exists a sequence m_k of Galerkin parameters and corresponding sequences $\boldsymbol{u}_k = \boldsymbol{u}_{m_k}$ and $\Theta_k = \Theta_{m_k}$ of solutions to the Galerkin approximations (5.4.61) and (5.4.62) such that the following convergences are valid:

$$\begin{aligned}
\dot{\boldsymbol{u}}_k &\rightharpoonup \dot{\boldsymbol{u}} \text{ in } \boldsymbol{H}^{1/2,1}(Q_T) \text{ and strongly in } L_2(S_T), \\
\Theta_k &\rightharpoonup \Theta \text{ in } H^{s,1}(Q_T), \text{ in } L_{2+\gamma+4/N}(Q_T) \\
&\qquad\qquad \text{and strongly in } L_2(Q_T), \\
[\Theta_k]_+ &\to \Theta_+ \text{ in } L_{r_R}(S_T), \\
\boldsymbol{u}_k(T,\cdot) &\rightharpoonup \boldsymbol{u}(T,\cdot) \text{ in } \boldsymbol{H}^1(\Omega) \\
\dot{\boldsymbol{u}}_k(T,\cdot) &\rightharpoonup \dot{\boldsymbol{u}}(T,\cdot) \text{ in } \boldsymbol{L}_2(\Omega) \text{ and} \\
\Theta_k(T,\cdot) &\rightharpoonup \Theta(T,\cdot) \text{ in } L_2(\Omega).
\end{aligned} \tag{5.4.63}$$

Due to the strong convergence of Θ_k in $L_2(Q_T)$ and the weak convergence in $L_{2+\gamma+4/N}(Q_T)$ we also have strong convergence of $c_{ij}(\Theta_k)$ in $L_r(Q_T)$ for $r < 1 + 2/\gamma + 4/(N\gamma)$. This yields the strong convergence of $c_{ij}(\Theta_m) \to c_{ij}(\Theta)$ in $L_2(Q_T)$, provided

$$\gamma < 2 + 4/N$$

holds.

In the limit $k \to +\infty$ it is necessary to ensure the convergence of the quadratic viscous heat term in the heat equation. Therefore, the limit is performed first in the Galerkin equation (5.4.61) of the contact problem. This is done after an integration of this equation with respect to $t \in I_T$ for a fixed test function $\boldsymbol{v} \in C^1(I_T, \mathscr{U}_m)$. Employing a partial integration of some terms with respect to time and the weak lower semicontinuity of positive definite bilinear forms it is easy to verify that the limit functions $(\boldsymbol{u}, \Theta)$ satisfy problem (5.4.58). Due to the density of $C^1\big(I_T; \cup_{m=1}^{\infty} \mathscr{U}_m\big)$ in $\mathfrak{U}$ this is true for all test functions $\boldsymbol{v} \in \mathfrak{U}$. Here the uniform boundedness of $\dot{\boldsymbol{u}}$ in $\boldsymbol{H}^{1/2,1}(Q_T)$ is essential. Then, in a next step, it is verified that the convergence of $\dot{\boldsymbol{u}}_k$ is in fact strong in $L_2\big(I_T; \boldsymbol{H}^1(\Omega)\big)$. This is done by adding the term $A^{(1)}(\dot{\boldsymbol{u}}, \dot{\boldsymbol{u}} - \dot{\boldsymbol{u}}_k)$ to both sides of the time integrated version of (5.4.61) with test function $\dot{\boldsymbol{u}}_k - \pi_{\mathscr{U}_{m_k}} \dot{\boldsymbol{u}}$. Then the limit $k \to \infty$ in the obtained relation proves

$$\lim_{k \to \infty} A^{(1)}(\dot{\boldsymbol{u}} - \dot{\boldsymbol{u}}_k, \dot{\boldsymbol{u}} - \dot{\boldsymbol{u}}_k) = 0.$$

From this and the ellipticity of $A^{(1)}(\cdot,\cdot)$ there follows the strong convergence $\dot{\boldsymbol{u}}_k \to \dot{\boldsymbol{u}}$ in $L_2\big(I_T; \boldsymbol{H}^1(\Omega)\big)$. Then the convergence can be performed in the heat equation for a test function $\varphi \in C^1(I_T, \mathscr{V}_m)$. This proves that $(\boldsymbol{u}, \Theta)$ solves also the heat equation (5.4.59). Due to the density of $C^1(I_T, \mathscr{V}_m)$ in $H^1\big(I_T; H^\ell(\Omega)\big)$ this is true for all test functions $\boldsymbol{v}$ from that space. □

The just described procedure also enables the limit procedure for the smoothing parameter $\eta_k \to 0$. Thus the following lemma is verified:

5.4.14 Lemma. *Under assumptions of Lemma 5.4.13 there exists a couple $[\boldsymbol{u}, \Theta]$ that satisfies all requirements of this lemma and the variational relations* (5.4.56), (5.4.57).

In order to perform the next limit procedure $M \to +\infty$ it is necessary to establish more sophisticated *a priori* estimates independent of this parameter. The basis of the following considerations is the regularity estimate (5.4.47) for the contact problem with given temperature. Moreover, the *non-negativity* of the temperature field will be crucial. It is proved by inserting the test function $\varphi = \Theta_- \chi_{t_0}$ with $\Theta_- \equiv \min\{0, \Theta\}$ and the characteristic function χ_{t_0} of the time interval $(0, t_0)$ into the heat equation (5.4.57). Due to $b_{ij} \mathsf{h}_M(\Theta_+)\Theta_- = 0$, $R(\Theta)\Theta_- = 0$, the non-negativity of the coefficient K, the function Υ, the frictional heat, the viscous heat and the initial temperature Θ_0 we obtain

$$\|\Theta_-(t_0)\|^2_{L_2(\Omega)} + \|\nabla \Theta_-\|^2_{L_2(Q_{t_0})} \le 0 \text{ for all } t_0 \in I_T. \tag{5.4.64}$$

Using this result it is easy to derive an *energy estimate*. We subtract inequality (5.4.56) with test function $\boldsymbol{v} = (1 - \chi_{t_0})\dot{\boldsymbol{u}}$ from equation (5.4.57) with test function $\varphi = \chi_{t_0}$. This yields

$$
\begin{aligned}
&\tfrac{1}{2}\|\dot{\boldsymbol{u}}(t_0)\|^2_{\boldsymbol{L}_2(\Omega)} + \tfrac{1}{2}\left\|a^{(0)}(\boldsymbol{u}(t_0), \boldsymbol{u}(t_0))\right\|_{L_1(\Omega)} + \|\Theta(t_0)\|_{L_1(\Omega)} \\
&\qquad + k_0\|\Theta\|_{L_1(S_{t_0})} + c_1 c_R\|\Theta\|^{r_R}_{L_{r_R}(S_{t_0})} \\
&\quad \le \tfrac{1}{2}\|\dot{\boldsymbol{u}}_0\|^2_{\boldsymbol{L}_2(\Omega)} + \tfrac{1}{2}\left\|a^{(0)}(\boldsymbol{u}_0, \boldsymbol{u}_0)\right\|_{L_1(\Omega)} + \|\Theta_0\|_{L_1(\Omega)} \\
&\qquad + \langle \mathscr{L}, \dot{\boldsymbol{u}}\rangle_{Q_{t_0}} + k_1\|\Upsilon\|_{L_1(S_{t_0})}
\end{aligned}
$$

with the lower and upper bounds k_0 and k_1 of K and c_1 of the function R from condition (5.4.50). This inequality is a consequence of the energy conservation of the system. The total energy, composed of kinetic energy $\frac{1}{2}\|\dot{\boldsymbol{u}}\|^2_{\boldsymbol{L}_2(\Omega)}$, elastic deformation energy $\frac{1}{2}\left\|a^{(0)}(\boldsymbol{u}, \boldsymbol{u})\right\|_{L_1(\Omega)}$ and thermal energy $\|\Theta\|_{L_1(\Omega)}$ at time t_0, is equal to initial energy plus energy flux across the boundary plus work done by the external forces. Application of the Gronwall lemma yields

$$
\begin{aligned}
&\|\dot{\boldsymbol{u}}\|^2_{L_\infty(I_T;\boldsymbol{L}_2(\Omega))} + \|\boldsymbol{u}\|^2_{L_\infty(I_T;\boldsymbol{H}^1(\Omega))} + \|\Theta\|_{L_\infty(I_T;L_1(\Omega))} \\
&\qquad + \|\Theta\|_{L_1(S_T)} + c_R\|\Theta\|^{r_R}_{L_{r_R}(S_T)} \le c
\end{aligned}
\tag{5.4.65}
$$

with a constant *independent* of the approximation parameters M, ε. The boundedness of Θ in $L_\infty(I_T; L_1(\Omega))$ is only a very weak result. However, in the estimates done below there appear non-monotone coupling terms with high order growth in lower order Sobolev spaces (compared to the monotone diffusion terms). In order to estimate them we employ the result of the energy estimate and interpolation results of Chapter 2 and thereby lower the order of growth at the expense of a higher order norm. This makes it possible to weaken the requirements for the growth exponent γ while still controlling those non-monotone terms. The calculations are rather technical, and in order to simplify the presentation we first prove three auxiliary estimates needed later:

5.4.15 Lemma. *(i) For* $\varphi \in H^{3/4,3/2}(Q_T)$ *there holds*

$$
\|\nabla\varphi\|_{L_p(I_T;L_q(\Omega))} \le c\|\varphi\|_{H^{3/4,3/2}(Q_T)},
$$

if $p, q \ge 2$ *and* $\dfrac{4}{p} + \dfrac{2N}{q} \ge N + 1$.

(ii) Let $\varphi \in H^{\beta,2\beta}(S_T)$, $0 < \beta < \dfrac{N+1}{4}$ *and* $p, q \ge 2$ *with* $\dfrac{2}{p} + \dfrac{N-1}{q} \ge \dfrac{N+1}{2} - 2\beta$. *Then*

$$
\|\varphi\|_{L_p(I_T;L_q(\Gamma))} \le c\|\varphi\|_{H^{\beta,2\beta}(S_T)}.
$$

(iii) For $\alpha \in (0,1)$ *and* $\varphi \in W_p^1(\Omega)$ *with* $\dfrac{2N}{N+2-2\alpha} \le p < N$ *there holds*

$$\|\varphi\|_{H^\alpha(\Omega)} \le c\|\varphi\|_{L_1(\Omega)}^{\lambda}\|\varphi\|_{W_p^1(\Omega)}^{1-\lambda}$$

with $\lambda < \dfrac{1}{2} - \dfrac{N-p+2p\alpha}{2(N(p-1)+p)}$.

Proof. (i): Using imbedding theorems for Sobolev spaces there follows

$$\begin{aligned}\|\nabla\varphi\|_{L_p(I_T;L_q(\Omega))} &\le c_1\|\nabla\varphi\|_{H^\alpha(I_T;H^\beta(\Omega))} \le c_2\|\varphi\|_{H^\alpha(I_T;H^{1+\beta}(\Omega))}\\ &\le c_3\|\varphi\|_{H^{3/4,3/2}(Q_T)},\end{aligned}$$

with $\alpha \ge 1/2 - 1/p$, $\beta \ge N/2 - N/q$ and $\alpha r = 3/4$, $(\beta+1)r' = 3/2$ for $r, r' \ge 1$ satisfying $1/r + 1/r' = 1$. From these conditions relation $\dfrac{4}{p} + \dfrac{2N}{q} \ge N+1$ follows easily.

(ii): In an analogous way as in the proof of part (i) just above, it is verified

$$\|\varphi\|_{L_p(I_T;L_q(\Gamma))} \le c_1\|\varphi\|_{H^\lambda(I_T;H^\mu(\Gamma))} \le c_2\|\varphi\|_{H^{\beta,2\beta}(S_{CT})}$$

with $\lambda \ge 1/2 - 1/p$, $\mu \ge (N-1)/2 - (N-1)/q$ and $2\beta = 2\lambda + \mu$. The latter equation leads to the requirement $\dfrac{2}{p} + \dfrac{N-1}{q} \ge \dfrac{N+1}{2} - 2\beta$.

(iii): The assertion is proved by the series of interpolation and imbedding results

$$\begin{aligned}\|\varphi\|_{H^\alpha(\Omega)} &\le c_1\|\varphi\|_{L_2(\Omega)}^{\mu}\|\varphi\|_{H^\beta(\Omega)}^{1-\mu} \le c_2\|\varphi\|_{L_1(\Omega)}^{\mu\nu}\|\varphi\|_{L_q(\Omega)}^{\mu(1-\nu)}\|\varphi\|_{H^\beta(\Omega)}^{1-\mu}\\ &\le c_3\|\varphi\|_{L_1(\Omega)}^{\mu\nu}\|\varphi\|_{W_p^1(\Omega)}^{1-\nu\mu}\end{aligned}$$

with $\beta = 1 - N/p + N/2$, $\mu = 1 - \alpha/\beta$, $q = Np/(N-p)$, $\nu = \big(1/2 - 1/q\big)/\big(1-1/q\big)$. Here the parameters β and q are calculated such that the imbeddings $W_p^1(\Omega) \hookrightarrow H^\beta(\Omega)$ and $W_p^1(\Omega) \hookrightarrow L_q(\Omega)$ are valid. The index λ is given by $\lambda = \nu\mu = 1/2 - N/\big(2(N(p-1)+p)\big)$.

□

Using the just established relations and the regularity result for the solution of the contact problem a new *a priori* estimate for the temperature is proved that is independent of the approximation parameters M and ε. For the problems with and without boundary heat dissipation the approach is slightly different. Let us first consider the case $R \ne 0$.

We assume first $\gamma < 1 - 1/N$. Let us take the test function $\varphi = \Theta^{1/N}$ in equation (5.4.57). From the heat diffusion term and the ellipticity condition

(5.4.48) we obtain the expression

$$\left\langle c_{ij}(\Theta)\partial_j\Theta, \partial_i\Theta^{1/N}\right\rangle_{Q_T} \geq c\int_{Q_T} \Theta^{\gamma+1/N-1}|\nabla\Theta|^2 dx\, dt.$$

For $p = 2/\big(2 - 1/N - \gamma\big)$ there holds

$$\begin{aligned}\|\nabla\Theta\|^2_{L_2(I_T;L_p(\Omega))} &= \int_{I_T}\left(\int_\Omega |\nabla\Theta|^p\Theta^{p/2-1}\Theta^{1-p/2}\, dx\right)^{2/p} dt \\ &\leq \int_{I_T}\left(\int_\Omega |\nabla\Theta|^2\Theta^{1-2/p}\, dx\right)\left(\int_\Omega \Theta\, dx\right)^{2/p-1} dt \\ &\leq \|\Theta\|^{2/p-1}_{L_\infty(I_T;L_1(\Omega))}\int_{I_T}\int_\Omega |\nabla\Theta|^2\Theta^{\gamma+1/N-1}\, dx\, dt.\end{aligned}$$

Hence, using the test function mentioned above the first two terms in (5.4.57) and the boundary heat dissipation term yield the expression

$$\|\Theta(T,\cdot)\|^{1+1/N}_{L_{1+1/N}(\Omega)} + \|\Theta\|^2_{L_2(I_T;W^1_p(\Omega))} + \|\Theta\|^{r_R+1/N}_{L_{r_R+1/N}(S_T)}$$

on the left hand side of an inequality to be established and all the remaining terms must be estimated by this.

For the viscous heat we get

$$\int_{Q_T} |\nabla\dot{\boldsymbol{u}}|^2\Theta^{1/N}\, dx\, dt \leq \|\nabla\dot{\boldsymbol{u}}\|^2_{L_2(I_T;L_{2N/(N-1)}(\Omega))}\|\Theta\|^{1/N}_{L_\infty(I_T;L_1(\Omega))}.$$

The norm of the velocity is estimated by the application of parts (i) and (iii) of Lemma 5.4.15 and the regularity result (5.4.47) for the solution of the contact problem,

$$\begin{aligned}\|\nabla\dot{\boldsymbol{u}}\|_{L_2(I_T;L_{2N/(N-1)}(\Omega))} &\leq c_1\|\dot{\boldsymbol{u}}\|_{\boldsymbol{H}^{3/4,\,3/2}(Q_T)} \\ &\leq c_2\|\Theta\|_{L_2(I_T;H^\alpha(\Omega))} + c_3 \\ &\leq c_4\|\Theta\|^\lambda_{L_\infty(I_T;L_1(\Omega))}\|\Theta\|^{1-\lambda}_{L_2(I_T;W^1_p(\Omega))} + c_5\end{aligned}$$

with $\alpha > 1/2$, index $p = 2/\big(2 - 1/N - \gamma\big)$ from above and λ from part (iii) of Lemma 5.4.15. Here condition $p > 2N/(N+1)$ is required in order to have $\lambda > 0$ for α sufficiently close to $1/2$. This condition is equivalent to

$$\gamma > 1 - \frac{2}{N}.$$

Then the viscous heat is bounded by

$$c_1\|\Theta\|^{2(1-\lambda)}_{L_2(I_T;W^1_p(\Omega))} + c_2. \tag{5.4.66}$$

For the deformation heat a similar procedure yields

$$\int_{Q_T} |\nabla \dot{\boldsymbol{u}}| \Theta^{1+1/N}\, d\boldsymbol{x}\, dt \leq \|\nabla \dot{\boldsymbol{u}}\|_{L_2(I_T;L_{2N/(N-1)}(\Omega))} \|\Theta\|^{1/N}_{L_\infty(I_T;L_1(\Omega))} \cdot$$

$$\cdot \|\Theta\|_{L_2(I_T;L_{2N/(N-1)}(\Omega))} \leq c_1 \|\Theta\|^{2-\lambda}_{L_2(I_T;W^1_p(\Omega))} + c_2$$

with $\lambda > 0$ from above. Here in the last step the imbedding

$$H^\alpha(\Omega) \hookrightarrow L_{2N/(N-1)}(\Omega)$$

for $\alpha \geq 1/2$ has been used.

The frictional heat is estimated by

$$\int_{S_{CT}} \tfrac{1}{\varepsilon}[\dot{u}_n]_+ |\dot{\boldsymbol{u}}_{\boldsymbol{t}}| \Theta^{1/N}\, ds_{\boldsymbol{x}}\, dt$$

$$\leq \left\| \tfrac{1}{\varepsilon}[\dot{u}_n]_+ \right\|_{L_2(S_T)} \|\dot{\boldsymbol{u}}\|_{L_q(S_T)} \|\Theta\|^{1/N}_{L_{r_R+1/N}(S_T)}$$

with $q = (2Nr_R + 2)/\big(Nr_R - 1\big)$. The norm of the displacement velocities is bounded by

$$\|\dot{\boldsymbol{u}}\|_{L_q(S_T)} \leq c_1 \|\dot{\boldsymbol{u}}\|_{H^{\beta,2\beta}(S_T)} \leq c_2 \|\dot{\boldsymbol{u}}\|^\nu_{H^{1/4,\,1/2}(S_T)} \|\dot{\boldsymbol{u}}\|^{1-\nu}_{H^{1/2,\,1}(S_T)}$$

with $\beta = \big(1/2 - 1/q\big)(N+1)/2$ and $\nu = 2 - 4\beta = \big(2N(r_R - 1)\big)/\big(Nr_R + 1\big)$. Using once more Assumption (5.4.47) and the standard *a priori* estimate for the contact problem

$$\|\dot{\boldsymbol{u}}\|_{\boldsymbol{H}^{1/2,\,1}(Q_T)} \leq c_1 \|\Theta\|_{L_2(Q_T)} + c_2,$$

the frictional heat is further estimated by

$$c\|\Theta\|^{2-\nu}_{L_2(I_T;H^\alpha(\Omega))} \|\Theta\|^\nu_{L_2(Q_T)} \|\Theta\|^{1/N}_{L_{r_R+1/N}(S_T)}.$$

Employing Lemma 5.4.15, part (iii) and the series of imbeddings and interpolations

$$\|\Theta\|_{L_2(\Omega)} \leq c\|\Theta\|^\mu_{L_s(\Omega)} \|\Theta\|^{1-\mu}_{L_1(\Omega)} \leq \|\Theta\|^\mu_{W^1_p(\Omega)} \|\Theta\|^{1-\mu}_{L_1(\Omega)} \tag{5.4.67}$$

with indices $s = Np/(N-p)$ (from the imbedding $W^1_p(\Omega) \hookrightarrow L_s(\Omega)$) and $\mu = s/(2s-2) = Np/\big(2(Np - N + p)\big)$ and Hölder's inequality yields an estimate of the frictional heat by

$$c(\eta)\|\Theta\|^\varkappa_{L_2(I_T;W^1_p(\Omega))} + \eta\|\Theta\|^{r_R+1/N}_{L_{r_R+1/N}(S_T)}$$

valid for every $\eta > 0$ with exponent

$$\varkappa = (2 - \nu + \nu\mu)\left(1 + \frac{1}{Nr_R}\right).$$

The condition $\varkappa < 2$ leads, after some calculation, to the requirement

$$r_R > \frac{p(2N-1) - 2N}{N(p(N+2) - 2N)}.$$

Since $p > \dfrac{2N}{N+1}$ holds, this is satisfied for all $r_R \geq 2 + \dfrac{1}{N}$.

Collecting all these estimates we realize that the inequality

$$\|\Theta\|_{L_2(I_T;W_p^1(\Omega))} + \|\Theta\|_{L_{r_R+1/N}(S_T)} \leq c \tag{5.4.68}$$

with constant c independent of the approximation parameters M and ε is established. Moreover, due to (5.4.47) the norms of $\dot{\boldsymbol{u}}$ in $\boldsymbol{H}^{3/4,3/2}(Q_T)$ and $\boldsymbol{H}^{1/2,1}(S_T)$ are uniformly bounded with respect to M and ε. This result is now verified for $1 - 2/N < \gamma < 1 - 1/N$.

From this first estimate we can obtain a better one by taking the test function $\varphi = \Theta^{1-\gamma}$ in the heat equation (5.4.57). Then, on the left hand side of the new estimate we have

$$\|\Theta(T,\cdot)\|^{2-\gamma}_{L_{2-\gamma}(\Omega)} + \|\nabla\Theta\|^2_{L_2(Q_T)} + \|\Theta\|^{r_R+1-\gamma}_{L_{r_R+1-\gamma}(S_T)}.$$

In the estimate of the remaining terms it is now possible to use the just proved uniform bounds for $\dot{\boldsymbol{u}}$ in $\boldsymbol{H}^{3/4,3/2}(Q_T)$ and in $\boldsymbol{H}^{1/2,1}(S_T)$. For the viscous heat the inequality

$$\int_{Q_T} |\nabla\dot{\boldsymbol{u}}|^2\Theta^{1-\gamma}\,dx\,dt \leq \|\nabla\dot{\boldsymbol{u}}\|^2_{L_{4/(1+\gamma)}(I_T;L_{2N/(N-\gamma)}(\Omega))}\cdot$$

$$\cdot\|\Theta\|^{1-\gamma}_{L_2(I_T;L_{N(1-\gamma)/\gamma}(\Omega))} \leq c\|\nabla\dot{\boldsymbol{u}}\|^2_{\boldsymbol{H}^{3/4,3/2}(Q_T)}\|\Theta\|^{1-\gamma}_{L_2(I_T;H^1(\Omega))}$$

is valid. Here the imbedding $H^1(\Omega) \hookrightarrow L_{N(1-\gamma)/\gamma}(\Omega)$ valid for $\gamma > 1 - 2/N$ has been employed. For the deformation heat

$$\int_{Q_T} |\nabla\dot{\boldsymbol{u}}|\Theta^{2-\gamma}\,dx\,dt \leq c_1\|\nabla\dot{\boldsymbol{u}}\|_{L_4(I_T;L_2(\Omega))}\|\Theta\|^{2-\gamma}_{L_{4(2-\gamma)/3}(I_T;L_{2(2-\gamma)}(\Omega))}$$

$$\leq c_2\|\dot{\boldsymbol{u}}\|_{\boldsymbol{H}^{3/4,3/2}(Q_T)}\|\Theta\|^{(2-\gamma)\nu}_{L_\infty(I_T;L_1(\Omega))}\|\Theta\|^{(2-\gamma)(1-\nu)}_{L_2(I_T;H^1(\Omega))}$$

with $(4-N)/(2N+4) \leq \nu \leq 4/(N+2)^2$ and $\nu > 0$. For the frictional heat the estimate

$$\begin{aligned}\int_{S_T} \tfrac{1}{\varepsilon}[\dot{u}_n]_+|\dot{\boldsymbol{u}}_{\boldsymbol{t}}|\Theta^{1-\gamma}\, ds_{\boldsymbol{x}}\, dt \ &\leq \ \left\|\tfrac{1}{\varepsilon}[\dot{u}_n]_+\right\|_{L_2(S_T)} \cdot \\ &\cdot \|\dot{\boldsymbol{u}}\|_{L_{2/\gamma}(I_T;\boldsymbol{L}_{2(N-1)/(N-1-2\gamma)}(\Gamma))}\|\Theta\|^{1-\gamma}_{L_2(I_T;L_{(N-1)(1-\gamma)/\gamma}(\Gamma))} \\ &\leq c\left\|\tfrac{1}{\varepsilon}[\dot{u}_n]_+\right\|_{L_2(S_T)} \|\dot{\boldsymbol{u}}\|_{\boldsymbol{H}^{1/2,1}(S_T)}\|\Theta\|^{1-\gamma}_{L_2(I_T;H^1(\Omega))}\end{aligned}$$

is true. Here the imbedding $H^{1/2}(\Gamma) \hookrightarrow L_p(\Gamma)$ valid for $p = (N-1)(1-\gamma)/\gamma < (2N-2)/(N-2)$ for $N \geq 3$ and $p < +\infty$ for $N = 2$ is employed. Hence, the collection of all these new estimates proves

$$\|\Theta\|_{L_\infty(I_T;L_{2-\gamma}(\Omega))} + \|\Theta\|_{L_2(I_T;H^1(\Omega))} + c_R\|\Theta\|_{L_{r_R+1-\gamma}(S_T)} \leq c \quad (5.4.69)$$

with constant c independent of any approximation parameter.

This estimate will now be derived for the case $1 - 1/N \leq \gamma < 1$. Here, the test function $\varphi = \Theta^{1-\gamma}$ is employed directly. Observe that the power $1-\gamma$ of the test function is in fact not bigger than $1/N$. Hence the estimates made above for the viscous heat, the deformation heat and the frictional heat can be easily reformulated. For the viscous heat we obtain

$$\begin{aligned}&\int_{Q_T} |\nabla\dot{\boldsymbol{u}}|^2\Theta^{1-\gamma}\, dx\, dt \leq \|\nabla\dot{\boldsymbol{u}}\|^2_{L_2(I_T;L_{2/\gamma}(\Omega))}\|\Theta\|^{1-\gamma}_{L_\infty(I_T;L_1(\Omega))} \\ &\leq c_1\|\Theta\|^2_{L_2(I_T;H^\alpha(\Omega))} \leq \ c_2\|\Theta\|^{2\lambda}_{L_\infty(I_T;L_1(\Omega))}\|\Theta\|^{2(1-\lambda)}_{L_2(I_T;H^1(\Omega))}\end{aligned} \quad (5.4.70)$$

with $\lambda < 1/(N+2)$. Here part (iii) of Lemma 5.4.15 is used with $p = 2$ in the case $N > 2$ and $4/3 < p < 2$ in the case $N = 2$. The deformation heat yields

$$\begin{aligned}\int_{Q_T} |\nabla\dot{\boldsymbol{u}}|\Theta^{2-\gamma}\, dx\, dt &\leq c_1\|\nabla\dot{\boldsymbol{u}}\|_{L_2(I_T;L_{2N/(N-1)}(\Omega))}\|\Theta\|^{2-\gamma}_{L_{2+2/N}(I_T;L_2(\Omega))} \\ &\leq c_2\|\Theta\|_{L_2(I_T;H^\alpha(\Omega))}\|\Theta\|^{(2-\gamma)\lambda}_{L_\infty(I_T;L_1(\Omega))}\|\Theta\|^{(2-\gamma)(1-\lambda)}_{L_2(I_T;H^1(\Omega))} \\ &\leq c_3\|\Theta\|^{1-\nu+(1-\lambda)(2-\gamma)}_{L_2(I_T;H^1(\Omega))}\|\Theta\|^{\nu+(2-\gamma)\lambda}_{L_\infty(I_T;L_1(\Omega))}\end{aligned}$$

with $\alpha > 1/2$, $1/(N+1) < \lambda < 2/(N+2)$ and $\nu < 1/(N+2)$. Here, in the last step part (iii) of Lemma 5.4.15 was again employed. If the parameters are chosen big enough, then the exponent satisfies $1-\nu+(1-\lambda)(2-\gamma) < 2$.

The frictional heat is estimated by

$$\int_{S_{CT}} \tfrac{1}{\varepsilon}[\dot{u}_n]_+|\dot{\boldsymbol{u}}_{\boldsymbol{t}}|\Theta^{1-\gamma}\, ds_{\boldsymbol{x}}\, dt \le \left\| \tfrac{1}{\varepsilon}[\dot{u}_n]_+ \right\|_{L_2(S_T)} \cdot$$

$$\cdot \|\dot{\boldsymbol{u}}\|_{\boldsymbol{L}_{2(N+1)/(N-1)}(S_T)} \|\Theta\|^{1-\gamma}_{L_{(N+1)(1-\gamma)}(S_T)}$$

$$\le c_1 \left\| \tfrac{1}{\varepsilon}[\dot{u}_n]_+ \right\|_{L_2(S_T)} \|\dot{\boldsymbol{u}}\|_{H^{1/2,1}(S_T)} \|\Theta\|^{1-\gamma}_{L_2(S_T)}$$

$$\le c_2 \|\Theta\|^{3-\gamma}_{L_2(I_T;H^\alpha(\Omega))} \le c_3 \|\Theta\|^{(3-\gamma)\lambda}_{L_\infty(I_T;L_1(\Omega))} \|\Theta\|^{(3-\gamma)(1-\lambda)}_{L_2(I_T;H^1(\Omega))}$$

with $\alpha > 1/2$ sufficiently close to $1/2$ and $0 < \lambda < 1/(N+2)$. Hence, for a suitable choice of λ the exponent $(3-\gamma)(1-\lambda)$ is again smaller than 2. Observe that we do not need the boundary heat dissipation term for the estimate here. Collecting all the estimates, we obtain again inequality (5.4.69).

In the case $\gamma \ge 1$ we use the test function $\varphi = \Theta^{1/(N+1)}\chi_{t_0}$. Due to the ellipticity relation

$$c_{ij}(\Theta)\, \partial_j\Theta\, \partial_i\big(\Theta^{1/(N+1)}\big) \ge c_1 \Theta^{1/(N+1)-1}\big(1+\Theta^\gamma\big)|\nabla\Theta|^2 \ge c_2|\nabla\Theta|^2$$

this is sufficient to obtain an estimate for $\|\Theta\|_{L_2(I_T;H^1(\Omega))}$. In fact, we will get an estimate for

$$\|\Theta\|_{L_\infty(I_T;L_{1+1/(N+1)}(\Omega))} + \|\Theta\|_{L_2(I_T;H^1(\Omega))} + \left\|\Theta^{(1+1/(N+1)+\gamma)/2}\right\|_{L_2(I_T;H^1(\Omega))}. \tag{5.4.71}$$

Since the $H^1(\Omega)$–norm is considerably stronger than the $W^1_p(\Omega)$-norm in the previous cases and the power $1/(N+1)$ of Θ is smaller, the estimates are much easier than before. For the viscous heat we obtain

$$\int_{Q_T} |\nabla\dot{\boldsymbol{u}}|^2\Theta^{1/(N+1)}\, d\boldsymbol{x}\, dt$$

$$\le c_1 \|\nabla\dot{\boldsymbol{u}}\|^2_{L_2(I_T;L_{2N/(N-1)}(\Omega))} \|\Theta\|^{1/(N+1)}_{L_\infty(I_T;L_1(\Omega))}$$

$$\le c_2 \|\Theta\|^2_{L_2(I_T;H^\alpha(\Omega))} + c_3$$

with α arbitrarily close to $1/2$. For the $H^\alpha(\Omega)$–norm of Θ the interpolation

$$\|\Theta\|_{H^\alpha(\Omega)} \le c_1 \|\Theta\|^{1-\lambda}_{H^1(\Omega)} \|\Theta\|^{\lambda}_{L_1(\Omega)}$$

with $\lambda < \dfrac{1}{2} - \dfrac{N-2+4\alpha}{2(N+2)} > 0$ is valid. Hence the viscous heat is bounded by

$$c_1 \|\Theta\|^{2(1-\lambda)}_{L_2(I_T;H^1(\Omega))} + c_2.$$

The deformation heat is bounded by

$$\int_{Q_T} |\nabla \dot{\boldsymbol{u}}| \Theta^{1+1/(N+1)} \, dx \, dt$$
$$\leq c_1 \|\nabla \dot{\boldsymbol{u}}\|_{L_2(I_T; L_{2N/(N-1)}(\Omega))} \|\Theta\|^{1/(N+1)}_{L_\infty(I_T; L_1(\Omega))} \|\Theta\|_{L_2(I_T; L_{2N/(N-1)}(\Omega))}$$
$$\leq c_2 \|\Theta\|^2_{L_2(I_T; H^\alpha(\Omega))} + c_3$$

which is further estimated as in the case of the viscous heat. For the frictional heat we have

$$\int_{S_T} \mathfrak{F} |\sigma_n(\boldsymbol{u}, \Theta)| |\dot{\boldsymbol{u}}_t| \Theta^{1/(N+1)} \, dx \, dt$$
$$\leq \|\mathfrak{F}\|_{L_\infty(S_T)} \|\sigma_n(\boldsymbol{u}, \Theta)\|_{L_2(S_T)} \|\dot{\boldsymbol{u}}_t\|_{L_{2+2/N}(S_T)} \|\Theta\|^{1/(N+1)}_{L_2(S_T)}.$$

The norm of $\dot{\boldsymbol{u}}$ is bounded by

$$\|\dot{\boldsymbol{u}}\|_{L_{2+2/N}(S_T)} \leq c_1 \|\boldsymbol{u}\|_{H^{1/4, 1/2}(S_T)} \leq c_1 \|\Theta\|_{L_2(Q_T)} + c_2;$$

this leads to an estimate of the frictional heat by

$$\|\Theta\|^{1+1/(N+1)}_{L_2(I_T; H^1(\Omega))} \big(c_1 \|\Theta\|_{L_2(Q_T)} + c_2\big).$$

The $L_2(\Omega)$–norm here is further estimated by

$$\|\Theta\|_{L_2(\Omega)} \leq c_1 \|\Theta\|^\lambda_{H^1(\Omega)} \|\Theta\|^{1-\lambda}_{L_1(\Omega)}$$

with $\lambda < \dfrac{N}{N+2}$. The frictional heat is thus bounded by

$$c_1 \|\Theta\|^\mu_{L_2(I_T; H^1(\Omega))} + c_2$$

with exponent $\mu = 1 + \dfrac{1}{N+1} + \lambda < 2$. This finishes the proof of

$$\begin{aligned} \|\Theta\|_{L_\infty(I_T; L_{1+1/(N+1)}(\Omega))} &+ \|\Theta\|_{L_2(I_T; H^1(\Omega))} \\ &+ \|\Theta^{(1+1/(N+1)+\gamma)/2}\|_{L_2(I_T; H^1(\Omega))} \leq c. \end{aligned} \tag{5.4.72}$$

Observe that no heat dissipation at the boundary is necessary for these estimates.

With this information, we derive a dual estimate for the temperature. For a test function $\varphi \in L_\infty\big(I_T; \mathring{H}^\beta(\Omega)\big)$ with $\beta > 1 + N/2$ in the heat equation we estimate all terms except the time derivative $\langle \dot{\Theta}, \varphi \rangle_{Q_T}$. The index β is chosen such that $W^1_\infty(\Omega)$ is imbedded into $H^\beta(\Omega)$. For the heat diffusion term and $\gamma \le 1$ we use the estimate

$$\int_{Q_T} |\Theta|^\gamma |\nabla\Theta||\nabla\varphi|\,d\boldsymbol{x}\,dt \le \|\Theta^\gamma\|_{L_2(Q_T)}\|\nabla\Theta\|_{L_2(Q_T)}\|\nabla\varphi\|_{L_\infty(Q_T)}.$$

In the case $\gamma > 1$ we have an additional bound for $\|\Theta^{(1+\gamma)/2}\|_{L_2(I_T;H^1(\Omega))}$ that is available from estimate (5.4.72). Then the heat diffusion term can be estimated by

$$\begin{aligned}\int_{Q_T} &|\Theta|^\gamma |\nabla\Theta||\nabla\varphi|\,d\boldsymbol{x}\,dt \\ &\le c_1(\gamma)\|\Theta^{(1+\gamma)/2}\|_{L_2(Q_T)}\|\nabla\Theta^{(1+\gamma)/2}\|_{L_2(Q_T)}\|\nabla\varphi\|_{L_\infty(Q_T)} \\ &\le c_2(\gamma)\|\Theta^{(1+\gamma)/2}\|^2_{L_2(I_T;H^1(\Omega))}\|\nabla\varphi\|_{L_\infty(Q_T)}.\end{aligned}$$

The viscous heat is bounded by

$$\begin{aligned}\int_{Q_T} |\nabla\dot{\boldsymbol{u}}|^2|\varphi|\,d\boldsymbol{x}\,dt &\le \|\nabla\dot{\boldsymbol{u}}\|^2_{L_4(I_T;\boldsymbol{L}_2(\Omega))}\|\varphi\|_{L_2(I_T;L_\infty(\Omega))} \\ &\le c\|\dot{\boldsymbol{u}}\|^2_{\boldsymbol{H}^{3/4,\,3/2}(Q_T)}\|\varphi\|_{L_2(I_T;L_\infty(\Omega))}.\end{aligned}$$

The deformation heat term is estimated by

$$\int_{Q_T} |\nabla\dot{\boldsymbol{u}}|\Theta|\varphi|\,d\boldsymbol{x}\,dt \le \|\nabla\dot{\boldsymbol{u}}\|_{L_2(Q_T)}\|\Theta\|_{L_2(Q_T)}\|\varphi\|_{L_\infty(Q_T)}.$$

Hence the dual estimate yields a uniform bound for $\dot{\Theta}$ in $L_\infty\big(I_T; H^\beta(\Omega)\big)^*$ with arbitrary $\beta > 1 + N/2$. This space is imbedded in $H^\alpha\big(I_T; H^\beta(\Omega)\big)^*$ with $\alpha > 1/2$. Interpolation with $\Theta \in L_2\big(I_T; H^1(\Omega)\big)$ yields a bound for $\|\Theta\|_{H^s(I_T;L_2(\Omega))}$ with $s < 1/(2\beta + 2) < 1/(4 + N)$.

In the case without boundary heat dissipation, it is only necessary to change the first estimate (5.4.68) done for $\gamma < 1 - \dfrac{1}{N}$. The heat dissipation term $R(\Theta)$ in the boundary condition (5.4.49) was required there in order to get an estimate for the frictional heat. The requirement for the growth coefficient γ was calculated from the necessity to estimate the viscous heat term. If the boundary heat dissipation is neglected, then the estimate of the frictional heat requires the slightly worse lower bound

$$\gamma > 1 - \frac{1}{N} - \frac{1}{N+1}.$$

Then estimate (5.4.68) can be recalculated in a modified way for the test function $\Theta^{1/(N+1)}$. The index p in the $W^1_p(\Omega)$ norm is chosen to be $p =$

$2/(2-1/(N+1)-\gamma)$; it is then again bigger than $2N/(N+1)$ and the crucial imbedding $W_p^1(\Omega) \hookrightarrow H^\alpha(\Omega)$ for every $\alpha > 1/2$ is valid. The estimates of the viscous heat and the deformation heat can be easily rewritten for the test function $\Theta^{1/(N+1)}$. In fact, they are valid with $\|\Theta\|^{1/N}_{L_\infty(I_T;L_1(\Omega))}$ replaced by $c\|\Theta\|^{1/(N+1)}_{L_\infty(I_T;L_1(\Omega))}$. The estimate for the frictional heat is

$$\begin{aligned}\int_{S_{CT}} \mathfrak{F}\frac{1}{\varepsilon}[\dot{u}_n]_+|\dot{\boldsymbol{u}}_{\boldsymbol{t}}|\Theta^{1/(N+1)}\,ds_{\boldsymbol{x}}\,dt &\le \left\|\frac{1}{\varepsilon}[\dot{u}_n]_+\right\|_{L_2(S_T)}\cdot \\ &\quad\cdot\|\dot{\boldsymbol{u}}\|_{\boldsymbol{L}_{2(N+1)/(N-1)}(S_T)}\|\Theta\|^{1/(N+1)}_{L_1(S_T)} \\ &\le c_1\|\Theta\|^2_{L_2(I_T;H^\alpha(\Omega))} \le c_2\|\Theta\|^{2(1-\lambda)}_{L_2(I_T;W_p^1(\Omega))} + c_3\end{aligned}$$

with $\lambda > 0$ given in part (iii) of Lemma 5.4.15. Hence the estimate

$$\|\Theta\|_{L_\infty(I_T;L_{1+1/(N+1)}(\Omega))} + \|\Theta\|_{L_2(I_T;W_p^1(\Omega))} \le c$$

with c independent of the approximation parameters ε and M is derived. From this, the better estimate (5.4.69) can be obtained as well; its derivation did not require the boundary dissipation term.

The relations proved are collected in the following Lemma:

5.4.16 Lemma. *Let Assumption 5.4.11 be valid. Then every solution of the approximate thermo-viscoelastic contact problem* (5.4.56), (5.4.57) *satisfies the* a priori *estimate*

$$\begin{aligned}\|\dot{\boldsymbol{u}}\|_{\boldsymbol{H}^{3/4,3/2}(Q_T)} + \|\Theta\|_{L_\infty(I_T;L_{1+r}(\Omega))} + \|\Theta\|_{H^{s,1}(Q_T)} \\ + c_R\|\Theta\|_{L_{r_R+r}(S_T)} \le c\end{aligned}$$

with $r = 1-\gamma$ *for* $\gamma < 1$, $r = 1/(N+1)$ *for* $\gamma \ge 1$ *and* $0 < s < 1/(4+N)$. *The constant* c *is independent of the approximation parameters* ε *and* M.

From these estimates, for every fixed penalty parameter $\varepsilon > 0$ there follows the existence of a sequence of parameters M_k tending to $+\infty$ and of a corresponding sequence of solutions $(\boldsymbol{u}_k, \Theta_k)$ to the approximate problem

(5.4.56), (5.4.57) such that the following convergences are valid:

$$\begin{aligned}
&\dot{\boldsymbol{u}}_k \rightharpoonup \dot{\boldsymbol{u}} && \text{in } \boldsymbol{H}^{3/4,\,3/2}(Q_T) \text{ and} \\
& && \text{strongly in } L_2\big(I_T; \boldsymbol{H}^1(\Omega)\big) \cap \boldsymbol{L}_p(S_T) \\
& && \text{for all } p < 2 + 4/(N-1), \\
&\Theta_k \rightharpoonup \Theta && \text{in } H^{s,1}(Q_T) \text{ with } s < 1/(4+N) \\
& && \text{and strongly in } L_2(Q_T), \\
&\boldsymbol{u}_k(T,\cdot) \rightharpoonup \boldsymbol{u}(T,\cdot) \text{ in } \boldsymbol{H}^1(\Omega), \\
&\dot{\boldsymbol{u}}_k(T,\cdot) \rightharpoonup \dot{\boldsymbol{u}}(T,\cdot) \text{ in } \boldsymbol{L}_2(\Omega), \\
&\Theta_k(T,\cdot) \rightharpoonup \Theta(T,\cdot) \text{ in } L_1(\Omega).
\end{aligned} \tag{5.4.73}$$

In the case with boundary heat dissipation there holds moreover

$$\Theta_k \rightharpoonup \Theta \text{ in } L_{r_R+r}(S_T) \text{ and strongly in } L_{r_R}(S_T)$$

with the r mentioned in the Lemma. Consequently, passing to the limit $k \to +\infty$ in the approximate problem with test functions $\boldsymbol{v} \in \boldsymbol{H}^{1/2,\,1}(Q_T)$ and $\varphi \in H^1\big(I_T; H^\ell(\Omega)\big)$ with $\ell > N/2+1$ it is easily verified that the limit functions $(\boldsymbol{u}, \Theta)$ are a solution of the penalized problem, that is (5.4.56), (5.4.57) with h_M replaced by the identity operator $\mathsf{h}_M(s) = s$.

This convergence procedure can be repeated for penalty parameter tending to zero. For a suitable sequence $\varepsilon_k \to 0$ and for a corresponding sequence of solutions $(\boldsymbol{u}_k, \Theta_k)$ of the penalized problem and its limit $(\boldsymbol{u}, \Theta)$ the convergences mentioned in (5.4.73) are valid. Moreover, for the normal traction there holds

$$\sigma_n(\boldsymbol{u}_k, \Theta_k) = -\frac{1}{\varepsilon_k}\big[(\dot{u}_k)_n\big]_+ \rightharpoonup \sigma_n(\boldsymbol{u}, \Theta) \text{ in } L_2(S_T).$$

The limit $k \to +\infty$ is done for test functions $\boldsymbol{v} \in \mathscr{C}$ and $\varphi \in W^1_\infty(Q_T)$ in the usual way. Here, due to the requirement $v_n \leq 0$ the penalty term is bounded from above by 0. Hence the limit functions $(\boldsymbol{u}, \Theta)$ are a solution of the original thermo-viscoelastic contact problem (5.4.53), (5.4.54). Thus, Theorem 5.4.12 is proved.

5.4.2.2 A problem with heat conductivity dependent on the temperature gradient

We still solve the problem (5.4.5), (5.4.6) with the boundary and initial conditions (5.4.7)–(5.4.12). Instead of the temperature-dependent tensor of thermal conductivity treated in the previous model we assume here that $\{c_{ij}\}$ depends on the temperature gradient and is locally Lipschitz with respect to $\nabla\Theta$. It may also depend on the space variables such that the Carathéodory conditions hold. Instead of the earlier assumption (5.4.48) the following conditions will be required: the growth condition

$$c_1\left(1+|\nabla\Theta|^\gamma\right)\xi_i\xi_i \le c_{ij}(\nabla\Theta)\xi_i\xi_j \le c_2\left(1+|\nabla\Theta|^\gamma\right)\xi_i\xi_i \qquad (5.4.74)$$

for $\boldsymbol{\xi}\in\mathbb{R}^N$, the strong monotonicity condition

$$\begin{aligned}&\langle c_{ij}(\nabla\Theta)\partial_j\Theta - c_{ij}(\nabla\Psi)\partial_j\Psi, \partial_i\Theta-\partial_i\Psi\rangle_{Q_T}\\ &\quad\ge c_3\|\nabla(\Theta-\Psi)\|^{2+\gamma}_{\boldsymbol{L}_{2+\gamma}(Q_T)} + c_4\|\nabla(\Theta-\Psi)\|^2_{\boldsymbol{L}_2(Q_T)}\end{aligned} \qquad (5.4.75)$$

for each $\Theta,\Psi\in L_{2+\gamma}\big(I_T;W^1_{2+\gamma}(\Omega)\big)$, and the continuity relation

$$c_{ij}\big(\nabla\Theta^{(k)}\big)\partial_j\Theta^{(k)} \to c_{ij}(\nabla\Theta)\partial_j\Theta \text{ in } L_{(2+\gamma)/(1+\gamma)}(Q_T) \qquad (5.4.76)$$

for $i=1,\dots,N$ and $\Theta^{(k)}$ tending strongly to Θ in $L_{2+\gamma}\big(I_T;W^1_{2+\gamma}(\Omega)\big)$. Here, γ is again a non-negative number.

An example for a matrix-valued function satisfying (5.4.74), (5.4.75) and (5.4.76) is

$$c_{ij}(\boldsymbol{x};\Xi) = \delta_{ij}\left(d_0(\boldsymbol{x}) + d_1(\boldsymbol{x})|\Xi|^\gamma\right)$$

with the Kronecker symbol δ_{ij} and measurable functions d_0 and d_1 taking their values in $[q_1,q_2]$, $i=0,1$, for some positive real constants q_1,q_2. For such matrix functions the relations (5.4.74) and (5.4.76) are obviously satisfied. In order to check (5.4.75) observe that

$$\begin{aligned}&\left(c_{ij}(\cdot,\boldsymbol{y})y_j - c_{ij}(\cdot,\boldsymbol{z})z_j\right)(y_j-z_j)\\ &\quad= d_0|\boldsymbol{y}-\boldsymbol{z}|^2 + d_1\left(|\boldsymbol{y}|^\gamma\boldsymbol{y} - |\boldsymbol{z}|^\gamma\boldsymbol{z}\right)\cdot(\boldsymbol{y}-\boldsymbol{z}).\end{aligned}$$

Therefore it suffices to find a lower bound of the form $k|\boldsymbol{y}-\boldsymbol{z}|^{2+\gamma}$ with $k>0$ for the factor of d_1. Without loss of generality, $|\boldsymbol{z}|\le|\boldsymbol{y}|$ and $\boldsymbol{y}\ne\boldsymbol{0}$. There exist orthogonal vectors $\boldsymbol{a}$ and $\boldsymbol{b}$ with norm 1 such that $\boldsymbol{y}=|\boldsymbol{y}|\boldsymbol{a}$ and $\boldsymbol{z}=|\boldsymbol{y}|(c\boldsymbol{a}+d\boldsymbol{b})$. Then an elementary calculation gives

$$\begin{aligned}C &\equiv \left(|\boldsymbol{y}|^\gamma\boldsymbol{y} - |\boldsymbol{z}|^\gamma\boldsymbol{z}\right)\cdot(\boldsymbol{y}-\boldsymbol{z})/|\boldsymbol{y}|^{\gamma+2}\\ &= \left(1-c\left(c^2+d^2\right)^{\gamma/2}\right)(1-c) + d^2\left(c^2+d^2\right)^{\gamma/2},\\ D &\equiv |\boldsymbol{y}-\boldsymbol{z}|^{2+\gamma}/|\boldsymbol{y}|^{2+\gamma} = \left((1-c)^2+d^2\right)^{1+\gamma/2}\end{aligned}$$

and $c^2+d^2\le 1$. If $c\ge 0$, then obviously $C\ge(1-c)^2+d^{2+\gamma}\ge(1-c)^{2+\gamma}+d^{2+\gamma}$. On the other hand, thanks to Hölder's inequality for sums $D\le 2^{\gamma/2}((1-c)^{2+\gamma}+d^{2+\gamma})$ so that the ratio of C to D is at least $2^{-\gamma/2}$ for

$c \geq 0$. If $c \leq 0$ we observe that $C \geq 1-c$ and $D \leq \left(2(1-c)^2\right)^{1+\gamma/2}$; therefore the quotient of C by D is at least equal to $2^{-1-\gamma/2}(1-c)^{-1-\gamma} \geq 2^{-2-3\gamma/2}$. Therefore we find that

$$\left(|\boldsymbol{y}|^\gamma \boldsymbol{y} - |\boldsymbol{z}|^\gamma \boldsymbol{z}\right) \cdot (\boldsymbol{y} - \boldsymbol{z}) \geq 2^{-2-3\gamma/2}|\boldsymbol{y} - \boldsymbol{z}|^{2+\gamma}.$$

Let us introduce the sets of admissible functions employed in the variational formulation of the problem:

$$\begin{aligned} &\mathscr{C} := \left\{\boldsymbol{v} \in \boldsymbol{H}^{1/2,1}(Q_T)\,;\ \boldsymbol{v} = \dot{\boldsymbol{U}} \text{ on } S_{U,T},\ v_n \leq 0 \text{ on } S_{C,T}\right\}, \\ &\mathfrak{U} := \boldsymbol{H}_0^{1/2,1}(Q_T) \equiv \left\{\boldsymbol{v} \in \boldsymbol{H}^{1/2,1}(Q_T);\ v = 0 \text{ on } S_{U,T}\right\}, \end{aligned} \tag{5.4.77}$$

$$\mathfrak{V} := L_{2+\gamma}\left(I_T; W^1_{2+\gamma}(\Omega)\right). \tag{5.4.78}$$

The definitions of $\mathscr{C}$ and $\mathfrak{U}$ are already introduced in (5.4.13) and (5.4.22), respectively, but the definition of $\mathfrak{V}$ is new. It is related to the studied model.

A weak solution of the thermo-viscoelastic contact problem is a pair of functions $[u, \Theta]$ *with* $\dot{\boldsymbol{u}} \in \mathscr{C}$, $\ddot{\boldsymbol{u}} \in L_2(\boldsymbol{I}_T; \boldsymbol{H}^1(\Omega)^*)$, $\Theta \in \mathfrak{V}$, $\dot{\Theta} \in \mathfrak{V}^*$ *such that the initial conditions* (5.4.12), *the variational inequality of the contact problem and the variational equation of the heat transfer problem are satisfied:*

$$\begin{aligned} &\left\langle \ddot{\boldsymbol{u}}, \boldsymbol{v} - \dot{\boldsymbol{u}}\right\rangle_{Q_T} + A^{(0)}(\boldsymbol{u}, \boldsymbol{v} - \dot{\boldsymbol{u}}) + A^{(1)}(\dot{\boldsymbol{u}}, \boldsymbol{v} - \dot{\boldsymbol{u}}) \\ &\qquad - \langle b_{ij}\Theta, e_{ij}(\boldsymbol{v} - \dot{\boldsymbol{u}})\rangle_{Q_T} + \langle \mathfrak{F}|\sigma_n(\boldsymbol{u}, \Theta)|, |\boldsymbol{v}_t| - |\dot{\boldsymbol{u}}_t|\rangle_{S_{CT}} \\ &\quad \geq \langle \mathscr{L}, \boldsymbol{v} - \dot{\boldsymbol{u}}\rangle_{Q_T} \end{aligned} \tag{5.4.79}$$

$$\begin{aligned} &\langle \dot{\Theta}, \varphi\rangle_{Q_T} + \langle c_{ij}(\nabla\Theta)\partial_j\Theta, \partial_i\varphi\rangle_{Q_T} + \langle b_{ij}\Theta\, \partial_j \dot{u}_i, \varphi\rangle_{Q_T} \\ &\qquad + \langle K(\Theta - \Upsilon), \varphi\rangle_{S_T} \\ &\quad = \left\langle a^{(1)}_{ijk\ell}\, e_{ij}(\dot{\boldsymbol{u}}) e_{k\ell}(\dot{\boldsymbol{u}}), \varphi\right\rangle_{\Omega} + \langle \mathfrak{F}|\sigma_n||\dot{\boldsymbol{u}}_t|, \varphi\rangle_{S_{CT}} \end{aligned} \tag{5.4.80}$$

for all $(\boldsymbol{v}, \varphi) \in \mathscr{C} \times \mathfrak{V}$.
Here the standard notation of the bilinear forms $A^{(\iota)}$, $\iota = 0, 1$ and of the linear form $\mathscr{L}$ from previous sections is preserved.

5.4.17 Assumption. *Let Assumption 5.1.6 be fulfilled*[1] *with* $\Omega_C = \Omega$. *The given data are assumed to satisfy* $\boldsymbol{\tau} \in L_2(S_{\tau T})$, $\boldsymbol{U}, \dot{\boldsymbol{U}} \in \boldsymbol{H}^{3/4,3/2}(Q_T)$, $0 \leq \Upsilon \in L_2(S_T)$ *and* $0 \leq K \in L_\infty(\Gamma)$. *The initial temperature* $\Theta_0 \in L_2(\Omega)$ *shall be positive. The diffusion coefficients* c_{ij}, $i, j = 1, \ldots N$, *satisfy the assumptions* (5.4.74)–(5.4.76), *the coefficients of thermal expansion* $b_{ij} \in$

[1] For a solution-dependent $\mathfrak{F}$ the discontinuous dependence in the sense of the Remark at the end of Section 5.1 is allowed.

$C^\alpha(\Omega)$ for $i, j = 1, \dots N$ with $\alpha > 1/2$ are bounded and symmetric, the possible boundary heat dissipation R satisfies the growth condition (5.4.50) *with $r_R > 0$ or, if $r_R = 0$, then $R = 0$ is assumed. We also assume that R is convex. The assumption* (5.4.47) *is fulfilled.*

The result to be proved is the following

5.4.18 Theorem. *Let Assumption 5.4.17 be satisfied. Let the dimension N and the growth exponent γ fulfil the inequality*

$$\gamma \geq \frac{\sqrt{12.2} - 3}{4} \text{ for } N = 2 \text{ and } \gamma \geq \frac{N-2}{2} \text{ for } N \geq 3. \tag{5.4.81}$$

Then there exists at least one solution to the coupled problem (5.4.79), (5.4.80).

5.4.19 Remarks. 1. The presented result is better compared to the preceding model in the case $N = 2$ only. In the case without boundary heat dissipation we have the restriction $\gamma > \left(\sqrt{12.2} - 3\right)/4 \sim 0.123$ compared to $\gamma > 1/6$ for the preceding model.

2. The problem can be also solved by the fixed-point method, see [81], but this method requires essentially stronger assumptions than that in (5.4.81); the requirement there is $\gamma > 4$ for both dimensions $N = 2$ and $N = 3$.

Proof. The scheme of this proof is the same as that for Theorem 5.4.12. We start with an approximate problem that is obtained by penalizing the contact condition, smoothing the Euclidian norm in the friction term and limiting the growth of the thermal sources with the help of the cutoff function h_M. The only difference to the system (5.4.56), (5.4.57) is the fact that $c_{ij} \equiv c_{ij}(\cdot, \nabla\Theta)$ here. The solvability of the approximate problem is again based on the Galerkin technique. The space $\mathscr{U} = \boldsymbol{H}_0^1(\Omega)$ has a basis $\{\boldsymbol{v}_1, \boldsymbol{v}_2, \dots\}$ such that $\bigcup_{m \in \mathrm{N}} \mathscr{U}_m$ with $\mathscr{U}_m = \operatorname{span}\left\{\boldsymbol{v}_1, \dots, \boldsymbol{v}_m\right\}$ is dense in $\mathscr{U}$. This basis can be chosen such that both the $\boldsymbol{L}_2(\Omega)$–orthogonal projection and the $\boldsymbol{H}^1(\Omega)$–orthogonal projection onto $\mathscr{U}_m$ are bounded uniformly with respect to m. For the space $\mathscr{V} = W^1_{2+\gamma}(\Omega)$ there is no such basis, because it is not a Hilbert space. Since $\mathscr{V}$ is separable, there is a family of normed bases $\left\{\varphi_{1m}, \dots, \varphi_{mm}\right\}$, $m \in \mathrm{N}$, and an associated family $\mathscr{V}_m = \operatorname{span}\left\{\varphi_{1m}, \dots, \varphi_{mm}\right\}$ of spaces such that $\varphi_{1m}, \dots, \varphi_{m-1m} \in \mathscr{V}_{m-1}$ and $(\varphi_{mm}, \varphi_{km})_\Omega = 0$ for $k < m$. The union $\bigcup_{m \in \mathrm{N}} \mathscr{V}_m$ is again dense in $\mathscr{V}$.

With this change the definition of the Galerkin problem is the same as that used before:

Find functions

$$\boldsymbol{u}_m(t,x) \equiv \sum_{i=1}^{m} p_i(t)\boldsymbol{v}_i(\boldsymbol{x}) \in L_\infty(I_T, \mathscr{U}_m) \text{ and}$$

$$\Theta_m(t,x) \equiv \sum_{i=1}^{m} q_i(t)\varphi_{im}(\boldsymbol{x}) \in L_\infty(I_T, \mathscr{V}_m)$$

that satisfy the approximate initial conditions

$$\big\langle \boldsymbol{u}_m(0,\cdot),\boldsymbol{v}\big\rangle_\Omega = \langle \boldsymbol{u}_0,\boldsymbol{v}\rangle_\Omega \text{ and } \big\langle \dot{\boldsymbol{u}}_m(0,\cdot),\boldsymbol{v}\big\rangle_\Omega = \langle \boldsymbol{u}_1,\boldsymbol{v}\rangle_\Omega$$

for all $v \in \mathscr{U}_m$ *and*

$$\big\langle \Theta_m(0,\cdot),\varphi\big\rangle_\Omega = \langle \Theta_0,\varphi\rangle_\Omega \text{ for all } \varphi \in \mathscr{V}_m$$

such that for all $\boldsymbol{v} \in \mathscr{U}_m$*, all* $\varphi \in \mathscr{V}_m$ *and all* $t \in I_T$ *the following equations hold:*

$$\begin{aligned}\langle \ddot{\boldsymbol{u}}_m,\boldsymbol{v}\rangle_\Omega &+ \big\langle a^{(0)}_{ijk\ell}e_{k\ell}(\boldsymbol{u}_m) + a^{(1)}_{ijk\ell}e_{k\ell}(\dot{\boldsymbol{u}}_m), e_{ij}(\boldsymbol{v})\big\rangle_\Omega \\ &- \big\langle b_{ij}\mathsf{h}_M\big((\Theta_m)_+\big), e_{ij}(\boldsymbol{v})\big\rangle_\Omega + \Big\langle \tfrac{1}{\varepsilon}\big[(\dot{u}_m)_n\big]_+, v_n\Big\rangle_{\Gamma_C} \\ &+ \Big\langle \mathfrak{F}\tfrac{1}{\varepsilon}\big[(\dot{u}_m)_n\big]_+ \nabla\Phi_\eta\big((\dot{\boldsymbol{u}}_m)_{\boldsymbol{t}}\big), v_{\boldsymbol{t}}\Big\rangle_{\Gamma_C} = \langle f,\boldsymbol{v}\rangle_\Omega + \langle \boldsymbol{\tau},\boldsymbol{v}\rangle_{\Gamma_\tau}\end{aligned} \tag{5.4.82}$$

and

$$\begin{aligned}\big\langle \dot{\Theta}_m,\varphi\big\rangle_\Omega &+ \big\langle c_{ij}(\nabla\Theta_m)\partial_i\Theta_m, \partial_j\varphi\big\rangle_\Omega \\ &+ \big\langle b_{ij}\mathsf{h}_M\big((\Theta_m)_+\big)e_{ij}(\dot{\boldsymbol{u}}_m),\varphi\big\rangle_\Omega \\ &+ \big\langle K(\Theta_m - \Upsilon) + R(\Theta_m),\varphi\big\rangle_\Gamma \\ &= \Big\langle \mathsf{h}_M\big(a^{(1)}_{ijk\ell}e_{ij}(\dot{\boldsymbol{u}}_m)e_{k\ell}(\dot{\boldsymbol{u}}_m)\big),\varphi\Big\rangle_\Omega \\ &+ \Big\langle \mathfrak{F}\mathsf{h}_M\Big(\tfrac{1}{\varepsilon}\big[(\dot{u}_m)_n\big]_+\Phi_\eta\big((\dot{\boldsymbol{u}}_m)_{\boldsymbol{t}}\big)\Big),\varphi\Big\rangle_{\Gamma_C}.\end{aligned} \tag{5.4.83}$$

The Galerkin equations are again equivalent to an initial value problem for a system of ordinary differential equations whose solvability follows from the theory of ordinary differential equations.

For the limit procedure $m \to \infty$ we derive *a priori* estimates, by taking $\boldsymbol{v} = \dot{\boldsymbol{u}}_m$ and $\varphi = \Theta_m$, as in the preceding model. The derivation is again substantially simplified by the boundedness of the thermal sources in the heat equation and the thermal stress in the viscoelastic system. The estimate for the displacement has the same form

$$\|\dot{\boldsymbol{u}}_m\|^2_{L_\infty(I_T;\boldsymbol{L}_2(\Omega))} + \|\boldsymbol{u}_m\|^2_{L_\infty(I_T;\boldsymbol{H}^1(\Omega))} + \|\dot{\boldsymbol{u}}_m\|^2_{L_2(I_T;\boldsymbol{H}^1(\Omega))} \le c$$

as in the preceding model, while condition (5.4.74) leads to

$$\|\Theta_m\|^2_{L_\infty(I_T;L_2(\Omega))} + \|\Theta_m\|^{2+\gamma}_{L_{2+\gamma}(I_T;W^1_{2+\gamma}(\Omega))} \\ + c_R\|(\Theta_m)_+\|^{r_R+1}_{L_{r_R+1}(S_T)} \le c. \tag{5.4.84}$$

We also need some time regularity of the Galerkin solutions uniform with respect to the Galerkin parameter m. A uniform bound for $\dot{\boldsymbol{u}}_m$ in $\boldsymbol{H}^{1/2,1}(Q_T)$ is proved in the same way as for the preceding model. From the equation (5.4.83) and the estimate (5.4.84) it is easy to see that

$$\|\dot{\Theta}_m\|_{\mathfrak{V}^*_m} \le c \tag{5.4.85}$$

with $\mathfrak{V}_m \equiv L_{2+\gamma}(I_T; \mathscr{V}_m)$ and a constant c independent of m.

Besides this dual estimate for the time derivative of the temperature we need an estimate for $\Theta \in H^\alpha(I_T; L_2(\Omega))$ with α sufficiently big. It is derived by a standard procedure. We take the test function $\varphi = \Theta_m(s_2) - \Theta_m(s_1)$ for $s_1, s_2 \in (0,t)$, $s_1 \ne s_2$ in the Galerkin equation (5.4.83) at time t, we multiply the result by $|s_2 - s_1|^{-1-2\alpha}$ with a parameter $\alpha \in (0, 1/2)$ and integrate the result both with respect to t from s_1 to s_2 and with respect to $s = [s_1, s_2]$ over I_T^2. Then we obtain the identity

$$\begin{aligned}
&\int_{I_T^2}\int_{s_1}^{s_2} \frac{\langle \dot{\Theta}_m(t), \Theta_m(s_2) - \Theta_m(s_1)\rangle_\Omega}{|s_2 - s_1|^{1+2\alpha}}\, dt\, ds \\
&= -\int_{I_T^2}\int_{s_1}^{s_2} |s_2 - s_1|^{-1-2\alpha}\Big(\big\langle c_{ij}(\nabla\Theta_m(t))\partial_j\Theta_m(t), \\
&\quad \partial_i\big(\Theta_m(s_2) - \Theta_m(s_1)\big)\big\rangle_\Omega \\
&\quad + \big\langle b_{ij}\mathsf{h}_M\big((\Theta_m)_+(t)\big)e_{ij}(\dot{\boldsymbol{u}}(t)), \Theta_m(s_2) - \Theta_m(s_1)\big\rangle_\Omega \\
&\quad + \big\langle K(\Theta_m - \Upsilon)(t) + R(\Theta)(t), \Theta_m(s_2) - \Theta_m(s_1)\big\rangle_\Gamma \\
&\quad - \Big\langle \mathfrak{F}\mathsf{h}_M\left(\tfrac{1}{\varepsilon}[\dot{u}_n(t)]_+ \Phi_\eta(\dot{\boldsymbol{u}}_{\boldsymbol{t}}(t))\right), \Theta_m(s_2) - \Theta_m(s_1)\Big\rangle_{\Gamma_C} \\
&\quad - \big\langle \mathsf{h}_M\big(a^{(1)}_{ijk\ell}\, e_{ij}(\dot{\boldsymbol{u}}(t))e_{k\ell}(\dot{\boldsymbol{u}}(t))\big), \Theta_m(s_2) - \Theta_m(s_1)\big\rangle_\Omega\Big)\, dt\, ds.
\end{aligned} \tag{5.4.86}$$

After performing the integration with respect to t one observes that the left hand side of (5.4.86) is equivalent to the square of the seminorm of Θ_m in the space $H^\alpha(I_T; L_2(\Omega))$. The right hand side is constituted of expressions of the type

$$\int_{I_T^2}\int_{s_1}^{s_2} \frac{|f(t)||g(s_2) - g(s_1)|}{|s_2 - s_1|^{1+2\alpha}}\, dt\, ds,\ \ f \in L_p(I_T),\ \ g \in L_q(I_T), \\ 1/p + 1/q = 1.$$

Such expressions are bounded by $const.\|f\|_{L_p(I_T)}\|g\|_{L_q(I_T)}$ with the constant independent of φ and g if $\alpha < 1/2$. Hence for $\|\Theta_m\|^2_{H^\alpha(I_T;L_2(\Omega))}$ we obtain

$$\|\Theta_m\|^2_{H^\alpha(I_T;L_2(\Omega))} \le c \tag{5.4.87}$$

with constant independent of $m \in \mathbb{N}$.

The above derived estimates allows finding a sequence $m_k \to +\infty$ and a limit $[u, \Theta]$ such that

$$\begin{aligned} &\dot{\boldsymbol{u}}_{m_k} \rightharpoonup \dot{\boldsymbol{u}} \text{ in } \mathfrak{U},\ \Theta_{m_k} \rightharpoonup \Theta \text{ in } \mathfrak{V} \cap H^\alpha(I_T; L_2(\Omega)),\\ &\dot{\boldsymbol{u}}_{m_k}(T) \rightharpoonup \dot{\boldsymbol{u}}(T) \text{ in } \boldsymbol{L}_2(\Omega),\ \Theta_{m_k}(T,\cdot) \to \Theta(T,\cdot) \text{ in } L_2(\Omega),\\ &[\Theta_{m_k}]_+ \rightharpoonup \Theta_+ \text{ in } L_{r_R+1}(S_T) \end{aligned} \tag{5.4.88}$$

and a limit C in $\boldsymbol{L}_{(2+\gamma)/(1+\gamma)}(\Omega)$ such that $c_{ij}(\cdot, \nabla\Theta_{m_k})\partial_j\Theta_{m_k} \rightharpoonup \mathsf{C}$ there. The space $\mathfrak{V}_0 \equiv \bigcup_{m\in\mathbb{N}} \mathfrak{V}_m$ is dense in $\mathfrak{V}$. Using the standard diagonal process we prove the existence of a $\Lambda \in \mathfrak{V}_0^*$ and a subsequence denoted by the indices m_k again such that $\langle\dot{\Theta}_{m_k}, \varphi\rangle_{Q_T} \to \Lambda(\varphi)$ for every $\varphi \in \mathfrak{V}_0$. Since $\|\Lambda\|_{\mathfrak{V}_0^*} \le c$ holds with c from (5.4.85), Λ has a continuous extension to a linear form on $\mathfrak{V}$, that is denoted by Λ again, such that $\|\Lambda\|_{\mathfrak{V}^*} \le c$ with the same c. Integration by parts with respect to the time variable easily shows $\Lambda = \dot{\Theta}$. By compact imbedding theorems we obtain the convergences $\Theta_{m_k} \to \Theta$ in $L_2(Q_T)$ and almost everywhere on Q_T. With the help of interpolation theorems for Sobolev spaces we also obtain the strong convergences $\Theta_{m_k} \to \Theta$ first in $L_{2+\gamma}(I_T; L_2(\Omega))$, then in $L_{2+\gamma}(Q_T)$ and finally in $L_{2+\gamma}(I_T; W^{1-\lambda}_{2+\gamma}(\Omega))$ with $\lambda > 0$ arbitrarily small. We first pass to the limit $k \to +\infty$ in the Galerkin approximation of the contact problem, after an integration with respect to time. Then, as described in the proof of Theorem 5.1.2, we prove the strong convergence $\dot{\boldsymbol{u}}_{m_k} \to \dot{\boldsymbol{u}}$ in $\boldsymbol{H}^{1/2-\lambda,1}(Q_T)$ for an arbitrarily small $\lambda > 0$. With this additional strong convergence it is easy to pass to the limit $k \to +\infty$ in the Galerkin-approximation of the heat equation and thus show that the limit functions $[\boldsymbol{u}, \Theta]$ satisfy the relations (5.4.58) and

$$\begin{aligned} &\langle\dot{\Theta}, \varphi\rangle_{Q_T} + \langle\mathsf{C}_i, \partial_i\varphi\rangle_{Q_T} + \langle b_{ij}\mathsf{h}_M(\Theta_+)e_{ij}(\dot{\boldsymbol{u}}), \varphi\rangle_{Q_T}\\ &\qquad + \langle K(\Theta - \Upsilon) + R(\Theta), \varphi\rangle_{S_T}\\ &\quad = \left\langle \mathsf{h}_M\left(a^{(1)}_{ijk\ell}e_{ij}(\dot{\boldsymbol{u}})e_{k\ell}(\dot{\boldsymbol{u}})\right), \varphi\right\rangle_{Q_T}\\ &\qquad + \left\langle \mathfrak{F}\mathsf{h}_M\left(\tfrac{1}{\varepsilon}[\dot{u}_n]_+\Phi_\eta(\dot{\boldsymbol{u}}_t)\right), \varphi\right\rangle_{S_{CT}}. \end{aligned}$$

It remains to prove $\Theta_{m_k} \to \Theta$ in $\mathfrak{V}$ and $\mathsf{C} = (c_{ij}(\cdot, \nabla\Theta)\partial_j\Theta)$. In fact, the monotonicity relation (5.4.75) yields

$$\liminf_{k\to+\infty} \langle c_{ij}(\cdot, \nabla\Theta_{m_k})\, \partial_j\Theta_{m_k}, \partial_i\Theta_{m_k}\rangle_{Q_T} \geq \langle \mathsf{C}_i, \partial_i\Theta\rangle_{Q_T}.$$

Simultaneously from the Galerkin equation (5.4.83) for Θ_{m_k}, the just derived equation for Θ and the strong convergences in the thermal sources and thermal stresses, we obtain

$$\limsup_{k\to+\infty} \langle c_{ij}(\cdot, \nabla\Theta_{m_k})\, \partial_j\Theta_{m_k}, \partial_i\Theta_{m_k}\rangle_{Q_T} \leq \langle \mathsf{C}_i, \partial_i\Theta\rangle_{Q_T}.$$

Here the weak lower semicontinuity

$$\begin{aligned}&\liminf_{k\to+\infty} \left(\left\langle\dot\Theta_{m_k}, \Theta_{m_k}\right\rangle_{Q_T} + \left\langle R(\Theta_{m_k}), \Theta_{m_k}\right\rangle_{S_T}\right)\\ &\quad\geq \left\langle\dot\Theta, \Theta\right\rangle_{Q_T} + \left\langle R(\Theta), \Theta\right\rangle_{S_T}\end{aligned}$$

is crucial; the relation for the term with R follows from the convexity of R. Hence we have $\lim\limits_{k\to\infty} \langle c_{ij}(\cdot, \nabla\Theta_{m_k})\, \partial_j\Theta_{m_k}, \partial_i\Theta_{m_k}\rangle_{Q_T} = \langle \mathsf{C}_i, \partial_i\Theta\rangle_{Q_T}$. From the limit in the monotonicity equation

$$\begin{aligned}&c_0\, \|\Theta_{m_k} - \Theta\|^{2+\gamma}_{L_{2+\gamma}(I_T;W^1_{2+\gamma}(\Omega))}\\ &\leq \left\langle c_{ij}(\cdot, \nabla\Theta_{m_k})\, \partial_j\Theta_{m_k} - c_{ij}(\cdot, \nabla\Theta)\partial_j\Theta, \partial_i\Theta_{m_k} - \partial_i\Theta\right\rangle_{Q_T}\end{aligned}$$

the convergence $\Theta_{m_k} \to \Theta$ in $\mathfrak{V}$ follows; therefore $\mathsf{C} = c_{ij}(\cdot, \nabla\Theta)\partial_j\Theta$. A similar convergence procedure can be carried out for the limit $\eta \to 0$ of the smoothing parameter. Hence we obtain a solution to the penalized thermo–viscoelastic contact problem, that is the variant of (5.4.56), (5.4.57) with $c_{ij}(\Theta)$ replaced by $c_{ij}(\nabla\Theta)$.

Now the energy estimate is derived. First of all, by taking the test function $\Theta_- = \chi_{t_0} \min\{0, \Theta\}$ with the indicator function χ_{t_0} of the time interval $(0, t_0)$ in (5.4.57) we prove $\Theta \geq 0$ on Q_T, because this yields

$$\|\Theta_-(t_0)\|^2_{L_2(\Omega)} + \|\nabla\Theta_-\|^{2+\gamma}_{\boldsymbol{L}_{2+\gamma}(Q_{t_0})} \leq 0 \text{ for } t_0 \in T.$$

Next, we choose $\boldsymbol{v} = (1 - \chi_{t_0})\dot{\boldsymbol{u}}$ in (5.4.56), $\varphi = \chi_{t_0}$ in (5.4.57) and add both relations. As in the preceding model, the thermal sources from the viscosity, friction and deformation disappear and we obtain the crucial *energy estimate*

$$\begin{aligned}&\|\dot{\boldsymbol{u}}\|_{L_\infty(I_T;\boldsymbol{L}_2(\Omega))} + \|\boldsymbol{u}\|_{L_\infty(I_T;\boldsymbol{H}^1(\Omega))}\\ &\quad + \|\Theta\|_{L_\infty(I_T;L_1(\Omega))} + \|\Theta\|_{L_1(S_T)} + c_R\|\Theta\|^{r_R}_{L_{r_R}(S_T)} \leq c\end{aligned} \tag{5.4.89}$$

with c independent of the parameters M and ε.

As in the previous model further estimates are developed by means of test functions $\varphi = \chi_{t_0}\Theta^\beta$ for an appropriate exponent $\beta \in (0,1]$. With this a new estimate of the expression

$$\|\Theta\|^{\beta+1}_{L_\infty(I_T;L_{\beta+1}(\Omega))} + \|\Theta\|^{2+\gamma}_{L_{2+\gamma}(I_T;W^1_p(\Omega))} + c_R\|\Theta\|^{r_R+\beta}_{L_{r_R+\beta}(S_T)} \tag{5.4.90}$$

with $p = \dfrac{2+\gamma}{2-\beta}$ is obtained. Indeed, for this value of p we have

$$\begin{aligned}\|\nabla\Theta\|^{2+\gamma}_{L_{2+\gamma}(I_T;\boldsymbol{L}_p(\Omega))} &= \int_{I_T}\left(\int_\Omega |\nabla\Theta|^p\,dx\right)^{2-\beta} dt \\ &\le c_1\int_{Q_T}\Theta^{\beta-1}|\nabla\Theta|^{2+\gamma}\,dx\,dt\,\sup_{I_T}\left(\int_\Omega \Theta\,dx\right)^{1-\beta} \\ &\le c_2\int_{Q_T} c_{ij}(\nabla\Theta)\,\partial_i\Theta\,\partial_j\left(\Theta^\beta\right)dx\,dt.\end{aligned} \tag{5.4.91}$$

The imbedding $W^1_p(\Omega) \hookrightarrow L_q(\Omega)$ holds with

$$\begin{aligned} &q \le q_0^{-1}, && \text{if } q_0 > 0,\\ &q < +\infty, && \text{if } q_0 = 0,\\ &q \le +\infty, && \text{if } q_0 < 0\end{aligned} \tag{5.4.92}$$

with $q_0 = \dfrac{2N - \beta N - 2 - \gamma}{N(2+\gamma)}$. By interpolation

$$\|\Theta\|_{L_{2+\gamma}(I_T;L_p(\Omega))} \le c\|\Theta\|^{1-\lambda}_{L_\infty(I_T;L_1(\Omega))}\|\Theta\|^{\lambda}_{L_{2+\gamma}(I_T;L_q(\Omega))}$$

$$\text{with } \lambda = \begin{cases} \dfrac{N(\beta+\gamma)}{N(\beta+\gamma)+2+\gamma}, & q_0 > 0,\\ \dfrac{q(\gamma+\beta)}{(q-1)(2+\gamma)}, & q_0 = 0,\\ \dfrac{\gamma+\beta}{2+\gamma}, & q_0 < 0.\end{cases}$$

Here for $q_0 = 0$, $q > p$ must hold. Using this interpolation in (5.4.91) yields, after some obvious calculation,

$$\|\Theta\|^{2+\gamma}_{L_{2+\gamma}(I_T;L_p(\Omega))} \le c_2\int_{Q_T} c_{ij}(\nabla\Theta)\,\partial_i\Theta\,\partial_j\left(\Theta^\beta\right)dx\,dt + c_1$$

with c_1 from (5.4.89) and the value of p from (5.4.90).

The estimates to be carried out make use of relation (5.4.47). It is therefore necessary to have an upper bound for $\|\Theta\|_{L_2(I_T;H^\alpha(\Omega))}$ with $\alpha >$

1/2. Such a bound can only come from $\|\Theta\|_{L_{2+\gamma}(I_T;W^1_p(\Omega))}$; this leads to the condition $W^1_p(\Omega) \hookrightarrow H^\alpha(\Omega)$. The corresponding condition for β, calculated with the value of p from (5.4.90), is

$$\beta > \beta_N(\gamma) = 1 - \frac{1}{N} - \frac{\gamma}{2} - \frac{\gamma}{2N}. \tag{5.4.93}$$

Since $\beta_N(\gamma)$ shall be positive, we confine ourselves to $\gamma < \gamma_N \equiv 2(N-1)/(N+1)$ for the first step. The limit values of p and q that correspond to $\beta = \beta_N(\gamma)$ are

$$\begin{aligned} p = p_N &= \frac{2N}{N+1} \text{ in (5.4.90) and} \\ q = q_N &= \frac{2N}{N-1} \text{ in (5.4.92).} \end{aligned} \tag{5.4.94}$$

Observe that these values do not depend on γ.

In order to derive an estimate for the norms in (5.4.90) it is necessary to bound the thermal sources by (5.4.90) with either a power lower than one or the power one but a sufficiently small factor. These estimates must be uniform with respect to the approximation parameters M and ε. In particular they must disregard the artificial limitation of the sources by the function h. The estimates are the easier, the lower β is taken. Hence we choose $\beta > \beta_N(\gamma)$ sufficiently close to $\beta_N(\gamma)$. We start with the estimate of the viscous heat. Using (5.4.47) and appropriate interpolation and imbedding theorems we obtain

$$\begin{aligned} \int_{Q_T} |\nabla \dot{\boldsymbol{u}}|^2 |\Theta|^\beta \, d\boldsymbol{x}\, dt &\le c_1 \|\nabla \dot{\boldsymbol{u}}\|^2_{L_2(I_T;L_{q_N}(\Omega))} \|\Theta\|^\beta_{L_\infty(I_T;L_{1+\lambda\beta}(\Omega))} \\ &\le \left(c_2 \|\Theta\|^2_{L_2(I_T;W^1_p(\Omega))} + c_3\right) \|\Theta\|^{\beta\mu}_{L_\infty(I_T;L_{1+\beta}(\Omega))} \|\Theta\|^{\beta(1-\mu)}_{L_\infty(I_T;L_1(\Omega))} \\ &\le \eta \|\Theta\|^{2+\gamma}_{L_{2+\gamma}(I_T;W^1_p(\Omega))} + c_4(\eta) \|\Theta\|^\kappa_{L_\infty(I_T;L_{1+\beta}(\Omega))} + c_5(\eta) \end{aligned}$$

with $\mu = \dfrac{\lambda(1+\beta)}{1+\lambda\beta}$ and $\eta > 0$ arbitrarily small. Here the first inequality is valid for $\dfrac{2}{q_N} + \dfrac{\beta}{1+\lambda\beta} \le 1$; this is equivalent to $\lambda \ge N - \dfrac{1}{\beta}$. Observe that q_N is just the parameter that permits the estimate $\|\nabla \dot{\boldsymbol{u}}\|_{L_2(I_T;L_{q_N}(\Omega))} \le c\|\dot{\boldsymbol{u}}\|_{\boldsymbol{H}^{3/4,3/2}(Q_T)}$. As to the second inequality we need $\lambda \le 1$ and thus $N - \dfrac{1}{\beta} \le 1$. This leads to the condition

$$\gamma > \frac{2(N^2 - 3N + 1)}{N^2 - 1} \equiv \gamma_{v,1}(N). \tag{5.4.95}$$

The exponent κ is $\kappa = \dfrac{2+\gamma}{\gamma}\beta\mu$. The necessary condition for κ is $\kappa < 1+\beta$. After some short calculation this leads to

$$\gamma > \frac{2N-4}{N+2} \equiv \gamma_{v,2}(N). \tag{5.4.96}$$

This bound is smaller than that in (5.4.95) for $N \geq 5$.

The deformation heat is bounded by

$$\int_{Q_T} |\nabla \dot{\boldsymbol{u}}||\Theta|^{1+\beta}\, dx\, dt \leq c\|\nabla \dot{\boldsymbol{u}}\|_{L_2(I_T;L_{q_N}(\Omega))}\|\Theta\|^{1+\beta}_{L_{2+2\beta}(I_T;L_s(\Omega))}$$

with $\dfrac{1+\beta}{s} + \dfrac{1}{q_N} = 1$ and thus $s = \dfrac{2N(1+\beta)}{N+1}$. The limit value of s for $\beta = \beta_N(\gamma)$ is $s_0 = \dfrac{4N-2}{N+1} - \gamma$. The norm of Θ is further estimated by

$$\|\Theta\|_{L_{2+2\beta}(I_T;L_s(\Omega))} \leq c\|\Theta\|^{\lambda}_{L_{2+\gamma}(I_T;L_{q_N}(\Omega))}\|\Theta\|^{1-\lambda}_{L_\infty(I_T;L_1(\Omega))}$$

with $\dfrac{1}{2+2\beta} \geq \dfrac{\lambda}{2+\gamma}$ and $\dfrac{1}{s} \geq \dfrac{\lambda}{q_N} + 1 - \lambda$. The latter condition is equivalent to $\lambda \geq \dfrac{2N}{N+1}\left(1 - \dfrac{1}{s}\right)$. The limit value of λ for $s = s_0$ here is $\lambda_0 = \dfrac{2N}{N+1}\dfrac{3(N-1)-(N+1)\gamma}{4N-2-(N+1)\gamma}$. For the condition $\dfrac{1}{2+2\beta} \geq \dfrac{\lambda}{2+2\gamma}$, the corresponding limit value with $\beta = \beta_N(\gamma)$ is $\lambda_1 = \dfrac{N(\gamma+2)}{4N-2-(N+1)\gamma}$. From the requirement $\lambda_0 < \lambda < \lambda_1$ follows

$$\gamma > \frac{4(N-2)}{3(N+1)};$$

this condition is weaker than (5.4.96). Another condition is $\lambda_0 < 1$. This is satisfied for

$$\gamma > \frac{2N^2-8N+2}{N^2-1};$$

this condition is also weaker than (5.4.96). Hence, under the condition (5.4.96) the deformation heat can be estimated by

$$c_1\|\Theta\|^{\kappa}_{L_{2+\gamma}(I_T;W^1_p(\Omega))} + c_2$$

with $\kappa = 1 + \lambda(1+\beta)$. The condition for κ that allows the derivation of the *a priori* estimate is again $\kappa < 2+\gamma$. After some calculation this leads to the requirement

$$\gamma > \frac{N-2}{N+1}$$

which is again weaker than (5.4.96).

The frictional heat is estimated by

$$\begin{aligned} \int_{S_T} \mathfrak{F}|\sigma_n(\boldsymbol{u},\Theta)||\dot{\boldsymbol{u}}_{\boldsymbol{t}}|\Theta^\beta \, ds_{\boldsymbol{x}} \, dt \le \|\mathfrak{F}\|_{L_\infty(S_{\mathfrak{F}T})} \cdot \\ \cdot \|\sigma_n(\boldsymbol{u},\Theta)\|_{L_2(S_{\mathfrak{F}T})} \|\dot{\boldsymbol{u}}\|_{L_{r_1}(I_T;\boldsymbol{L}_{s_1}(\Gamma))} \|\Theta\|^\beta_{L_{\beta r_2}(I_T;L_{\beta s_2}(\Gamma))} \end{aligned} \tag{5.4.97}$$

with $\dfrac{1}{r_1} + \dfrac{1}{r_2} = \dfrac{1}{s_1} + \dfrac{1}{s_2} = \dfrac{1}{2}$. The values of r_1 and s_1 are chosen such that the imbedding $L_{r_1}(I_T;\boldsymbol{L}_{s_1}(\Gamma)) \hookrightarrow \boldsymbol{H}^{1/2,1}(S_T)$ holds, according to Lemma 5.4.15 that leads to the requirement $\dfrac{2}{r_1} + \dfrac{N-1}{s_1} = \dfrac{N-1}{2}$. This can be translated to the condition $\dfrac{2}{r_2} + \dfrac{N-1}{s_2} = 1$ for r_2 and s_2. It remains to estimate the norm $\|\Theta\|_{L_{\beta r_2}(I_T;L_{\beta s_2}(\Gamma))}$. This is done by the interpolation

$$\|\Theta\|_{L_{\beta r_2}(I_T;L_{\beta s_2}(\Gamma))} \le c\|\Theta\|^\lambda_{L_{r_3}(I_T;L_{s_3}(\Gamma))}\|\Theta\|^{1-\lambda}_{L_1(S_T)}$$

that is valid for $\dfrac{\lambda}{r_3} + 1 - \lambda \le \dfrac{1}{\beta r_2}$ and $\dfrac{\lambda}{s_3} + 1 - \lambda \le \dfrac{1}{\beta s_2}$. The first norm here is further estimated by the trace theorem and another interpolation,

$$\begin{aligned} \|\Theta\|_{L_{r_3}(I_T;L_{s_3}(\Gamma))} &\le c_1\|\Theta\|_{L_{r_3}(I_T;W^\mu_{s_4}(\Omega))} \\ &\le c_2\|\Theta\|^\mu_{L_{2+\gamma}(I_T;W^1_p(\Omega))}\|\Theta\|^{1-\mu}_{L_\infty(I_T;L_1(\Omega))} \end{aligned}$$

with $\dfrac{1}{s_4} < \mu \le 1$, $-\dfrac{N-1}{s_3} \le \mu - \dfrac{N}{s_4}$, $\dfrac{\mu}{2+\gamma} = \dfrac{1}{r_3}$ and $\dfrac{\mu}{p} + 1 - \mu < \dfrac{1}{s_4}$. Here the first and the fourth condition can be satisfied, if

$$\mu > \frac{2N}{3N-1}; \tag{5.4.98}$$

this condition is calculated with the limit value p_N of p. Expressing all the indices in terms of μ and inserting the obtained expressions in the condition $\dfrac{2}{r_2} + \dfrac{N-1}{s_2} = 1$ yields, after some calculation,

$$\lambda > \frac{N+1-\dfrac{1}{\beta}}{\mu\left(\dfrac{N+1}{2} - \dfrac{2}{2+\gamma}\right)+1} \equiv \lambda_1.$$

The total exponent of $\|\Theta\|_{L_{2+\gamma}(I_T;W_p^1(\Omega))}$ that is obtained in the estimate of the frictional heat by the described procedure is $\kappa = 2 + \beta\lambda\mu$. It has to be smaller than $2 + \gamma$. This leads to the requirement $\beta\lambda\mu < \gamma$. From

$$\beta\lambda_1\mu = \frac{(N+1)\beta - 1}{\dfrac{N+1}{2} - \dfrac{2}{2+\gamma} + \dfrac{1}{\mu}}$$

it is easy to see that the optimal choice of μ is as small as possible. Since condition (5.4.98) must be satisfied, we choose μ as close as necessary to $\dfrac{2N}{3N-1}$ and obtain, after some extended calculation, the condition

$$N(N+3)\,\gamma^2 + \big(N^2 + 5N + 1\big)\,\gamma - 2\big(N^2 - N - 1\big) > 0.$$

Hence γ has to be larger than both the roots of this polynomial,

$$\begin{aligned}\gamma &> \frac{\sqrt{9N^4 + 26N^3 - 5N^2 - 14N + 1} - N^2 - 5N - 1}{2N(N+3)} \\ &=: \gamma_{\mathfrak{F}}(N).\end{aligned} \tag{5.4.99}$$

This lower bound is bigger than that in (5.4.96) for $N = 2$, but not for $N \geq 3$. For $N = 2$ it is given by

$$\gamma_{\mathfrak{F}}(2) = \frac{\sqrt{12.2} - 3}{4}.$$

In the case $\gamma \geq \gamma_N \equiv 2(N-1)/(N+1)$ the suitable test function is $\Theta^{1/N}$. This yields an estimate for

$$\begin{aligned}&\|\Theta\|^{2+\gamma}_{L_{2+\gamma}(I_T;W_p^1(\Omega))} + \|\Theta\|^{1+1/N}_{L_\infty(I_T;L_{1+1/N}(\Omega))} \\ &\qquad + c_R\|\Theta\|^{r_R+1/N}_{L_{r_R+1/N}(S_T)}\end{aligned} \tag{5.4.100}$$

with

$$p \equiv p(\gamma) = \frac{N(2+\gamma)}{2N-1} > \widetilde{p}_N = p(\gamma_N) = \frac{4N^2}{(2N-1)(N+1)}. \tag{5.4.101}$$

Observe that $W_p^1(\Omega) \hookrightarrow L_q(\Omega)$ with $q \leq \widetilde{q}_N \equiv (4N^2+N)/(2N^2-3N-2) > 2$ for $N \geq 3$ and $q < +\infty$ arbitrarily large for $N = 2$. The viscous heat is now estimated by

$$c_1\|\nabla\dot{\boldsymbol{u}}\|^2_{L_2(I_T;L_{q_N}(\Omega))}\|\Theta\|^{1/N}_{L_\infty(I_T;L_1(\Omega))} \leq c_2\|\Theta\|^2_{L_{2+\gamma}(I_T;W_p^1(\Omega))}. \tag{5.4.102}$$

The deformation heat is bounded by

$$c_1 \|\nabla \dot{\boldsymbol{u}}\|_{L_2(I_T; L_{q_N}(\Omega))} \|\Theta\|^{1+1/N}_{L_{2+2/N}(I_T; L_2(\Omega))}. \tag{5.4.103}$$

For $N \geq 3$ this is already the required estimate, because $2 + 2/N < 2 + \gamma_N$. In the case $N = 2$ the $L_3(I_T; L_2(\Omega))$–norm of Θ is further estimated by the interpolation

$$\|\Theta\|_{L_3(I_T; L_2(\Omega))} \leq c \|\Theta\|^{1/6}_{L_\infty(I_T; L_1(\Omega))} \|\Theta\|^{5/6}_{L_{5/2}(Q_T)}.$$

The frictional heat has the estimate

$$\|\mathfrak{F}\|_{L_\infty(S_{CT})} \|\sigma_n(\boldsymbol{u}, \Theta)\|_{L_2(S_{\mathfrak{F}T})} \|\dot{\boldsymbol{u}}_{\boldsymbol{t}}\|_{L_{(2N+2)/(N-1)}(S_T)} \cdot \|\Theta\|^{1/N}_{L_{(N+1)/N}(S_T)}. \tag{5.4.104}$$

Since the trace theorem $W^1_{\tilde{p}_N}(\Omega) \to L_{\tilde{r}_N}(\Gamma)$ holds with

$$\tilde{r}_N = \frac{4N^2 - 4N}{2N^2 - 3N - 1} > \frac{N+1}{N} \text{ for } N \geq 3 \text{ and } \tilde{r}_N = +\infty \text{ for } N = 2$$

and $\gamma_N > 1/N$ holds, the frictional heat is bounded by

$$c_1 \|\Theta\|^{2+1/N}_{L_{2+\gamma}(I_T; W^1_p(\Omega))} + c_2$$

which is sufficient due to $2 + 1/N < 2 + \gamma$.

For the limit in nonlinear terms we will need an estimate for

$$\|\Theta\|^{2+\gamma}_{L_{2+\gamma}(I_T; W^1_{2+\gamma}(\Omega))} + \|\Theta\|^2_{L_\infty(I_T; L_2(\Omega))} + c_R \|\Theta\|^{r_R+1}_{L_{r_R+1}(S_T)}. \tag{5.4.105}$$

Such an estimate is proved with $\varphi = \chi_{t_0}\Theta$ as the test function. Then the elliptic term and the time derivative yield the expression (5.4.105) on the left hand side. In the estimate of the remaining terms we can use the imbedding

$$W^1_{2+\gamma}(\Omega) \hookrightarrow L_r(\Omega) \text{ with } r = \frac{N(2+\gamma)}{N-2-\gamma}$$

and the trace theorem

$$W^1_{2+\gamma}(\Omega) \hookrightarrow L_r(\Gamma) \text{ with } r = \frac{(N-1)(2+\gamma)}{N-2-\gamma},$$

both valid for $\gamma < N - 2$. In the case $\gamma \geq N - 2$ these imbeddings are valid with $r = +\infty$, for $\gamma = N - 2$ with $r < +\infty$. Moreover, we also use the imbeddings

$$\nabla\dot{\boldsymbol{u}} \in \boldsymbol{H}^{1/4,1/2}(Q_T) \hookrightarrow L_r(I_T, \boldsymbol{L}_s(\Omega)) \ \text{ for } \frac{2}{r} + \frac{N}{s} \geq \frac{N+1}{2} \quad \text{and}$$
$$\dot{\boldsymbol{u}} \in \boldsymbol{H}^{1/2,1}(S_T) \hookrightarrow L_r(I_T; \boldsymbol{L}_s(\Gamma)) \ \text{ for } \frac{2}{r} + \frac{N-1}{s} \geq \frac{N-1}{2}$$

with $r, s \geq 2$ in both cases. The difference to the preceding estimates is that we need not take care about the power of the norms of Θ, because all the employed norms of $\boldsymbol{u}$ are already estimated and the powers of the norms of Θ do not exceed 2. However, the requirements for the order of the norms are rather high. They contribute to the restrictive assumption of our Theorem. The viscous heat is estimated by

$$\int_{Q_T} |\nabla\dot{\boldsymbol{u}}|^2 \Theta \, dx \, dt \leq \|\nabla\dot{\boldsymbol{u}}\|^2_{L_{r_1}(I_T;\boldsymbol{L}_{s_1}(\Omega))} \|\Theta\|_{L_{2+\gamma}(I_T;L_{s_2}(\Omega))}$$

with $\frac{2}{r_1} = \frac{1+\gamma}{2+\gamma}$, $\frac{2}{s_1} = 1 - \frac{1}{q_1}$. The restriction $\frac{2}{r_1} + \frac{N}{s_1} \geq \frac{N+1}{2}$ leads to $\frac{N}{s_2} \leq \frac{\gamma}{2+\gamma}$. With the value $s_2 = \frac{N(2+\gamma)}{N-2-\gamma}$ valid for small γ we obtain the condition

$$\gamma \geq \frac{N}{2} - 1. \tag{5.4.106}$$

This condition is stronger than (5.4.95) and (5.4.96), with the only exception of the case $N = 2$ where (5.4.96) requires $\gamma > 0$. The deformation heat is bounded by

$$\int_{Q_T} |\nabla\dot{\boldsymbol{u}}| \Theta^2 \, dx \, dt \leq \|\nabla\dot{\boldsymbol{u}}\|_{L_2(I_T;\boldsymbol{L}_{2N/(N-1)}(\Omega))} \|\Theta\|^2_{L_4(I_T;L_{4N/(N+1)}(\Omega))}.$$

The norm of Θ is further estimated by interpolation

$$\|\Theta\|_{L_4(I_T;L_{4N/(N+1)}(\Omega))} \leq c\|\Theta\|^{\lambda}_{L_r(I_T;L_2(\Omega))} \|\Theta\|^{1-\lambda}_{L_{1+N/2}(I_T;L_s(\Omega))}$$

with $s = +\infty$ for $N = 2$ and $s = \frac{N(N+2)}{N-2}$ for $N \geq 3$. The indices in the last norm here satisfy $1 + \frac{N}{2} \leq 2 + \gamma$ and $s \leq \frac{N(2+\gamma)}{N-2-\gamma}$, hence

$$\|\Theta\|_{L_{1+N/2}(I_T;L_s(\Omega))} \leq c_1 \|\Theta\|_{L_{2+\gamma}(I_T;W^1_{2+\gamma}(\Omega))}.$$

The parameter λ here is chosen as $\frac{N+1}{4N} = \frac{\lambda}{2} + \frac{(1-\lambda)(N-2)}{N(N+2)}$; this is $\lambda = \frac{N^2 - N + 10}{2(N^2+4)}$. This value satisfies $\frac{1-\lambda}{1+N/2} < \frac{1}{4}$; hence the parameter

r is smaller than $+\infty$ and the estimate is done. For the frictional heat we use the estimate

$$\int_{S_T} \mathfrak{F}|\sigma_n(\boldsymbol{u},\Theta)||\dot{\boldsymbol{u}}_{\boldsymbol{t}}|\Theta\, ds_{\boldsymbol{x}}\, dt \le \|\mathfrak{F}\|_{L_\infty(S_T)}\|\sigma_n(\boldsymbol{u},\Theta)\|_{L_2(S_T)}\cdot$$
$$\cdot\|\dot{\boldsymbol{u}}_{\boldsymbol{t}}\|_{L_{2+4/\gamma}(I_T;\boldsymbol{L}_r(\Gamma))}\|\Theta\|_{L_{2+\gamma}(I_T;L_s(\Gamma))}$$

with $s = \dfrac{(N-1)(N+2)}{N-2}$ for $N \ge 3$, $s = +\infty$ for $N = 2$, and $r = \dfrac{2(N+2)(N-1)}{N^2-N+2}$. The parameter s should allow the trace imbedding $W^1_{2+\gamma}(\Omega) \hookrightarrow L_s(\Gamma)$; this is true due to $2+\gamma > 1 + \dfrac{N}{2}$. The norm of $\dot{\boldsymbol{u}}_{\boldsymbol{t}}$ is bounded by

$$\|\dot{\boldsymbol{u}}_{\boldsymbol{t}}\|_{L_{2+4/\gamma}(I_T;\boldsymbol{L}_r(\Gamma))} \le c_1\|\dot{\boldsymbol{u}}_{\boldsymbol{t}}\|_{\boldsymbol{H}^{1/2,1}(S_T)},$$

because $\dfrac{2(4+2\gamma)}{\gamma} + \dfrac{N-1}{r} > \dfrac{N-1}{2}$. This finishes the estimate for the frictional heat, and the relation (5.4.105) is established.

It remains to prove an estimate for a fractional time derivative of Θ. This is done as described above in the derivation of (5.4.87). It shows

$$\|\Theta\|_{H^\alpha(I_T;L_2(\Omega))} \le c$$

for $\alpha < \frac{1}{2}$ with constant c independent of the approximation parameters.

We are now in the same situation as in the convergence proof in the Galerkin approximation. There exist sequences of approximation parameters $M_k \to +\infty$, $\varepsilon_k \to 0$ and corresponding sequences of solutions $(\boldsymbol{u}_k, \Theta_k)$ of the approximate problems such that

$$\begin{aligned}
&\dot{\boldsymbol{u}}_k \rightharpoonup \dot{\boldsymbol{u}} \text{ weakly in } \boldsymbol{H}^{3/4,3/2}(Q_T),\\
&\ddot{\boldsymbol{u}}_k \rightharpoonup \ddot{\mathrm{u}} \text{ weakly in } L_2\big(I_T;\boldsymbol{H}^1(\Omega)^*\big),\\
&\Theta_k \rightharpoonup \Theta \text{ weakly in } \mathfrak{V}, \text{ and, for the case } R \not\equiv 0, \text{ also in } L_{r_R+1}(S_T)\\
&\qquad\qquad \text{and strongly in } H^\alpha(I_T;L_2(\Omega)) \text{ for } \alpha < \tfrac{1}{2},\\
&\dot{\Theta}_k \to \dot{\Theta} \text{ weakly-}* \text{ in } \mathfrak{V}^*,\\
&\Theta_k(T,\cdot) \rightharpoonup \Theta(T,\cdot) \text{ weakly in } L_2(\Omega).
\end{aligned}$$

Due to compact imbedding theorems the convergence $\Theta_k \to \Theta$ is strong in $H^{\alpha,1/2}(Q_T)$ for every $\alpha < \frac{1}{2}$. As in the proof of Theorem 5.1.8 for the contact problem without heat transfer, $\dot{\boldsymbol{u}}_k$ tends to $\dot{\boldsymbol{u}}$ strongly in $H^{1/2,1}(Q_T)$. Moreover, as in the convergence of the Galerkin method we prove that Θ_k

tends to Θ strongly in $L_{2+\gamma}\left(I_T; W^{1-\lambda}_{2+\gamma}(\Omega)\right)$ for an arbitrarily small positive λ. With these convergence properties it is possible to pass to the limit $k \to +\infty$ in the heat equation and to prove $\Theta_k \to \Theta$ strongly in $\mathfrak{V}$. This is sufficient to pass to the limit $k \to +\infty$ in the corresponding version of the approximate problem (5.4.58), (5.4.59) and to prove that (u, Θ) is a solution of the thermo–viscoelastic contact problem (5.4.79), (5.4.80). $\square$

Bibliography

[1] Adams R. A. *Sobolev Spaces.* Academic Press, New York 1975.

[2] Adams R. A. Reduced Sobolev inequalities. *Canad. Math. Bull.* **31**, 159–167, 1988.

[3] Agmon S., Douglis A., and Nirenberg L. Estimates near boundary for solutions of elliptic partial differential equations satisfying general boundary conditions. I. *Comm. Pure Appl. Math.* **12**, 623–727, 1959. Part II. *Comm. Pure Appl. Math.* **17**, 35–92, 1964

[4] Amanov T. I. *Spaces of Differentiable Functions with Dominating Mixed Derivatives* (in Russian). Nauka Kaz. SSR, Alma-Ata 1976.

[5] Amerio L. and Prouse G. Study of the motion of a string vibrating against an obstacle. *Rend. Mat.* Ser. 6, **8**, 563–585, 1975.

[6] Amann H. *Linear and Quasilinear Parabolic Problems.* Vol. **1**, Birkhäuser, Basel 1995.

[7] Amann H. Operator-valued Fourier multipliers, vector-valued Besov spaces, and applications. *Math. Nachr.* **186** (1997), 5–56.

[8] Amann H. Vector-valued distributions and Fourier multipliers. Preprint, Zürich 1997.

[9] Amann H. and Escher J. *Analysis III.* Birkhäuser, Basel 2001.

[10] Andersson L. E. A quasistatic frictional problem with a normal compliance penalization term. *Nonlinear Anal., Theory Methods Appl.* **37A**, 689–705, 1999.

[11] Andersson L. E. A quasistatic frictional problem with normal compliance. *Nonlinear Anal., Theory Methods Appl.* **16** (4), 347–369, 1991.

[12] Andersson L. E. Existence result for quasistatic contact problem with Coulomb friction. *Appl. Math. Optimiz.* **42**, 169–202, 2000.

[13] Andrews K. T., Shillor M., and Wright S. Dynamic evolution of an elastic beam in frictional contact with an obstacle. **In**: *Contact Mechanics* (Proc 2nd Cont. Mech. Int. Symp. held at Carry-le-Rouet, M. Raous and E. Pratt, Eds.), 49–56, Plenum Press, 1995.

[14] Andrews K. T., Kuttler K. L., and Shillor M. On the dynamic behaviour of a thermoviscoelastic body in frictional contact with a rigid obstacle. *Eur. J. Appl. Math.* **11**, 1–20, 1996.

[15] Bamberger A. and Schatzman M. New results on the vibrating string with a continuous obstacle. *SIAM J. Math. Anal.* **14**, 560–595, 1983.

[16] Ballard P. A counterexample to uniqueness of quasistatic elastic contact problems with small friction. *Int. J. Engrg. Sci.* **37** (2), 163–178, 1999.

[17] Ballard P. Elastic problems with arbitrarily small Coulomb friction at the boundary admit, in general, a continuum solutions. *In preparation.*

[18] Barbu V. *Nonlinear Semigroup and Differential Equations in Banach Spaces.* Noordhoff International, 1976

[19] Baiocchi C. and Capelo A. *Variational and Quasivariational Inequalities.* Wiley, Chichester 1984.

[20] Benedek A. and Panzone R. The spaces L^p with mixed norms. *Duke Math. J.* **28**, 301–324, 1961.

[21] Bhattacharya T. and Leonetti F. On improved regularity of weak solutions of some degenerate, anisotropic elliptic systems. *Ann. Mat. Pura Appl.* **IV**, Ser. 170, 241–255, 1996.

[22] Bennett C. and Sharpley R. *Interpolation of Operators.* Academic Press, Boston 1988.

[23] Bergh J. and Löfström J. *Interpolation Spaces. An Introduction.* Springer-Verlag, Berlin 1976.

[24] Besov O. V. On a family of function spaces (in Russian). *Dokl. Akad. Nauk SSSR* **126**, 1163–1165, 1959.

[25] Besov O. V. On a family of function spaces in conection with embedings and extensions (in Russian). *Trudy Mat. Inst. Steklov* **60**, 42–81, 1961.

[26] Besov O. V., Il'in V. P., and Nikol'skii S. M. *Integral Representation of Functions and Embedding Theorems* (in Russian). Nauka, Moskva 1975. English transl.: Halsted Press, V. H. Winston & Sons, New York 1978/79.

[27] Brézis H. *Opérateurs maximaux monotones et semi-groupes de contraction dans les espaces de Hilbert.* North-Holand & Elsevier, Amsterdam & New York 1973.

[28] V. I. Burenkov. On one method of extension of functions (in Russian). *Trudy Mat. Inst. Akad. Nauk SSSR* **140**, 27–67, 1976.

[29] Butzer P. L. and Berens H. *Semi-groups of Operators and Approximation.* Die Grundlehren der mathematischen Wissenschaften in Einzeldarstellungen. 145. Springer-Verlag, Berlin 1967.

[30] Cabannes H. and Citrini C. (eds.). *Vibrations with Unilateral Constraints.* Proceedings of the Colloquium Euromech 209 held at Como 1986. Tecnoprint, Bologna 1987.

[31] Calderón A. P. Lebesgue spaces of differentiable functions and distributions. *Proc. Symp. Pure Math.* **5**, 33–49, 1961.

[32] Citrini C. Vibrating strings with obstacles: The analytic study. **In**: Proc. conf. Advances in kinetic theory and continuum mechanics, R. Gatignol and Soubbaramaya, Eds., Springer-Verlag, Berlin 1991.

[33] Citrini C. and Marchionna C. Some unilateral problems for the vibrating string equation. *Eur. J. Mech. A/Solids* **8**, 73–85, 1989.

[34] Citrini C. and Marchionna C. On the motion of a string vibrating against a glueing obstacle. *Ren. Accad. Naz. Sci., Mem. Mat.* **107** (XIII/8), 141–164, 1989.

[35] Cocu M. and Rocca R. Existence results for unilateral quasistatic contact problems with friction and adhesion. *ESAIM Math. Modelling Num. Anal.* **34**, 981–1001, 2000.

[36] Coddington A. and Levinson N. *Theory of Ordinary Differential Equations.* McGraw–Hill, New York 1955.

[37] Coulomb C. A. de. Théorie des machines simples, en ayant égard au frottement de leurs parties et la raideur des cordages. (Prix double Acad. Sci. 1781). *Mémoirs Savants Etrang.* **X**, 163–332, 1785 (reprinted by Bachelier, Paris 1809).

[38] Dacorogna B. *Weak Continuity and Weak Lower Semicontinuity of Nonlinear Functionals.* Lect. Notes in Math., Vol. 922, Springer-Verlag, Berlin 1982

[39] Diestel J. and Uhl J. J. *Vector Measures.* Amer. Math. Soc. Surveys 15, Providence, RI 1977.

[40] Duvaut G. Equilibre d'un solide elastique avec contact unilateral et frottement de Coulomb. *C. R. Acad. Sci. Paris*, Série A **290**, 263–265, 1980.

[41] Duvaut G. and Lions J. L. Un problème d'elasticité avec frottement. *J. Mécanique* **10**, 409–420, 1971.

[42] Duvaut G. and Lions J. L. *Inequalities in Mechanics and Physics* (in French). Dunod, Paris 1972 (English translation Springer, Berlin - Heidelberg - New York, 1976).

[43] Eck C. Existenz und Regularität der Lösungen für Kontaktprobleme mit Reibung. (Existence and regularity of solutions to contact problems with friction, in German). *Thesis*, University of Stuttgart, 1996.

[44] Eck C. and Jarušek J. Existence results for the static contact problem with Coulomb friction. *Math. Models Methods Appl. Sci.* **8** (3), 445–468, 1998.

[45] Eck C. and Jarušek J. Existence results for the semicoercive static contact problem with Coulomb friction. *Nonlinear Anal., Theory Methods Appl.* **42** A(6), 961–976, 2000.

[46] Eck C. and Jarušek J. Existence of solutions for the dynamic frictional contact problem for isotropic viscoelastic bodies. *Nonlin. Anal., Theory Meth. Appl.* **53** (2), 157–181, 2003.

[47] Eck C. and Jarušek J. The solvability of a coupled thermo-viscoelastic contact problem with small Coulomb friction and linearized growth of frictional heat. *Math. Methods Appl. Sci.* **22**, 1221–1234, 1999.

[48] Eck C. Existence of solutions to a thermo-viscoelastic contact problem with Coulomb friction and heat radiation. *Math. Models Methods Appl. Sci.* **12** (10), 1491–1511, 2002.

[49] Edmunds D. E. and Triebel H. Logarithmic Sobolev spaces and their applications to spectral theory. *Proc. London Math. Soc.* **71**, 333–371, 1995.

[50] Edmunds D. E. and Triebel H. *Function Spaces, Entropy Numbers and Differential Operators*. Cambridge Univ. Press, Cambridge 1996.

[51] Ekeland I. and Temam R. *Convex Analysis and Variational Problems* (in French). Dunod, Paris 1974 (English translation North-Holland, Amsterdam 1976).

[52] B. L. Fain. On extension of functions from Sobolev spaces for non-regular domains with preservation of the smoothness parameter (in Russian). *Dokl. Akad. Nauk SSSR* **32**, 296–301, 1985. (English translation *Soviet Math. Dokl* **32**, 688–692, 1985.)

[53] Fichera G. Problemi elastostatici con vincoli unilaterali: il problema di Signorini con ambique condizioni al contorno. *Mem. Acc. Naz. Lincei* **VIII** (7), 91–140, 1964.

[54] Fichera G. *Existence Theorems in Elasticity. Boundary Value Problems of Elasticity with Unilateral Constraints*. Springer-Verlag, Berlin

[55] Franciosi M. and Moscariello G. Higher integrability results, *Manuscripta Math.* **52**, 151–170, 1985.

[56] Gagliardo E. Caratterizzazioni delle tracce sulla frontiera relative ad alcune classi di funzioni in n variabili. *Rend. Sem. Mat. Univ. Padova* **27**, 284–305, 1957.

[57] Gagliardo E. Proprietà di alcune classi di funzioni in più variabili. *Ricerche Mat.* **7**, 102–137, 1958.

[58] García-Cuerva J. and Rubio de Francia J. L. *Weighted Norm Inequalities and Related Topics*. North-Holland, Amsterdam 1985.

[59] Genebashvili I., Gogatishvili A., Kokilashvili V., and Krbec M. *Weight Theory for Integral Transforms on Spaces of Homogeneous Type*. Addison-Wesley, Pitman Monographs and Surveys in Pure and Applied Mathematics 92, Harlow 1998.

[60] Gol'dshtein V. M. and Reshetnyak Yu. G. *Introduction to the Theory of Functions with Generalized Derivatives and Quasiconformal Mappings* (in Russian). Nauka, Moscow, 1983.

[61] Grisvard P. Espaces intermediaires entre espace de Sobolev avec poids. *Ann. Scuola Norm. Sup. Pisa* **III**, Ser. 17, 255–296, 1963.

[62] Grisvard P. Commutativité de deux foncteurs d'interpolation et applications. *J. Math. Pures. Appl.* **45**, 143–290, 1996.

[63] de Guzmán M. *Differentiation of Integrals in* $\mathbb{R}^N$. Lecture Notes in Math., Vol. 481. Springer-Verlag, Berlin 1975.

[64] de Guzmán M. An extension of Sard's theorem. *Bol. Soc. Brasileira Mat.* **3**, 133–136, 1972.

[65] Hardy G. H., Littlewood J. E., and Pólya G. *Inequalities*. Cambridge Univ. Press, Princeton 1951.

[66] Haslinger J., Hlaváček I., Nečas J. and Lovíšek J. *Solution of variational inequalities in mechanics*. Springer, New York 1988.

[67] Haslinger J., Hlaváček I. and Nečas J. Numerical methods for unilateral problems in solid mechanics. **In:** *Handbook of Numerical Analysis* (P.G. Ciarlet and J.-L. Lions, Eds.) Vol. IV, Elsevier, Amsterdam 1996.

[68] Heinz E. An elementary analytic theory of the degree of mapping in n-dimensional space. *J. Math. Mech.* **8**, 231–247, 1959.

[69] Hestenes M. Extension of the range of a diferentiable function. *Duke Math. J.* **8**, 183–192, 1941.

[70] Hild P. Non-unique slipping in the Coulomb friction model in two-dimensional linear elasticity. *J. I. Mech. Appl. Math* **57** (2), 225–235, 2004.

[71] Hille E. *Functional Analysis and Semi-groups*. Amer. Math. Soc. Coll. Publ. 31, New York 1948.

[72] Ionescu I. R. and Paumier J.-C. Instabilities in slip dependent friction. *ESAIM, Proc. 2*, 99-111, 1997

[73] Jarušek J. Contact problems with bounded friction. Coercive case. *Czechoslovak Math. J.* **33** (108), 237–261, 1983.

[74] Jarušek J. Contact problems with bounded friction. Semicoercive case. *Czechoslovak Math. J.* **34** (109), 619–629, 1984.

[75] Jarušek J. Contact problems with given time-dependent friction force in linear viscoelasticity. *Comment. Math. Univ. Carolinae* **31** (2), 257–262, 1990.

[76] Jarušek J. Solvability of unilateral hyperbolic problems involving viscoelasticity via penalization. *SAACM* **3** (2), 129–140, 1993.

[77] Jarušek J. Solvability of the variational inequality for a drum with a memory vibrating in the presence of an obstacle. *Boll. Un. Mat. Ital.* **7** (8-A), 113–122, 1994.

[78] Jarušek J. Dynamic contact problems for bodies with a singular memory. *Bol. Un. Mat. Ital.* **7** (9-A), 581–592, 1995.

[79] Jarušek J. Dynamic contact problems with given friction for viscoelastic bodies. *Czechoslovak Math. J.* **46** (121), 475–487, 1996.

[80] Jarušek J. Remark to dynamic contact problems for bodies with a singular memory. *Comm. Math. Univ. Carolinae* **39**, 545–550, 1998.

[81] Jarušek J. Solvability of nonlinear thermo-viscoelastic contact problem with small friction and general growth of the heat energy. *Preprint* Nr. 273, Inst. Angew. Math, Univ. Erlangen-Nürnberg, Erlangen 2000.

[82] Jarušek J. and Eck C. Dynamic contact problems with friction in linear viscoelasticity. *C. R. Acad. Sci. Paris* **322**, Sér. I, 497–502, 1996.

[83] Jarušek J. and Eck C. Dynamic contact problems with small Coulomb friction for viscoelastic bodies. Existence of solutions. *Math. Models Meth. Appl. Sci.* **9** (1), 11–34, 1999.

[84] Jarušek J. and Eck C. Existence of solutions for a nonlinear coupled thermoviscoelastic contact problem with Coulomb friction. **In:** *Applied Nonlinear Analysis* (in honour of the 70th birthday of Prof. J. Nečas; A. Sequiera, H. Beirão da Veiga, J. H. Videman, Eds.) Kluwer Acad./Plenum Publ., 1999, pp. 49–65.

[85] Jarušek J. and Eck C. On the solvability of unilateral dynamic frictional contact problems of isotropic viscoelastic bodies in three and more space dimensions. *Preprint* Nr. 287, Institute for Applied Mathematics, University Erlangen-Nürnberg 2002.

[86] Jarušek J., Málek J., Nečas J., and Šverák V. Variational inequality for a viscous drum vibrating in the presence of an obstacle. *Rend. Mat.*, Serie VII **12**, 943–958, 1992.

[87] Jones P. W. Quasiconformal mappings and extendability of functions in Sobolev spaces. *Acta Math.* **147**, 71–88, 1981.

[88] Kakutani S. A proof of Schauder's theorem. *J. Math. Soc. Japan* **3**, 228–231, 1951.

[89] G. A. Kalyabin. Theorems on extension, multipliers and dipheomorphisms for generalized Sobolev-Liouville class in domains with Lipschitz boundary (in Russian). *Trudy Mat. Inst. Akad. Nauk SSSR* **172**, 173–186, 1985.

[90] Kawohl B. *Rearrangements and Convexity of Level Sets in PDE.* Lecture Notes in Math., Vol. 1150, Springer-Verlag, Berlin 1985.

[91] Kim J. U. A boundary thin obstacle problem for a wave equation. *Comm. Part. Diff. Equations* **14**(849), 1001-1025, 1989.

[92] Kinderlehrer D. and Stampacchia G. *An Introduction to Variational Inequalities and Their Applications.* Academic Press, San Diego, 1980.

[93] Klarbring A., Mikelić A., and Shillor M. Frictional contact problems with normal compliance. *Int. J. Engrg. Sci.* **26** (8), 811–832, 1988.

[94] Klarbring A., Mikelić A., and Shillor M. On friction problems with normal compliance. *Nonlinear Anal., Theory Methods Appl.* **13** (8), 935–955, 1989.

[95] Klarbring A. Example of non-uniqueness and non-existence of solutions to quasistatic contact problems with friction. *Ingen. Arch.* **60**, 529–541, 1990.

[96] Kolyada V. I. On embeddings of certain classes of functions of several variables. *Sib. Mat. Zh.* **14** (4), 766–790, 1973.

[97] Krbec M. and Schmeisser H.-J. Limiting imbeddings—the case of missing derivatives. *Ricerche Mat.* **XLV**, 423–447, 1996.

[98] Krbec M. and Schmeisser H.-J. Imbeddings of Brézis-Wainger type. The case of missing derivatives. *Proc. Roy. Soc. Edinburgh, Sect. A* **131**, 667–700, 2001.

[99] Kufner A., John O., and Fučík S. *Function Spaces.* Academia, Prague 1977.

[100] Lebeau G. and Schatzman M. A wave problem in a half-space with a unilateral constraint at the boundary. *J. Diff. Eq.* **53**, 309–361, 1984.

[101] Lieb, E. H. and Loss, M. *Analysis.* 2nd ed., Graduate Studies in Mathematics, Vol. 14. AMS, Providence, RI 2001.

[102] Lions J. L. and Magenes E. *Non-homogeneous Boundary Value Problems and Applications* (in French). Dunod, Paris 1968 (English translation Springer, Berlin - Heidelberg - New York, Vol. I and Vol. 2 1972, Vol. III 1973).

[103] Lizorkin P. I. and Nikol'skii S. M. Classification of differentiable functions on the basis of spaces with a dominating mixed derivative (in Russian). *Trudy Mat. Inst. Steklov* **77**, 143–167, 1965.

[104] Luenberger, D. G. *Introduction to Linear and Nonlinear Programming.* Addison-Wesley, 1973.

[105] Marchaud, A. Sur les dérivées et sur les differences des fonctions des variables réeles. *J. Math. Pure Appl.* **6**, 337–425, 1927.

[106] Martins J. A. C. and Oden J. T. Models and computational methods for dynamic friction phenomena. *Comp. Meth. Appl. Mech. Engrg.* **52**, 527–634, 1985.

[107] Martins J. A. C. and Oden J. T. Existence and uniqueness results for dynamic contact problems with nonlinear normal and friction interface laws. *Nonlinear Anal., Theory Methods Appl.* **11** (3), 407–428, 1987.

[108] Maruo K. Existence of solutions of some nonlinear wave equations. *Osaka Math. J.* **22**, 21–30, 1984.

[109] Maz'ya V. G. *Sobolev Spaces.* Springer-Verlag, Berlin 1985.

[110] Mendez O. and Mitrea M. The Banach envelopes of Besov and Triebel-Lizorkin spaces and applications to partial differential equations. *J. Fourier Anal. Appl.* **6**, 503–531, 2000.

[111] Meyers N. and Serrin J. $H = W$. *Proc. Nat. Acad. Sci.* **51**, 1055–1056, 1964.

[112] Nagumo M. Degree of mapping based on infinitesimal analysis. *Amer. J. Math.* **73**, 485–496, 1951.

[113] Nečas J., Jarušek J., and Haslinger J. On the solution of the variational inequality to the Signorini problem with small friction. *Boll. Un. Mat. Ital.* **5** (17-B), 796–811, 1980.

[114] Nečas J. *Les méthodes directes en équations elliptiques.* Academia, Prague 1967.

[115] Nečas J. and Hlaváček I. *Mathematical Theory of Elastic and Elasto-Plastic Bodies. An Introduction.* Studies in Appl. Math. 3, Elsevier, Amsterdam 1981.

[116] Panagiotopoulos P. D. A nonlinear programming approach to the unilateral contact- and friction boundary value problem. *Ingenieur-Archiv* **44**, 421–432, 1975.

[117] Peetre J. Espaces d'interpolation et théoréme de Soboleff. *Ann. Inst. Fourier (Grenoble)* **16**, 279–317, 1966.

[118] Peetre J. Sur la transformation de Fourier des fonctions à valeurs vectorielles. *Rend. Semin. Math. Univ. Padova* **42**, 15–26, 1969.

[119] Peetre J. *New Thoughts on Besov Spaces.* Duke Univ. Math. Series I, Duke University, Durham 1976.

[120] Peetre J. A counter-example connected with Gagliardo's trace theorem. *Commentat. math., spec. Vol. II, dedic. L. Orlicz*, 77–282, 1979.

[121] Petrov A. and Schatzman M. A pseudodifferential linear complementarity problem related to a one dimensional viscoelastic model with Sigorini conditions. *To appear*

[122] Reshetnyak Yu. G. Integral representations of differentiable functions in domains with non- smooth boundary (in Russian). *Sib. Mat. Zh.* **21**, 108–116, 1980. English transl in: *Sib. Math. J.* **21**, 833–839, 1981.

[123] Rocca R. Existence of a solution for a quasistatic problem of unilateral contact with local friction. *C. R. Acad. Sci. Paris* **328**, Sér. I, 1253–1258, 1999.

[124] Rocca R. Analyse mathématique et numérique de problèmes quasi statiques de contact unilatéral avec frottement local de Coulomb en élasticité (Mathematical and numerical analysis of quasi-static problems with unilateral contact and local Coulomb friction in elasticity, in French). *Thesis*, Laboratoire de Mécanique et d'Acoustique de Marseille 2000.

[125] Rychkov V. S. On restrictions and extensions of the Besov and Triebel-Lizorkin spaces with respect to Lipschitz domains, *J. London Math. Soc.* **60** 237–257, 1999.

[126] Sadosky C. *Interpolation of Operators and Singular Integrals. An Introduction to Harmonic Analysis.* M. Dekker Inc., New York 1979.

[127] Schatzman M. A hyperbolic problem of second order with unilateral constraints: the vibrating string with a concave obstacle. *J. Math. Anal. Appl.* **73**, 183-191, 1980.

[128] Schatzman M. The penalty method for the vibrating string with an obstacle. **In:** *Analytical and Numerical Approaches in Asymptotic Problems in Analysis* (Proc. Conf. Univ. Nijmegen, Nijmegen 1981). North-Holland, Amsterdam 1981.

[129] Schatzman M. Un problème hyperbolique du 2ème ordre avec contrainte unilatérale: la corde vibrante avec obstacle ponctuel. *J. Diff. Equations* **36**, 295–334, 1980.

[130] Schauder J. Der Fixpunktsatz in Funktionalräumen. *Studia Math.* **2**, 171–180, 1930.

[131] Schmeisser H.-J. On spaces of functions and distributions with mixed smoothness properties of Besov-Tiebel-Lizorkin type. I. Basic properties. *Math. Nachr.* **98**, 233–250, 1980.

[132] Schmeisser H.-J. On spaces of functions and distributions with mixed smoothness properties of Besov-Triebel-Lizorkin type. II. Fourier multipliers and approximation representations. *Math. Nachr.* **106**, 187–200, 1982.

[133] Schmeisser H.-J. Vector-valued Sobolev and Besov spaces. *Semin. Analysis*, Berlin 1985/86, Teubner-Texte Math. **96**, 4–44, 1987.

[134] Schmeisser H.-J. and Sickel W. Traces, Gagliardo-Nirenberg inequalities and Sobolev type embeddings for vector-valued function spaces. Jenaer Schriften zur Math. and Inf., Mat/Inf/24/01.

[135] Schmeisser H.-J. and Triebel H. *Topics in Fourier Analysis and Function Spaces.* Geest & Portig, Leipzig 1987; Wiley, Chichester 1987.

[136] Seeley R. T. Extension of C^∞-functions defined in a half space. *Proc. Amer. Math. Soc.* **15**, 625–626, 1964.

[137] Sickel W. and Triebel H. Hölder inequalities and sharp embeddings in function spaces of $B^s_{p\,q}$ and F^s_{pq} type. *Z. Anal. Anwendungen* **14**, 105–140, 1995.

[138] Slobodetskii L. N. Sobolev spaces of fractional order and their application to boundary problems for partial differential equations (in Russian). *Dokl. Akad. Nauk SSSR* **118** (2), 243–246, 1958.

[139] Slobodetskii L. N. Sobolev spaces of fractional order and their application to boundary problems for partial differential equations (in Russian). *Uch. Zapisky Leningrad. Ped. Inst. im. A. I. Gercena* **197**, 54–112, 1958.

[140] Sobolev S. L. On a theorem in functional analysis (in Russian). *Mat. Sb.* **4**, 471–479, 1938. English transl. in: *Am. Math. Soc., Transl., II,* Ser. 34, 39–68, 1963.

[141] Souček J. Spaces of functions on domain Ω, whose k-th derivatives are measures defined on $\overline{\Omega}$. *Časopis Pěst. Mat.* **97**, 10–46, 1972.

[142] Sparr G. Interpolation of several Banach spaces. *Ann. Mat. Pura Appl.* **99**, 247–316, 1974.

[143] Stein E. M. *Singular Integrals and Differentiability Properties of Functions,* Princeton Univ. Press, Princeton, New Jersey 1970.

[144] Talenti G. Inequalities in rearrangement invariant function spaces. **In:** *Nonlinear Analysis, Function Spaces and Applications, Vol. 5* (M. Krbec et al., Eds.), Prometheus Publ. House, Prague 1994, pp. 177–230.

[145] Tartar L. Compensated compactness and applications to partial differentila equations. **In:** *Research Notes in Mathematics, Heriot–Watt symposium,* Vol. **4** (R.J. Knops, Ed.), Pitman Press, London 1979.

[146] Tikhonov A. Ein Fixpunktsatz. *Math. Ann.* **111**, 767–776, 1935.

[147] Torchinsky A., *Real-Variable Methods in Harmonic Analysis.* Pure and Applied Mathematics, Vol. 123, Academic Press, San Diego 1986.

[148] Triebel H. *Interpolation Theory, Function Spaces, Differential Operators.* VEB Deutsch. Verl. Wissenschaften, Berlin 1978. Sec. revised ed.: North-Holland, Amsterdam 1978.

[149] Triebel H. *Theory of Function Spaces.* Geest & Portig K.-G., Leipzig, Birkhäuser, Basel 1983.

[150] Triebel H. *Function spaces in Lipschitz domains and on Lipschitz manifolds. Characteristic functions as pointwise multipliers. Rev. Mat. Complutense* **15**, 475–524, 2002.

[151] Troisi M. Teoremi di inclusione per spazi di Sobolev non isotropi. *Ricerche Mat.* **18**, 3–24, 1969.

[152] Troisi M. Ulteriori contributi alla teoria degli spazi di Sobolev non isotropi. *Ricerche Mat.* **20**, 90–117, 1971.

[153] Vodop'yanov S. K., Gol'dshtein V. M. and Latfullin T. G. A criterion for extension of functions in the class L_2^1 from unbounded planar domains. *Sib. Mat. Zh.* **20**, 416–419, 1979.

[154] Yosida K. *Functional Analysis.* Springer-Verlag, Berlin 1965.

[155] Zeidler E. *Nonlinear Functional Analysis and its Applications. I: Fixed-point theorems.* Springer-Verlag, New York 1993.

[156] Zeidler E. *Nonlinear Functional Analysis and its Applications. II B: Monotone Operators.* Springer-Verlag, New York 1990.

[157] Ziemer W. P. *Weakly Differentiable Functions.* Springer-Verlag, New York 1989.

[158] Zygmund A. *Trignometric Series.* Vols. I, II, Cambridge Univ. Press, New York 1959.

This is a list of the most important symbols used in the book. The number (or the first of numbers) usually refers to the page, where the notation was encountered for the first time.